APPLIED GEOMORPHOLOGY

Geomorphological Surveys for Environmental Development

H.Th. Verstappen

International Institute for Aerial Survey and Earth Science (I.T.C.)
Enschede, The Netherlands

ELSEVIER
Amsterdam – Oxford – New York 1983

ELSEVIER SCIENCE PUBLISHERS B.V.,
Molenwerf 1
P.O. Box 211, 1000 AE Amsterdam, The Netherlands

Distributors for the United States and Canada:

ELSEVIER SCIENCE PUBLISHING COMPANY INC.
52, Vanderbilt Avenue
New York, N.Y. 10017

Library of Congress Cataloging in Publication Data

Verstappen, H. Theodoor (Herman Theodoor)
Applied geomorphology.

Includes bibliographies and index.
1. Geomorphology. I. Title.
GB401.5.V47 1983 551.4 83-1663
ISBN 0-444-42181-5 (U.S.)

ISBN 0-444-42181-5

Printed in The Netherlands

.... Because natural phenomena reveal to us their full dimensions only when places in the human context ....

- To Grace, my wife and to our children Chandra, Rosanne and Armand -

PREFACE AND ACKNOWLEDGEMENTS

Surveying for development is a complex, interdisciplinary subject and this book is certainly not meant to claim that a geomorphologist - or any other lone scientist - can, in isolation, come to satisfactory results. The content rather proves the contrary! However, geomorphology is a key element when in the early phases of the work an adequate break-down of the land into terrain units is aimed at and, together with other ecosciences, 'environmental geomorphology' plays a part in the subsequent phases of surveys for environmental resources. In the ultimate assessment of the land for specific purposes also socio-economic, cultural and other factors have to be considered and then one may, with justification, speak of fully 'integrated geomorphology'. In all cases, factual and functional information is required about landforms, geomorphological processes, their morphogenetical situation and their environmental context, which together represent the four major aspects of geomorphology.

The book is composed of three parts. The first part begins with a general introduction to the subject and thereafter concentrates on the position of geomorphology among environmental sciences and on the importance of this science in monodisciplinary resource surveys related to geological, soil, water and vegetation/forest resources. The second part deals with the utilization of environment by man and shows how on one hand geomorphological factors affect the works of man and how, inversely, man has become a factor changing geomorphology and environment to an important degree. The subjects covered range from rural land use with its wide-spread, diffused but not always very intensive impact, to spatially more limited, more concentrated and often more violent impacts caused by urban land use, engineering works and mining. It concludes with the evident need for planned land utilization as to optimally use the available resources and to prevent or to stop environmental degradation.

The third part gives survey methodologies. These comprise of analytical geomorphological survey, which is largely monodisciplinary and emphasizes matters such as landforms, processes, genesis and chronology, of synthetic (holistic) surveys which are characteristically multidisciplinary and emphasize the environmental land qualities, and of pragmatic, problem-oriented surveys such as flood susceptibility surveys, erosion surveys, drought susceptibility surveys, etc.

Geomorphology has benefited substantially from the increasing potentialities of earth observation methodology in recent decades. This field has been covered by the author in an earlier book entitled 'Remote Sensing in Geomorphology' in which the contribution of aerospace technology to the advancement of geomorphology has been elucidated. The effect of these technological innovations has been most marked where geomorphological surveying is concerned and it has also become one of the major forces giving momentum to the rapidly growing field of applied geomorphology. Therefore this book on the contribution of geomorphology to surveying for development should be considered as a logical follow-up of the earlier book mentioned above. Together these two books lead the reader from the advancing technology through the evolving scientific discipline to the application of knowledge to resource development and environmental management.

The author is grateful for the numerous fruitful exchanges of views with his colleagues of the former Commission on Applied Geomorphology and of the Commission/Working Group on Geomorphological Survey and Mapping of the International Geographical Union, which he served as a Secretary and (Vice)President for many years.

The book is a reflection of the author's lectures at the International Institute for Aerial Survey and Earth Sciences (ITC), Enschede, The Netherlands. The author expresses appreciation to the Rector of this institute for the opportunity granted to develop his concepts on applied geomorphology.

Special thanks are due to the collaborators of the Geomorphology Department of ITC: Drs. N.H.W. Donker, Dr. A.M.J. Meijerink, Dr. R.A. van Zuidam and Drs. R.P.G.A. Voskuil for their valid comments and suggestions. Several of the former M.Sc. students, whose research and thesis work was often concerned with applied geomorphology, contributed to the examples in this book.

My gratitude is expressed to all those who have been directly involved in the preparation of the book. This applies in the first place to Miss Marlies Simmons who corrected the English typescript and prepared the camera-ready copy. Also the work of Mrs. Laura Flamand, who typed the lengthy reference lists, Mr. A.M. van Leeuwen, who made numerous drawings, Mrs. Anne van der Linde, who typed the manuscript, Mrs. Edith Hoschtitzky-Dantas Lic., who assisted in various ways and Mr. P.H. Hofsteenge B.Sc., who did the proof reading and final editing, is gratefully acknowledged.

It is an essential prerequisite for the successful application of geomorphology to surveys for development to acquire a pragmatic, society-oriented appreciation of this discipline, because natural phenomena reveal to us their full dimensions only when placed in the human context.

H.Th. Verstappen

PART A. GEOMORPHOLOGY AND THE SURVEY OF ENVIRONMENTAL RESOURCES

GEOMORPHOLOGY AND ENVIRONMENTAL RESOURCES

1.1 Pure and Applied Geomorphology

For a long time, geomorphology was considered a subject comprising mainly academic studies of the types and origins of landforms. Only relatively recently has it found many applications, particularly in the various fields of resource and environmental surveys. As a result, geomorphology has received much attention from growing numbers of scientists and planners. Swelling these numbers, there are clearly many non-geomorphologists, including soil scientists, engineers, geologists, regional and urban planners, etc. The development of the subject has therefore gained new momentum with respect to its scientific foundations and its methods of research. At the same time, the scope of geomorphology has gradually evolved; the emphasis is now placed on several diverse aspects which formerly received only limited attention. This development in turn caused further expansion of the field of applications.

It is, in fact, rather surprising that applied geomorphology has developed so late (Tricart, 1962, 1978; Hails, 1977), considering that mankind survived by successfully applying his knowledge of the world around him since the dawn of his existence (Marsh, 1964, 1965). Early man lived in caves and 'abris sous roche' for reasons of self-defence and also to protect himself from adverse weather conditions.

At a later stage, he constructed his settlements, fortresses and castles as true strongholds on mountain peaks and other inaccessible places. But his choice was also influenced by the availability of water and the location of suitable grounds for hunting or agriculture. With the advance of science and technology, he gradually learned to make efficient use of his environment to a great extent. Clearly the terrain configuration of the land surrounding him has always been a challenge to mankind, although his appreciation of it has evolved in the course of time, depending on his attitude and needs.

Several factors account for the long time which applied geomorphology took to evolve. First there is the discrepancy between the human and the geological time scales. The relief of the earth's crust tends to change at a very slow rate and therefore the geological time scale is needed to record the evolution of landforms, the gradual transformation of one type into another. Emphasis was therefore normally put on long-term landform evolution. Although the scientific information so obtained may also be of practical importance, it is logical that for purposes of applied research, emphasis should be placed on the situation as it is here and on the changes expected within the next few years, or generations. This often requires a different type of study and a different attitude on the part of the geomorphologist (Mabbutt and Steward, 1963; Mitchell, 1979, Ryabehikov, 1964; Shchukin, 1960; Young, 1968).

Second is the fact that the application of the subject could only evolve after geomorphology was firmly established as a science with its own methods of investigation. Generalities such as the theory of the geomorphological cycle had to be replaced by more adequate concepts, and qualitative/quantitative studies on processes and landform development under different climatic conditions. Thereafter an improved evaluation of the structural, climatological and other impacts on processes and landform development became possible.

Among the methods of investigation that have strengthened geomorphology as a subject, and have indirectly increased its capacity for application, mention should be made of the furthering of laboratory methods of research and the diversified techniques and tools now available for field measurement of various processes. Recently, more stress has been put on the techniques of surveying and the mapping of geomorphological features. Methods and legends have been devised for such maps in large, medium and small scales. An important factor in this respect has been the application of aerial survey methods to geomorphology. The interpretation of photographic and other images taken from aircraft or orbiting satellites (Verstappen, 1977) is nowadays a condition sine qua non in geomorphological survey. The more sophisticated methods of research mentioned above have resulted in three inseparable theatres of modern geomorphological investigations: airphoto, field and laboratory. Observations have become more reliable and the study of problems related to the derivation of one landform type from another over geological periods of time, has become much more reliable.

The third factor explaining the recent rise of applied geomorphology is the socio-economical and political context of the present world (Matley, 1966; Tricart, 1956). Millions of people live near or on the minimum subsistence level and their numbers are rapidly increasing. At the same time there is a universal drive, particularly in the new nations of the Third World, to raise the standard of living (Ehrlich, et al., 1977). Never before has mankind been confronted with such an immense problem, the solution of which must be found within a few decades. There is also a great demand for geomorphologists in countless projects for economic development aiming at an optimal, balanced utilisation of the natural environment.

It is not easy to give a completely satisfactory definition of applied geomorphology as the distinction between applied and pure research is not always clear and general agreement does not exist concerning the limits of geomorphology. A certain type of pure research may suddenly become of great practical importance due to entirely new developments in adjacent fields of science and technology. Differences exist in the degree of applicability of scientific research as well. Some types of basic study may be essential for applied research, though they may not be geared directly to practical uses. Other types of studies may be of general interest for applied research, whereas some are executed specifically for particular projects in the fields of engineering, conservation, etc. It may even be that a certain kind of study, e.g. relating to ravines or landslides, is considered as pure research and only of academic interest in one situation, whereas it would have been regarded as applied research in the true sense of the word if it had been carried out in another context and the investigator had been asked specific questions by engineers, hydrologists or others (Brunsden, 1981; Derbyshire and Sperling, 1981; Dixey, 1962; Jones, 1980; Sidorenko, 1972).

The geomorphologist engaged in applied research should be aware that it is essential that the information provided is both factual and functional. All facts about geomorphological features should be as exact as possible. Their precise spatial situation and interrelations should be given in the geomorphological map and the quantification of data on forms, processes, etc. and their effects should be sought. The study should be geared to the purpose of the work, avoiding unnecessary sidesteps. Ultimately, only functional information provided at the right time will be of use for those who requested the study, have to take decisions and make plans on the basis of the reports and maps produced. An early request, a good definition of the task, continuous contact with the principal-commissioner (engineer, planner, decision maker, etc.) and consultation of experts in neighbouring disciplines, are among the essential pre-requisits for obtaining optimum results.

A useful side-effect of the execution of applied geomorphological research is found in the frequent contacts with and growing appreciation of, scientists working in neighbouring disciplines. Teamwork is essential in most cases. These contacts lead to a broader knowledge of subjects on the fringe of geomorphology. The de-specialization thus achieved often opens new and interesting vistas. It is clear that geologists, geographers, soil scientists and also hydrologists, photogrammetrists and civil engineers, all have their own specific requirements where geomorphology is concerned. The aspects in which they are most interested vary with their aims and attitudes, but they all require factual and functional information.

The vital importance of pure research does, of course go without saying. This is the source on which applied research depends. We should also bear in mind the numerous, purely scientific investigations which sometimes quite unexpectedly and much later prove to be of immense practical value. It would be superfluous to give examples of this. However, applied research, launched as a result of well-defined needs of society, has often given great impetus to the furthering of pure science. Thus we may justly speak of the symbiosis between pure and applied research. It is certainly not true that applied work is a lower class derivate of pure science, a view still held in some academic circles. On the contrary, only the best, more well-evolved concepts yield reliable, practical conclusions (Dury, 1972; Gellert, 1968; Joly, 1977; Klammer, 1965; Mensching, 1979; Semmel, 1979; Tricart, 1968).

1.2 Major Aspects of Geomorphology and their Applicability

The diverse fields of science covered by geomorphology encompass many subjects, particularly since the recent development of the last few decades. Basically, four major aspects of geomorphology can be distinguished:

1. Static geomorphology, concerned with actual landforms.
2. Dynamic geomorphology, concerned with processes and the short-term changes so caused in landforms.
3. Genetic geomorphology, concerned with long-term development of relief.
4. Environmental geomorphology, concerned with the landscape ecological links between geomorphology and neighbouring disciplines or elements (parameters) of the land.

In applied geomorphological research, one inevitably touches upon one or more of these aspects, although the importance of each of them and thus their most appropriate combination varies with the type and purpose of the study at hand. Geomorphological surveying and mapping being an essential prerequisite for almost any kind of applied geomorphological work, should therefore incorporate information of each of these four major aspects which are further defined below.

When seen in a historical context it appears that emphasis has been shifted from one major aspect to another in the course of time; it was not until recently that the role and importance of each of them was fully understood. It should be clear from the onset that scientific work, including that in geomorphology, should be rooted in observed facts. A theory can subsequently be formulated and become the starting point for deductive reasoning which may eventually lead to the prediction of events or can be used to account for situations of the past. Checking the predictions or the assumed previous situations is the ultimate proof of the validity of the conclusions. It is only in this way that any trustworthy contributions to applied geomorphology can be expected.

The study of present landforms, disregarding the hypothetical shapes of the land millions of years in the past or in the future, and leaving aside the causative processes of the past as well as the processes operative at the moment is the purpose of static geomorphology (Verstappen, 1963, 1968). Investigations of this type are often classed under the heading of 'morphometric' studies. Although the aim of such studies is basically to arrive at a non-committing classification of the relief, unbiased by any theory, it is seldom purely descriptive. The author therefore prefers the term static geomorphology. The studies may encompass matters such as slope steepness, slope form in profile and in plan (concave, straight, convex), degree of dissection or valley density, relief amplitude, etc.

This approach can be useful, but it should be understood that geomorphology, as a pure and applied science, has much more to offer. It will be clear that a realistic and reliable landform classification is only possible once there is a proper understanding of the nature of the units and features distinguished (such as gravelfans and deltaic lowlands). Both the genesis and the interrelation between the various geomorphological phenomena have to be incorporated in the study. Also a thorough knowledge of the physical properties of each unit and feature is essential in order to render the classification useful for practical purposes such as the evaluation of the hydrological conditions, agricultural potential and the engineering properties. It is evident that landform studies of this kind must also be shouldered by investigations in other major aspects of geomorphology (dynamic, genetic, environmental).

It is of interest to note that up to the early Eighteenth Century the concept of the permanence of landforms prevailed. They were believed to have resulted from a cataclysm in the past and were there to remain unchanged forever. Nothing but a description of landforms could be given in those days and geomorphology could only develop after this concept had been abandoned. Previously only a few isolated voices, such as those of some Greek and Roman scientists, the Arab scientist Ibn-Sina also known as Avicenna (980-1037) and of Leonardi da Vinci (1452-1519), were heard concerning changes of landforms, valleys being formed by rivers, etc. Their ideas, however, were only generally accepted centuries later. Man has also given rise to new, anthropogenic landforms, either involuntarily or deliberately. The stereopair of Figure 1.1 gives an example.

The study of geomorphological processes, classified under the field of dynamic geomorphology, which was formerly referred to as physiological geomorphology, gradually evolved in the early Eighteenth Century, when matters such as the role of rivers in valley formation, the effects of marine abrasion on cliff recession and the impact of glacial erosion and deposition became subjects of study. These investigations are deliberately designed for a limited scope: emphasis is on active processes of erosion, sedimentation, etc. and on the minor changes in the land caused by these processes in short periods of time, for instance, years, decades or centuries. No attempt is made towards unravelling the evolution of landforms over longer periods of time. The human time scale rather than the geological time scale is used as a term of reference.

Occasional occurrences which may be abrupt and sometimes even catastrophic such as landslides, volcanic eruptions or sudden changes in the position of river courses are the more spectacular processes. Their effects in particular confront us with the obvious incorrectness of the permanence of landforms. However, the less spectacular, slower changes provoked by everyday processes of weathering and mass movements, the work of rivers and wind, glacier and wave action, which can only be demonstrated by careful observations and precise measurements, are in the long run, more effective than the rare abrupt changes. Investigations concerning these processes and rapid changes in the landforms obviously have ample scope for applied geomorphological research also.

Important early contributions to the study of geomorphological processes come from agriculturists and foresters who sought to control soil erosion, which was recognized as a problem in Europe in the early Eighteenth Century. Farmers saw their land threatened by slope processes, particularly by sheet, rill and gully erosion due to inadequate agricultural practices and a growing population. The traces of this are still noticeable in several parts of Europe. Another contribution from the practical side came from engineers who, when constructing roads or bridges, became aware of river work and of matters such as cliff recession due to wave action, etc. An early (1884) example in this field is given by Lamblardie in a study in Normandy of cliff recession under the influence of the surf. These early factual and functional studies were already clearly demonstrating the applicability of geomorphological research.

Simultaneously, there was a growing academic interest

in geomorphological processes such as the role of glacial erosion and deposition, as illustrated for example by the studies by De Saussure (1740-1799). It was vital to the advancement of geomorphological knowledge that in those days the old doctrine of catastrophism and permanence of landforms was gradually replaced by the uniformitarian concept of their slow but continuous development under the influence of everyday processes. The names of Hutton (1726-1797), Playfair (1748-1819) and particularly Lyell (1797-1895) are associated with this concept. They were also the ones who claimed that the present (i.e. the study of actual processes) was a key to the past.

The interest in processes and their short-term effects, was then neglected for quite some time, but it again became a major focus of interest among geomorphologists in the last few decades.There is a strong tendency in modern geomorphology to study active processes from both a qualitative and a quantitative point of view, and also processes of the past. The awareness that processes vary in type and intensity with climate, and that climates of the past differ substantially from those prevailing today, has led to the development of climatic geomorphology, and this has contributed substantially to this tendency.

At present much more reliable information about processes is available. Exact field and laboratory methods, both simple and sophisticated, have led to quantifiable data of many processes and to a better understanding of their mode of operation and geomorphological effects. Processes caused or accelerated by human activities are of wide-spread occurrence (Sherlock, 1922; Rathjens, 1979; Goudie, 1981) and in numerous localities, endanger further human occupance. Figure 1.2 gives an example.

The long-term development of landforms, requiring extrapolation in both the past and future, is treated in genetic geomorphology. This subject became a focus of interest especially in the later part of the Nineteenth Century. It is evident that the extremely slow and almost imperceptible change of the landforms through geological times represents an aspect of geomorphology which deviates largely from that which was the main concern of the previously mentioned early agriculturalists and engineers. It is the concept of synthesised landform development versus the detailed analysis

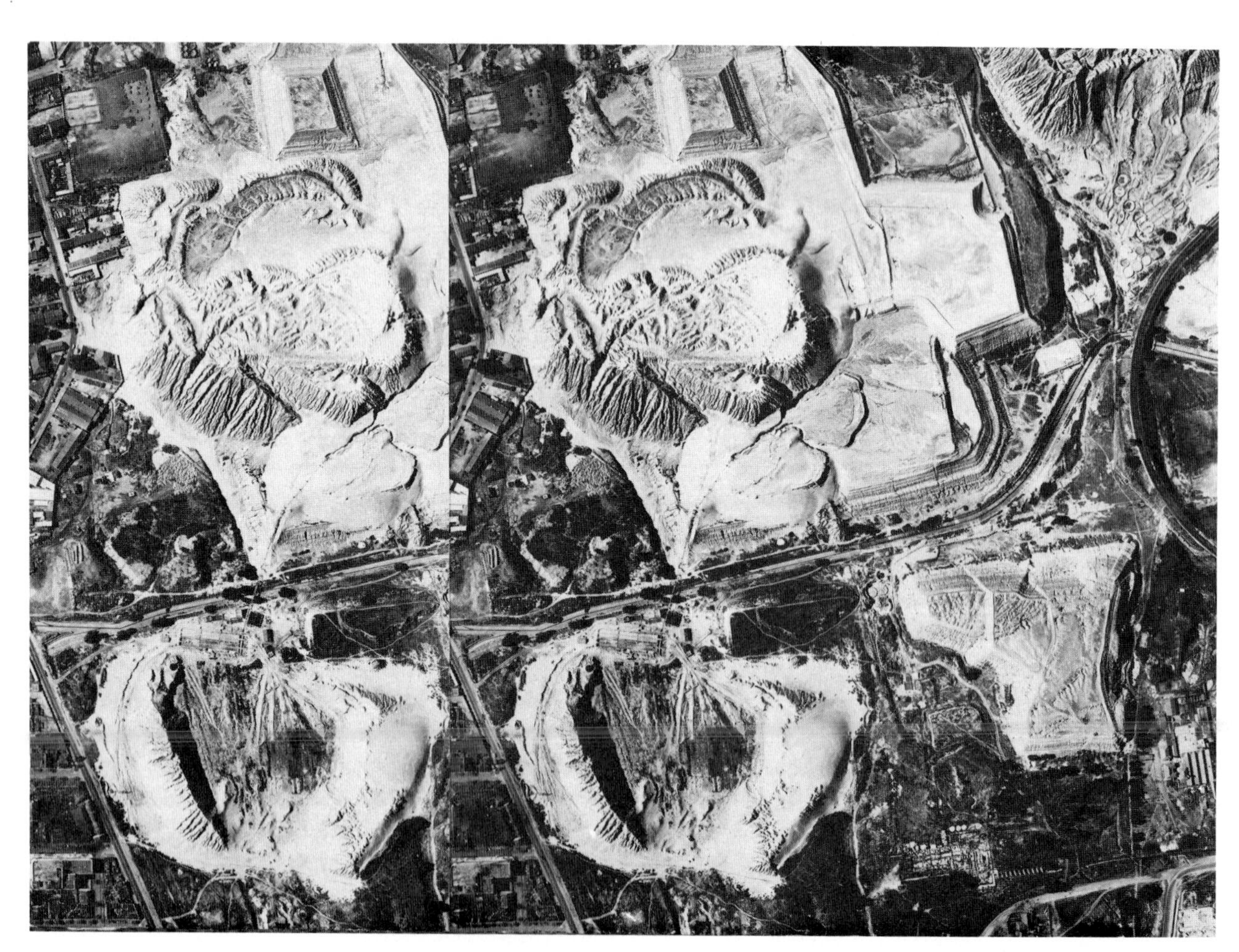

Fig. 1.1 Anthropogenic landforms: stereopair showing gold mine tips, Johannesburg, R.S.A., scale 1:7,000

of the various processes (dynamic geomorphology); which requires an entirely different attitude towards geomorphology and uses the geological, and not the human time scale as a term of reference. This sometimes abstract type of study seeking to establish a sequence of landform development covering periods of millions of years, is sometimes labelled as 'physiognomic' geomorphology. Both structural and climatological factors have to be taken into consideration in such attempts to recontruct the landforms from which the present situation derived, or to visualize the landforms which will evolve in the future. However, the multitude of processes which make up the mechanism of these changes has often been neglected, particularly in the earlier studies of this kind. Many lacked adequate academic foundation due to the insufficient knowledge of the mechanism and effects of the diversified geomorphological processes or those of climatic factors, etc.

Fig. 1.2 Man-induced and acceleratec processes: total erosion of soil as a result of excessive deforestation and harmful agricultural practices. Carius Area (Paraiba), Brazil. Courtesy: UNESCO- Aubert de la Rue.

At present genetic geomorphology is still an important and essential branch of geomorphology. It has now emerged as a subject of study supported by a sound scientific framework and numerous appropriate methods of research. The broad views on synthesised landscape development under the influence of both endogenic and exogenic factors were advocated by W.M. Davis (1850-1934) and others in the field of earth sciences, as a sort of logical follow-up to Darwin's views on the evolution of life. They were, however, severely critisized.

The endogenic factors such as crustal movements, geological structure and lithology are now carefully considered together with the exogenic factors such as weathering, mass movements, erosion and aggradation under different climatic conditions. More realistic concepts of landscape evolution have thus been developed and the field has become more factual, although in some respects, such as planation, it is not yet always as solidly based on observed facts as one would like it to be. It was only after the concept of the 'geomorphological cycle' had been replaced by more adequate concepts such as the open system approach with a defined in- and output, that genetic geomorphology also became useful in applied research. In modern genetic geomorphology, detailed analysis of forms, processes, soils and sediments replaced the synthesised approach and hypothetical deductions of classical geomorphology. Whether or not the vast amount of factual information that can be collected on long-term geomorphological development is always functional in the context of a certain type of applied research is a different matter. If correct, it is functional in a geological context where the landform development represents the younger part of the geological history. In many types of applied geomorphological research one is inclined, however, to replace the geological time scale by the human time scale and it may thus be more appropriate to put more emphasis on the aforementioned static and dynamic aspects of geomorphology or on the environmental context of the geomorphological framework. Nevertheless, the information provided by genetic geomorphology has proved useful on many occasions for various practical purposes, as will be evident in the forthcoming sections of this book.

Landscape ecological links of geomorphological phenomena are treated in environmental geomorphology. This study of the interrelation between geomorphology on the one hand and neighbouring earth science disciplines on the other and also including the role of geomorphology as an element of the physical environment of the human group, is the newest major branch of geomorphology. (Barsch (Ed.), 1979; Coates (Ed.), 1971, 1972/4; Cooke and Doornkamp, 1974; Dawson and Doornkamp, 1973; Flawn, 1970; Hall, 1974; Leser, 1976; Morgan and Moss, 1965; Tank, 1976; Tricart, 1973; Tricart and Kilian, 1979). It is remarkable how the interdisciplinary concept of the environment has affected geomorphology in recent years. The relation between landforms and processes with land(scape) elements such as soils, ground- and surface water, vegetation through the landscape ecological situations and even including man as an agent through his use of the land, is a fascinating field of study. This study was stimulated some decades ago by the use and interpretation of aerial photographs, showing the landforms in their environmental context, and developed rapidly during the Fifties when it was felt that the land as a whole deserved more attention than it had previously received and that the relation between geomorphology and other environmental elements or parameters should therefore be studied. Comprehensive investigations of this type may either reveal the impact of the natural environment on human activities or inversely, the effect of human activities on the environment. It is obvious that the geomorphologist cannot claim the whole field of environment to be his. Scientists from various disciplines can make great contributions into finding and understanding the environment in all its complexity. Teamwork is almost always required. The contribution to be made by the geomorphologist is an essential one, and fully justifies the concept of environmental geomorphology.

It goes without saying that whatever the approach to a certain applied geomorphological study and regardless of the particular aspect of geomorphology to be emphasised, the investigations should always be factual and at the same time functional. One will often find it necessary to concentrate on the human rather than on the geological time scale; in other words, mainly studying the present landforms (static geomorphology) and the active processes and rapid changes (dynamic geomorphology). Figure 1.3 gives an example.

The evaluation of these data in terms of genetic geomorphology will lead to the necessity of accurately plotting all observations in the form of a map. Geomorphological maps, compiled using analytical principles, have become an important tool in applied geomorphology. (Kienholz, 1980). It is the task of the environmental geomorphologists to fit this information into a landscape ecological context. This can be realised either by pairing this map with a landscape ecological map compiled using syntheticism or by adding the various environmental parameters to the map in a tabular form providing a solid base for terrain classification and evaluation.

Fig. 1.3 Misuse of environment: stereotriplet showing landforms and active processes in an area of derelic land in the western USA. Scale 1:20,000. Innumerable gullies and ravines provide excessive bedload to braided main river. Problems should be studied in a morphogenetic as well as in an environmental context.

1.3 Main Fields of Applied Geomorphology

The results of geomorphological survey find application in two basic groups, first, in the various adjacent fields of (mainly) earth sciences and resource surveying, and second in the assessment of the land for the numerous specific needs of mankind.

The first group mainly involves bilateral scientific contacts whereas the second area, at least in part, involves multilateral scientific contacts. Both the relations of geomorphology with other earth sciences and its role in matters relating to a planned, optimal utilization of the land by man, will be elaborated upon in the forthcoming sections of this book.

The relations between geomorphology and geology are so close that some even consider geomorphology a part of geology. One of the most outspoken adherents of this 'geological' geomorphology was Penck (1977), who interpreted geomorphological observations mainly in terms of tectonic movements of the earth's crust. More recent studies of this kind have led to the detection of minor uplift and/or subsidence in alluvial plains, where, for lack of outcrops the geologist has to rely on borings, geophysics and geomorphological indications. Kurashkowskaja (1971) is an advocate of 'geological' geomorphology in the USSR and divides geology into three major parts: petro-mineralgenesis, tectogenesis and morphogenesis. Geomorphology is thereby fully incorporated into geology. It is evident that a good knowledge of geology is essential for any geomorphologist and that a geologist will benefit considerably from a thorough knowledge of geomorphology. However, geomorphology also covers a sizeable scientific field outside the sphere of geology.

A geographer, for instance, will find studies of the type indicated above insufficient for most of his requirements. Russel (1949) stressed the need for what he called 'geographical' geomorphology some decades ago. He pointed out in his criticism that 'geomorphologists have been unrealistic, too geological in their interests and have failed to cover their field'.

It is fortunate that the attitude of many geomorphologists since then has fundamentally changed. What the geographer wants of geomorphology is factual distributional patterns. He is furthermore interested in accurate information about processes which have led to measurable changes in the human time scale, or which have been the most decisive factors influencing the existing patterns of soils, vegetation, etc.

More recently, geomorphological applications have found a place in the field of soil science, comparable to those existing in geology and geography. The somewhat belated start of this development is due to a great extent to the concentration of many geomorphologists on erosional forms and mountainous areas and their neglect of areas of deposition. By the time geomorphologists began to take an interest in soils and their significance in the study of landforms, soil scientists had already come to recognize the role of landforms in the study of soils. Interesting studies, including some relating to catenary soils, have resulted from the contact between geomorphology and soil science. In the Netherlands Edelman (1957) successfully introduced the physiographic method of soil mapping, thus establishing the much required link between these two sciences.

What the soil scientist needs from geomorphology is: information on the classification and distribution of landforms and second, information on their significance to the geomorphological environment in which soils develop. This means that a mere descriptive system of physiographic or landform classification, as developed in some countries, is not satisfactory.

Thorough geomorphological knowledge is required to make best use of the physiographic approach to soil survey. Since both soil science and geomorphology have developed rapidly in the last few decades, it is out of the question that both these fields can be mastered sufficiently by an individual; geomorphologists and soil scientists must cooperate effectively to obtain optimum results.

Interesting applications of geomorphology are also found in the borderlands between geomorphology and hydrology (Chorley, (Ed.) 1969). It has been established that there is a general link between hydrological and geomorphological variables. It is therefore logical that the hydrologist engaged in the evaluation and appraisal of ground and surface water resources should take an interest in geomorphology in the widest sense of the word. Geomorphological studies, which may include data on size, shape and altitude of the drainage basins concerned, become of paramount importance when hydrological investigations are carried out in areas where data on river discharge and other parameters are insufficiently available. Morphometric aspects rank high in geomorphological studies orientated towards hydrology. Through them a quantitative analysis of drainage basins can be carried out. This is fundamental to the study of matters such as surface runoff and infiltration and it is also useful for calculating the time of flood concentration, sediment yield and other characteristics of rivers. Environmental and other major fields of geomorphology are also of considerable interest to the hydrologist.

There are links between geomorphological and hydrological variables also in the lower reaches of the rivers. In deltaic areas they are so all-encompassing that one may speak with some justification of an amalgamation of the two sciences. Our modern insights into fluvial geomorphology are the invaluable fruits of many years of investigations in the borderland of geomorphology and hydrology.

Aerial photographs provide the geomorphologist with unprecedented possibilities for studying landforms from great altitudes and for discovering patterns and anomalies of various types. Photographs taken from spacecraft and sophisticated remote sensing techniques such as thermal infrared imagery or side-looking radar have further increased the arsenal of aerial survey techniques available to geomorphological research as already briefly mentioned in the previous section.

Geomorphology has gained considerably in prestige through its applications in aerial survey techniques (Verstappen 1963, 1977).

Quite naturally in applied geomorphology, great advantage can be taken of the powerful tools of aerial survey. Aerial photographs are a means of making objective terrain observation possible. Because the photographs depict the diversified landform elements or units and their spatial relationship, they are an excellent starting point for geomorphological survey. Quantitative analysis of slopes, drainage density and numerous other landform elements can be carried out most accurately and rapidly on the basis of aerial photographs. Aerial photographs can also be successfully used as an aid to the study of dynamic geomorphology, because former positions of coastlines and river courses are often clearly depicted on them. Sometimes a few additional data are sufficient to date such traces, making quantitative studies possible. In the case of rapid, recent changes, sequential air photography at regular intervals of days, months or years, becomes specially interesting. Most important of all is the airphoto, however, as this is an unique means of visualizing the whole complex of natural environment and landscape ecology in the context of the intricate interrelationships which exist between landforms, rocks and soils, ground- and surface water, climatic conditions, vegetation, animal life and land utilization.

The unravelling of the interrelationships between these various elements of the natural environment and of their impact on society requires a thorough analysis of all elements concerned and thereafter a synthesis giving the evaluation of their interdependence. To achieve this the whole landscape ecological situation of the environment has to be studied. The idea of landscape ecology is not new in itself; it was established as an academic subject decades ago, but came into its own when the concept was recently accepted as a means of estimating the economic potentialities (resources) in surveys for regional development. It is evident that in these surveys geomorphology plays an important part in understanding the environment in areas where soils develop, ground and water circulate, vegetation grows and man tills his fields.

Planners and governmental or non-governmental decision-makers at all levels involved in promoting an optimum use of the land for the benefit of the inhabitants, taking into consideration the economic welfare and the well-being of the people and the importance of long-term safeguarding of the environmental resources, will inevitably have to consult geomorphologists for matters related to landforms, processes and/or environmental geomorphology. This is increasingly so because of the growing emphasis on land development in almost every part of the world. When the aim is to raise the standard of living and to allow for growing numbers of inhabitants, new areas have to be opened up, and other, more intense kinds of use have to be introduced in areas which are already heavily populated. Thus there is a growing need to master adverse environmental factors and and to establish a solid scientific basis for physical planning in various scales of magnitude.

Summarizing, it can be said that though diversified, applications of geomorphology can be grouped under a limited number of headings:

1. Applications in the field of earth sciences (geology, soil science, hydrology), vegetation science, etc. including the topographic and thematic mapping related to the study of natural resource development.
2. Applications in the field of environmental studies and surveys, either of a general nature or geared to specific natural hazards, such as landslides, avalanches, earthquakes, volcanism, land subsidence, flooding or drought.
3. Applications in the field of rural development and planning. Emphasis is often on agriculture, herding or other types of rural land utilization. Improvements of the rural land use by erosion control and conservation and/or by river basin development also fall under this heading.
4. Applications in the field of urbanization. Here the impact of man on the land is usually more intense and limited in areal extent and consequently the problems involved are often of a different nature. The studies may relate to matters such as urban extension, site selection for settlements or industry, or mining activities.
5. Applications in engineering. These are manifold and in relation to urbanization and industrialization they are frequently called for. Broadly speaking, engineering applications can be divided into two parts: highway engineering (and related fields such as railway engineering and airfield construction) and river and coastal engineering.

Together these five groups cover the two areas mentioned

at the beginning of this section, namely the applications in earth science and resource surveying and the assessment of the land for specific uses by mankind. They are systematically discussed in the first two parts of this book. For all these applications, analytical geomorphological mapping as well as synthesised mapping of terrain types and terrain classification and evaluation for specific purposes is essential. These techniques are elaborated upon in the third part of the book and illustrate their importance in applied geomorphology geared to surveys for development.

References

Barsch, D., 1979. The geomorphological approach to environment. Geo-Journal, 3(4): 329-416.

Brunsden, D., 1981. Geomorphology in practice. Geogr. Mag., 53(8): 531–533.

Chorley, R.J., 1969. Water, Earth and Man. A synthesis of hydrology, geomorphology, and socio-economic geography. Methuen & Co. Ltd., London: 1-588.

Coates, D.(Editor),1971. Environmental Geomorphology. Proc. Symp. State Univ. New York, Binghampton: 1-262.

Coates, D.(Editor),1972–1974.Environmental Geomorphology and Landscape Conservation. Hutchinson and Ross Inc., Benchmark papers in geology, 3 vols.: 1-485; 1-454; 1-483.

Cooke, R., and Doornkamp, C., 1974. Geomorphology in Environmental Management. Clarendon Press, Oxford: 1-413.

Dawson, J.A., and Doornkamp, J.C., 1973. Evaluating the Human Environment. Edward Arnold, Ltd., London: 1-288.

Derbyshire, E., and Sperling, C.H.B., 1981. Geomorphology in Practice. Geogr. Mag., 53(7): 464–467.

Dixey, F., 1962. Applied Geomorphology. S. Afr. Geogr. J., 44:3-24.

Dury, G.H., 1972. Some current trends in geomorphology. Earth Sci. Rev., 8:45-72.

Edelman, C.H., 1957. De betekenis van de geomorfologie voor de bodemkunde. Tijdschr. Kon. Nat. Aardr. Gen., 74:257-262.

Ehrlich, P.R., et al., 1977. Ecoscience:population, resources, environment, Freeman, San Francisco.

Flawn, R., 1970. Environmental Geology. Conservation, Land Use Planning and Resources. Harpers & Row, New York, Evanston, London: 1-313.

Gellert, J., 1968. Zum Wesen der angewandten Geomorphologie. Petermanns Geogr. Mitt., 112:256-264.

Goudie, A., 1981. The Human Impact. Man's Role in Environmental Change. Basil Blackwell Publ., Oxford: 1-316.

Hails, R., 1977. Applied Geomorphology. Elseviers Sci. Publ., Amsterdam, 418 pp.

Hall, V., 1974. Environmental geology: a selected bibliography. Geoscience Inform. Soc., 4:55-74.

Joly, F., 1977. Point de vue sur la géomorphologie. Ann. Géogr. 86:522-541.

Jones, D.K.C., 1980. British applied geomorphology: an appraisal. Ztschr. f. Geom., Suppl. Bd. 36:48-73.

Kienholz, H., 1980. Beurteilung und Kartierung von Naturgefahren: mögliche Beiträge der Geomorphologie und der geomorphologische Karte 1:25,000. Berliner Geogr. Abh. 31:83-90.

Klammer, G., 1965. Geomorphologie und erdwissenschaftliche Praxis. Ztschr. f. Geom., 9:115-129.

Kurashkowskaja, J.A., 1971. Das geologische materielle System und die Gesetzmässigkeiten seiner Entwicklung. Urania, Berlin.

Leser, H., 1976. Landschaftsökologie, UTB 521, Stuttgart, 1-432.

Mabbutt, J.A., and Steward, G., 1963. The application of geomorphology in resources surveys in Australia and New Guinea. Rev. Géom. Dyn., 14:97-109.

Marsh, G.P., 1965. Man and Nature. Schribner, New York, 1864. Re-ed. Lowenthal, D. Belknap Press, Cambridge, Mass., U.S.A.

Matley, I.M., 1966. The Marxist approach to the geographical environment. Ann. Assoc. Amer. Geog., 56:97-111.

Mensching, H., 1979. Angewandte Geomorphologie. Beispiele aus den Subtropen und Tropen. Abh. 42.D. Geogr. Tag., Göttingen: 25-34.

Mitchel, B., 1979. Geography and Resource Analysis. Longman Ltd., London: 1-399.

Morgan, W.B., and Moss, R.P., 1965. Geography and ecology: the concept of the community and its relationship to environment. Ann. Assoc. Amer. Geogr., 55(2):339-350.

Rathjens, C., 1979. Die Formung der Erdoberfläche unter dem Einfluss des Menschen, Stuttgart.

Russel, R.J., 1949. Geographical Geomorphology. Ann. Assoc. Amer. Geogr., 39:1-11.

Ryabehikov, A.M., 1964. On the interaction of the geographical sciences. Soviet Geogr., 5(10):45-60.

Semmel, A., 1979. Geomorphologie als geowissenschaftliche Disziplin-praktische Erfahrungen, theoretische Möglichkeiten. Stuttgarter Geogr. Stud., 93:23-32.

Shchukin, E.S., 1960. The place of geomorphology in the system of natural sciences and its relationship with integrated physical geography. Soviet Geogr., 1(9):35-43.

Sherlock, R.L., 1922. Man as a geological agent. Witherby, London.

Sidorenko, A.V., 1972. Geomorphology and the national economy. Soviet geogr., 13:344-355.

Tank, R., 1976. Focus on environmental geology. 2nd Ed. Oxford Univ. Press, New York, London, Toronto:1-538.

Tricart, J., 1956. La géomorphologie et la pensée marxiste. La Pensée, 69:3-24.

Tricart, J., 1962. L'Epiderme de la terre. Esquisse d'une géomorphologie appliquée. Masson et Cie., Paris:1-167.

Tricart, J., 1968. Aspects méthodologiques des études de ressources pour le développement. O. Tulippe Mém. Vol., Duculot, Gembloux:345-361.

Tricart, J., 1973. La géomorphologie dans les études integrées d'aménagement du milieu natural. Ann. Géogr., 82:421-453.

Tricart, J., 1978. Géomorphologie applicable, Masson et Cie., Paris:1-167.

Tricart, J., and Kilian, J., 1979. L'Ecogéographie. F. Maspero/ Hérodo, Paris: 1-326.

Verstappen, H.Th., 1963. The role of aerial survey in applied geomorphology. Rev. Géom. Dyn., 10:237-252.

Verstappen, H.Th., 1968. Geomorphology and Environment. Inaugural address, Waltman, Delft: 1-23.

Verstappen, H.Th., 1977. Remote sensing in geomorphology. Elsevier Sci. Publ. Co. Amsterdam, London, New York: 1-214.

Whyte, A.V.T., 1977. Guidelines for field studies in environmental perception. MAB Techn. Notes, 5, UNESCO.

Wiggers, A.J., Functionele fysische geografie. Geogr. Tijdschr. 4:296-302.

Young, A., 1968. Material resources surveys for land development in the tropics. Geography, 53:229-248.

GEOMORPHOLOGY IN SURVEYING AND MAPPING

2.1 Introduction

The application of geomorphology in surveying and mapping is a matter of long-standing. The reason is obvious: both the geomorphologist and the mapmaker are interested in landforms; the former from a genetic or environmental point of view, and the latter because he has to represent the landforms adequately in the form of a map. (Hoffmann and Louis, 1968-1975). Nevertheless, the importance of geomorphological knowledge for mapping purposes is worth elaborating upon, since not all mapmakers have a clear view of the benefits, nor are all geomorphologists aware of what they may contribute where the interest of the mapmaker is concerned.

The ground surveyor, photogrammetrist or cartographer, in representing the earth's surface - and thus landforms - in his maps, is engaged in mapping both planimetric detail: rivers, coastlines, etc. and relief features. A knowledge of the characteristics of the landforms involved and the effect of seasonal variations, secular changes, tidal range, etc. will enable the mapmaker to produce a map of better quality. This means he will acquire a better understanding of what he is actually doing, he maps more specifically as he is familiar with the characteristics of the features to be mapped and able to evaluate their importance and nature.

It will be evident that the use and interpretation of topographic maps, though definitely less revealing than aerial photo interpretation, is an essential part of geomorphological studies. Therefore, accurate and vivid representation of landforms on maps is of the utmost importance for the geomorphologist. Good contacts between mapmakers and geomorphologists are, therefore, mutually profitable. (Verstappen, 1982).

2.2 The Terrain as seen in the Field and on Aerial Photographs

The representation of landforms on topographic maps is not as simple as initially envisaged. Inaccuracies inherent to the methods of survey and mapping have a distinct effect on the precision of the ultimate cartographic results, whereas, a measurement which is too precise may lead to unsatisfactory visualisation of the relief which is detrimental to the legibility of the map. (Phillips, et al., 1975; Potash, et al., 1978). The matter is elaborated upon in section 2.3.

The fact that the impression of the relief gained in the field depends on the position of the observer and his angle of view, and that the relief seen in aerial photographs, either stereoscopic verticals or obliques, also deviates considerably from reality, complicates the mental exercise of imagining the precise terrain configuration on the strength of the picture given in the map (Imhoff, 1950, 1965/82). Further this section deals specifically with the discrepancies between the landforms as they are in reality and as they present themselves to an observer in the field or in the air (Jenschor, 1978).

In plainlands, perspective distortion may bias the terrain appreciation of the observer on the ground. For example, it may obscure the parallelism of certain features such as low dune ridges. It also causes visual compression of the distant terrain, thus impeding the estimation of distances in these parts, particularly in the line of view. The estimation of distances is facilitated in the middle and foreground by the apparent size of known objects, such as houses and trees. This factor lessens in importance as the distance increases and the size of these objects diminishes. These perspective distortions also affect sinuous features, such as rivers and indented coastlines which appear unduly curved when viewed lengthwise, and scarcely reveal their sinuosity when viewed at a right angle. Vegetation and other groups of vertical objects may hide substantial parts of the terrain and sparse vegetation may give the impression of a fair denseness. This is because the foremost trees join up with the more remote ones and their different distances to the observer are not readily seen. A higher viewpoint, e.g., from a tower or a low-flying aircraft gives a more realistic picture of the ground cover.

In areas of considerable relief slope angles tend to be overestimated when faced from below, and underestimated when viewed downslope. Only those slopes which are perpendicular to the line of sight can be correctly appreciated - at least when the atmosphere is transparent. In case of hazy conditions, the various slopes seen in the same direction and partially covering each other cannot be properly distinguished and are consequently visually merged into one single slope of less gradient. Another common mistake is that sub-horizontal crestlines seen in the mid or far distance are considered flat, such as terraces, planation surfaces and structural surfaces. The recognition of such surfaces, if occurring approximately at the same altitude as the observer and thus perspectively shortened, is sometimes facilitated by the greater apparent density of the vegetation occurring there as compared with that on the slopes. Other terrain objects, such as roads, fences and telephone lines may accentuate relief differences which may have otherwise escaped attention. Fig. 2.1 gives an example from Botswana.The field pattern contributes substantially to

Fig. 2.1 Track accentuating the scarp marking an ancient (920 m) lake shore in northern Botswana.

the visualisation of the relief. Variable factors, such as the angle of incidence of sun rays, snow cover and seasonal changes in vegetation also affect the relief perception of the observer in the field. The visibility of the horizon, which marks the altitude of the observer, is, in mountainous terrain, a great help in estimating the height of mountain tops. When obscured by clouds or haze, the height of nearby peaks is often underestimated as compared to those at a greater distance.

When the terrain is seen vertically from an aircraft, a broad overview is obtained and almost no parts are hidden behind others as is normal in the case of terrestrial observation. However, the view angle combined with the great distance is unfavourable for a proper appreciation of the relief. The relief impression obtained from vertical aerial photographs is enhanced, however, by the effect of shadows, and in many instances also through (sub)horizontal linear features of natural or cultural origin acting as form lines and by drainage lines, etc. The airphoto of fig. 2.2 at a scale of 1:9,000 demonstrates the effect on the relief visualization brought about by the subparallel bunds of paddy fields in a hilly area near the town of Lembang, Western Java, Indonesia. The vertical airphoto of fig. 2.3 (1:10,000) illustrates the enhancement of relief impression in a part of Cyprus, caused by outcropping sub-horizontal geological beds with the associated banding of the vegetation and by the pattern of field boundaries of the non-irrigated fields characterizing that area. The (near) absence of fields on the steeper slopes or scarps where scrub prevails also adds to the vivid relief impression as do the shadows cast by scarps, stone walls and hedges.

When partly overlapping, vertical aerial photographs are available they can be studied stereoscopically and relief can be seen three-dimensionally. However, the apparent slopes seen in the stereomodel deviate substantially and are usually considerably steeper than the true terrain slopes. Further distortion is caused by the radial displacement resulting from the fact that an aerial photograph is a central projection of the earth's surface. Radial displacements increase towards the edges of the photograph and give rise to an oversteepening or even overturning and shortening of the slopes turned away from the centre, whereas the stereoscopic oversteepening is slightly reduced for slopes facing the central point of the photograph if their vertical dimension surpasses the so-called stereo threshold. This threshold only precludes the detailed study of micro relief if aerial photographs are used, but becomes a severely limiting factor when images taken from orbital altitudes are used.

Optical relief illusion and even apparent relief inversion may result from the illumination of the terrain and by the colours of its photographic image. Shadows enhance the relief perception when they are cast towards the lower right (as if the sun were in the upper left). If the shadows point in the opposite direction, a pseudo-relief is impressed upon the observer which may lead to the perception of an erroneous relief inversion. Colours also may suggest height differences: reddish tints giving the impression of higher ground than green and particularly blue colours. This is a long-established technique used by cartographers in the selection of colours for altitude zoning. It may complicate a correct perception of minor relief differences particularly when coloured infrared aerial photographs are used. The relief impression of digital (e.g. Landsat) imagery can be optimally enhanced for the purpose of topographic mapping (Donker and Meijerink, 1977).

It will be evident from the aforesaid that even a well-trained observer will not always find it easy to 'translate' the terrain configuration as observed from his viewpoint and angle into the relief actually occurring. This adds to the difficulties in comparing this relief with its presentation on a topographic map.

2.3 The Terrain as depicted in Topographic Maps

It is obvious that the aim of producing a good topographic map can only be achieved by a proper depiction of the terrain which results from skilled and dedicated measuring, combined with a well-developed feeling for cartographic realization. (Peucker, 1970; Bosse (Ed.), 1978). The cartographic image presented in a topographic

Fig. 2.2 Vertical aerial photograph, scale 1:9,000, illustrating the role of the low paddy field bunds in enhancing the relief impression in a hilly part of Western Java, near the town of Lembang.

map should not be simply an abstract geometric one, but should give a concise and legible picture of the terrain while maintaining the accuracy of the map content and an adequate degree of detail (Brandstaetter, 1941, 1957; McDonald, 1963).

An advantage of photogrammetric techniques is that all parts of the terrain are equally visible and can be accurately mapped irrespective of the accessibility. This results e.g. in more reliable contours and in a better representation of the terrain in swamps and marshy areas than terrestrial survey permits.

The vertical dimensions of terrain forms are visualized by contour lines at a selected interval. Their spacing is a measure for slope steepness: divergence of contours indicates a decrease of slope angle and convergence an increase. Where the spacing of contour lines of a slope varies in the slope direction, concave, straight and convex parts of the slope can be pointed out.

In order to facilitate the legibility of the pattern of contour lines the most important ones (usually every tenth contour) are thicker. Minor irregularities of a slope, the vertical amplitude of which is less than the contour interval and which would thus escape cartographic visualization, may be indicated by using intermediate (dashed) contour lines wherever appropriate (see e.g. the map of fig. 2.7). Isolated small hills may be marked on the map by the same means, but only a limited and well-contemplated use of this should be made.

A tiny downslope arrow is sometimes added to contour lines of shallow closed depressions to avoid mistaking them for minor rises of the terrain. Distinct breaks of slope may also be mapped separately by way of a dashed line. Modern mathematical approaches to contouring have problems of their own (Peucker, 1980).

On the basis of the contour line pattern only, it is not always easy to tell whether a contour line is higher or lower than the adjacent one. Therefore, their altitude is indicated at several suitable places, and spot heights may be added to facilitate map reading (Baldock, 1971; Brod, 1979 and Toepfer, 1974).

Rapid visualization of the relief all over a map sheet is only possible by the juxtaposition of the contour lines and other topographic information, particularly the drainage pattern. Hill shading is also important in this respect (Castner and Wheat, 1979). The merit of contour lines for relief representation lies not only in their being a systematic and geometrically satisfactory procedure, but even more important, that they result in a good visualization of the natural terrain by way of their coherent patters (Engelbert, 1963; Erb, 1950 and Toepfer, 1964). Any sizeable curve or indentation of a contour line can be traced in the adjacent contour lines because it is associated with a geomorphological phenomenon the vertical dimension of which largely exceeds the contour interval. Furthermore, this geomorphological phenomenon has often evolved under the influence of processes operating in the slope direction, either upslope (e.g. headward erosion of valleys) or downslope (e.g. the deposition of a scree slope or alluvial fan).

The wide range of terrain forms occurring in nature and depicted on topographic maps can be explained by the geomorphologist from the variety of endogenous and exogenous, causative factors and their different degree and period of activity. The differences in terrain form

Fig. 2.3 Vertical aerial photograph, scale 1:10,000 showing outcropping geological strata, vegetation and field patterns enhancing the relief visualisation in a part of Cyprus.

thus produced, are often so noticeable that conclusions can be drawn from a good topographical map concerning landform types, lithology, etc. Inversely, it is possible, once a proper idea of the nature of the landforms has been obtained by the geomorphologist, to assist the mapmaker in producing a better picture of the terrain configuration. (Carlberg, 1958; McDonald, 1963; Mietzner, 1964; Pannekoek, 1946, 1962; Elvhage, 1980).

The characteristic shapes and profiles of terrain slopes, the pecularities of river valleys, etc. in a given morphological situation can be more properly mapped once the cartographer's attention has been drawn to these characteristics. Examples are: the forms resulting from the alternation of horizontal hard and soft beds in canyons and the variations in valleywidth where a river cuts through alternating hard and soft (folded) beds.

In lowland areas where contour lines are scarce, an occasional scarp (terrace edge or limit of a riverbed) may occur, but the representation of the terrain configuration in plan becomes dominant. The phenomena concerned are often mainly of fluviatile, deltaic, littoral or other origin and have gone through numerous changes since the formation of these plains. The actual terrain configuration still is strongly affected by former positions of riverbeds, coastlines, etc. (Keller, 1975; Paris, 1974; Polcyn and Lyzenga, 1975), as evident from the distributional pattern of vegetation, hydrographical features and land utilization. Proper mapping of this situation substantially adds to the value of the topographic maps, not only from a geomorphological point of view but also for the legibility of the map and the assessment of the terrain.

Two fields of mapping problems inherent to many lowland areas merit special attention. First, there is the matter of the rapid changes of many fluvial forms and lowland coasts (Emplaincourt and Wielchowsky, 1974). The recording of continuously changing river channels requires repeated measurements. Sequential maps are used when it comes to the execution of engineering works such as bridge construction or harnessing the river for purposes of flood control. The same applies to the rapid changes of the littoral zone, the mapping of which is not only of prime importance for navigation in the coastal waters, but which is of great concern to geomorphologists and coastal engineers involved in coastal protection, harbour construction, etc. Secondly, there is the difficulty encountered in plotting clear-cut boundaries in natural terrain which is by far the most serious problem in areas of faint relief. When mapping a riverbed and associated topographic details, the mapmaker should be able to distinguish between a) those parts next to the actual bed which are more or less stable, b) the areas which are flooded in the wet season, where seasonal and secular changes are important, and c) those (e.g. braided) parts of the floodplain where the changes are so continuous that he should content himself with mapping the characteristics of the pattern rather than with the details actually observed during the survey. Therefore, he should be familiar with concepts such as bankful stage and minor riverbed and be able to distinguish them in the field and/or on aerial photographs. In shallow coastal areas, particularly in tidal flats and estuaries, the effect of the tidal range on the position of the waterline in plan is considerable. If the mean high tide level is considered to be the coastline, as is common practice in many countries, the mapmaker should know how to recognize and map it. (Lobeck/Tellington, 1944; Madden, 1978; McCurdy, 1947; Gierloff-Emden and Wieneke, 1978). Precise levelling must be carried out, and bench-marks have to be established along the shore in order to determine the tidal planes. Such bench-marks are also of use when it comes to precise mapping of coastal changes.

Fig. 2.4 illustrates how a geomorphological input in the training of surveyors (and photogrammetrists) may contribute to an improved accuracy of mapping shoreline features which are often poorly defined. The example relates to the sandy cay (Njamuk Kecil Island) situated on top of a tiny coral reef in the Bay of Jakarta. The geomorphologist indicated to the trainee what had to be mapped; i.e. (i) the high tide mark on the reef platform with isolated coral blocks and mangroves at a greater distance from the island; (ii) the border of the sandy cay partly covered by grass and scrub; and (iii) the higher, older part of the sandy cay with a tree vegetation and partly bordered by a distinct scarp. This map (A) at a scale of 1:8,000 was prepared using a compass and a pedometer. The trainee made a map using a plane table depicted in (B). His measured lines, stations and bearings are indicated in (C) and the precision thereby achieved is demonstrated in (D) in which the two maps are superimposed. The geomorphologist then mapped the geomorphological features of the island (E) indicating the reef platform, the shingle rampart at the exposed northern extremity of the island (which is formed and subsequently modified by high winds/waves), the gradual 'ringwise' growth of the sand cay at the southeast leeward side, and the older, higher core of the island. The trainee then mapped these features on his plane table map (F). He was able to identify not only the main features to be mapped on this particular sand cay, but also understood their mode of origin which facilitated the mapping of other coral islands of the same type.

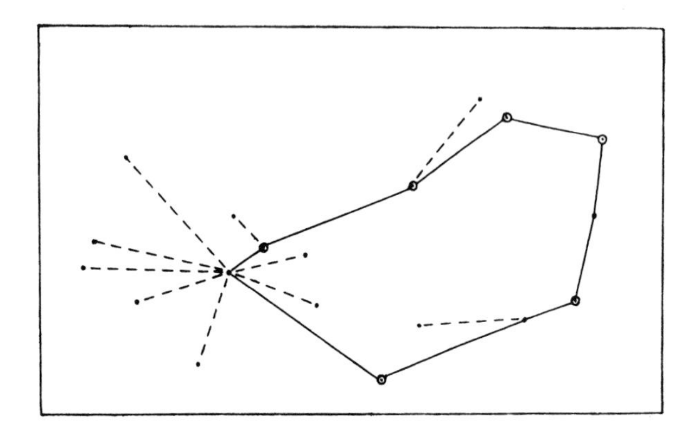

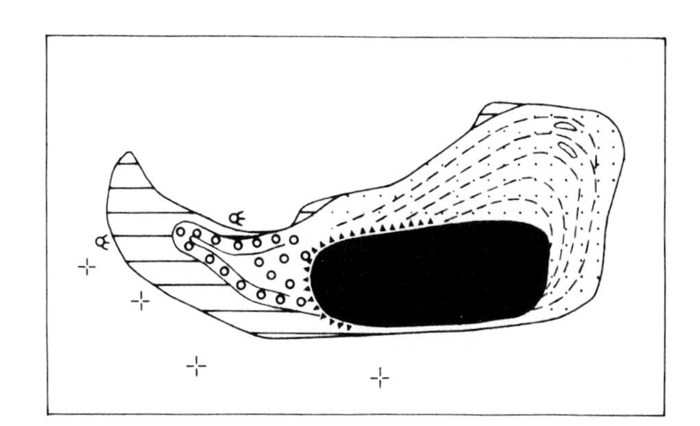

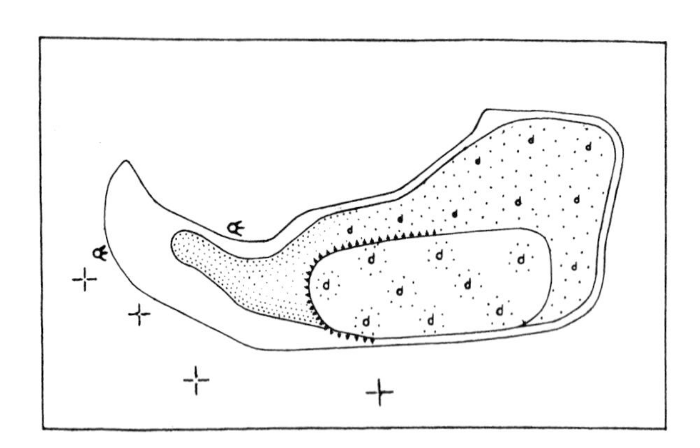

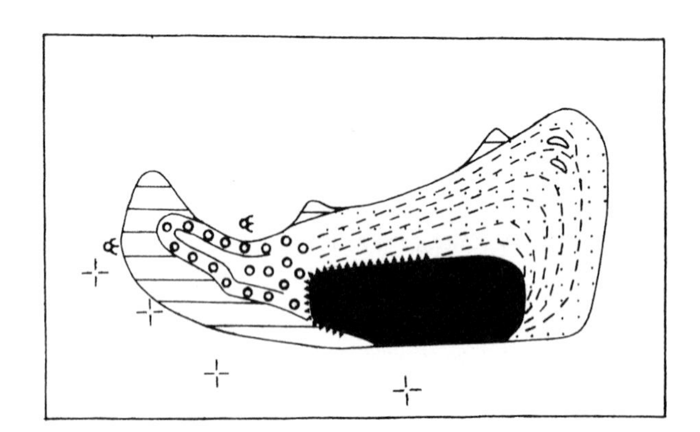

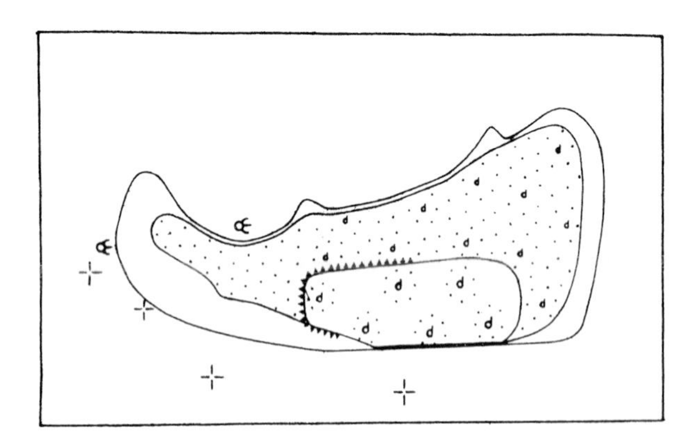

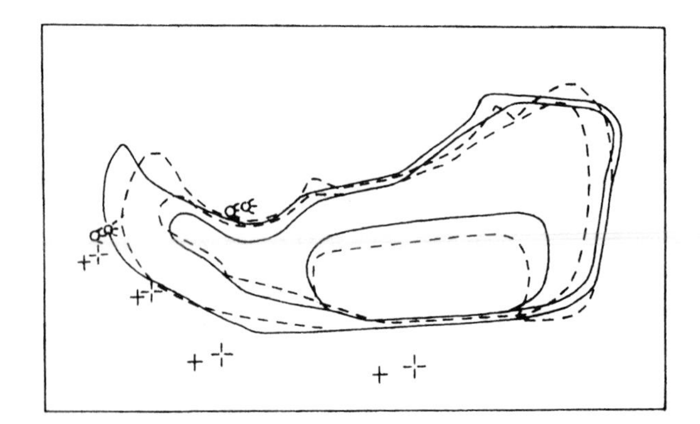

Fig. 2.4 The coral cay Njamuk Besar island in Jakarta Bay, Indonesia at a scale of 1:8,000. An example of cooperation between geomorphologist and ground surveyor. For explanation see text.

2.4 Geomorphology in different Types of Mapping

2.4.1. Generalities

The representation of the factual terrain configuration realised by a mapmaker is based on a firm footing of measurement facts. The question is therefore, how to represent the situation in the most convenient and reliable way, and in case of insufficiently mapped areas, how to represent it best schematically. (Karaszewska and Lopatto, 1964; Mietzner, 1964; Pietkiewicz, 1964; Sarkisjan, 1963; Schmidt, 1964; Seeler, 1964).

This section deals with the importance of a sound knowledge of the geomorphology of the terrain, in ground survey, photogrammetric survey and the cartographic phase of the work. It should be understood from the onset that in cases of detailed mapping as well as small-scale mapping, the interest of the mapmaker is mainly in the present landform and further in rapid changes, which either render mapping of certain features a futile effort, or have to be taken into consideration during map revision.

2.4.2 The Ground Surveyor

The ground surveyor interpolates the location of contour lines on his map between the traverses that he has measured, or sketches them e.g. during plane table mapping. A large part of the lines he draws are not actually measured. The result is greatly influenced by the density of the traverses, although much depends on his drawing abilities, see fig. 2.5.

It is sometimes possible to trace the areas mapped by the individual surveyors through different drawing styles. In some extreme cases difficulties are even encountered in fitting the contour lines of two adjacent areas. A conception of landforms is in this case a definite advantage to the surveyor. The common practice of indicating ridges by more or less rounded shapes of contour lines, and valleys by angular contours, though in many cases correct, is not always justified.

The slope of an active strato-volcano is often characterized e.g. by undissected, rather flat strips, separated from each other by steep and narrow gullies. In such a case, the contour lines should be drawn as fairly straight lines, the slope only being interrupted by minor indentations, the narrow gullies. Another example of this type is the mapping of minor circular depressions, such as sinkholes in a limestone area. Geomorphology can be especially helpful when mapping areas such as swamps, etc., which can only be entered with great difficulty. The sketched map can then be greatly improved by geomorphological considerations.

2.4.3 The Photogrammetrist

The photogrammetrist maps the contour lines to their full extent irrespective of accessibility, with the aid of the floating mark. So there seems little or no room for the inaccuracies inherent to ground survey. However, there are other aspects to the problem. In some extreme cases, the photogrammetric map cannot be fully accurate. Narrow, steep-sided river gorges and rocky slopes of extreme steepness cannot be adequately represented by contour lines, so rock symbols replace them or are added. Another extreme case is formed by areas of very gentle slopes (up to 3° or 4°). Then the pattern of contours can only be accurately depicted if the prevailing landforms are clearly understood. Furthermore, it should be realised that an accurate map does not always give a correct impression of the terrain forms. This is the case where small conical (e.g. limestone) hills or shallowly incised river valleys occur on a gently downsloping surface. A very irregular relief is suggested at those localities where a contour line intersects the slope, whereas areas with exactly the same type of topography but not intersected by a contour appear flat and non-dissected. This type of terrain is often misinterpreted, even by cartographers: hill shading is then added to the map but only in the zones adjacent to the contour lines.

Quite another aspect is the fact that the great detail of photogrammetric maps makes them less easy to interprete by the average map user. It is therefore, usually necessary to generalize the contours to some extent in order to get a more coherent pattern of contour lines, and thus achieve a better legibility of the map.

2.4.4 The Cartographer

The cartographer is confronted with the latter problem to a far greater extent when compiling small-scale maps from existing topo-sheets (Habel, 1971). The reduction of scale necessitates him to simplify the lines on the map and to omit or combine certain features. This generalization (choice and simplification) should be carried out in such a way, that the characteristics of the landforms, coastlines, etc. remain clearly visible. Fig. 2.6 gives an example of generalization of two rivers in France. Cartographic generalization is thus not merely the haphazard elimination of details, but rather the selection of relevant information and the omission of trivialities. In simplifying line patterns, care should be taken that only useful characteristics are preserved or even emphasized. This is a criterion of good cartography at any scale: neither the inclusion of abundant irrelevant information, which makes the map illegible; nor undue smoothing or rounding off of topographic forms.

When generalizing natural terrain forms in plan and in relief, the guidance of a trained geomorphologist will

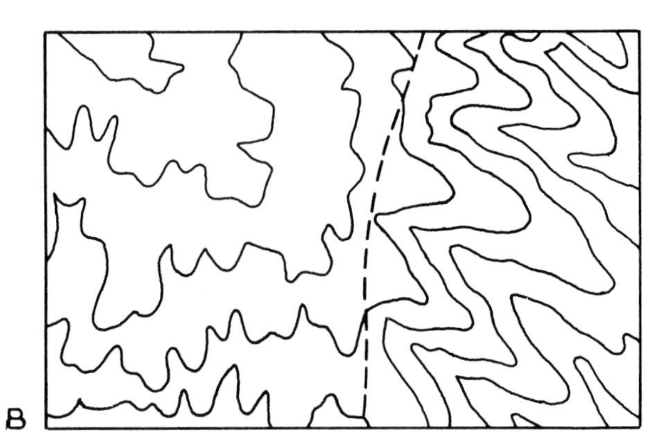

Fig. 2.5 Patterns of contour lines obtained by terrestrial survey as adversely affected by an unduly low density of measurements and over-generalization (A) and by differences in cartographic style and ability of two surveyors (B). Courtesy: A.J. Pannekoek.

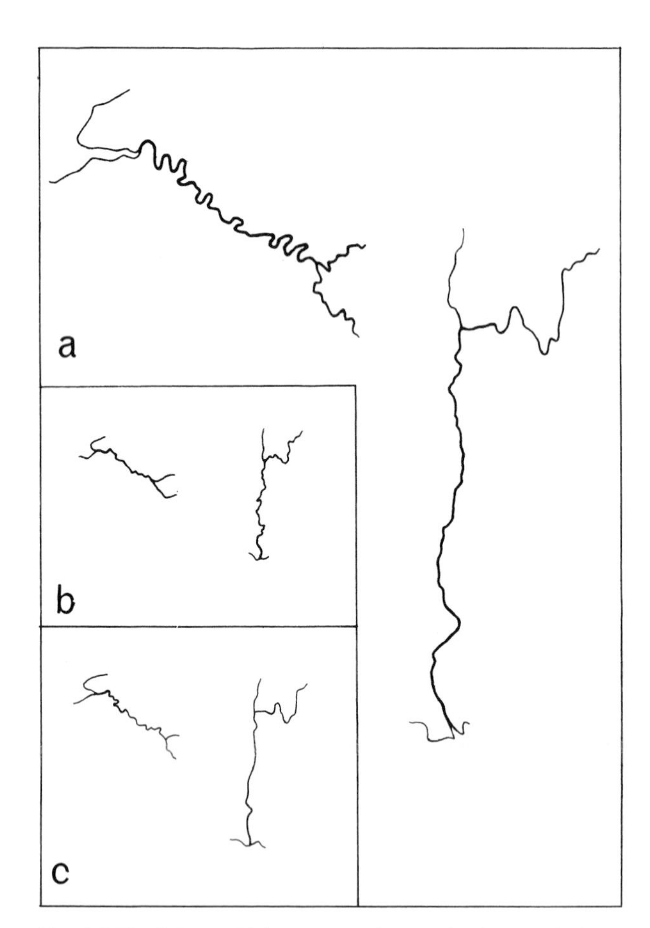

Fig. 2.6 The Seine and Rhone rivers characterized respectively by meandering and straight forms (A) as depicted improperly (B) and properly (C) generalized in existing atlasses. Courtesy: A.J. Pannekoek.

improve the ultimate result (Knorr, 1965). It is important to the cartographer whether the small-scale manuscript map is obtained by photographic reduction or from small-scale (e.g. Landsat) imagery, or by means of a pantograph.

In the first case generalization must be completely effectuated whereas in the second case part of the generalization is done mechanically by pantographing, though not necessarily in the most appropriate way. A simple but characteristic cartographic representation is always preferable to one showing too many minute details. This fact is also mentioned in an English instruction book for draughtsmen and artists: 'By jumping in and out of promontories and dipping into every estuary and little bay, you are apt to make a most wobbly caricature of our island. If, on the other hand, you draw only the chief curves and straight lines, noting how they differ from one another, you will get an exaggerated but unmistakable likeness'.

Geomorphological knowledge is indispensable for cartographic generalization because a conception of the characteristics of the features concerned (coasts, rivers, mountains, etc.) is essential. The same is true when preparing hill shading, when choosing contours and their intervals, and when making sketch maps.

The importance of the study of landforms for topographic mapping may be sufficiently evident from the above consideration. But since not every mapmaker, whether ground surveyor, photogrammetrist or cartographer can be expected to have the thorough knowledge of these specific aspects of geomorphology which are relevant to his work, some of the larger survey departments may require the full-time assistance of a geomorphologist to ensure proper representation of landforms on the maps which they produce. The geomorphologist should draw the mapmaker's attention to the characteristics of the features occurring in the area to be mapped, emphasizing their form and the validity and permanence of certain natural boundaries such as riverbeds and coastlines. In order to be effective, the geomorphologist should study the area concerned and subsequently instruct the mapmaker prior to the beginning of the mapping programme. This is applicable if detailed mapping is concerned. In the case of small scale maps produced by the cartographer by way of scale reduction and generalization of detailed maps, some advice on landform characteristics might be given before the work starts, but more important in this case is a thorough scrutinizing of the final manuscript before it is prepared for final cartography and printing. The geomorphologist should carry out or at least supervise the final generalization, when the possibilities of terrain representation have become clear. It is, of course, essential for him, that the mapmaker has a basic knowledge of geomorphology in order to understand the 'language' of the geomorphologist and to know the nature of the problems involved. Equally important is that the geomorphologist understands the problems the mapmaker has to face and the mapping instructions he has to apply.

2.5 Geomorphological Processes and Map Revision

It will be evident from the aforesaid that in the application of geomorphology in mapmaking, emphasis is on the correct representation of landforms both in plan and in relief by means of contour lines and/or hill shading. Geomorphological processes play a role only through the forms which they have created. The U-shaped cross-section of valleys in alpine regions, for example, reflects glacial scouring of the past. Similarly, an abandoned riverbed in an alluvial plain illustrates the process of river diversion following overtopping or breaching of the natural levees which low-land rivers tend to build up. The asymmetry displayed in river bends, where characteristically the outward, concave bank is steep and the inward bank gently sloping, indicates lateral sapping by the river water. Many other examples of this kind could be given, but in all these cases a correct representation of the landforms produced will suffice to elucidate the processes that have caused them - which are irrelevant to the mapmaker.

When revising a map, one has to consider the possibility that the accumulative effect of the continuous or discontinuous processes during a given interval may have resulted in significant changes in landforms which makes them mappable, and thus mapworthy on topographical maps even at scales of 1:50,000 or smaller. However, the progressive effect of the various geomorphological processes is normally slow and almost imperceptible and the changes thus brought about are mostly within mapping accuracy. As a result the contour lines and the configuration in plan of natural features, as depicted on the earlier map edition, can often be used for the revised map. This is not the case if drastic changes such as a major landslide or a volcanic eruption, have occurred. Mappable changes of configuration in plan may result from matters like coastal accretion, fluvial erosion, etc. In his appreciation of the topographic situation, the mapmaker should be aware of the danger of being misled by, for example, tidal differences existing in the case of a coastal area, or by differences in discharge and thus, of water level and width of the river when mapping a riverbed. Thorough instructions given by a trained geomorphologist, strict adherence to the mapping prescriptions in use, and correct, consistent interpretation of these will assist the mapmaker in discriminating between apparent and actual changes of terrain forms of mappable magnitude. This applies both to the ground surveyor and the photogrammetrist. If the changes are very rapid, for example, as is the case with some braided rivers, it is more important to indicate the characteristics of the riverbed and its main, possibly somewhat more stable, parts than to map all minor, variable details. If, however, minor details are more or less stable, in position, such as some glacier crevasses which are the reflection of irregularities on the ice-covered rock surface, they may merit precise mapping. Thus the surveyor should be properly instructed about the spatial (in)stability of the terrain which he is mapping.

Fig. 2.7 examplifies how major changes in terrain configuration may occur following a volcanic eruption and how these changes are reflected in the pattern of contour lines. An example is the Rinjani Volcano (3726 m) on the island of Lombok, Indonesia. The volcanic activity of the mountain is not at its top but near the shore of a fairly large lake of tectonic origin situated to the west of the main cone at an altitude of 2008 m. Fig. 2.7a is a part of the 1:25,000 topographical map dating from 1925 showing the G. Baru (now known as G. Baru I) cinder cone with solfatara in its crater and the lava flows to the west which emerged from it and terminate in the lake. The depth of the lake is also indicated in meters. Eruptions (probably) centered in the G. Baru I date from 1884, 1900-1901, 1906 and 1915. A new cinder cone, G. Baru II to the northwest of G. Baru I, arose during the eruptions of 1941-1944 and lava flows spread west and north of it. The resulting topographic changes are depicted in fig. 2.7b which is a part of a 1:25,000 photogrammetric map based on aerial photographs of

1946. Comparison of the two maps with the aerial photograph of fig. 2.8 and careful study of the contour lines (interval 25 m; in the area of the new lava flows 12,5 m) gives a clear picture of the volcanic events and how they have affected the terrain configuration and thus, the contour pattern (Milius, 1954).

Fig. 2.9 is another example of a major change in terrain configuration, leading to revision of the contour lines when reprinting the topographic map concerned. It relates to a catastrophic landslide that occurred in the Buonamico Valley, Calabria, Italy in February, 1972. The old contour lines as appearing on the existing topographic map 1:25,000 based on aerial photographs of 1955 are indicated by dashed lines and the new contours (aerial photographs of 1975) by full; (courtesy Prof. Dr. P.Ergenzinger, F.U. Berlin). The new position of the river and the lake ponded up by the slide are also indicated in the figure. The photograph of fig. 2.10 pictures part of the slide.

Changes in the position of rivers and coastlines are common and numerous examples can be found by careful comparison of older and newer editions of topographic map sheets. Rivers are usually more stable upstream than in the lowland and braided rivers may change so rapidly that substantial changes may occur in the period elapsed between the photographic mission and the plotting of the photogrammetric map. Mapping of the braided characteristics rather than the detailed position of various forms should then be attempted. Rocky coasts, of course, change considerably less than alluvial, lowland coasts. Apparent coastal accretion may occur where vegetated coasts (for example, mangrove) occur because in such cases the seaward limit of closed vegetation is considered the coastline, regardless of its position with respect to the mean high tide level, which is considered the coastline in many countries where there are no mangroves. A seaward spread of the mangrove may suggest coastal accretion, which in reality is not, or not to this extent, substantiated by sedimentation.

2.6 Generalization of Settlement, Land Use and Other Topographic Patterns

In many areas a significant and distinct correlation exists between geomorphological features, the distributional pattern of agricultural and other land uses, the location of settlements and roads, the natural vegetation, etc. In numerous tropical lowlands, for example, it can be observed that settlements and homestead gardens have a strong afinity to sandy beach ridges and loamy or silty natural levees of rivers. The same applies to the roadnetwork. Paddy fields are often associated with the swales between the beach ridges and the basins (back swamps) between the present and former natural levees. The lowest parts of these areas may be occupied by swamps or even shallow lakes or lagoons. In other types of terrain, different but likewise significant correlations of this kind often exist. The natural or semi-natural vegetation in mountainous areas may show the effect of vertical zoning and of exposure to the sun. It may also clearly show the effect of avalanches by way of breaches (and other destructions) in the vegetative cover.

A good topographical map shows settlements and other features and reveals their relation to the terrain configuration (Sandy, 1977). In case of detailed maps, settlements can be shown true to scale in their proper form and situation. In such cases geomorphologists can do nothing for the mapmaker who simply maps the situation as he finds it. In case of medium to small-scale mapping, where generalization plays a more important part, the mapmaker should be taught how to generalize in order to emphasize the relation: settlement, etc. and terrain configuration. This is particularly rewarding if the settlements are depicted in their true form and not by dots or small circles. The matter is elaborated upon in sections 6 and 7. Fig. 7.1 gives an example of Central Java where the settlement patterns alone clearly indicate the lowlands along the south coast with numerous beach ridges, interrupted in the centre by the Karangbolong limestone hills and bordered to the north by the more thinly populated south Serayu Mountains. In the north and west settlements are preferably located on the levees of the Serayu River and its tributaries.

Fig. 2.11 gives an example of the east coast of Sumatra where mangroves and other forests border the coast of the Malacca Straits and are succeeded further inland by a belt of fish ponds, which in turn gives way to paddy fields where the salinity decreases. Villages and homestead gardens are found along the main rivers; rubber and other crops on the flat interfluves, and forest, tall grass and shifting cultivation in the hills.

The settlement and land use patterns mentioned above are dynamic features that change with the environmental conditions occurring in the area. Fig. 6.25 a,b and c, show the land use and settlement patterns in the delta of the Cimanuk River, Western Java, Indonesia, as they appear on topographical maps of 1857, 1940 and 1969 respectively. It is evident from the changes depicted that settlements and paddy fields are abandoned when sea water penetrates, thus increasing salinity,when the discharge of a distributary decreases (1940, 1969; northwest) and immediately following the sedimentation and freshwater conditions around a new distributary (1969; east).

Fig. 16.7, a vertical aerial photograph of a valley in the Alps illustrates how the upper forest limit is attacked and lowered in the course of the years, shaped by avalanches,

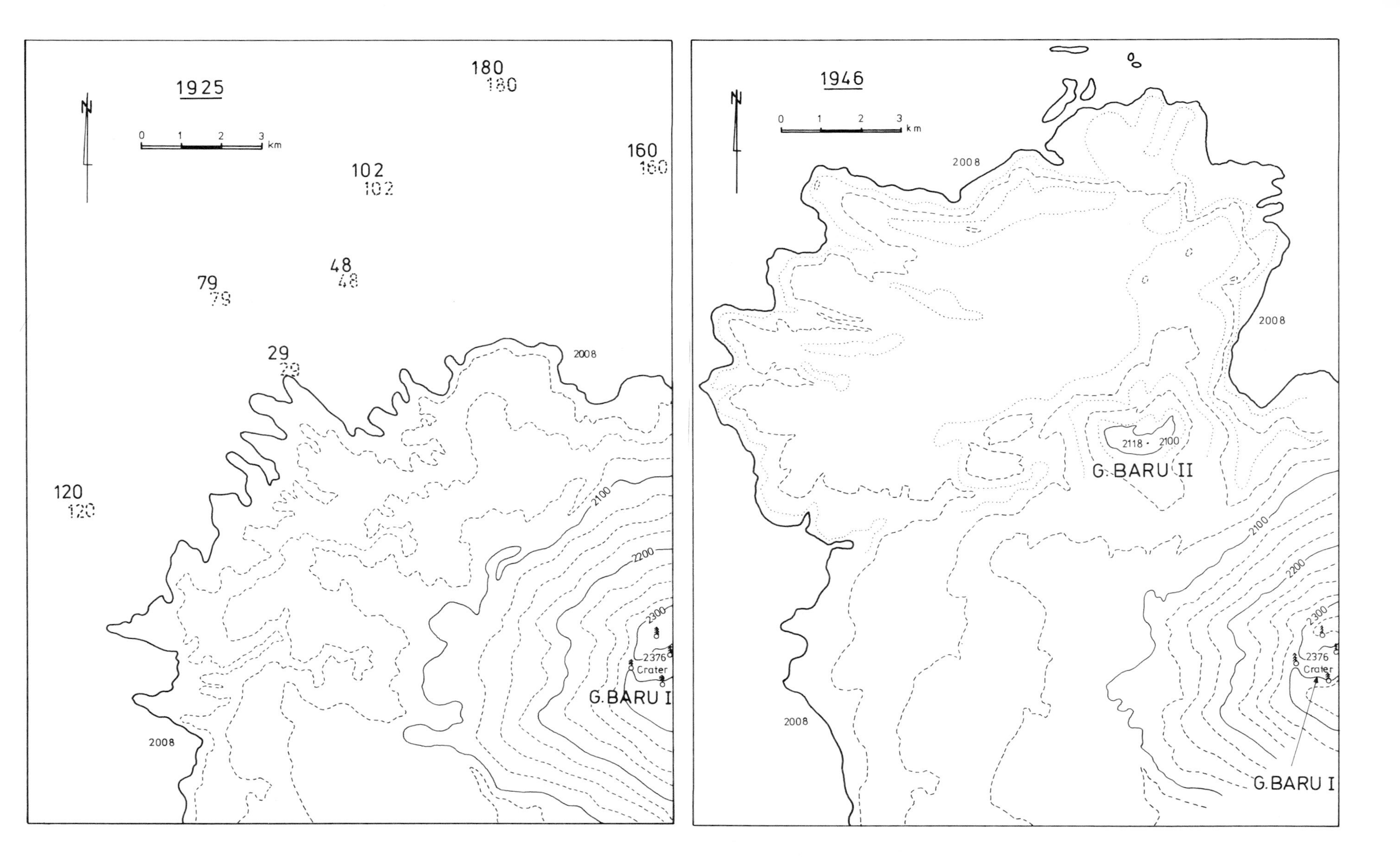

Fig. 2.7 Changes in terrain configuration due to a volcanic eruption of the Rinjani Volcano, Lombok, Indonesia as evident from a comparison of 1:25,000 contour maps dating from 1925 and 1946 respectively. Compare with fig. 2.8. Courtesy: G. Milius.

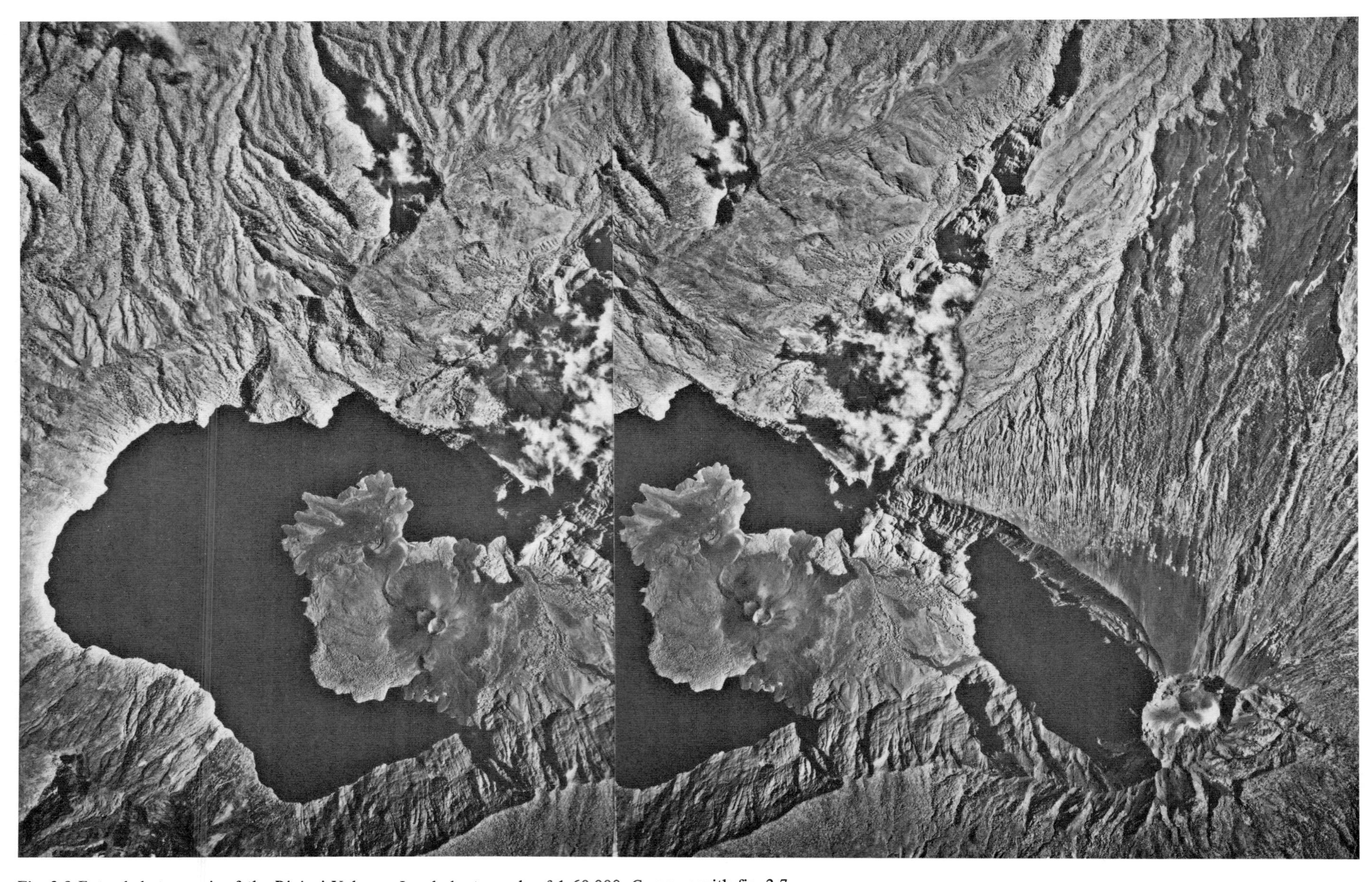

Fig. 2.8 Extended stereopair of the Rinjani Volcano, Lombok at a scale of 1:60,000. Compare with fig. 2.7.

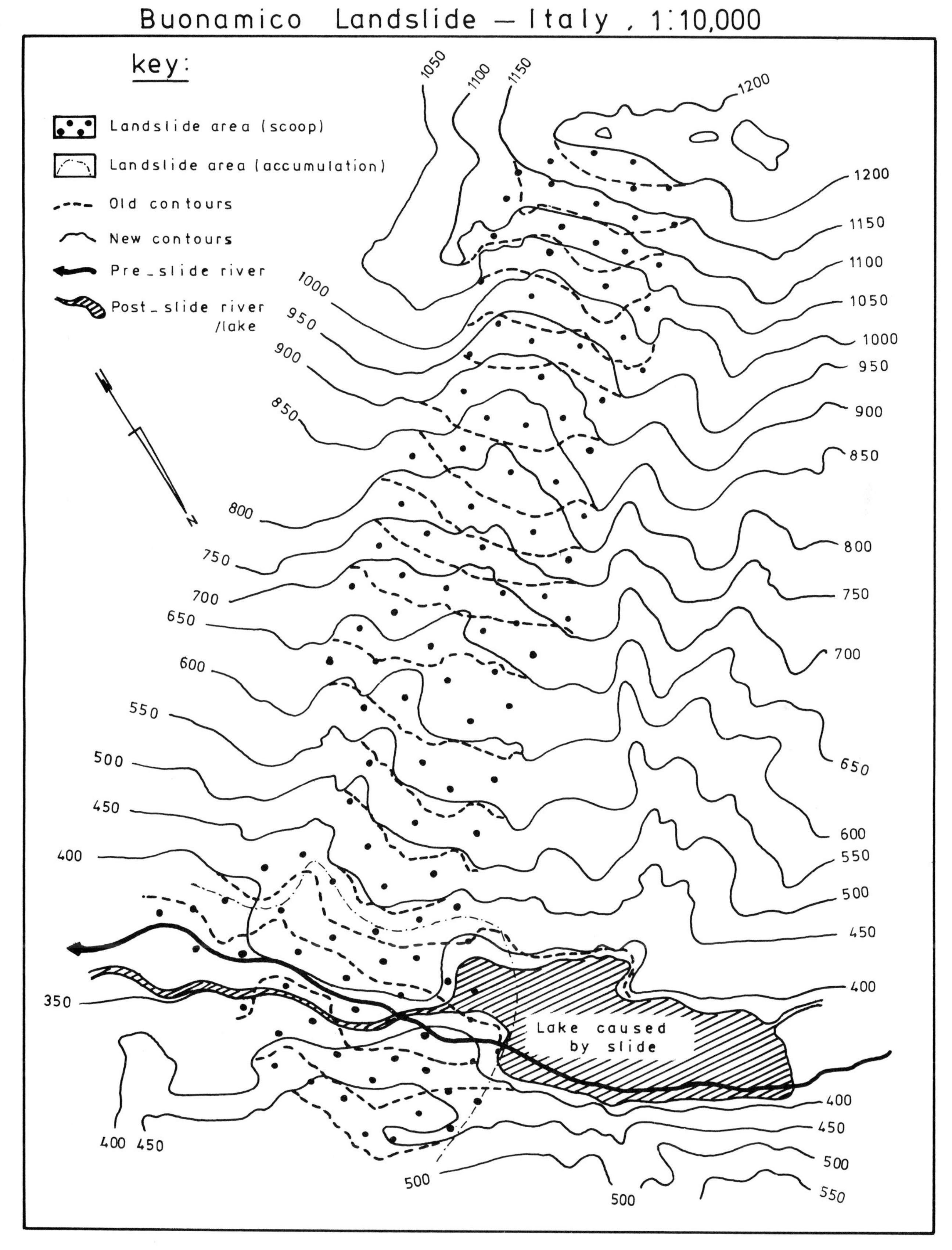

Fig. 2.9 Changes in terrain configuration due to a major landslide in the Buonamico Valley, Italy, as evidenced by a comparison of photogrammetric maps of 1955 and 1975. Based on data provided by P. Ergenzinger.

most of which have fairly well-established tracks.

2.7 Thematic (resource) Mapping

Geomorphology also plays a part in thematic maps (Verstappen, 1964), although in geological, soil, hydrological, vegetation and land use maps the topography, and thus landform, serves simply as a basis for the thematic information and should not lead to overloading of the map. Correct interpretation of a geological map does not only depend on a good conception of the geological structures depicted, but also requires an insight in the topography, which may have a substantial influence on the patterns of outcropping rocks. Likewise, the interpretation of a soil map requires understanding of the relief to readily understand the patterns of deep soil occurring on plateaus, truncated soils on the higher parts of the slopes, and colluvial soils further downslope, etc. Hydrological maps, apart from supplying information on surface and groundwater, geological formations occurring, etc. also may include data on the shape of river basins, density of dissection, and other matters related to relief. Vegetation maps clearly reflect the vertical zoning of various types of vegetation and the effects of the aspect of slopes in connection with their exposure to sun, rain or wind. Land use maps often show clear relations with the major landforms of an area. Therefore, landform and other geomorphological characteristics may be visualized in thematic maps, if not directly through contours, then at least by way of the patterns of the thematic information in the map.

While doing survey work and before the cartographic elaboration of the thematic maps, geomorphology is a factor to be considered. One has to be careful, however, that the relief observed, is correctly interpreted. In ground survey a fact to be considered is, that the relief impression gained is affected by the perspective of view and that when using aerial photographs, the stereoscopic relief exaggeration might result in an incorrect evaluation of relief amplitude and steepness of slope. (see section 2.2). When photo-interpretation is involved, as is usually the case in resource surveys, the role of geomorphology becomes particularly evident. Since landforms are among the most conspicuous features seen in the three dimensional image of the terrain obtained through stereoscopic study and their interpretation is thus a very direct, straightforward one, geomorphology has become a valuable aid in other earth sciences such as geology and soil science the subject of which is less readily seen in aerial photographs. Geomorphology has therefore, become a solid link between the various earth sciences.

Geological photo-interpretation is largely based on geomorphology, as the geological features such as rock types and structures, are usually not directly visible from the air. In most cases, their presence has to be deduced from related geomorphological phenomena. Although greytone (density) and vegetation elements are sometimes an indication for certain geological details, relief and drainage pattern are the most used indicators of geological features. Photogeology was once defined as 'the geological interpretation of the morphological outlook of a landscape as appearing on aerial photographs'.

The soil scientist engaged in photo-interpretation is also largely dependent on geomorphology as his object of study, due to the fact that the soil is usually covered by vegetation and the soil profile cannot be seen at all from the air. It is very fortunate that the boundaries of soil associations often coincide with, or have simple relations to, geomorphological limits.

There are, of course, other indications, but landform study certainly is basic to pedological aerial photograph interpretation. It is generally recognized in relation to soil survey that photo-interpretation becomes much easier if the terrain can be analysed and classified in physiographic units.

The hydrologist, when applying photo-interpretation will also heavily depend on geomorphology as will many others when engaged in thematic mapping on the basis of aerial photograph interpretation. The emphasis is always on landform, process, genetic or environmental factors or more commonly, on a combination of these four major aspects of geomorphology.

Fig. 2.10 Ground view of the site portrayed in the map of fig. 2.9 as it was in 1976 and showing the slide and the lake. Courtesy: K. Sijmons.

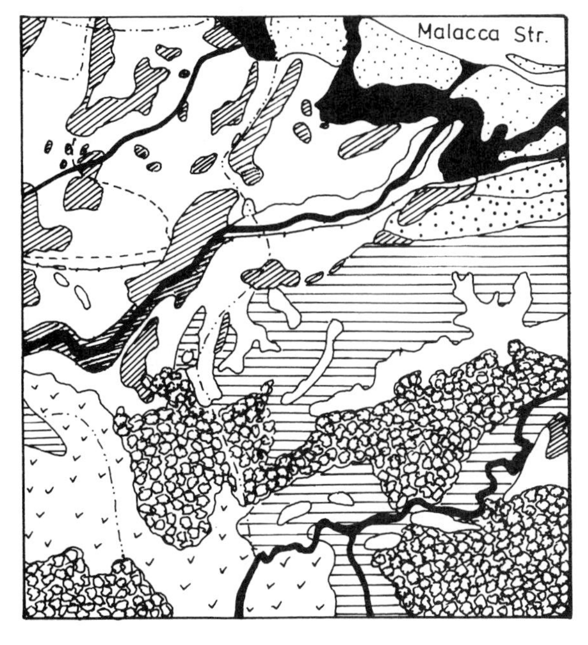

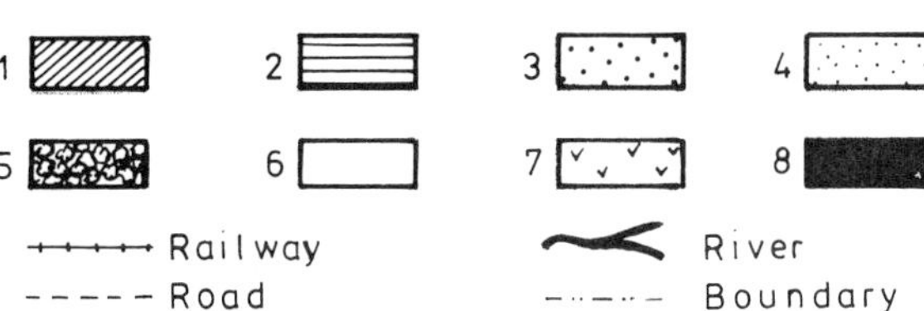

Fig. 2.11 Land use patterns and geomorphological terrain conditions in southern Sumatra, scale 1:50,000. Key: 1. settlements, 2. dry fields, 3. agricultural estates, 4. mangrove swamps, 5. secundary forests, 6. paddy fields, 7. sedges/grasses, 8. fish ponds.

References

Baldock, F.D., 1971. Cartographic relief portrayal. Int. Yrb. Cart.: 75-78.

Bosse, H. (Editor), 1978. Probleme der Geländedarstellung. Verh. 11., Arbeitskurs der D.K.V., Niederdollendorf, 401 pp.

Brandstätter, L., 1941. Das Geländeproblem in der Hochgebirgskarte 1:25,000. Jrb.f.Kartogr.: 5-23.

Brandstätter, L., 1957. Exakte Schichtlinien und topographische Geländedarstellung. Österr. Ztschr.f.Vermessungsw., Sonderheft 18, 94 pp.

Brod, R., 1979. Generalization–a fundamental process in cartography. Bull. Illinois Geogr. Soc., 21(1): 24-29.

Carlberg, B., 1958. Geographisch-morphologische Forderung an die Geländedarstellung. Kartogr. Nachr., 8:24-32.

Castner, H.W., and Wheate, R., 1979. Reassessing the role played by shaded relief in topographic scale map. Cart.J., 16: 77-85.

Donker, N.H.W., and Meijerink, A.M.J., 1979. Digital processing of Landsat imagery to produce a maximum impression of terrain ruggedness. ITC Journal, 4: 683-704.

Elvhage, C., 1980. An experimental series of topo-geomorphological maps. Geogr. Annaler, 62A: 105-111.

Emplaincourt, J.L.S., and Wielchowsky, C.C., 1974. Detection of shoreline changes from ERTS-1 data. Southern Geographer, 14(1): 38-45.

Engelbert, W., 1963. Studie zur Geländedarstellung in grossmassstäblichen Karten. Ztschr.f.Vermessungswesen, 88(1): 453-457.

Erb, G., 1950. Kaarteringsvoorschrift voor het topografisch opnemen en kaarteren. Publ. Top. D. Jakarta, 76 pp.

Finsterwalder, R., 1962. Topographisch-morphologische Kartenproben 1:25,000. Grundlagenforschung auf dem Gebiet der deutschen Originalkartographie. Tagungsber. u. Wiss. Abh. d. Deutschen Geogr. Tages, Köln: 259-275.

Gierloff-Emden, H.G., and Wieneke, F., 1978. Anwendung von Satelliten- und Luftbildern zur Geländedarstellung auf topographischen Karten und zur bodengeographische Kartierung. Münchener Geogr. Abh., 20: 1-33.

Habel, R., 1971. Zum Problem der Kombination von Reliefzonen mit anderen phys. geogr. Erscheinungen in kleinmassstäblichen Karten. Int. Yrb. Cart.: 90-92.

Hofmann, W., and Louis, H., 1969-1975. Topographisch-morphologische Kartenproben 1:25,000. G. Westermann Verlag, Braunschweig.

Imhoff, E., 1950. Gelände und Karte. Eugen Rentsch Verlag, Zürich, 255 pp.

Imhoff, E., 1965. Kartographische Geländedarstellung. W. de Gruyter Verlag, Berlin, 256 pp.

Imhoff, E., 1982. Cartographic relief presentation. 384 pp.

Jenschor, A. 1978. Gelände, Karte und Luftbild. Walhalla und Praetoria, Regensburg, 2nd Ed., 448 pp.

Keller, M., 1975. Aerial photography in the NOS coastal mapping division. Ph. Eng. R.S., 41(8): 1005-1011.

Knorr, F., 1965. Die Herausarbeitung der Landschaftsformen in der neuen topographischen Übersichtskarte 1:200,000 von Deutschland. D. Geod. Komm. Bayer. Akad. Wiss., Bd. 118.

Lobeck, A.K., and Tellington, W.J., 1944. Military maps and airphotographs: their use and interpretation. McGraw-Hill, New York, London, 253 pp.

Madden, J.D., 1978. Coastal delineation by aerial photography. Austral. Surveyor, 29(2): 76-82.

McCurdy, P.G., 1947. Manual of coastal delineation from aerial photographs. U.S. Hydrogr. Office, publ., 592, 143 pp.

McDonald, D., 1963. The presentation of relief on maps. Austral. Surveyor, 19(5): 303-306.

Mietzner, H., 1964. Die kartographische Darstellung des Geländes unter besondere Berücksichtigung der geomorphologische Kleinformen. Inst. f. angew. Geodäsie, Frankfurt, 121, 39 pp.

Milius, G., 1954. De jongste eruptie in het Rindjani gebergte. Publ., Geogr. Inst. DJATOP, Jakarta: 3-12.

Monkhouse, F.J., and Wilkinson, H.R., 1966. Maps and diagrams. Univ. paperbacks, 75, Methuen, London, Ch. 2 relief maps and diagrams, 432 pp.

Pannekoek, A.J., 1946. Enige aantekeningen over indische kaarten. Tijdschr. Kon. Ned. Aardrijksk. Gen., 63: 627-639.

Pannekoek, A.J., 1946. Geografische factoren in de kartografie. Tijdschr. Kon. Ned. Aardrijksk. Gen., 63: 779-788.

Pannekoek, A.J., 1962. Generalization of coastlines and contours. Int. Yrb. Cart., 2: 55-75.

Paris, J.F., 1974. Coastal zone mapping from ERTS-1 data. 2nd Can. Symp. Remote Sensing, Guelph: 515-528.

Peucker, K., 1970. Beiträge zur Geschichte und Theorie der Geländedarstellung. Meridian Publ., Amsterdam, 132 pp.

Peucker, T.K., 1980. The impact of different mathematical approaches to contouring. Cartographia, Monogr. 25: 73-95.

Phillips, R.J., et al., 1975. Some objective tests of the legibility of relief maps. Cartogr. J., 12: 39-46.

Pietkiewicz, S., 1964. Moraines and dunes on small scale maps. Geogr. Polonica, 2: 257-259.

Polcyn, F.C., and Lyzenga, D.R., 1975. Nearshore coastal mapping. Proc. NASA earth res. survey symp., Vol 1-C: 2075-2086.

Potash, L.M., et al., 1978. A technique for assessing maprelief legibility. Cartogr. J., 15: 28–35.
Sandy, I.M., 1977. Penggunaan tanah (land use) di Indonesia. Dir. tata guna tanah, 75, 115 pp.
Sarkisjan, G.N., 1963. On the representation of some volcanic landforms on topographical maps (in Russian). Sbornik naucn. Erevansk. Inst. Pedol., Ser. Geogr. Nauk., 1: 123–155.
Schmidt, W., 1964. Die Darstellung von Steinriedeln und Steinaufhäufung in der topographischen Karte 1:10,000. Vermessungstechn., 12: 443–445.
Seeler, A., 1964. Beiträge zur Morphologie norddeutscher Dünengebiete und zur Darstellung des Dünenreliefs in topographischen Karten. Ztschr. f. Vermessungsw., 12(3): 112–113.
Töpfer, F., 1964. Zufallsschnitt und Niveauveränderung bei der topographischen Höhenliniengeneralisierung. Vermessungstechn., 12(4): 137–140.
Töpfer, F., 1964. Untersuchung qualitativer Kriterien für die Höhenliniendarstellung der Bodenvormen. Vermessungstechn., 12(8): 329–333.
Töpfer, F., 1974. Kartographische Geländedarstellung. Petermanns Geogr. Mitt., Erg. Heft 276.
Verstappen, H.Th., 1964. Geomorphology as an essential element in aerial survey. UN. Reg. Cart. Conf. Manilla. Abbreviated version, ITC Publ. B. 32, 12 pp.
Verstappen, H.Th., 1981. Kartograaf en Explorateur. Bull. vakgr. kartografie, Dept. Geogr. Utrecht State Univ., 14: 5–16.

Chapter 3

THE ROLE OF GEOMORPHOLOGY IN GEOLOGICAL AND SOIL SURVEY

3.1 Introduction

Among the earth sciences, geomorphology takes a distinct intermediate position between geology on one hand and soil science on the other, thus bridging these two fields of science. The relations between geomorphology, geology and soil science are discussed in their totality in this section. Both geologists and soil scientists have much to gain from geomorphological information, though each of these groups of investigators, and particularly those engaged in thematic (geological and soils) mapping, has its own specific interests and needs with respect to geomorphology. However, an overly restricted interest should be avoided. In this case neither geology nor soil survey would benefit from the potential contribution of this science to the solution of scientific and survey problems.

In fact, both geologists and soil scientists tend to look at geomorphology from a different and often rather opposite angle. Broadly speaking, the existing geological situation, including both lithology and structure, together with the geological development of the past, more or less sets the scene for geomorphological evolution under the influence of a variety of exogenous factors. It is the geomorphological situation and development which, in its turn, is the environment in which pedogenesis occurs (Veen, 1970). The evolution of the landforms prevailing in an area, determine to a large extent the distributional pattern of the unconsolidated materials as created by weathering agents and transportation processes. It will also affect the stage of development of the soils produced.

It goes without saying that inversely, geomorphological investigations can only be optimally executed if supported by adequate geological information and data on soil development and profiles. The three sciences are thus intricately interwoven although each of them has its own distinct objectives. Nevertheless, some confusion still exists in certain circles and sometimes unfounded claims are made that geomorphology is part of geology and even that soil science is part of geomorphology. Seen within the framework of one of these sciences this may be correct; Sidorenko (1971), therefore, justly considers lithogenesis, tectonogenesis and geomorphogenesis as the three major parts of geology. It should be kept in mind, however, that geomorphology has a wider scope and has more to offer in other fields of science, such as soil science and hydrology, and can contribute to many practical problems which have no bearing on geology.

It is important to note that soil science, geomorphology, geology and also geophysics form a sequence whereby each subject exceeds the previous one in terms of space and time. Generally speaking, soil development proceeds more rapidly than landform development, pedological mapping units tend to be smaller than geomorphological units. Geological developments cover larger periods and result in larger units than geomorphological units and geophysiscists think in a even larger framework of space and time. This situation, affecting the relations between sciences and matters such as mapping scales and time required for surveys, will be explained later.

3.2 Relevance of the Various Aspects of Geomorphology for Geological Purposes

The close links between geomorphology and geology which have led some to consider the study of landforms and their development, as merely a part of geology, merit further elaboration. It is particularly interesting to know what aspects of geomorphology are most relevant in geological studies. The various processes sculpturing rock types and geological structures, create characteristic forms by means of which these rocks and structures can be recognized. Landform is thus an important aspect of geomorphology for geological purposes. It is generally beneficial to the study of many kinds of geological phenomena to relate the observations carried out at rock outcrops to the landforms associated with these phenomena (Semmel, 1980).

The effectiveness of this procedure varies from one place to another. In areas of degradation, selective erosion due to lithological differences plays a role of prime importance. It becomes less revealing in areas where rocks of similar resistance outcrop, or where the effects of selective erosion are obliterated. The latter situation occurs, for example, in rugged, mountaineous terrain where the drainage pattern is largely governed by gravity, regardless of the geological conditions and in areas where the relief is carved by glacial erosion and does not necessarily relate to the strike of beds.

A sedimentary cover, of fluvial, glacial or aeolian origin, may interfere with geological observations. Geomorphological studies of such areas are often quite useful. Geomorphological analysis of such terrain may reveal peculiarities in drainage pattern, variations in types, shapes or distribution of deposits which can give a clue to geological problems.

In areas covered for example, by flood plains, aeolian sands, alluvial fans or lake deposits, geomorphological study for geological purposes is indicated. In areas where thick glacial deposits, such as morainic ridges, till or outwash plains cover the bedrock, a good geomorphologist may be able to find some indications concerning the underlying rocks on the strength of patterns and other characteristics of the covering deposits.

In some cases it may suffice to merely study the surface appearance, but more often elaborate geomorphological reasoning and detailed geomorphological studies are required, in order to optimally benefit from contributions of this science. The study of geomorphological processes may be useful for certain types of geological study, for example, the presently active exogenous processes may be of use in connection with geochemical investigations. Also processes of the past may be relevant in this context.

Maurice and Meyer (1975) found in Ireland that weathering and erosion during the Tertiary, resulted in secondary dispersion near exposed mineralisations. The patterns of this are still partly traceable under the glacial cover and give rise to important geochemical anomalies. A surfacial lead-zinc anomaly was proved to be formed over a buried sinkhole, probably of Tertiary age, and filled with iron hydroxide. Unraveling dispersion patterns led to the detection, under the glacial cover, of two residual anomalies with secondary zinc enrichment. This example shows that the study of various exogenous processes may also be important for mining purposes, as will be further explained in section 9.

The study of present-day processes such as accumulation in fluvial, deltaic and certain marine environments are appropriate in order to arrive at a better understanding of the sedimentological patterns of similar environments of the past. This may also be of considerable economic importance, for example, in the context of oil exploration. For example, Moore (1969) pointed out that sizeable petroleum reserves are known to exist in Cenozoic deep water sediments, whereas such occurrences are less probable in Pleistocene sediments. He suggests that the different fluvial regimes of that period are responsible for this having transported large quantities of sediment to the foot of the continental slope where it formed large abyssal fans. These fans, according to Moore, may have considerable oil potential.

The geologist often has a special interest in the geomorphological indications for active endogenous processes. This applies both to localized movements, for example, along faults and more generalized endogenous processes such as subsidence or uplift, synclinal downwarping versus anticlinal upwarping which affects larger tracts of land. His interest stems from two factors: first, the geological methods of research may be inadequate compelling him to rely on geomorphological and/or geophysical evidence. Secondly, in many cases the recent tectonic movements reflect similar processes of the past at those localities and have thus considerable structural importance (Hills, 1961). In section 3.3 these matters are discussed in more detail.

Morphogenesis and morphochronology are of particular significance in studies of areas dominated by Quaternary accumulation where traditional geological research methods are of limited value. The correlation with Quaternary sculpturing in the hinterland is also an essential aspect.

Full-fledged geomorphological studies supported by pedological, sedimentological and other data, are essential to unravel the younger part of the geological history. The long-term morphogenetic development is however, not exclusively interesting when restricted to the Quaternary. It is often essential to trace back in the deeper past, for example, where Tertiary or older planation surfaces, or other old landforms are concerned. A practical field application here is, for example, the exploration for weathering ores (section 9). The old landforms may be tectonically deformed, partly eroded or buried, which may complicate the investigations. The palaeogeomorphological situation is often an important subject of study.

Miller (1961) specified the role of geomorphology in geological investigations and distinguished four categories:

a) Elementary geomorphology. This relates to the numerous erosional and depositional landforms, such as volcanic cones, dunes and shorelines which can be primarily identified on the basis of their form alone.
b) Supplemental geomorphology. This relates to geomorphological evidence, which may contribute to the solving of geological problems. This method of applying landform characteristics if often, and almost unconsciously used during geological photo-interpretation, where areas of different rock type are delineated on the strength of differential erosion, as explained above.
c) Complemental geomorphology. This concerns information of geological interest revealed through geomorphological studies and having a tendency to escape pure geological observations. Anomalies in stream deposition and/or erosion may give a clue to tectonic or epirogenic movements and peculiarities in the drainage pattern may reveal a fault structure.
d) Independent geomorphology. This is the application of geomorphology by the geologist in areas where no mappable outcrops and no easily discernable structures occur, such as in areas of extensive glaciation, major alluvial plains, but also where small patches of alluvium or colluvium occur in otherwise well-exposed territory. Thorough study of the geomorphological situation and history may then give useful information.

Although these four types may often tend to intermingle, their proper distinction is useful in clarifying the position of geomorphology in geological sciences.

Fig. 3.1 Oblique aerial view of the Cisarua waterfall on the southwest slopes of the Tangkuban Prahu Volcano, Java, Indonesia, a drainage anomaly, caused by a buried lava flow.

The reader will find it easy to evaluate the meaning of Miller's concept of elementary geomorphology and the same applies to supplemental and independent geomorphology. Several photographs in this section further clarify these distinctions. Confusion may exist concerning complemental geomorphology; in this case the following example may serve as a clarification. It relates to the south-western slopes of the Tangkuban Prahu Volcano, Java, Indonesia and is illustrated by the oblique aerial photograph of fig. 3.1 (Verstappen, 1955).

The Cimahi River formed a 80-metre high waterfall near the village of Cisarua which is situated approximately 4 kms north-west of the town of Lembang. The road connecting the village with this town passes around the gorge and crosses the river by way of a tiny bridge where the river is shallow, upstream of the waterfall. Further upstream, the river is much deeper than at the site of the bridge and another, smaller, waterfall is visible in the extreme left. A small tributary enters the Cimahi River upstream of the bridge and is shallowly incised. The explanation for this anomalous fluvial form is the presence of a basaltic lava flow of eruption period B, which acts as a local base level of erosion. It is buried by the young volcanic tuffs of eruption period C and followed, during its outflow, the palaeo relief carved in the tuffaceous breccias of eruption period A. The fragment of the geological map 1:100,000, (sheet Bandung, fig. 3.2) gives the situation and exactly covers the area pictured in the oblique aerial photograph. The lava flow is the causative factor of the waterfall and of the difference of depth of incision, upstream of the bridge, between the Cimahi River (situated aside the flow) and the small tributary which runs on the flow.

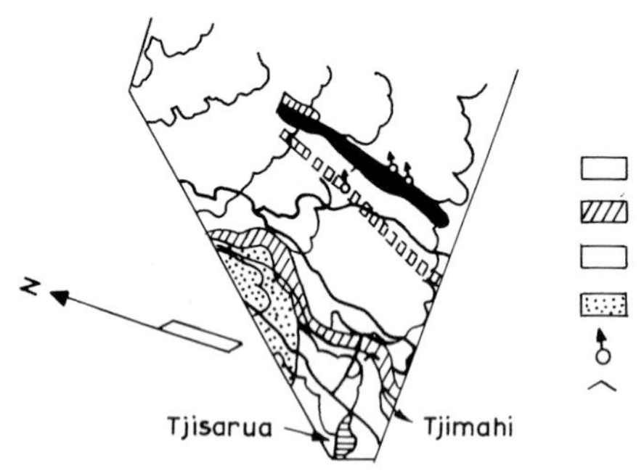

Fig. 3.2 Part of the geological map of Java 1:100,000 depicting the area covered by the photograph of fig. 3.1 and showing the lava flow(s). Key : 1.tuffs of eruption phase C (white); basalt flows and sheets of eruption phase B (hachured); tuffaceous breccia and block flows of eruption phase A (black); old quaternary volcanic complex (stipples); hotspring; waterfall.

3.3 Landforms and their Geological Evaluation

In geological applications of geomorphology, emphasis is often on landforms reflecting lithological conditions and geological structures. Structural geomorphology even receives undue attention. The photogeologist, clearly benefits from a thorough study of the relief produced by selective erosion and thus, reflecting features of geological interest. A common error among geologists is, however, that they conceive landforms as the straightforward result of the geological setting. Evidently, this concept is incorrect, since other factors related to the prevailing climatological conditions and the stage of terrain development also have a profound effect on the landforms found in certain geological situations. The following examples may clarify this.

The Appalachian relief developed in the Condroz Region on the planation surface of the Ardennes, Belgium, has long been considered by geologists a classic example of the direct reflection of the geological conditions on the relief. The psammites of the Famennien are major ridge-forming features in large parts of the Condroz. The gently downsloping ridge caused by this formation near Durnal was considered the morphological expression of a pitching anticline. A deep road cut during improvements of the road Namur-Marche, clearly revealed that the anticline as much steeper than formerly assumed and that its top had been removed presumably by deep tropical weathering during the Miocene. Altiplanation processes here resulted in a terrain configuration giving the erroneous impression of a moderately steep, pitching anticline. It is, therefore, clear in this case that the morphology is not a mere direct reflection of the lithology.

The above example highlights the effect of palaeoclimatic conditions. Contrarily, the following example emphasizes the effects of stages of development. It concerns vast surfaces of planation, dating from the Tertiary, or older geological periods. These have been studied in Africa in particular, but are also of widespread occurrence in other parts of the world. These surfaces of planation are bounded by high and usually steep scarps, the erosion of which is believed by many to result in parallel scarp retreat. A staircase of these extensive surfaces often develops. A granitic area in central Angola is partly incorporated in such a planation surface which is characterized by a low relief amplitude and a widely-spaced drainage pattern. The other part, however, has recently been affected by erosion due to its greater proximity to the main drainage lines and as a consequence, the planation surface has disappeared. This area, which is much lower and more deeply dissected, shows a greater drainage density. Its geomorphological appearance is thus completely different, although the lithology in both parts is identical, i.e. granites (personal communication by Dr. De Vries Lapido Loureiro).

Comparable examples can be given from areas of (Quaternary) aggradation. In the northern part of the Porali Plain, near Lasbela, Pakistan a radial pattern of broad, gravelly channels occurs on a fan structure. The gravelly channels are separated from each other by narrow, elongated 'islands' of silt loams which are under cultivation. It can be observed that the gravels of the channels continue at the same level under the silt loams of the 'islands' which are about 3 metres thick. The gravels were obviously deposited during a more humid period of the past when the Porali River had a greater transporting capacity and prior to the deposition of the silt loam. The gravel was re-exposed due to renewed river incision following a decrease in bed load. Vertical erosion was checked once the river incision reached the gravels and from then on lateral erosion began. Excessively wide channels were thus formed, which are still occasionally flooded. No sizeable transport of grave occurs and thus, many miles of unmetalled trunk road to Kalat could be constructed within these gravelly channels without risk of undue gravel deposition by the river. Increased aridity clearly affected the texture of transported and deposited material, hence the shape of the river channels, and ultimately, the potentialities of the latter for highway location (Verstappen, 1965).

Structural geomorphology is a particular aid in unraveling geological structures and tectonic situations. However the rocks have developed a certain appearance under the influence of exogenous processes which vary particularly with the climatic conditions of the past and present. Climatic geomorphology is thus important in considering the variety of types and intensities in geomorphological processes. Minor climatic differences may produce substantial changes in form which are characteristic for certain types of rock. Minor lithological differences may, (though not always) be clearly reflected in landforms prevailing in areas with comparable climatic conditions.

Structural geomorphological observations are particularly important in areas of sedimentary rocks. Not only can major structural units, such as table mountains and cuesta belts be distinguished, but also structural elements of smaller dimensions. Asymmetry of hills, true dipslopes, etc. are useful criteria for revealing the geological dip and strike direction, particularly since the actual outcrops of exposed rocks play a minor role in many areas and are insufficient for establishing outcrop patterns, which are required for an adequate evaluation of the geological structure.

Further indications can be found in the drainage pattern and particularly in matters such as: variations of the drainage density, amount of geological control on it, and integration and homogeneity of the pattern. Dendritic patterns may indicate the occurrence of horizontal beds or possibly very weak beds of any dip direction. The length of tributaries may be indicative for dip direction in certain cases, as usually the longer ones are found on dip slopes and the shorter ones on the face slope of the beds. Exceptions to this rule may, however, occur when the beds dip steeply. The crestline pattern is also a feature to be considered in conjunction with drainage analysis. It is especially useful in areas where non-resistant beds of gentle to moderate dips occur. The dip slopes are then largely eroded but the U- or V-shaped divide lines may still indicate certain structural elements.

In metamorphic areas structural geomorphology is usually considerably less revealing and often only scattered features can be observed, the visual coherence and geological relations of which may remain obscure. Foliation and cleavage are rather well-reflected in the terrain configuration. Contact zones, veins, and plugs, etc. are structural elements to be mapped in igneous areas where cleavage can also be observed. Dykes may stand out in relief or may form depressions, depending on the resistance to erosion of the dyke material as compared to that of the surrounding rocks. The metamorphosed belts adjacent to the dykes are ridge forming. Fig. 3.5 gives an example.

Features of particular interest in mapping igneous and certain sedimentary rocks are joint patterns. Major aspects to be considered are: the frequency distribution of joint directions (Verstappen, 1959) the regional distribution of certain patterns and the delineation of areas of comparable joint density characteristics. It should be noted that the number of joints and faults visible in the field and on aerospace imagery is not always a good indication for the degree of jointing and/or faulting occurring. In unconsolidated materials faults may be poorly visible, whereas in hard rock they show up clearly. Sometimes major dislocations are difficult to trace, whereas even the smallest joints can be seen in quartzites. In areas where the land has been covered by continental glaciation, the ice has usually carved out very dense patterns of small joints which are reflected in the alignment and pattern of residual hills, drainage lines, etc.

Even though it may be comparatively simple to detect structurally controlled landforms, particularly when aerial photographs are consulted, it is often indispensable to take the entire, often complex morphogenetical history of the region into consideration to arrive at reliable conclusions of geological interest. In those areas where erosion surfaces are deformed and partly buried due to more recent endogenous forces, or where Appalachian structures are crowned by lateritic caps, etc. thorough analysis of the geomorphological development is essential. Climatic geomorphology then becomes almost invariably important. The presently prevailing climatic conditions have a profound influence on the existing landforms. The fact that similar lithological situations in different climatic conditions may result in substantially deviating landforms, is well-known. This is very outspoken for granites and related rocks (Verstappen, 1968). The climatic influences on karst landforms developed in limestone areas in different climatic belts are also quite outspoken.

These matters are not always easy to unravel, since minor differences in mineralogical composition, physical properties of the rocks, relief conditions, groundwater level, etc and other factors in landform development have various effects on each other, which are often difficult to quantify. True humid tropical karst relief such as conical and tower karst hills are best developed where porous (e.g. coral) limestones are concerned and considerably less in case of impure and compact limestones (Verstappen, 1960, 1964). In non-resistant chalks, containing a considerable amount of clay-like particles, the joints opened by the dissolution of lime are gradually filled with clay. Thus, they will no longer affect the infiltration of rainwater and become indistinct in the surface configuration. Small sinkholes and shallow, dry valleys then occur, and the hill slopes tend to be smoothly rounded. However, quite different landforms will evolve if less clay is present: jointing will then become the leading

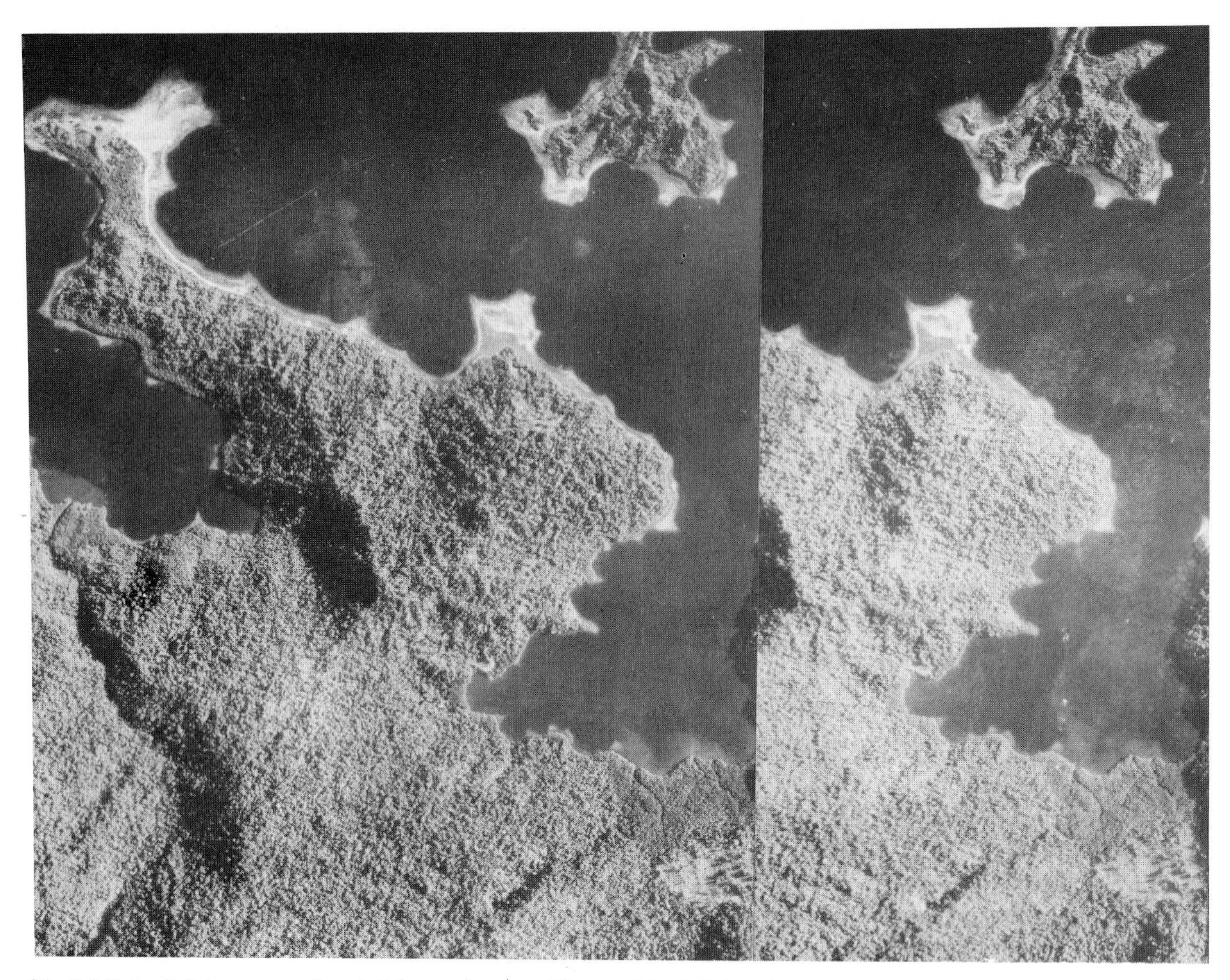

Fig. 3.3 Extended stereogram of part of the north coast of Batanta Island, Indonesia, showing asymmetric hills formed in seaward dipping limestones. Scale 1:40,000.

element in the alignment of steep-sided karst hills.

The examples given illustrate on one hand the use of landform characteristics and structural geomorphology revealing lithology and structure and on the other, the importance of climatic geomorphology. The extended stereogram of fig. 3.3 shows a part of the northern coast of the island of Batanta, Indonesia, covered by dense tropical rainforest. Notwithstanding the vegetative cover, the northward dip of the Miocene limestones occupying most of the area pictured is clearly indicated by the asymmetry of the hills. The limestones are fairly compact and no tropical karst is developed. A few major river valleys can be seen. They originated largely as underground rivers but the roof of their subterraneous course has collapsed long since. The landforms of the small island near the lower photo edge are quite different. More intense dissection, concave slopes, and sharp-crested ridges are characteristic to the (andesitic) volcanic breccias occurring here, where lineaments can be further observed. The lithological differences are thus clearly reflected in the terrain configuration. The northward dip of the limestones, and particularly the deeply penetrating bays at either side of the limestone peninsula in the centre, point to subsidence of the coastal zone. Since the volcanic breccias found on the island to the north of the northward dipping Miocene limestone beds are Pre-Miocene, the occurrence of a WNW-ESE running dislocation is evident in the water south of the peninsula.

The vertical aerial photograph of fig. 3.4 showing a part of the hills on the east flank of the Barisan Mountains in northern Sumatra, illustrates how faults may affect the geomorphological situation even when their existence may be difficult to prove by conventional geological means. The situation of the alluvial plains and of the fan pictured, together with the straight sector of the river and the rectilinear boundary of the grass-covered hill in the left are converging geomorphological evidence, leading to the detection of the fault(s) indicated. In the terms of Miller (1961), this is a case of complemental geomorphology.

The stereotriplet of fig. 3.5 shows an area in Lesotho, where basalt dykes clearly stand out in relief and are situated in the weak Red Beds and the Transition Beds forming the lower part of the stratigraphic sequence, and outcropping in the valleys and lower hill slopes respectively. In the more resistant, overlying Cave sandstones of the

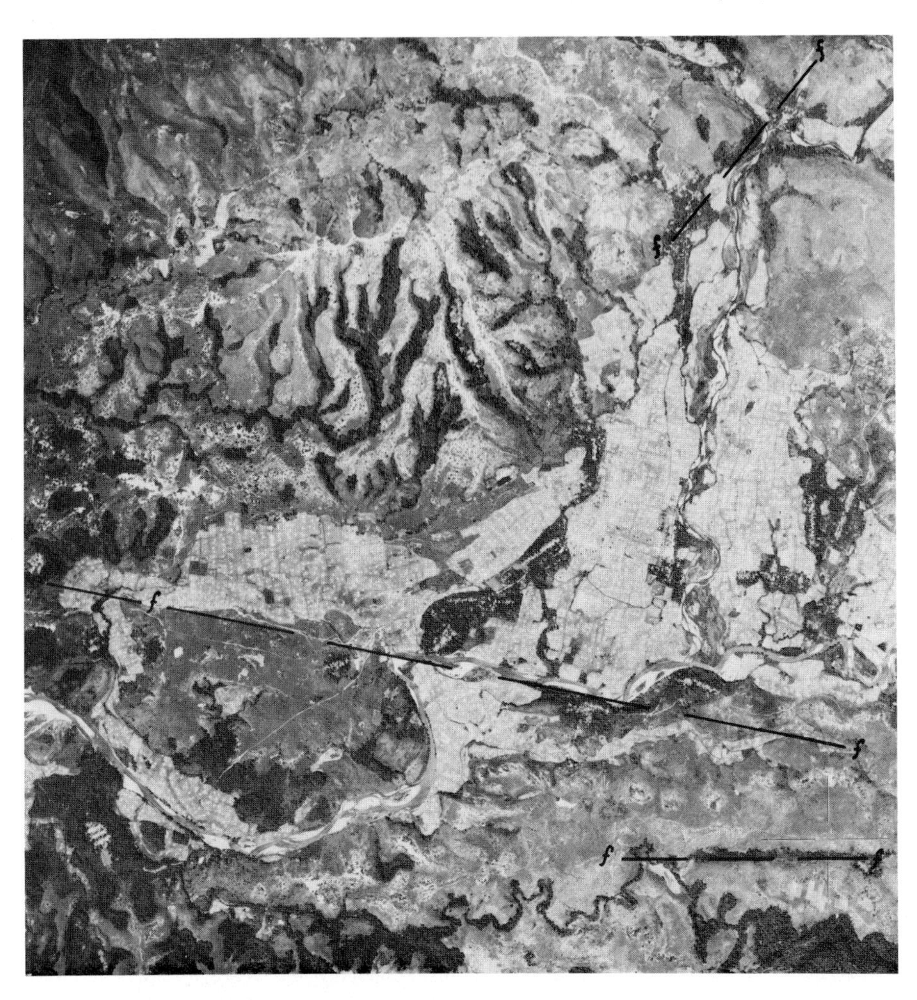

Fig. 3.4 Vertical aerial photograph (scale 1:40,000) showing faults affecting the terrain configuration along the eastern side of the Barisan Mountains in North Sumatra, Indonesia.

Clarence Formation, the dykes are partly ridge forming and may be traced in many cases by differences in tone in the photographic images. Locally, they are almost indistinguishable from depressions between ridges made up of contact metamorphosed 'baked' adjacent sandstones or other country rock. In the basalts of the Drakensberg Formation, dykes are geomorphologically developed as distinct and narrow trenches. Obviously, the landforms sculptured in the dykes largely depend on the differential resistance between dyke and adjacent rocks. Offsets of dykes and the anomalies in the drainage pattern caused by dykes are other geomorphological indications that assist in their correct geological evaluation. An interesting geomorphological phenomenon of a completely different nature is the erosion process of 'scarp collapse'. This is found particularly near the contact between basalts and Cave sandstones where the thin soil is often soaked by emerging groundwater and subsequently slides downslope, forming minute scarps and barren sandstone surfaces (light toned on the photographs). The matter will be dealt with in section 15 on erosion surveys.

Fig. 3.6 is a vertical aerial photograph at the scale of 1:40,000 of an area in the island of Sumba, Indonesia. Weak horizontally bedded sedimentary rocks outcrop in the entire area, which is characterised by a dense dendritic drainage pattern. The hills are mainly grass-covered but it seems that the dissection and erosion of the terrain has been more intense in the past than at present. The valleys pictured and even some of the major ravines have been partially filled with eroded unconsolidated materials and apparently the rivers are presently uncapable of evacuating these from the area. The effects of past climates on the landforms are clearly demonstrated. This is even more pronounced on the extended stereogram of fig. 3.7 picturing a semi-arid area in northern Africa on a scale of 1:30,000. The glacis relics depicted in this stereogram are basically an erosion glacis formed by beveling of the easily eroding, underlying Miocene, marly siltstones, and shales (the dip and strike of which can be clearly seen in those parts where the glacis has been removed). The distinctly radial pattern of rather shallow drainage channels on the glacis relics, classes them as cone-shaped erosion glacis.

The veneer of detritus that has been deposited on top of the glacis prior to dissection and further destruction, is very thin and amounts to 1 metre at the maximum. The formation of the erosion glacis over the weak under-

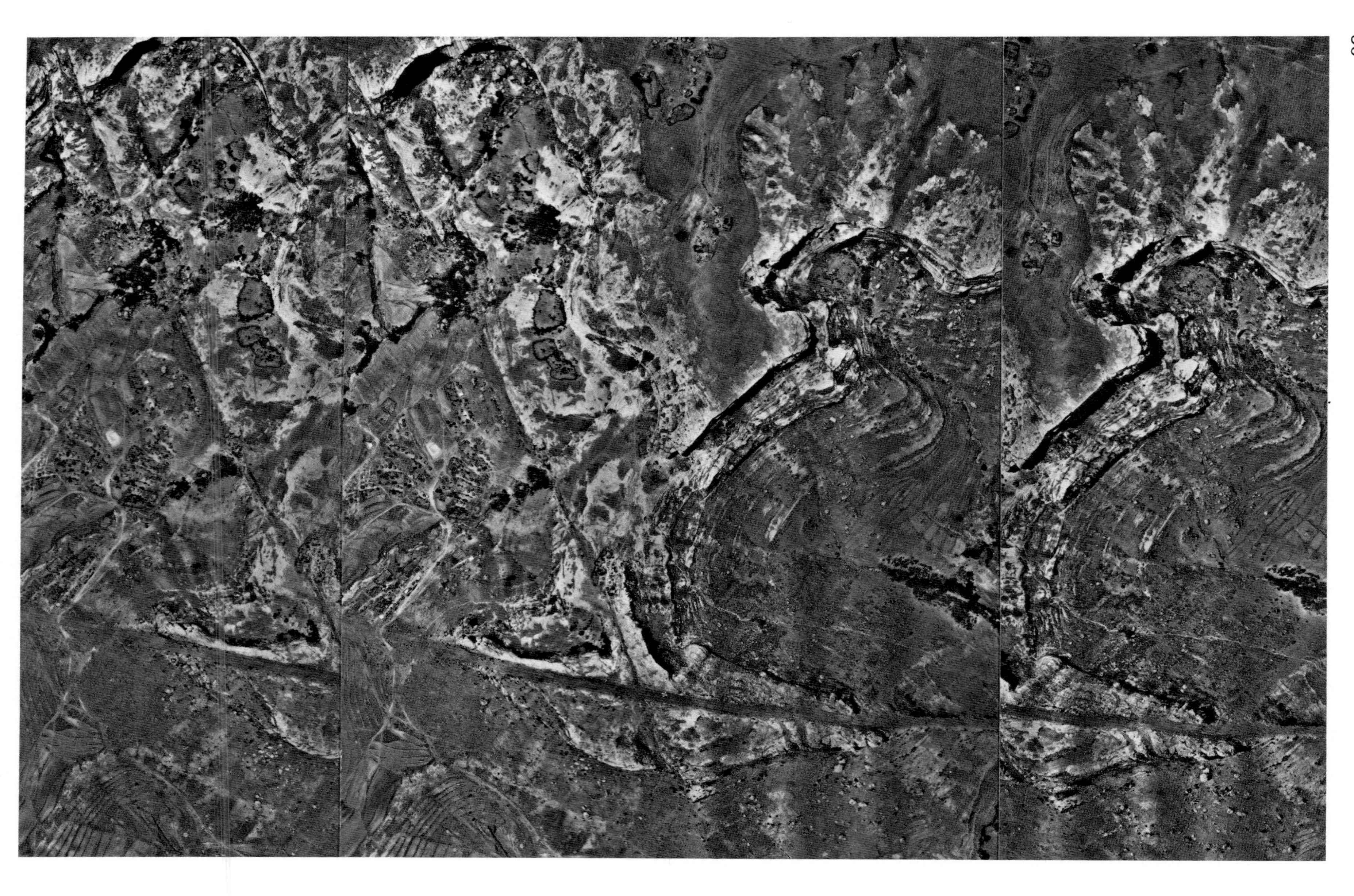

Fig. 3.5 Stereotriplet of part of the Thaba Bosiu Area, Lesotho. Scale 1:20,000. Basalt dykes are ridge-forming in the soft Red Beds and Transition Beds but usually form depressions in the Cave sandstones (and the Basalt plateau, outside the stereotriplet).

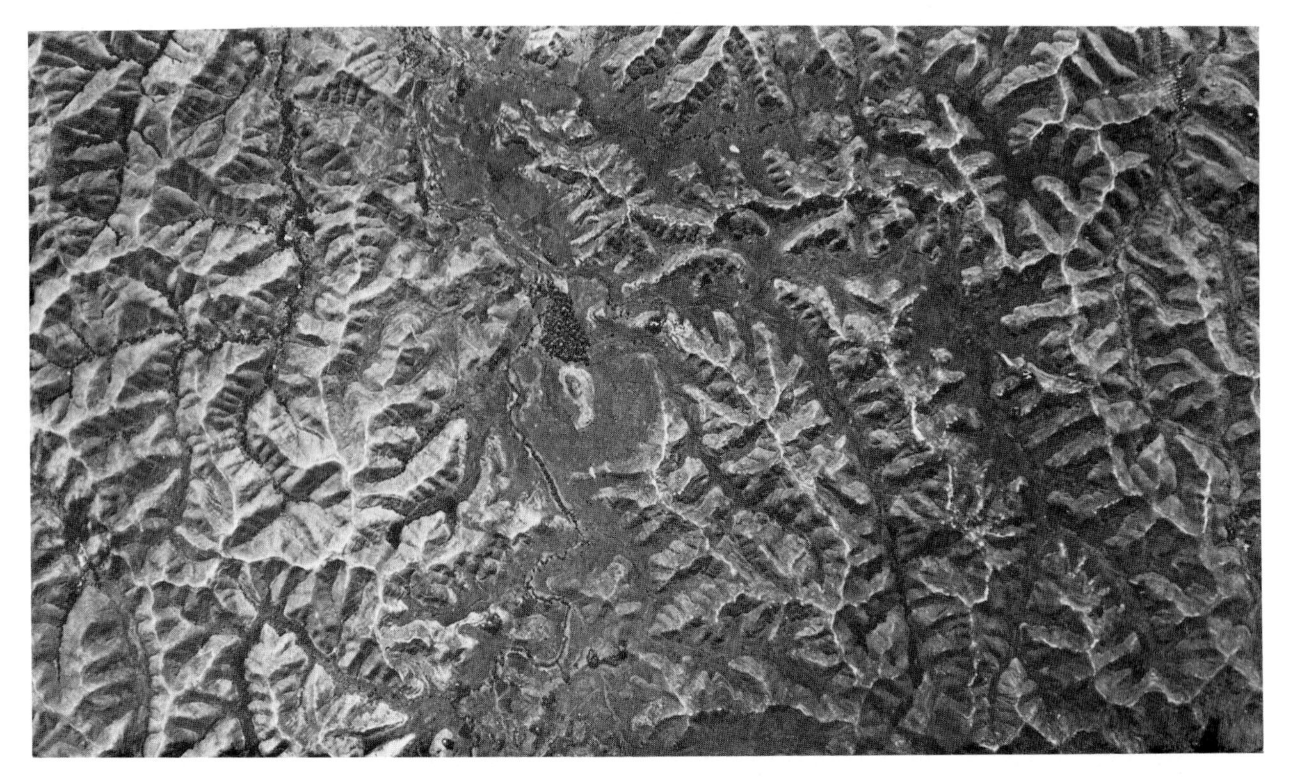

Fig. 3.6 Vertical aerial photograph 1:40,000 of part of Sumba Island, Indonesia. Intense dissection of soft rocks is, at least in part, a phenomenon of the past.

lying rocks can be attributed to a relatively humid period. The deposition of debris on top followed when the rivers became incompetent with increasing aridity. Calcrete development occurred in a relatively dry period until finally, dissection ended the sequence of events which in many cases has repeated itself several times. The darkish tone of the glacis surface cover is due to the desert varnish of the debris.

Active dissection of the weak rocks underlying the glacis, presently proceeds at a high rate and the waste is deposited in the form of slightly concave mud slopes, or cones. When transported to the ephemeral rivers, the material is carried off to form alluvial fans, or cone-shaped accumulation glacis. In this case, the alternation of planation and dissection of the rocks, and thus their morphological expression, is governed in the first place by climatic fluctuations of the past and not by structural factors.

3.4 Geomorphological Evidence for recent Tectonic Movements

Another point of interest to the geologist concerning geomorphology is directed towards indications for young (recent) tectonic movements. Type, number and clarity of such indications vary from one area to another, but reliable conclusions can often be arrived at, as has been demonstrated by many authors. The subject is of considerable geological interest due to the fact that the actual, recent or sub-recent crustal movements in the majority of cases, are related to older, and possibly buried structural trends; though they may also be merely Holocene continuations of e.g. subsidence, uplift, tilting or warping which have characterised the areas concerned, during several consecutive geological periods. Their analysis is not only important for the study of the Quaternary, but also for an understanding of older geological structures and tectonic movements. Geomorphological indications of neotectonics are frequently used also in matters such as oil exploration (Aristarkhova, 1968; Yonekura, 1972; Zvonkova, 1972), and the search for groundwater. It is essential in such cases to clearly establish the relationship between recent and older tectonic movements. They may show the same trends and lateral displacements; or even it may occur that the recent trend is opposite to that of earlier situations (Berlyant, 1969).

The geomorphologist engaged in these studies unravels endogenous processes, which themselves are invisible and hard to trace, when using purely geological methods of research. Clues are sought concerning type and configuration of landforms, certain exogenous processes of erosion and sedimentation triggered by these endogenous processes which are reflected in the landforms and the general geomorphological situation. The data gathered show great diversity in type and reliability, and often a convergence of evidence will be conclusive.

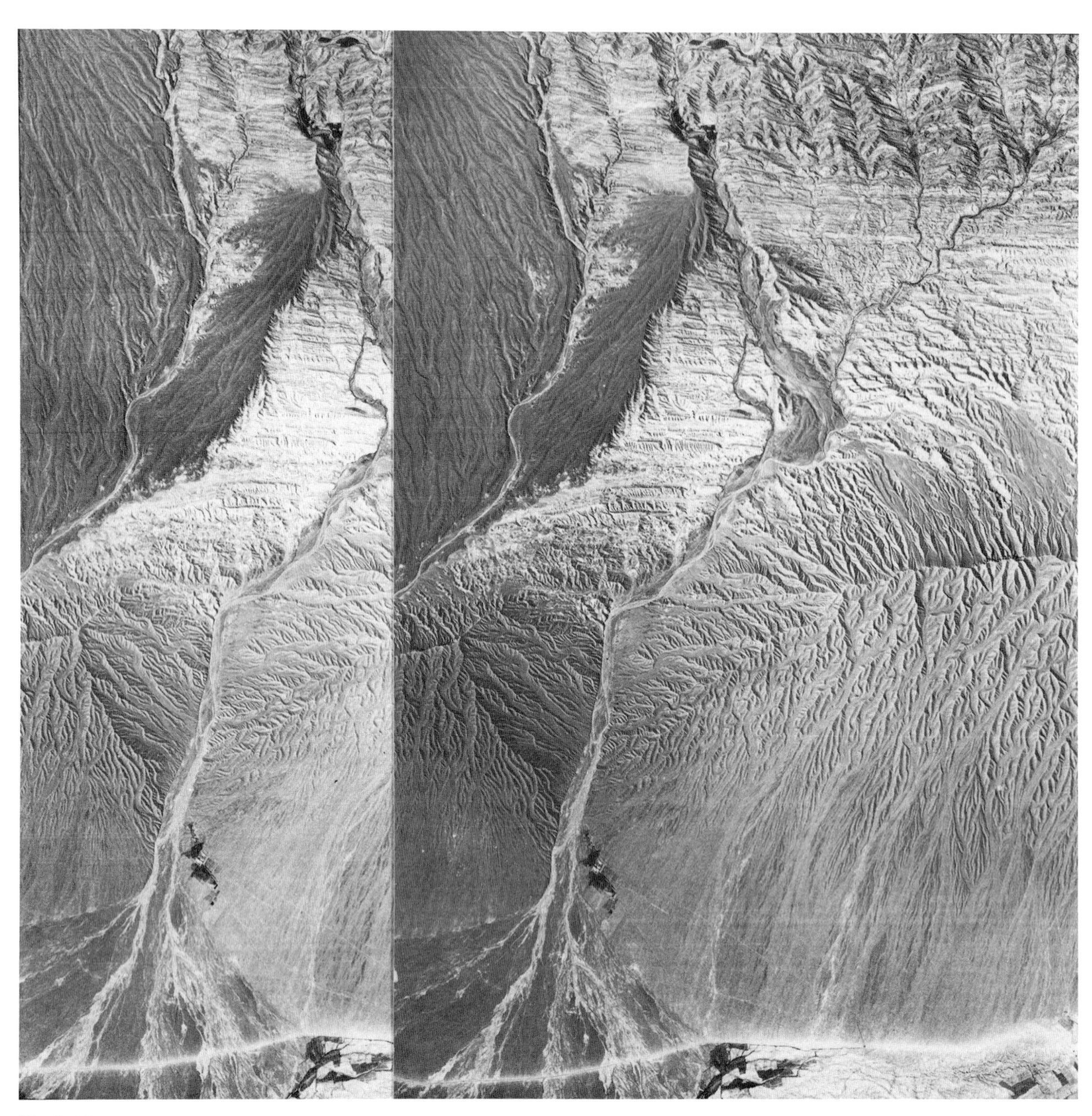

Fig. 3.7 Extended stereogram (1:30,000) of a dissected erosion glacis in northern Africa. Palaeoclimates played a leading part in its formation as is explained in the text.

Precise information from seismological observations (Yonekuda, 1972) on the frequency of earthquakes and their epicentres, as well as borehole data is useful for the location of major faults. Comparison of these data with geomorphological indications is essential. Geodetic information has also proven its usefulness in many cases. Lateral movements along fault lines have been ascertained and quantified on certain occasions as an outcome of triangulation operations where the trig points were situated at opposite sides of the fault. An example is mentioned by Verstappen (1973) from the rift zone in northern Sumatra.

Recent vertical displacements along faults are usually clearly traceable in the terrain by way of minor fault scarps, emerging groundwater and cracks in the soil. Fig. 3.8, a groundphoto showing a young faultscarp in a gravelfan in the Magdalena Valley, Colombia, exemplifies this. If, however, regional uplift or subsidence is concerned, the direct effect in the landforms may be less distinct and then comparison of older and more recent precise levelling may be extremely useful. In the Netherlands, for example, this method of research has been successfully applied (Edelman, 1954). Some complications are for example, that it is not easy to discriminate between tectonic influences, sea level changes, and anthropogenous factors such as compaction

Fig. 3.8 Active fault in a Young Pleistocene gravelfan, Magdalena Valley, Colombia.

of peat following drainage improvements. The patterns of measured subsidence need not by consistant with time, as illustrated in fig. 3.9 showing a gradual concentration of the subsidence in a narrow coastal strip of part of eastern Hokkaido, Japan. The isolines showing the subsidence of the mostly swampy area in the periods 1910-1955, 1955-1970, 1970-1975, and based on precise levelling in the four years mentioned, reveal a distinctly changing trend (personal communication by Dr. I. Yokoyama, Hokkaido University).

The traceability of young tectonic movements by geomorphological means depends on various factors. First, one should clearly distinguish between localized movements (e.g. along faults), and regionalized movements, (e.g. subsidence). The indications for the former are, in many cases, relatively clear even if the general terrain configuration is not favorable for the detection of minor differential tectonic movements e.g. due to the dominance of active erosional processes in areas of high relief. The regionalized crustal movements are generally more difficult to trace where the geomorphological situation is complex. Three broad physiographic environments which respond differently to the geomorphological expression of the above mentioned localized and regionalized tectonic movements, can be distinguished: tectogene areas generally of high degradational relief, kratogene areas of low degradational relief and large areas of aggradation.

Localized young tectonic movements along faults in tectogene areas of high relief can be easily traced, particularly if they have a vertical component. In order to date events

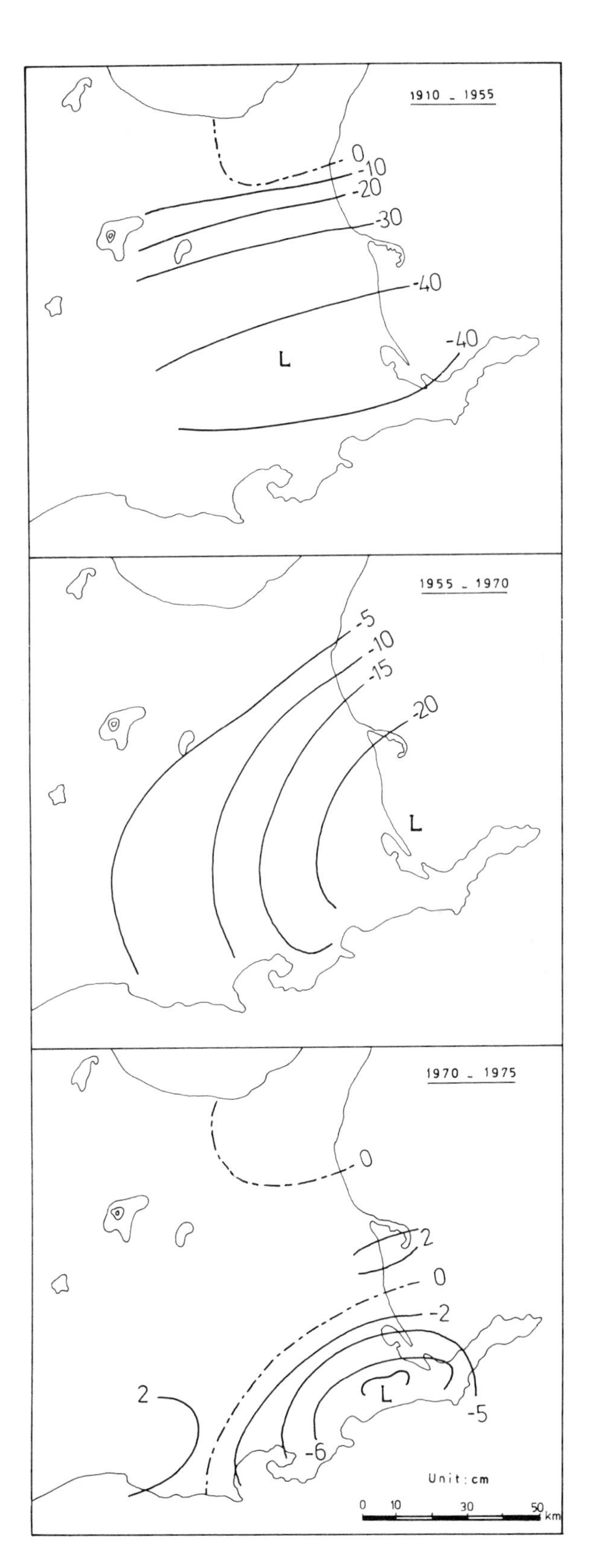

Fig. 3.9 Subsidence patterns in a coastal area in East Hokkaido, Japan, changing with time (1910-1955; 1955-1970; 1970-1975). Courtesy: Dr. I. Yokoyama.

one must make a clear distinction between older movements (and forms created by them) and younger events or forms. There is often a gradual transition between these because tectonic movements are generally continuous or at least occur repeatedly at relatively short intervals. In such cases, there is no methodological difference from the detection of structural, linear forms, mentioned in the previous section. Criteria for truely neotectonic features are: the age of the youngest beds affected by these movements and the freshness of the forms created. Indications include matters such as abrupt and non-dissected scarplets, well-preserved facets, a linear base of the scarps, ponded-up drainage, deviating gradients of rivers where they traverse a fault line or fault zone. Stream off-sets and poorly matching landforms at either side of the fault suggest recent lateral movements. The faults indicated in fig. 3.4 of northern Sumatra are a good example. Another good example is the fault scarp occurring in the surroundings of the town of Lagunillas in the Venezuelan Andes. Here a high fault scarp has been formed in the gravels of one of the higher terraces of the Chama River. A lake (Laguna de Urao) has been ponded up at the downthrown side of the fault scarp and the drainage pattern is strongly influenced by faulting movements.

Generalized crustal movements in areas of high relief are usually difficult to detect by geomorphological means. Nevertheless, various indications may exist. River valleys (see fig. 3.10) may change in width and general character and deformations of river terraces or anomalous zones of aggradation may occur. Fig. 3.11 shows warped lake terraces along the eastern side of Lake Toba in northern Sumatra. Differential tectonic movements of about 450 metres within a distance of about 35 km are reflected here by the deformation of the terraces, which are only a few thousand years old (Verstappen, 1973). Marine terraces and particularly coral terraces or reef caps provide precise registrations of uplifting movements of various types. Where absolute dating of the coral polyps can be effectuated, quantification of the neotectonic movements becomes feasible (Tjia, 1970, 1975).

Verstappen (1957) proved the general WNW-ward tilting of the island of Muna, Indonesia on the strength of the reef rings and reef terraces, the absolute height of which had been determined by photogrammetric means. At first only a few, low coral islands existed, from which a platform reef evolved. Reef rings concentrated along the eastern side of the island originated and now reach a height of 445 metres above sea level. They downslope towards the WNW as does the reef platform as a whole. Fig. 3.12 is an oblique aerial photograph of part of the Tukangbesi Islands, Indonesia showing how two coral reefs merged into the present island by uplift, and how subsidence is evident in the coastal area in the lower right.

Subsidence may also be evidenced in coastal areas of high relief, by the bays and larger indentations of the coastline produced by transgression of the sea. Care should be taken to separate (glacio)eustatic sea level movements from tectonic causes. Since the former are universal or regional, such distinction is possible in many cases. Coral reefs are a classic example of the potentialities of geomorphological indications for subsiding coastal or oceanic areas. The almost-barrier reef surrounding the north-east side of the Ujung Kulon Peninsula, W. Java pictured in fig. 3.13 is a clear example of this (Verstappen, 1956). Also in other parts of the tectogene island arcs of Indonesia, such examples are numerous. Subsidence of the east coast of the non-volcanic island festoon situated to the West of Sumatra has been recorded by navigators in the early part of this century. Comparison of old charts with more recent ones supports these observations as do geomorphological studies of these coastal zones (Verstappen, 1973).

Sometimes regionalized crustal movements are accompanied by localized faulting movements particularly in rigid covering strata. These may be also reflected in the geomorphology of uplifted coral reefs. Fig. 3.14 shows part of the reef cap of Muna Island, Indonesia with several edges of coral terraces indicating a general uplift and tilt and two approximately parallel fault zones that affect the living reef. The stereo-triplet of fig. 3.15 is another example. This figure depicts part of the reefcap of the Schouten Islands, Indonesia near Cape Lumbee on south-eastern Biak at a scale of approximately 1:20,000. Basically, the reefcap is composed of several subhorizontal reef terraces which are separated by steep scarps and higher ridges. These terraces reflect the gradual emergence of the area as a result of regional tectonic movements and of sea level changes.

An interesting feature is the sudden interruption of the reef edges and their continuation at a lower level along a SW-NE line which marks a fault scarp, post-dating most of the young Quaternary reef development. It runs parallel to the major Sorendidori fault which separates Supiori Island from Biak. Tectonics, either pre-dating or post-dating the coral growth, can often be traced on the strength of the morphology of the reefcap (Verstappen, 1960).

No surface drainage occurs due to the porosity of the coral limestones. Sinkholes can be seen in the more compact and impure limestones of the uplifted lagoons. The pure and highly permeable limestones of the ridges are more apt to develop conical karst hills, which characterize many karst formations in the humid tropics. The tropical forest covering the larger part of the reefcap

Fig. 3.10 Tilted dry valley of a former tributary of the E-Digul River (top centre). Upwarping of an anticline situated at the foot of the limestone scarp has disrupted the drainage system in this part of the Central Range, forming the backbone of the island of New Guinea.

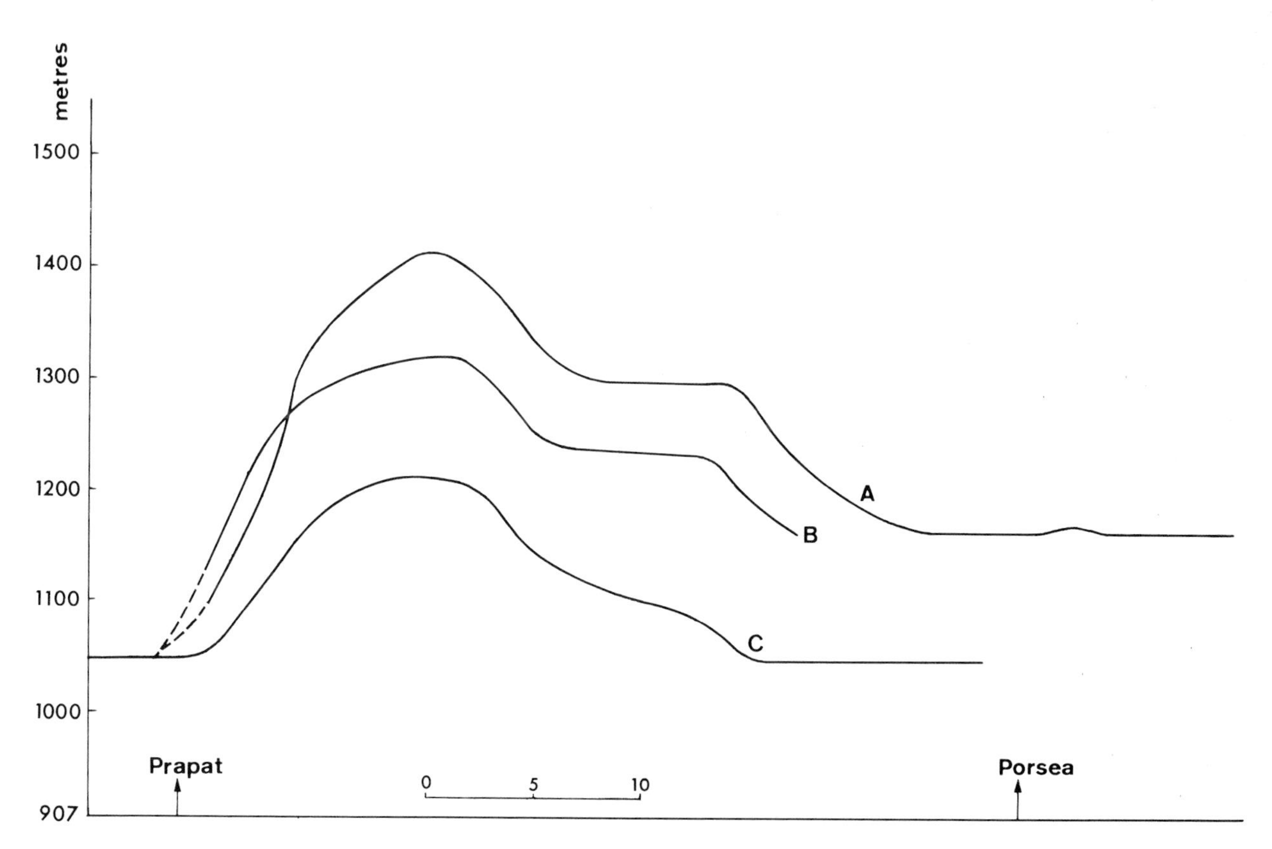

Fig. 3.11 Warped lake terraces along the eastern side of Lake Toba, Sumatra, Indonesia.

Fig. 3.12 Oblique aerial view of the raised coral reefs of WangiWangi, Tukangbesi Archipelago, Indonesia.

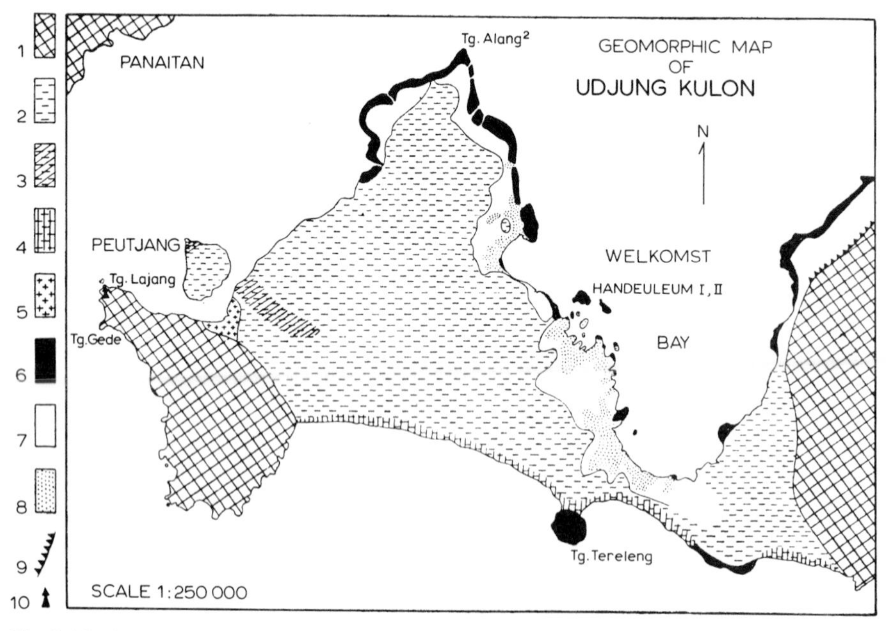

Fig. 3.13 Almost barrier reef (black) along the northern and eastern sides of the Ujung Kulon Peninsula and surrounding areas of Western Java, Indonesia indicating northward tilt of the area. Key: 1. hills and mountains; 2. low plateau; 3. higher ridges of low plateau; 4. low sandstone cliffs and dunes of the south coast; 5. alluvial plain; 6. almost-barrier reef; 7. lagoon; 8. mangrove; 9. fault scarp; 10. lighthouse.

Fig. 3.14 Vertical aerial photograph (1:40,000) of the terraced reef cap of Muna Island, Indonesia, with post-reef faulting.

is best developed on the ridges.

The sub-recent reefcap has an estimated thickness of approximately 100 metres and unconformably overlies the older rocks (basalts, tuffs, serpertines) which emerge from it at certain localities. ^{14}C dating indicates a rate of uplift of the reef of approximately 0.7 mm/yr (Tjia et al, 1975). The 25 m level of the reef pictured in the photograph has an approximate age of 36.370 B.P.

In cratogene areas of low relief, one might not have at first sight, high expectations for geomorphological indications of the neotectonic movements occurring there, which are of relatively small magnitude. Because of the low relief amplitude that characterises most of these areas, important geomorphological consequences may nevertheless result and assist in the analysis of neotectonics. Major deviation of rivers, the formation of lakes and swamps in depressed zones, and even vegetation patterns allowing for differentiation between dry and marshy land, may serve for indications. A good example is the Ob River basin in the USSR where anticlinal and synclinal axes were found in lowland areas reflecting recent movements along older structural trends in the underlying rocks. Comparable studies from the USSR are from Vilenkin (1968) and Shumilov (1969).

Sri Lanka, with its staircase of old planation surfaces, is normally considered a tectonically stable zone, where Quaternary crustal movements cannot be expected. Nevertheless, when Herath (1962) studied the coastal development and protection near the town of Negombo, situated along the west coast of the island, it became clear that the geomorphology of the coastal plain and the configuration of the coastline in that area were profoundly affected by differential Holocene tectonic movements along the E-W faultline or zone which is covered by alluvial deposits and sea water, but the existence of which was proven by airborne geophysical survey. As a consequence, a lagoon-barrier type of coast is found to the south of Negombo, whereas to the north of the town, a recession of the coastline has occurred since the last few thousand years, as evidenced by the pattern of cut-off beach ridges and the present coastal forms and processes (see also section 8).

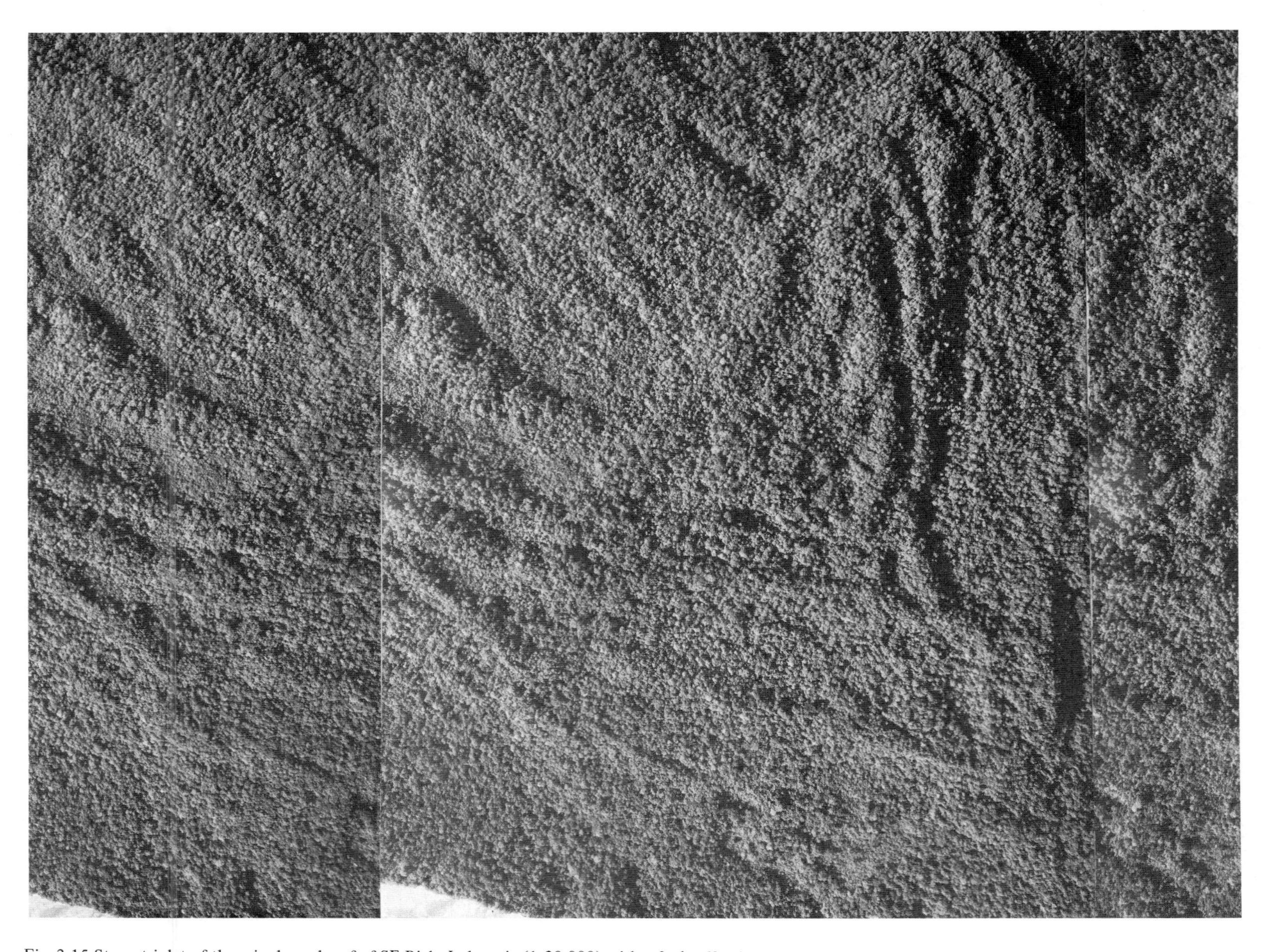

Fig. 3.15 Stereotriplet of the raised coral reef of SE Biak, Indonesia (1:20,000) with a fault affecting all but the youngest reef edges both as far as altitude and pattern is concerned.

Another clear example of how geomorphological features may reveal neotectonic movements in cratogene areas is the incipient south-eastward extension of the East African rift first discovered on Landsat images in northern Botswana. Diversion of the river water by these recent fault systems in a north-eastern and also in a south-western direction affects the whole hydrological situation in the Okavango Delta and adjacent areas of Botswana. Further to the east a complex and closed tectonic depression became the site of a Quaternary lake, the size of which fluctuated under the influence of climatic variations and the tectonic developments mentioned above. Fig. 14.8 illustrates the situation. (Verstappen and Cook (Eds.), 1981).

Further to the north, other geomorphological indications can be found which aid in clarifying the tectonical history of the East African rift zone. In Rwanda, for example, rifting and accompanying tilting has resulted in drastic changes in the drainage pattern. The rivers, formerly draining towards the rift, have been diverted to form the head waters of the Nile. The head waters have been largely converted into papyrus-covered lakes by way of tectonic damming (see fig. 4.4).

Localized neotectonic movements are usually difficult to trace in lowland areas, although features such as aligned mud volcanoes may be indications. The indications for regionalized neotectonic movements, however, are numerous and diversified. Zones of subsidence are often clearly marked by for example, broad flood plains with extensive inundations, swamps, and drowned forests.

The map of fig. 3.16 pictures such a situation in Sumatra and the oblique airphoto of fig. 3.17 gives an example of the lowlands near the Fly River, southern New Guinea. Slightly higher grounds, with tropical rain forests, indicate alternating zones of upheaval where flood plains tend to be somewhat narrower. The distributional pattern of finer and coarser sediments and their thickness are other good indicators in such lowland areas. Geophysical data may provide further evidence.

Slow, overall subsidence of a delta under the influence of its own weight or due to tectonic movements, is not always easily separated from a rise in sea or lake level. Young differential crustal movements in the deltaic area, however, can often be readily traced and have a great impact on the development of numerous deltas. Recent accretion near the mouth of the Digul River, New Guinea, for example, is of little importance, as shown by a comparative study of old and new maps and of recent air photos. The main outlets of this huge river draining a large part of southern New Guinea have some unpronounced estuary characteristics. Drowned forests covering extensive areas further inland, indicate that recent subsidence is common to considerable parts of the southern lowland of the island. Most of the river load apparently is deposited there, which explains the stationary situation at the river mouth. It is astonishing that from a study of the same sources, it is clear that rapid seaward displacement of the coastline occurs several tens of kilometres further to the south, near the western side of Kolepom Island. The changes of the coast, evident from maps dating from 1903 and airphotos of 1945 are indicated in fig. 3.18. The growth there has no direct relation to the sedimentation of the Digul River, but is due to a zone of recent uplift which passes through Cape Valsch and which can be traced from the air across Kolepom Island (Frederik Hendrik, Dolak) at a short distance from the coast and further towards the southern lowlands of Irian/New Guinea. The larger part of Kolepom Island is scarcely above sea level and is covered by reed swamps.

The Digul River, farther to the east, is forced to alter its southern course and to take a western direction to the north of the well-known 'Merauke ridge'. The Princess Marianne Straits which stretch in a south-southwest direction to the east of the vast Kolepom Island is to be considered a lower course of the Digul River, which was in use when the river erosion was still capable of adequately counteracting the upwarping of the latter ridge. The straits are still navigable due to the scouring action of tidal currents.

Another example of a delta where tectonic movements enter the scene and counteract aggradational processes, is the Mamberamo Delta, located at the northern tip of Irian Barat, Western New Guinea. The huge Mamberamo River system drains the larger part of the island, north of the Central Range. Its main branches are the Tuaral (Rouffaer)-Tariku (van der Willingen) River and the Taritatu (Idenburg) River, both of which flow in the low lying and rather swampy belt, called 'Meervlakte' and stretching in an east-west direction to the north of the Central Range. Downstream of their confluence, the Mamberamo River takes a northern direction and breaks through the east-westerly stretching Rouffaer - Van Rees Mountains to reach the Pacific Coast at Cape d'Urville.

Recent accretion of the delta (fig. 3.19) is unimportant which is easily explained by the fact that most of the river load is deposited further upstream, primarily in the 'Meervlakte' zone, but to some extent also in an east-west zone of subsidence occurring in the deltaic area. This situation explains why the river quietly flows between its natural levees and has no connection with its former branches. It also clarifies the existence of an important saline intrusion as observed from the broad mangrove belts stretching far inland. A beach ridge

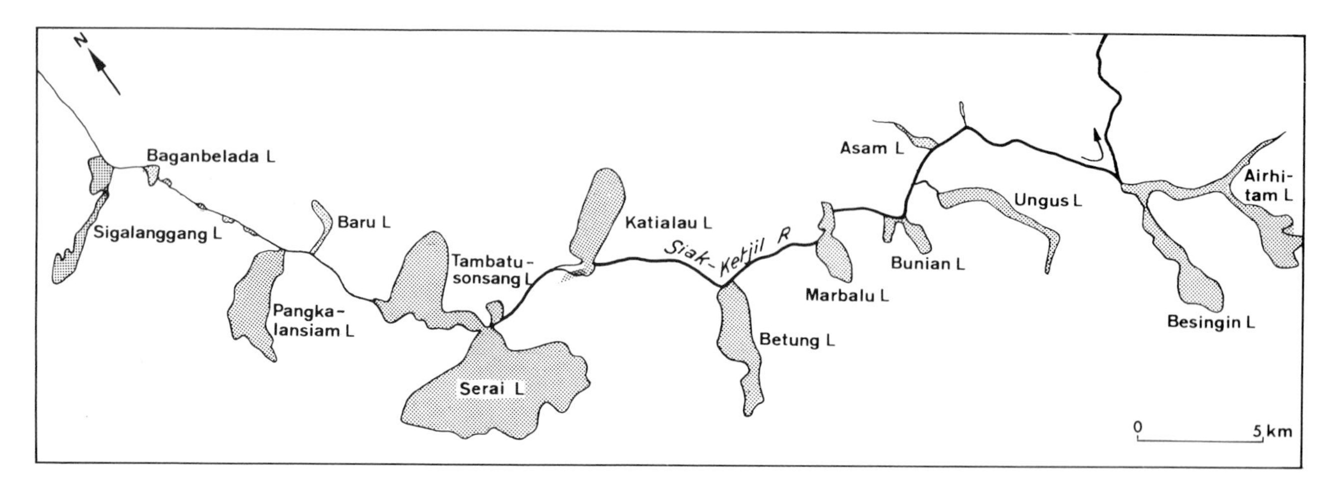

Fig. 3.16 The Siak-kecil River with permanently flooded back swamps in a zone of active subsidence in Sumatra, Indonesia.

Fig. 3.17 Oblique aerial view of the drowned forests and backswamps along the Fly River, New Guinea; a situation comparable to that of fig. 3.16.

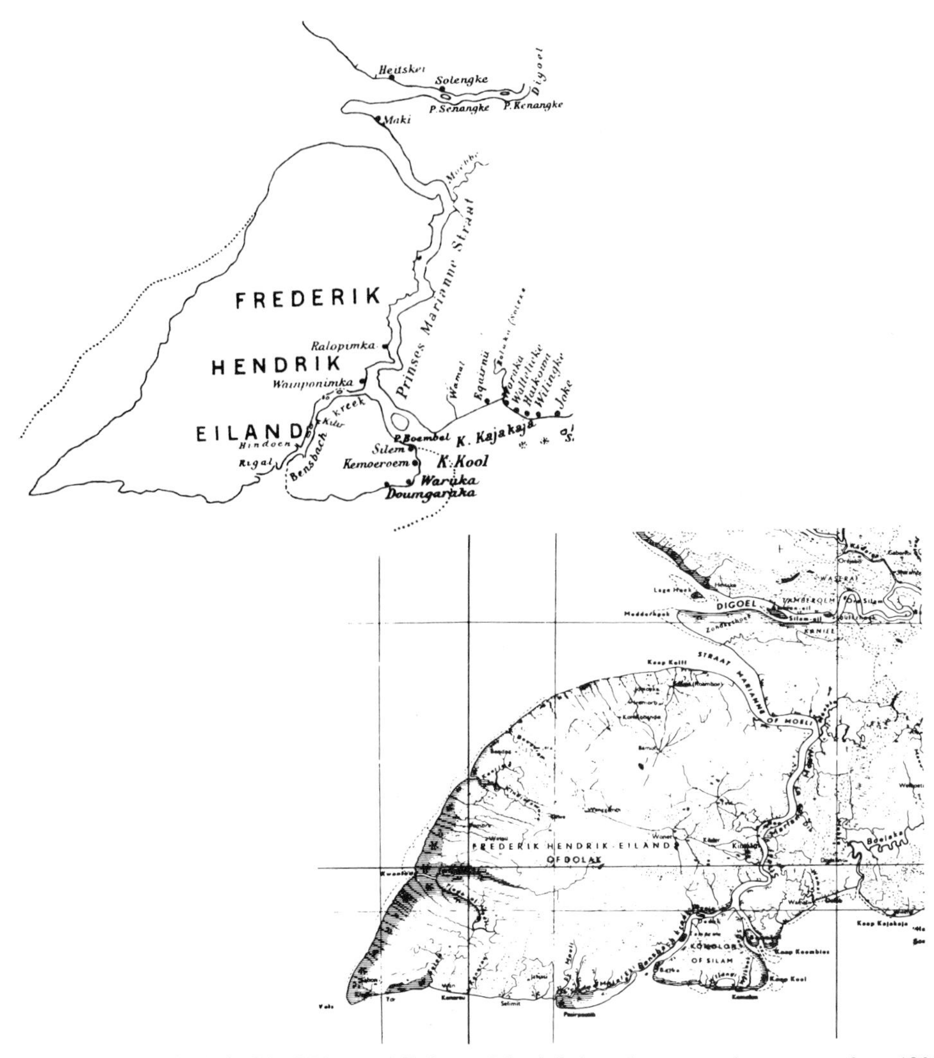

Fig. 3.18 The mouth of the Digul River and Kolepom Island, Indonesia as appearing on maps from 1903 (top) and 1945 (bottom). The seaward displacement of the coastline is due to recent upheaval.

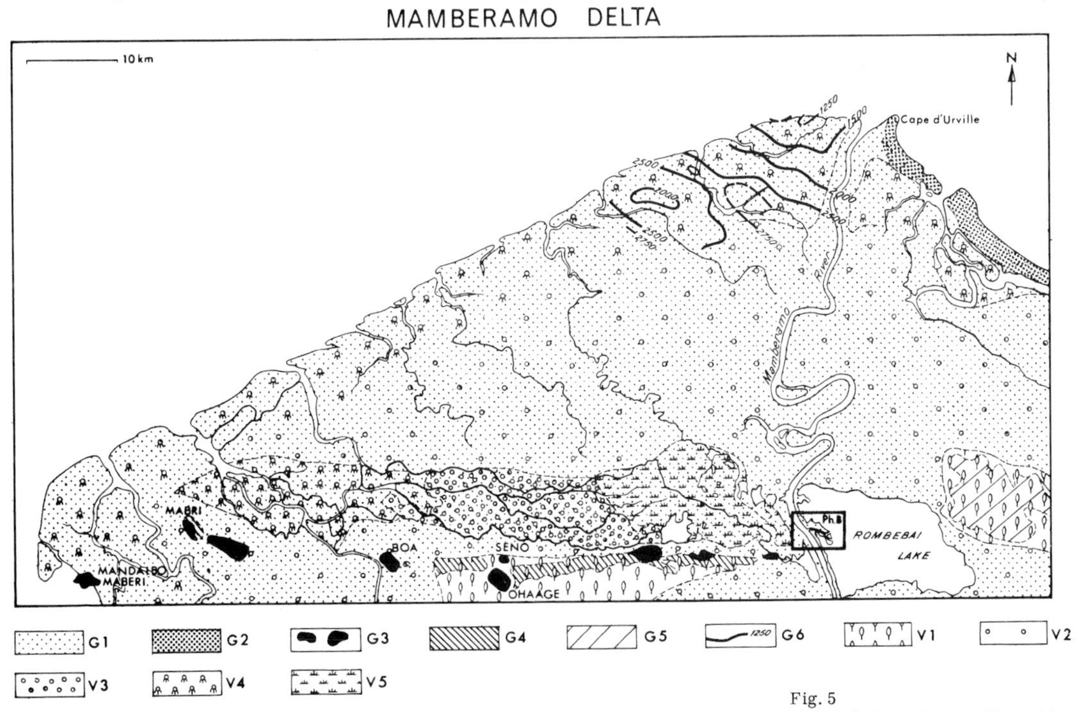

Fig. 3.19 Map showing an EW stretching zone of subsidence in the Mamberamo Delta, Irian, Indonesia as reflected in geomorphological/geological (G) and vegetation (V) features. Key: G 1. alluvium, 2. beach ridge, 3. mud volcanoes, 4. Mamberamo formation, 5. Kukunduri formation, 6. depth contours of crystalline basement (m). V: 1. hill forest, 2. swamp forest, 3. low swampy vegetation, 4. mangrove vegetation, 5. grass and rush marshes.

bordering the delta at its seaward side, and dead tree stumps found on the beach, make some degree of subsidence likely, probably due to compaction of the sediments. The total picture obtained is one of inactive deltaic environment.

A first glance at the map reveals the delta as roughly triangular in shape and characterised by a rather straight shore line. However, a closer inspection of the area reveals that the situation is a more complex one and that structural phenomena determine the outline of the delta to a considerable degree. The location of Cape d'Urville, on the northern tip of the delta, is certainly related to the occurrence of a structural high where crystalline basement rocks are present underneath the various members which unconformably underlay the recent and sub-recent alluvial deposits. The depth of the crystalline basement is indicated in fig. 3.19. The minimum depth known is 1227 metres (W.A. Visser and J.J. Hermes, 1962). Another structural feature of importance is the occurrence of an east-west ridge rising from the plain and approximately 100 metres high, where the Mamberamo formation is exposed. According to Visser and Hermes, a fault occurs along its northern edge. The ridge is crowned by a row of mud volcanoes, the most important of which are from West to East: Mandaibo/Maberi, Mabri, Boa, and Seni/Ohaage. Parts of the ridge, particularly in the west, are submerged below the alluvium. Its continuation can be found in the island of Kurudu located to the west of Dombo straits.

The western limit of the delta is thus also due to structural phenomena. The alluvial plains measuring approx. 40.000 sq. km. to the north of the above mentioned ridge also cover large areas further south, where the Andaiwaren River flows westward and debouches in the Waropen Bay (to the south of fig. 3.19). The outline of the deltaic area is thus strongly influenced by young crustal movements and older structural phenomena.

The interior of the delta also is affected by recent crustal movements and the above mentioned ridge appears to be accompanied by a swampy zone of subsidence at its northern side. This zone is indicated by numerous lakes, the largest of which is the Rombebai Lake, located directly east of the Mamberamo River. The east-west trend of the creeks occurring within this zone is another characteristic. The distribution of vegetation types in the delta is largely influenced by the subsidence of this zone and is evidence of its very wet conditions. This fact even appears on the rather sketchy 1:250,000 topographic map of the area, dating from 1944, compiled from trimetrogon aerial photography. A 'grassy swamp' area is indicated to the west of the Rombebai Lake and further westward 'low swampy vegetation' and mangroves occur. More detailed information on the vegetation types of an approx. 10 km wide belt at either side of the river can be derived from the 1:100,000 Schematic Vegetation Map of the Mamberamo prepared from vertical air photos by Ir. D.A. Boon and Ir. D.A. Stellingwerf of the ITC Forestry Branch in 1953.This map shows that the mangrove vegetation is much more extensive than indicated in the above mentioned 1:250,000 map; which is in line with the statements of Visser and Hermes. Near the Mamberamo mouth a 8-10 kilometre broad mangrove zone occurs. Swamp forest is developed further to the south but river bank forest occurs on the natural levees of the Mamberamo and also in some other localities pointing to former courses of this river. Grass and rush marshes are mapped to the west of the Rombebai Lake, in the 'grassy swamp' area of the 1:250,000 map. Hill forests occur on the east-west running ridge to the south of this zone of recent subsidence and furthermore to the south of Rombebai Lake.

The subsidence of this zone also affects the Mamberamo River itself. As its natural levee cannot escape the subsidence, it breaches when the river is in spate, which occurs preferably in this part of its course. Several small deltas formed in the Rombebai Lake as a result of these circumstances. The formation of these 'deltas in a delta' may finally result in the silting up of the lake. However, it is also possible, that some day the river will take a new course as a result of these tectonic movements.

It is evident from the aforesaid that geomorphology contributes considerably to the unraveling of neotectonics.

3.5 Geomorphology and Soil Development

3.5.1 General

As outlined in section 3.1, the relations of geomorphology with soil science differ essentially from its linkages with geology. Although geomorphological evolution may be considered to be embedded in a geological framework the landforms and relief evolution form in a general sense the environment in which the soils develop. Geomorphological factors of various kinds strongly influence the distributional pattern of the soils in an area and also play an important role in the stage of soil development as evidenced by the soil profiles. Often the soils are not directly derived from the underlying bed rock, but develop in slope deposits or other covering materials that pre-date soil formation. Such deposits may be related to earlier geomorphological situations when relief, climatic and other situations differed from the present ones (Riezebos, 1979). In certain cases sedimentation continues up to the present and then precludes maturing of the soils.

Soil formation, in fact, is a complex matter and depends on various factors. The disintegrated rock forms the in situ and/or allochthonous parent material of the soil. This is affected by the combined soil forming factors, which come under the main headings: climate, relief, vegetation, time and man. The ultimately developed soil is a function of these factors (Buringh and Vink, 1965). In order to clarify the role of geomorphology in this context, it is important to emphasize that the factors mentioned above are of interest, due to their effect on the geomorphological processes related to weathering, removal, transportation and deposition of parent material and soil. The role of climate is, in this respect, partly direct and partly due to its control of vegetation, (micro) organisms, etc. The role of man also is of importance, mainly because of the changes in slope and other processes (Morgan, 1979) brought about by deforestation, agricultural practices and engineering works. With these changing processes, changes in soil development will also be generated.

Among the changes which occur gradually, the accumulation of organic matter in the upper soil horizon should be mentioned. Organic and inorganic compounds evolve and mix with the primary mineral particles to form soils of distinct structure, porosity, etc. Soil horizons of various colour and structure develop under the influence of biotic and abiotic factors. Even the mineral content may change to a certain extent: new clay minerals, not initially present can often be traced. Leaching in one horizon and accumulation in another are other forms by which stratification of soils is effectuated. Among the biotic factors affecting the physical and chemical changes brought about, micro- and macro organisms, and root development should be mentioned; the abiotic factors involved englobe internal factors such as humidity conditions, soil climate and external factors which are the various geomorphological processes.

The relief in itself is also an important geomorphological factor in soil formation (Bruce King, 1975; Bushnell, 1928; Curtis, 1965; Walker et al., 1968; Walker and Ruhe, 1968; Wall, 1964). Rain water tends to collect in depressions where also shallow groundwater may occur. Hydromorphic soils with gley development are common there. On the higher, drier parts of the relief, such as on slopes and plateaus other soil characteristics prevail and a good correlation between land form and soil patterns often exists. The same applies to areas where aeolian relief, composed of deflation forms and those produced by wind accumulation, dominate. It should be understood, however, that linking soil patterns to relief only, is unsatisfactory because the correlation between soils and land form is usually rooted in spatial variations in type and intensity of geomorphological processes and in the age and chronology of land forms. A study of the whole morphogenetic development of an area will ultimately lead to results which are useful for the soil scientist.

In areas where, for example, a staircase of fluvial or coastal terraces is developed the higher terraces are usually markedly older, and aging, or the maturing of soils is much more advanced there than on the lower, more recent terraces. One may say that the factor 'time' plays a role here but this is rather a simplification as one must bear in mind that climatic changes of the past are often involved in the associated variations in kind and density of vegetation and type and intensity of geomorphological processes. The importance of morphogenesis, morphochronology and palaeogeomorphology also applies to periods pre-dating soil development: only in relatively rare cases does the parent rock in situ provide the parent material for soil development. In many cases this is a Quaternary (or older) sediment, the distributional pattern, characteristics and origin of which can only be revealed by thorough geomorphological analysis. To most geologists this material is a thin unconformable cover of no particular interest. It may be a glacial till or an aeolian cover, but it may also consist of various kinds of reworked, old weathering products.

It is evident from the aforesaid that geomorphology may interest the soil scientist in many ways. The lithological factor in soil formation can be studied with the aid of geomorphological methods, because of the linkage between the type of parent rock in situ and the land forms and processes developed there. Where allochthonous parent materials are concerned, the role of geomorphology becomes even more explicit, as outlined above. Of the biological factors in soil formation, vegetation and the associated soil forming processes are in particular, subject to geomorphological interest. Geomorphological processes are distinctly affected by the vegetative cover and its influence on matters such as surface run-off, infiltration and incision of ravines. They have, inversely, an effect on the type and distributional pattern of plant associations developed. (Tricart, 1952, 1965). The influence of geomorphology on soil formation (Blackburn, 1968; Furley, 1968; Goosen, 1961, 1967, 1971; Hodgsen and Catt, 1976; Knott, 1980), is so important, that it is justified to classify the geomorphological factor in soil formation under a separate heading although it is intricately interwoven with the two soil forming factors mentioned. The role of land forms, processes (Kleiss, 1970; Moss, 1965), and the morphogenetical evolution of the area of study, should be adequately incorporated in geomorphological studies for pedological purposes (Leveque, 1969). This also applies if specific soil characteristics, e.g. salinity is concerned. The map of fig. 3.20 shows part of the Ebro Valley, Spain where salinity occurs in zones bordering alluvial fans where groundwater from the fans emerges and evaporates. Fig. 3.21 shows saline patches near a

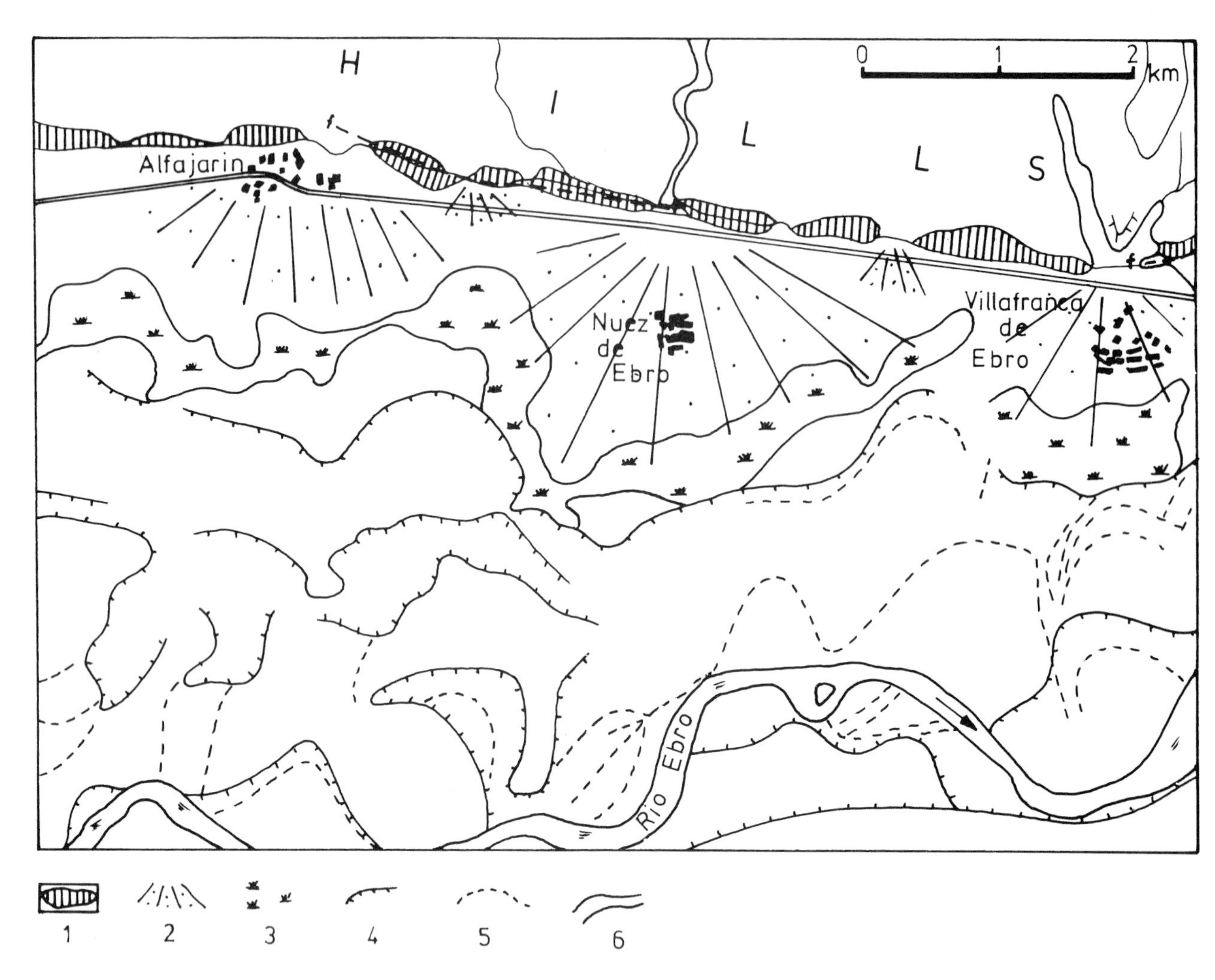

Fig. 3.20 Ebro Valley, east of Zaragoza, Spain showing saline areas associated with lower parts of fan structurs. Key: 1. scarps, 2. fans, 3. saline areas, 4. terrace edges, 5. former river courses, 6. Ebro River. After: van Zuidam, 1976.

Fig. 3.21 Salinity due to evaporation of moist zone around spring, Quetta, Pakistan.

spring in a comparable situation in Pakistan.

Soil scientists have long since recognized this situation. Ruhe (1956) e.g. stated that the study of parent materials, the changes which have taken place in these materials in the course of time and relief or topography of the earth's surface are important factors in the study of soils and soil conditions. It should be stated here, however, that the role of geomorphological processes and morphogenesis have often been underestimated among soil scientists, whereas land form has been relatively overemphasized as a geomorphological factor.

3.6 Some special Geomorphological Influences on Soils

Some aspects of the relation between morphogenesis and pedogenesis deserve further mention. They relate to the velocity and intensity of morphogenetic processes and to their variations in space and time. In many areas of denudation, the geomorphological slope processes are diffuse and generalised in nature (Lietzke, 1975; Malo, 1974; Pandy et al., 1967), and comprise of diffuse run-off, surface wash, creep, etc. Normally the evolution caused here is comparatively slow, thus soil development can keep pace with it. As a consequence, a distinct catenary soil sequence is formed along such slopes. Deep, well-developed soils are found on the flat or gently rolling uplands, where infiltration dominates over surface run-off. Truncated soil profiles may occur on the higher parts of the slopes, moderately deep soils usually characterise the middle parts and immature colluvial soils without much profile development tend to be formed in the low areas surrounding the hills. Countless variations to this general picture exist, depending on the local soil forming conditions, as discussed in the next sub-section.

Geomorphological processes which are localised, linear or concentrated, are in one way or another, usually more violent and proceed at a rate many times higher than the formation of soils. Pedogenesis is then interrupted, if it ever had a chance to start at all. This is the case when ravines and badlands invade an area, either resulting from natural causes or, more frequently, due to accelerated erosion initiated by land misuse. In areas of rapid accumulation, for example, a flood plain (Michel, 1978) a similar situation may exist: the rate of deposition may be such that soil development is precluded, with the exception of the hydromorphic soil processes which proceed more rapidly than other aspects of the pedogenesis. In natural levees composed of fairly pervious material, a hydromorphic soil is often found at some depth, with a gley horizon in the profile. The lower lying backswamps where both external and internal drainage is normally poor, gley development is much more intense and found at and near the surface. Such areas are furthermore, characterised by peat formation and swamps, as well as by an absence of A and B horizons in the profile because of the absence of leaching.

Spatial variations in type and intensity of geomorphological processes are common and their interrelations with soil development are often complex. They strongly affect the distributional patterns of soils in many areas as evidenced by the example of hydromorphic soils, previously mentioned. In fact, morphogenesis and pedogenesis are usually so interwoven that geomorphologists and soil scientists benefit from each others work and have much to gain by joint research and surveying (Jungerius, 1964). The following example may serve to further elucidate the role of spatial variations in geomorphological processes.

The steep and sparsely vegetated mountains in semi-arid areas are often surrounded by a gently sloping and slightly concave foot slope or glacis, from which they are separated by a distinct break of slope. These glacis are generally characterized by the occurrence of calcrete formation, the thickness of which varies from place to place. Depending on the climatological and lithological conditions affecting the migration of calcium, iron and silicium, ferricretes and silcretes may also occur. Their formation is associated with certain well-defined phases of the morphogenetical development of the areas concerned (Mulcany, 1961). When mapping the areal extent of these duricrusts (Goudie, 1973; Martin, 1967; McFarlane, 1976; Wright, 1963) and studying their various characteristics, the normal pattern is for the duricrusts to be the thickest and hardest at some distance from the hills and mountains forming the hinterland. This is explained by the fact that the diffuse surface run-off and imperfect sheet flood phenomena characterizing the foot slopes, diminish downslope; whereas infiltration and evaporation proportionally gain in importance in this direction. If the footslopes come under the attack of linear erosion, this process tends to be more violent on the somewhat steeper glacis slopes near the hinterland. This stronger erosion, in combination with the relatively weak duricrust formation in these parts, may easily lead to the development ment of a depression separating the glacis form the hinterland. A pattern of calcrete 'cuestas' facing the hills of the hinterland may then result.

Temporal variations in type and intensity of geomorphological processes can result from diversified causes, such as tectonic movements, changes in groundwater level or vegetation, and above all, climatic variations. The soil forming factors change accordingly and traces of this persist up to the present. Relic soils may occur at places where the profile is comparatively unaffected by later soil development. Truncated soils may occur where periods of violent morphogenetic developments have caused the removal of parts of the soil profile. Soils covered by more recently developed soil profiles and polygenetic soils developed during several phases of different climatic morphological and pedogenetical conditions have also been reported from many areas.

Well-known relic soils are the red-yellow podzol soils found in isolated pockets of western and central Europe with about 45-50% kaolinite in the clay fraction and dating from the lower Pliocene when apparently a warm monsoonal climate prevailed in Europe. In the same area yellow-red podzolic soils and bleached sediments are also found which contain 75-85% illite in the clay fraction and are believed to have formed under drier, Mediterranean climatic conditions from the middle Pliocene to the earlier Pleistocene Interglacials. (Bakker, 1960). Thereafter, grey-brown and brown soils with illite dominance developed.

An interesting example of palaeogeomorphology and Pleistocene climates as affecting type and distributional patterns of soils is given by Tricart (1962), from the cuesta area of NW France, in particular the Champagne crayeuse. The soil here is developed on superficial deposits formed during periglacial conditions as a result of solifluction to which process chalk is very susceptible. Three geomorphological subdivisions of the cuesta scarp can be clearly indicated, viz:

1. The top of the cuesta where the chalk outcrops with many joints. The soil is largely removed and the permeability is low. After rain, the surface becomes muddy whereafter it dries out and becomes hard. It is unsuitable for agriculture and is largely under forest and grass.
2. The solifluction has been active in the middle portion of the slopes below zone 1, at least during certain parts of the upper Pleistocene. Though chemically the same, the soil is lighter in texture (sandy) and much better for agriculture.
3. The best fields, however, occur at the foot of the slope where the fines are completely washed away and where the terrain is flat enough to permit mechanical farm practices. It is evident from this example that climatic events of the past and the morphogenetic developments generated by them have a great influence on the soils occurring at present and also on the agricultural utilization of the land.

Comparable effects of palaeogeomorphological and climatological situations have also been reported from other parts of the world. Lime concretions and soft calcretes at shallow depth, locally known as 'kankar' are of widespread occurrence in the lowlands of India. They are clearly phenomena inherited from the past. The same applies to the calcretes and ferricretes found in numerous semi-arid and semi-humid regions in Africa, Latin America Asia and Australia, (Pujos and Raynal, 1960; Litchfield and Mabbutt, 1962). Pedogenetic aspects such as leaching and depth of various soil constituants, colour, enrichment of fine-textured materials, etc. are affected there by palaeogeomorphological conditions.

In Swaziland the author observed that the humid, tropical weathering in the 'Middle Veld' pediplain which in the past resulted in strong alteration of granites and in deep dark-red latosols where the weathered granitic corestones can hardly be recognized, must be largely a fossil phenomenon. In parts, the latosol is completely or largely removed, leaving granitic block fields and giving rise to extensive sedimentation in the valleys where at least one major accumulation terrace in the allochthonous red material has been formed. In other localities, a capping of thin stone lines separates the old latosols from the overlying greyish soils which are formed under the conditions presently prevailing. Characteristic profiles and further evidence can be found in many places. The dating of the morphogenetic crisis that must have occurred here in the past still waits further investigation.

In SE Asia Pleistocene climatic variations have given rise to considerable variations in the morphogenetical development and consequently, to the occurrence of palaeosols of various kinds. Deep red latosols were found to underly considerably younger brown latosols in Java. Laterites and latosols were proven to be relic features in Malaysia (Verstappen, 1974, 1980). New and interesting data have been recently gathered in Central Java by Van der Linden (1978). It will be clear from the aforesaid that palaeogeomorphology and palaeoclimates may render the relation between morphogenesis and pedogenesis a very complex matter. Precise geomorphological investigations, in cooperation with soil scientists, are required to satisfactorily unravel these relations and to enable the geomorphologist to optimally contribute to soil science and soil survey.

3.7 Geomorphology and Soil Survey

It is evident that it is a great advantage for a soil surveyor if ample geomorphological information about the area to be surveyed is available. Due to the close relationship between these two environmental sciences soil indications in turn may be of great use for the analysis of the geomorphological situation and development (Jungerius, 1964). It is generally beneficial for the geomorphological survey and the soil survey to be executed in good coordination (Morariu, 1976; Soil Survey Manual, 1951; Vink, 1962). How this is to be realised, depends on the pre-existing data of the area, the professional skills of the soil surveyor and geomorphologist concerned and the purpose and scale of the survey.

Simultaneous execution of geomorphological and soil survey by the two surveyors is often a good solution, though it may be useful if the geomorphological survey is a few steps ahead of the soil survey, in order to work efficiently from the general to the specific.

The 'chain' of earth sciences: geophysics-geology-geomorphology-soil science, in this order, leads to a subdivision of the land into units of decreasing spatial and temporal dimensions. A consequence in the field of surveying is, that when making a separate geomorphological map to support soil mapping, it usually suffices to prepare that map in a smaller scale than the soil map, e.g., a 1:50,000 geomorphological survey is usually adequate for a 1:25,000 scale soil survey.

The mapping scale has a rather important effect on the type of geomorphological input in soil survey and on the mode of cooperation between geomorphologist and soil surveyor (Tricart, 1978). In case of detailed mapping, the geomorphologist will direct his attention towards geomorphological processes, whereas the soil surveyor concentrates on the study of the soil profiles by augering and the study of pits and natural outcrops using pedological methods of research. In this way, the two work on equal footing; the contribution of the geomorphologist being, in the first place, to interprete the effects of geomorphological processes in characteristics of the profile. In semi-detailed and, particularly in reconnaissance survey, the task of the geomorphologist is first to establish a breakdown of the land into geomorphological units. The soil surveyor then attempts, in cooperation with the geomorphologist, to correlate the boundaries of the soil associations, etc. classified by him, with the general geomorphological patterns.

In an attempt to define the various factors influencing soil formation and thus the spatial distribution of types of soil, Vink (1964) listed a number of such elements which he classified under six headings as follows:

1. Elements related to the soils themselves
2. Elements related to the general morphology of the terrain
3. Elements related to special aspects of the terrain
4. Elements related to the vegetation cover
5. Elements related to specific human aspects
6. Inferred elements or elements based on converging evidence

Although this classification is to be used in photo interpretation for soil survey purposes and one may debate about details, it clearly shows the importance attached by numerous soil surveyors to geomorphological characteristics and related parameters of the land. It is interesting to note in this context that Vink uses the term 'physiographic systems' thus underlining that what matters are not merely the topographic patterns, but that the mapping units are dynamic, landscape ecological phenomena.

In many cases the geomorphological analysis for purposes of soil mapping concentrates mainly on the classification of land forms and is guided by the concept of the toposequence, which links the distribution of certain soils and soil characteristics to their topographic situation (Adams and Walker, 1975; Hanna, 1975). In section 3.5 it was already stated that deep soils are often associated with areas of gentle relief, such as plateaus, whereas truncated soil profiles are more common on the higher parts of slopes and colluvial soils on the lower parts and the foot of slopes. It is evident that this approach is only valid if, apart from the land form, also the slope processes responsible for these characteristics are taken into consideration. The fact that soils develop in an organised manner on any given land form and that a close association exists between the different elements of landscapes and soils along the slope from the hill tops to the valley floor is now generally recognized. Wooldridge (1939), observed that the geomorphological evolution of Chalk country was reflected in variation in soil type (and vegetation). Twidale (1961), expresses the view that a close association exists between pedogenesis and geomorphological evolution of the land surface.

The mechanism by which the toposequence is brought about is often clearly unraveled. Biswas and Gawande (1967, 1964), mention the occurrence of red lateritic soils on the ridges of part of Madhya Pradesh, India, and black regur soils in the valley bottoms where calcification occurs. The marked effect of drainage on the chemical composition and the clay fraction of these soils is emphasized and four main soil types are recognizable within the toposequence (see also: Gawande and Biswas, 1967). Biswas et al (1966), have made comparable observations also in Andhra Pradesh, India, and elaborate on the changes in colour, texture and chemical composition of the soils along the slopes. They also make a point of the role of topography in diversifying the external and internal drainage conditions and ultimately, the characteristics developed from the soils.

Closely associated with the toposequence is the site-catena concept. The main differences being that sites and catenas are mapping units of a certain areal extent, rather than profiles and that emphasis is not solely on the relation between landform and soil, but on the land as a whole, therefore including other elements of the land such as the hydrological conditions. A site can be defined as a small, fairly uniform land surface, a subdivision of the terrain, which occurs repeatedly within a region. Approximate equivalents are, for example, facet and ecotope. A catena can be defined as the assemblage of a number of sites, characteristic for a certain part of the terrain. Approximate equivalents are pattern and region. Milne (1935), developed the catena concept to indicate a regular repetition of a certain sequence of soil profiles, together with a certain topography. This definition could also be used for toposequence. The uniformity of parent material was considered by him to be of subsidiary interest. It is now generally understood that in any region within a limited climatic range, similar land forms of the same geomorphological origin have similar groupsof soil and soil moisture relationships (Christian, C.S., 1958, 1968; Webster, 1962).

It is evident from this quotation that in the course of the years emphasis has gradually moved away from the topographic aspect, to incorporate the morphogenetic factors in the site-catena concept. The matter is elaborated upon in section 12.

The essence of the geomorphological contribution to soil surveying is in the recognition that soils are landscapes (Pitty, 1979; Pullan, 1970), as well as profiles. This understanding was firmly established in early soil science and particularly among early Russian soil scientists but it has been neglected for several decades. Emphasis was shifted to the study of soil profiles, without paying much attention to the physiographic context from which they were obtained and the subsequent details of these soil samples in the laboratory. This approach is strongly advocated by soil scientists of the USA where soil surveys are mainly based on the results of systematic augering from which a complex taxonomic system of soil classification has been formulated. The value of this method is not challenged, but it should be clear that optimum results can only be expected when a balance between the landscape aspect and the soil profile aspect is maintained. A revival of the landscape approach to soil science and survey came in the nineteen thirties, particularly in the Netherlands, where the late Professor Edelman established what is often referred to as the 'Dutch physiographic school of soil survey' (Edelman, 1957, Buringh, 1954; Vink, 1964, 1976; Goosen, 1967). The taxonomic and landscape approaches to soil survey are not necessarily contradictory. An ideal situation is having the geomorphological survey provide physiographic data and breakdown of the land into units of geographical importance i.e. mappable areas of acceptable taxonomic homogeneity. Fig. 3.22 gives an example. Thereafter, the taxonomic classification of the soils characterizing these units is based on the profile descriptions and laboratory results. Geomorphology and photo interpretation logically are considered important by the adherents of this school.

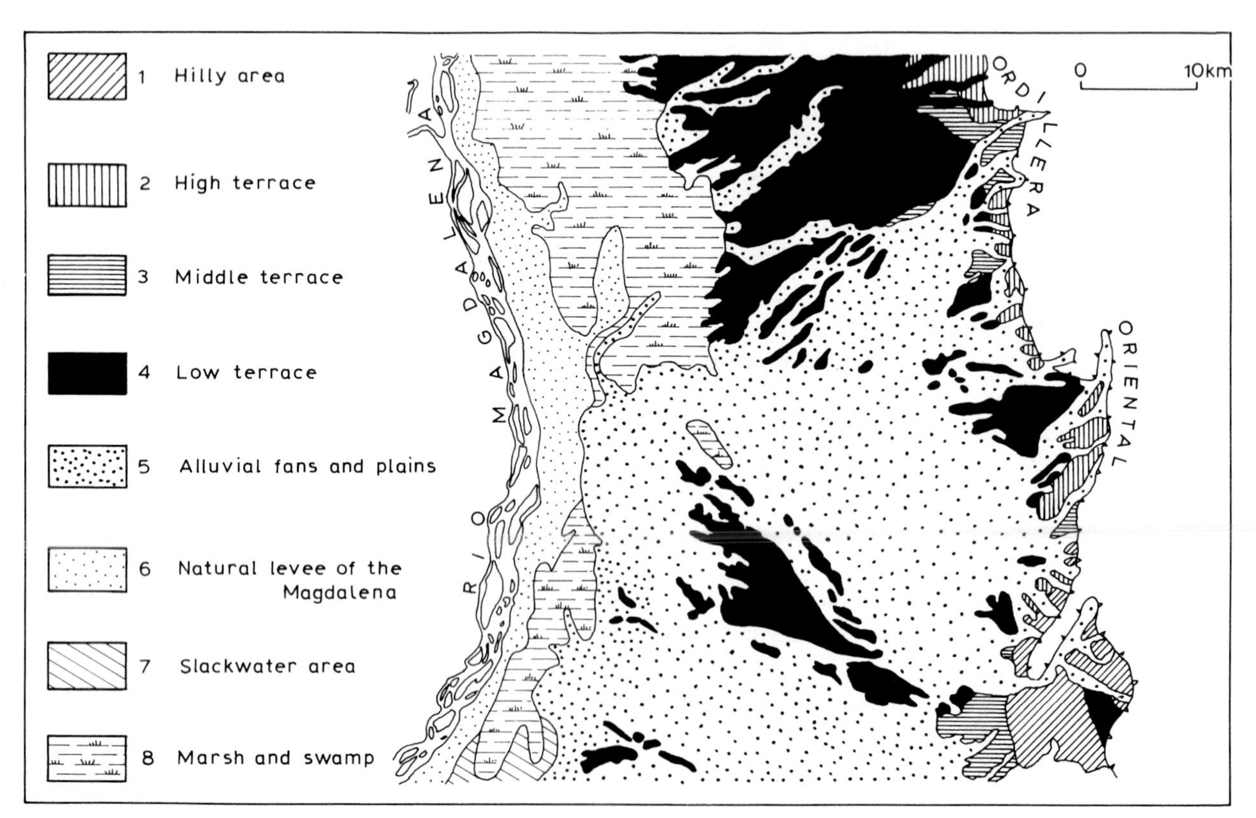

Fig. 3.22 Geomorphological terrain units in part of the middle Magdalena Valley, Colombia for soil survey purposes. After: Goosen, 1961.

References

Adams, J.A., and Walker, T.W., 1975. Some properties of a chrono-toposequence of soils from granite in New Zealand. Geoderma, 13(1): 41-52.

Aristarkhova, L.B., 1968. A rational complex of methods of detailed structural geomorphological studies in the salt-dome region of the Sub-Urals plateau (in Russian). Sbornik. Geomorfologisch. analiz. pri geologich. is issled v. Prikaspijksvpadine: 5-21.

Bakker, J.P., 1960. Some observations in connection with recent Dutch investigations about granite weathering and slope development in different climates and climatic changes. Ztschr.f.Geom., Suppl. Bd. I: 69-92.

Berlyant, A.M., 1969. Regularities in the connection between the most recent and the oldest tectonic structures in the N. Perchskaya Depression (in Russian). Sovetskaya geol., 1: 115-120.

Biswas, T.D., and Gawande, S.P., 1962. Studies in genesis of catenary soils on sedimentary formation in Chhatishgarh Basin of Madhya Pradesh. I. Morphology and mechanical composition, Vol 10.

Biswas, T.D., and Gawande, S.P., 1964. Relation of manganese in genesis of catenary soils. I. Indian Soc. Soil Sci., 12: 261-267.

Biswas, T.D. et al, 1966. Characteristics of catenary soils on granite-gneiss parent rock in the Kurnool District of Andhre Pradesh. I. Indian Soc. Soil Sci., 14: 183-195.

Belkhodja, K., 1972. Sols de Tunisie. Origine, évolution et caractères de la salinité dans les sols de la plaine de Kairouan (Tunisie Central). Contribution a l'étude de leur mise en valeur.

Blackburn, G., 1968. Soil distribution and geomorphology of constructional coastal lowland. Trans. 9th Intern. Congr. Soil Sc., Adelaide, Australia: 623-630.

Bruce King, R., 1975. Geomorphic and soil correlation analysis of land systems in Northern and Luapula Provinces of Zambia. Trans. Inst. Br. Geogr., 64: 67-76.

Bruce King, R.A., 1975. A comparison of information theory interdependence and product-moment correlation analyses as applied to geomorphic and soils data. Ztschr.f.Geom., 19(4): 393-404.

Buringh, P., 1954. The analysis and interpretation of aerial photographs in soil survey and land classification. Neth.J. Agric.Sci, 2(1): 16-26.

Buringh, P., and Vink, A.P.A., 1965. The importance of geology in airphoto interpretation for soil mapping. ITC Publ. B 33: 24 pp.

Bushnell, T.M., 1928. To what extent should location, topography or physiography constitute a basis for differentiating soil into units or groups? Proc. 1st Int. Congr. Soil Sci., IV: 502-506.

Christian, C.S., 1958. The concept of land units and land systems. Proc. 9th Pacific Sci. Congress, 20: 74-81.

Christian, C.S., and Stewart, G.A., 1968. Methodology of integrated surveys. Proc. 1964 Toulouse Conf. Aerial Survey and Integrated Studies: 223-280.

Curtis, L.F., 1965. The description of relief in field studies of soils. J. Soil Sc., 16: 16-30.

Edelman, C.H., 1957. De betekenis van geomorfologie voor de bodemkunde. Tijdschr. Kon. Ned. Aardrijksk. Gen., 74: 257-263.

Edelman, T., 1954. Tectonic movements as resulting from the comparison of two precision levellings. Geol. en Mijnbouw, 16 : 209-213.

Furley, P.A., 1968. Soil formation and slope development. Ztschr.f.Geom., NS 12: 25-42.

Gawande, S.P., and Biswas, T.D., 1967. Studies in genesis of catenary soils on sedimentary formation in Chhatisgarh Basin of Madhya Pradesh, III. J. Indian Soc. Sci., 15: 158-172.

Gerrard, J., 1981. Soils and landform. An integration of geomorphology and pedology. G. Allan & Unwin, London, 238 pp.

Goosen, D., 1961. A study of geomorphology and soils in the middle Magdalena Valley, Colombia. ITC Publ. B 9, 24 pp.

Goosen, D., 1967. Aerial Photointerpretation in soil survey. FAO, Soils Bull., 6, 55 pp.

Goosen, D., 1971. Physiography and soils of the Llanos Orientales, Colombia, Thesis Univ. Amsterdam, ITC Publ. B 64: 1-198.

Goudie, A., 1973. The geomorphic and resource significance of calcrete. Progress in Geography, 5: 77-118.

Hanna, W.E., et al., 1975. Soil-geomorphic relationships in a first-order valley in Central New York. Proc. Soil Sci. Soc. Am., 39(4): 716-722.

Herath, L., 1962. Shoreline development and protection of Negombo Beach, Ceylon. An aerial photographic approach. Proc. 1st Symp. Comm. VII, I.S.P., Delft: 453-460.

Hills, E. Sherbon, 1961. Morphotectonics and the geomorphological sciences with special reference to Australia. Quat. J. Geol. Soc., London, 117: 77-89.

Hodgson, J.A., and Catt, J.A., 1976. Soils and geomorphology of the Chalk in south-east England. Earth Surface Processes, 1(2): 181-193.

Jungerius, P.D., 1964. De betekenis van de bodem voor de geomorfologie. Public Lecture. Univ. Amsterdam, 15 pp.

Kleiss, H., 1970. Hillslope sedimentation and soil formation in Northeastern Iowa. Proc. Soil Sci. Soc. Amer., 34: 287-290.

Knott, P.A., et al., 1980. The relationship between soils and geomorphological mapping units. A case study from Northamptonshire. Bull. Ass. Eng. Geol., 21: 186-193.

Levèque, A., 1969. Les principaux évènements géomorphologiques et les sols du socle granito-gneissique du Togo. Cahiers ORSTOM, Pédol., (2): 203-224.

Lietzke, D.A., et al., 1975. Relationship of geomorphology to origin and distribution of a high charge vermiculitic soil clay. Proc. Soil Sci. Soc. Amer., 39(6): 1169-1176.

Linden, P. van der, 1978. Contemporary soil erosion in the Sanggreman River basin related to the Quaternary landscape development. A pedogeomorphic and hydro-geomorphological case study in middle Java, Indonesia, Thesis Univ. Amsterdam, 110 pp.

Litchfield, W.H., and Mabbutt, J.A., 1962. Hardpan in soil of semi-arid Western Australia, J. of Soil Sci., 13: 148-159.

Malo, D.D., et al., 1974. Soil-landscape relationships in a closed drainage system,. Proc. Soil Sci. Soc Amer., 38(5): 813-818.

Martin, D., 1967. Géomorphologie et sols ferrallitiques dans le Centre-Cameroun. Cahiers ORSTOM. Pédol, V(2): 189-218.

Maurice, Y.T., and Meyer, W.T., 1975. Influence of preglacial dispersion on geochemical exploration in County Offaly, central Ireland. J. Geochem. Explor., 4(3): 315-330.

McFarlane, 1976. Laterite and landscape. Acad. Press, London, New York: 1-151.

Michel, P., 1978. La vallée alluviale du Sénégal. Catena, 5(2): 213-225.

Miller, V.C., 1961. Photogeology. Mc.Graw-Hill, New York, 248 pp.

Milne, G., 1935. Composite units for the mapping of complex soil associations, Trans. 3rd. Congr. Soil Sci., I: 345-347.

Moore, G.T., 1969. Interaction of rivers and oceans.

Pleistocene petroleum potential. Bull. Amer. Soc. Petrol. Geol., 53(12): 2421-2430.
Morariu, T., 1967. Les cartes Pédo-géomorphologiques. Leur importance pour les travaux hydroamélioratifs des plaines inondables. Symp. Int. Géom. Appliquée, Bucarest.: 1-4.
Morgan, R.P.C., 1979. Soil erosion topics in applied geography. Longman Green Ltd., London, 113 pp.
Moss, R.P., 1965. Slope development and soil morphology in a part of South-West Nigeria. J. Soil Sci. 16: 192-209.
Mulcahy, M.J., 1961. Symp. Geochronology, landsurfaces and soils. Adelaide, Australia: 142-151.
Pandey, S., et al., 1967. Geomorphic Influence on soil genesis in semi-arid and arid environments. . J. Indian Soc. Soil Sci., 15: 163-174.
Pitty, A.F., 1979. Geography and soil properties. Methuen & Co., London, 287 pp.
Polynov, B., 1928. Der Boden als landschaftselement. Proc. 1st Int. Congr. Soil Sci., IV: 502-506.
Pujos, A., and Raynal, R., 1959. La géomorphologie appliquée au Maroc. Revue Géom. Dyn., 10: 103-105.
Pullan, R.A., 1970. The soils, soil landscapes and geomorphological evolution of a metasedimentary area in northern Nigeria. Dept. of Geogr., Univ. Liverpool. Res. paper 6, 144 pp.
Riezebos, H.Th., 1979. Geomorphology and soils of the Sipalwini savannah, Suriname. Thesis, Utrecht State Univ., 168 pp.
Semmel, A., 1980. Die geomorphologische Karte als Hilfe bei der geologischen Landesaufnahme. Berliner Geogr. Abh., 31: 67-73.
Shumilov, Y.V., 1969. Some signs of neotectonic movements in the basin of the Malyi Anyui (in Russian). Kolyma, I: 36-39.
Sidorenko, A.W., 1971. Geomorphologie und Volkswirtschaft. Fragen der praktischen Geomorphologie. Ztschr.f.Angewandte Geol., 7: 257-263.
Soil Survey Manual, 1951. U.S. Dept. Agric. Handbook, 18, 503 pp.
Tjia, M.D., et al., 1975. Additional dates on raised shorelines in Malaysia and Indonesia. Sains Mal., 4(2): 69-84.
Tricart, J., 1952. Climat, végétation, sols et géomorphologie. Vol., jub. E. de Martonne, Rennes: 225-239.
Tricart, J., 1962. L'épiderme de la terre. Esquisse d'une Géomorphologie appliquée. Masson et Cie., Paris: 1-160.
Tricart, J., 1965. Morphogénèse et pédogénèse. Science du sol, (1): 69-85.
Tricart, J., 1978. Géomorphologie applicable. Masson, Paris, 204 pp.
Twidale, R.T., 1961. Symp. Geochronology, land surfaces and soils. Adelaide, Australia: 167-174.
Veen, A.W.L., 1970. On geogenesis and pedogenesis in the Old Coastal Plain of Surinam. Thesis Univ. Amsterdam, 120 pp.
Verstappen, H.Th., 1955. Geomorphologische Notizen aus Indonesiën. Erdkunde, 9: 134-144.
Verstappen, H.Th., 1956. Landscape development of the Udjung Kulon Game reserve. Penggemar Alam, 36: 37-51.
Verstappen, H.Th., 1957. Eén en ander over het rifpantser van het eiland Muna (Celebes). Tijdschr. Kon. Ned. Aardrijksk. Gen., 74: 441-449.
Verstappen, H.Th., 1959. Geomorphology and crustal movements of the Aru Islands in relation to the Pleistocene drainage of the Sahul shelf. Amer. J. of Sci., 257: 491-502.
Verstappen, H.Th., 1960. On the geomorphology of raised coral reefs and its tectonic significance. Ztschr.f.Geom., 4: 1-28.
Verstappen, H.Th., 1964. Karst morphology of the Star Mountains (Central New Guinea) and its relation to lithology and climate. Ztschr.f. Geom., 8: 40-49.
Verstappen, H.Th., 1965. Geomorphology. An integrated survey of the Porali Plain, Lasbela, Publ. 2, Arid Zone Res. Inst., Pak. Met. Service, Karachi: 18-34.
Verstappen, H.Th., 1968. Fundamentals of photo-geology/geomorphology (V Vols), ITC Textbook of Photo Interpretation, VII.1, 101 pp.
Verstappen, H.Th., 1973. A geomorphological reconnaissance of Sumatra and adjacent islands. Wolters/Noordhoff Publ. Comp. Groningen: 1-182.
Verstappen, H.Th., 1974. On Palaeo climates and landform development in Malaisia. J. Quat. Res. in SE Asia, I: 3-35.
Verstappen, H.Th., and Cooke, H.J. (Editors), A drought susceptibility pilot survey in northern Botswana. Final Report ITC Publ., 237 pp.
Vilenkin, V.L., 1968. The problem of the tectonic landscapes of Levoberezhnaya (left bank), Ukraine (in Russian). Materialy Khar'kovsk otd Geogr. o-van Ukrainy, 6: 61-70.
Vink, A.P.A., 1968. Planning of soil surveys in land development. Publ. I.L.R.I., Wageningen, 10: 45 pp.
Vink, A.P.A., 1962. Practical soil surveys and their interpretation for practical purposes. ITC Publ. B 16, 20 pp.
Vink, A.P.A., 1968. Aerial photographs and the soil sciences. Proc. Conf. Aerial photographs and integrated studies, Toulouse, UNESCO, Paris: 81-141.
Visser, W.A., and Hermes, J.J., 1962. Geological results of the exploration for oil in Neth. New Guinea. Verh. Kon. Ned. Geol. Mijnb. Gen., 20; 1-265.
Walker, P., et al., 1968. Relation between landform parameters and soil properties. Proc. Soil Sci. Soc. Amer. , 32: 101-104.
Walker, P., and Ruhe, R., 1968. Hillslope models and soil formation: II, closed systems. Journal Paper J - 5737, Iowa Agric. and Home Econ. Experiment station Ames (Iowa). Project 1250.
Wall, J.R.D., 1964. Topography-soil relationships in lowland Serawak. J. Trop. Geogr., 18: 192-199.
Webster, R., 1962. The use of basic physiographic units in air photo interpretation. Trans. First Symp. Photo interpretation, Comm. VII, Waltman, Delft: 143-148.
Wooldridge, S.H., 1939. The cycle of erosion and the representation of relief. Scott. Geogr. Mag., 48: 30-36.
Wright, R.L., 1963. Deep weathering and erosion surfaces in the Daly River basin, N. Territory, Australia. J. Geol. Soc. Austral., 10: 151-164.
Yonekura, N., 1972. A review on seismic crustal deformation in and near Japan. Bull. Dept. Geogr., Univ. Tokyo, 4: 17-50.
Zuidam, R.A. van, 1976. Geomorphological development of the Zaragoza Region, Spain. Thesis, Utrecht State Univ, 211 pp.
Zvonkova, T.V., 1970. Geomorphic methods in oil and gas prospecting. Soviet Geography, 13(6): 353-363 (translated from: Prikladnaya geomorfologiya, 1970: 99-114).

GEOMORPHOLOGY IN HYDROLOGICAL SURVEYS

4.1 Introduction

The relationship between geomorphology and hydrology is diversified and complex. The four major aspects of geomorphology mentioned in section 1: morphographic and particularly morphometric landform data, fluvial and other processes, genetic landform development and environmental geomorphology have to be considered, although emphasis varies with the type of study and with the terrain characteristics of the area concerned. The importance of geomorphology for hydrological purposes has been increasingly appreciated among hydrologists in the last few decades when they became engaged in the development of water resources in developing and other countries where the available hydrological data were found to be inadequate for their aims. Observations on discharge, sediment load, etc. over a sufficient period of time, are often few in number or even totally absent. Thus the hydrologist was obliged to use different, unconventional approaches to evaluate the characteristics of rivers and drainage basins and to get a proper idea of the order of magnitude of the various aspects of the water problem (Kuiper, 1971).

The study of fluvial geomorphology and the morphometrical and environmental geomorphological analysis of drainage basins have become increasingly important. In fact, geomorphology was found to have very close links with both surface and subsurface water conditions. A specific asset of geomorphology is that it aids to describe and evaluate the environment in which the water circulates thus providing the hydrologist working in areas where essential data are lacking, with information enabling him to understand the situation and to make the proper decisions. (White, 1975; Zaporozec, 1979). Schumm (1964), emphasized this role of geomorphology and maintains that a general relationship exists between hydrological and geomorphological variables. Once these relations are established, the probable hydrological characteristics of other areas, which are geomorphologically similar, can be estimated. This applies to both surface and groundwater resources. Meijerink (1974), elaborated on this and came to similar conclusions. Popov (1960), states that whereas the topographer is only interested in the actual situation, the hydrologist tries to grasp the natural processes to forecast future developments. Thus, it appears that geomorphological criteria do not only assist the hydrologist in evaluating the hydrological characteristics of his area of study but also facilitate their extrapolation in space and time (Lohman and Robinove, 1964). In this context a distinction should be made between surface water, water associated with alluvial/colluvial deposits and water occurring in older geological formations.

It follows from the aforesaid that it is justifiable to carry out geomorphological and hydrological surveys jointly or to integrate them into a more comprehensive study of the land potential (Schumm, 1974). In Japan hydrographical maps form part of the nation-wide landform classification survey (Nakano, 1962), and in Poland a systematic, detailed hydrographical map at the scale of 1:50,000 forms a counterpart of the detailed geomorphological map on the same scale (Klimaszewski, 1956). Considering that hydrology is actually the combined effect of climate, landforms, geology, soils and vegetation on the natural regime of water and the effect of man on the water balance, it is evident that geomorphological studies for hydrological purposes should not only stress the genetical aspect of landform development, including processes, but also the environmental aspect. The geomorphological approach to hydrological studies then becomes a counterpart of the mathematical concept of hydrological models. It may also provide data for the elaboration of such mathematical models.

The surface water present in rivers, lakes and swamps is, in principle, readily observable but the important seasonal and secular variations of its availability necessitate careful study of the characteristics of form, gradient and roughness of the riverbed. This variability and the fact that water may be required at localities other than those where it is abundant, may necessitate engineering structures of various kinds so as to make optimum use of the surface water resources.

Groundwater generally escapes direct observation, except where it emerges in springs or is tapped by wells. An indirect approach to the evaluation of groundwater resources is, therefore, common. Landforms may give a clue to subsurface water conditions, for example by the occurrence of sinkholes or fractures. The relative importance of surface run-off and infiltration can be estimated from comparative drainage density studies. A complete terrain classification is often required to evaluate the environmental situation and for this, geomorphological surveying is essential. It should be understood that in practice it is not always feasible to separate the links between geomorphology and subsurface water resources completely from ties with surface water conditions, as these two components are strongly interwoven (Hesters, 1981).

The importance of geomorphology in hydrology becomes even more outstanding when aerial photograph interpretation or other remote sensing techniques are applied, as is common practice in many kinds of hydrological surveys. (Lohman and Robinove, 1964; Radai, 1969; Robinove, 1968; Schumm, 1969; Tulio Benavides, 1976). Geomorphological factors also play a role in photo-geological, or -pedological surveys, as an aid in establishing lithology, structure, aquifers and many other matters of hydrological relevance. Data on surface water features can also be easily and accurately observed and measured on these kinds of imagery. It is evident from the afore said that geomorphological studies for hydrological purposes should stress the role of landforms and other geomorphological factors in the broadest sense and should ultimately lead to an appreciation of the environmental situation in which the water occurs.

Such studies may encompass the following main fields:

- A survey of the whole watershed emphasizing the morphometric characteristics of size, shape, etc., and the environmental situation on the interfluves.
- The analysis of the drainage network in the degradational part of the basin.
- The geomorphology of the river channel and the mode of development of the aggradational part of the basin.
- The geomorphological-hydrological relationship in tidal areas.
- The discharge characteristics of the main river, its tributaries and the sediment yield of the basin or parts of it.

Each of these main fields is elaborated upon in a separate sub-section following the discussion of the two main components: surface and groundwater resources.

4.2 Surface Water Resources

The geomorphological contribution to the study of surface water features and the evaluation of surface water resources includes straightforward mapping of surface drainage, such as rivers, lakes, swamps, springs, floodplains and highwater marks. The divide lines of various orders should also be mapped, indicating size, shape and further characteristics of the (sub) drainage basins concerned, including their drainage density. Hill slopes also merit attention (Kirkby, 1978). Springs can be classified according to their yield, but the geomorphologist can add to the water resource situation by indicating their geological-geomorphological type and situation. Streams should be grouped into permanent and intermittent (periodic or ephemeral) categories and furthermore, according to size/discharge. Water-logged areas, either permanent or periodic, should be mapped and areas susceptible to flooding should be indicated. In addition, minor and major riverbeds, indications of river dynamics (cutting, lateral sapping, aggradation), the occurrence of sand- and gravelbanks, rapids, etc. have to be indicated.

It is clear that aerial photographs (e.g. black and white IR) and other remote sensing images are of great use in this phase of the work. Drainage lines are usually visible from the air and even if this is not the case there may be enough indications in the vegetation: type, height, crown diameter, etc., to plot the situation of the drainage lines with sufficient accuracy. In the Taiga zone of northern Europe and Siberia, homogeneous pine stands gradually give way to birch and spruce towards the rivers; in Canada similar indications exist as described in section 5; in the humid tropics sago swamps are found in back swamp areas; nipah palms are characteristic for brackish water swamps and for riverine tracks in the brackish zone. Geomorphology becomes of particular interest when the extent, nature, origin and the (often obscure) drainage direction of swamps have to be investigated. Landform and specifically micro relief then play a leading role. Where rivers of more distinct gradient occur, it is generally easier to establish the current direction, especially if larger areas are studied. Nevertheless, geomorphological indicators, such as angle of confluence, shape of sandy shoals, etc. are of use. If large-scale aerial photographs are available, qualitative data on current speed can also be gathered (structureless water surface: current speed$<$ 1,5 m/sec.; foam streaks downstream of obstacle diverging: current speed approx. 2 m/sec.; foam streaks converging: current speed$>$4 m/sec.).Gospodinov (1961), elaborates on the application of photo interpretation for the study of the hydrographic situation and Bogomolov (1963) gives interesting information on the relation between slope steepness, slope form, drainage, snow/silt retention and vegetation in the tundra zone of the USSR as indicated in the tables 4.1 and 4.2.

Table 4.1 Steepness of slope and tundra vegetation (after Bogomolov, 1963).

SLOPE	DRAINAGE; SNOW/ SILT RETENTION	VEGETATIVE COVER
very gentle $< 2^{o}$	Drainage absent, silt/snow retained	Mossy and sedge swamps
gentle 2-15°	Solifluction current of silt; snow is retained	All types of tundra vegetation predom. meadow and swampy tundra
average 15-25°	Solifluction of silt; weak shifting of snow	All types altern. with stony surface and alluvial deposit
steep 25-45°	Silt partly retained; snow not retained	Stony surface and alluvial deposit & sparce vegetation
very steep$>$ 45°	Silt/snow not retained	Lichen-shrubbery cover absent

Table 4.2 Shape of slope and tundra vegetation (after Bogomolov, 1963).

SLOPE SECTOR	EXPECTED VEGETATIVE COVER
Convex	Lichens and shrubbery; gravel, peat, sand and stony alluvial deposits encountered
Transitional (upper part)	Lichens and shrubbery; moss and sedge encountered
Transitional (lower part)	Mossy, sedge and shrubbery; shrubbery and lichen encountered
Concave	Swamp (sedge and mossy), meadow and shrubbery

For the evaluation of surface water resources, (Cooke and Doornkamp, 1974; Douglas and Crabb, 1972), spatial distribution as well as seasonal and/or secular variations are of the utmost importance, as has already been briefly mentioned. This applies to both humid areas where floods may occur when the river is in spate, and (semi) arid areas where flash floods may occur but where a shortage of surface water resources in dry months or years may be particularly harmful. Surface water resources in (semi) arid areas fall under three headings: ephemeral streams; natural water holes in alluvium, sometimes feeding a short streamlet and natural water holes in bed rock. Emphemeral streams give a seasonal supply only and water holes in alluvium or in bedrock may also dry out after a prolonged drought. The existing groundwater resources have to be relied upon, for example, where pervious materials rest upon impervious bedrock. Care should be taken, however, to prevent excessive water use, as the storage may be partly fossil (Dixey, 1966).

Apart from seasonal variations in surface water supply, secular changes may occur. An example of particular interest is the Okavango Inland delta in Botswana, pictured in the NOAA satellite image of fig. 14.6. This delta is bounded at its lower end by a NE-SW directed fault system which represents an incipient, southerly continuation of the East-African rift. As a result of upwarping part of the waters of the Boteti River, which drains the delta SE wards to the Makgadikgadi Pans, at times is diverted SW wards to the Ngami Depression and NE wards to the Mababe Depression (and the Zambesi River). Farther upstream the drainage of the delta is affected by faulting and part of its water is drained towards the Chobe and the Zambesi. The flow direction of the water of the delta is determined by the discharge generated by rains in the head waters, change in position and relative importance of the various channels in the deltaic zone and by neotectonic movements along faults.

For a full understanding of the surface water resources, it is essential to investigate the overland flow of the rainwater before it reaches the first order channels. Generally speaking, it is difficult to estimate the overland flow, but its magnitude can be ascertained by substracting the infiltration capacity from the effective rainfall. Detention on the slopes will temporarily store the water and thus retard the flow. Storage during longer periods of time will occur in depressions acting as retention reservoirs. The density of the vegetation (canopy of trees, scrub, grass) and the thickness of organic material (litter) on the ground are factors in this respect. Geomorphological factors are steepness and length of slope and in particular surface roughness. Other factors are duration and intensity of the rainfall. The infiltration capacity of the soil has already been mentioned as a factor in overland flow estimation. Various geomorphological indications exist for its estimation and its measurement in the field is comparatively easy.

It follows from the above that geomorphological study and a thorough terrain classification are important approaches to the evaluation of surface water resources. These studies may emphasize the general hydrological conditions in a given area or river basin and then assist in a better definition of the effective precipitation, surface run-off, infiltration and evapotranspiration. The surface area for which existing rain gauging stations are representative can, for example, be outlined on the strength of the relief (altitude, exposure to rain bringing winds, etc.). Also the sites of rain gauging stations to be installed to form a representative network can be indicated. Isohyetal maps can be improved by taking the relief factor into consideration and patterns of snow melt can yield information of hydrological relevance. The siting of gauging stations for meausring discharge in various parts of a basin can be facilitated by a geomorphological approach. Surface run-off and infiltration are both strongly affected by the steepness, length and roughness of the slopes. Drainage density also gives useful indications.

Geomorphological analysis and terrain classification are also of use in flood protection. The flood susceptibility survey discussed in section 13 illustrates this. Flood control measures can be precisely planned and the rate of silting-up of reservoir lakes can be estimated. The last mentioned application is based on the fact that areas where soils and rock types occur which are particularly susceptible to erosion can be outlined and a better estimate of the quantities involved can thus be made. This is illustrated in fig. 4.1 of the Haputale Area, Sri Lanka given by Herath. (Verstappen, 1967).

A number of landform units can be distinguished in the Haputale area, each having different lithological, morphometrical and vegetational characteristics. Initially, little attention was paid to the rock types because the drainage pattern, -density and terrain dissection are usually easier to establish on airphotos than lithology and thus more valid in hydrological aerial reconnaissance. Since fracturing is also of paramount importance for appraising infiltration, surficial run-off, etc. all detectable fault zones and joints were plotted on the map, a simplified version of which is given at a strongly reduced scale in fig. 4.1. The following four landform units were mapped and their hydrological characteristics evaluated.

1. The northernmost and highest part of the area. This high-lying zone is characterized by well-rounded hills, a network of broad and rather shallow valleys with moderate drainage density and poor vegetation and grasses. Most of the terrain slopes are towards the north and beyond the drainage basin under consideration.
2. A steep escarpment, 250-300 metres high, formed in very resistant rocks. The dissection of the escarpment is only slight, in spite of an annual precipitation of nearly 3,000 mm. recorded in these parts. Dense natural vegetation, well-kept terracing and contouring in the numerous tea gardens favourably affect the hydrological conditions in this steep zone.
3. A low-lying zone of low hills. High drainage density and intense dissection indicate low permeability of the comparatively weak rocks. The unit is broadest near the centre of the map. It tapers out to the east and extends southwards in the west. Homestead gardens occur in great numbers and accelerated erosion is obvious at many localities.
4. A broad zone of fairly resistant charnokites in the south. These rocks have a characteristic grey tone on the airphotos and a moderate to low drainage density.

A striking feature of the map is a major fault zone running across the entire area in a NNE-SSW direction. Distinct drag was observed. Renewed movements along the fault are not likely to occur, but it certainly can be considered as a zone of weakness where seepage can also be expected.

It was concluded from this reconnaissance survey that: 1) the proposed dam should be located within unit 4, but beyond the fault zone; 2) soil conservational practices will be necessary in zone 3 in order to reduce the rate of silting up of the reservoir; 3) little danger of erosion exists in unit 1 despite the high relief, because of lithology, vegetative cover and estate management.

Other examples of surface water resources and their utilization are given in the extended stereopairs of figs. 4.2 and 4.3. Fig. 4.2 depicts an area in Tunisia at the scale of 1:50,000, where a number of ephemeral streams, coming from the mountains (right side of the photo), are collected by two small subsequent rivers which are situated in a low-lying, NS zone of shales. After joining, they break through a likewise NS stretching ridge of westward dipping marly sandstones, in a narrow gorge to the left of which a fan is formed (largely out of stereo). The rather limited surface water resources of the tiny basin are concentrated towards the gorge and small amounts of water which percolate in the materials of the riverbed are brough to the surface where the river traverses the hard rocks of the ridge. A small village occurs at this strategic site where drinking water is available and where a short irrigation canal to the north of the river provides water for a few fields. The site is by no means safe from flood waters although in recent times they have been drained towards the SW (lower left corner of the photo). The hazard is limited by the small size of the basin although this is gradually growing by way of headward erosion (upper right corner) as a result of its low position. The fields situated on the lower part of the fan are mainly rain dependent, although some moisture may be derived from emerging groundwater. The latter seems to be rather unimportant in this fan and only one chain well exists, having its origin near the gorge and stretching SW ward between the flood channel and the mountain front. In areas where the lower parts of the fan are dissected by ravines (upper left) no land use occurs.

An entirely different situation is pictured in fig. 4.3 of an area in Montana, USA (scale 1:20,000). Here the ephemeral waters of the streams are not used and the water needed for irrigation of the fields is derived from a nearby storage reservoir and transported to the fields by way of a sub-horizontal irrigation canal with aquaducts crossing the ephemeral streams. Considerable erosion as a result of overgrazing occurs in the hills where these streams have their headwaters. Obviously, there is a reliable water supply for irrigation at places and abuse of the land in other parts. The example is of interest for comparison with the technologically less advanced situation pictured in fig. 4.2.

An example of an area where the development of rural water supply depends mainly on surface water resources, is the Bugesera area, straddling the boundary between Rwanda and Burundi.

The area is located to the east of the East African rift zone at about 1.500 metres a.s.l. The gently undulating and comparatively low central part, where granitic rocks occur, is surrounded by quartzites and shales of the Urundi Series. The quartzites form distinct ridges, thus a somewhat more pronounced relief is found near the edges of the area. The annual precipitation amounts to 1,100mm and vegetation is of the savanna type. Deep weathering and important colluviation characterize the large, old erosion surfaces making up the area.

The main divide is located in the east and most of the area is drained towards the west and northwest by broad valleys with flat valley bottoms. These valleys have extremely low gradients and their downstream portions

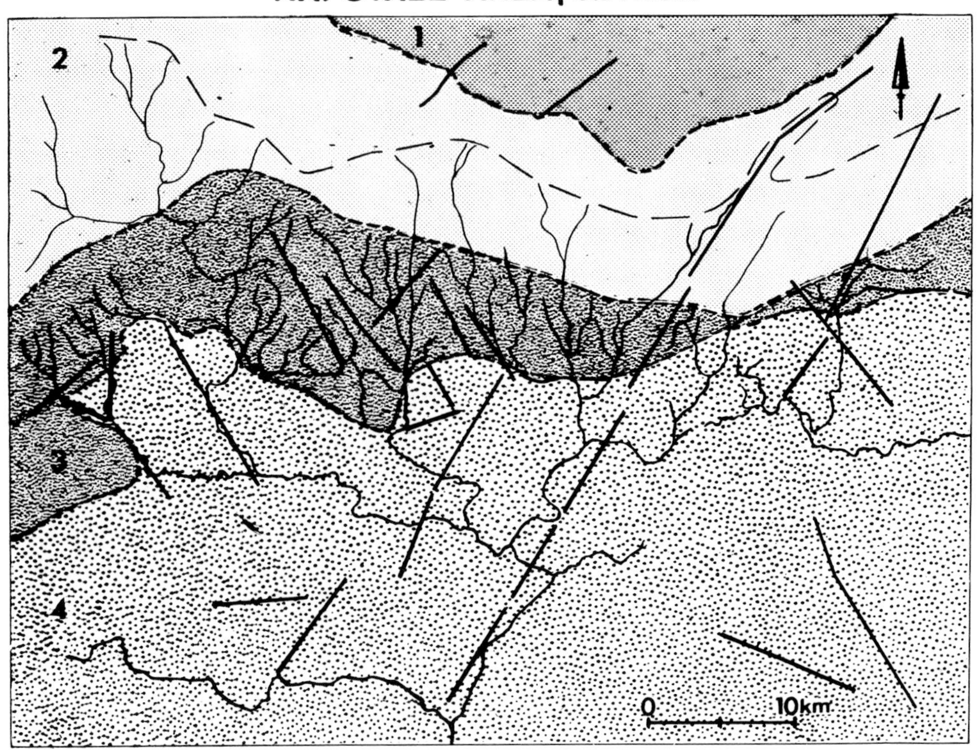

Fig. 4.1 Photo geomorphological sketchmap of the Haputale Area, Sri Lanka, showing four major terrain units, each with distinct lithological and relief characteristics. The divide line (long dashes) and several major faults are also indicated.

Fig. 4.2 Extended stereopair (1:50,000) showing an area in Tunesia where both the ground and the ephemeral surface water resources converge towards a narrow gorge, where a settlement is located, and subsequently diverge on the fan situated downstream.

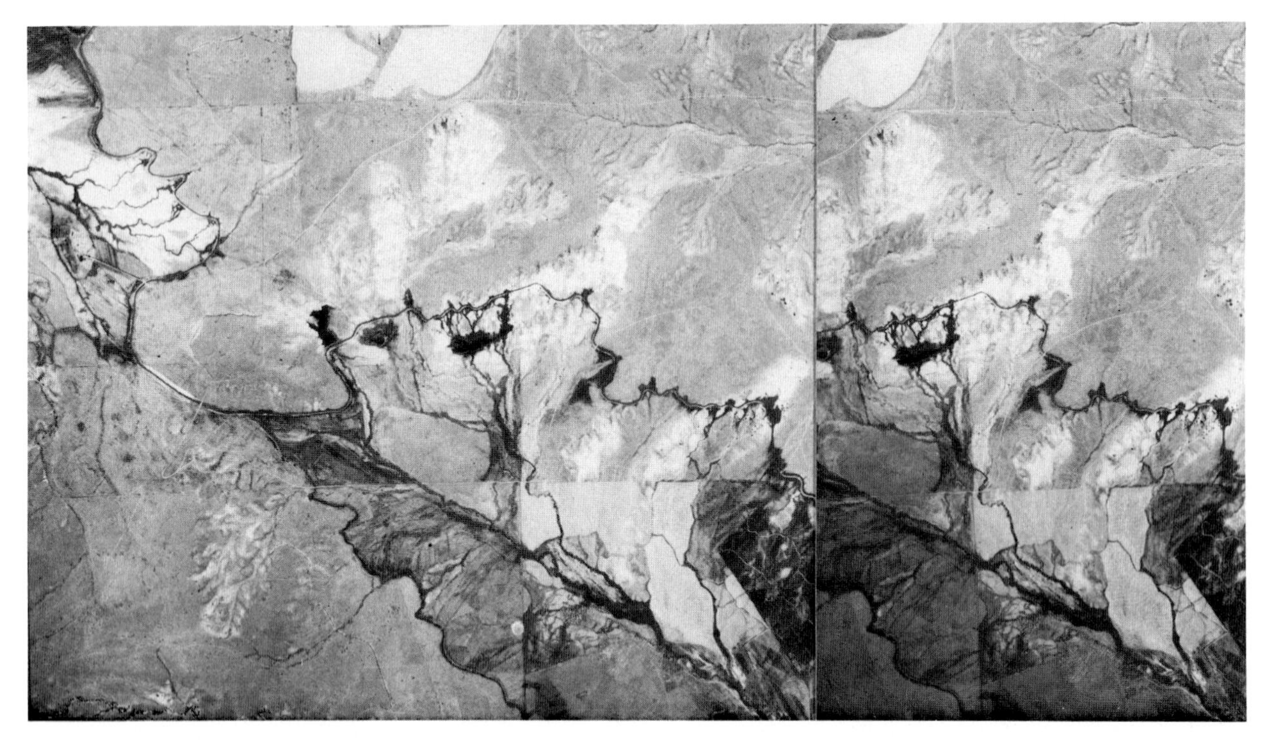

Fig. 4.3 Extended stereopair (1:20,000) showing an area in Montana, USA, where the ephemeral waters are not used and irrigation water is obtained from a reservoir lake.

are occupied by lakes and marshes containing floating Papyrus vegetation. Their upper portions are dry and do not contain water even after the occasional heavy rain showers. It is evident that the drainage system is a very old one and predates the present tectonic and climatological situation. The rivers entering the northward stretching Akanyaru Valley in the west, once drained northwestwards into the East African Graben. At present the waters of the Akanyaru join the Nyabarongo River (one of the headwaters of the Nile), which loops round the northern and eastern sides of the Bugesera (fig. 4.4). This complete change in drainage direction and the formation of the above-mentioned lakes is due to the combined effects of a general eastward tilt of the area during the development of the rift structure and an important fluvial sedimentation in the Nyabarongo Valley near the confluence of the Akanyaru (Verstappen, 1963, 1967).

The absence of running water in the upper parts of the fossil valleys and the low relief of most of the geomorphological units distinguished, render it likely that most of the rainwater falling on the gentle slopes is absorbed by the soil and subsequently largely evaporates. Only some slight infiltration may occur. No construction of (earthen) dams in these valleys for the purpose of collecting surface water seems justified. Only near the Akanyaru Valley, where more relief occurs, do active gullies and valleys occur. They are rather short and important surficial run-off is unlikely. Cattle raising is only possible in broad zones around the lakes and marshes. The water in the wells near the lake shores is somewhat lower than the lake levels and saline, due to evaporation. A number of small pools, occurring locally on quartzites or lateritic crusts may be improved to some extent and could contribute to the poor surface water conditions of the higher parts of the area.

It is obvious that the surface water resources of the area need further development. This is particularly so because the existing rock types are not very favourable to groundwater development and the few water borings executed had little success. Water circulating in colluvial materials may be of some use.

4.3 Groundwater Resources

For the evaluation of groundwater resources, a geomorphological terrain classification leading to the delineation of hydromorphological units is useful, taking both morphological and lithological factors into consideration (Anon.,1973; Fisk, 1951; Freers, 1970). The aquiferous areas, zones of groundwater replenishment and areas where groundwater emerges from springs should be mapped. In fact, many factors mentioned in the previous sub-section also affect the groundwater situation (Williams, 1970). The infiltration should be estimated, using as main criteria matters such as: permeability of rocks, fracturing,

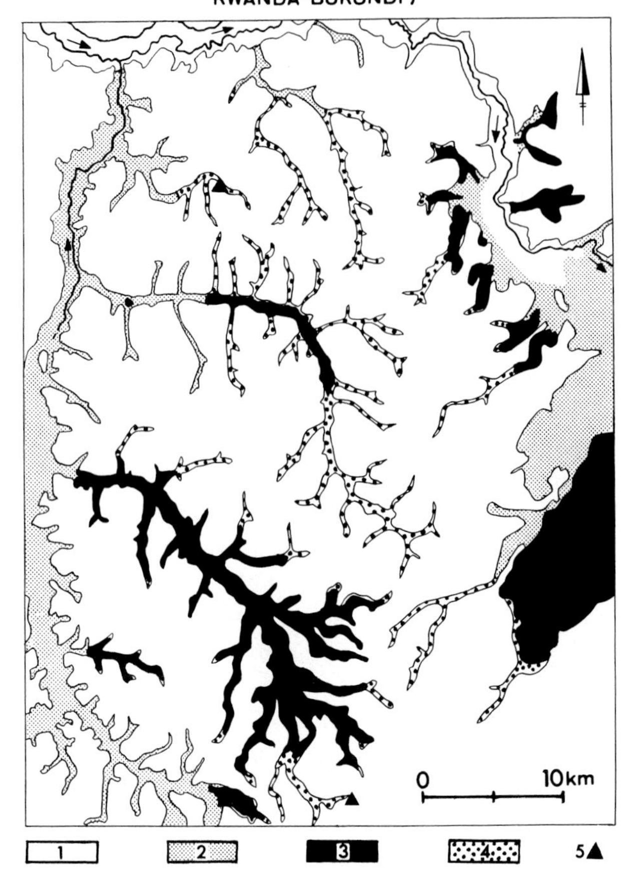

Fig. 4.4 Drowned river valleys in the Bugesera Area (straddling the boundary between Rwanda and Burundi) due to a westward tilt of the area. Key: 1. alluvial plain of Akanyaru River, 2. lake covered by floating papyrus vegetation, 3. open water, 4. dry, colluviated valleys, 5. springs.

permeability of soil cover, steepness and regularity (retention) of slopes and the effect of vegetative cover.

Infiltration measurements or estimates thereof are important not only for evaluating the surface run-off conditions on the interfluves and determination of the time of flood concentration of basins, but also for the study of groundwater conditions. Surface roughness and vegetative cover of the slopes, together with the texture and other characteristics (structure, organic matter content, pH, etc.) of the soil are important matters. The moisture content of the soil is also a major factor; the infiltration rate of dry soil is low because the capillary tension impedes the penetration of the water. Once the soil is slightly moist, however, infiltration reaches maximum values, thereafter decreasing as the soil gradually becomes saturated. As long as the rainfall intensity is less than the infiltration capacity, no surface run-off will occur.

The angle of slope, together with the length, shape in profile and plan and the roughness are geomorphological variables affecting the infiltration. Slopes which are convex in plan tend to spread the overland flow and thus favour infiltration, whereas slopes which are concave in plan promote concentration of flow and linear run-off. (Kirkby,1978). Apart from its role in the recharge of a groundwater reservoir, relief is also a major factor in the circulation of water by way of interflow at a shallow depth, above the groundwater level. The interflow is fed by the soil moisture storage, which in turn, depends on infiltration and evapotranspiration. The infiltration capacity which depends on the permeability of the bedrock, characteristics of soil and sediment cover, relief and vegetation, can be estimated by terrain classification using aerial photographs and/or other remote sensing imagery. If rainfall data are available, the surface run-off can then be approximated. A breakdown can be made of the land into units according to their relative infiltration characteristics. A rating is given, distinguishing 3 to 5 classes. During fieldwork, when site investigations yield data on the nature of the soil material, quantification becomes feasible, either on the basis of infiltration measurements, or using the infiltration characteristics known from comparable materials in other areas.

The following example of Syria (Voûte, 1958), illustrates the combined effect of geological factors such as permeability and fracturing and geomorphological factors, e.g. slope steepness, on the groundwater potential of two adjacent limestone areas situated to the W and E of the Ghab graben respectively.

Fractured Cenomanian limestone occurs east of the graben and acts more or less as a single underground reservoir with major recharge areas SE and secondary ones NE of a number of springs. The fracturing and karstification of the Jurassic limestones outcropping W of the graben are much less and the water circulation there is often restricted to rather shallow and isolated reservoir areas. It thus appears that the two limestone areas, although closely spaced, have entirely different reservoir characteristics and have reached a different degree of karstification in spite of the fact that they have been exposed to weathering for about the same length of time and are comparable with regard to the degree of tectonic deformation. The reasons for this are both lithological and morphological: the rugged topography of the higher Dj. Ansarieh in the west not only provokes more important rainfall, but also promotes surface run-off and restricts infiltration. The plateau-like and rolling areas of the Dj. Zawiye, the Massyaf region and the Naarret/Naamane regions are, on the contrary, more favourable for infiltration. The extensive alluvial fans developed along the foot of the Dj. Ansarieh and only locally present along the scarp of the Dj. Zawiya, add to the different hydrological potential of these two areas.

It is obvious that in the hydrological evaluation of limestone areas, particular attention should be paid to primary as well as secondary permeability. The former depends on the initial intergrain voids and may be low in compact limestones during the early stages of erosion. The latter results from solution and weathering processes and is a major factor in the yield of springs. Secondary permeability increases with time where pure limestones are concerned and tends to decrease in impure limestones, chalks, etc. Distinct geomorphological indications for such differences exist, providing valuable hydrological information.

The geomorphological situations may also give a clue to the hydrologist in other environments. Springs may occur in areas where a capping of resistant and pervious rocks, (such as limestone beds and ferricretes) is found on top of less resistant impervious material, particularly where the contact is in a topographically low position, for example due to a slight inclination of the beds. Slope retreat is then not due to retrogressive erosion, but is governed in particular by denudation of the supporting soft material and the subsequent collapse of blocks of the capping formation. Recementation of those fragments by lime or iron containing water may then occur and further affect the hydrological situation.

Comparatively shallow groundwater may occur where alluvial and/or colluvial deposits cover the bedrock. In areas of Pleistocene glaciation, for example, extensive areas of glacial till may occur which are poor in groundwater and where water sufficient for local use is restricted to the occasional sand lenses. Contrarily, the coarser textured outwash plains, valley trains, intertill gravels, etc. usually make good aquifers which can be located through geomorphological survey and mapping. Actually, most of the groundwater in such areas is derived from buried pre- and interglacial valleys, particularly in the USA (Watt, 1958). In the lowlands or northern Europe the fill of proglacial valleys which once bordered the receding continental ice sheet, is a major source of groundwater. The yield of such buried valleys is affected by the texture and permeability of the fill. A study of glacial geomorphology is clearly of considerable hydrological relevance.

Generally speaking, one can justly maintain that geomorphology contributes to the location of recharge or intake areas of groundwater and the prediction of recharge of various groundwater zones (Knisel, 1972). The further sub-division of a groundwater reservoir into confined and unconfined zones and into compartments separated by barriers such as dykes and faults impeding the horizontal transportation of groundwater, is facilitated by a thorough geomorphological analysis of the area concerned. An example is given in the stereopair of fig. 3.5 depicting some near-vertical dolerite dykes occurring in the sub-horizontal Basalts, Transition beds and Cave sandstones in Lesotho where they act as groundwater traps, particularly in the basalts. The dykes form conspicuous ridges in the Transition beds because of their greater resistance to erosion. They are topographically depressed in the basalts where only the contact zone may stand out in relief. The geophysical exploration for groundwater can be guided by geomorphological observations previously carried out. The extent of confined zones of groundwater is basically determined by the sub-surface geology (Le Grand, 1970). Storage capacity follows from the volume of the various formations and the recharge rates and transmissibility depend on their permeability and thickness. The importance of perched and artesian water tables is demonstrated in many parts of the world, for example, the water supply in the Tertiary Basin of Paris and the date cultivation in Algeria. Among major groundwater basins are: the Dakota Sandstone Basin to the east of the Rocky Mountains, USA,which carries the groundwater eastwards over a distance of 500 km and the basin underlying the Murray and Darling river basins in SE Australia which yields approx. 8.10^6cbm daily from 400 - 1.500 m depth and was discovered largely by deductive reasoning based on rainfall and discharge data. Groundwater reserves in limestone areas are a particularly interesting field of study. (Brucker, et al., 1972; Ion, 1970; Knisel, 1972; Marketik, 1969; Roglic, 1969; Sweeting et al., 1973). Where shallow karst occurs, damming of underground rivers may be feasible (see fig. 4.5).

It is evident from the examples given that a breakdown of the land into geomorphological units having distinct characteristics with respect to the groundwater problem is valuable for the assessment of sub-surface water resources. This applies both to areas of degradation where lithology, fracturing, weathering and slope play a role, and to areas of fluvial, aeolian and glacial aggradation, where emphasis is on permeability, topographical situation and size. Recharge areas and groundwater losses can often be estimated and matters of access, retention and release of sub-surface water have to be considered. Aerial photograph interpretation and other remote sensing techniques have also proved useful.

The role of geomorphology and more specifically the use of stereoscopic aerial photographs and photo mosaics for evaluating the groundwater situation have been recognized for several decades. Lohmann and Robinove (1964) mentioned the importance of landform evaluation in terms of lithology as a clue to the prevailing water-bearing properties. They emphasize the mapping of river alluvium and fluvial terraces and use the height of the terrace above the river as a measure for groundwater depth. The discharge of springs is estimated on the basis of vegetal growth around them and its length along the streams they feed. More recently, Meijerink (1974, 1977) followed up this aerial photographic approach. Other remote sensing

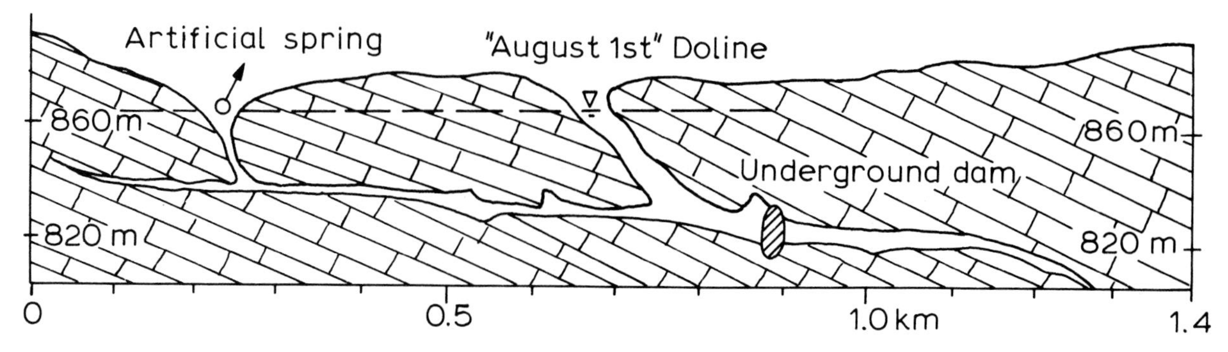

Fig. 4.5 Damming of an underground river near Lutoa, Hunan Province, P.R. China. After: Zhang, 1980.

techniques, especially Landsat imagery have also been successfully used for this purpose (Robinove, 1968), always employing the same basic principles. Apart from the vegetation indicators, the acreage of irrigated fields can also be used to estimate the yield of the shallow open wells from which the water is derived. The acreage is an index of the sub-surface storage at shallow depths. The author applied the method of irrigated fields watered by chain wells near Quetta, Pakistan (sub-section 6.41) and indicated localities where the groundwater was inefficiently used. Meijerink (1974) compared irrigated areas with the corresponding recharge areas in the Cuddapah Area, Southern India, where partly pediplaned Pre-Cambrian rocks with local unconsolidated deposits in semi-arid climatic conditions only provided scanty water resources. Local differences in shallow sub-surface water conditions could be correlated with the size of the recharge area. Fig. 4.6 gives an example and shows the irrigated areas (hachures), recharge areas (dashed lines) and the wells (open circles, black dots when sampled). Fig. 4.7 is a graph showing the irrigated area as a function of the recharge area for all zones investigated.

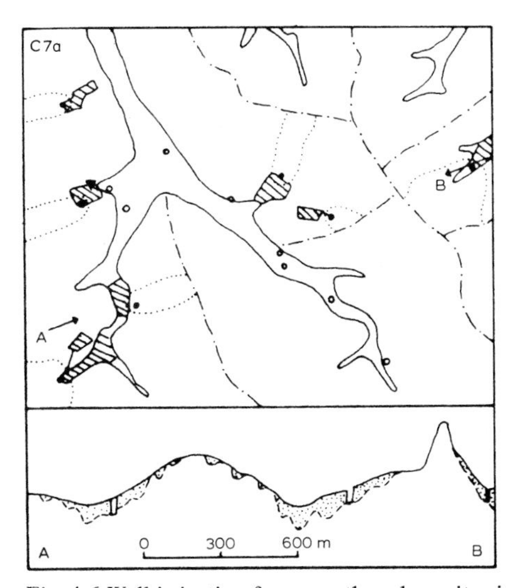

Fig. 4.6 Well irrigation from weathered granites in the Cuddapah area, S. India. The map shows the wells (open circles), sampled wells (dots) with the corresponding recharge areas (dotted lines) and irrigated zones (hachures) Meijerink, 1974.

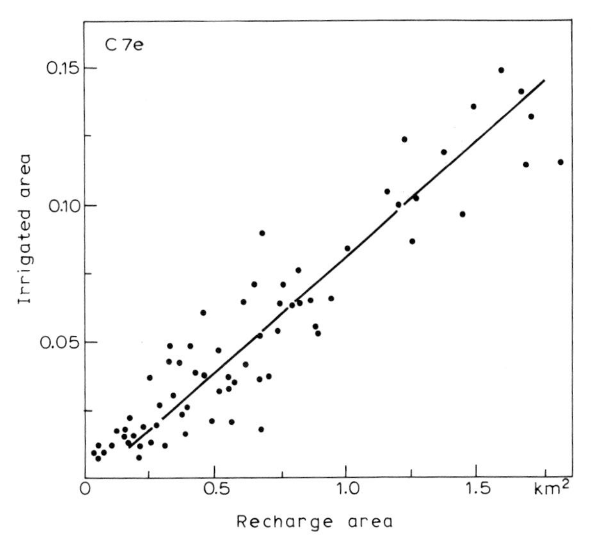

Fig. 4.7. Graph giving irrigated area as a function of the recharge area for the sampled areas of fig. 4.6.

4.4 The Geomorphology of Drainage Basins

This sub-section deals with the geomorphological studies of watersheds or parts thereof, for hydrological purposes (Chorley, et al., 1969; Cooke and Doornkamp, 1974). Emphasis is on the degradational, upstream parts of the basins as the lowland geomorphology will be treated separately. Both a qualitative study and a quantitative analysis of basins and sub-basins are involved.

The qualitative study of geomorphological and other watershed properties for hydrological purposes can be applied to evaluate the water resources of the basin studied (Chiang and Peterson, 1970; Popp, 1969), and can also be applied for extrapolation of data from one basin to another, although the latter approach has been mainly used in surface run-off estimates and much less for purposes of groundwater. Leaving apart the mapping of surface water features, in particular the drainage of swamps, the melting of snow, the identification of salinity encroaching the land, etc., it is basically a multi-disciplinary approach, in which geomorphology has a leading role, particularly when aerial photographs and other remote sensing imagery are used. It is especially valuable for reconnaissance studies and requires increasing support of ground truth when (semi) detailed surveys are concerned. The approach has proved relevant for various matters of hydrological importance such as: the relative magnitude of annual discharge, the peak discharge of certain recurrence intervals, the areas susceptible to inundation (with the depth), potential flood damage, areas of high evapotranspiration and areas of excessive sediment production characteristics and recharge of shallow and deep aquifers. The survey results contribute not only to pure hydrological studies but assist in the assessment of surface and groundwater resources of the basin or the area investigated (Maitre, 1968). As such they may be used for matters such as the selection of sites for open wells for irrigation, sites for tubewells, storage lakes for agricultural and other purposes, etc. The engineer, faced with cross-drainage problems, may benefit from this kind of physiographic survey and analysis of drainage basins (Tricart, 1962, 1978). Invariably both surface water and shallow or deep groundwater resources have to be considered. From a quantitative hydrological point of view, shallow groundwater resources are often not interesting, though on the village level they are of utmost importance for many subsistence farmers (ITC, 1969). Even small amounts of shallow groundwater may suffice to overcome short periods of rainfall deficiency and also for the growing of cash crops.

The assessment of the groundwater recharge potential is facilitated by a study of the texture and permeability of the surficial materials and deposits in the riverbed. A careful study of the soil profile and the evaluation of the internal drainage may give a clue to infiltration capacity. Soil erosion and vegetation characteristics are other useful indices. The depth of the water table and its fluctuations in open and tube wells of known topographic height as well as their yield is needed for assessing shallow and deep groundwater resources respectively. Discharge measurements of small, low-order streams in the dry season may give representative data on base flow characteristics and on medium to long storage in watersheds. If the measurements are carried out at a number of properly selected sites in representative physiographic units, an insight can be gained in the regional differences of long term storage in the drainage basin as a whole. Channel geometry is important for the estimation of the bank-full discharge, using the Manning equation.

Since the bank-full discharge is theoretically related to the short-term storage of the watershed, channel geometry also gives a clue to surface and near-surface storage conditions. Low and brief storage is accompanied by high, bank-full discharges and vice versa.

The depression storage in swamps and lakes can be evaluated by a measurement of the fluctuations in their level and extent. Soil mottling and gley layers are other indicators of water-logging and poor internal and external drainage.

The map of fig. 4.8. shows an example of how a geomorphological breakdown of the land into major units, an evaluation of their hydrological characteristics, and the mapping of relevant surface water features may contribute to an assessment of ground and surface water potential. It relates to the arid and semi-arid Borunda Area, Rajasthan in India. (Meijerink, 1974). It is evident that the limestone area has the best groundwater resources, particularly where fractures occur. Recharge of this aquifer is by infiltration from local rains/sheetwash and by sheetwash from adjoining areas (for example, North of Gagrana). The conglomerates and sandstones of unit 2 contribute by subsurface flow. Water losses of the aquifer may be to the Join Nadi fracture and to the Luni deposits in the lower right. The only place where surface water could possibly be stored seems to be the Jojri Nadi downstream of the sandstones.

The study on the rural development potential of the Jalor Area, Rajasthan, given in sub-section 6.4 may serve as an example of geomorphological evaluation of water resources.

4.5 Morphometric Analysis of River Basins

In these quantitative studies of river basins, morphometric parameters are emphasized (Morisawa, 1959) and can be grouped into three main categories: size, shape and relief (angle and aspect of slopes, vertical dimension). These categories will be discussed in this sub-section and in the above-mentioned order. Emphasis is on quantitative approaches and whenever possible the relevance of parameters for hydrological purposes (discharge, sediment, etc.) will be mentioned. When calculating the surface area of a watershed, the entire area between the divide line and the outfall with all sub and inter-basin areas should be considered. In case of substantial infiltration and underground drainage, inflow from adjacent basins or outflow towards them may occur. Care should be taken to investigate, on the strength of the geological and geomorphological situation, whether the groundwater divide coincides with the topographical divide. If this is not the case one should be aware that only part of the precipitation in a zone (the width of which is difficult to indicate with certainty) contributes to the water-circulation in the basin. The percentage depends on the ratio between infiltration and surface run-off, since the former may drain to one basin and the latter to the other.

It has been established by Schumm (1964) that the average basin size of a certain order has an exponential relation to the average size of basins of a higher order.

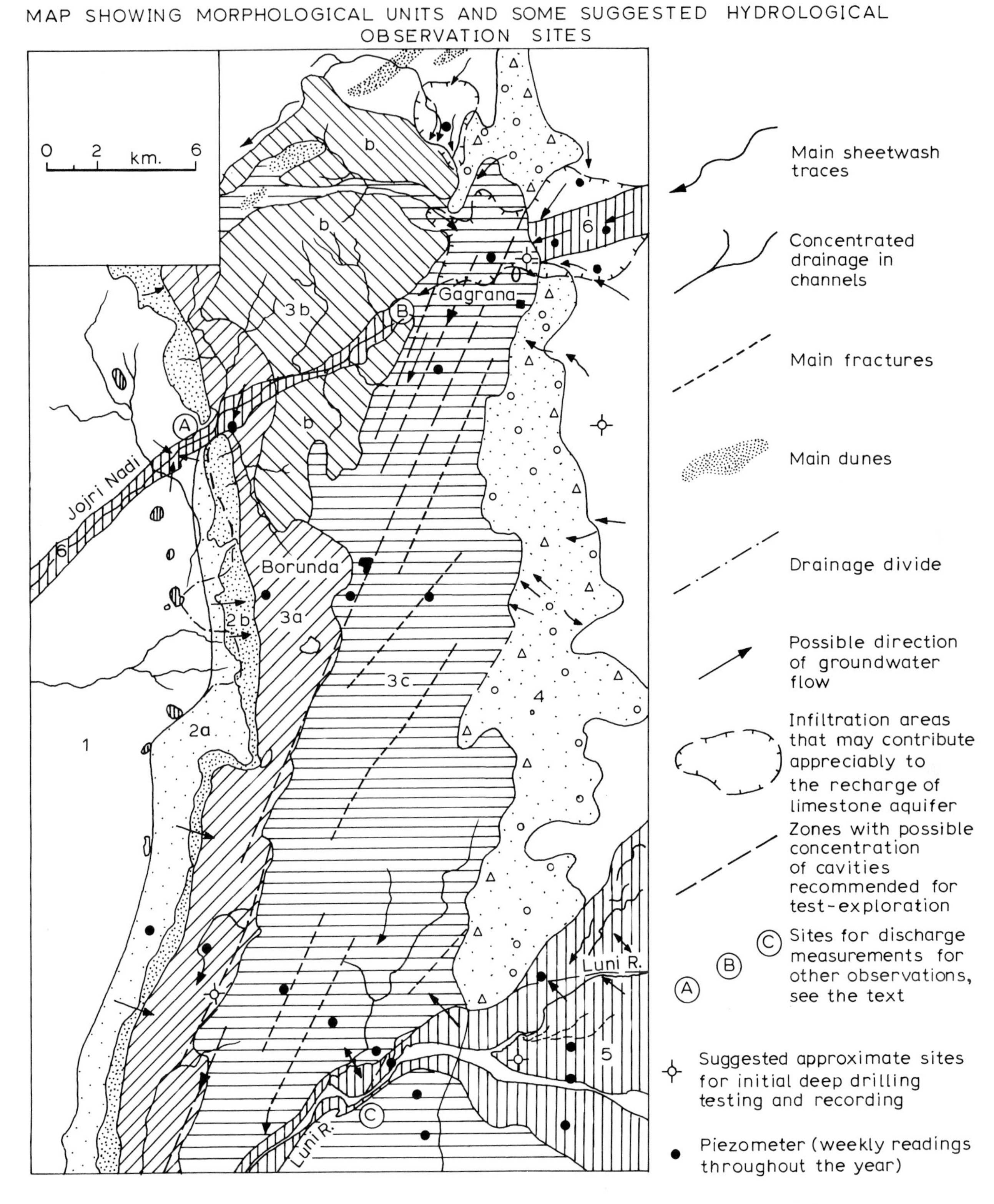

Fig. 4.8 Geomorphological units of the Borunda Area, Rajasthan, India with their hydrological characteristics and water resources. The units distinguished are: 1. basement and sand cover; 2.a. conglomerates and sand cover; 2.b. sandstone excarpment; 3.a. flat sandy area; 3.b. limestone area, continuous exposures; 3.c. limestone hillocks and sand covered depressions; 4. low hills, gravel deposits; 5. Luni River deposits, cover sands; 6. Jojri Nadi deposits. (Meijerink, 1974).

If stream order and basin size are plotted on double log paper, straight lines result, the inclination of which depends on a constant R_a = rate of area.

The larger the size of a basin, the less irregularities in discharge. Peak discharge becomes proportionally less, whereas the base flow increases. As a result, the hydrograph of large basins tends to be smoother than that of comparable basins of smaller size. There are several reasons for this situation. First, the overall rainfall intensity is less when a larger basin is considered, whereas the storage capacity and the time of flood concentration increase. Empirical formulae have been developed showing that the discharge at the outlet of a basin is exponentially related to the basin area upstream. When peak discharges are considered, the exponent evidently has a negative value since the importance of peak flow decreases with the increase in basin size as already stated above. The empirical formula $q = C \cdot A^{-0,5}$ applies for this. (q = peak discharge; A = basin size and C = empirical constant).

Since peak discharge depends on many physical characteristics of the basin concerned, including vegetative cover and anthropogenous impact, the applicability of the formula to basins is limited. A pragmatic approach based on the estimated effect of various basin parameters and not solely on basin size is preferred, although exact quantitative data cannot always be given. As peak discharge is of considerable importance, for example, in deciding upon the size of culverts, bridges or causeways during highway construction, the matter is elaborated upon in sub-section 4.7, using the Trans Sumatra Highway as an example.

The shape of the basin is another morphometrical factor affecting the discharge characteristics of a river as reflected in the curve of the hydrograph. This is demonstrated in fig. 4.9 showing an elongated basin and an approximately circular one of equal size with their respective unit hydrographs. The curve of the elongated basin is smoother which is explained by the greater time lag for the water from the upper catchment to reach the outlet. In case of a more circular basin, water from the lower, middle and upper catchment reaches the outlet in less time and causes higher discharges during a shorter period. It should be noted that these differences in shape are also expressed in the bifurcation ratio (see sub-section 4.6) which is substantially higher for elongated basins.

A crude indication of basin shape can be obtained by using linear measure, such as the length of the main channel or the distance to the most remote point of the perimeter. However, a variety of indices is used which involve parameters such as width, length and perimeter of the basin. Aerial photographs may be used with advantage to their determination. Some indices follow:

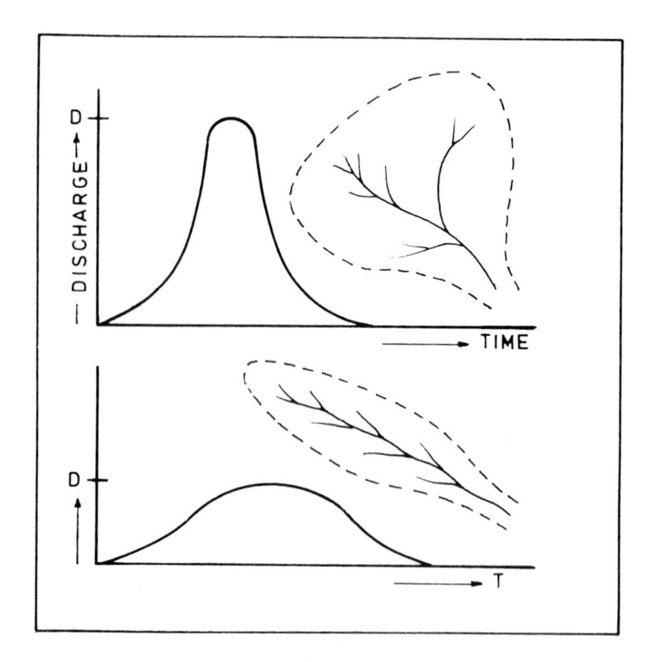

Fig. 4.9 Two differently shaped drainage basins with their respective unit hydrographs. The water produced by a cloudburst in the elongated basin (below) does not reach the outfall at the same time (T) and the discharge thus is less flashy than in the watershed at the top.

The Shape Index:

$$S_w = \frac{L}{W} = \frac{L^2}{A}$$

in which L is the length of the watershed along the main stream in km. (measure the valley length and not the length of the meandering river), W is the average width in km. (=A/L), and A is the watershed size (area) in km^2 (Horton, 1945).

The Gravelius Index:

$$K_c = 0,28 \frac{P}{1/2A}$$

in which K_c is Gravelius shape index, A the area of the basin in square kilometres and P the basin circumference in kilometres. A and P can be measured on aerial photographs by means of a planimeter and a curvimeter, respectively.

The Langbein Index:

This amounts to the summation of the areas of many small subdivisions of the drainage basin, each multiplied by its stream distance to basin outlet or gauging point (a.l.). The Index, $\frac{a.l.}{A}$ is a measure of the effective length of water travel.

The Circularity Ratio:

$$R_c = \frac{A_d}{A_c}$$

in which A_d is the basin area, and A_c is the surface of a circle, possessing the same perimeter of the basin in question. The maximum value of R_c is unity.

The Elongation Ratio:

$$R_e = \frac{D}{L}$$

in which D is the diameter of a circle having the same area of the basin and L represents the maximum length of the basin, measured parallel to the main drainage line. The maximum value of R_e is unity (Schumm, 1956).

The latter shape parameters are comparatively stable numbers and are only slightly influenced by geologic conditions. Using the above-mentioned shape factors, no differentiation is made between watersheds with a normal drainage network and watersheds having an unusual drainage system resulting from structural influences. The shape factors assist in the prediction of streamflow intensity in river basins where data are scarce.

The slope angle and further relief characteristics of a basin are a third group of morphometrical factors of hydrological relevance. (Speight, 1980). Steep slopes generally have high surface run-off values and low infiltration rates. Consequently, they add to the steepness of the hydrograph and lead to relatively high peak discharges. The high proportion and velocity of the overland flow easily leads to sheet, rill and ultimately gully erosion. Sediment production thus tends to be high except when largely barren slopes are concerned. The matter is elaborated upon in section 15 Erosion Surveys. Apart from slope angle, length of slope associated with relief amplitude should also be considered and in this way, terrain dissection is expressed in drainage density and the gradient of the channels. Slope angle can be determined in the field, but contour maps or aerial photographs are more commonly used. Because the measurements required are laborious, slope determinations are limited and random or grid sampling techniques are applied.

Strahler (1956) developed a method for the preparation of a slope map by using a good photogrammetrical map at a scale of 1:25,000 or larger, depending on the relief and lengths of slope. Topographical maps are not necessary since steepness of slope can be directly measured from aerial photographs.

In order to reduce the rather laborious preparation of a slope map, a slight modification of the method may be applied:

1. The slopes of the terrain are divided into several gradient classes. The width of a class depends on the difference of the maximum and minimum value of slope steepness in the terrain and on the desired accuracy.
2. Terrain-facets having the same slope steepness are determined and delineated from the stereoscopic image.
3. Within each delineated terrain-facet, several measurements of slope steepness are made in order to check the accuracy of the delineations and to calculate the average slope steepness.

The results of the slope measurements can be presented in the form of a slope map or as frequency distribution histograms. When contour maps are used, the surface area between two adjacent contour lines can be measured and a cumulative hypsometric curve can be constructed.

The average slope of the watershed can be determined using contour maps made from aerial photographs by applying the following formula:

$$S = \frac{MN}{A} \times 100$$

in which M is the total length of contours within the watershed in metres, N is the contour interval in metres, and A is the size of the watershed in m^2. For small watersheds the average slope can be taken as the ratio of the difference in elevation between the watershed outlet and the most distant ridge to the approximate average length of the watershed. The formula for the relief ratio then reads:

$$R_r = \frac{H_{max} - H_{min}}{L \text{ average}}$$

In case of larger basins, more elaborate means of quantification are required, encompassing various parameters. The ruggedness number (HD) is another interesting relief parameter. It is the product of local relief and drainage density. This number suggests steepness of slope implicitly: if drainage density increases and local relief (R) remains constant, slope steepness will also increase. If R increases and D_d remains constant, the slopes will be steeper and longer. For high relief and high drainage density, the ruggedness number attains high values.

4.6 Drainage Network

The qualitative and quantitative analysis of the drainage network prevailing in an area may also yield information of hydrological interest although the relations between network and water resources are not always straightforward and at best give only general indications. Nevertheless, these may be of use for drainage basins which are short on data. This is due to fracture systems or other structural factors which are reflected in them but also because of the effect of soil conditions and in particular, permeability of these patterns.

Factors to be considered are: density, which also has a bearing on the permeability of the rocks, the amount of geological control on the drainage pattern and the integration and homogeneity of the patterns (Schumm, 1972;

Thomas and Benson, 1970).

A major distinction exists between patterns showing negligible and those showing distinct influence of the geological structure. Among the former the dendritic or arborescent pattern is by far the most common. Branches make acute angles with the main river. This coherent pattern is best developed in areas of homogeneous lithology and sub-horizontal beds. The basins tend to be tear-shaped and have a bifurcation ratio of 3,0 - 3,7. A parallel pattern may occasionally be the result of structural control but more commonly is the outcome of a distinct general slope of the terrain. The basins in this case tend to be more elongate, with bifurcation ratios of over 4. A radial pattern is often not directly influenced by structure. It may occur in a centrifugal form, for example, on volcanic cones, in a centripetal form in depressions where evaporation and/or infiltration is high.

Slight effects of geological structure on major drainage lines may result in sub-dendritic and sub-parallel patterns. The effect of structure becomes dominant when beds of different resistance to erosion are exposed. In a dome structure an annular pattern will then result. It is composed of circular or ellipsoidal, subsequent drainage elements located in the less resistant rocks, which debouch in consequent, radial streams of higher order. The trellis pattern is comparable but originates where the strike lines of the beds are more rectilinear. Numerous small, first order tributaries join the major streams; therefore, the bifurcation ratio and drainage density are often high. In areas where faults and/or fractures are numerous, angulate and rectangular patterns are common, and are characterized by straight sections connected by sharp, acute or right angles. Only the main types are briefly mentioned here and given in fig. 4.10. Many complex and incoherent patterns such as deranged (in glacial till) and shallow hole patterns (in limestone terrain) occur, each providing clues to the specific hydrological situation of their environment.

The importance of quantitative analysis of drainage patterns has been increasingly emphasized since the early investigations of Horton (1945), Strahler (1964) and others. Their approach is centered on a considerable number of basin parameters, the major ones of which are listed below. These parameters of the drainage network have been mainly applied by hydrologists in correlation models or in multiple regression studies. It has, however, proved difficult in many cases to establish clear-cut relations between these morphometric parameters on one hand and the characteristics of the hydrograph and the sediment production on the other. This is partly explained by the fact that it is often limited parts of a basin, such as narrow strips along the river which are almost continuously saturated, that deliver most of the flood waters. The same applies to the sediment load of a river for which specific parts of limited extent are mainly responsible.

Stream order is a major aspect of quantitative analysis of drainage network. Horton (1945) introduced a system for it which was modified by Strahler (1964) and this latter system is at present commonly used. Designating the hierarchy of the channels of a river basin is the basis of various formulae used. Horton found three major laws governing the organisation of the drainage network. The number of streams of each order in a given basin form an inverse geometric series in which the first term is unity and the ratio is the bifurcation ratio R_b. The latter can be defined as the value with which the number of streams of a certain order has to be multiplied to get the number of streams of the next-lower order. Similarly, the average length of streams of each order in a given basin form a geometric series (Law of stream lengths) in which the first term is unity and the ratio is the stream length ratio (R_l). Horton's third law links general gradient of streams with stream order and reveals that with respect to this parameter, an inverse geometric series relation exists (Law of stream slopes).

The bifurcation ratio:

$$R_b = \frac{N_u}{N_{(u+1)}}$$

has shown to be of particular interest. Although its value will not precisely be the same from one order to the next, because of change variations in watershed geometry, it tends to be constant throughout the series. This is the basis of the Law of stream numbers. When the logarithm of number of streams is plotted against order, most drainage networks show a linear relationship with a small deviation from a straight line. Bifurcation ratios characteristically range between 3.0 and 5.0 for watersheds where the influence of geological structure on the drainage network is negligible. The theoretical minimum possible value of 2.0 is rarely approached. The relative constancy of R_b is due to the dimensionless property of R_b and the fact that drainage systems in homogeneous materials tend to display geometrical similarity.

Abnormal bifurcation ratios usually have a marked effect on maximum flood discharges, for example, basins with a high R_b yield a low but extended peak flow. In areas of active gulleys and ravines, the bifurcation ratio between first and second order streams may be considerably higher than the R_b of higher order streams. This is indicative for a state of accelerated erosion. High bifurcation ratios are common where the effect of geological structure is dominant, but the shape of the basins also has an important effect. The bifurcation ratio tends to reach high values where elongated river basins are concerned. Interesting is also the relation to the groundwater situation which is particularly clear in the low order streams in the upper reaches of the basin. If the ground-water is deep and the infiltration high, surface run-off will be relatively low and less channels and a lower bifurcation ratio can be expected.

The drainage density in areas of comparable climatic conditions basically depends on geological factors (particularly lithology and the resistance of rocks to erosion) and on infiltration capacity. The latter is partly affected by the lithological factor of permeability, but relief and soil moisture are also factors.

Drainage density is another quantitative parameter of the drainage network. It is expressed in kilometres of channel length per sq. km. The drainage density of a drainage basin is written in the form:

$$D_d = \frac{L}{A_d}$$

in which L represents the total length of all channels and A_d the area of the basin. It is a measure for the degree of dissection of the basin and can be determined from topographical maps using a planimeter and a curvimeter. If aerospace imagery is used, care should be taken for geometrical correction of the measurements which may then be required. Stream density is defined in a similar way, but only the accumulated lengths of channels, in which stream flow actually occurs are taken into account. Drainage densities may vary from zero, for example, in dune areas, to more than a thousand km/sq.km. in highly dissected badlands on impermeable shales. Heavy rainfall, annual or episodic and seasonal, seems to be largely responsible for high values of drainage densities.

The drainage density is in fact influenced by numerous factors, among which resistance to erosion of the rocks, infiltration capacity of the land and climatic conditions rank high. Low drainage densities and large basins with comparatively low-order main streams are common in resistant rocks, whereas high drainage densities and higher order basins are common in soft rocks such as shales. The infiltration capacity, as affected by permeability, slope steepness and soil moisture conditions (see sub-section 4.3), is another major factor as is the accompanying climate and vegetation cover. Humid climates tend to have a lower drainage density than arid areas of comparable lithological composition resulting from the regulating effect of vegetation on streamflow and erosion. Areas of low relief and good infiltration capacity generally show lower drainage density values than zones of higher relief or lower permeability.

It is important, however, to remember that present day drainage density may partly represent the long-term effects of changes in rainfall characteristics, climatic conditions or land management in the past. In Western Europe, for example, many presently dry valleys were formed during the time of Pleistocene glacials. Overgrazing or deforestation may result in the development of gullies, which persist as conspicuous channels for a very long time. New drainage lines cut as the result of peak run-off conditions would appear to be the most persistent and even self-perpetuating features.

Although the hydrological evaluation of drainage density data is not always an easy matter, several interesting aspects can be recognized. In well-dissected areas this parameter gives a clue to the distance between drainage lines, thus the average length of overland flow is 1/2 drainage density. In poorly dissected areas a low drainage density may indicate important storage capacity.

Among the other morphometric parameters of the drainage network, drainage frequency (F) and the constant of channel maintenance (C), should be mentioned. Drainage frequency is defined as the number of rivers or more precisely river segments, of all orders in a basin divided by the surface area of that basin. It is closely associated with the bifurcation ratio. The constant of channel maintenance is a parameter introduced by Schumm (1956, 1964) and is defined as the area in sq.m. to maintain one meter of channel length. It is thus the reciprocal of drainage density ($C = 1/D_d$). Since the number of streams, length of streams and surface area of basins are all related to stream order, they are interrelated and the same also applies to most other drainage network parameters. It usually suffices to select one or two of them which are most appropriate for the study at hand.

4.7 The Hydrological Significance of Lowland Geomorphology and Channel Shape

The characteristics and temporal changes of the river-bed form another group of geomorphological data of hydrological relevance (Allan, 1968; Carey, 1969; Coleman, 1969; Martinec, 1967), in particular with respect to discharge and sediment transportation. Channel forms can be classified into meandering, braided and straight types, the roughness of the bed can be determined, the slope of the channel established, geomorphological traces of high water marks detected and the cross-section of the river-bed measured. (Lacey, 1929/1930; Orme and Bailen, 1970; Tricart and Vogt, 1967). These observations then assist in estimating peak discharges, etc. Aerial photograph interpretation and radar and landsat images also play an important role here (Abdel-Hady, M. et al., 1970). General indications can be obtained and provisional conclusions arrived at. Rapid changes in width to depth ratio and in sinuosity of a river for example, point to differences in bank materials and a rapid increase in the width of a channel downstream usually results from aggradation caused

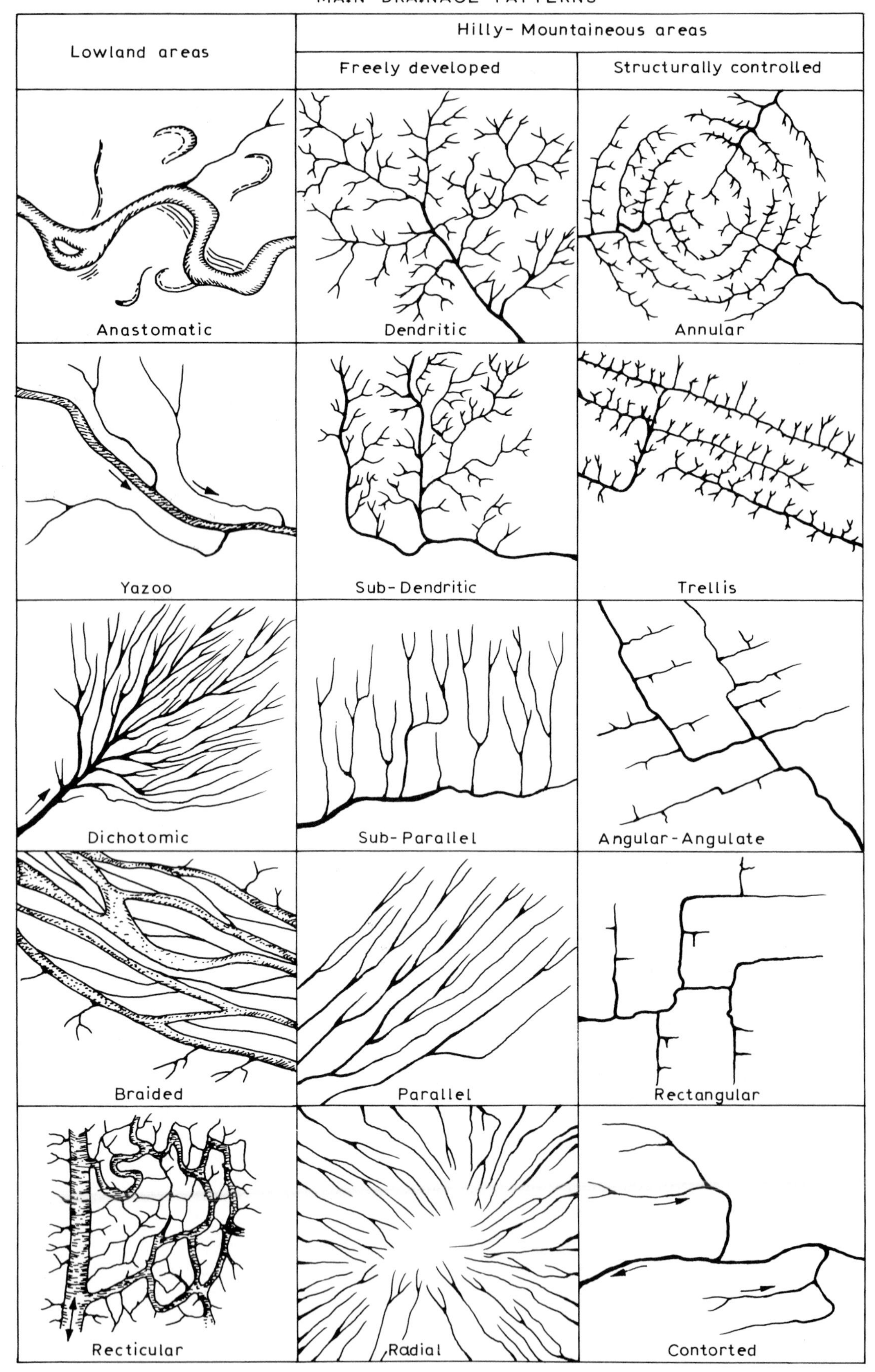

Fig. 4.10 Main drainage patterns of accumulation and denudation zones (Verstappen, 1972).

by severe accelerated erosion upstream. Braided characteristics generally indicate fairly high fluctuations in discharge, high sediment load and a fairly steep gradient and meandering rivers usually have lower peak flows because of the overbank storage in backswamps. Various types of rivers occurring within one area indicate hydrological differences among the basins (Leopold, 1953; Leopold and Wolman, 1957; Leopold et al., 1964).

The map of fig. 4.11 shows the lower reaches of the Palu River, Sulawesi, Indonesia, based on aerial photographs and demonstrates the hydrological importance of observed changes in river morphology. A considerable increase in sediment yield of the Palu River basin is evident from increased braiding.

The 1924 topographical map used in this study shows the river channel in operation then and also several older, meandering courses dating from 1880-1900 or earlier. Most of these are visible in detail on the 1972 aerial photographs. Comparison of the pre-1900 channels with the more recent ones shows that the latter are less sinuous and wider. They also have a longer wave length and probably a somewhat steeper gradient. These changes concord perfectly with those mentioned by Schumm (1972) and Gregory (1977) as criteria for increased sediment discharge. The 1924 situation shows that most of these changes pre-date topographic mapping. Since then, sediment discharge has remained high and has possibly even increased as the average width of the river has slightly increased. The channel now tends to braid and contrasts significantly with the earlier meandering characteristics. Sobur (1980) has studied the changes in morphology of the Lower Serayu River in detail and has mapped the resulting land use dynamics.

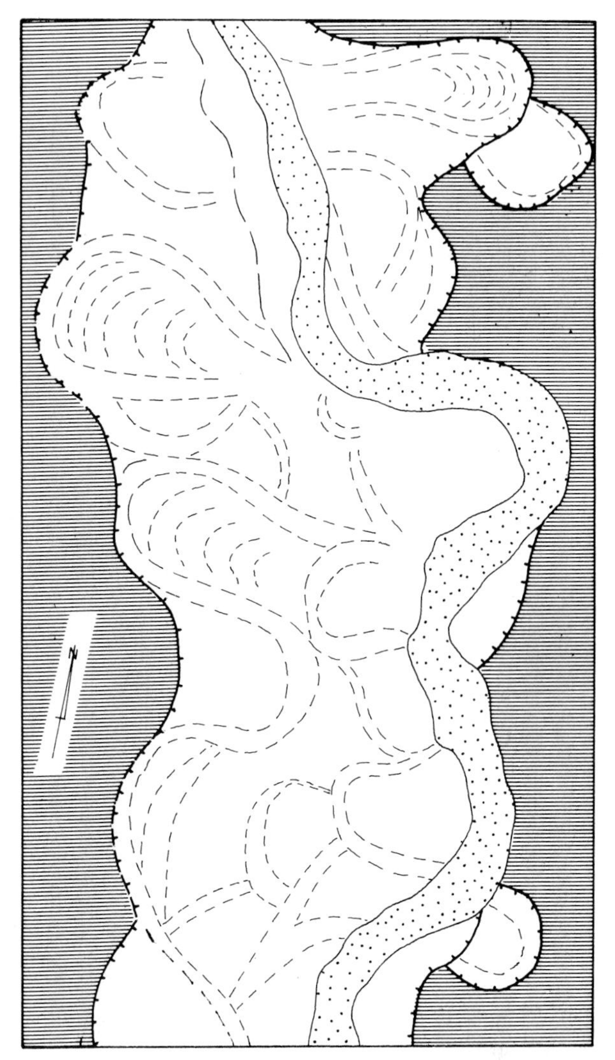

Fig. 4.11 Map, based on aerial photographs, showing the transition from meandering to braiding of the Palu River, Sulawesi, Indonesia as a result of increased bedload caused by accelerated erosion upstream (1:13,000).

The diversity of applications of the geomorphological approach to the hydrological evaluation of lowland areas is evident from a summary review given by Verstappen (1972). The hydrological characteristics of some salient geomorphological features of lowland rivers are given below:

Braided rivers, such as those from Seran, Indonesia, depicted in the oblique aerial view of fig. 4.12 may occur in a variety of circumstances but they normally indicate an important bedload of the rivers and a rather rapid deposition of coarse material. Their formation is favoured by situations such as accelerated erosion in the head waters, a pronounced seasonality of flow, a sudden decrease in gradient where the river emerges from the mountains, a rock sill interfering with river cutting and furthering lateral river work. The example given clearly demonstrates that braiding is restricted to the zone of rapid sedimentation; it does not occur in the mountains and towards the lowlands free meanders become characteristic (lower left). Apart from rapid deposition, other hydrological implications are: important fluctuations in discharge, frequent changes of the channel pattern and groundwater flow in the sandy/gravelly material. In areas of flooding, this is usually of short to medium duration.

Alluvial fans, formed where tributaries enter the main stem of a river of less gradient or where mountains streams or torrents enter the plain, are another important geomorphological feature resulting from fluvial accumulation. The oblique aerial view of fig.4.13 gives an example and depicts a torrent debouching in the Balim Valley in the Central Range of Irian Barat, Indonesia. The rapid erosion in the catchment, the collecting trunk of the torrent and the alluvial fan (lower right) are clearly visible. Two phases of activity can be distinguished on the fan: an older, presently inactive and vegetated part and a narrower, barren, active zone. These phases reflect two erosion phases in the catchment of the torrent. Other hydrological implications of alluvial fans are: the rapid deposition of usually coarse

Fig. 4.12 Oblique aerial view of braided rivers on Seran Island, Indonesia, when leaving the mountains and entering the plains.

Fig. 4.13 Aerial view of a torrent in the Balim Valley, Central Range of Irian, Indonesia, showing eroding headwaters, transport channel and zone of deposition.

materialwhere active fans are concerned, the radial pattern of partly abandoned channels and the occurrence of imperfect sheet flood in (semi) arid areas. Gravity flow of groundwater can be expected when the water table is low, when the fan is gravelly, or when perched watertables occur with silty/loamy beds. Springs are often found in areas where the fan tapers out down slope. Flooding is usually shallow and of short duration but disasters may result from a sudden change in the position of channels or from the occurrences of upstream debouching in the trunk rivers. Generally the flood hazard is greatest along the active channel, less in the upper and lower parts of the fan and minimum in its central portion (fig. 4.14).

Natural levees gradually built up during bankful discharges of a river, are common along present and former channels in areas where more quiet sedimentation conditions exist. The oblique aerial view of fig. 4.15 gives an example and depicts the birdsfoot delta of the Arabu River protruding into the Panai Lake (Wissel Lakes), Central Range, Irian, Indonesia. It clearly shows the levees at either side of the river and the swamps and lakes in the low-lying backswamps formed farther from the river between recent and older levee ridges. The hydrological implications of a levee-back swamp situation include a high position of the riverbed with respect to the adjacent parts of the plain,which may lead to seepage or to breaches in the natural levee. Tributaries may be prevented from directly entering the trunk river due to its high position and are forced to flow parallel to it before debouching. The areas between natural levees often have impeded external drainage and difficult internal drainage due to the fine texture of the soil material. Waterlogging and gley formation in the soil profile are common in backswamps. Flooding of natural levees is of minor importance although some overtopping occasionally occurs. (Tada and Oya, 1957; Oya, 1973). Lateral sapping of the river may, however, bring disaster to settlements situated on the levee, lead to the formation of breaches in the levee (fig. 4.16) or create splay deposits between levee and backswamp zone. Inundation of the backswamps is characteristically deep and of long duration but may serve for overbank storage. The acid-sulphate 'cat-clays' frequently occurring in backswamps pose problems for reclamation as improved drainage easily leads to their toxication.

The study of river meanders and their cut-offs is of considerable hydrological relevance. Their occurrence generally indicates a comparatively quiet aggradation environment where mainly medium to fine textured material is deposited (Inglis, 1967). A qualitative estimate of the bedload situation can often be made on the strength of the change rate of the meanders and the surface area of point bars. The width of the meander belt is a function of the bankful discharge of the streams concerned.

The mode of the cut-offs also is significant. Complete meander development ending in the formation of a neck cut-off generally indicates a fairly regular discharge, a less important bedload and a fairly great stability of the channels. If chute cut-offs are dominant, the discharge of the river is subject to greater fluctuations and flashy floods may occur. The reason is that when the river is in spate and the water level high, the river will be straightened and

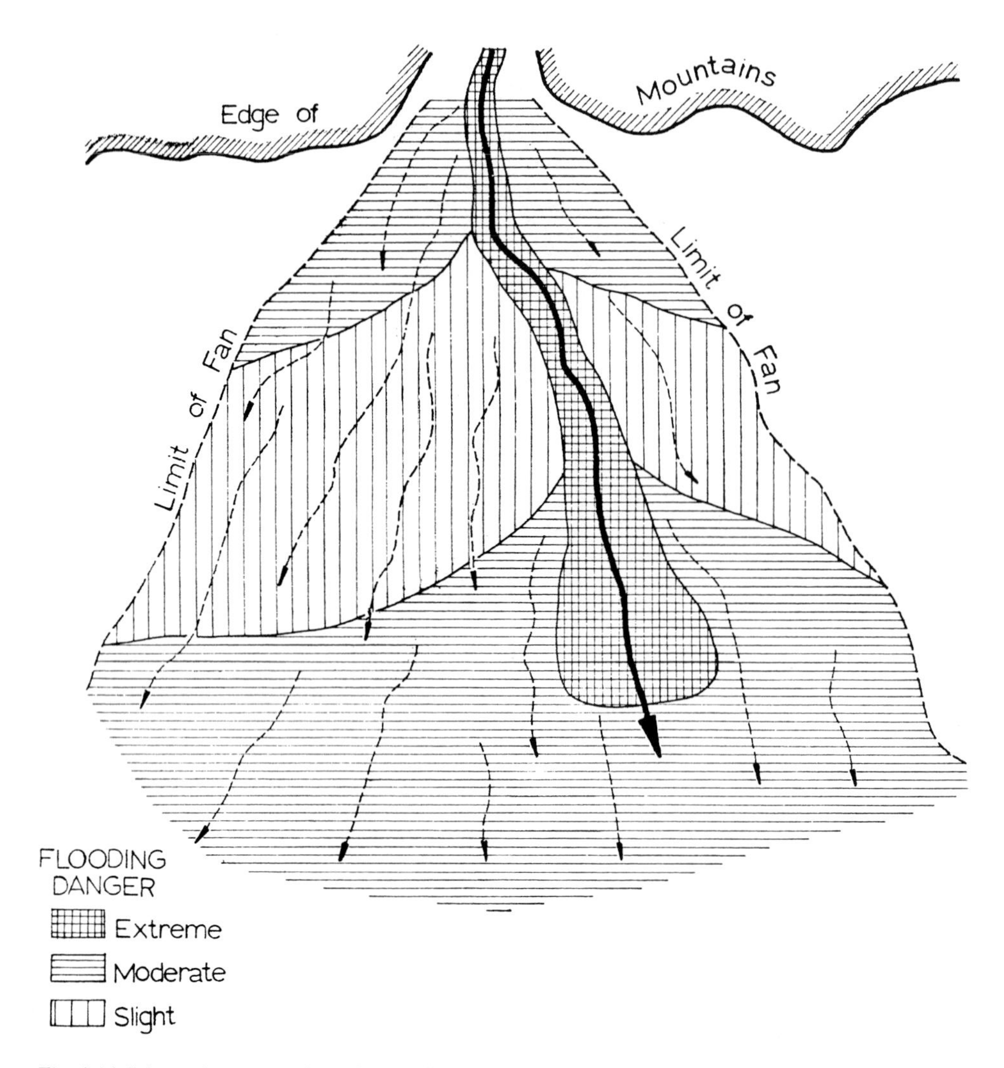

Fig. 4.14 Schematic presentation of major flood hazard zones on a (semi) arid fan. (Schick, 1971).

Fig. 4.15 Oblique aerial view of the typical bird's foot delta of the Arabu River where it enters Paniai Lake; Central Range, Irian, Indonesia.

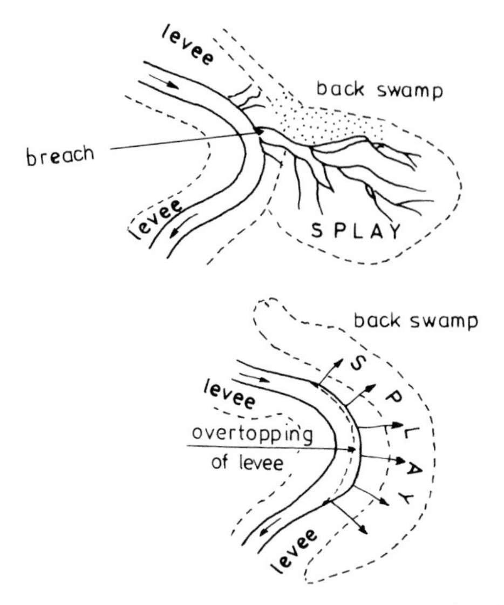

Fig. 4.16 Sketch of a natural levee and a splay deposit on the edge of the back swamp.

scouring will then occur in the shallowly inundated area in the convex curve of the meander where the gradient is steeper and the flow velocity of the water larger than in the minor riverbed. Fig. 4.17 elucidates this.

4.8 Deltaic Areas

There is probably no other natural environment where geomorphology and hydrology are so interwoven as in deltaic areas. Hydrological factors having a profound impact on deltaic geomorphology are: fluvial factors such as frequency and duration of floods, rate of rise and fall in river stage, flood volume, minimum dry period flow and the amount of transported sediment as well as marine factors such as astronomical tides and occurrence and magnitude of storm surges. The landforms produced in turn affect the hydrological conditions of the delta. Three parts of deltaic areas are distinguished: the upper part where the river flow is the dominant factor, the middle part where both river and sea play a role and the lower part where the influence of the sea is most marked. It is found in many deltas that the middle portion poses the greatest technical problems for delta development and flood protection. A detailed geomorphological survey in which aerial photo interpretation plays a particularly important role (as it offers the best means to establish the exact location of former channels), may be of considerable practical value for flood protection and delta development schemes. This is mainly because insight in minor relief differences and the type of deposits occurring adds to an understanding of flood conditions.

The natural levees in the lower zone of the delta build up rapidly particularly if the tidal range is large. This is due to the fact that the bankfull stage is frequently reached

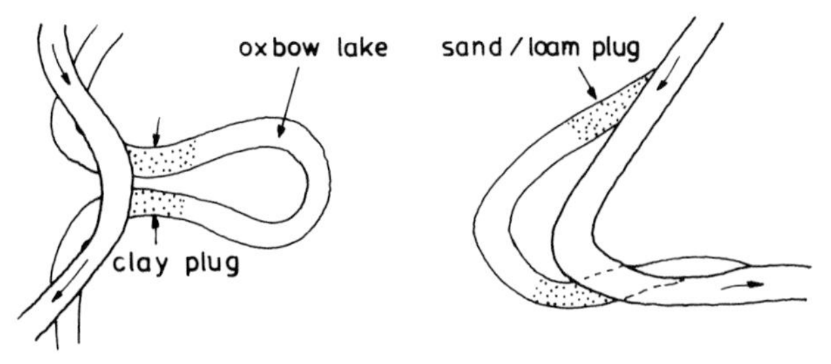

Fig. 4.17 Sketch of a neck cut-off and a chute cut-off.

under the influence of the daily tides. Under such conditions high and broad levee ridges are common and the whole coastal zone of the delta tends to be higher than the areas farther inland, where a zone of shallow lakes and swamps may consequently occur. Farther upstream where the river builds the levees at a rate of a few times a year, when in spate or at bankfull discharge, the levee ridges are much less developed. This is particularly so in the middle part of the delta where textural difference of the sediments forming the levees and backswamps is minimum. This explains why these parts are the most susceptible to flooding.

The gradient of a delta diminishes during delta growth, therefore, the riverbed is raised as an adaptation to the new longitudinal profile. This results in more frequent flooding, which forms part of the natural development of deltas and results in the building up of top sets. The delta slope also affects the distributary pattern of the channels. Volker (UNESCO, 1966) states that few large channels usually occur if the slope is between 5×10^{-5} and 5×10^{-4} whereas numerous branches are common in cases of more gentle gradient. At the same time the size of the interdistributary basins diminishes and the levees become lower and gradually transitional to the basins.

In areas where a new distributary is formed, due for example, to a breach in the levee, a new flow pattern is initiated. Whether the old or the new branch will silt up or remain in operation depends on various circumstances and particularly on the direction and curvature of both branches at the bifurcation. If the new branch is straight and departs from the old course in a concave meander bend, the tendency will be for the old channel to be put out of use as the water diverted into it will get an upward impulse due to the curvature of the bed and much silt from the bottom will be deposited. This is a fairly common situation owing to the strong lateral sapping in these concave river stretches. If however, the new branch is curved and the old channel straight at the bifurcation, it is likely that the new channel will be temporarily used only. If the curvatures of both arms are comparable they may remain simultaneously in use provided that other circumstances are equal. Differences in length of the two branches may also be decisive: the branch having the shortest distance to the outfall will have a stronger gradient and thus be in a favorable position at least until further seaward growth has eliminated the advantage.

Mikhailov (UNESCO, 1966) pointed out that the current velocity in estuaries decreases both parallel and perpendicular to the current direction. The decrease in velocity is most rapid in the central part of the estuary and bars are preferably formed there. This changes the current pattern and other shoals are formed. This mechanism explains why in case of an estuary with many branches, the central ones tend to silt up whereas the two extreme positions tend to become dominant. Longshore current, beach drifting and wave action are processes which tend to close the outfalls or to deviate them sidewards. Sandy beach ridges may be formed, but where they are absent the near-coastal backswamps are surrounded by horse-shoe shaped natural levees and drained off toward the sea. Thus they are to some extent protected from flooding by the rivers but fully exposed to storm surges at their open, seaward side.

The importance of the tides in building natural levees in the lower part of deltas has already been mentioned. However, the tides have other important geomorphological effects and contribute to the integration of geomorphology and hydrology in deltaic areas. In areas of great tidal range, such as the Ganges delta, creeks have an aggressive, eroding character and on aerial photographs the broad transitional zone between the low and high tide marks is striking. The tides assist in keeping seaward channels open and straightening their lowest parts. Farther upstream increased curvature and reticular patterns of interconnected channels may result from the ponding up of river water during high tide.

The effects of tides are complicated especially when not simultaneously occuring throughout along the coast, but proceeding from one side to another. Along the Dutch coast the tides proceed from south to north and as a result the SW channels in the estuaries have more gradient, stronger current velocity and less sedimentation than the NW directed channels.

The effects of the tides are diversified and many different situations exist, for example, when the tidal movements result in strong and counteracting ebb and flood currents, a variant of the levee building process can be observed. Hook or horse-shoe shaped initial levees of sand and silt are formed and ebb and flood currents meet, thus forming an interlocking system of channels and shoals. Vertical accretion may result in the formation of islets, known as 'opwas' in the Netherlands and as 'dwip' in the Sunderbans zone of the Ganges Delta, forming a contrast with the lateral accretion (Dutch: 'aanwas') found along certain parts of channels or coast.

Other complications in the lowest reaches of the rivers result from the contact of saline/brackish and fresh water. Mixing these waters results in the formation of an inclined interface, the inclination of which may range from near vertical to almost horizontal. The lower current speed near this plane and particularly the flocculation process occurring will result in important sedimentation in these river parts, especially of very fine particles (< 0,0002 mm). Deepening of harbours,lowering river courses or canals for purposes of navigation for seagoing vessels may result in saline intrusions and as a consequence contributes to increased sedimentation by way of flocculation in the formerly fresh waters that are turned brackish. This process has been observed in the Durme River, a branch river of the Schelde, after the deepening of the latter for improving the access to the harbour of Antwerp.

The Coriolis force which tends to deviate the rivers of the northern hemisphere to the right and those of the southern hemisphere to the left has hydrological as well as geomorphological effects as both salinity and sedimentation are influenced by it. It has been observed that in straight parts of larger rivers in the northern hemisphere lower salinities occur near the right bank as compared to the left bank. As a result sedimentation stretches farther upstream near the left bank. The effect is particularly clear in estuaries. If an important tidal range occurs in such areas, the Coriolis force may substantially contribute to bank erosion on both sides during rising as well as falling tides.

It is evident that it is important for hydromorphological studies of deltaic areas to take the characteristics of the whole river basin into consideration as they will affect the regime of the lowland portion of the river and the geomorphological situation prevailing there. The latter will,in turn, have an influence on the hydrological situation. A good example is given by Tada and Oya who compared the Kiso and Chikugo rivers in Japan (see also section 13). The Kiso River carries a high bedload as a result of its mountainous hinterland, steep gradient and absence of basins where the load could be deposited. As a result large natural levees have been built which often lead to long-lasting inundations in extensive areas. The Chikugo River on the contrary, has a very limited bedload and the levee development along its lower course is negligeable. Its branch rivers entering the lowland plain have steeper gradients and have formed fans. As a result, the Chikugo River is in a low position and inundations by the Chikugo are limited in extent and of short duration.

The nature of the hinterland (together with velocity and turbulence during sedimentation) also determine the texture of the sediments. Stratification of the sediments is explained by the fact that a higher current velocity is required for the erosion of a sediment than for its sedimentation. Thus fine-textured sediments deposited during quiet conditions are covered by sandy material during a period of greater turbulence, without being removed. The hinterland also has an effect on the fertility of the delta deposits and of the suspended load. The heavily weathered hinterland of humid tropical rivers usually produces rather senile material and the lime content of tropical river water is often low. Water passing through cat clay areas may become toxic and brackish water may be a factor in potassium enrichment if the sodium and chlorine concentrations can be reduced. These matters play an important role in irrigation projects where fertility depends not only on soil quality but also on the quality of the irrigation water.

The combined effects of factors such as river discharge, sediment load, littoral drift, tides, storms, size of catchment and crustal movements, result in a great diversity in delta morphology. Each delta of the major SE Asian rivers has its own identity. The Irrawaddy delta, where flooding is due to upper reach discharge and heavy local rains, shows very little vertical accretion but grows seaward at a rate of approximately 50 metres annually. Erosion is common due to the great tidal range. The Chaophya Delta in Thailand has four major levee ridges with intervening backswamps where stagnant water is at places 3 metres deep and floating rice is cultivated. The gradient of the delta is very low and floods are gentle. As a consequence the deposits are very fine textured. The seaward growth of the delta has led to a decreased gradient and an increased frequency of flooding in the plains. The Ganges delta is composed of an extensive system of interconnected distributaries, particularly in the Sundarbans. The most prominent natural levee extends from Calcutta to the SE which reflects a major shift of the Ganges from west to east due to crustal movements. In the west the levees are higher, the banks of the creeks more pronounced, the drainage lines deeper incised, the estuaries wider and the fresh water influence more limited.

4.9 Estimating River Discharge and Sediment Yield

It has been stated on various occasions in previous sub-sections how and when geomorphological observations and parameters can be used for making qualitative estimates of river discharge, peak flow, sediment yield, etc. (Gottschalk, 1964). In this sub-section reference to these earlier sections will be made.

A first approximation of the discharge characteristics of a river can be obtained by a purely qualitative assessment of the various geomorphological units of the basin. (Gagoshidze, 1976; Ghanem, 1972; Holeman, 1975). Somewhat greater precision may be achieved by applying semi-empirical methods and giving ratings to these units such as is used in the Cook method (Ven Te Chow, 1964). Straightforward morphometrical parameters of a watershed such as drainage density, basin shape and slope may be successfully applied to the estimation of peak discharges using multiple regression techniques. As mentioned earlier, spacing of the drainage lines gives qualitative information on the surface run-off as compared to infiltration and evaporation, if climatological factors are known, in particular, rainfall. Drainage density may serve as a useful indicator. Surface run-off depends on the permeability of the rock type and soil cover as well as on the morphological characteristics of the terrain. The classification of the terrain in hydro-morphological units is based on factors such as dissection of the area, and the steepness and form of slopes. There is usually a close relation between these factors and hydrological matters such as peak run-off. Because lithology is one of the factors influencing the geomorphology of the basin, hydromorphological units may or may not coincide with lithological boundaries. The evaluation of these units in terms of peak run-off has to be adapted to the wet season, when the soils are saturated and a rainstorm of considerable intensity and duration results in relatively high run-off rates.

In fact the run-off of a river and the shape of its hydrograph giving the variations in discharge with time is determined by the rainfall intensities and their temporal distribution as well as characteristics of the watershed. Among these characteristics, relief and shape rank high as geomorphological variables; other importnat environmental parameters are: matters of permeability and fracturing of the bedrock, texture and thickness of unconsolidated covering materials, soil type and moisture conditions. In small basins quick run-off response to rainfall usually occurs whereas large basins have a slow response. This is understandable because the perennial flow of large basins is sustained by the groundwater discharge and the effect of peak run-off produced in sub-basins by brief intense local rainfall, is dampened.

The time of flood concentration is interesting from a hydrological point of view. This is the time it takes for water to travel from the most distant point of a watershed to the watershed outlet. Nomographs exist for estimating the time of concentration. The formula is:

$$T_c = 6.95 \frac{L^{1.15}}{H^{0.385}}$$

where L is the length of the watershed along the main stream in m and H is $H_{max} - H_{min}$ in m.

However, differences in elevation due to waterfalls, rapids or other sudden drops should be substracted from the value of H before using the formula. All necessary measurements and checks can be made on aerial photographs. Slopes occurring in the river basin (steepness, frequency distribution, etc.) are of great hydrological interest, as they affect both run-off and sediment transport along the slopes.

Several practical applications of morphometrical approaches are elaborated upon by Hirsch (1975). This author claims that the ratio between the drainage densities of episodic and ephemeral rivers can be used as an indication for the torrentiality and that anthropogenic influences cause a rapid increase of drainage density and in the occurrence of a larger number of higher-order streams.

Another interesting hydrological application of relief factors, briefly mentioned in sub-section 4.2, is the determination of the rainfall in a catchment on the basis of the data gathered at the sites of rain gauging stations. The simplest method, using the arithmetic average of observations,is normally unsatisfactory because of the gauges being non-uniformly distributed. More satisfactory results are obtained by weighing the gauging data according to the areal distribution of the stations. This can be done using the so-called Thiessen method of drawing perpendicular bisectors to the connecting lines of the station network (fig. 4.18). The best results are reached with a map showing the isohyetes drawn by interpolation in that zone. The shapes of the isohyetes are often strongly affected by the relief situation. In the case of orographic precipitation rainfall increases with height, except at great altitudes where a decrease is common. The exposure to winds bringing rain is also important. Considerable differences may exist between windward and leeward slopes and the crestline separating them then is marked by a sudden drop in precipitation.

The assessment of the sediment production and transportation within the various parts of a watershed (fig. 4.19) and the ultimate sediment yield at the outfall of the river is a delicate matter (Allan and Barnes, 1973; Bedito and Wieber, 1967; Bowie et al., 1975; Glymph, 1972; Hopp, 1972).

The geomorphological contribution to the problem is two-fold: a geomorphological classification of the land into genetical types and a thorough qualitative and quantitative study of the various erosion processes of the present and the past. Furthermore, pedological data on the erodibility of soils and the stage of erosion reached as evident from truncation of profiles, for example, is essential. Evidently, there is a relation between erodibility of the soil and erosion processes. The universal soil loss equation, developed in the USA,gives results of reasonable accuracy for small watersheds. The matter is elaborated upon in section 15 dealing with erosion surveys.

One of the difficulties in estimating existing and po-

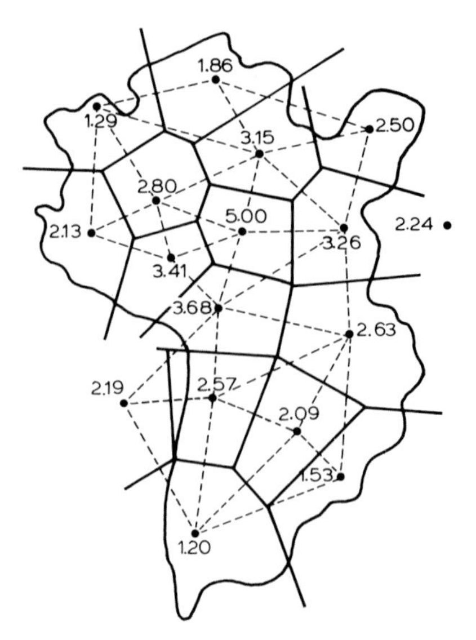

Fig. 4.18 Example of Thiessen network: average precipitation is 2,54 in.).

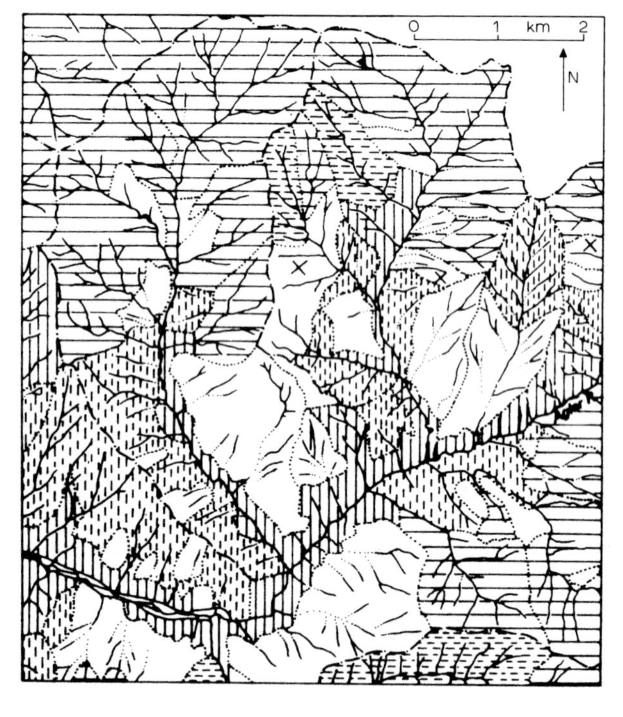

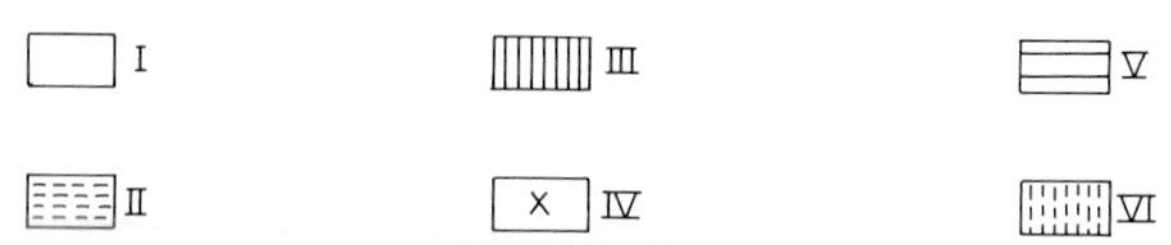

Fig. 4.19 Map showing part of the Aglar River basin, N. India, with classes of relative sediment production. Drainage channels are divided into those which are well-defined and incised (fully drawn lines) and those which disappear in periglacial deposits (dashed). Six classes of relative sediment production are distinguished:
Non-forested slopes: I - low (5), II - moderate (30), III high - very high (77,5).
Forested slopes: IV - low (2,5), V - moderate (15), VI moderate - high (45) (Meijerink, 1974).

tential sediment production from the interfluves is that sporadically occurring excessive rainfall intensities may change the characteristics of a basin overnight and that changes in the vegetative cover such as deforestation and other works of man may have a similar effect in the course of several years, decades or centuries. Another difficulty is that various sediment sources exist, some of which are not readily detectable or quantifiable unless elaborate and detailed investigations are carried out. Linear erosion, such as the development of gullies, ravines and badlands can be mapped comparatively easily, particularly when aerial photographs are used,although quantification may be difficult either in the absence of sequential imagery or because the photo scale does not permit the precise measurement of the (increased) incision of the gullies and ravines. The degree of sheet and rill erosion can be determined from the truncation of soil profiles and the different, usually lighter tone of the areas where the topsoil has been removed. The effect of mass movements is considerably more difficult to assess and bank erosion is not always easily established.

The vegetative cover and landuse patterns may add to the difficulties mentioned above. (Walling, 1974). A third difficulty in the assessment of sediment yield is that one has to consider which proportion of the sediment produced on the interfluves is deposited before reaching the riverbed and which part remains (temporarily) in the aggrading riverbed or is deposited in depressed zones and other areas where the water spreads, dropping part of the load.

In spite of these difficulties, the sub-division of a watershed into classes of equal sediment production using aerial photographs and a subsequent attempt at quantification by field observation in each of these classes is a valid approach. The studies carried out by Meyerink (1977) in the Serayu River Basin, Java, Indonesia give a good insight into the methods involved and the results obtained when applying this geomorphological approach.

When a basin is investigated for the estimation of sediment yield, many different factors have to be taken into consideration. (Schumm, 1961, 1969; Scott and Williams, 1974; Somogyl, 1967). It may be important here to make make an estimate of the rate of silting up of reservoir lakes on the basis of the rock types occurring, their erosion susceptibility and the quantification of processes operating.

The universal soil loss equation may be used to make an estimate of the gross erosion occurring within a basin (Williams, 1975). Rates of sediment yield of watersheds are immediately comparable only if their gross erosion rates per unit area are approximately the same. Otherwise their delivery rates must be compared. Once the causal factors of the sediment delivery rate have been evaluated for a certain hydromorphological unit within a climatic region, it is possible to predict sediment yields rather precisely from gross erosion data and the delivery rate characteristics of individual watersheds within that region.

Adjustments to the sediment delivery ratio are necessary because of the large variations in gross erosion rates and in the percentage of eroded material transported to the point of the measurements. It has been shown that as little as 5% and as much as nearly 100% of the material eroded in a watershed may be delivered to a downstream point in the plain. It should also be considered that relatively small parts of a watershed can sometimes produce the majority of the gross erosion. In this case an arbitrary size of net-contributing area is a better indicator of the sediment yield than the total watershed area. Several geomorphological parameters are of importance in sediment yield estimates. Steep watersheds with well-defined channels attain higher values than watersheds of low relief and poorly defined channels. Drainage density or rather the density of incised channels is also important in many cases (Rees, 1967; Renard and Lane, 1975; Rhoades, 1975).

It has been stated that a catastrophic event may completely change the river regimen overnight. This happened in the Guil Valley, France, during a very severe flood in 1957 (Tricart, 1962). Tree trunks and improperly constructed bridges caused a terribly pulsating floodflow. As a result, huge blocks were transported which had been in rest since early Post-glacial times. Many slides added to the river load and deposits up to 10 metres thick were formed in 48 hours in the main riverbed behind obstacles. Deposition took place where the riverbed widened (due to lesser current speed occurring there). The river was then diverted around its deposits and caused lateral sapping at both sides. The transported material was again deposited against the next alluvial fan downstream. Erosion was observed where the river had a narrow passage at the lower end of a side-fan. The material eroded was again deposited in broader parts farther downstream. It proved of particular importance that the angle of confluence of the side rivers on the fans with the trunk river points downstream. Otherwise, a decrease in current velocity will be produced and excessive deposition and ponding-up will easily be provoked during a freshet. Other examples of similar phenomena are, for example, rapid deposition in the so-called Wara triangle in Java and in the Brantas River basin where the volcanoes Merapi and Kelud were the cause of excessive river load (see section 6).

The river regimen is usually of a more quiet nature than that described above. A good understanding of horizontal and vertical accretion of the river and the lateral and vertical erosion produced is essential in all cases. A change in the course of a (lowland) river, a meander cut-off, for example, immediately influences the current and effects the distribution of lateral sapping and sedimentation downstream. The material forming the river sides is of the greatest importance here: if fine textured material occurs, it will probably be removed and transported over considerable distances. If however, coarse grained material such as pebbles are concerned, they will most likely be deposited after a comparatively short distance. These deposits then form a narrow part in the river where erosion is likely to occur during the next flood. River dynamics have to be studied in detail; the geomorphologist does not look at the sediments as 'deposits' only, but tries to understand the dynamics that have produced them. For example, gravel deposited during a flash flood has non-oriented long-axes, whereas the majority of long-axes of gravel transported by quiet saltation is perpendicular to the current direction.

Changes occurring in river courses can often be traced by comparing old and new topographical maps or old maps with airphotos. The period covered is at the most 100-150 years. However, on aero space imagery much older river courses can often be traced without difficulty. Therefore, areas where changes frequently occur can be distinguished from more stable parts of the river; this is of great practical importance, for example, for bridge-construction. Approximate dating of the changes can be attempted by estimating the annual rate of accretion or the total time needed to form the alluvial plain. Free meanders tend to enlarge their curves and to slowly shift downstream, however, their centre may remain at the same place, particularly in places where the space for meandering is limited due, for example, to rock outcrops. It should be understood that differences in river load up- and downstream of a river section are not always indicative for the number of changes of the bed, but only point to the general direction of the changes. The continuous exchange existing between the riverbed and the valley and the suspended and bottom deposits, forming the total river load,should receive due attention. After ample observation on several rivers in the USA, Leopold and Wolman (1957) came to the conclusion that lateral erosion in meanders is mainly due to the average-sized floods which fill the minor bed completely. Larger floods overflow the levees and are therefore less effective in eroding the bed. They also pointed out that the great majority of floodplain deposits is of the channel type and result from lateral sapping, whereas overbank deposits due to vertical accretion are of minor importance. Every deposition by the river is associated with the existence of a circulatory motion of helical flow which affects every river bend. The opposite bank (concave) is eroded, thus the width of the channel remains the same. It should be remarked that the textbook-type cut-offs are much less frequent than a break through of the floodwaters over the the gently downsloping convex side of the meanders. Maximum sapping usually occurs after thorough wetting of the bank and secondly, cold periods seem to be effective.

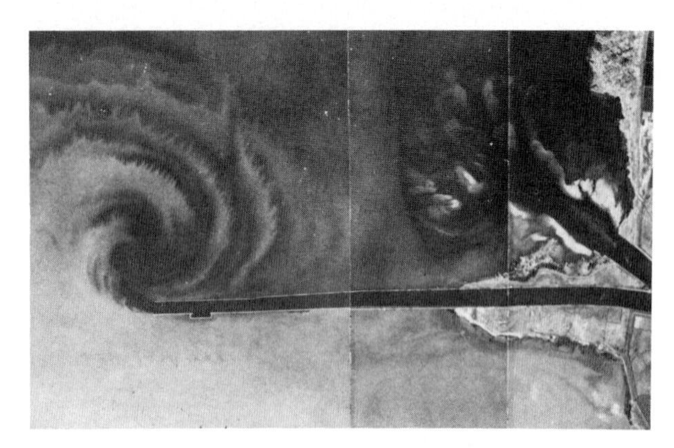

Fig. 4.20 Turbidity patterns at the outfall of the Keteldiep, the Netherlands. Scale: 1:25.000.

Turbidity patterns produced by clay brought by the rivers and carried in suspension into the coastal waters are useful indicators of flow patterns and the spatial variations in sedimentation rates so caused. This is particularly marked in the humid tropical zone though it also occurs elsewhere. Black and white infrared aerial photographs show the turbidity patterns at the surface of the water only because their water penetration is negligible. If the filters used in the survey also let the red and possibly part of the orange light pass through, they will at best show the turbidity of a very shallow layer of water. Panchromatic emulsions show a thicker slice of water and usually give more satisfactory results. The vertical aerial photograph of fig. 4.20 gives an example and shows the outfall of the Keteldiep, the Netherlands at a scale of 1:40,000. A gradual build-up of natural levees is characteristic for the distributary in the right half of the picture, whereas a pronounced eddy current developed at the mouth of the other.

As the green part of the spectrum gives the best water penetration, the turbidity in the thickest possible slice of water can then be recorded; however, this does not necessarily mean that optimum results will be obtained (Verstappen, 1976). It may be that the turbidity at the time of exposure indiscriminately covers large parts of the shallow waters and that the patterns stand out better if the observation is limited to a thinner slice of water. Furthermore, the recording in the green part of the spectrum is also rather easily adversely affected by atmospheric conditions, resulting in an image of inferior quality. It may be advantageous to record the turbidity patterns simul-

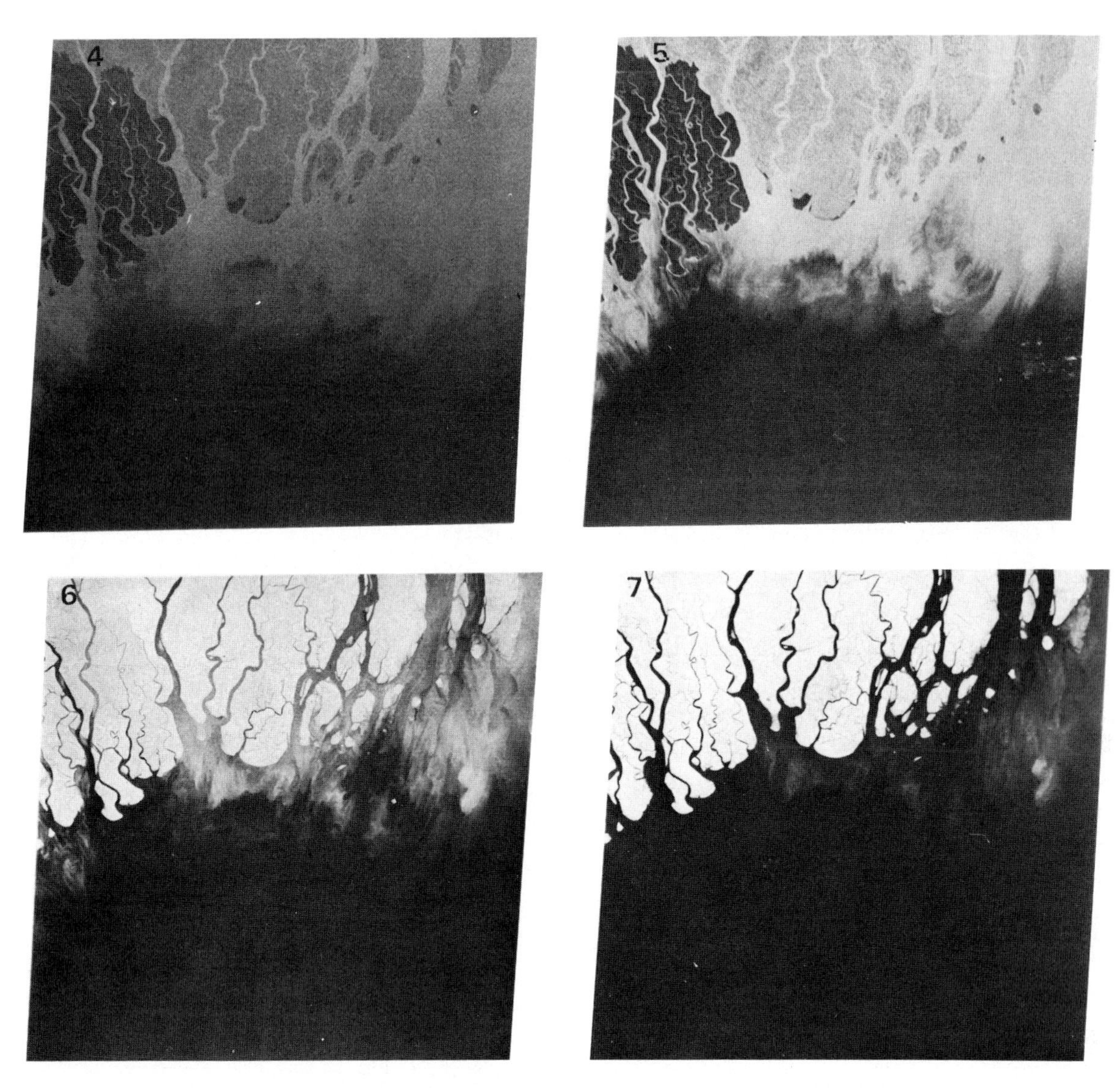

Fig. 4.21 Landsat image (Bands 4, 5, 6 and 7) of the lower Ganges Delta showing decreasing turbidity pattern visibility and increasing aptness for coastal deliniation from band 4 to 7.

taneously in various wave lengths and to analyse them separately afterwards. Landsat imagery is ideal for this purpose, particularly where broad patterns in small scale surveys are concerned. Fig. 4.21 depicts the Sundarban part of the Ganges Delta in the four MSS bands of Landsat-2 and shows how the turbidity patterns decrease in size and detail from band 4 (green) to 5 (red), 6 and 7 (both in the infrared).

The matters of erosion, sedimentation rates in river basins, sediment yield and silting up of reservoirs will be further elaborated upon in section 15, Erosion Surveys.

References

Abdel-Hady, M. et al., 1970. Subsurface drainage mapping by airborne infrared imagery techniques. Proc. Oklahoma Acad. Sci., 50: 10–18.

Allan, P.B., and Barnes, B.B., 1972. Total sediment load by the extrapolated data procedure. Proc. Sediment-yield Workshop, USDA Sedimentation Lab., Oxford (Miss.), Nov. 1972. USDA, Agric. Res. Service, ARS-S-40: 100–108.

Allan, W.H. Jr., 1968. Pedo-geomorphic elements of the channel and floodplain of a small Nebraska watershed. Proc. Iowa Acad. Sci., 75: 243–252.

Anon., 1973. Géomorphologie et eaux souterraines: présentation de la carte géomorphologique de la pampa del Tamarugal (Désert nord chilien). Bull. Inst. d'Etudes Andines, 2(2): 39–57.

Bakker, A.J., 1951. The hydrology of the Anten River, W. Java

Bedito, G. and J.C. Wieber, 1967. Quelques observations sur l'hydrologie et les transports solides dans le bassin-versant de la haute Dordogne. (Some observations on the hydrology and transport of solids in some drainage basins of the Haute Dordogne). Revue Géogr. de l'Est, 7(4): 427–459.

Bleys, C., 1953. Enkele aantekeningen over Surinaamse ferrietkappen en hun waterafvoer. Geol. en Mijnbouw, 15: 175–179.

Bogomolov, L.A., 1963. Topografičeskoe deshifrirovaniiye prirodnogo landšafta na aẙrosnimkow. (Topographical interpretation of aerial photographs of a natural landscape). Moscow, Gosgeoltekhizdat, JPRS: 17–771.

Bowie, A.J. et al., 1972. Sediment yields related to characteristics of two adjacent watersheds. Proc. Sediment-yield Workshop, USDA Sedimentation Lab., Oxford (Miss.), Nov. 1972, USDA, Agric. Res. Service, ARS-S-40: 89-99.
Boyce, C., 1975. Sediment routing with sediment-delivery ratios. Proc. Sediment-yield Workshop, USDA Sedimentation Lab., Oxford (Miss.), Nov. 1972. USDA, Agric. Res. Service, ARS-S-40: 61-65.
Brucker, R.W. et al., 1972. Pole of vertical shafts in the movement of ground water in carbonate aquifers. Ground Water, 10(6): 5-13.
Bührer, M., et al., 1981. Hochwasserschutz und Wassernutzung im Gebirge. Vermessung, Photogramm., Kulturtech., 79(3): 81-91
Bureau of Reclamation, 1975. Procedures for predictive sediment yield. Proc. Sediment-yield Workshop, USDA Sedimentation Lab., Oxford (Miss.), Nov. 1972. USDA, Agric. Res. Service, ARS-S-40: 10-15.
Carey, C., 1969. Formation of flood plain lands. Proc. Amer. Soc. Civil Eng., J. Hydraul. Div., 95: 981-994.
Chiang, S.L.,and G.W. Petersen, 1970. Soil catena concept for hydrologic interpretations. J. Soil and Water Conserv., 25(6): 225-227.
Chorley, R.J., et al., 1969. Water, Earth and Man, Methuen, London, 588 pp.
Chow, V.T. (Editor) et al., 1964. Handbook of applied hydrology. McGrawHill, New York, 1418 pp.
Coleman, J.M., 1969. Brahmaputra River: Channel processes and sedimentation. Special Issue Sedimentary geology: 129-239.
Cooke, R.U., and Doornkamp, J.C., 1974. Geomorphology in Environmental Management. Clarendon Press, Oxford, 412 pp.
Diacony, C., 1967. La modification des lits sablonneux pendant les crues et ses conséquences pour l'activité hydrométrique. Int. Ass. Sci. Hydrology, 75: 250-254.
Dixey, F., 1966. Water supply, use and management. In: E.S. Hills (Editor) Arid Lands, a geographical appraisal. London and Paris: 77-102.
Douglas, I., and Crabb, P., 1972. Conservation of water resources and management of catchment areas in Upland Britain. Biol. Conserv., 4: 109-116.
Fisk, H.N., 1951. Mississipi River Valley geology in relation to river regime. Proc. Amer. Soc. Civil Eng., 77: 1-30.
Freers, T.F., 1970. Geology and ground water resources, Williams County, North Dakota, Part I, Geology. Bull. N. Dakota Geol. Survey, 48(1): 55 pp.
Gagoshidze, M.S., et al., 1976. Criteria of stability and transportation capacity of different kinds of mountain streams. Int. Ass. Sci. Hydrol, 75: 209-214.
Ghanem, H., 1972. A study on the morphological changes in part of the Zeroud River Basin (Central Tunisia), induced by the flood of September-October 1969. M.Sc. thesis, ITC, 120 pp.
Glymph, L.M., 1975. Evolving emphases in sediment-yield predictions. Proc. Sediment-yield Workshop, USDA Sedimentation Lab., Oxford (Miss.), Nov. 1972. USDA, Agric. Res. Service, ARS-S-40,: 1-4.
Gospodinov, G.V., 1961. Deshigrirovaniiye Aerosnimkow/Interpretation of aerial photographs. Univ. Moscow, 186 pp. (Translated from the Russian by US Dept. Commerce).
Gottschalk, L.C., 1964. Reservoir sedimentation. In: V.T. Chow (Editor). Handbook of applied Hydrology, McGraw-Hill, New York.
Gregory, K., 1977. River channel changes. Wiley, New York, 281 pp.
Happ, S.C., 1975. Valley sedimentation as a factor in sediment-yield determinations. Proc. Sediment-yield Workshop. USDA Sedimentation Lab. Oxford, (Miss.), Nov. 1972. USDA, Agric. Res. Service, ARS-S-40: 57-60.
Hesters, M.E.S., 1981. Praktijkvoorbeeld van winterbeheersing in Vlaanderen; De Zwalm vallei. Cultuurtechn. Tdschr., 14: 1-5.
Hirsch, F., 1975. Application de la morphométrie à l'hydrologie. Revue Géom. dyn., 15: 172-175.
Holeman, J.N., 1975. Procedures used in the Soil Conservation Service to estimate sediment yield. Proc. Sediment-yield Workshop, USDA Sedimentation Lab., Oxford (Miss.), Nov. 72. USDA, Agric. Res. Service, ARS-S-40: 5-9.
Horton, R.E., 1945. Erosional development of streams and their drainage basins; hydrophysical approach to quantitative morphology. Bull. Geol. Soc. Amer., 56: 275-370.
Inglis, C.C., 1967. Meanders and their bearing on river training. Proc. Inst. Civ. Eng., Maritime and waterways, paper 7: 71-85.
Ion, D.I., 1970. Carstul din nordul Olteniei şi citeva probleme de geografie aplicatǎ (North Oltenia's karst and some problems of applied geography); Anal. Univ. Bucuresti, Geogr., 19: 105-112.
ITC, Proc. ITC-UNESCO seminar on integrated surveys for river basin development. 1969, 110 pp.
Kirkby, M.J. (Editor), Hillslope hydrology. J. Wiley, New York, 389 pp.
Klimaszewski, M., 1956. The detailed hydrological map of Poland. Przeglad Geogr., 28, suppl: 42-47.
Knisel, W.G., 1972. Response of karst aquifers to recharge. Hydrol. Papers, Colorado State Univ., Fort Collins, 60, 51 pp.
Kuiper, E., 1971. Water Resources development. 3rd Ed. Butterworths, London.
Lacey, G., 1930. Stable channels in alluvium. Proc. Inst. Civil Eng., 1929-1930, 229: 259-292.
Legrand, H.E., 1970. Comparative hydrogeology: an example of its use. Bull. Geol. Soc. Amer., 81(4): 1243-1248.
Leopold, L.B., 1953. The hydraulic geometry of stream channels and some physiographic implications. U.S. Prof. Paper, 252, 57 pp.
Leopold, L.B., and Wolman, H.G., 1957. River channel patterns—braided, meandering and straight. U.S. G.S. Prof. Paper, 282-B, 22 pp.
Leopold, L.B., et al., 1964. Fluvial processes in geomorphology. Freemans Co., San Francisco, London, 522 pp.
Lohman, S.W., and Robinove, C.J., 1964. Photographic description and appraisal of water resources. Int. Archives of Photogrammetry, XIV, Lisbon. (also in: Photogrammetria, 19(3), 40 pp.
Maitre, G., 1968. Etude hydrogéomorphologique comparée de cinq bassins-versants D'Allier, Thesis, Strasbourg Univ., 85 pp.
Markotic, M., 1969. Jezero. Prilog poznavanju problema polja u kršu. (Jezero. A contribution to the understanding of poljes in karst). Geografski Glasnik, 31: 155-170.
Martinec, J., 1967. The effect of morphological processes in alluvial channels on flow conditions. Int. Ass. Sci. Hydrol., 75: 243-249.
Meijerink, A.J.M., 1974. Photohydrological reconnaissance surveys. Thesis Free Reformed University Amsterdam, 372 pp.
Meijerink, A.J.M., 1977. A photohydrological reconnaissance of the Serayu River basin. ITC Journal, 4: 646-674.
Morisawa, N.H., 1959. Relation of quantitive geomorphology to stream flow in respective watersheds of the Appalachian Plateau Province. Techn. Report 29, Dept. Geol., Columbia Univ., New York, 94 pp.
Müller, R., 1943. Theoretische Grundlagen der Fluss- und Wildbachverbauung, Mitt. 4, Versuchsanstalt f. Wasserbau, ETH, Zürich.
Nakano, T., 1962. Landform type analysis on aerial photo-

graphs. Its principles and techniques. Trans. Symp. Comm. VII, I.S.P. , Delft: 149-152.
Orme, A.R., and Baileu, R.G., 1970. The effect of vegetation conversion and flood discharge on stream channel geometry: the case of southern California watersheds. Proc. Ass. Amer. Geogr., 2: 101-106.
Oya, M., 1973. Relationship between geomorphology of the alluvial plain and inundation. Asian Profile, 1(3): 479-538.
Popov, I.V., 1960. Aerofotos - emka i izučenic vos suŝi (Aerial survey and the study of inland waters. Hydr. Meteor. Publ. House, Leningrad) Translated from russian. ITC Netherlands, 60 pp.
Popp, N., 1969. Les conditions géomorphologiques dans l'aménagement hydroaméliorätif des basses plaines alluviales en Roumanie. Symp. Int. Géom. Appliquée, Bucharest 191-199.
Radai, Ö, 1969. Legifotó-értelmezós alkalmazása karsztuízföldtani térkópezéshez (Aero photographic interpretation and hydrogeological mapping of karstic areas), with english summary, Budapest, 82 pp.
Rees, D.C., 1967. A study of the movements of bed sediment along Azusa River, Japan. Geogr. Reports, Tokyo Metropolitan Univ., 2: 29-40.
Renard, K.G., and Lane, L.J., 1975. Sediment yield as related to a stochastic model of ephemeral runoff. Proc. Sediment-yield Workshop, USDA Sedimentation Lab., Oxford (Miss.), Nov. 1972. USDA, Agric. Res. Service, ARS-S-40: 253-263.
Rhoades, E.D., et al., 1975. Sediment-yield characteristics from unit source watersheds. Proc. Sediment-yield Workshop, USDA Sedimentation Lab., Oxford (Miss.), Nov. 1972. USDA, Agric. Res. Service, ARS-S-40: 125-129.
Robinove, Ch.J., 1968. The status of remote sensing in hydrology. Arch. Photogrammetry, 16: 827-831.
Roglič, J., 1967. L' analyse morphologique et la solution des problèmes hydrologiques dans les poljes du Karst dinarique. Symp. Int. Géom. Appliquée. Bucarest.
Schick, A.P., 1971. A desert flood: physical characteristics, effects on man,Geomorphic significance, Human adaption. Jerusalem Studies in Geogr., 2: 91-155.
Schumm, S.A., 1961. Effect of sediment characteristics on erosion and deposition in ephemeral stream channels. USGS Prof. Paper, 352-C.
Schumm, S.A., 1964. Airphotos and water resources. Trans. UNESCO Symp. Aerial Surveys and Integrated studies, Tolouse. UNESCO: 70-80.
Schumm, S.A., 1969. River metamorphosis. Amer. Soc. Civil Eng. Proc., 95. J. Hydrol. Div., HY1: 255-273.
Schumm, S.A. Variability of river patterns: Nature (Phys. Sci.), 237: 75-76.
Schumm, S.A., 1974. Geomorphic thresholds and complex response of drainage systems. In: Fluvial Geomorphology; Proc. 4th Ann. Symp. Geom. Sunj.-Binghampton, N.Y.: 299-310.
Scott, K.M., and Williams, R.P., 1974. Erosion and sediment yields in mountain watersheds of the Transverse ranges, Ventura and Los Angeles Counties, California. Analysis of rates and processes. Final report. USGS, Menlo Park, Cal., Water Res. Div., WR1-47, WRD 75/019.
Somogyl, S., 1967. Relationships between morphology and sediment transport in river beds. Int. Ass. Sci. Hydrol., 75: 151-161.
Speight, J.G., The role of topography in controlling through flow generation: a discussion. Earth Sci. Proc., 5(2): 187-191.
Strahler, A.N., 1964. Quantitative geomorphology of drainage basins and networks. In: Ven Te Chow (Editor). Handbook of Applied Hydrology, New York.
Sweeting, M., et al., Some results and applications of karst hydrology: a symposium. Geogr. J., 139(2): 280-310.
Tada, F., and Oya, M., 1957. The flood type and the classification of topography. Proc. I.G.U. Reg. Conf., Tokyo: 192-196.
Thomas, D.M., and Benson, M.A., 1970. Generalization of streamflow characteristics from drainage basin characteristics. U.S.G.S. Water-Supply Paper, 55 pp.
Tricart, J., 1962. Mécanismes normaux et phénomènes catastrophiques dans l'évolution des versants du bassin du Guil (Htes-Alpes, France). Ztschr.f.Geom., 5(4): 277-301.
Tricart, J., and Vogt, H., 1967. Quelques aspects du transport des alluvions grossières et du façonnement des lits fluviaux. Geogr. Annaler, 49A: 351-366.
Tricart, J., 1978. Géomorphologie applicable. Masson, Paris, 204 pp.
Tulio Benavides, S., 1976. Introducción a la fotointerpretación en estudios del terreno y aplicaciones en investigaciones hidrológicas. Revista CIAF, 3(1): 103-138.
UNESCO, 1966. Proc. Symp. Scientific problems of the humid tropical zone deltas and their applications, Dacca, 1964, UNESCO Humid Tropics Research Publ. 63, 422 pp.
UNESCO, 1971. Design of water resources projects with inadequate data. Nat. Sci. Div., Publ., UNESCO, Paris, 166 pp.
Verstappen, H.Th., 1961. Etude hydrogéologique du Bugesera (Ruanda-Burundi); unbl. EEC Report, 23 pp.
Verstappen, H.Th., 1964. Geomorphology in delta studies. ITC Publ. B 20, 20 pp.
Verstappen, H.Th., 1967. Geomorphology using airphotos in hydrological studies. Symp. Int. Géom. Appliquée, Bucharest: 159-168.
Verstappen, H.Th., 1972. Geomorphology. In: Veldboek voor land en water deskundigen. Publ. I.L.R.I.: C. 1-16.
Verstappen, H.Th., 1972. Purnavarman's river works near Tugu. Bijdr. Taal, Land en Volkenk., 128(2, 3): 1-10.
Voûte, C., 1958. A comparison between some hydrological observations made in the Jurassic and Cenomanian Limestone mountains, situated to the West and to the East of the Ghab Graben (U.S.R., Syria). Publ. I.A.S.H., 57: 160-166.
Walling, D.E., 1974. Suspended sediment and solute yields from a small catchment prior to urbanization. In: K.J. Gregory and D.E. Walling (Editors), Fluvial processes in instrumented watersheds, Inst. Brit. Geogr., Spec. Publ. 6: 169-192.
Watt, A.K., 1968. Pleistocene geology and groundwater resources, Township of Etobicoke, York County. Geol. Report Ontario Dept. of Mines, 59, 50 pp.
Williams, J.R., 1970. Ground water in the permafrost regions in Alaska. U.S.G.S., Prof. Paper, 696.
Williams, J.R., 1975. Sediment-yield prediction with universal equation using runoff energy factor. Proc. Sediment-yield Workshop, USDA Sedimentation Lab., Oxford (Miss.), Nov. 1972. USDA, Agric. Res. Service, ARS-S-40: 244-252.
White, G.F., 1975. Trends in River Basin development. Law and Contemporary Affairs, 22: 167-187.
Zaporozec, A. (Editor), 1979. Water: availability, demand and use. GeoJournal, 3(5): 417-508.
Zhang, Zhi-Gan, 1980. Karst types of China. GeoJournal, 4(6): 541-570.

THE ROLE OF GEOMORPHOLOGY IN VEGETATION SURVEYS

5.1 Introduction

The relation between the sciences of geomorphology and vegetation differs essentially from that existing between geomorphology and the earth sciences, elaborated upon in sections 3 and 4. The reason for this is that a landscape ecological relation is concerned between two elements of the land, both of which are visible in the field as well as in aerospace imagery: landform and (vegetal) land cover. As a result, the two sciences mutually support each other when thematic mapping is at hand and in various degrees, depending on the environmental situation. The same applies to agricultural landcover in cultural landscapes, as will be discussed in section 6.

To many ecologists (Gaussen, 1946; Boon, 1968; Morrison, 1981; Peterson and Billings, 1978; Rey, 1968 and Zonneveld, 1970, 1980), engaged in vegetation survey, a map indicating the distributional pattern of the various vegetation types or plant communities is not the ultimate aim, but is meant to serve as a basis for resource inventory. The philosophy behind this concept is that the vegetation occurring at a certain locality reflects the effects of all environmental parameters combined. A study of the vegetation offers a clue to the role of each of these environmental parameters in the ecological situation prevailing at that locality, and thus yields information of practical importance for the study and development of an ecosystem. To avoid confusion, it is relevant in this context to properly define the term 'ecosystem' as follows: 'a system of interaction encompassing living creatures and their abiotic habitat' (Tansley, 1935). It does not simply include the interrelation complexes within the living world, but englobes all interacting factors of the physical environment. One may distinguish plant (animal) and landscape ecology, within the field of ecology, although these two (biotic and abiotic) elements are in fact so intricately interrelated that their separation is hardly feasible in reality (Montoya Maquin, 1966).

Vegetation may be labelled as the main visible landscape parameter in the biotic sector in much the same way as geomorphology, and landform in particular is the main visible landscape parameter in the abiotic sector. Therefore, both may serve as a starting point for the study of environmental resources, particularly when aerospace imagery is involved in arriving at an adequate breakdown of the land into mapping units of acceptable homogeneity. Obviously, proper attention should be paid to all other ecological parameters when it comes to evaluating the ecological situation and the economic potential of the units outlined.

The advantage of the geomorphological approach is in the relative stability of landforms as well as in the effectiveness of their 3-dimensional representation on stereoscopic aerial photographs. The geomorphological situation sets the scene for the development of the ecosystem (Howard and Mitchell, 1980). For the vegetation approach, the third dimension is less effective because of the smaller size of the objects under study. Emphasis is on the reflection and emission characteristics of the vegetation which are reflected in the image density and colour (Boon, 1964; Versteegh, 1966, 1974). An advantage here is the ease with which reflection data is obtained from digital approaches, for example, from Landsat satellites. According to the author, the weakest point of the vegetation approach is the discrepancy existing almost everywhere in the world between vegetation theoretically present as a consequence of environmental conditions and the vegetation patterns actually occurring as a result of intense human interference with the environment in general and with the vegetative cover in particular.

Nevertheless, valid contributions can be made by applying the vegetation approach. This is, however, neither the subject of this section nor the specialisation of the author. This section deals with the role of geomorphology in vegetation science and survey and not vice versa. Therefore, matters such as the effect of the vegetative cover on geomorphological processes, and thus on geomorphological development, are purposely left out of consideration, thereby not diverting the reader's attention from the main theme of this book. Some practical aspects of reafforestation and the protection of natural vegetation for purposes of stabilization of wind-blown sands, combating accelerated erosion and river basin development are dealt with in the relevant sections on specialized geomorphological surveys.

5.2 Relief and Major Vegetation Zones

The study and mapping of major vegetation zones in a latitudinal as well as in an altitudinal sense is among the major foci of interest in vegetation science and may also be referred to as vegetation geography. The influence of climatic factors such as rainfall, humidity, temperature, etc. is so dominant in this respect that many vegetation maps may, with justification, be considered climatic documents (Kuchler, 1963). Landform, and particularly its vertical component, relief, is also an important factor. Broad vegetation patterns often show a close relationship with the topographic situation. The reason for this is that landform and particularly elevation, affect the climatic and/or microclimatic situation and influence the environmental conditions for plant growth. The straightforwardness of observing the terrain configuration in the field, its clear representation on topographic (contour) maps and aerial photographs, renders landform an important aid in delineating vegetation zones which are more directly linked to 'invisible' climatic factors.

The effect of landforms on vegetation patterns and their effectiveness as an aid to vegetation mapping, varies with the topographic situation, the mapping scale, etc. In the case of small-scale mapping of areas of comparatively low relief, the global distribution of land masses affects the continentality, which is reflected in the patterns of latitudinal vegetation zones. When smaller areas of low relief are mapped at large or medium scales, the climatic conditions tend to be rather homogeneous and the effect of relief on vegetation patterns is, in most cases, more limited. Even so, the terrain configuration may affect matters such as the flow of cold air to topographically depressed areas by way of valleys. Micro relief may become crucial for certain types of vegetation in such circumstances, particularly when critical climatic values are reached (for example, freezing or dew point). The role of landform is most outspoken in mountainous areas where the vertical zoning of vegetation and the exposure of the slopes to the sun are major issues. Terrain conditions also affect forest exploitation (Remeijn, 1978). It should be added here that the considerations outlined above also apply to the agricultural vegetal cover and further aspects of rural land utilization, dealt with in section 6. This aspect has not been studied as intensely, at least on a global scale, as the topo-climatic relations of natural vegetation zones.

The most important climatic factors which vary with altitude are temperature and rainfall/humidity. Temperature decreases with increasing elevation, whereas rainfall and humidity usually increase with elevation until a certain altitude after which a marked decrease occurs. Snow and ice may be factors at higher altitudes, and the seasonal and perennial snow line merits consideration in conjunction with the upper vegetation limit, the forest line and the altitude of various vertical vegetation zones. A variety of other climatic factors may also exert considerable influence on the distributional pattern and altitudinal situation of the vegetation zones occurring in an area as evident from the examples given in figs. 5.1a and 5.1b. The botanical composition of these zones varies with the region, but the same factors and principles apply everywhere. Together with the height of the stands, the number of species decreases with altitude as is shown by the example of fig. 5.1 from Brunei (Jacobs, 1980).

As an example, the vertical vegetation zoning in the northern part of the Mediterranean zone is elaborated upon and illustrated by the stereopair of fig. 5.2, which depicts parts of the Herault Valley in southern France. Three major vegetation zones can be distinguished in the image, viz.:

i. *Quercus ilex* (green oak) in the rather warm, dry lower parts of the slopes;
ii. *Quercus pubescens* on the somewhat cooler middle portion of the slopes and
iii. *Fagus silvatica* (beech) on the higher parts of the slopes which are relatively cool and moist due to the frequent occurrence of clouds and fog.

When these zones are studied in detail, it appears that the lower zone, with *Quercus ilex,* is used for agriculture and vineyards. The lower portion of the *Quercus pubescens* zone has chestnut trees *(Castanea sativa)* and the highest part of the *Fagus silvatica* zone is, at places, reafforested with conifers. The effect of sun exposure is visualized by an altitudinal rise of *Quercus* on south-facing slopes and by a lower altitude of *Fagus* on north-facing slopes.

Variations in this general north-Mediterranean situation are numerous. In Calabria, southern Italy, for example, a *Laurus* zone up to approximately 600 metres a.s.l. is followed by a *Castanea* zone, where Quercus also occurs and reaches approximately 1200 metres a.s.l. The upper portion of the latter, overlaps with the Fagus zone which starts at approximately 1000 metres a.s.l. In this area a marked difference in altitude of the various vegetation zones exists between the west-facing slopes bordering the Tyrrhenian Sea, which are exposed to the rain bringing westerly winds and the slopes of the drier interior, for example the Crati Valley, situated to various degrees in the rain shadow. Coniferous trees have been planted in the highest areas such as the Sila Plateau.

In the Alps, the altitudinal vegetation zoning and the effect of exposure to sun are very pronounced. The stereopair of fig. 5.3, shows a part of the northern Limestone Alps at a scale of approximately 1:15,000 and gives a clear picture of the general situation. Broad leaf forest is practically absent on north-facing slopes. Here spruce *(Picea, Abies)* and fir *(Pinus silvestris*, etc.) are common on the lower slopes and with the gradually increasing altitude, give way to larch *(Larix decidua)* and pine *(Pinus mugo).* On many south-facing slopes, beech *(Fagus silvatica)* is found and may form almost pure stands at places, particularly on the lower mountain slopes. Higher up spruce, larch and pine gradually gain in importance.

The latitudinal and altitudinal vegetation zoning has been studied in many parts of the world by scientists such as C. Troll (1952, 1959, 1962) who specifically mentions the cordilleras of North and South America. This major mountain range stretches in a N-S direction almost from pole to pole and at most places is of sufficient height to show well-developed altitudinal zoning. Among the investigators in the humid tropics in SE Asia, Van Steenis (1964, 1965) should be mentioned. Particularly interesting effects on topo-climatic vegetation

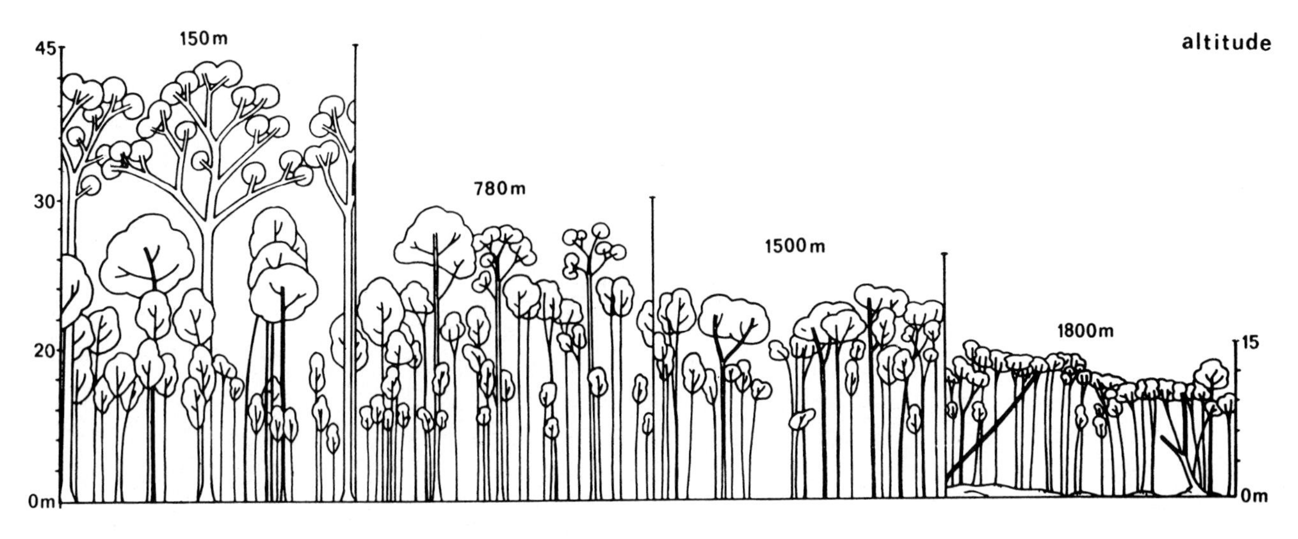

Fig. 5.1.a Decrease in forest stature in Malaya (Jacobs, 1980).

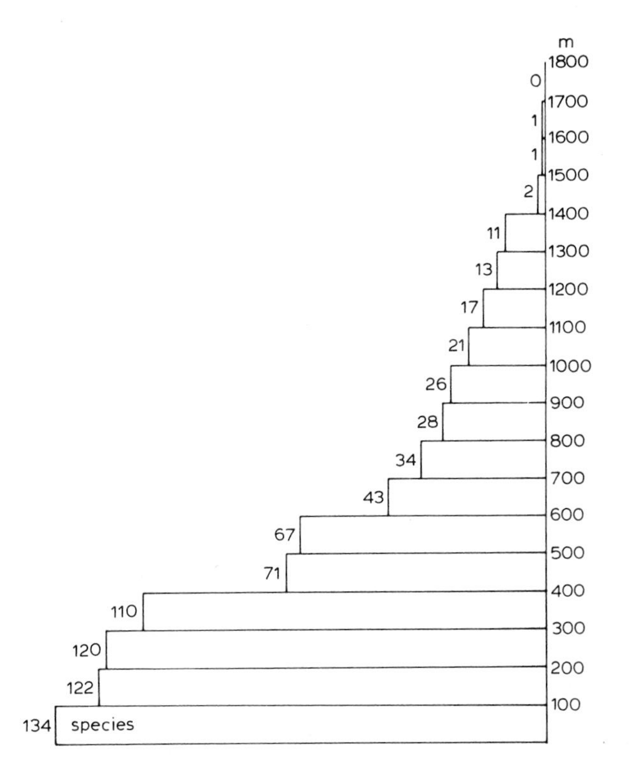

Fig. 5.1.b Decrease in number of species in Brunei, with increasing elevation (Jacobs, 1980).

zoning can be observed where relatively dry, intramontane valleys occur and where strong mountain winds, funneled through transverse wind-gaps cause increased dryness. (Troll, 1952). The föhn of the Alps and the Bohorok winds of northern Sumatra, Indonesia, are among the classic examples of this phenomenon. Extensive, high-altitude land masses, such as those occurring, for example, in Central Asia, have a tendency to increase the general temperature and affect the vegetation zoning accordingly. Smaller land masses have a comparable effect and deviate in altitudinal zoning from isolated peaks of the same height.

An interesting phenomenon of species occasionally occurring at below-normal altitudes has been described by Van Steenis (1965), who labelled this as 'elevation effect'. Certain species pertaining to the montane zone are sometimes found even at very low altitudes, where they are, however, sterile. The condition for this phenomenon is that the mountain reaches up to the montane zone where the species is fertile. In areas where this condition is not fulfilled, adjacent slopes are devoid of such anomalous low occurrences of the species (for example, *Albizia montana* Bth.).

Vegetation zoning and particularly the height of the forest line may be affected by other, non-climatic factors. The oblique aerial photograph of fig. 5.4, showing the Slamet Volcano, Central Java, examplifies this. Volcanic activity here determines the height of the forest line, which is lowest at the side of the main active crater (right-hand side of the photo). Sometimes the climatic conditions of the past have an effect on the distribution of the present vegetation, particularly of certain species that have

Fig. 5.2.Stereopair (1:25,000) showing an example of altitudinal zoning of vegetation in southern France.

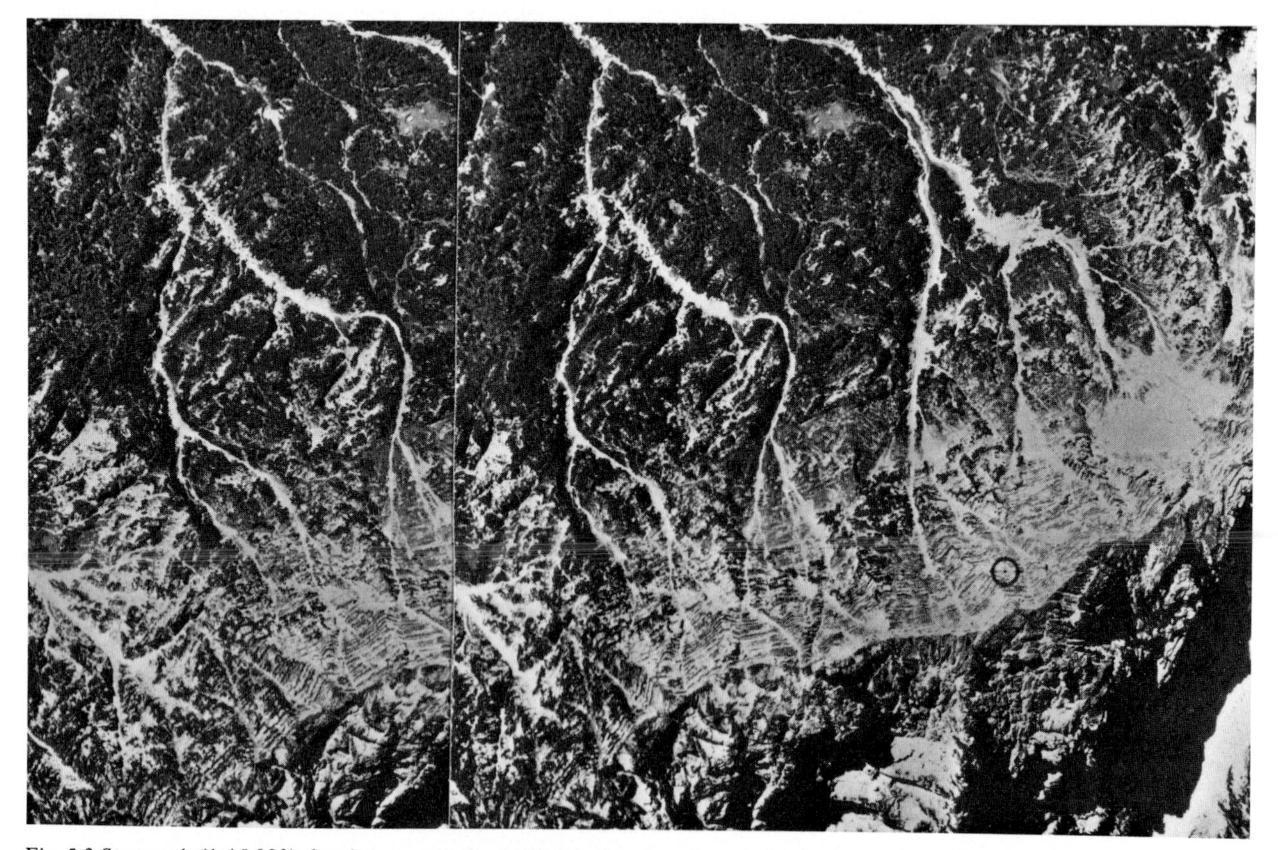

Fig. 5.3 Stereopair (1:15,000) showing an example of altitudinal zoning of vegetation in the northern Limestone Alps.

Fig. 5.4 Oblique aerial view of the Slamet Volcano, Central Java, Indonesia, showing a well-defined forest limit related to volcanic activity.

survived climatic changes, usually of Pleistocene, but also of greater age. The disconnected occurrence of such species is indicative of this phenomenon. Examples are the disjunctions in the occurrence of teak *(Tectona grandis L.f)* and sandalwood *(Santalum album L.)* between Burma and eastern Java, Indonesia (Verstappen, 1975, 1980).

5.3 Geomorphology in Forest Inventory and Range Land Survey

In case vegetation mapping concerns climatologically homogeneous areas of smaller size and of limited altitudinal range, thereby minimizing topo-climatological effects, other environmental factors such as internal and external drainage conditions, salinity, texture and type of surficial deposits, degree of weathering and mineral content of the soil, play a decisive role in the distributional patterns of herbaceous and arborescent vegetation. Since all these parameters are in one way or another associated with the geomorphological situation, a break-down of the land into morphogenetic units is a valid approach to the inventory of forest and range lands. Landform, being a visible parameter both in the field and in aerospace imagery, take a key position here. The importance of the genesis of landforms, often revealed by certain relief characteristics, should be stressed.

Within a certain region landforms of comparable origin are usually covered by surficial deposits of comparable texture, soils of comparable type and well-defined vegetation types or complexes. The average composition and even productive capacity of forest and range lands can be evaluated on this basis.

One may argue that since vegetation itself is a visible terrain parameter, the vegetative cover can be analysed directly, for example, with the aid of large-scale vertical aerial photographs and that there is, therefore, no need for the indirect approach via the study of landform. One may also claim that the evaluation of the potential of forest and grazing land is effectuated in a positive way through the study of soil profiles and hydrological conditions on the spot. Indeed, the ground truth gathering of vegetation and the use of large-scale verticals is the most reliable approach but can only be successfully applied if relatively small and readily accessible areas are concerned. Where large areas of difficult access have to be evaluated, small-scale aerial photographs or other images, for example, SLAR, may have to be used, on which the individual trees are barely visible and the species occurring cannot be directly established, in combination with a limited field check only (Molina, 1973). Under such conditions the approach of geomorphological classification of forest and grazing land becomes optimally efficient.

It is not surprising, therefore, that the foresters engaged in the nation-wide inventory of Canadian forest resources were among the first to appreciate geomorphological terrain classification for forest inventory and developed a methodology for it in the nineteen-sixties (Gimbarzevsky, 1964, 1966). Apart from the forester's professional competence, this also requires a comprehensive knowledge of geomorphology (in the Canadian context, glacial geomorphology in particular), and of related surface materials, soils and hydrological conditions. The objectives of the classification were:

i. an assessment of the land for setting-up forest management units,
ii. the selection of particularly productive areas for intensification of forestry practices,
iii. the delineation of logging areas, forest roads, etc.

The task was carried out in two steps, namely a general classification of the forest land into major units with the aid of aerial photographs at a scale of 1:40,000 and subsequently a detailed classification into five classes of expected productivity, using 1:16,000 scale aerial photographs. In the first phase, main units such as the following were distinguished:

- alluvial terrain
- organic terrain
- aeolian deposits
- glacial till (deep, shallow, bedrock controlled, compacted)
- glacio-fluvial deposits (outwash, esker),
- glacio-lacustrine deposits (kames)
- lake deposits
- colluvial deposits
- exposed bedrock

Drainage conditions, erosional features and land use which were associated with landforms, served as criteria for establishing the nature of the surficial deposits on which the forest developed.

In the second phase a more detailed capability map was compiled using local relief, angle and aspect of slope, moisture regime, texture, mineral composition and thickness of unconsolidated surficial materials and soil type as indicators. The volume/age relationship of mature and fully-stocked forest stands is considered a fairly reliable indicator of land productivity and expected per-acre volumes at the maturity age of 100 years. An estimation for each of the five classes distinguished has been made, which ranges from 5-10 cords for class five to more than 60 cords for class 1 (Gimbarzevsky, 1964).

The stereo grams of fig. 5.14 exemplify the method and show how closely type, density and productivity of forest land in Canada is associated to the (glacio)-geomorphological situation. Some characteristic species are specifically marked in the stereogram to elucidate the various correlations.

The philosophy behind this geomorphological approach is that the forester, having a knowledge of the natural range of tree species, is able to predict, with reasonable accuracy, which species are likely to occur in an area, knowing that within such an area the distribution of these species is strongly influenced by the physiographic situation (Zsilinszky, 1964). Similar approaches have been developed for tropical areas and have been successfully applied by Sicco Smit (1969, 1974, 1975) in Colombia, who extended the approach to side-looking airborne radar (SLAR) images. In other countries the geomorphological approach to forest inventory has found various forms of application, for example, by Giordano (1964) and Leven et al., (1974).

In the field of range land evaluation, a similar geomorphological approach exists. Verboom (1965), in a grassland survey in eastern Zambia, subdivided the area into four major terrain units: the valley floor of the Luangwa River, the Escarpment slopes, the Plateau and the Nyika Mountains; and correlated the herbaceous vegetation accordingly. He found that many grasses occur in all four units whereas others are restricted to specific environments. An example of the latter is the Mopane forests with their annually occurring grasses. These grasses show a preference for alkaline environments. The occurrence and distributional pattern of many grasses were found to be largely governed by water/soil relationships, and can be classified into two groups: the wet-land grasses of the so-called Dambo Valley bottoms and the dry-land grasses of the higher areas.

In a comparable range land survey in the northern province of Zambia, carried out in the framework of a rural development scheme, Verboom (1961) mapped the Chambesi flood plain, terraces and adjacent higher ground. He correlated the geomorphological position, texture, and organic matter content of the surficial material with the vegetation and ultimately arrived at a grassland quality rating, an estimation of the carrying capacity of animals/acre per year and the determination of the optimal season for grazing of every geomorphological unit. Table 5.1 gives a summary of the results and the aerial photograph of fig. 5.5 illustrates the general conditions in the flood plain and the mode of its break-down into geomorphological units with distinct herbaceous vegetation types.

It is interesting to note that in these savanna areas the tsetse fly, and in particular, the common *Glossina morsitans*, is restricted to certain well-defined habitats, namely the riverine belt (A), the dambos (B) and the dambo fringes (C). Fig. 5.6 and the aerial photograph of fig. 5.7 illustrate this convincingly. Campaigns to combat Trypanosomiasis (sleeping sickness) affecting humans and cattle and caused by the tsetse fly, can be optimalised by geomorphological terrain classification using aerial photographs (Verboom, 1965).

Investigations by Thalen (1978, 1979) particularly in Iraq, clearly show the general importance of geomorphological terrain classification and geosyntaxa for the evaluation of range lands. Reafforestation projects in (semi) arid lands also have to consider geomorphological and other environmental factors (Petrov, 1970; Raheja, 1963).

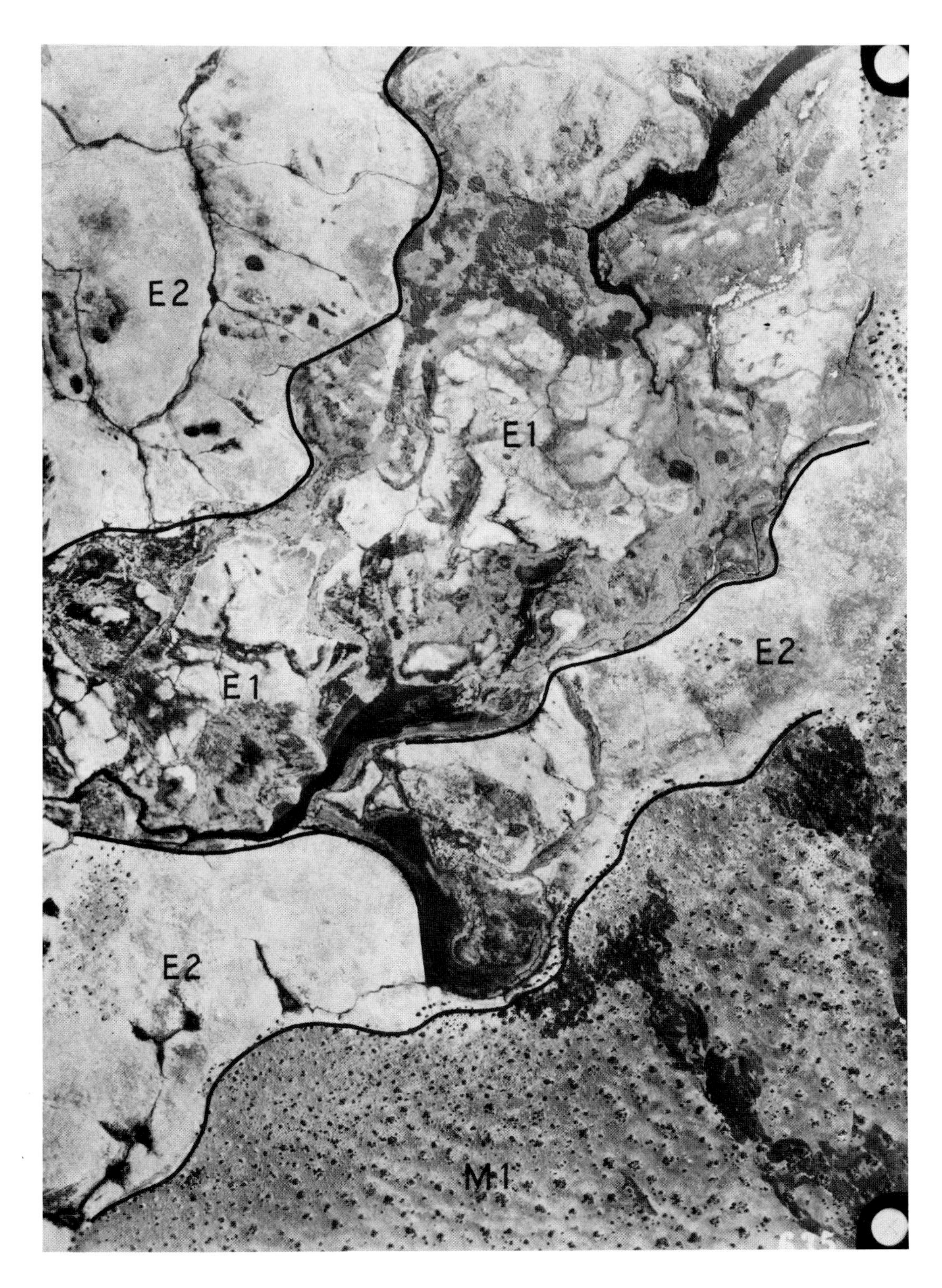

Fig. 5.5 Vertical air photo of part of the Chambesi flood plain, Zambia, showing valuable winter grazing areas near the flood channels (El, in the flood plain (E2) and on the terrace (M1). (Verboom, 1961).

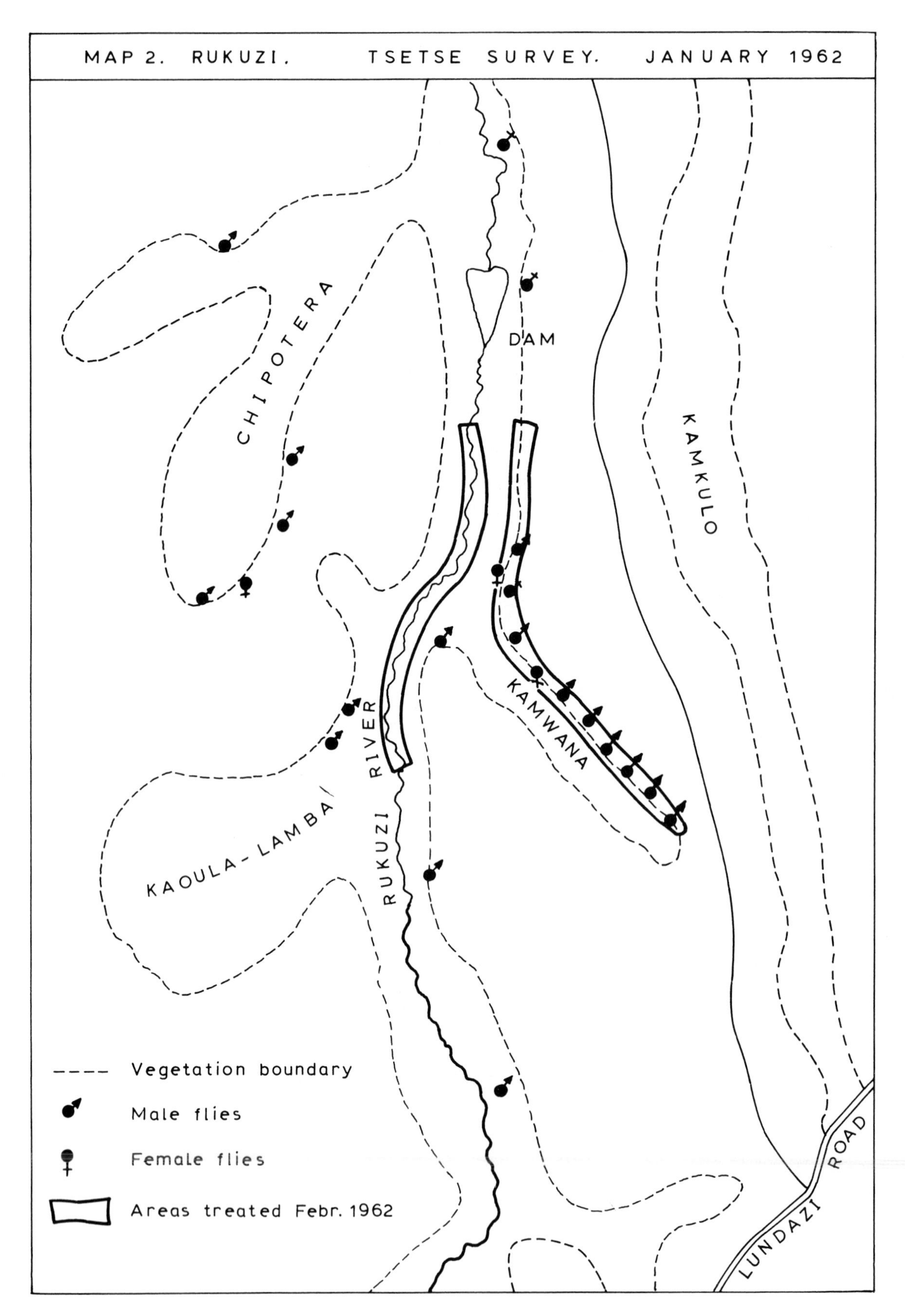

Fig. 5.6 Occurrences of male and female tsetse flies near Rukuzi, Zambia, associated with distinct vegetation and landform types. (Verboom, 1965).

SYMBOL	Geomorphological position	SOIL			VEGETATION ASSOCIATION	GRASSLAND QUALITY RATING Based on river deposits	ANALYSES grasses % C. P. at Grazing time	CARRYING CAPACITY acres per beast per year (Estimated)	LAND USE AND TIME OF GRAZING	REMARKS	% of AREA ACREAGE % of ERROR
		TEXTURE	ORGANIC C%	TOTAL N%							
E 1	Flood Channel	Clay	6.01	0.498	Vossia cuspidata Echinochloa haploclada Aeschynomene sp. near A. schimperi ("Mau" in Chibemba)	EUTROPHE active silt deposit by seasonal flooding	12.17	1	Late end of dry season	Dose for worms	10.4% (40,690 ± 3.4%)
E 2	Flood Plain	Clay	3.48	0.243	Acroceras macrum (Nile grass) Echinochloa haploclada Saccolepis sp. Entolasia imbricata Sesbania kapangenus	EUTROPHE occasional silt deposit at high flood level.	5.90	5	Early end of dry season	Dose for worms	6% (23,290 ± 4.5%)
L 1	Levee recent	Sandy Clay Loam			Cynodon dactylon Acroceras macrum Mimosa pigra Pictus mundtii	MESOTROPHE	-	8	Villages and village gardens	Fence off.	2.4% (9,320 ± 7.3%)
M 1	River Terrace	Clay Loam	1.36	.090	Paspalum commeresonii Digitaria scalarum Loudetia simplex Hyparrhenia gazensis Themeda triandra Cymbopogon sp.	MESOTROPHE	4.87	15	Early middle dry season	Consider mowing & other improvements. Adequate water supply.	19.5% (76,340 ± 2.3%)
OR 1	Old River Bed.	Clay Loam with Peat A o	2.27	.177	Entolasia imbricata Acroceras macrum Saccolepis sp. Setaria sphacelata Digitaria sp.	MESOTROPHE	5.7	12	Late middle dry season	Consider mowing & other improvements. Adequate water supply.	19.5% (76,240 ± 2.3%)
L 2	Old Levee or Old Terrace Remnants	Sandy Loam			Danthoniopsis sp. Hyparrhenia nyassae Trees on anthills	OLIGOTROPHE	-	15	Kraals for Cattle and shelter belts	Tsetse Control	2.2% (8,440 ± 7.5%)
0 1	Pediment Overlaying River Terrace	Loamy Sand	0.90	0.072	Tristachya tholonii Brachiaria sp. Digitaria gayana Digitaria monodactyla Panicum fulgens Eragrostis chalcanta Setaria sphacelata Alloteropsis semialata Eragrostis hispida	OLIGOTROPHE	4.8	20	Late beginning dry season	Dip for ticks. Water supply.	11.1% (43,395 ± 3.4%)
0 2	Dambo margin	Loamy Sand			Loudetia superba Sedges Faurea sp. (Tree) Monotes sp. (Tree) Swartzia madagascariensis (Trees) Cryptosepalum maraviense (Shrublet)	OLIGOTROPHE	-	25	Early beginning dry season	Dip for ticks	5.6% (21,790 ± 4.7%)
W 1	Up land	Loamy Sand	1.34	.088	Isoberlinia spp.) Woodland Brachystegia spp.) Hyparrhenia nyassae) Sub-climax Andropogon gayanus) grasses	OLIGOTROPHE	5.1	20	Rainy season	Dip for ticks. Thin out trees & other improvements.	15.0% (59,160 ± 2.7%)
W 2	Rocky slopes and Hills	Loamy sand mixed with rubble			Pterocarpus sp.) Isoberlinia spp.) Woodland Brachystegia spp.)	OLIGOTROPHE	-	30	Woodland late rainy season grazing if needed.	Dip for ticks.	7.1% (28,060 ± 4.1%) Question on Area 1.2% (4,600 ± 10.4%) Total: 391,000 Acres

Table 5.1 Geomorphology, soils, vegetation and grazing capacity in Eastern Zambia (Verboom, 1965).

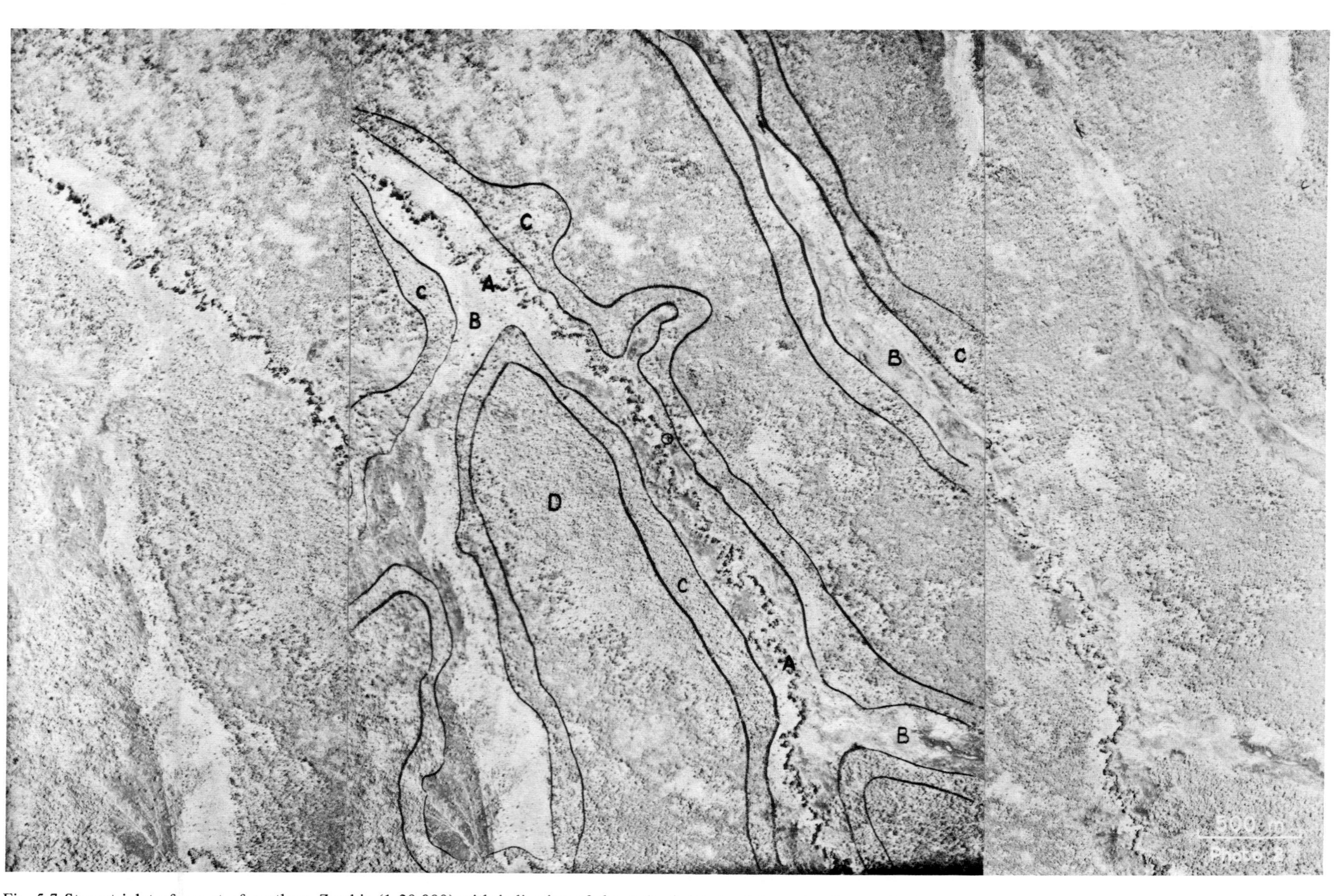

Fig. 5.7 Stereotriplet of a part of northern Zambia (1:20,000) with indication of the major habitats of the tsetse fly: A. riverine belt, B. dambos, C. dambo fringes (Verboom, 1965).

5.4 The Correlation between Geomorphological Variables and Vegetation Patterns

The usefulness of geomorphology, for purposes of vegetation survey, initially appears optimal in areas where relief is of sufficient height to be observed on stereoscopic aerial photographs and can therefore be adequately studied. The vegetation species and associations cannot be directly identified in the field because of lack of time or from the air because of limited vertical dimension and untypical spectral characteristics, for example, of heterogenous stands. The situation is not so simple, however, for two main reasons. First, a change in landform does not necessarily provoke a change in vegetative cover. In the humid tropics (Demangeot, 1976), subdivision of the tropical rain forest for purposes of forest exploitation is fraught with difficulties due to the intermingling of valuable and less valuable hardwoods. Differentiation on the basis of relief is of little use in this context. Only where the altitude range of the moss forest is reached or where processes such as leaching have resulted in poor edaphic conditions, is sensible subdivision of the forest land feasible (Boon, 1964; Giordano, 1964).

Secondly, the vegetative patterns are, in most cases, more influenced by hydric, salinity and plant nutritive patterns which have evolved within the geomorphological environment, than by the amplitude of the prevailing relief. The aerial photographs of figs. 5.8 and 5.9 illustrate this. Photo 5.8 shows an area of tropical karst hills in Irian Barat (New Guinea), Indonesia, where the tops and the higher slopes of the limestone hills are almost completely devoid of shrub and trees as a result of the low water table. Tree vegetation is concentrated in the irregular depressions surrounding the hills. Photo 5.9 shows an intricate vegetation pattern associated with a shallowly incised drainage net in Kalimantan (Borneo), Indonesia. Rather low, dark-toned, wet-land forest is found in the valley bottoms and narrow strips of higher, dry land forest border on these valleys, and grasslands are dominant on the higher grounds. It is evident that under such conditions a knowledge of the geomorphological situation is as important to the forester as the study of the vegetation pattern is to the geomorphologist as an indicator of relief and geomorphological environment. A study of the two visible landscape parameters, landform and vegetation, is mutually beneficial.

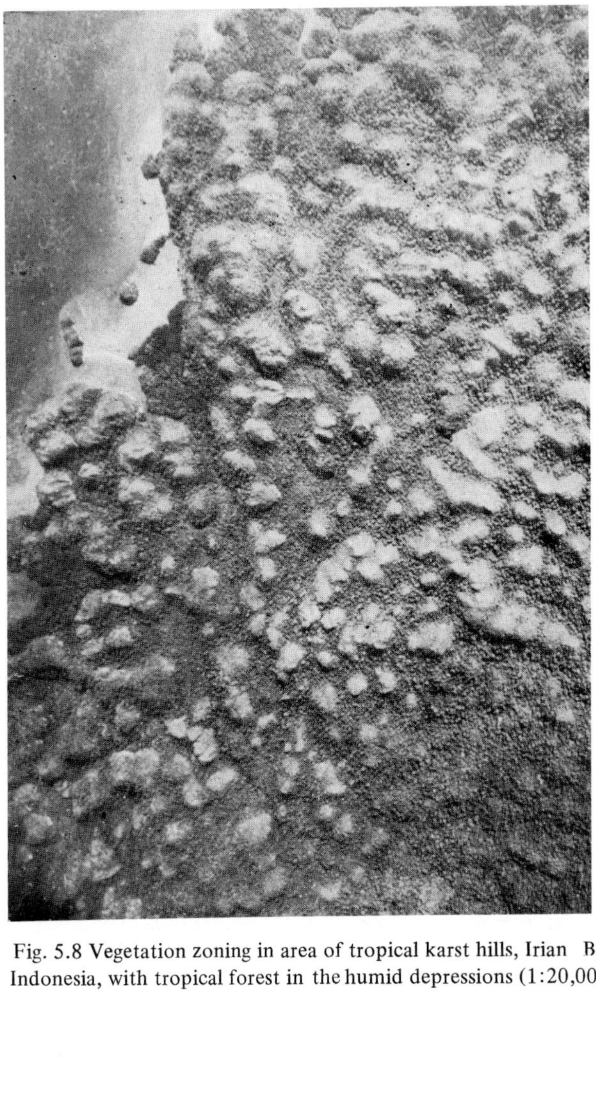

Fig. 5.8 Vegetation zoning in area of tropical karst hills, Irian Barat, Indonesia, with tropical forest in the humid depressions (1:20,000).

Fig. 5.9 Vegetation zoning in a shallowly incised area in Kalimantan, Indonesia, with low swamp forest in the valley bottoms, dry land trees on the edges and mostly grass and scrub on the hills (1:40,000).

The closest correlation between geomorphology and vegetation exists in swampy lowland areas of stereoscopically imperceptible relief. Outspoken patterns are often found in saline to brackish mangrove swamps and in tropical freshwater swamps (Richards, 1957). Among the first to stress the relation between terrain configuration and mangrove vegetation zones in tropical lowland coast environments is Kint (1935),who gained his experience during a photographic survey of the island of Banka, Indonesia. More recent comparable studies are numerous, such as the study by Allen (1965) on the vegetated tidal flats and beach ridge islands in the deltaic zone of the Niger River. Three examples are given to illustrate the relations between geomorphology - in particular the patterns of creeks, natural levees and back swamps - and the associated hydric, salinity and edaphological conditions on the patterns of mangrove and fresh water swamp vegetation.

The stereotriplet of fig. 5.10 gives a typical example of a humid tropical lowland coast, bordered by a wide belt of diversified mangrove vegetation. Mangroves grow particularly well on mud coasts, but they also flourish on sandier ground such as along the coast of the Tanimbar Islands pictured. The dry areas are covered by tropical rain forests, with numerous tall trees and large crowns. On the seaward side of it a largely unvegetated, barren zone occurs where neither the rain forest nor the mangrove swamp vegetation find a suitable environment. This zone is whitish in the photos, due to sand and salt encrustations. The zone is best developed in mangrove areas with a sandy bottom. Saline water occasionally reaches this zone.

The area between this whitish, barren zone and the coastline is under the influence of the tides. Seawater penetrates here by way of the sinuous creeks during high tide. A narrow strip, bordering on the coast, is under the immediate influence of the tides and accretion occurs here. *Rhizophor stylosa* flourishes and forms a belt of medium-high and dark-toned trees. A sandy ridge with Casuarina, casting spiry shadows, borders the coast at several places. On more exposed coasts *Sonneratia alba* may grow. Some *Avicennia marina* may be found directly to the landward side of the *Rhizophora.* The dark *Rhizophora* can also be observed along the creeks and particularly around their headward ends, where seawater spreads out over the low swampy terrain.

The landward limit of this dark *Rhizophora* belt is the average high tide mark and may thus be considered the coastline. However, it is common practice in surveying and mapping of mangrove coasts to map the seaward limit of the coherent vegetation as the coastline. A broader and lighter-toned zone of lower, mixed mangrove vegetation can be observed to the landward side of the Rhizophora and on the levees bordering the creeks. These are brackish areas, giving way in turn to darker areas in the lowest parts of the swamp, where conditions are less saline and where freshwater swamps occur. Vegetation in the transition area of these two vegetation belts is absent which indicates that the hydrological and salinity conditions there are unsuitable for both vegetation types. The mangrove vegetation of the neo-tropics is of a different species composition; nevertheless, a distinct zoning can also be observed.

The vertical aerial photograph of fig. 5.11 shows another pattern of swamp vegetation zoning in the Masamba Area, Sulawesi (Celebes), Indonesia, at a scale of 1:15,000. Mangroves are found in the shallow coastal waters, mostly forming a coherent zone but also as isolated occurrences. Landward coconut growths (light-toned) have been planted on a sandy beach ridge and are separated by a knife-edge boundary from the fine-textured and medium-toned nipa palm *(Nipa fruticans)* with their strongly reflecting leaves. Nipa palms are lower than most mangroves and are less resistant to salinity. They occur in brackish water farther inland and upstream, where the tides are still effective. The roots are usually inundated at high tide. Farther inland the large rosets of the sago palms *(Metroxylon sagu)* with their white flowering crowns can be seen in the photograph. This palm is typical for freshwater backswamps, particularly in eastern Indonesia; it is 10-13 m high and grows on a muddy substratum. Another typical fresh water swamp palm occurring mainly in a narrow strip in the transition zone of freshwater swamp and tropical (dry land) rain forests, is the nibung palm *(Oncosperma filamentosum)* with its characteristic whitish crowns (not pictured). The vertical aerial photograph of fig. 5.12 clearly pictures the difference in height, tone and texture of fresh water swamp forest with distinct white-barked slender *Melaleuca leucadendron,* as compared to tropical rain forests of the Masamba Area, Sulawesi (Celebes). In the lowest areas open water surfaces can be seen; under such conditions floating vegetation may develop (Polak, 1951).

A vegetation expert may attempt to single out environmental factors which are favourable or unfavourable in varying degrees to the growth of certain plants and represent the results in the form of a map (Legris, 1963; Tansley, 1963). Various factors may even be subjected to multivariate analysis (Phipps, 1968). From a geomorphological point of view, it suffices in most cases to establish a vegetative toposequence within a region, keeping in mind the causes of this sequence. Such a sequence has been established for the semi-natural vegetation around Madrid, Spain,by Spiers (1978), who gives full weight to the geomorphological situation.

In humid tropical regions the following generalized vegetational toposequence often occurs: in saline and

Fig. 5.10 Stereotriplet (1:40,000) showing diversified mangrove vegetation on the Tanimbar Islands, Moluccas, Indonesia.

Fig. 5.11 Vertical aerial photograph showing a coastal swamp area in the Masamba area, Sulawesi, Indonesia (1:15,000). Dark-toned *Rhizophora,* lighter-toned coconut growths, fine-textured nipa palm areas can be recognized and are associated with distinct geomorphological zones.

brackish water swamps with mangroves and related vegetation (pictured in the aerial photographs accompanying this sub-section). The toposequence then continues in the natural levee - point bar - fresh water (back) swamps of the flood plains and from there to the lowest terraces which are still, or have been until recently, periodically or episodically flooded. These are comparatively rich in minerals and capable of supporting a well-developed forest and some agricultural land use. The higher, older terraces are not subjected to flooding and have often been leached to such an extent that only poor forest grows and land use is absent.

A narrow marshy zone may separate these higher terraces and foot plains from hills and other landforms of the hinterland. Tropical rain forest occurs there, unless densely populated areas are concerned. Poor vegetation, for example savanna vegetation, may be locally developed where poor hydrological or edaphological conditions prevail (cappings of ferricrete). Fig. 5.13 pictures this strongly generalized vegetative toposequence, the variations of which are numerous. One example will be described in detail, namely the vegetative toposequence as it presents itself in northern Suriname. From the coast to the interior, four main geomorphological zones can be distinguished, each with a clear vegetative association, namely:

1. The young coastal plain consisting of rich alluvial clays covered by brackish and fresh water swamps bearing grassy vegetation, mangrove and swamp forests, occasionally traversed by low ridges of fine sand or shells covered with dry-land forest.
2. The old coastal plain consisting of poorer clay and sands, gently undulating towards the interior and inter-

Fig. 5.12 Vertical aerial view (1:15,000) of a fresh water forest in the Masamba area, Sulawesi, Indonesia surrounded by dry land forest. Note the characteristic light tone and fine texture caused by the slender and low trees growing in the swamp.

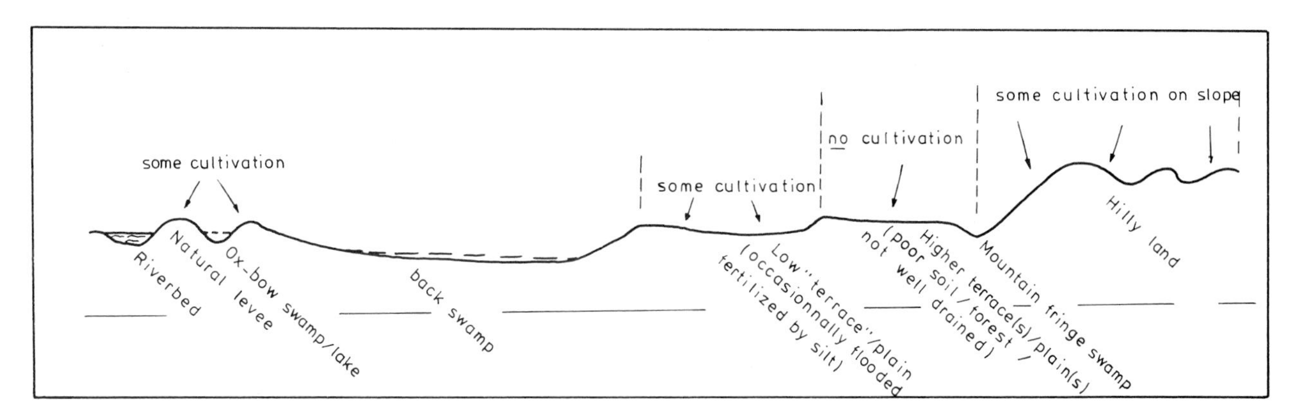

Fig. 5.13 Sketch of a common toposequence in humid tropical areas with poor and strongly leached soils on the higher terraces and a humid depression separating the terraces from the hills and mountains.

spersed with fresh-water swamps. The higher sites bear dry land forest while in the transition to the fresh water swamp forest, marsh or creek forests occur.

3. The savanna zone, an area where the oldest unconsolidated sediments of coarse white and loamy brown sands occur, with poor vegetation of grasses and shrubs or, on the better sites, savanna forests of mostly small-sized trees.
4. The residual soils of the basement complex of pre-Cambrian age, bearing high, dry land forest - the proper tropical rain forest - and creek forest on the colluvial marshy sites along the creeks. (Versteegh, 1966).

Distinct vegetation patterns associated with the geomorphological situation can be distinguished in the young coastal plain, in both the saline/brackish mangrove zone and in the fresh water swamps of zone 1. In zone 2 alternations between fresh water swamp and dry land forest are common, whereas in zone 3 differentiation of savanna vegetation on a geomorphological basis is evident. The most productive zone for forestry purposes is the dry land forest between the savanna belt in the north and the navigability/trafficability limit in the south, which precludes economic exploitation of natural forest stands in the interior.

The swamp vegetation in Suriname shows a distinct relation to the geomorphological situation. Geyskes (1948) stated that the aerial photographs 1:40,000 available for the area can be successfully used to map the extent and distribution pattern of the various vegetation types. For determination of species, further field checking is required. The botanical composition of the swamps changes with decreasing salinity from the coast in the north to the interior in the south. The swamp belt borders inland on the savanna belt. Farther inland, only isolated small swampy or marshy areas occur, which are characterized by the occurrence of the palissade palm *(Euterpe oleracea)*.

Cohen and Van der Eijk (1953) have made a genetic classification of the savannas of Suriname and classified them according to the landscapes to which they belong. In most landscapes only one savanna type occurs, but in some, several types can be observed. In total ten different types of savanna have been distinguished, spread over eight landscapes. Maurisi palms *(Mauritia flexuosa)* often mark the water courses in the savannas. The causes of savanna formation vary with type, but in general three main causes or combinations of causes can be given (Riezebos, 1979). One of the main causes is poor edaphic conditions and in particular, a low content in nutritive minerals due to strong leaching. The strongly leached areas often have rather sharp boundaries and the vegetation changes abruptly from dry land forest to savanna forest where the brownish sand gives way to the white leached sandy soil. In other types of savanna, impervious layers in the soil profile, in combination with the resulting marshy conditions are considered a main causative factor. Podzolization accounts for the savanna occurring on sandy beach ridges while climatic factors play a role in the formation of some savannas occurring at greater altitudes in the interior.

Dillewijn (1957) published a most useful guide with stereoscopic terrestrial and aerial photographs of the major vegetation types and the main characteristic species occurring in the various landscapes of northern Suriname, such as the coastal zone, swamps, ridges, marshy areas, savannas, dry land zones and riverine belts. Similar landscape ecological approaches to vegetation mapping also exist for many other parts of the world and prove the universal applicability of the method (Wyatt - Smith, 1963; Troll, 1939, 1962; Richards, 1957).

5.5 Vegetation Elements as Geòmorphological Indicators

It has been stated in the preceding sub-section that vegetation patterns are sometimes as important for the geomorphologist as geomorphological patterns are for the plant ecologist and forester. Indicator vegetation associations and species may form valuable convergent evidence where the geomorphological phenomena themselves can be readily observed and mapped, in which case, however, they are not indispensable. For example, it is of scientific, landscape-ecological interest to note that *Araucaria cunninghamii* trees often mark the terrace edges in Central New Guinea but since the terrace edges themselves are easily mapped, this information is not absolutely required for the geomorphologist. The same applies to several of the vegetation types and species previously mentioned.
phological situation is difficult to unravel by geomorphological means, for example due to inconspicuous topographic expression of the landforms, certain vegetation indicators may be of crucial importance in geomorphological survey and for environmental evaluation.

The Ontario Department of Highways (Anderson et al., 1954) has successfully applied vegetation indicators in establishing engineering soil conditions for highway location in Canada and Tomlinson and Brown (1962) have done the same for the interpretation of surface material. They found that in northern Canada, where the forest tundra (taiga) prevails, Black Spruce and Balsam Fir associations occur on steep scree slopes, whereas Willow and Alder show distinct affinity to wet depressions filled with muck. Open stands of Black Spruce are found on the sand and on coarser materials such as of

TREE SPECIES	ASSOCIATED TREES		PHOTO RECOGNITION		GROUND FEATURES	
	Swamp Species	Upland Species	Shape & Texture	Tone	Site	Soil
A. SOFTWOODS (CONIFEROUS) :						
1. BALSAM FIR	b.spruce,w.birch birch, tamarack	hard maple,beech y. birch	Fairly conical; low crown density	Dark to black	Swamps to slopes	Various moist to wet soils
2. CEDAR	b.spruce, b. ash tamarack	w. & j. pine w. spruce	Small,round crown; symmetrical	Dark to black	Mainly swamps & bottomlands	Wet, organic soils
3. PINES	aspen, balsam w.birch,b.spruce	beech,maple,birch hemlock,w.spruce	Irreg.crown shape; medium crown density	Medium to black	Lowlands to hilltops	Dry, granular soils
4. SPRUCES	balsam, tamarack w.birch, cedar	aspen, pines hemlock	Very conical shape; low crown density	Dark to black	Swamps (b.sp) hillsides(w.sp)	Wet,organic or moist,mineral
5. TAMARACK	b.spruce, cedar balsam, b.ash		Smooth, round tip; low crown density	Light to white	Swamp margins & bottomlands	Imperfectly drained organic
B. HARDWOODS (DECIDUOUS):						
1. BEECH & SUGAR MAPLE		aspen,elm,ash,w. pine,balsam,hem.	Large, full crowns; high crown density	Medium to light	Upland sites mainly	Rich,well-dr. mineral soils
2. BIRCHES & POPLARS	b.spruce,balsam alder, willow	maple, beech hemlock	Rounded crowns; med.crown density	Medium to light	Slopes & lake & swamp marg.	Moist mineral soils
3. WHITE ELM		most hardwoods	Large, flat, fan-shaped;high density	Medium to light	High water table sites	Rich, moist granular soils
4. SCRUB WILLOW & ALDER	usually pure		Low continuous bush; high crown density	Light	Bottomlands	Mineral soils Water-saturate
5. ORCHARD TREES		pure	Uniform,even spaced; low crown density	Medium to dark	Well-protected slopes,terrace	Well-drained mineral soils

Table 5.2 Indicator trees of engineering soil conditions in Ontario, Canada. (Anderson et al., 1954).

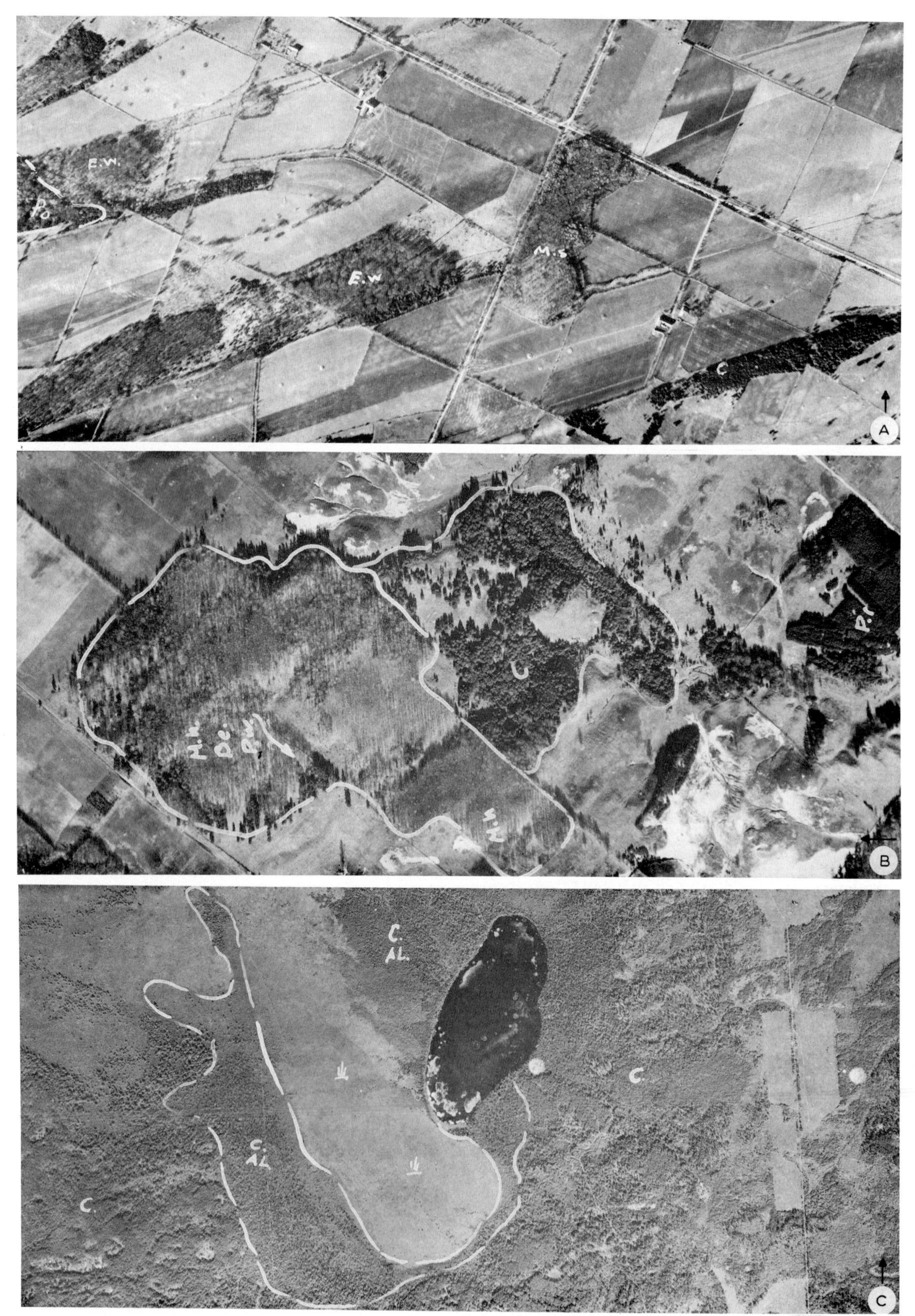

Fig. 5.14 Six examples (A-F) of tree species indicative for distinct geomorphological situations in Ontario, Canada, as depicted on vertical aerial photographs, 1:20,000. A: Drumlins with elm (Ew), maple (M), cedar (C) and Poplar (Po). B: Interlobate (Kames) moraine with Maple (Mn), Elm (Em), Red Pine (Pr), Cedar (C). C: Water-saturated clay/silt in swamp, with Cedar (C) and Alder (Al).

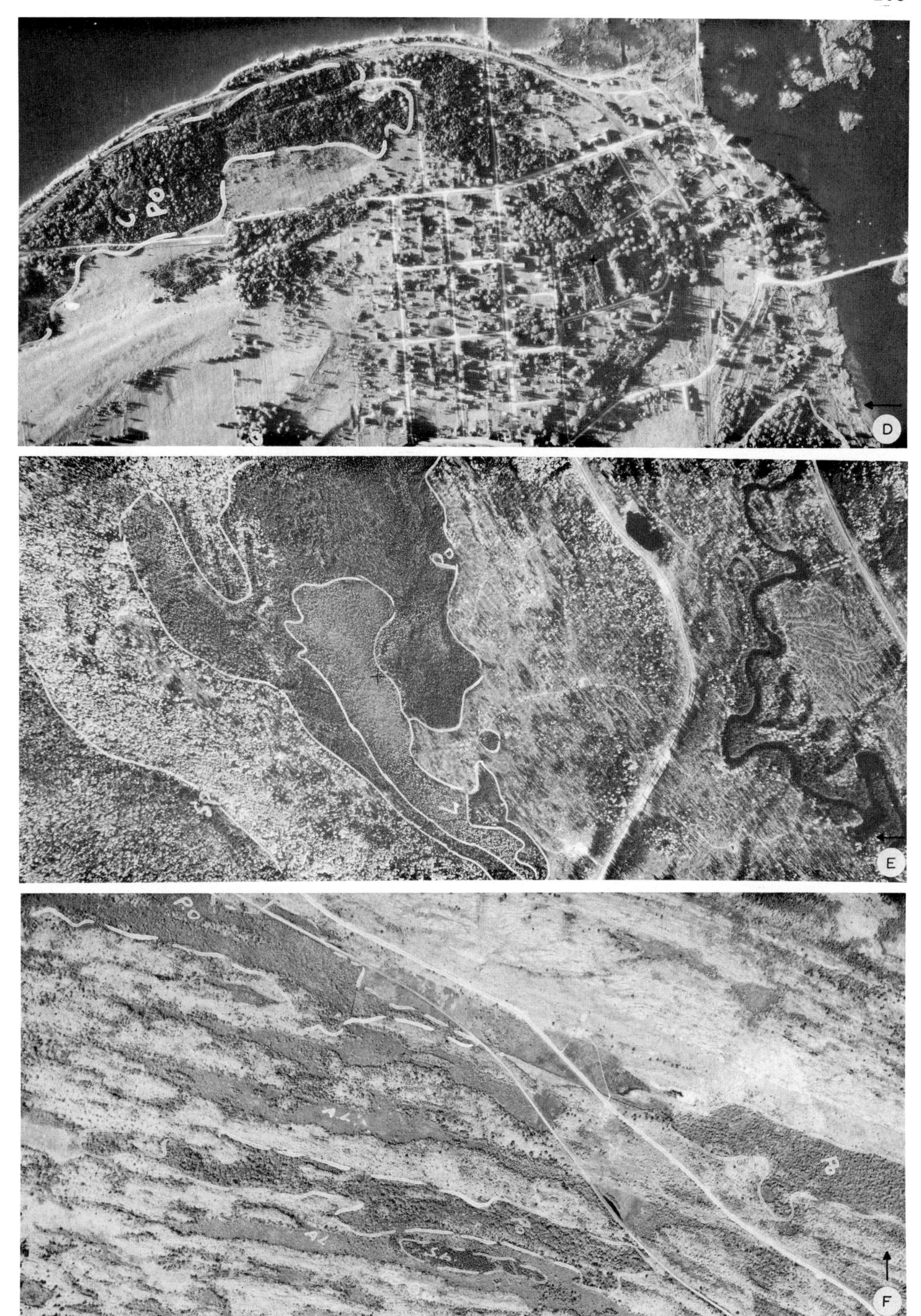

D: Raised benches with Poplar (Po), Willow (W) and Cedar (C). E: Rock and thin overburden with Spruce (Sp), Tamarack (L) and Poplar (Po). F: Metamorphic rock/soil with Poplar (Po), Alder (Al) and White Birch (Sb).

eskers and sandy drumlins. Dwarf Birch and Willow often occupy slopes composed of medium textured glacial, till materials, sometimes in combination with Spruce. Fine-textured surface materials are usually under grass and sedge.

Further south, in the boreal forest zone, different species appear and there rock and talus is usually covered with Black Spruce mixed with Jack Pine and White Birch. White Birch often marks boulder slopes, wetted by seepage water. Sand and coarse material, often an important source of construction material, is sometimes covered with pure stands of Jack Pine. The density of the stand and the height of individual trees is affected by hydrological and edaphic factors which relate to the texture and mineral content of the material. In the east, Balsam Fir is an indicator. White Spruce is common on the medium-textured, silty-clayey glacial drifts and White Birch indicates glacial till. Aspen Poplar (with White Spruce) is found on fine-textured material and Alder has a preference for soils with a high content of organic matter. Larch, sometimes mixed with Black Spruce, indicates peat.

The northern hardwood zone of southernmost Canada shows different species on the kinds of sites and surface materials mentioned. Red and White Pine, White Birch and White Cedar characterize the rocks and talus slopes; sand and coarser materials are usually covered by Yellow Birch and Sugar Maple,but are replaced by Poplar and White Birch in glacial till. The hydrological condition of these sands is reflected in the vegetation: wet sands are often under Hemlock, moist sands under White Pine and dry sand under Red Pine. Sugar Maple prefers the medium materials in this zone,in combination with other tree species depending on depth of bedrock, etc. Elm indicates clay. Soft Maple is indicative for lacustrine or marine clays, which are distinctly avoided by the Sugar Maple. White Cedar, sometimes in combination with Black Ash, White Elm or Swamp Maple, indicates boggy and peat conditions.

Table 5.2 lists the indicator trees used by Anderson et al. (1954) for establishing engineering soil conditions in Ontario, Canada,and is self-evident. The sterograms of figs. 5.14 depict some of the major indicator-species and associations and should be studied together with table 5.2,taking the geomorphological situation into consideration.

References

Allen, J.R.L., 1965. Coastal geomorphology of Eastern Nigeria: beach-ridge barrier islands and vegetated tidal flats. Geol. en Mijnbouw, 44(1): 1–21.

Anderson, J., et al., 1954. Aerial photogrammetry as applied to highway location. Ontario Dept. Highways, 122 pp.

Boon, D.A., 1964. Some aspects of plant-ecology in the tropics in connection with the use of aerial photography. Proc. U.N. Cartogr. Conf. Asia and Far East, Manilla: 1–4.

Boon, D.A., 1968. Hydrology and vegetational influences. Unpubl. ITC lecture notes, 20 pp.

Cohen, A., and Eijk, J.J. van der, 1953. Klassificatie en ontstaan van savannen in Suriname. Geol. en Mijnbouw, 15: 202–214.

Demangeot, J., 1976. Les espaces naturels tropicaux. Masson, Paris, 190 pp.

Dillewijn, F.J. van, 1957. Sleutel voor de interpretatie van begroeiingsvormen uit luchtfoto's 1:40,000 van het noordelijk deel van Suriname. Landsbosbeheer, Paramaribo, 45 pp.

Gaussen, H., 1946. La connaissance du monde végétal par la photographie aérienne. Atomes, 7: 26–28.

Geijskes, D.C., 1948. Luchtfotografie en zwampbegroeiing in Suriname. Tijdschr. Kon. Ned. Aardrijsk. Gen., 65(4/5): 665–668.

Gimbarzevsky, Ph., 1964. The significance of landforms in the evaluation of forest land. Canadian Pulp and Paper Magazine, Hinton (Alb.), Canada: 1–18.

Gimbarzevsky, Ph., 1966. Land inventory interpretation. Ph. Eng., 32(6): 967–976.

Giordano, G., 1964. Forest inventory in the Province of Misiones, Argentine. Trans. 10th I.S.P. Conf.: 1–25.

Howard, J.A., and Mitchell, C.W., 1980. Phyto-geomorphic classification of landscape. Geoforum, 11(2): 85–106.

Jacobs, M., 1980. Significance of the tropical rainforests on twelve points. Bio Indonesia, 7: 73–94.

Kint, A., 1935. De luchtfoto en de topografische terreingesteldheid in de mangrove. Jaarverslag Top. Dienst, N.O.I.: 173–189.

Küchler, A.W., 1963. The vegetation map as a climatic document. Paper Reg. IGU Conf.,Kuala Lumpur, 5 pp.

Legris, P., 1963. Botanical and ecological cartography in India. Paper Reg. IGU Conf., Kuala Lumpur, 8 pp.

Leven, A.A., et al., 1974. Land response units – an aid to forest land management. Proc. Soil Sci. Soc. Amer., 386: 140–143.

Molina, I. de., et al., 1973. SLAR en la mapificación de los bosques humidos tropicales de Colombia. Publ. CIAF, 6, Bogotá, 20 pp.

Montoya Maquin, J.M., 1966. Ecosystème et photo-interprétation: fondement théorique et pratique des inventaires de ressources intégrées. Trans. 2nd. Symp. Comm. VII, I.S.P., Paris, IV: 2–37.

Morrison, R., et al., 1981. Ecological land survey as a basis for environmental assessment. Newsletter Can. Comm. Ecol. Land classif., 10: 1–10.

Peterson, N.M., and Billings., W.D., 1978. Geomorphic processes and vegetational change along the Meade River sand bluffs in northern Alaska. Arctic, 31(1): 7–23.

Petrov, M., 1970. The U.S.S.R., In: Afforestation in arid zones, (Editor R.N. Kaul): 210–267.

Phipps, M., 1968. Analyse d'une structure régionale de modèles bio-géographiques. Vie et Milieu, 19 (2.C): 303–330.

Polak, B., 1951. Construction and origin of floating islands in the Rawa Pening (Central Java). Contr. Gen. Agr. Res. Sta. Bogor, 121: 1–11.

Raheja, P.C., 1963. Shelter-belts in arid climates and special techniques for tree planting. Annals of Arid Zone, 2(1): 1–13.

Remeijn, J.M., 1978. Forest road planning from aerial photographs. ITC Journal, 3: 429–444.

Rey, P., 1968. Photographie aérienne et végétation. Proc. Toulouse Conf. on aerial surveys and integrated studies. UNESCO, Paris: 187–207.

Richards, P.W., 1957. The tropical rain forest. Cambridge Univ. Press: 1–450.

Riezebos, H.Th., 1979. Geomorphology and soils of the Sipalwini savanna (S. Suriname). Thesis Utrecht State Univ., 169 pp.

Sicco Smit, G., 1969. Sistema de fotinterpretacion recomendado para los bosques humedos tropicales de Columbia, CIAF Publ. B 1, Bogotá: 1–27.

Sicco Smit, G., 1974. Practical applications of radar images to tropical rainforest mapping in Colombia. Mitt. Bundesanstalt f. Forst- u. Holzwirtsch., Rheinbeck bei Hamburg, 99: 51–64.

Sicco Smit, G., 1975. Will the road to the green hell be paved by SLAR? ITC Journal, 2: 245–266.

Sicco Smit, G., 1978. SLAR for forest type classification in a semi-deciduous tropical region. ITC Journal, 3: 385–401.

Spiers, B., 1978. A vegetation survey of semi-natural grazing lands (dehasas) near Merída, Spain. ITC Journal, 4: 649–679.

Steenis, C.G.G.J. van, 1964. Plant geography of the mountain flora of Mt. Kinabalu. Proc. Roy. Soc. B.161: 7–38.

Steenis, C.G.G.J. van, 1965. Concise Plant Geography of Java. In: Flora of Java. Vol. 2 (Editors Backer and Bakhuizen, v.d. Brink): 1–72.

Tansley, A.G., 1965. The use and abuse of vegetational concepts and terms. Ecology, 16(3): 284-307.

Thalen, D.C.P., 1978. Complex mapping units, geosyntaxa and the evaluation of grazing areas. Proc. Int. Symp. Rinteln. Cramer Verlag, Vaduz: 491–514.

Thalen, D.C.P., 1979. Ecology and utilization of desert shrub rangelands in Iraq. Thesis, Groningen Univ., 428 pp.

Tomlinson, R.F., and Brown, W.G.E., 1962. The use of vegetation analysis in the photo interpretation of surface material. Ph. Eng., 28(4): 584–592.

Troll, C., 1939. Luftbildplan und ökologische Bodenforschung. Ztschr.d.Gesellsch.f.Erdkunde, Berlin, (7/8): 1–58.

Troll, C., 1952. Das Pflanzenkleid der Tropen in seiner Abhängigkeit von Klima, Boden und Mensch. Ber. D. Geogr. Tag, 28, Frankfurt, 1951. Verlag d. Amtes f. Landesk., Remagen: 35–67.

Troll, C., 1952. Die Lokalwinde der Tropengebirge und ihr Einfluss auf Niederschlag und Vegetation. Bonner Geogr. Abh., 9: 124–182.

Troll, C., 1959. Die tropischen Gebirge. Ihre dreidimensionale klimatische und pflanzen geographische Zonierung. Bonner Geogr., Abh., 25: 1–93.

Troll, C., 1962. Die dreidimensionale Landschaftsgliederung der Erde. Hermann von Wissman-Festschrift, Tübingen: 54–80.

Verboom, W.C., 1961. Planned rural development in N. Rhodesia. ITC Publ., B 12/13, 52 pp.

Verboom, W.C., Tzetze control and grassland surveys. ITC Publ., B. 28, 21 pp.

Verstappen, H.Th., 1975. On palaeo climates and landform development in Malesia. Modern Quaternary Research, 1: 3–36 (Dr. H.R. v. Heekeren Memorial Vol, Balkema, Rotterdam).

Verstappen, H.Th., 1980. Quaternary climatic changes and natural environment in SE Asia. GeoJournal, 4(1): 45–54.

Versteegh, P.J.D., 1966. Tropical forestry development and aerial photo-interpretation. ITC Publ. B34, 21 pp.

Versteegh, P.J.D., 1974. Short introduction to forest hydrology. Unpubl. ITC lecture notes, 22 pp.

Wyatt-Smith, J., 1964. Some vegetation types and a preliminary schematic vegetation map of Malaya. J.Trop. Geogr., 18: 200–213.

Zonneveld, I.S., 1970. The contribution of vegetation science to the exploration of natural resources. Misc. Papers L.H. Wageningen, 5: 31–44.

Zonneveld, I.S., et al., 1980. Aspects of the ITC approach to vegetation survey. Doc. Phytosociologiques: 1029–1063.

Zonneveld, I.S., 1982. Principles of indication of environment through vegetation. In: Steubing, J., and Jäger, H.J. (Editors), Monitoring of air polluants by plants. Junk Publ., The Hague: 3–17.

Zsilinszky, V.G., 1962/64. The practice of photo interpretation for a forest inventory. Trans. 9th Congress I.S.P., Lisbon, Photogrammetria, 19(5): 1–5.

PART B. GEOMORPHOLOGY AND APPROPRIATE USE OF THE NATURAL ENVIRONMENT

Chapter 6

GEOMORPHOLOGY AND RURAL LAND USE

6.1 Introduction

A distinct relationship exists between the geomorphological situation and land utilization of rural areas, although the type of relationship may vary from one area to another. In some cases landform is a key factor, although it should be understood that only in relatively few instances is relief in itself the main causative factor of the observed distributional patterns of land utilization. More commonly it serves as a vehicle for detecting and surveying the distributional pattern of other environmental parameters, for example related to soil conditions or hydrological situations which are more directly responsible for the land use patterns. It is thus essential for the geomorphologist to study both the mode of development or the genesis and the landscape ecological relations characterizing the environment.

The genesis of landforms is often of improtance, although in many cases emphasis should be on the human rather than on the geological time scale. In other words the short-term geomorphological and environmental changes which have a noticeable effect within the span of a man's life or within a few generations should receive particular attention. Therefore, active geomorphological processes of various kinds and intensities are often of crucial importance. These processes frequently tend to escape direct observation during surveys unless on-the-spot measurements are taken during several years. Mapping of the landforms, easily visible in the field and on aerospace imagery, may serve as a guide to the distributional patterns of geomorphological processes (Verstappen, 1977) and thus provide the surveyor with a blue-print of the morpho-dynamical situation as subjected to - and often modified by - the various types of rural land utilization.

Every environmental parameter,by itself or in combination with others, may favour, limit or preclude certain kinds of land utilisation (Dylik, 1954). Therefore, it is of great importance to the geomorphologist engaged in surveys for rural development to place his studies in the environmental context of what some refer to as integrated geomorphology. Unraveling the relations between land utilization and physical environment is essential, particularly if the aim is to improve the existing conditions in order to arrive at a more efficient use of the land in the future.

Therefore, it is useful to carry out a land use survey in preparing for the study of the physical environment, because the local population has used the land since generations and is familiar with the pecularities of every part of the terrain. As a consequence of this aspect of ethnoscience, superb modes of adaption to the environmental conditions or a (partial) mastering or improvement of these conditions can be observed in many parts of the world. One is often led to a deep esteem for the achievements of the simple peasants, which unfortunately not always graces the technocrat's mind. Some examples of such achievements are discussed below.

A systematic land use survey executed in this context may be beneficial in two ways: first, the distributional pattern and the type of land utilization may be indicative of certain environmental conditions and could offer clues to geomorphological, pedological, hydrological and other investigations. Inversely, however, once the environmental factors are unraveled in more detail, it is possible to come to conclusions concerning the most efficient land utilization of the various parts of the land in the future. The effort is thus beneficial both ways and as a consequence the mapping of the present land utilization should always be an integral part of development surveys (Vink, 1975, 1980).

It should be emphasized that the geomorphological and related environmental parameters never completely dictate the prevailing land use. Much depends on cultural heritage, technological level and socio-economic conditions of the people concerned. It is possible that in similar geomorphological situations, an entirely different mode of utilization evolves. For example, in the lowlands of southern Irian Barat, Indonesia, which are characterized by natural levees along the rivers separating extensive backswamp areas and bordered on the sea side by a series of sandy beach ridges alternating with elongated swampy swales, one is confronted with a situation where the staple food is the flour of the stems of sago palms growing in the backswamps and fish of the rivers are the main source of protein. In other, more developed, humid tropical areas of comparable kind, for example on Java, paddy fields are commonly found in the backswamps and in the swales, whereas coconut palms are planted on the sandy beach ridges. An important protein source is the fish ponds laid out in low areas which are too saline for paddy cultivation.

Similarly, many arid and semi-arid areas are often considered the domain of nomadic herdsmen, but sedentary groups of population and even forms of agriculture which have been cleverly adapted to the scarcity of water, have been developed in some areas. Examples are given below.

The cultural and socio-economic conditions may modify or, in technologically advanced areas, obliterate to some extent the effects of physical environment, but they are seldom completely antagonistic to the natural patterns. When land improvement programmes are to be implemented these patterns (and also processes) should be given full weight so as to reach optimal results with lowest cost.

6.2 Landforms and their Effect on Land Utilization

As outlined in the previous sub-section, the relations between landform and land use are diversified. Figures 6.1 and 6.2 illustrate this and relate to Malaysia and Senegal (Michel, 1978) respectively. An agricultural pattern of paddy fields in the broad valley bottoms, sago palms along the edges and rubber plantations in the hills are characteristic for large tracts in Malaysia (fig. 6.1). The stagnant water collected in the flood channels and other depressions of the Lower Senegal Plain in normal years provides the animals of the nomadic herdsman of the Ferlo Plateau with water even in the late part of the dry season and thus forms a crucial part of their seasonal migratory movements (fig. 6.2).

In some cases the relations may be direct but more frequently land utilization is affected by a specific environmental parameter (or a combination of such parameters) or by certain geomorphological processes. In these two cases the landforms act as an intermediate, a vehicle, for the detection of less easily studied factors, particularly when aerospace imagery is incorporated in the survey work. Even in cases where it can be established that the land use patterns largely coincide with or have a simple relation to the prevailing landform patterns, one should be aware of the fact that this situation may be affected by parameters other than relief. (Cotet, 1969; Stärkel, 1967). Two examples, from a lowland and a hilly terrain respectively, may illustrate the subtility of these relations.

Fig. 6.3 shows part of the Indogangetic Plain, near Patiala, India,as seen on ERTS imagery (Meijerink, 1974). Apart from the present Sutlej River Plain, a number of abandoned river courses marked by their natural levees with intervening backswamp basins are clearly indicated. Salt efflorescence has adversely affected extensive basin areas and even rendered their agricultural utilization impossible, whereas the adjacent natural levees are still under extensive use. Although the direct result of the river work has been the formation of levee ridges, it is the accompanying patterns of soil (texture) characteristics and of inherent internal and external drainage situation that together with the relief, have conditioned the land which deteriorated by increased salinity following irrigation. Menon (1969) describes similar matters from the Indus Plain in Pakistan and relates the landforms,such as meander floodplains and bar deposits, to the effects of a rising water table, such as water logging and salinity and to the agricultural use of the land. He found that of the area studied, 38% was severely affected by salinity, the remainder being moderately affected; a high water table affected 60%. Better drainage may improve the situation. See also Ghose (1962) and Jacobsen and Adams (1958).

Similar patterns of natural levees and backswamps have a profound effect on the land utilization in humid tropical lowlands. Paddy fields often characterize the backswamps whereas homestead gardens and coconut plantations are normally found on the levee ridges. Here the relief factor proves to be important,but soil texture, hydromorphic soil development, gley formation and the internal and external drainage associated with the levee-backswamp situation under those climatic conditions cannot be separated from the relief factor proper.

Fig. 6.4 is a stereotriplet depicting a portion of the north coast of western Java, Indonesia and is particularly interesting as it shows the effect of relief on land utilization in hilly terrain. The isolated plateau-like marly limestone hills with steep sides and a rather broken-up topography should be classified as waste land. The gently sloping footslopes surrounding them are, however, fully occupied by rain-dependent fields, the topographic situation precluding irrigation. These fields can be readily distinguished from the smaller paddy fields that occupy the entire lowland where irrigation is possible. Only where salinity is high in some near-coastal parts (for example, in the lower left corner) do fish ponds replace rice fields. The villages and homestead gardens are, in the interior, located along the lower limit of the non-irrigated slopes. In the lowlands they are situated outside the precious irigable land, viz. on natural levees and beach ridges.

In areas of more considerable relief a vertical zoning of land utilization can often be observed, comparable to the vertical zoning of natural vegetation discussed in section 5. The effect of relief in such areas is very pronounced,although it functions in this case with climate as a causative factor. The lower temperatures at greater altitudes increase the danger of frost damage and reduce the productivity of the higher parts of the mountains where the slopes are normally steeper. The differences in vertical zoning and the altitude ranges of these zones between sun-exposed slopes and on differently orientated slopes are often very outspoken and are brought about by the different insolation in various topographic situations. The economic importance of this factor may be considerable, for example, in vineyards. Although climate is the causative factor, it cannot be denied that the relations between relief and land use are rather 'direct' when matters of vertical zoning and exposure are concerned.

The importance of relief in land utilization is particularly outspoken when (canal) irrigation is concerned. The fields which are to be irrigated should be horizontal and when relief occurs, they have to form terraces separated by bunds and be provided with overflows to transport the water from one field to the next. The irrigation canals are usually sub-horizontal and therefore, have to be adapted to the topographic situation. The example of

Fig. 6.1 A characteristic land use - land form relationship from Malaysia : paddy fields on the flat valley bottoms; sago palms in a narrow strip along the edges; rubber in the hills.

Fig. 6.2 Stagnant water in depressions of the alluvial plain of the Senegal River is vital to the herdsmen of the Ferlo Plateau during the dry season.

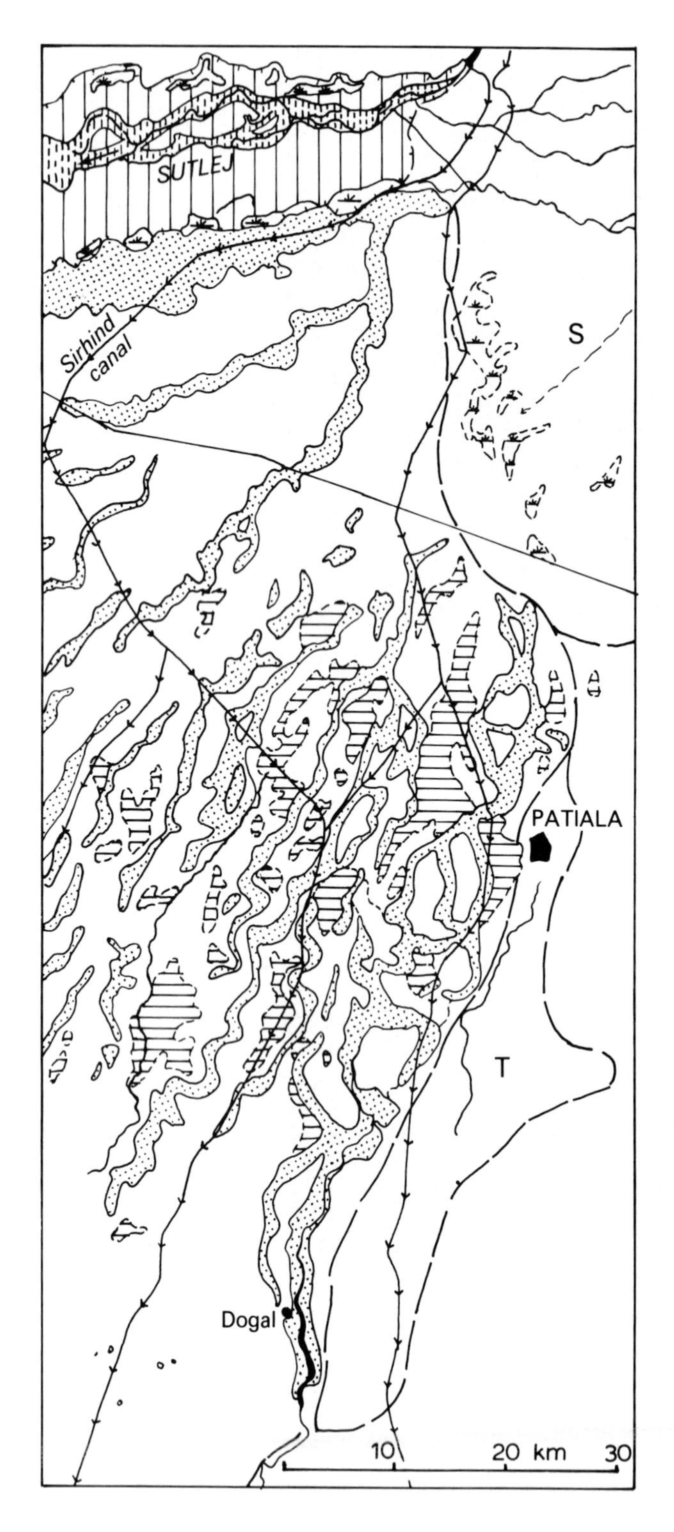

Fig. 6.3 Natural levees, basins and areas with salt efflorescence, Punjab, India (Meijerink, 1974).

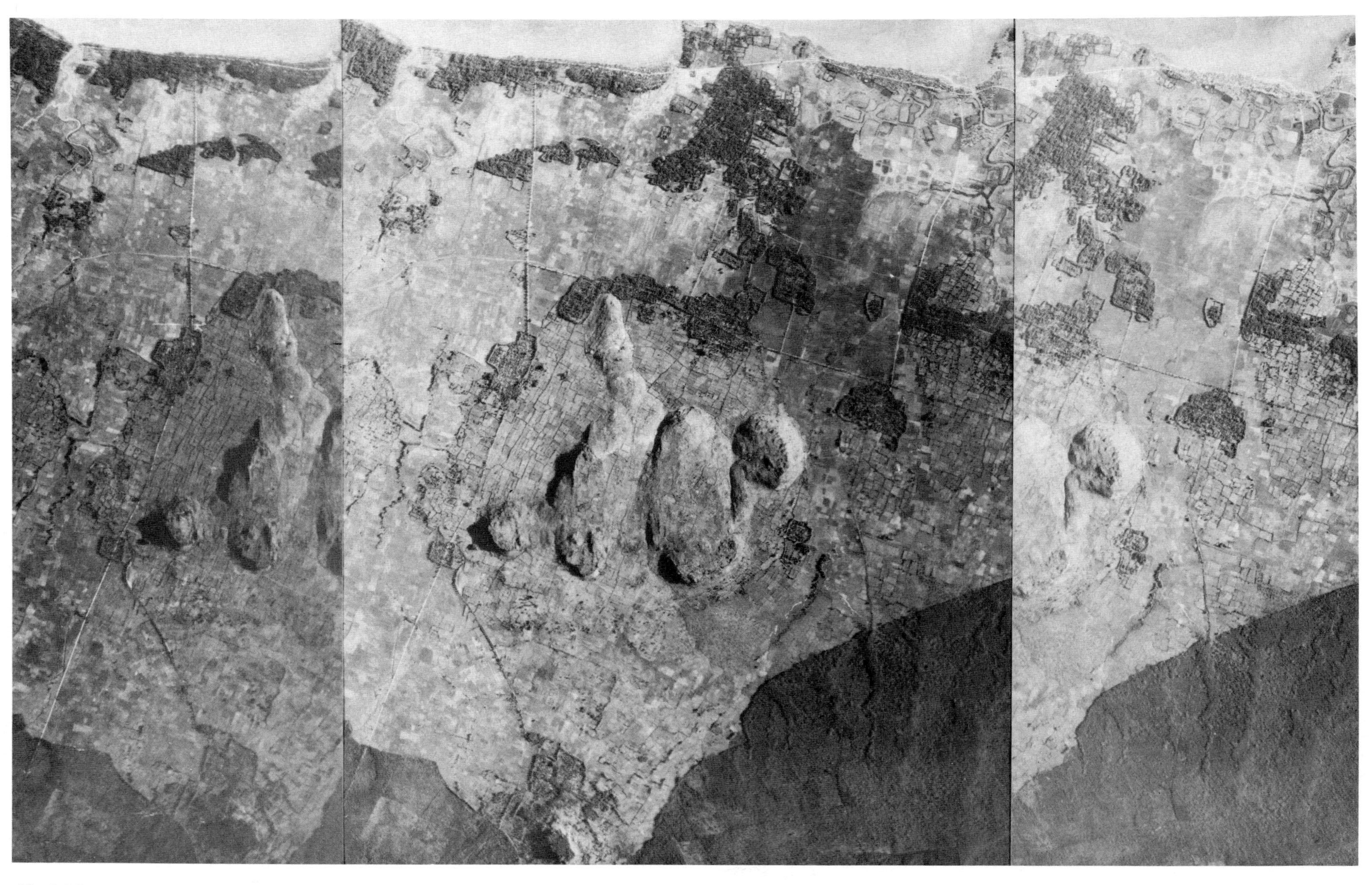

Fig. 6.4 Stereotriplet (1:40,000) showing the relation between geomorphology and land utilization in part of the north coast of western Java, Indonesia.

fig. 6.5 relates to tidal irrigation in the coastal lowlands of Kalimantan (Borneo) near Banjarmasin, Indonesia. The principle is to construct intakes for primary irrigation canals in the lower reaches of the main rivers where the fresh river water is ponded up at high tide and upstream of the lowest parts of the river course which are then invaded by saline or brackish water. These primary canals stretch from the natural levee into the backswamp and there secondary and tertiary irrigation canals are constructed for irrigation of the paddy fields to be laid out here. The primary canals may (Kalimantan; see example) or may not be connected with the adjacent river. Tidal irrigation is, particularly in S.E. Asia, an increasingly common mode of irrigation, effective and comparatively simple. Normally, the irrigation canals are constructed as straight lines in view of the fact that the relief of the backswamps is usually faint. Nevertheless, it appears to the author that if some considerations were given to the relief of the backswamps when planning the lay-out of the canal system (and consequently the villages, etc.) the water in the paddy fields could be more efficiently controlled and productivity could be increased. Other examples of agricultural adaptation to the surface conditions in humid tropical lowlands have been mentioned in earlier publications by the author related to the tidal zone in Kalimantan (Borneo) (Verstappen, 1977) and and the Solo Delta, E. Java (Verstappen, 1964). See also: Andriesse, 1974. Fig. 6.6 is a Landsat image of a coastal area of the Lampung district, Southern Sumatra, showing technical, government-sponsored, tidal irrigation plots (large rectangles) as well as Buginese tidal small holdings (1:300,000; July 1978).

In irrigation projects situated outside the tidal zone, relief plays a leading role in the location of the irrigation canals and consequently in the land use patterns in the irrigated zone (Anderson et al., 1970). The annotated aerial photograph of fig. 6.7 pictures the intramontane plain of Los Andes in the Chilean Andes at a scale of 1:90,000. The Aconcagua River has built a fan which is largely under irrigated agricultural use. The area is situated in the Central Chilean Depression between the Central and the Coastal Ranges and as a result is rather dry. Great fluctuations occur from year to year; the average annual precipitation in Los Andes amounts to 319 mm with a distinct maximum in June.

The Aconcagua River emerges from the mountains slightly east of the town of Los Andes near the upper limit of the annotated aerial photograph and flows in a WNW direction from the apex of the fan towards the town of San Felipe, which is situated just outside the lower photo edge. The riverbed, ranging in width from a few tens of metres near the two towns mentioned, to several hundreds of metres elsewhere on the fan, is largely composed of boulders and gravels. The river is

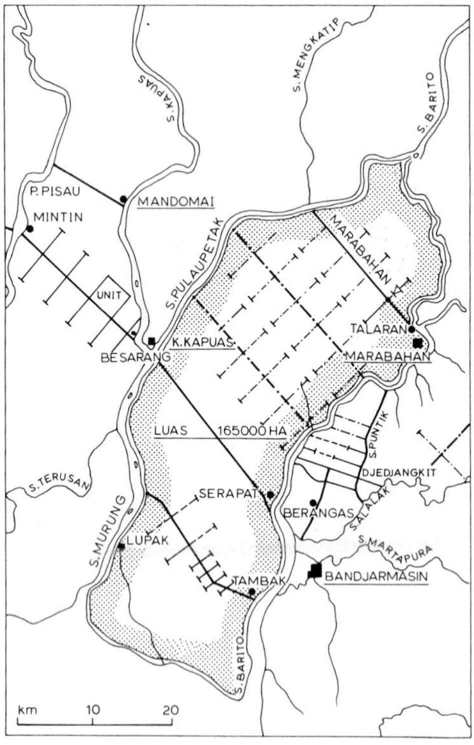

Fig. 6.5 Primary and secondary irrigation canals of the tidal irrigation scheme to the West of Banjarmasin, Kalimantan, Indonesia. Canals under construction are indicated by dashed lines; those planned, by dotted lines.

Fig. 6.6 Tidal irrigation scheme in Southern Sumatra as pictured on Landsat image of 22nd July, 1978 (Band 4, 5 and 7). Both the large fields of government-sponsored and the small tidal fields of the Buginese fishermen can be seen.

Fig. 6.7 Vertical aerial view of Aconcagua fan near Los Andes, Chile,at an approximate scale of 1:90,000. The pattern of irrigation canals indicated clearly reflects effective adaptation to the terrain configuration.

characterized by important fluctuations in discharge: there may be just a trickle of water in the dry season, whereas the bed may be filled from bank to bank when the river is in spate. The river is very shallowly incised, about 2 m, in the fan.

The thickness of the fan deposits is 45-56 metres near Los Andes and thicker farther downstream near San Felipe. The topography of the rocky substratum is irregular, however, and rocky hills rise above the fan at several places. The thickness of the deposits and the good permeability of some of them accounts for the ample groundwater supply. Borings indicate that the composition of the fan differs substantially from that of the riverbed: apart from gravel, fine sands and clays also frequently occur in alternating beds and lenses. This explains why irrigation is possible and has been practiced for over 300 years. This prolonged irrigation has resulted in an increase of the argillaceous and colloidal components of the soil and has led to a reduced permeability.

Most of the irrigation water is derived from the Aconcagua River and distributed over the fields by a network of radiating larger and smaller irrigation canals, some of which pass over minor rivers. Others flow parallel to the surrounding mountains for quite some distance and irrigate the more remote parts of the fan. The most important intakes are situated upstream of Los Andes, which explains the site of this town. It is also well-protected from flooding by the two hills directly upstream at either side of the river.

After use the irrigation water is drained into the Estero Pocuro River, which marks the boundary between the lowest portions of the fan in the south and west and the surrounding mountains. This river enters the Aconcagua River near San Felipe,where the basin narrows and where the concentration of ground and surface water results in water-logged, swampy areas, unsuitable for agriculture. Irrigation water is insufficient in particularly dry years, which adds a risk to agriculture in this area.

The example of the fan of Los Andes, Chile, is an excellent example of how the geomorphological situation affects the irrigation potential and the lay-out of the system of irrigation canals. In this case the settlers have cleverly made use of the natural environment in the organisation of their agricultural activities and location of settlements. (Wilgat, 1972).

6.3 Geomorphological Processes and Land Utilization

As already stated in sub-section 6.1, the influence of geomorphology on land utilization is often effectuated through active geomorphological processes of various kinds (Niculescu, 1967). These may be exogenous or endogenous, almost imperceptibly slow or spectacularly instantaneous, but invariably the processes themselves can only be studied and quantified after patient research and observation covering considerable periods of time, if they can be adequately studied at all. For the purpose of surveying for development it is important that every process in a given geomorphological environment leaves characteristic traces in the landforms which enable us to identify the process which caused them. Another clue directing further studies is obtained where the processes clearly affect the pattern and type of land utilization.

The processes concerned may range from surface wash, lateral sapping of rivers and cutting of ravines, to tectonism and volcanism (Pewe, 1954; Tricart, 1962, 1966, 1967; Young, 1973). It is important in this context to distinguish between the effects of the more gradual and diffusily occurring natural processes with which the agriculturist can live without much trouble or which he may even use to his advantage, and the more violent processes which are often difficult to master and which may upset a whole agricultural system. The latter are often of a localized and spasmodic occurrence. Some examples are given below. It should be stressed that only natural geomorphological processes are considered in this sub-section. Processes induced by man and which may drastically deteriorate his environment, will be dealt with in sub-section 6.5.

Leaching empoverishes the soils of areas of high rainfall such as the humid tropics. The stage of leaching reached and thus the productivity of the land, depends on the time elapsed since the surface was first subjected to the rain and thus to the process of leaching. In a valley where the floodplain is bounded by a series of river terraces of different height and age, the soils of the younger, lower terraces are less intensively leached and thus richer in nutritive minerals than those of the older, higher terraces. Consequently, the best yield can be expected on the younger terraces. Agriculture may even be completely confined in the latter. In the uplands, the deepest and best developed soils normally occur in the (sub)-horizontal and gently sloping parts where leaching of the topsoil has reached an advanced stage. On the steeper parts of the slopes, truncated soil profiles are common where the leached topsoil has been removed by slope processes of various kinds and where as a result the subsoil which is richer in minerals,is within the reach of the roots of the crops.

This situation explains the surprising fact that in many mountainous areas in the humid tropics the farmers, particularly shifting cultivators, have a distinct preference for the steeper slopes, in spite of the inconveniences often connected with such steeply-inclined fields (Verstappen, 1964). By cutting and burning the forest cover and tilling the fields, the shifting cultivators in fact introduce a certain degree of accelerated erosion which favors the process described and thus tends to temporarily increase the yields. Needless to say, these farming practices have a considerable inherent risk factor and become dangerous, particularly when the rotation period is reduced due to increasing population density. In the Sibil Valley, Central Range of Irian Barat, Indonesia, where the author first observed these phenomena, the land use is concentrated on the lowest terrace and the steeper parts of the mountain slopes.

In the example given above, the farmers merely passively adjusted themselves to the naturally existing distributional patterns of processes and the inherent soil fertility and field productivity. In the second example, which relates to surface wash in the extensive savanna areas of eastern Africa and in particular, Rwanda, man plays a more active role. He actually uses the natural process of unconcentrated surface wash to his advantage, although he does so without being fully conscious of the effects of his activities. The non-irrigated and permanently used agricultural fields in this area are commonly situated on the gentle colluvial footslopes surrounding the steep slopes of the high plateau-like erosion surfaces. This is understandable,taking into account that these thick and medium-textured colluvial soils are well moistened by the water running diffusely down the steep slopes after having fallen on the plateaus during the rainy seaosn.

Deposition of soil particles on the fields accompanies surface wash in the zone of colluvial accumulation. Much more important for the productivity of the fields is the fact that a considerable amount of ashes is deposited on the colluvial slopes during the wet season and acts as an annual gift of fertilizer. These ashes are derived from the burning of savannah vegetation on the plateaus during the dry season. Although the fires, occurring in all savanna areas in the world, have distinct detrimental effects on vegetation and environment, they are beneficial for the colluvial slope farmers mentioned above.

Imperfect sheet flood is a process efficiently used by farmers on (wheat) fields depending on the rare surface and rain water in many (semi-) arid areas. The extended stereopair of fig. 6.8 gives an example from the Near East. Such fields are commonly situated on the gentle slopes of alluvial fans where the occasional, spasmodic flow of the mountain torrents is dispersed and forms a thin film of water, the velocity of which gradually decreases. Low, horseshoe-shaped, earthen bunds open at the upslope side, collect this water on the fields where it infiltrates and in years of adequate rainfall produces sufficient moisture to grow crops. Depending on the discharge characteristics of the torrent, the lithology and morphology of its basin and the texture of the material of which the fan is composed, these fields are located near the apex of the fan as in this example (upper part of fig. 6.8), or in the lower parts of the fans.

The water which remains unused will gradually infiltrate but some of it may concentrate in channels where a network of low bunds is constructed for collecting it, provided that these valleys are sufficiently wide and flat-bottomed. The lower portion of fig. 6.8 depicts some of these fields. It should be noted, however, that it is not uncommon that these fields derive most of their water from the rain that falls on the fields themselves and on the immediately adjacent slopes. The matter is elaborated upon in sub-section 6.6.

The three examples given above concerning agricultural adaptation to and utilization of exogenous processes relate to rather gentle processes to which such adaptation is feasible with technologically simple means. It is the occasionally violent processes which are more difficult to handle and which are, therefore, usually destructive and disrupt agricultural land use. They are classified under the heading natural disasters of exogenous origin (see sections 10, 13, 15, 16 and 17). The lateral sapping of the Merguellil River, Tunisia, during the catastrophic flood of 1969 for example, caused the loss of numerous agricultural fields along its course. A detail of the damage is given in figure 6.9, where a strip of up to 200 metres wide was lost along the northern bank. The position of the northern river bank prior to the 1969 flood is indicated by a barbed line and the land lost during the flood is dotted. Several fields and farm roads were destroyed.

In case of a meandering river, the losses at the concave curves are generally counterbalanced by comparable gains of land in the convex parts of the meanders. However, the land lost is often situated higher and thus is more or less safe from flooding. Soil formation has made these areas agriculturally more valuable than the newly gained, lower lying and often coarser-textured sediments. The bed of a braided river will be widened on the occasion of every major flood until the width is properly adjusted to the largest floods. Much of the bed load in such cases is derived from bank erosion.

Endogenous processes may also be of a gradual nature, thereby enabling the rural population to adapt their land-use to the process. More frequently, however, endogenous processes of tectonic or volcanic origin operate shock-wise during earthquakes or volcanic eruptions. In this case, they fall under the heading of natural hazards of endogenous origin (see section 17). Two examples of tectonic and volcanic processes affecting land utilization follow.

Slight subsidence occurs in the intra-montane Kerinci Plain in Sumatra, Indonesia, on the occasion of almost every earthquake which has its epicentre along the fault

Fig. 6.8 Extended stereopair (1:40,000), showing rain-dependent fields, surrounded by horse-shoe-shaped low bunds and situated near the apex of a fan (top) and in shallow flat-bottomed valleys (bottom) in a semi-arid environment in the Near East.

scarp which borders the wedge-shaped graben to the NE (see fig. 6.10). The plain slopes down very gently and gradually widens towards the SSE, where a volcano-tectonic lake is situated, partly surrounded by a cone of tuffaceous deposits. The lake receives most of its water from the Siulak River and is drained to the east by the same river which crosses the fault scarp immediately downstream of the lake. Rice cultivation, the economic mainstay of the plain, is endangered by the subsidence which accompanies the earthquakes. The drainage of the Siulak River is then impeded where it enters the mountainous zone east of the fault scarp and the lake is thus ponded up. After the 1909 earthquake, the lake level rose slightly more than a metre and flooded the paddy fields along its shores. The aerial photograph of fig. 6.11 pictures the situation. Fortunately, so far it has always been possible to remedy the problem by repeatedly widening and deepening the riverbed downstream of the lake and by constructing an artificial outlet which provides a shortcut with increased gradient (fig. 6.12). Obviously, the situation is critical and paddy cultivation in the lower parts of the plain is vulnerable. External pro-

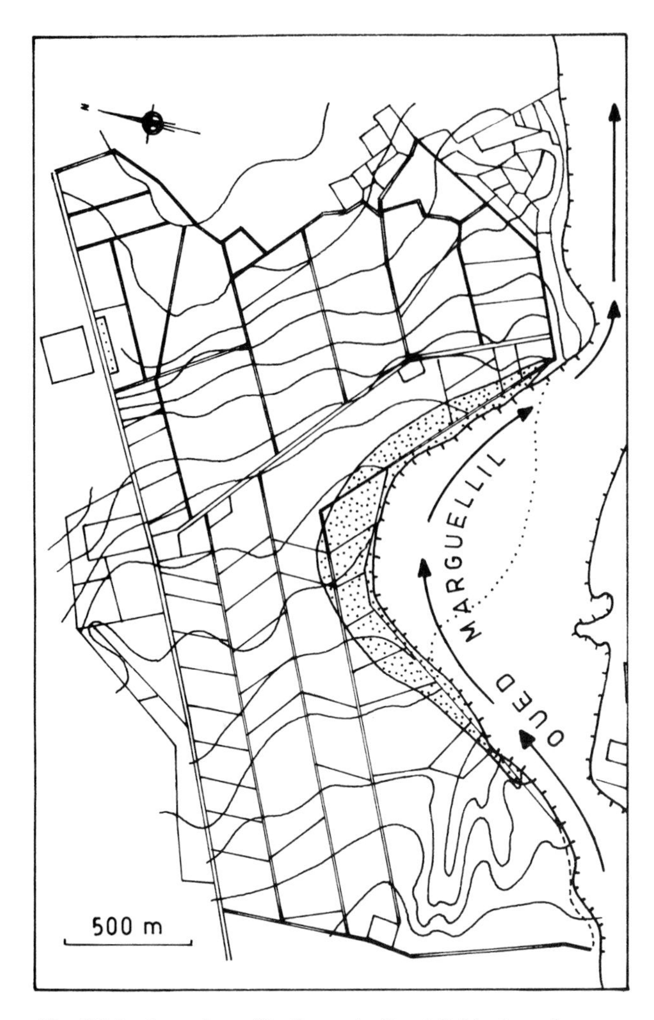

Fig. 6.9 Bank erosion, affecting agricultural fields along the Merguellil River, Tunisia, produced during a catastrophic flood in 1969.

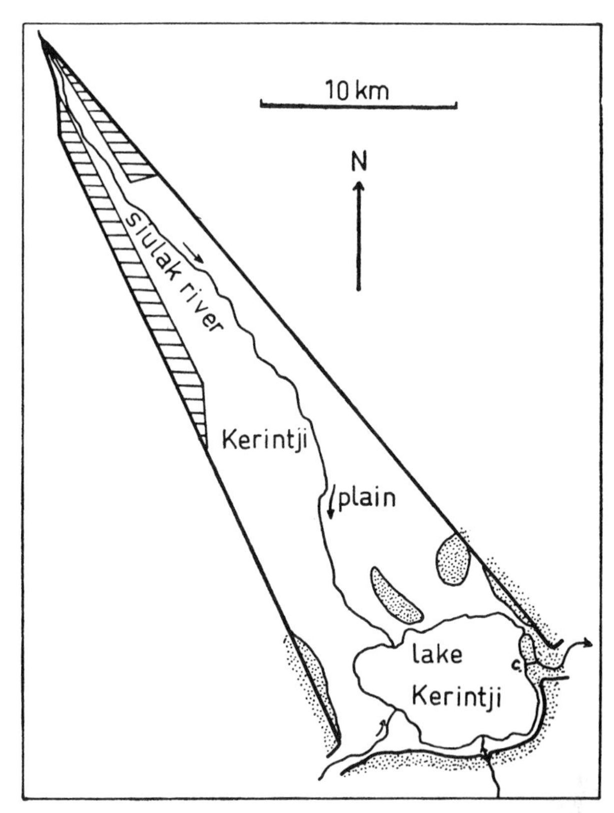

Fig. 6.10 Sketchmap of the intra-montane Kerinci graben, Sumatra, bordered at either side by rugged mountainous terrain. The plain is under paddy cultivation except the higher lying splinters/slivers (hachured) in the northwest. Tuffaceous deposits (dotted) occur around the volcanic lake in the southeast. The plain and the lake are drained across the active fault scarp which borders the graben to the northeast and drainage is impeded when earthquakes occur.

cesses have substantially added to the problem, especially in the past few decades. Deforestation of the steep fault scarps and adjacent parts of the mountains, mainly for growing cash crops, has resulted in an increased variability of the discharge of the Siulak River and in deposition of comparatively coarse material in the plain which can only be partly transported to the lake because of the faint gradient of the plain.

Characteristically, volcanic activity has beneficial as well as detrimental effects on land utilization. On one hand many volcanic areas are endowed with a high soil fertility as a result of the deposition of volcanic ashes, but more violent eruptions may occasionally bring disaster to densely populated volcanic slopes (van Bemmelen, 1956; Verstappen, 1963).

Shortly after such occurrences, the rural population, needing land and attracted by the soil fertility, reoccupies these slopes and even the sites known to be particularly dangerous. The natural hazards associated with volcanic activity are diversified in nature and may result from the endogenic, volcanic processes directly or from exogenous processes associated with the eruptions.

Volcanic mud flows or 'lahar' are an important phenomenon in many volcanic areas. Their mechanism is sparked by the volcanic activity and is usually spasmodic and violent. They may lead to major natural disasters resulting in a high death toll and loss of property (see, for example sub-section 6.6, Kelut Volcano). The mudflows may be initiated in various ways: the sudden melting of snow and ice on top of a volcano at the advent of an eruption, the emptying of a crater lake during an eruption, the crumbling of a lava plug or the deposition of ashes with subsequent downslope transportation by rain and running water. The latter process is common in the humid tropics where so-called 'cold' lahars occur during the rainy season following an eruption (Taverne, 1926; Thal Larson, 1926).

The Merapi Volcano, Java, Indonesia, is ill-famed, particularly because of its cold lahars that derive from an active lava plug at its top. Fig. 6.13 is a northward oblique view of the Merapi top area and clearly shows

Fig. 6.11 Vertical aerial view (1:12,000) of part of the shores of Lake Kerinci, Sumatra, showing several former positions of the shoreline each of which is related to earthquake-induced lake level changes.

the dark-coloured lava plug. A younger, active part of the plug with solfataras is slightly lower and to the left of the main plug. The viscous lava flows slowly downslope, disintegrates and ultimately contributes to the fragmentary material building up the steep and straight upper slope of the volcano below the plug where 'dry' downslope transportation of volcanic debris under the influence of gravity is a dominant process. The more gentle fluvio-volcanic slopes of the volcano can be seen near the upper photo edge and in particular in the upper left corner Fig. 6.14 gives a ground view of the lava plug as seen from the north. Cool (1931), Escher (1931) and Neumann van Padang (1933) have described the phenomena that occurred during the ill-famed Merapi eruption of 1930.

Fig. 6.15 is a vertical black and white infrared aerial view of the large slender cone of Mt. Merapi (2,911 m) as it rises from the plains of Central Java. The plug at the top and the steep upper slopes are shown in stereo; the lower and more gentle fluvio-volcanic slopes to the southwest of the volcano with numerous volcanic mudflows are pictured non-stereo, farther to the left. The photo dates from 1970 and the dark flows date from the eruption of 1969/1970. Most of the material is transported downstream by the ill-famed Batang and Blongkeng ravines. Interesting is the comparison with the panchromatic aerial photographs of fig. 6.16 which date from 1948 and cover part of the lower left corner of the 1970 photograph. Volcanic mudflows can also be seen on this photo but they date mostly from the 1930 eruption; agricultural fields and villages which were obliterated as a result of the 1969 eruption can be traced as well. The lahars even reach down as far as the road Yogyakarta-Magelang, endangering the village of Muntilan and making protection of roads and bridges necessary. The damage

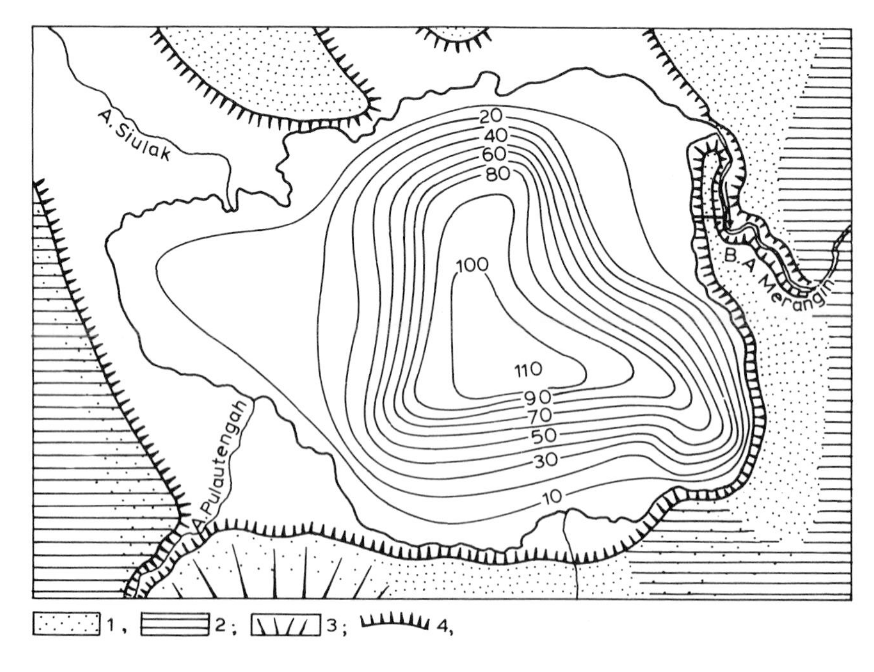

Fig. 6.12 Map of the southeastern part of the Kerinci graben, showing the lake (with depth contours in metres) and the artificial short cut made in the Merangin River to improve the drainage of the lake. Key: 1. tuffs, 2. Tertiary rocks, 3. Bambang Volcano, 4. escarpment.

Fig. 6.13 Oblique aerial view of the Merapi Volcano, Central Java, showing the active plug at the top, the disintegration of which feeds the volcanic mudflows on the lower slopes to the left. The slopes to the right are (temporarily) protected by a low crater rim. An old crater rim can be seen in the upper right and offers effective protection at that side.

Fig. 6.14 Ground view of the Merapi lava plug as seen from the north. Note the active flow towards the lower right corner of the photo.

to life and property in these parts, caused by the volcanic mudflows, can be dramatic at times.

A crater rim can be seen on the oblique photo of fig. 6.13,immediately to the right of the lava plug. It represents the rim of the lava plug of 1930. Farther downslope and near the right photo edge, another, older crater rim can be traced. It forms part of the 'Old Merapi' and is known as Mt. Pusunglondon. It protects the slopes below it from volcanic mudflows if the activity of the plug changes to that side of the mountain. In the beginning of this century there was activity near the top area and volcanic debris was transported downslope by the Wara Ravine, situated between the Merapi proper and Mt. Pusunglondon. Several rivers in the lowlands situated south of the Merapi were heightened rapidly in those years due to accumulation of volcanic ashes and sands. These 'sand rivers' are now several metres above the rice fields and impede drainage of the rice fields. Several irrigation canals now pass underneath them through small tunnels. The situation is depicted by figures 6.17 and 6.18.

6.4 Geomorphology, Environment and Rural Land Utilization

When discussing the effect of landform on rural land use in sub-section 6.2, it was pointed out that in many cases it is difficult to separate the effects of landform from those of other environmental parameters and that

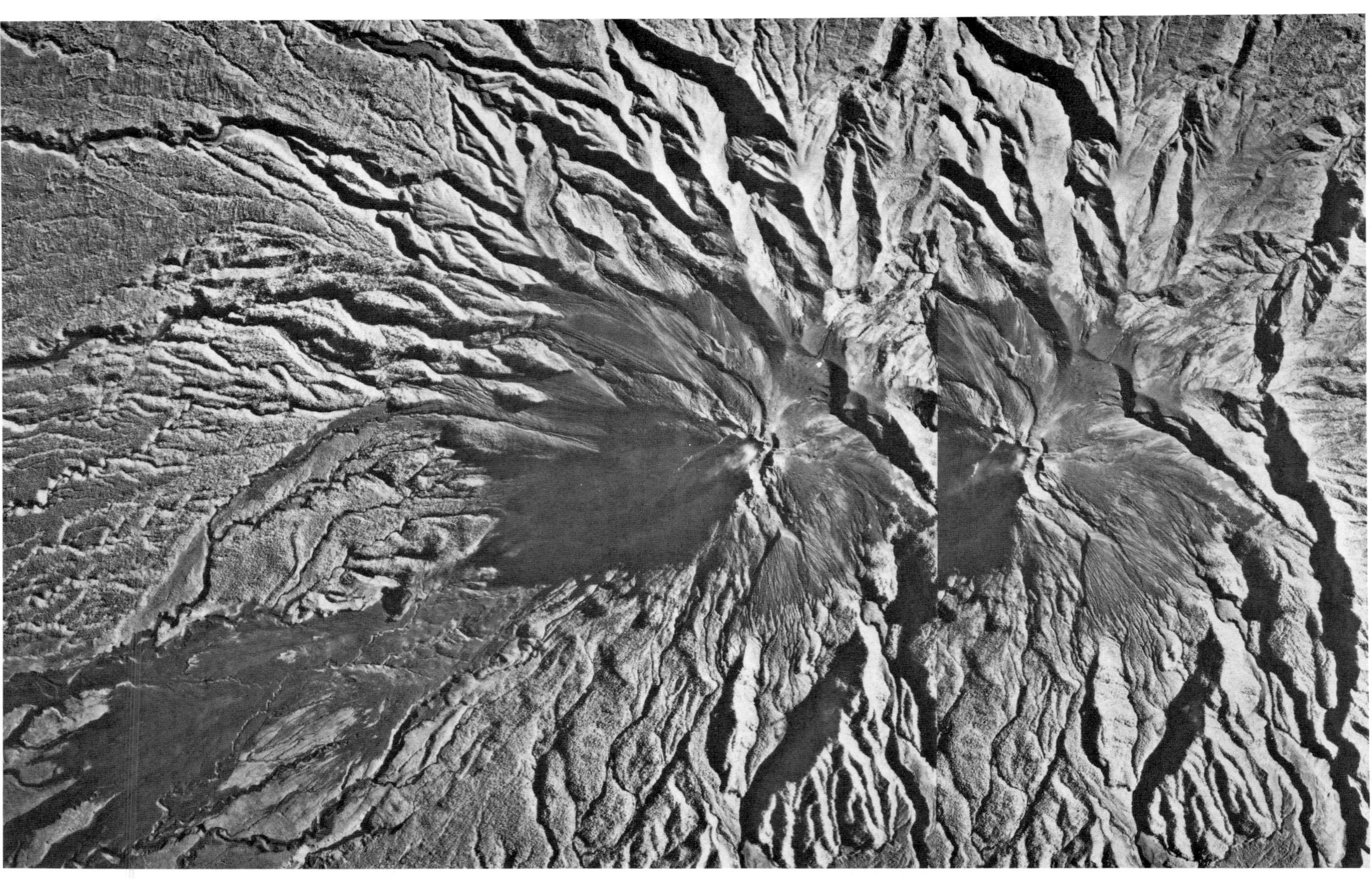

Fig. 6.15 Extended infrared stereopair (1:35,000) dating from 1970, showing the top of the Merapi volcano in stereo and the active volcanic mudflows (lahar) on the southwestern slopes in the extended non-stereo part. Comparison with the aerial photograph of

Fig. 6.16 Aerial photograph at approximately the same scale as that of fig. 6.15, showing the volcanic mud flows (lahar) on the southwestern slopes of the Merapi Volcano as they were in 1948. The limit of fig. 6.15 is indicated to facilitate comparison.

the emphasis on landform is justifiable mainly because of the convenient approach it offers to the survey of the invariably complex environmental situation. It may therefore be maintained that to arrive at an adequate contribution to the evaluation of this situation, the geomorphologist should emphasize those aspects of his scientific field which are directly relevant in the context of environment, i.e. environmental geomorphology (Calder, 1951; Kent et al., 1969; Kugler, 1976; Lidov, 1954; Martinine, 1967; Mateesen, 1967; Pecsi, 1968; Thomas and Whittington, 1967; Verstappen, 1956; Young, 1968).

Although this is undoubtedly correct, one should be aware that the present environmental situation often has certain elements which are inherited from the past and that genetic geomorphological studies may also yield relevant information. It is an established fact that processes and developments of the past commonly differ substantially from those prevailing at present. Climatic fluctuations during the Quaternary and Tertiary are of particular importance in this respect, but other factors, for example tectonics, also have played a role. Both environmental and genetic aspects of geomorphology should, therefore, be considered.

In fact geomorphological processes of the past may influence present-day land use in various ways. A particularly clear case has been described by the author and concerns the Porali Plain, near Lasbela, Pakistan (Verstappen, 1966, 1969). The agriculture in this plain is almost entirely based on flood waters of the Porali River and smaller streams entering the plain from the sides. Aerial photographs show that during the formation of the northernmost, older part of the plain, the river had braided characteristics. Numerous abandoned river channels can be recognized. At present the Porali River has a 10-metre incision in this part of the plain. Irrigation of this flat area is difficult and it is, therefore, almost completely devoid of agricultural land use. The younger parts

Fig. 6.17 'Sand river' of the Wara drainage basin, on the southern foot of the Merapi Volcano, flowing several metres above the paddy fields as a result of lahar activity at the beginning of this century.

Fig. 6.18 Same area as fig. 6.17, seen from above. An irrigation canal passes underneath the 'sand river'.

of the plain farther south are not braided, but show distinct levee and basin development. The river changed its course several times when entering this southern part as evidenced by abandoned levee ridges. The morphological differences between the northern and the southern parts of the plain result from a change in the river regime brought about by a climatic change which occurred about 5000 years ago and contributed to the decline of the Indus Valley Civilization. The change in climate is also reflected in the dune morphology near the coast where parabolic dunes crown an old offshore bar (which is associated with a strandflat 4 m above sea level) and barchans occur on the present offshore bar.

The somewhat higher position of the levees with respect to the basins renders parts of the younger river deposits suitable for irrigation. A simple, recently improved dam was constructed directly downstream of the point where the former levees depart from the present riverbed. The flood waters of the Porali River are thus ponded up and diverted to irrigation canals, - the former channels located on the abandoned levees for irrigation of the fields in the basins (back 'swamps') (see fig. 6.19). A change in the river regime from braided to levee-basin, caused by a climatic change of the past, still determines the present distributional patterns of the agricultural activities in the plain. It is evident that in most cases descriptive landform classifications will not suffice for purposes of rural development. It is necessary to unravel the genesis of the geomorphological phenomena occurring, their mutual relationship in space and time and particularly the interdependency of geomorphology, hydrological conditions, soil development, vegetation and land utilization.

In most areas where the environmental situation is fairly stable or only gradually changing, the patterns of rural land utilization also tend to be stable, provided the socio-economic and technological situations remain the same. If no deteriorating side effects of human occupance become manifest, man can live in balance with his environment using the accumulated know-how of generations. However, in some rapidly changing environments such as deltaic areas and river plains, the land use of the fields has to be continuously modified and land use patterns have to be adapted even to the slightest morphodynamical developments. The two examples given below illustrate how environment and geomorphology affect rural land use in case of a static situation (in India) and a dynamic situation (in Indonesia).

Extensive aggradational plains, so-called 'Old Alluvium' mostly sandy and silty loams, are a dominant feature in the arid to semi-arid realm of Central Rajasthan, India. Steep-sided, residual hills composed of granites, rhyolites or Vindhyan sandstones rise from these plains. The surroundings of the town of Jalor, pictured on the map of fig. 6.20, are fairly characteristic of the situation.

Jalor was built long ago, at the foot of a steep granitic residual hill (1), with rhyolites occurring farther to the west (2). The hills are surrounded by a broad Old Alluvial Plain (3) which, as everywhere in Rajasthan, is underlain by a thick layer of so-called 'kankar', a whitish calcrete at a depth of 30-50 cm which can be clearly seen in the profile pit pictured in fig. 6.21. This calcrete,dating from a more humid period of the past, does not exist in the Young Alluvial Plain (4) which is of a more recent origin and borders on the Jawai River in the east and north. The granitic residual hills and the Old and the Young Alluvial Plain can be clearly seen on the photograph of fig. 6.22, which was taken from the fortress that crowns the hilltop rising above Jalor. The southwest slopes of the hills are exposed to the southwestern monsoon winds which prevail during the dry season. A large, cone-shaped, windward obstacle dune (5), more than a hundred metres high near the hills, developed here, probably in the early Holocene and was dissected during the more humid conditions that prevailed about 500 years ago. Sands, washed out from the obstacle dune and redeposited by the southwestern monsoonal winds, now form a zone of low, inactive, longitudinal and parabolic dunes (6) to the northwest and the southeast of the hills.

Agriculturally, the most valuable land of the area is the rather narrow belt of Young Alluvium along the ephemeral Jawai River. The soil is mostly loamy sand to sandy loam but sand and gravel is common underneath. Because of the good permeability and the absence of calcrete, a free exchange of shallow groundwater exists between the Young Alluvium - which is tapped by wells - and the water percolating through the coarse deposits of the river channels. Water is lost to the channels in the dry season, but recharge of the wells occurs after rains and floods. The yield of the numerous wells is sufficient to irrigate the many fields scattered all over this zone.

The shallow calcrete characterizing the Old Alluvium has a profound effect on the hydrological and agricultural conditions of these extensive plains. Precipitation cannot contribute to the groundwater supply to any considerable extent due to the impervious nature of the calcrete, causing a fair proportion of it to evaporate, though near the Young Alluvium some water may drain off laterally to this slightly lower zone. As a consequence of these conditions, the few shallow wells dug into the calcrete have a low yield and salinity is often considerable. These areas are evidently of low agricultural value,as irrigation depends fully on the occasional rains. After a single crop, fallows of one to two years follow. The plains are almost completely devoid of surface drainage and few fossil drainage lines exist. Near the hills, however, some ephemeral channels can be traced

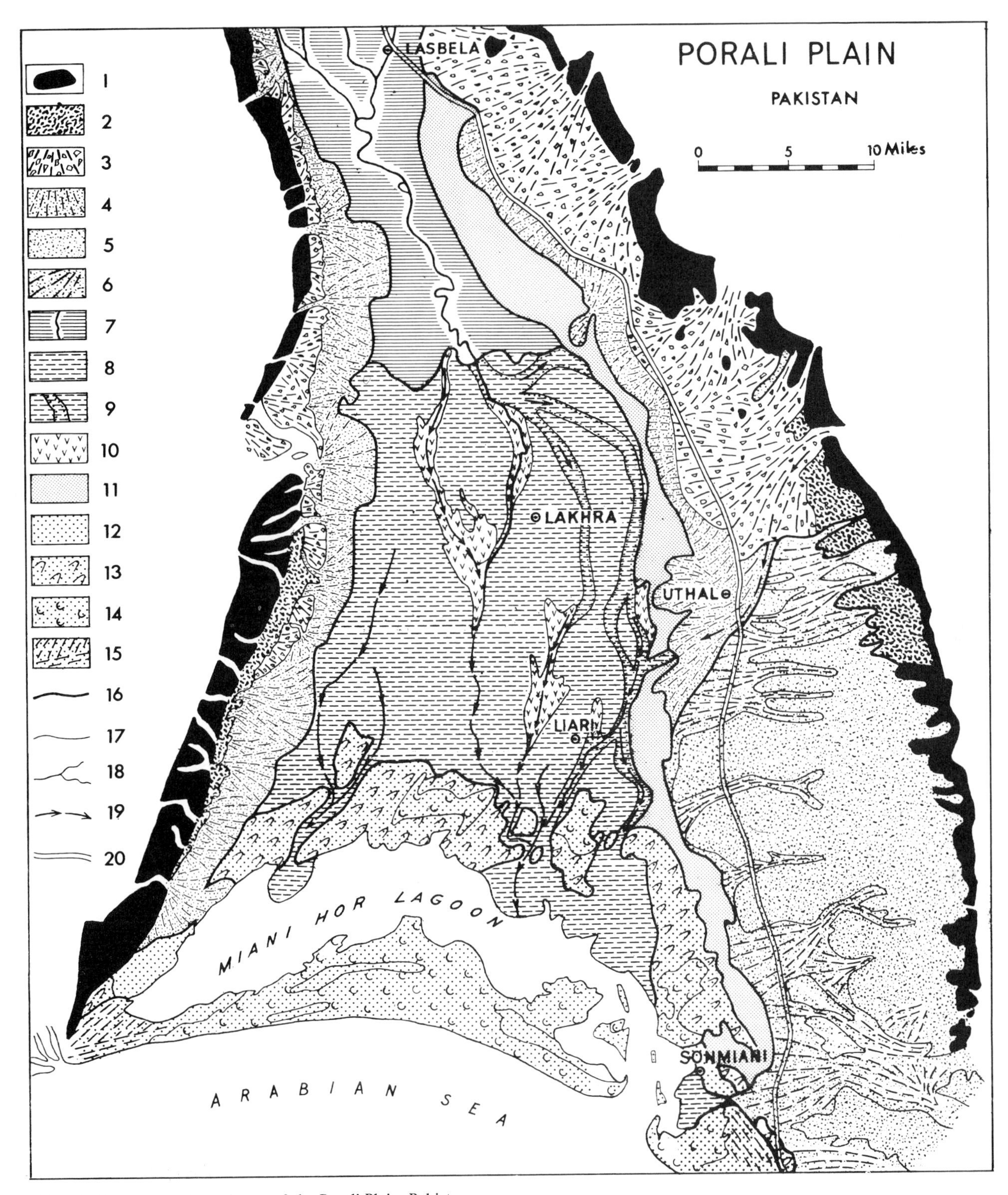

Fig. 6.19 Geomorphological map of the Porali Plain, Pakistan.
Key:
1. Mountains bordering the plain
2. Pediments developed at the foot of the mountains
3. Gravel fans
4. Lower, silty-loamy parts of fans
5. Sub-recent sandy deposits
6. Washed-out parts of same
7. Older, braided parts of plain
8. Younger parts of plain
9. Levees in same
10. Eroded parts of same
11. Poorly drained zone between plain and E-fans
12. Beach sands (barrier)
13. Fossil parabolic dunes in same
14. Active barchans in same
15. Beach ridges in same
16. Land system boundaries
17. Other boundaries
18. Rivers, mostly perennial
19. Ephemeral drainage lines
20. Trunk road Karachi-Lasbela-Kalat

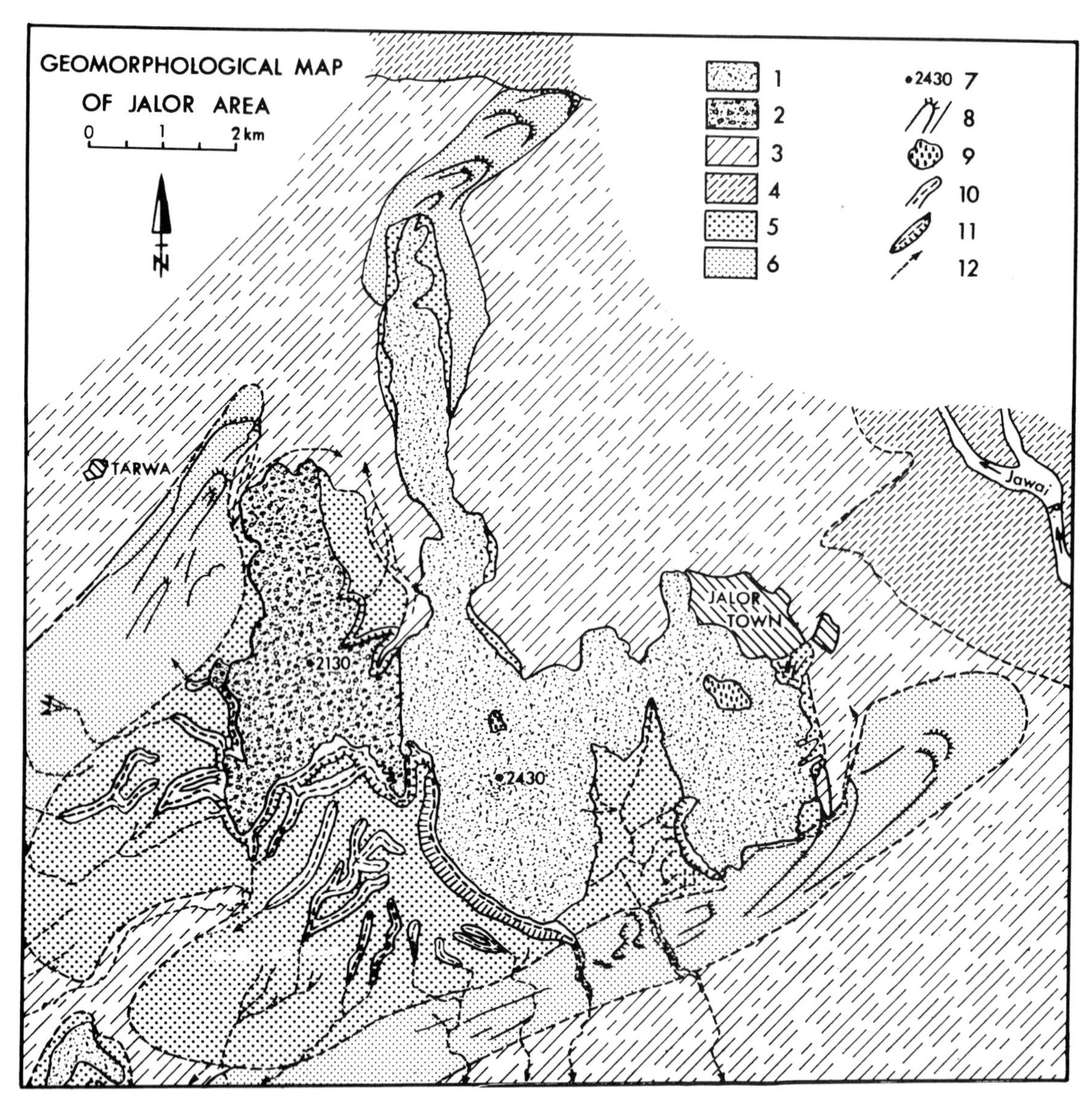

Fig. 6.20 Geomorphological map of the Jalor area, Rajasthan, India.
Key:
1. Granitic residual hills
2. Rhyolitic residual hills
3. Old alluvial plain
4. Young alluvial plain
5. Dissected dunes
6. Parabolic/longitudinal dunes
7. Spot heights (in ft.)
8. Dune ridges with slip-off slopes
9. Pockets of detritus
10. V-shaped ravines
11. Depression between dune and hillslope
12. Traces of flood run-off channels

which feed artificial storage lakes or 'tanks' (fig. 6.23), situated at topographically suitable sites and surrounded at three sides by a sizeable earthen bund. Because of the low permeability of the calcrete, surface water can be stored,providing drinking water for the animals until nearly the end of the dry season, at leas in 'normal' years. A major reservoir of this type can be seen to the north of Jalor. This was the main water supply of Jalor prior to the construction of a tube well.

Evidently, the best sites for the construction of 'tanks' are where the rainwater draining the steep and poorly vegetated slopes of the hills can reach the Old Alluvium unhampered. For this reason they are scarce to the south-west of the hills where extensive dune areas commonly occur. In these parts the rainwater is rather adequately retained in the dune sands, causing increased moisture content in the lower parts of the sand slopes and in inter-dune depressions, which are hydrologically more favourable. Though the conditions in the dune areas are unsuitable for agriculture, the possibilities of establishing areas for improved grazing are better than elsewhere.

The steep residual hills are economically the least valuable geomorphological units and only some extensive grazing is found, mainly by goats. The importance of these hills is however, strategic. The bowl-shaped top area of the residual hill rising above Jalor was turned into a fortress

Fig. 6.21 Calcrete ('kankar') occurring at shallow depth in the Old Alluvial Plain near Jalor.

where the population of the town could seek safety during turbulent days in the 1000 years of its existence. Drinking water is always available in the ancient fortress: a major and almost vertical dyke forms a depressed lineament within its confines. Water circulates in the weathered rock of the dyke, part of which was removed in the distant past to form a reservoir (fig. 6.24), supplying sufficient water to the fortress even in cases of prolonged besiege.

This example clearly demonstrates how the rural population properly evaluated the suitability of every part of its territory and used it in concordance with its cultural, socio-economic and technological background. The situation has not changed essentially since historical times and can be considered fairly static or stable. The effectiveness of the traditional adaptation (outlined above) to the environment must have fluctuated with dry and wet periods. The system is vulnerable, particularly in the case of drought.

A much more dynamic mode of adaptation is required of the rural populations living in rapidly changing geomorphological environments, such as the Cimanuk Delta, West Java, Indonesia. The growth of this delta and the accompanying changes in land utilization are given in fig. 6.25, picturing the situation in 1857, 1940 and 1969 as appearing on topographical maps of those years. The legend applies to the three maps (Sandy, 1977). In 1857 the main distributary of the Cimanuk River reached the sea slightly south-west of Cape Indramaju. A sizeable natural levee developed, the higher parts of which were

Fig. 6.22 General view of Jalor and the Old Alluvial Plain with reservoir lakes (tanks) as seen from the hilltop fortress. The Jawai River with the Young Alluvial Plain,where agriculture concentrates, can be seen in the distance.

Fig. 6.23 One of the reservoir lakes (tanks) in the Old Alluvial Plain near Jalor. The underlying calcrete ('kankar') interferes with infiltration (see fig. 6.21).

Fig. 6.24 The dyke in the top area of the Jalor hill and the part of it that has been excavated centuries ago to provide drinking water for the fortress in case of emergency.

the site of settlements. The lower parts between the villages, the mangrove swamps and the areas downstream of the villages were used for paddy fields. It appears from the map of 1940 that in the period 1857-1940, considerable accretion occurred in the north-west, where an entirely new outfall can be seen. As a result, fresh water advanced seaward and paddy fields replaced the mangroves near the lower margin of the map. Vertical accretion in the backswamp zones contributed to this development. In the north-east, onshore winds straightened the coastline and formed a low sandy beach ridge which was also put to use. An important change occurred in the interval 1940-1969, when a new outfall formed in the east. It broke through the above-mentioned beach ridge which was used in 1969, for the location of settlements. The paddy fields were immediately extended following this new delta growth and fishponds were laid out in the more saline parts. Some further gain of land and particularly an extension of villages marks the outfall that first appears on the map of 1940. The old outfalls of the river and particularly the former main channel near Cape Indramaju,lost much of their water as a result of these newer developments and saline water could penetrate there. Considerable loss of paddy fields was the result as can be clearly seen in the map of 1969. This example shows how the rural population, particularly in densely populated areas,adapts itself immediately and tries to make optimum use of newly accumulated land.

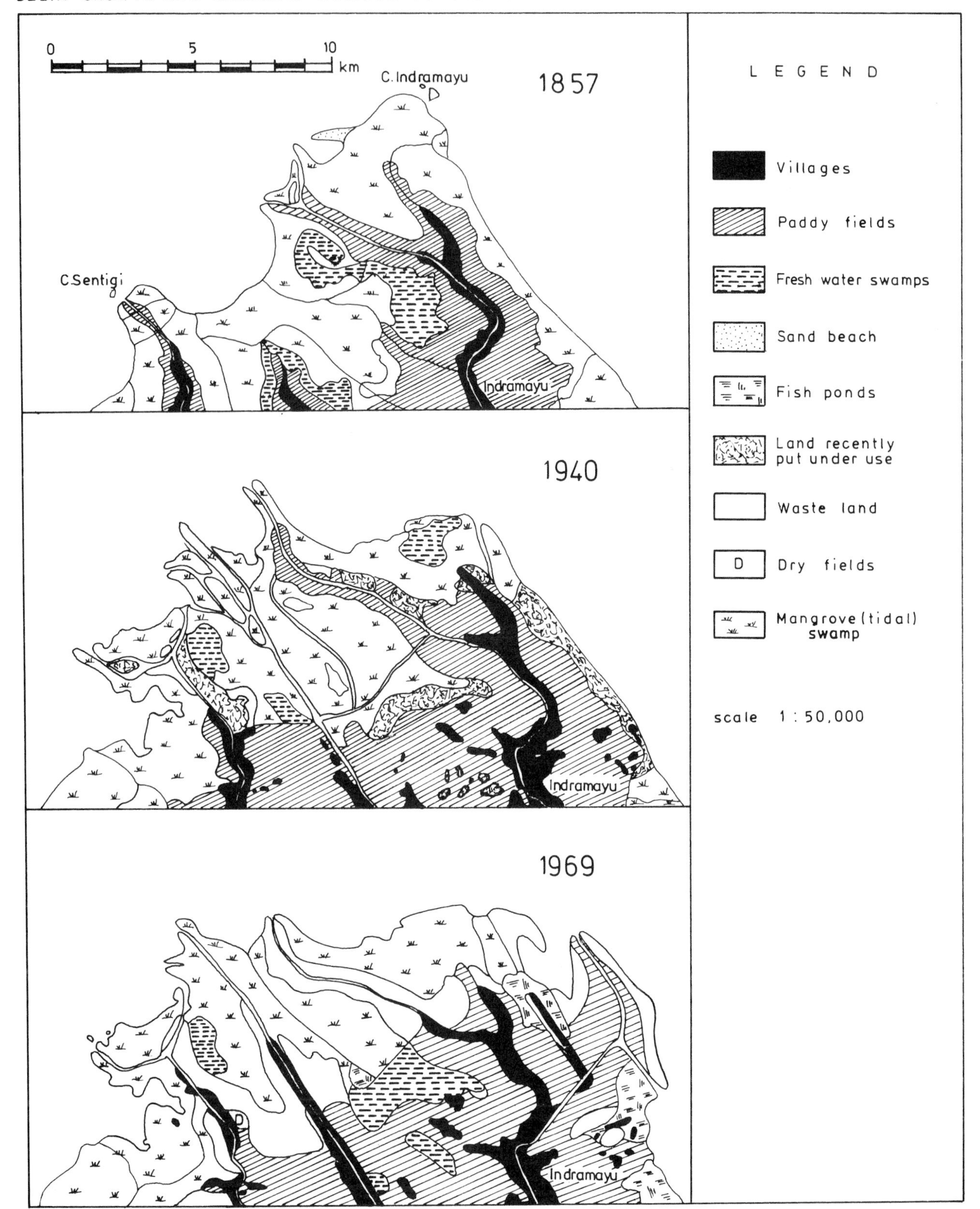

Fig. 6.25 Accretion and abrasion of the Cimanuk River delta, Western Java, Indonesia as pictured on topographic maps dating from 1857, 1940 and 1969. The land utilization of the delta is being constantly adapted to the changing morphological and hydrological situation and the resulting distribution of fresh, brackish and saline water (Sandy, 1977).

6.5 Man as a Geomorphological Factor deteriorating Rural Areas

To some extent, man has changed his environment in almost every part of the world by cutting forests, cultivation, irrigation and in many other ways. A recent review of the matter is given by Goudie (1981). In some areas, if not too densely populated and industrialised, the environmental effects of his activities are slight or even negligible, because he has skillfully adapted himself to the environment without causing much disruption or change. A state of equilibrium has been reached and one can rightly speak of a balanced environment in which man is one of the partaking elements. (Butzer, 1972; Erdosi, 1969; Gil and Slupik, 1972; Leser, 1975; Scott, 1975; Zapletal, 1973, 1975).

Shifting cultivators have basically established an equilibrium which is disrupted only when the rotation period becomes too short, for example, due to population increase. Otherwise the forest has an opportunity to restore itself and the soils are enriched with humus and minerals. In areas which are seasonally flooded, such as backswamps in many alluvial plains, other kinds of equilibrium often develop. A striking example is the Mekong River basin where the so-called 'floating rice' grows up with the rising floodwaters to a length of several metres. Another, equally striking mode of adaptation to such seasonal flooding conditions is found, for example, in broad river plains in (semi) arid Africa along the Niger and Senegal rivers, where the system of flood-retreat cultivation has been common practice for generations. The fields are planted shortly after they run dry due to the falling water level and therefore, have a good moisture content. Planting is carried out from the highest areas which have been submerged to the lower ones and the harvest is reaped before the water rises again during the next season. Most of the examples of land utilization affected by landforms, processes or the environmental situation 'dans son optique total' given in the previous sub-section, relate to various kinds of adaptation, making optimum use of the land without disturbing the subtle ecological equilibrium. Effective measures for protecting or even improving the quality of the land in historical periods are also on record (Knight, 1928; Lowdermilk, 1942; Ranere, 1970).

In many cases, however, man through his various actions unvoluntarily alters and often deteriorates the land (Brown and Sheu, 1975; Hempel, 1971; Lange and Pshenim, 1969; Tricart, 1956). He has done so repeatedly in the past and the adverse effects can still be felt. Man has also tried to change the land to his advantage. Modern technology has opened new possibilities in this respect. However, unexpected or underestimated side effects of such projects are numerous. Utmost care is required even when minor changes in the land are contemplated, otherwise the disadvantages may easily exceed the benefits envisaged.

In areas where land is put to use, natural vegetation is replaced by hydrologically less effective crops and the slopes may even be barren for a part of the year. As a result, the unconcentrated run-off on the slopes increases and a tendency towards concentrated run-off may develop. In cases of improper land use these processes are insufficiently controlled and abuse of the land may result in accelerated erosion in the form of sheet, rill or gully erosion, ultimately extensive badlands may be formed (Morgan, 1979). Serious erosion has been reported from various parts of the world in specific periods: after introduction of new crops, high pressure on the land, etc. (Lisitsina, 1976; van Zuidam, 1975). The Mediterranean region, for example, was seriously eroded following the large scale cultivation of wheat during the Roman Period. The fact that the wheat fields were barren during the rainy winter season was particularly harmful. A time of marked erosion occurred in North Africa and in Europe during the Neolithic period. Even more important in many parts of Europe are the traces left by the accelerated erosion which occurred in the Middle Ages and in the mid 18th to mid 19th century. The eroded areas as well as deposits (up to 5 m thick) such as coarse-textured low terraces, floodplains and other anthropogenous landforms formed in this period have been studied in detail. Anthropogenic river levees exist in lowland areas such as Flanders. In the 18th century, population pressure, revolutions and large land holdings were,factors but the introduction of potatoes as a staple crop was the major cause. Potatoes are a very poor protector of the soil and leave the land barren in the spring after the snow melts. The situation gradually improved with the introduction of small holdings and the planting of lucerne (Vogt, 1953; Pecsi and Stärkel, 1975). Also many coastal areas have been severely affected by land utilization (Gilberton, 1981; Verger, 1959).

In the east of the USA accelerated erosion started with the agricultural development by the settlers. As a result, accelerated deposition has occurred in the last 150 years, for example, in the Chesapeake Bay. In Central America the cultivation of mais as a staple crop resulted in serious erosion, in particular as it leaves the land barren in the rainy season. It has been claimed that for centuries the Maya Empire was forced to move from one place to another due to the erosion brought about by massive mais cultivation. Potatoes are an erosion factor in the Andes.

Numerous examples can be given of the Far East. Interesting is the case of the Padang Lawas Area, situated south of Lake Toba, Sumatra, Indonesia. It is an area of comparatively low rainfall where dry and hot 'föhn' winds frequently blow, descending from a narrow, depressed part of the Barisan Range, situated to the west.

The climatic pecularities caused by this topographic setting in combination with the easily eroded soil are the causes of the serious erosion which now characterizes this area, most likely triggered by intensive land utilization during the period when it was a centre of Hindu - Javanese civilization. Temple ruins are scattered over the devastated land.

It is very probable that the large scale cutting of tropical forests, extensively carried out in the equatorial belt, may lead to catastrophic erosion in the near future if no precautions are taken. There are numerous indications that the process has already begun. Many tropical soils are susceptible to erosion and in particular to splash erosion, which seals the surface of the soil, increasing surface run-off and linear cutting. In general, diffuse, areal processes such as surface wash proceed at approximately the same rate as soil formation; which is in most cases not harmful. If deforestation leads to the introduction of linear erosion, devastation starts and pedogenesis will be interrupted as explained in section 3.

Areas where the population is dense and rapidly increasing such as the island of Java, Indonesia are particularly vulnerable to environmental deterioration as demonstrated in the following example from Central Java.

The stereopair of fig. 6.26, its interpretation of fig. 6.26a and the figs. 6.27 and 6.28 show how man's interference with the natural equilibrium in a drainage basin gradually changes the various slope and other processes and ultimately results in a deterioration of the environment, unless protective measures are taken before it is too late. The stereopair depicts a part of the Sapi River basin, a tributary basin of the Serayu Basin, South Java, Indonesia, at the approximate scale of 1:20,000 as it was in 1973. The Sapi River runs from left to right near the lower edge of the photo. Its flood plain is bordered by an alluvial plain with irrigated rice fields. Farther to the north a broad terrace can be seen bordering a zone of comparatively low hills. The South-Serayu Mountains are situated in the upper portion of the photograph. These main geomorphological units are indicated in the interpretation of the stereopair given in fig. 6.26a.

Fig. 6.27 depicts the same areas as fig. 6.26 but as it was in 1946 and at a scale of approximately 1:50,000. It is known that the population density of this area rapidly increased in the period 1946-1973 and with it the expansion of the settlements and surrounding homestead gardens. Comparison of old and recent photographs shows that there has been almost no change in the distributional pattern of villages and paddy fields. The reason for this is obvious: the paddy fields are too precious to be used for extension of the villages. The houses are probably more closely spaced and each one accommodates more people than before, but this is not visible on the photographs. The relative permanence of the boundary between settlements/homestead gardens and paddy fields is particularly clear in the dissected terrace where the lower parts are irrigated and the slightly higher, non-irrigable areas are occupied by settlements.

The increasing population found room for expansion in the zone of low hills where in 1946 (light-toned) dry fields were situated and which were almost entirely covered by (darker-toned) settlements in 1973. The growing population's need for dry fields, land for grazing the animals and wood for purposes of cooking led to an almost complete deforestation of the mountainous part of the photograph; in 1948 this zone was still largely covered with scrub and forest and rather dark-toned on the photo, whereas in the photograph of 1973 it is light-toned, grass-covered and partially barren.

This development led to a drastic change of slope processes in the mountains. Formerly, creep resulted in a slow downslope movement of waste which over-steepened the ravines, dropped or slipped into the water due to lateral sapping and was ultimately removed from the area. These processes were slow, non-concentrated and represented a balanced environment. Deforestation reduced the the effect of creep which is now largely replaced by surface wash and by the formation of rills and gullies which cut into and gradually destroyed the grass cover. Much more waste is brought into the ravines and subsequently into the Sapi River. The processes have become predominantly linear and violent and are indicative for a deteriorating environment. Fig. 6.28 schematically depicts this change in morpho-dynamics which will result in a true morpho-genetic crisis if the slope processes in the mountains are not kept under control by measures such as terracing.

Figs. 6.29 and 6.30 depict an area where creep is still a dominant process, as demonstrated by the inclination of the trees in a rubber plantation and another area where linear erosion has started its destructive work.

Wind work is another complex geomorphological process easily initiated or reactivated by misuse of the land. It is a serious problem in many parts of the world, for example in the United States, where the ill-famed 'dust-bowl' of the 1930's marked the beginning of a country-wide soil conservation programme. Fine-textured, fertile soil particles are blown away from the fields, strongly reducing productivity and sometimes even precluding further agricultural use. Sandy deposits may be turned into dune fields and existing dune fields may be reactivated, for example, by cutting scrub and bushes for domestic and other purposes. The problem is not restricted to arid and

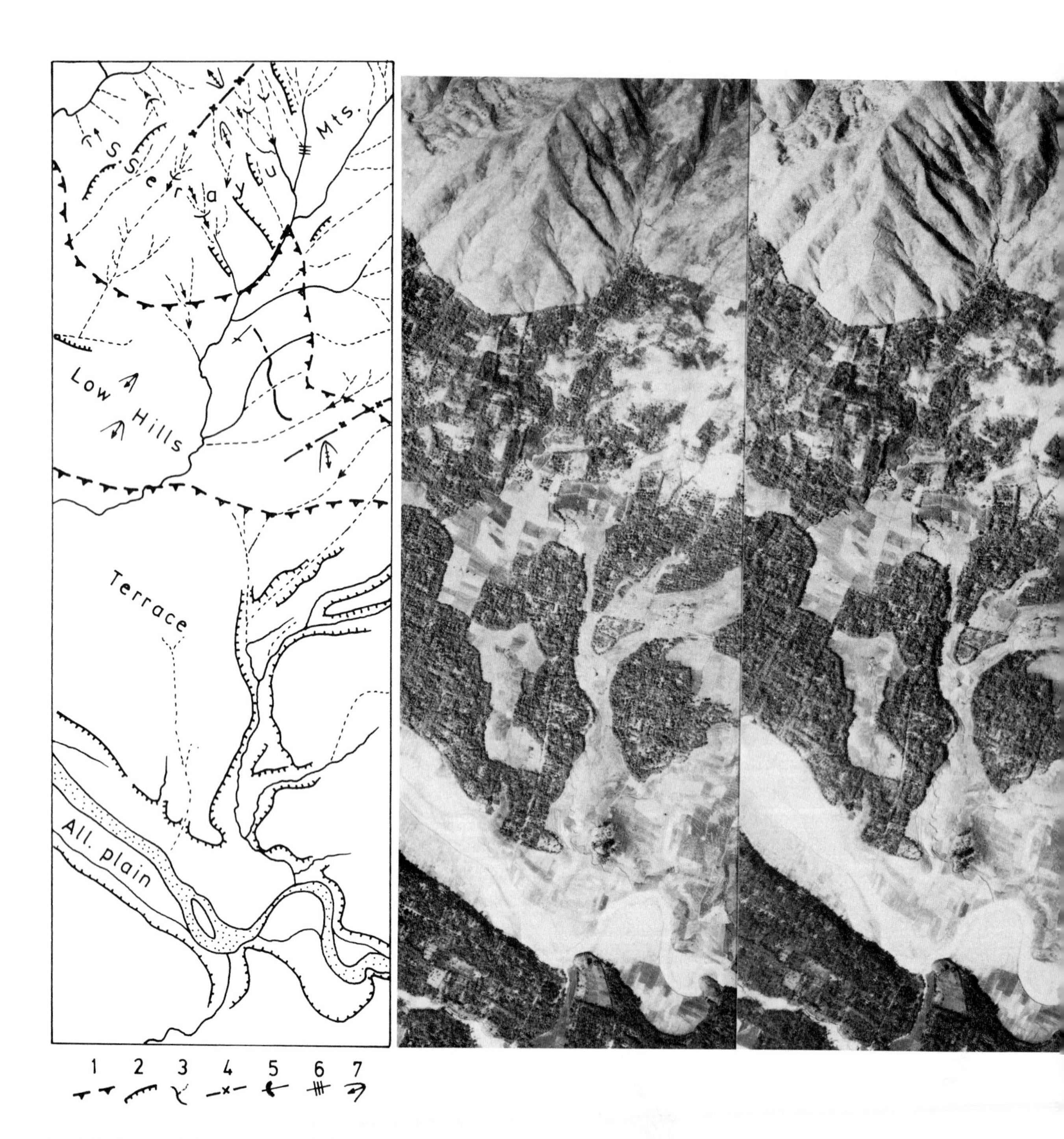

Fig. 6.26 Stereopair showing part of the Sapi River basin, Central Java, Indonesia, as it was in 1973; scale 1:20,000.
Key of interpretation (Fig. 6.26a):

1. limits between terraces, low hills and South Serayu Mountains
2. terrace edges
3. drainage lines
4. crest lines
5. active headward erosion
6. rapids
7. dip slopes

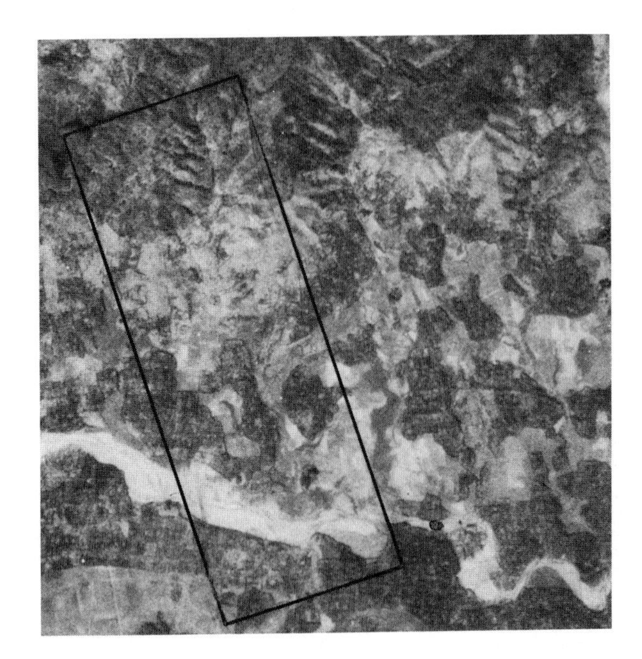

Fig. 6.27 (left) Vertical aerial view of part of the Sapi River basin dating from 1946, the area covered by the stereo pair of fig. 6.26 is indicated by the black frame.

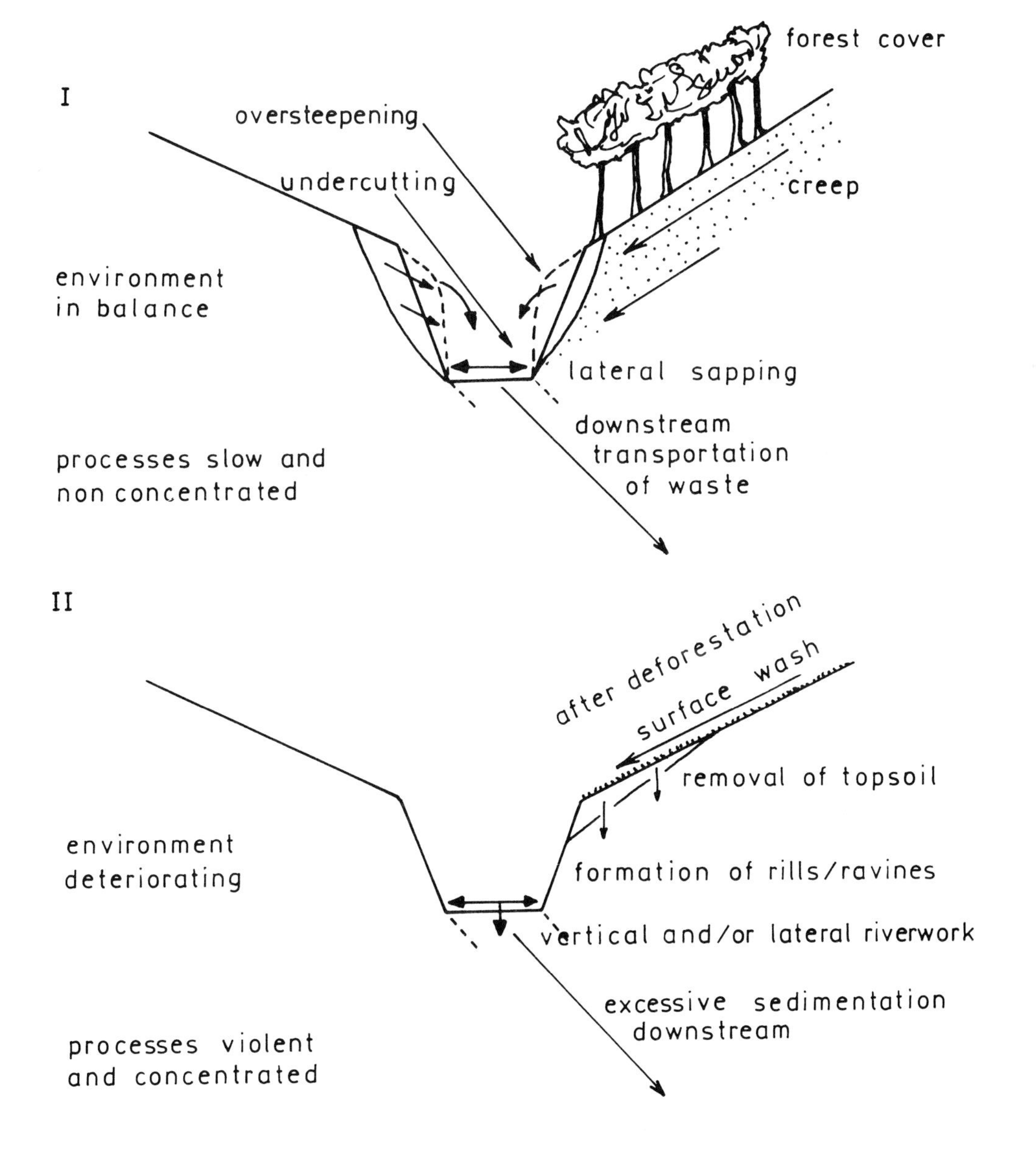

Fig. 6.28 (below) Morphodynamic changes in the deteriorating Sapi River basin: creep and other mild processes are being replaced by surface wash, linear cutting and other violent processes following deforestation.

Fig. 6.29 Down-slope inclined rubber trees in an area where creep is still dominant.

Fig. 6.30 Surface wash and ravines characterize this area where deforestation led to accelerated erosion.

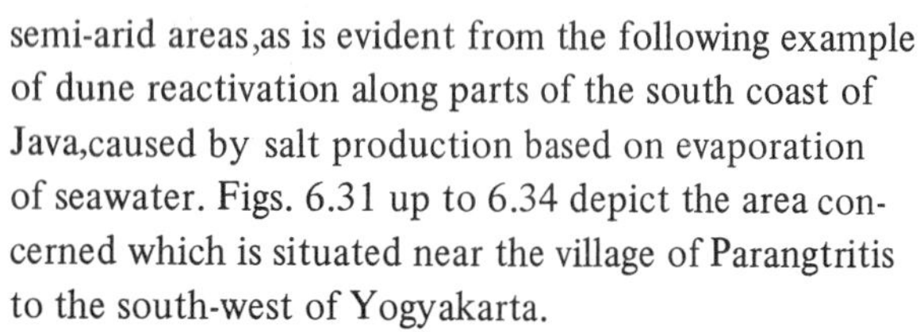

semi-arid areas,as is evident from the following example of dune reactivation along parts of the south coast of Java,caused by salt production based on evaporation of seawater. Figs. 6.31 up to 6.34 depict the area concerned which is situated near the village of Parangtritis to the south-west of Yogyakarta.

The dune belt is generally about 2 km wide and the maximum height of the dunes is roughly 15 metres. The dunes are formed on a number of fairly low beach ridges which are separated by swales or furrows. This alternation of ridges and swales can often still be found on the older, inland parts. Along the beach a foredune developed where active aeolian processes occur. Large quantities of sand are blown inland by the sea winds during the dry east monsoons. Constructional hillocks or tongues of all sizes are formed behind every clump of *Spinifex littoreus* the most important sand binder. Immediately landward of *Spinifex,* the following vegetation is found: *Calotropis gigantea R.Br., Ipomoia pes-Caprae* and *Pandanus* scrub. A rejuvenation ridge of fresh sands can be observed at the leeward slope of the foredune indicating active dune development. On the other hand gaps in the foredune are being constantly scoured by the wind, continuously destroying even the smallest vegetation. As a result of frequent wind scouring a broad belt of parabolic dunes formed behind the foredune. Barchans replace them where vegetation completely fails (Verstappen, 1957).

It is obvious that the dunes have started wandering and now move under the influence of the east monsoon winds with an azimuth of 325°. It is difficult to say when this dune reactivation started, but it is evident that since the compilation of the residency map of 1892, the process developed at an increasing pace. Increased wind work is due to the thorough destruction of the dune vegetation by the increasing population of the nearby villages. Even the thinnest twigs were used for firewood and Pandan leaves are commonly used as roofing. Since the abolishment of the salt monopoly, firewood is also used for accelerating the evaporation of seawater and increasing the salt production. Large areas are almost completely stripped of any vegetation. In some instances the villagers have constructed low hedges perpendicular to the east monsoon winds to reduce the drifting of sand. They are, however, insufficiently kept and thus not very effective. Abandoned, sand-covered fields can be traced everywhere. According to the 1892 residency map, several villages situated in the dune area have disappeared or were moved inland. A few cemeteries can still be traced and indicate the site of a former village. Evidently, the dunes were considerably better anchored at the turn of the century than they are today. Fig. 6.31 illustrates the damage done to an irrigation canal due to the invasion of dunes. It is based on the topographic maps of 1919, 1933 and on an aerial photography of 1946.

Fig. 6.32 is a panchromatic aerial photograph at the scale of 1:52,000 and depicts the coastal dune area near Parangtritis as it was in 1946. The foredune, wandering, parabolic dunes and barchans can be clearly seen and are evidence of active dune development which gradually penetrated into the agricultural fields situated landward. Fig. 6.33 depicts the same area in black and white infrared, at the same scale and in 1970. The landward penetration of the dunes is striking, as is the loss of valuable land. Fig. 6.34 is based on the residency map of 1892 and the aerial photographs of 1946 and 1970. It summarizes the landward growth of the dune belt and shows the seaward

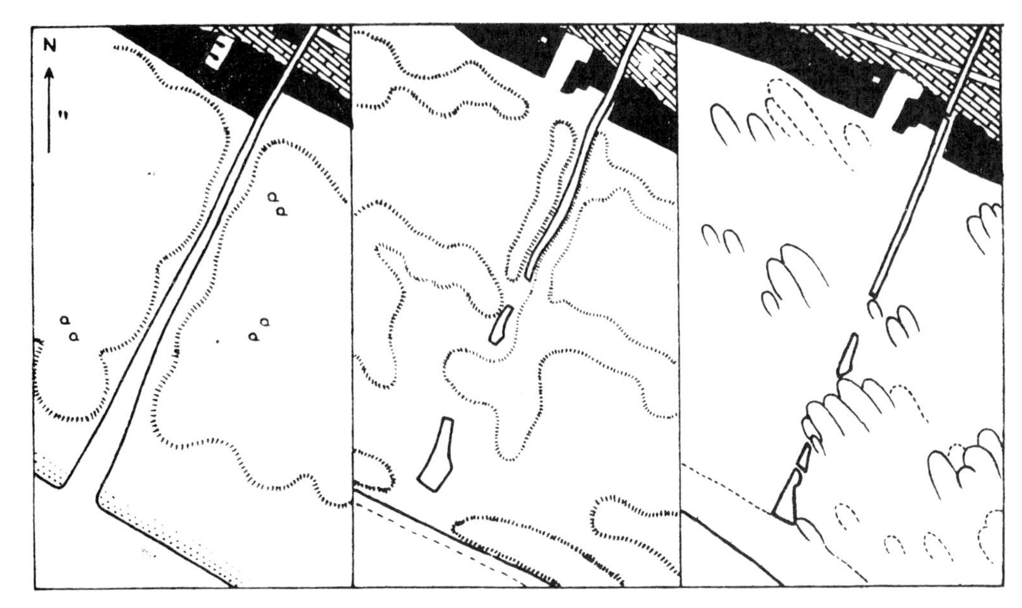

Fig. 6.31 The irrigation canal K.Cent as it appears on the 1919 topographic map (left), on the 1933 map (centre) and on an aerial photograph of 1946. The canal is filled up and obliterated by the restored fore dune and by wandering parabolic dunes.

Fig. 6.32 Panchromatic aerial photograph of the dunes near Parangtritis, Java, between the Opak River in the west and the Sewu Mountains in the east dating from 1946. Scale 1:52,000.
Key:
1. Foredune
2. Parabolic dunes
3. Barchanes
4. Swales
5. Fields
6. Village
7. Paddy fields

Fig. 6.33 Infrared aerial photograph of the same area as 6.32 and at approximately the same scale, dating from 1970. Reactivation of the dunes is evident. A large part of the older dunes (5 of fig. 6.32) have been covered by new dunes. Some seaward growth of the coast also can be observed.

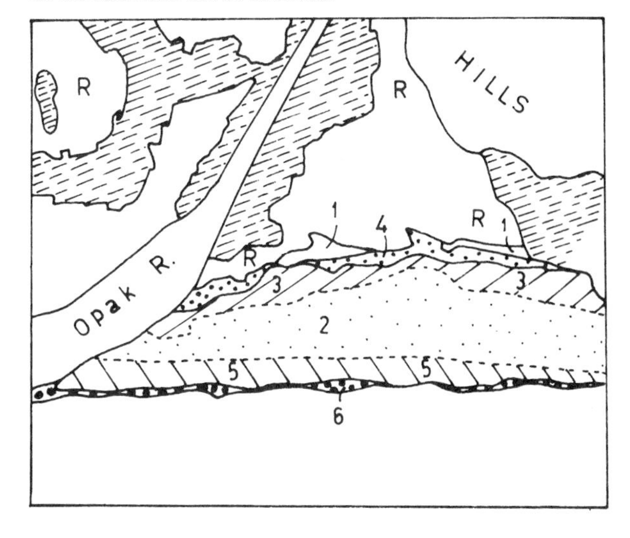

Fig. 6.34 Landward displacement of the dunes and seaward growth of the coast near Parangtritis in the periods 1892-1946 and 1946-1970. Sources: topographic map of 1892 and the aerial photographs of figs. 6.32 and 6.33.
1. dune area active before 1892
2. active dune area in 1892
3. landward dune displacement 1892-1970
4. landward dune displacement 1946-1970
5. coastal accretion 1892-1946
6. coastal accretion 1946-1970

displacement of the coastline,which contributes to the increased width of the dune belt. Obviously, the damage by the increased aeolian activity is considerable in the area and also diversified. Restoration of the vegetation or rather reafforestation of the dunes,seems to be the only satisfactory solution to prevent further deterioration of the environment,which was initiated by salt producers who destroyed the natural vegetation of the dunes, thus triggering their reactivation. Sequential maps and aerial photographs have recorded the changes of the past and signal the processes that are actively progressing at the present. It is up to the geomorphologist to properly interprete the data and to contribute to projects of dune stabilization which are to be effectuated in the future.

6.6. Man, Protecting or Improving the Rural Environment

Deliberate efforts to improve the land are as diversified as the modes of deterioration described above. Rivers may be harnessed by building dikes or dams, coastal abrasion may be checked and irrigation introduced, although the various side-effects of such measures are not always easily assessed. Dikes tend to result in a gradual rise of the river above the adjacent plain. Reservoir lakes change the river regimen and the groundwater level. They may cause erosion, sedimentation, landslides and even earthquakes. Irrigation may lead to salination if the drainage is poor and evaporation high. Harnessing a river and reducing its seasonal fluctuations in discharge for purposes of irrigation or navigation will adversely affect the flood-retreat cultivation system mentioned in sub-section 6.5. Artificial outlets of rivers will cause abrasion near the old outfall and the construction of dams will have comparable effects. The aerial photographs of figs. 6.35 and 6.36 show the accretion at the mouth of an artificial, new outfall of the Cidurian River in West Java and the abrasion of the coast near the old river mouth. Thorough studies have to be carried out before any land development measures are applied to avoid unexpected adverse side effects.

Successful attempts to improve the environmental situation are not new. The author studied an artificial river diversion in Java, commemorated by a Sanskrit inscription of 1500 years ago (Verstappen and Noorduyn, 1972). Classic examples are the irrigation works in ancient Egypt and in Mesopotomia. Jacobsen and Adams (1958) studied the development of ancient irrigation in part of what is presently Iraq. They found that the problem of salination was well known in Mesopotomia although it could not be combated with the modern means of deep drainage to lower the water table and the utilization of chemicals to restore the texture of the soils. In Mesopotomia an attempt to keep salination under control was made by avoiding over-irrigation and alternating productive years with weed fallow. Deep-rooted plants consumed the soil moisture at a certain depth and created dry zones in the soil profile which prevented the capillary rise of saline groundwater. Nevertheless, three distinct periods of salination occurred between 2400 BC and 1200 BC. The earlier one was the most serious and is thought to have resulted from the construction of an irrigation canal bringing water from the Tigris to areas which previously received much smaller quantities from the Euphrates. Over-irrigation, flooding and seepage thereafter introduced salination.

Siltation was also a problem in the irrigation of Mesopotomia. In the beginning, irrigation canals were short as only the areas near the river were used. Siltation was a minor problem but with the increasing population, the irrigation canals had to be extended over almost flat terrain. They silted up considerably and gradually formed ridges in the plains. Maintenance was increasingly required and negligence, for example, in times of turbulence, ultimately disrupted the irrigation system. Ancient irrigation schemes are also known from other parts of the world, such as the dry belt of Sri Lanka, but the environmental and other causes behind their growth and decline have not yet been firmly established.

Ancient agriculture is known to have existed in the Negev Desert and the ingenious modes of the early farmer's use of water in this harsh, arid environment was unraveled by Kedar (1957) and Evenari et al. (1951, 1961, 1963). Three different techniques were applied: the use of water collected on the slopes immediately adjacent to the fields, the use of water collected by short irrigation canals constructed in larger watersheds (up to 10,000 ha), and the use of groundwater by way of chain wells, used in many semi-arid parts of the world and mentioned in sub-section 6.3. The first method of 'run-off farming' is based on the sealing of the moist soils which increases surface run-off. To optimalise this sealing process, the surface had to be cleared of rock fragments, pebbles, etc. Evidence of this are regularly spaced gravel mounds. The ratio between catchment area and field acreage was found to range from 1:17 to 1:30. The fields situated in the valley bottom were surrounded by low bunds to collect the run-off water and received an estimated 150-200 cbm/ha water annually.

The oldest run-off farms found date from the 10th - 8th century BC (Evenari et al., 1959) and the system probably reached its peak in the Roman - Byzantine era. Run-off farming is still practised in some parts of the world, such as in the shallow, flat-bottomed vales dissecting the Quaternary glacis in the Ebro Basin, Spain. Here run-off water from the slopes is an essential addition to the annual rainfall and assures a reasonable wheat crop, particularly in dry years. The stereotriplet of fig. 6.37 pictures the situation. The annual precipitation in this area is such that

Fig. 6.35 Vertical aerial view of the artificial outfall of the Cidurian River, West Java, Indonesia. About 2 1/2 km accretion has occurred since it was brought into use. The new land is mainly under fish ponds. Scale: 1:12,500.

a dense scrub vegetation occurs on the slopes,which proportionally reduces the surface run-off. In wet years care should be taken that not too much water collects on the fields,which could damage crops.

The author observed that ravines occasionally develop at the foot of the hills, preventing the run-off water from reaching the adjacent fields, thereby providing fields,situated on the valley bottom farther downstream,with water. If the catchment area becomes too large, ephemeral, concentrated run-off will establish itself in the vales and gradually disrupt the run-off farming system.

The second technique of using water in the ancient Negev, mentioned by Evenari et al. (1959, 1961, 1968): the use of run-off from larger watersheds by deviating the river water into short irrigation canals, requires a higher technological skill and in particular the ability to

Fig. 6.36 Abrasion of the coast approximately 2 km to the East of the old mouth of the Cidurian River, West Java, Indonesia, following the artificial displacement of the outfall (Fig. 6.35). The mangrove belt is disappearing and many bunds bordering the fish ponds are destroyed and replaced by new ones situated farther inland. An ancient creek system is clearly visible. Vertical aerial view at a scale of 1:12,500.

construct intakes strong enough to withstand occasional flash floods. It should be understood that,whereas a rainfall of 3-6 mm will suffice for run-off farming, water flows in the channels of the Negev only after 10-15 mm precipitation. Surface run-off on the slopes may be 20-40% of the annual rainfall,but is in the order of 3-6% in the larger watersheds. The water diverted from the wadi by rising its water level slightly at the intake,was distributed over fairly large fields. Fig. 6.39 gives an example. Silting up of the channels and incision of the wadis forced the farmers to modify their irrigation system several times in the course of the centuries. Also from ancient Egypt remarkable achievements in this field are known (Murray, 1955). Traditional means of water harvesting in dry lands have received considerable scientific interest in recent years (Aschenbach, 1973; Kirkby, 1969; Boers and Ben-

Fig. 6.37 Stereotriplet showing the Ginel River basin and surroundings, draining towards the Ebro River in Northern Spain, at a scale of 1:33,000. The larger part of the area depicted is an intensely dissected but at present largely inactive badland area. The ravines are filled-in and have flat valley bottoms. The latter are provided with numerous low transverse bunds which serve to retain the scanty rainwater falling on the fields and the adjacent slopes in the summer season. Where excess water occurs it is drained off by usually dry ravines situated at the left and/or right margin of the valley bottom at the foot of the hills. Irrigated fields occur in the upper part of the photograph and derive their water from the permanent spring of Mediana (top centre).

Fig. 6.38 An international group of trainees being instructed on the construction of micro dams for run-off water catchment in desert areas. Courtesy: Prof. M. Evenary, Hebrew University.

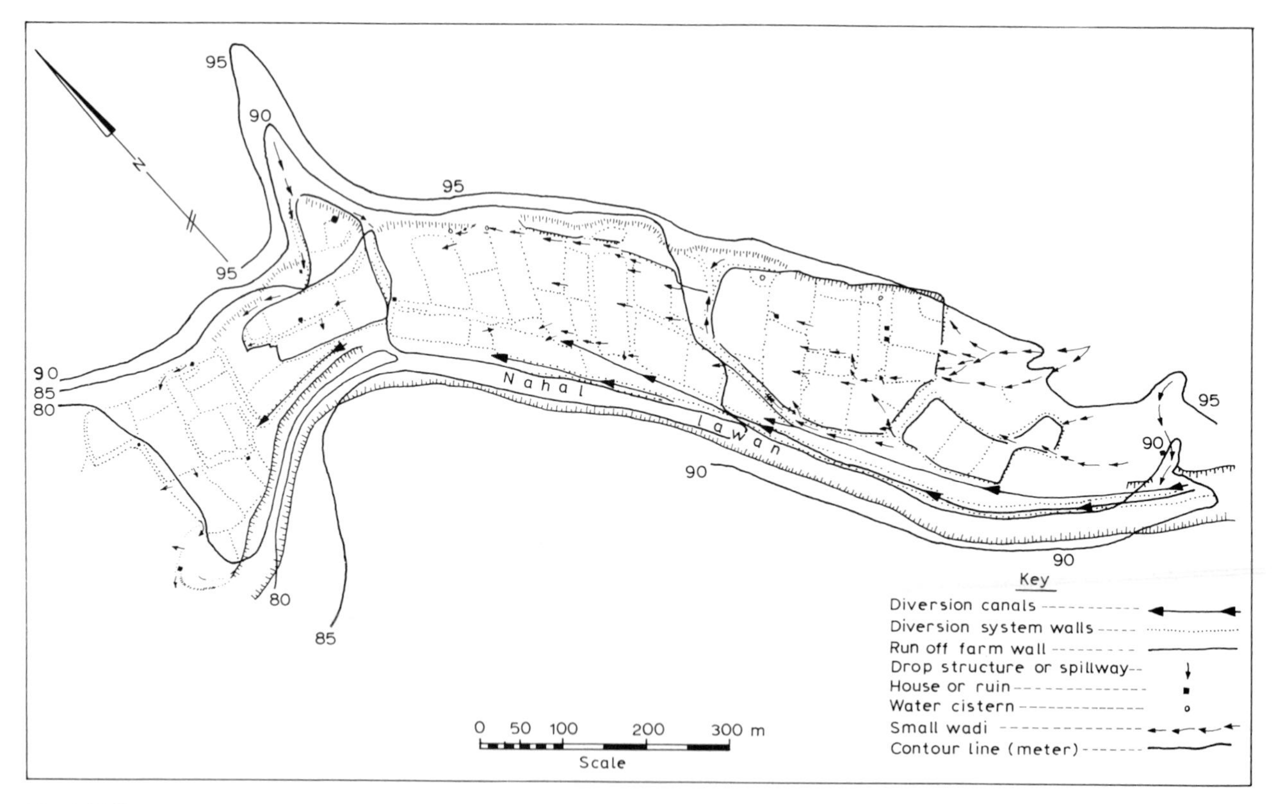

Fig. 6.39 Example of an ancient desert-irrigation scheme based on deviated flood waters from the Nahal Lavan wadi, Negev Desert (Evenari et al., 1961).

Asher, 1979; Hillel, 1967; FAO, 1967, 1976; Frasier et al., 1974; Shanan, 1970; Solignac, 1952/53; Tadmor, 1971; Troll, 1963). Revival of the technology is aimed at as shown in fig. 6.38.

Land improvement can sometimes be achieved by simple means, for example, in the area in Central Java pictured in fig. 6.40,where the groundwater level was too high for non-irrigated crops and the topographic situation too high for irrigation: simple, manual earthwork altered the area into a highly productive zone with irrigated paddy fields in the excavated strips and (non-irrigated) vegetables in the adjacent higher strips where the excavated material was deposited.

However, more sophisticated technology may offer solutions for improvement where simple means,carried out by the local rural population itself, are inadequate (Tricart, 1960). An example is given below of the extension of irrigated acreage around large gravel fans near Quetta, Pakistan. These gravel fans are situated where seasonal river channels enter the plain. They are characterized by a system of shallow radial channels which only drain off part of the occasional flood waters. The remainder of the flow drains off as a sheet of water covering the whole surface of the fan (Verstappen, 1966, 1969).

The surface water is then collected on the non-irrigated sailaba fields where wheat is grown. This is effectuated by the construction of low, horse-shoe shaped earthen bunds around the fields,which have their open ends at the upslope side to let the water in. The sailaba fields are concentrated in the lower parts of the gravel fans where silt loam occurs. The upper parts of the fans are gravelly and cannot be used for agricultural purposes. It is fortunate that the sheetflood is not ideal and that some concentration of flow in the shallow channels always remains. This enables the farmers to establish other sailaba fields downslope of the first ones. The bunds of these partly enter the channels and therefore, derive floodwater from them sideways. The map of fig. 6.41 pictures the situation.

Downslope of the interfan depressions, the radiating channels of two adjacent fans meet,causing concentration of the surface flow. This water can be used for other fields farther downslope,though flash floods may have damaging effects. The convexity of the fans causes a dispersion of the water over larger areas in other parts and less water is available here than downslope of the interfan depressions.

The groundwater of the fans is tapped by subhorizontal tunnels called chainwells and locally known as karezes. These tunnels are dug into the gravels and are provided with vertical shafts used for hoisting the material excavated during construction and also serving for the maintenance of the tunnels. The lower end of the karezes is open canals and from these the water is diverted and used on orchards, vegetable gardens, etc. Fields situated at a greater distance from the lower end of the karezes can only be provided with water after the rains, when the flow is at its peak. These fields are only suitable for growing wheat (Kirkby, 1969; Wadji, 1972; Rahman, 1981; Scholz, 1970).

Although this traditional use of the surface and the groundwater of the fans is quite ingenious, it should be understood that irrigation using chainwells is a wasteful use of groundwater: the flow cannot be regulated, it flows whether it is required or not. There is considerable seepage in the tunnels which increases with length and the unused groundwater often emerges farther downslope where it may cause either gullying or, if the water evaporates, salinity. The inefficiency of chainwells is exemplified in the area near Quetta where the largest fans have a lower ratio irrigated acreage/surface area of catchment (fan) than the medium and small fans. In other words, relative to their size, they irrigate less fields due to water losses.

Certain areas near the larger fans were indicated, where extension of irrigation by tube wells is feasible. The results of this study were confirmed by geophysical observations.

The use of the surface water can be improved if the terrain configuration permits. In fact,agriculture based on unreliable sheet flood is a rather risky matter. Famine may result in case of drought. The stereotriplet of fig. 6.42 pictures a fan in Arizona, USA, at the scale of 1:20,000 where the irrigation is effectuated by a sub-horizontal canal deriving its water from a small storage lake nearby, thereby decreasing the risk of water failure. The water which occasionally floods the fan is concentrated in channels and drains to the river which can be seen in the lower part of the photograph. This water may be collected in a storage lake and used for irrigation. In areas where pumping can be economically applied, it may even flow back through the irrigation canal and contribute to the irrigation of the fields situated on the fan.

In some instances only advanced technology can bring about urgently needed land improvements or harness a natural hazard, such as the volcanic hazard around the Kelut Volcano, East Java, Indonesia, which was ill-famed for its volcanic mudflows produced by the emptying of its crater lake (see fig. 6.43). In the past approximately 40.10^6 cbm lake water was washed downslope during every eruption. As a result the slopes are now almost completely deprived of ashes and other debris. All fragmentary material was transported in the form of mud flows and the lava of the volcanic body became exposed. The morphology of the Kelut is thus characterized by its strongly eroded slopes, rugged topography and jagged crater rim. The mudflows produced during the 1919 Kelut eruption

Fig. 6.40 Earthwork for agricultural purposes in an alluvial plain in Southern Java, Indonesia. The low strips are now suitable for irrigated rice cultivation and vegetables can be grown on the intervening raised strips.

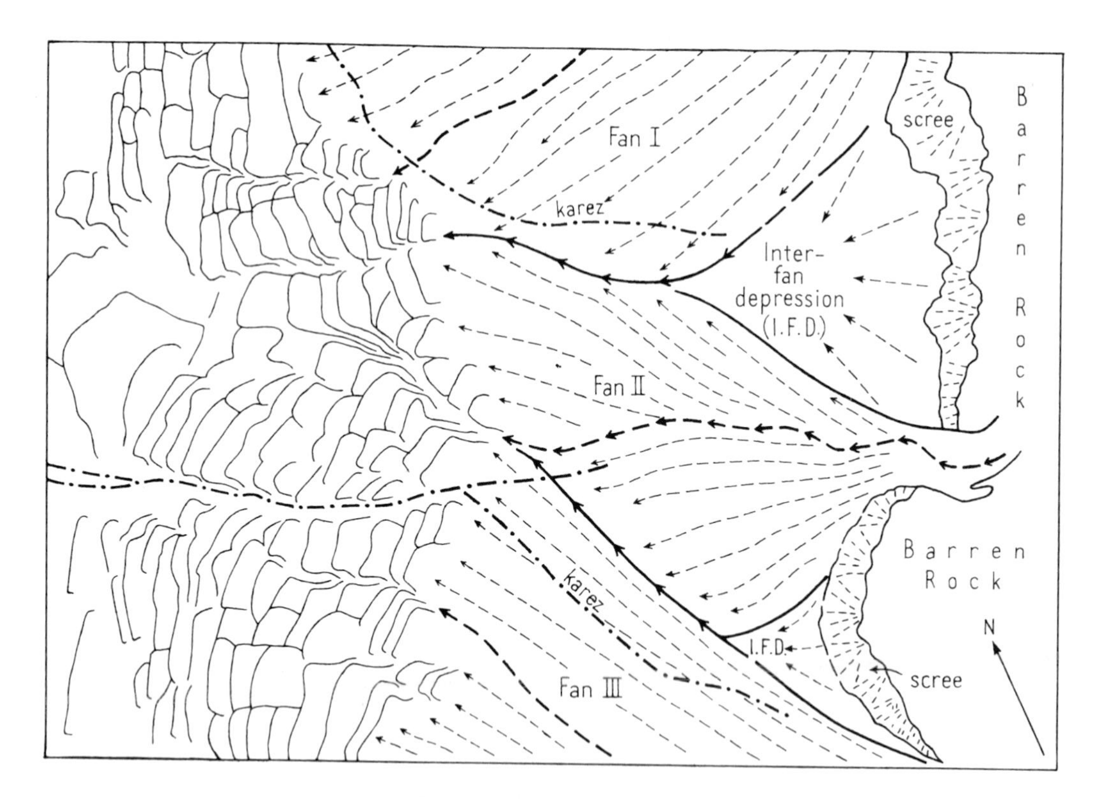

Fig. 6.41 Rain dependent wheat fields, locally known as 'sailaba' fields near Quetta, Pakistan. They are located on the lower, silty-loamy parts of gravel fans and they derive their water from surface wash and imperfect sheet flood. A system of low, horse-shoe shaped bunds is used for the purpose. Groundwater of the fans is tapped by several chainwells or 'karezes' and is used for irrigated vegetable and fruit gardens situated outside the left edge of the map.

Fig. 6.42 Extended stereopair showing a piedmont zone with fans in Arizona, USA, at a scale of 1:20,000. Irrigation here is based on a technologically more advanced system of sub-horizontal canals originating in a storage lake. The flood waters from the torrents pass underneath the canals.

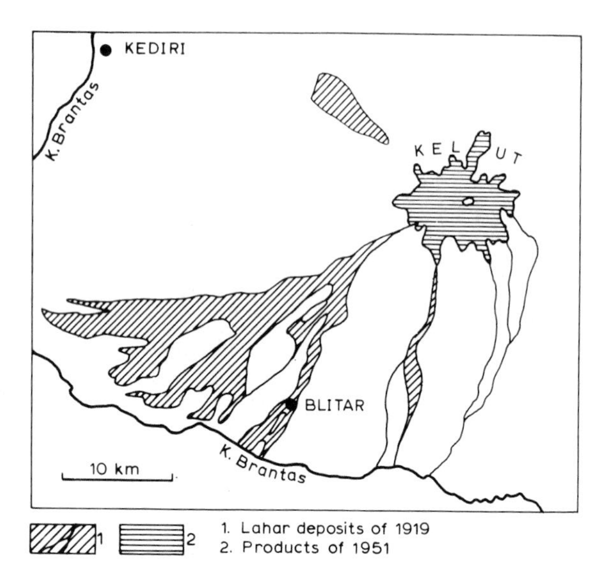

Fig. 6.43 The volcanic mudflows of the Kelut Volcano, East Java, produced in 1919 prior to the emptying of the crater lake by a system of tunnels and in 1951 when the tunnels had their protective role (IJzendoorn, 1952).

caused disaster in the area and particularly in the southwest section of the volcano where the town of Blitar is situated. About 30,000 people died as a result of the 35-40 km long lahars which covered 131 km^2 of agricultural land.

Soon after this tragic event the Kelut tunnel works were constructed, draining the lake almost entirely, leaving only 1.10^6 cbm of water. Fig. 6.44 gives a cross-section of the structure (IJzendoorn, 1952). The effectiveness of the structure was proved during the 1951 Kelut eruption: the volcanic products remained largely on the non-inhabited, higher parts of the slopes.

Fig. 6.43 demonstrates the difference between mudflow activity in 1919 and 1951 as a result of the artificial drainage of the lake by the tunnel works. The volcanic mud previously produced,caused serious problems of excessive sedimentation in the Brantas River which drains the area. This river, in fact, almost entirely circumferes the Kelut Volcano and has a very low gradient. Its transporting capacity is also low and it is now situated several metres above the surrounding rice fields in large tracts of its downstream course. Development projects have been launched in the Brantas River basin to improve the situation.

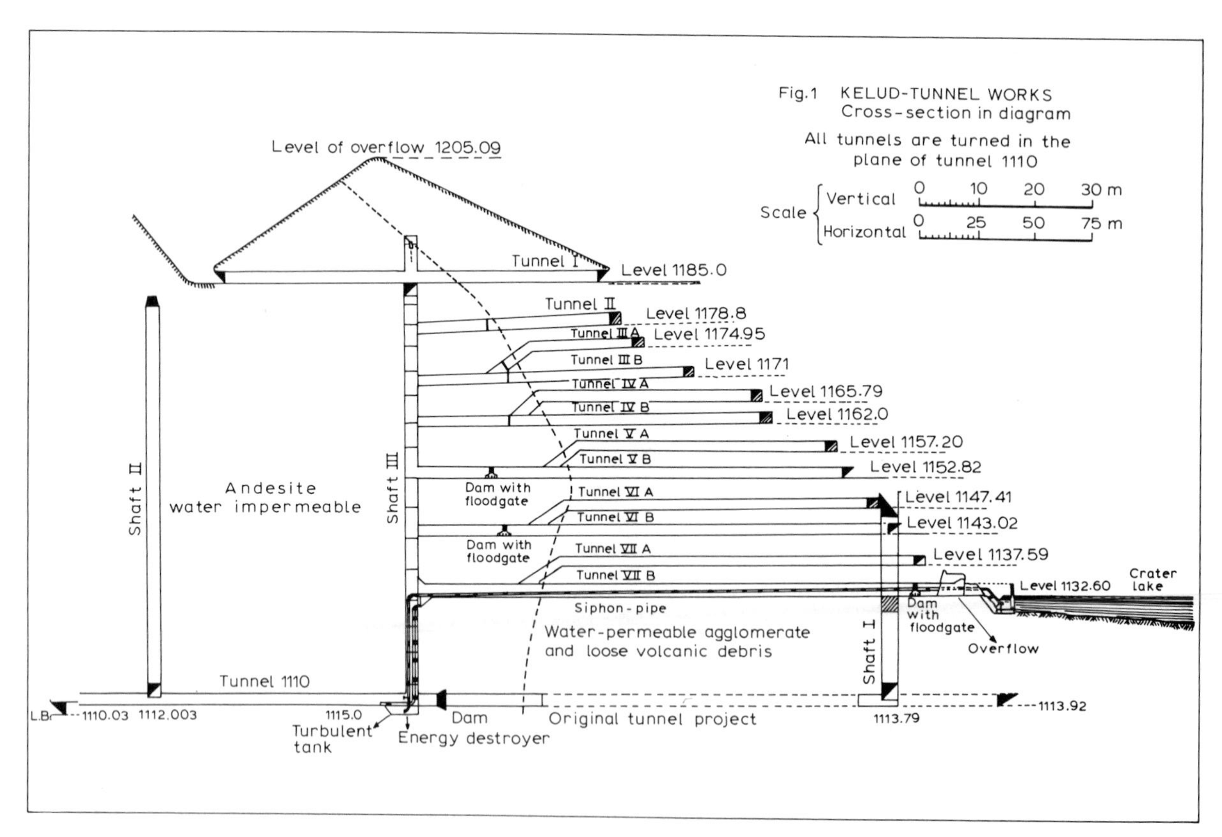

Fig. 6.44 A cross section of the Kelut tunnel works draining the crater lake, that used to cause catastrophic mudflows on the occasion of an eruption.

References

Anderson, J.U., et al., 1970. Land classification for irrigation, Roosevelt County. New Mexico Agric. Exp. Station, Univ. Park. Res. Report, 163: 1-13.

Andriesse, J.P., 1974. The characteristics, agricultural potential and reclamation problems of tropical lowlands peats in south-east Asia. Amsterdam, Royal Tropical Inst., 63 pp.

Aschenbach, M., 1973. Römische und gegenwärtige Formen der Wassernutzung im Sahara–Vorland des Aurès (Algerien), Die Erde, 104(2): 157-175.

Bemmelen, R.W. van, 1956. The influence of geologic events on human history. Verhand. KNGMG, Kon. Ned. Geol. Mijnbouwk. Gen., 16: 1-17.

Boers, Th.M., and Ben-Asher, J., 1979. Harvesting water in the desert. Annual Report I.L.R.I., Wageningen: 6-23.

Brown, C.E., and Sheu, M.S., 1975. Effects of deforestation of slopes. Geotech. Eng. Div., ASCE, 101(GT2), Proc. Paper 11141: 147-165.

Butzer, K.W., 1972. Environment and archaeology–an ecological approach to prehistory. Methuen, London, 524 pp.

Calder, R., 1951. Man against the desert. Allan and Unwin Ltd., 241 pp.

Cool, W., 1931. Het Merapi-gebeuren in Midden Java bij de jaarwisseling 1930-31. De Ingenieur, 46(37), A: 341-357.

Cotet, P., 1969. Prémisses géomorphologiques en marge des travaux pratiques dans la plaine Roumaine. Symp. Int. Géom. appliquée, Bucharest: 197-203.

Dylik, J., 1954. Geomorphological problems as related to agricultural needs (in polish). Przeglad Geograf., 26(4): 4-36.

Erdosi, F., 1969. Social effects in the development of the alluvial fans in the Pćs area (in hungarian). Földrajzi Ertesito, 17(3): 293-308.

Escher, B.G., 1931. Gloedwolken en lahars, vulkanische catastrophes in Nederlands Indië. Tropisch Nederland, 3(19/20): 291-320.

Evenari, M., et al., 1959. Ancient agriculture in the Negev. J. Israel Exploration, 8, 231 pp.

Evenari, M., et al., 1961. Ancient agriculture in the Negev. Science, 133 (3457): 304-321.

Evenari, M., et al., 1968. Run-off farming in the desert. Agronomy J., 60: 29-32.

FAO, 1967. U.N.D.P. pilot project in watershed management of the Nahal Shikma, Israel. FAO, Land and Water Use, 3, 281 pp.

FAO, 1976. Medina conservation in arid and semi-arid zones. Conservation Guide FAO, 3: 61-72.

Frasier, G.W. (Editor), 1975. Proceedings of the water harvesting symposium, Phoenix, Arizona, March 26-28. USDA Agric. Res. Service, Western region, 22: 1-329.

Galon, R., 1954. An experimental interpretation of the geomorphological map of Bydgoscz from the point of view of the regionalization of agricultural production. Przeglad Geograf., 26(4): 48-53.

Ghose, B., 1962. Salinity and salt basins in arid lands. Their origin and distribution, problem in irrigation–a geomorphic approach. Arid Zone Newsletter: 1-14.

Gil, E., and Slupik, J., 1972. The influence of the plant cover and land-use on the surface run-off and wash down during heavy rain. Studia Geom. Carpatho-Balcanica, 6: 181-190.

Gilberton, D.D., 1981. The impact of past and present land use on a major barrier system. Appl. Geogr., 1(2): 97-120.

Goudie, A., 1981. The human impact. Men's role in environmental change. Basil Blackwell, Oxford, 316 pp.

Hempel, L., 1971. Die Tendenzen anthropogen bedingter Reliefformung in den Ackerlandereien Europas. Ztschr.f.Geom., 15(3): 312-329.

Hillel, D., 1967. Run-off inducement in arid lands. Techn. Report USDA. Volcani. Inst. Agric. Res. and Hebrew Univ., Jerusalem, Rehovot, Israel, 142 pp.

IJzendoorn, M.J. van, 1953. The eruption of Gunung Kelud on August 31, 1951, proved the utility of the Kelud tunnel works. Berita Gunung Merapi, 3/4: 1-5.

Jacobsen, T., and Adams, R.M., 1958. Salt and silt in ancient Mesopotamian agriculture. Science, 128(3334): 138-145.

Kashtanov, A., 1980. Zum bodenschutzenden Ackerbau in der U.d.S.S.R. Int. Ztchr.d.Landwirtschaft, 3: 240-242.

Kedar, Y., 1957. Water and soil from the desert: some ancient agricultural achievements in the Central Negev. Geogr. J., 123(2): 323-333.

Kent, B.H., et al., 1969. Geology and land-use in eastern Washington County, Pennsylvania. Pennsylvania Geol. Surv. Bull. G 56, 31 pp.

Kirkby, A.V., 1969. Primitive irrigation. In: R.J. Chorley (Editor), Water, Earth and Man (chapter 4, III), Methuen & Co. Ltd., London,

Knight, M.M., 1928. Water and the course of empire in North Africa. Quart. J. Econ., 43: 44-93.

Kugler, H., 1976. Geomorphologische Erkundung und agrarische Landnutzung. Geogr. Berichte, 21(3): 190-204.

Lange, K.O., and Pshenin, G.N., 1969. Anthropogenic diluvial pediments of foothill loess plains of the Fergana and circum-Tashkent region (in russian). Izvestiya Vsesoyuznogo geogr. Obshch., 101(4): 354-358.

Leser, H., 1975/76. Anthropogene Beeinflussung des Faktors Boden in Ökosystemen der westlichen Kalahari (SW Afrika). J. S.W.A. Wiss. Gesellschaft, 30: 25-38.

Lidov, V.P., 1954. Geomorphological research for the needs of agriculture (in russian). Geogr. Studies, Geomorphology, Moscow: 30-39.

Lisitsina, G.N., 1976. Arid soils, the source of archaeological information. J. Archaeol. Sci., 3(1): 55-60.

Lowdermilk, W.C., 1942. Lessons from the old world to the americas in land-use. Proc. 8th Amer. Sci. Congr., (5): 147-161.

Martiniuc, C., 1967. Recherches géomorphologiques pour les projets des canaux d'irrigation. Int. Symp. Appl. Géom., Bucharest: 191-195.

Mateescu, F., 1967. Conditions géomorphologiques de l'utilisation des versants du côte danubien de la Dobroga. Int. Symp. Appl. Géom., Bucharest.

Meijerink, A.M.J., 1974. Photo hydrological reconnaissance surveys. Thesis Free Ref. Univ., Amsterdam, 372 pp.

Menon, M.M., 1969. Alluvial morphology of the lower Indus Plain and its relation to land-use. Pakistan Geogr. Rev., 24(1): 1-34.

Michel, P.,1978. La vallée alluviale du Sénégal. Catena, 5(2): 213-225.

Morgan, R.P.C., 1974. Soil erosion topics in applied geography. Longman Green Ltd., London, 113 pp.

Murray, G.W., 1955. Water from the desert: some ancient Egyptian achievements. Geogr. J., 121: 171-181.

Nadji, M., 1972. Kanale in der Ebene von Kashan, Iran. Erdkunde, 103(3/4): 209-215.

Neumann von Padang, M., 1933. De uitbarsting van de Merapi in de jaren 1930-1931. Ned. Ind. Vulk. Med., 12, 116 pp.

Niculescu, P., 1967. L'influence des processus géomorphologiques actuels sur l'utilisation des terrains dans les Carpates Méridionales. Symp. Géom. appl., Bucharest.

Pecsi, M., 1968. A Duna–ártéri szinterek kialakulása és fontosabb agráföldrajzi vonatkozásai. (Evolution of the floodplain levels of the Danube and their principal bearings on the geography of agriculture). Földrajzi Közlemények, 16(3): 215-222.

Pécsi, M., and Stärkel, L., 1978. Human impact on the physico-geographical processes. Proc. 2nd. Polish-Hungarian Seminar, Budapest 1975. Geogr. Polonica, 41: 1-87.

Pewe, T.L., 1954. Effect of permafrost on cultivated fields, Alaska. U.S. Geological Survey Bull., 989, F: 315-351.

Rahman, M., 1981. Ecology of Karez irrigation: a case of Pakistan. Geojournal, 5(1): 7-15.

Ranere, A.J., 1970. Prehistoric environments and cultural continuity in the western Great Basin. Tebiwa, 13(2): 53-73.

Sandy, I.M., Penggunaan tanah (land-use) di Indonesia. Publ. Dir. Tata Guna Tanah, Jakarta, 115 pp.

Scholz, F., 1970. Beobachtungen über künstliche Bewässerung und Nomadismus in Belutschistan. Beiheft, Geogr. Ztschr., 26: 55-79.

Scott, A.J., 1975. Soil profile changes resulting from the conversion of forest to grassland in the Montana of Peru. Great Plains - Rocky Mountain Geogr. J., 4: 124-130.

Shanan, L., et al., 1970. Run-off farming. Agronomy J., 62: 445-447.

Solignac, M., 1952/1953. Recherches sur les installations hydrauliques de Kairouan et des steppes tunisiennes du VIIe au XIe siècle. Ann. Inst. Etudes Orientales, Fac. Lettres Univ. d'Algiers, 10, 5-273; 11: 60-170.

Stärkel, L., 1969. L'influence du relief sur l'utilisation des sols sur les plateaux de Flysch. Int. Symp. Appl. Geom., Bucharest: 85-90.

Tadmor, J., 1971. Run-off farming. Agronomy J., 63: 91-95.

Taverne, N.J.M., 1926. Vulkaanstudiën op Java. Ned. Ind. Vulk. Meded., 7: 65-71.

Thal Larsen, J.H., 1926. Bestrijding van de zandbezwaren aan de voet van werkzame vulkanen op Java. De Ingenieur, 41(51): 1009-1023.

Thomas, M.F., and Whittington, G.W., 1967. Environment and land-use in Africa. Methuen & Co. Ltd., London: 103-145.

Tricart, J., 1956. Dégradation du milieu naturel et problèmes d'aménagement en Fouta-Djalon (Guinée). Revue Géogr. Alpine, 44: 7-36.

Tricart, J., 1960. Etude géomorphologique du projet d'aménagement du Lac Faguibine (Rép. du Mali). African Soils, 5(3): 207-289.

Tricart, J., 1962. L'epiderme de la terre, équisse d'une géomorphologie appliquée. Masson et Cie., Paris, 160 pp.

Tricart, J., 1966. Géomorphologie et aménagement rural (example du Venezuela). Cooperation Technique, 44-5: 1966-81.

Tricart, J. 1967. Rapport entre le milieu physique et la géographie agraire (cours inférieur du rio La Villa, SE de Chitré, Panama). Photo Interprétation, 67-3(2): 8-14.

Troll, C., 1963. Quanatbewässerung in der alten und neuen Welt. Mitt. Oesterr. Geogr. Ges., 105: 313-350.

Verger, F., 1959. Colmatage et endiguements sur les rivages de la baie de Bourg-neuf. Bull. C.O.E.C., 9: 179-188.

Verstappen, H.Th., 1955. Geomorphic notes on Kerintji, Central Sumatra, Indonesia. J.Nat. Sci., 111: 166-177.

Verstappen, H.Th., 1956. The physiographic basis of pioneer settlement in southern Sumatra. Publ. Geogr. Inst., Djakarta, 8, 15 pp.

Verstappen, H.Th., 1957. Short note on the dunes near Parangtritis (Java). Tijdschr. Kon. Ned. Aardrijksk. Gen., 74: 441-449.

Verstappen, H.Th., 1960. The role of aerial survey in applied geomorphology. Rev. Géom. Dyn., 10: 156-162.

Verstappen, H.Th., 1963. Geomorphological observations on Indonesian volcanoes. Tijdschr. Kon. Ned. Aardrijksk. Gen., 80: 237-251.

Verstappen, H.Th., 1964. Geomorphology of the Star Mountains (Central New Guinea). Sci. Results Neth. New Guinea Expedition 1959. Nova Guinea, Geol. Ser., 4/5:101-158.

Verstappen, H.Th., 1964. Geomorphology in delta studies. ITC Publ., B.24: 1-24.

Verstappen, H.Th., 1966. Landforms, water and land-use west of the Indus Plain. Nature and Resources (UNESCO Bull.), 3: 6-8.

Verstappen, H.Th., 1969. Aerial survey and rural development in South Asia. Jahrbuch Südasien Institut, Univ. Heidelberg 1968-69, 3: 1-6.

Verstappen, H.Th., et al., 1969. Landforms and resources in Central Rajasthan (India). ITC Publ., B.51, 20 pp.

Verstappen, H.Th., and Noorduyn, J., 1972. Purnavarman's river-works near Tugu. Bijdr., Taal-, Land- en Volkenkunde, 128(2/3): 1-10.

Verstappen, H.Th., 1977. Remote sensing in geomorphology. Elsevier's Publ. Comp., Amsterdam ,214 pp.

Vink, A.P.A., 1975. Landuse in advancing agriculture. Springer Verlag, Berlin, Heidelberg, New York, 394 pp.

Vink, A.P.A.,1980. Landschapsecologie en landgebruik. Bohn, Scheltema en Holkema, Utrecht, 151 pp.

Vink, A.P.A., (in print). The suitability of the soils in the Netherlands for arable land and grassland.

Vink, A.P.A., (in print). The role of physical geography in integrated surveys of developing countries.

Vogt, J., 1953. Erosion des sols et techniques de culture en climat tempéré maritime de transition (France et Allemagne). Rev. Géom. Dyn., 3: 157-183.

Wilgat, T., 1972. Mapa geologiczno-hydrograficzna dorzecza górnej Rio Aconcagua. Przeglad Geografizny, 44(4): 635-648.

Young, A., 1968. Natural resource surveys for land development in the tropics. Geography, 53(3): 229-248.

Young, A., Mapping Africa's natural resources. Geogr. J., 134 (2): 236-241.

Young, A., 1973. Rural land evaluation. In: J. A. Dawson and J.C. Doornkamp (Editors), Evaluating the human environment, essays in applied geography. Arnold Ltd., London: 5-33.

Zapletal, L., 1973. Indirect anthropogenic geomorphological processes and their influence on the earth's surface. Acta Univ. Palackianae Olomucensis, Rac.R.N., 42, Geogr.-Geol., 13: 239-261.

Zapletal, L., 1975. Nevratné antropogenni transformace reliefu slovenska. (Irreversible anthropogenic transformations of Slovakia's relief). Geografický Casopis, 27(2): 141-153.

Zuidam, R.A. van, 1975. Geomorphology and archaeology: evidence of interrelation at historical sites in the Zaragoza Region, Spain. Ztschr. f. Geom., 19(3): 319-328.

GEOMORPHOLOGY AND URBANIZATION

7.1 Introduction

The location of settlements and their internal structure is often influenced to a considerable degree by environmental factors and particularly by terrain configuration. This is particularly clear in the case of small settlements in rural areas which often form patterns reflecting those geomorphological factors (Grumazescu, 1967; Harjoaba, and Donisa, 1967; Nimigeanu, 1967; Sandy, 1977). These factors may also affect certain constructional details within a larger settlement, the urban morphology as a whole, general characteristics (concentrated, dispersed, etc.) and patterns of urbanization. Fig. 7.1 shows the patterns of villages with their homestead gardens near the south coast of Central Java, reflecting the general distribution of plains and hills in the area. The series of sandy beach ridges which have developed parallel to the coast and the natural levees of rivers such as the Serayu River are near the left hand side of the figure. The vertical aerial photograph of figure 7.2 illustrates how the location of a single settlement - in this case Dobo, Aru Islands, Indonesia - is dictated by the geomorphological situation: a small triangular sandy plain not only offers a suitable site for construction, but is also the only reasonable mooring for fishing boats and other small ships in the wide surroundings. The plain has been formed by littoral currents and beach drifting processes, which in turn are affected by the configuration of the sea bottom, following an ancient slide of the coastal cliff and subsequent coral growth.

The suitability of the location of a settlement does not only depend on the characteristics of the site itself, but also on the situation of the settlement within the area concerned, for example, along a caravan route, near the mouth of a river with an important hinterland, at a river crossing, at the junction of natural transportation routes, etc. (Chandler, 1976). Figure 7.3 shows the town of Tampere in Finland which has developed at the crossing of an important NNW-SSE stretching waterway, formed by glacial scouring during the Pleistocene and used for the transportation of logs and a land route located on an approximately WSW-ENE stretching ridge of ice-marginal deposits marking one of the recessional stages of the continental ice sheet.

In many cases, therefore, the importance of the situation is so dominant that eventual disadvantages associated with the site have to be accepted (Berry and Neils, 1969). In the selection of a site, one factor, for example defence or water supply,may be so dominant that other, less favourable aspects of the site are overruled.

It should be understood that both site and situation may change with time and that therefore, the environmental situation of existing settlements may be rather inadequate for the present needs (Hilpman, 1970). The situation may change either physically, for example, the silting up of a river mouth, general degradation of the land or socio-economically, for example, the changing of trade routes. The site itself may change by natural processes, by processes provoked or accelerated by man, or simply by a different appreciation or requirement such as recreation or industrialisation which was not initially considered. In numerous cases such changes have led to the decline and ultimate abandonment of the settlements concerned. Schmidt (1969) has vividly described ancient ports in Italy which he rediscovered and studied using aerial photographs. Folk (1975) has investigated the ancient town of Stobi in Yugoslavia that flourished in the Hellenistic-Byzantine period from about 400 BC to 600 AD and was gradually destroyed because of foundation problems, earthquakes and mudflows and was ultimately abandoned as a result of increasing desiccation leading to the deposition of flood deposits and windblown dust. However, many other settlements have survived and have even grown, notwithstanding considerable changes in situation and/or site, often utilizing other economic generating forces. In southern Italy, for example, most of the villages were founded in the long-drawn period of political unrest following the collapse of the Roman Empire and were situated on hilltops and other inaccessible,steep places for reasons of defence. Widespread badland development has since occurred and mudflows have repeatedly destroyed and still endanger many of these old settlements. Nevertheless these settlements are maintained,notwithstanding the environmental degradation of their surroundings, the shifting of the agricultural development towards the plains and the peaceful situation now prevailing.

In case of deterioration of the site of the settlement itself or of its surroundings, for example, deforestation of the higher slopes, problems related to the physical environment may concern the old core of the settlement.

In most cases, however, environmental problems and hazards occur in the extension of a settlement, beyond the safe confines of its initial and usually well-chosen site. The rapid urban sprawl characterizing many large conurbations nowadays also calls for careful studies of environment and terrain configuration, particularly in the sub-urban zones (McCulloch et al., 1969; Pestrong, 1968; Stauffer, 1974; Taylor, 1971; Wayne, 1969).

It is evident from the aforesaid that it is not always feasible to develop urban areas in a fully adequate natural environment. This is particularly the case where extension of existing settlements is concerned: one then has to make the best of the given situation. A terrain classification

Fig. 7.1 Settlement patterns and geomorphological situation in South Central Java, Indonesia. Settlements are located on beach ridges in the lowlands bordering the south coast and on the natural levee of the Serayu River in the west. Low densities characterize the limestone areas of the Karangbolong Hills (centre-south) and the South Serayu Range which stretches from west to east in the middle of the map (Sandy, 1977).

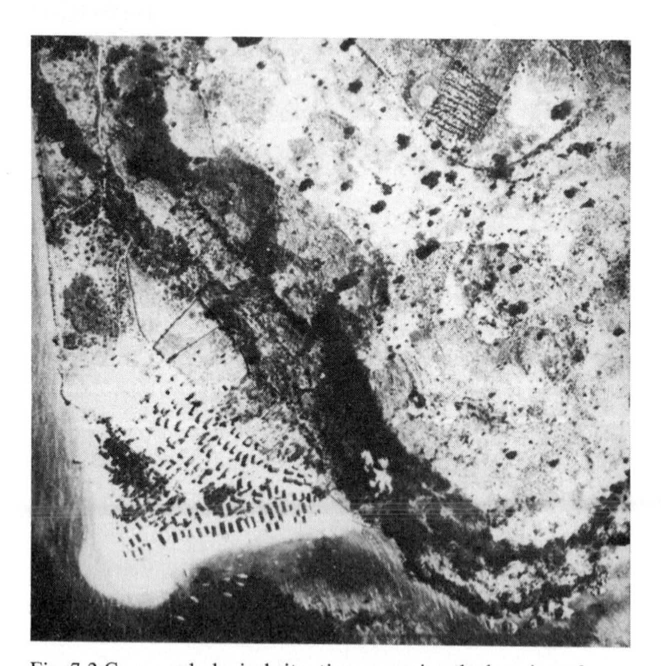

Fig. 7.2 Geomorphological situation governing the location of a settlement on a raised coral reef in the Moluccas, Indonesia. Scale 1:15,000.

geared to urban land use of the town and its surroundings is then appropriate. New settlements cannot always be located in ideal sites, as location is decided by other matters, such as the site of mineral deposits in case of a mining town or the possibility of obtaining drinking water. The availability of water from a storage lake for municipal uses was one of the major factors in siting the capital of Botswana, Gaborone. Other factors affecting the location may be the presence of mountains, an earthquake belt or widespread slope instability. Thorough study of the environmental situation and a terrain classification of the area concerned will contribute to the siting of future settlements in the most adequate and the least endangered parts.

Due to the present state of technology, many environmental factors do not present unsurmountable barriers to urban development. It is, however, usually advisable for reasons of economy to adapt the urban land use to the characteristics of the natural environment of the various parts of the terrain to be put under use. Multi-story buildings and industrial areas, for example, require a higher bearing strength of the soil than most residential areas, because otherwise the foundation cost will be unduly high. One may also consider some modifications of the

Fig. 7.3 The town of Tampere, Finland, situated at the junction of a glacially scoured waterway and a land route on ice-marginal deposits.

existing terrain configuration to improve the conditions, for example, filling poorly drained depressions where inundations and adverse hygienic conditions may otherwise occur. Some environmental factors are more difficult to master, such as deep-seated land sliding and earthquakes. Careful zoning of the urban land use (Anon. 1969) and appropriate building codes are then required to minimize the effects of these natural hazards.

This is, of course, of special importance in major conurbations (Abreu and Verhasselt, 1981) where large concentrations of people exist and large sums are invested in real estate. The various environmental factors and natural hazards affecting urban areas will be discussed in the following sub-sections.

7.2 Some Major Geomorphological Factors in Urban Development

7.2.1. Landforms

Among the geomorphological factors affecting the selection of optimum sites for settlements and their further development and urban structure, landform plays an important role. For example, the inclination of the land is to be considered as grading increases building cost. Densely dissected terrain will certainly not be the first choice for urban development since it adversely affects the communications and will require numerous bridges, culverts, etc. It is not so much the morphogenesis but more specifically the morphographical and morphometrical relief data that influence the general lay-out of the settlement and its subdivision into functional urban land use types. Even the occurrence and spatial distribution of smaller (meso) forms is of importance, particularly for construction work and landscaping of recreational areas. Kugler (1968) specifies that the morphographical characteristics of an area encompass the following four main aspects:

1. its geographical situation,
2. the spatial distribution of terrain forms,
3. the trend and aspect of relief elements such as altitudinal range, degree of dissection, orientation and angle of slope, etc.,
4. the size of the terrain forms

The morphographical situation affects the urban development in many different ways. The general characteristics of the urban structure will, for example, reflect the availability of flat or gently sloping land: congested settlements are common in narrow valleys whereas the most distinct urban sprawl with its inherent high cost for public facilities occurs around settlements in flat lands. The general lay-out of the settlement and the situation of major functional zones such as residential, service and industrial areas is normally governed, to a large extent,by matters of relief. The scenic quality of the settlement is also influenced by this and thus the price of the land in the various parts. The meso and micro climatic conditions associated with certain relief elements may add to this: insolated, sun-exposed slopes tend to be more valuable than, for example, depressed zones where temperature inversions may cause frequent fog conditions. It has already been mentioned that grading tends to increase building cost. Earthwork is thus often justified from an economic viewpoint and one can generally observe a decline in amplitude and complexity of meso relief forms as well as complete obliteration of micro relief in built-up areas. Sometimes however, artificial relief is added either to dispose of debris or urban waste or simply to increase the attractivity of recreational areas. Finally, landforms also affect the pattern of urban transportation lines, sewer systems, etc.

7.2.2 Geomorphological processes

Geomorphological processes of diversified nature and intensity form a second group of physical factors influencing urban development. Whenever possible, settlements should be free from flooding, land slides, avalanches and other adverse processes and natural hazards. In many localities, however, these processes pose problems to the efficiency of specific urban functions, either continuously or occasionally. They may also lead to the necessity of engineering designs which raise the construction cost and they may increase the cost of maintenance or decrease the durability of the structures. In extreme cases, they represent distinct natural hazards for which remedies may or may not exist (Burton et al., 1969; Blanc and Cleveland, 1971; Hutchinson, 1959).

A process of considerable magnitude and intensity, such as major land sliding, occurring continuously or at short intervals may force the inhabitants to abandon the settlements or part of them. If, however, the process operates at large and irregular intervals,such as is characteristic for earthquakes, floods and most volcanic processes, the tendency is that considerable concentrations of costly urban structures are built and large numbers of people continue to live in potentially dangerous sites. Hazard zoning of the urban and sub-urban areas is then an important task for the geomorphologist (Kollmorgan, 1953). This may serve to concentrate further urban development in the least endangered zones, to enforce, for example, building codes in certain zones and to guide insurance companies in determining the extra premiums for insuring real estate in the areas concerned (Kiersch, 1969; Ker, 1970; Leighton, 1972). One should bear in mind that the large number of people living in major towns and the great investments made in urban construction justify high expenditures to control a process and thus to reduce the risk of important loss of life and property. For humanitarian and economic reasons costly and drastic control measures can thus be often contemplated which could not be afforded elsewhere.

It is evident from the aforesaid that the study of geomorphological processes and the precise mapping of their spatial distribution and intensity is of considerable importance for urban planning and construction.This applies to generalized and often almost unperceptible processes such as creep and cryoturbation as well as to localised phenomena of greater magnitude and violence such as land slides and mudflows. Particularly for the latter group of phenomena it does not suffice to map the sites of actual processes; one should also indicate the areas where they may potentially occur in the future. Sites where these processes have occurred in the past may indicate zones of greater susceptibility where reactivation of the process is more probable than elsewhere. Thus both processes of the present and past should be considered to arrive at a proper assessment of the (risk) factors which affect hazard zoning.

It should be emphasized in this context that many geomorphological processes are substantially changed after urbanisation (Houghton, 1975). Usually the natural situation has deteriorated and Kotlov (1972) maintains that up to 70% of failures of buildings and other structures in Canadian cities are due to man-made processes and situations (Guy, 1975). For example, if a small drainage basin is the scene of urban extension, the infiltration gradually decreases and the surface run-off increases while the original natural cover of grass and forest is replaced by buildings and roads of asphalt or concrete. The hydrological regime of the basin changes, the peak discharges become higher and culverts and bridges which served their purpose perfectly prior to the building activities may become inadequate (Wolman and Schick, 1967). A marked increase in sediment yield often accompanies construction and adds to the problems caused by urbanization (Becker and Mulhern, 1975; Powell et al., 1975). The cost of combating such developments in an early stage are considerably lower than those of the damage already caused and subsequent remedial measures, but this fact is often ignored (Brandt et al., 1971).

A cutting made in a slope for purposes of construction may introduce instability in an otherwise stable slope (Hunter and Schuster, 1968) and the weight of the structure may aggravate the situation. The following example relating to the construction in 1974 of a new church in the small town of Gallicchio, Basilicata, Italy,illustrates this. A deep cut was made in the fine-textured Aliano sands of a moderately steep slope and the church was subsequently built on the excavated platform and the adjacent fill. Although a massive concrete wall was erected to support the undercut slope, sliding occurred during construction of the church when, after heavy winter rains, the beds were saturated with water. The fact that the beds dipped gently in the same direction as the slope and that water emerged at the contact of some slightly coarser textured beds, unfavourably affected the stability conditions of the site. Figure 7.4 depicts the situation with the cracked supporting wall and the landslide scar above it in the right-hand part of the picture. Figure. 7.5 shows the damage done to the church on that occasion.

Fig. 7.4 Newly constructed church in Gallicchio, Basilicata, Italy, severely damaged by a landslide in 1974.

Fig. 7.5 Detail of destroyed church of Gallichio.

The numerous bridges found in many urban areas, if improperly designed, may act as bottlenecks for the evacuation of flood waters. This is particularly so if the whole river basin is adversely affected by anthropogenic factors. Apart from river basin development, including reafforestation and the construction of retention reservoirs in the head waters, zoning ordinances in the urban areas affected by the floods, building codes, etc. may contribute to mitigation of the damage (Kates, 1962, 1965; Mittelstedt, 1976; White, 1968). Degradation of surrounding rural lands may also affect an urban site in hilly terrain, and ravines or badlands may, for example, invade the town by headward erosion. This is illustrated by the town of Montalbano Ionico, Basilicata, Italy. The badlands around the town are extensive and are developing dramatically (fig. 7.6). This has, among other things, resulted in the collapse of the far end of the street pictured in fig. 7.7 which is now closed off by a concrete wall to prevent casualities. Fig. 7.8 gives a 1:7,000 vertical aerial view of the site.

7.2.3 Soil and subsoil conditions

A third group of physical factors affecting urban planning and construction relates to soil and subsoil conditions. The distribution of rock outcrops, the depth of solid rock, the thickness of weathered material and of allochthonous covering materials are matters of great concern for the urban planner but particularly for the urban engineer (Calembert, 1958). The engineering properties of the soil and subsoil materials such as bearing strength, are especially critical where heavy, multi-story structures are to be erected or in locating industries, if foundation problems are to be avoided. Data on bearing strength are mainly obtained by way of laboratory tests. One should, however, be aware of the fact that the results so obtained usually relate to disturbed samples. They may, nevertheless, be applicable to conditions of flat terrain, but the data should be carefully considered where slopes are concerned.

Soil and subsoil conditions are also of importance, for matters such as earthquake hazard zoning (Cluff, 1968, 1969; Berlin, 1975), municipal and industrial water sup-

Fig. 7.6 Ground view of the badlands near Montalbano Ionico, Basilicata, Italy, threatening parts of the town by headward erosion and scarp collapse.

Fig. 7.8 Vertical aerial view of part of Montalbano Ionico, Basilicata, Italy with adjacent badlands. Scale 1:7,000.

Fig. 7.7 Dead-end street in Montalbano Ionico, Basilicata, Italy. The far end of the street fell victim to the development of the badlands and the escarpment is only a few metres past the concrete wall closing off the street.

ply (Mihailescu and Martiniuc, 1967), the disposal of urban waste waters including the proper functioning of septic tanks, etc. (Bouma, 1973; Loughry, 1973; Pavoni and Hagerty, 1973). Finally, surface deposits may be potential construction materials (Young, 1968) and therefore, one should avoid whenever possible, constructing the town on top of valuable building materials, particularly when they are scarce or expected to be so in the future. Problems of this kind exist in several urban areas (Davis and Meyer, 1972). In semi-arid and arid areas soil salinity may be a problem since these salts tend to be highly aggressive to concrete (Golani (Ed.), 1978; Cooke et al., 1979, 1980).

It is a task for the geomorphologist to study these surficial deposits and to classify them according to genesis, because usually each type has its own inherent characteristics and engineering properties. The interpretation of aerial photographs and other remote sensing techniques, supported by field studies,is a common approach to such studies (Johnson, 1970). Glacial deposits are of widespread occurrence in North America and northern Europe. Attention has been drawn to the fact that 50% of the US population lives on this kind of materials, some of which are inferior for purposes of foundation. Glacial till is particularly ill-famed in this respect. Lobdell (1970) studied the effect of weight on the settling of varved clays. Serious problems exist in zones of permafrost, to be dealt with in section 8 (Engineering). In almost every geomorphological setting foundation problems can be expected to some degree (Hewitt, 1970). In some cases swelling of clays rather than settling of soil and foundations is the problem. This depends on the clay percentage of the material, the clay mineral itself, the hydrological conditions, etc. (Demirev et al., 1974; Hamilton and Owens, 1972).

Subsidence in urban areas may relate to the weight of the structures erected but it is often also caused by the extraction of groundwater. This may have serious consequences for the drainage of the urban areas affected and may even lead to inundations. This has been observed in Tokyo and other cities in Japan (Matsuda, 1974; Nakano, 1970), in the London area (Wilson and Grace, 1942) and is particularly spectacular in Venice, where the extraction of large quantities of groundwater at the nearby port of Marghera is one of the major causes of subsidence and inundations (Gambolati et al., 1974).

Particularly poor sites for urban construction are fills, for example, in lakes and bays. The San Francisco area is particularly ill-famed in this respect: urban growth has led to the necessity of construction on such fills because of scarcity of other land around the bay. In the case of an earthquake these sites are, however, particularly hazardous; therefore a strict system of risk zoning and building codes has been enforced to minimize eventual losses. Subsidence and settlement, liquefaction and subaquous landsliding are among the other problems to be faced (Bolt et al., 1969; Goldman, 1969; Wiggington, 1969).

7.3 Categories of Geomorphological Studies for Urban Development

7.3.1 Generalities

There are four major groups of interested parties who benefit from studies of the physical environment of urban areas: the urban and sub-urban planners, the engineers involved in urban construction works, individual inhabitants or groups of inhabitants and insurance companies. Most survey work is, in fact,done for the planners and/or the engineers who use factual environmental data in three different phases of their work, namely (1) prior to construction during the planning of urban extension or renewal, (2) immediately before and during construction when preparing the area or the site for building and when deciding on the most appropriate foundation and further design and (3) after construction when it comes to maintenance for which site improvement (for example, drainage) may be required (Leser, 1976; Kolomensky, 1959).

The area included in the survey and the scale of mapping vary with the purpose of the study:

1. Geomorphological mapping of fairly large zones at scales of approximately 1:25,000 to 1:10,000 is often required for purposes of locating areas for urban extension, for deciding on the general lay-out of the new developments or for controlling urban sprawl.
2. Geomorphological mapping of the urban zone proper at scales of approximately 1:10,000 to 1:5,000 may serve for purposes of urban renewal, risk zoning, etc.
3. Geomorphological mapping at scales of approximately 1:5,000 to 1:1,000 is most appropriate for the study of selected construction sites.

Obviously, the planners have a particular interest in the first category of maps, whereas the engineers benefit in particular from the third category. The second category is of interest to both types of users.

Good co-operation between those parties studying the urban environment on one hand and the planners and engineers on the other,is essential for an adequate implementation of the recommendations made. The former should be provided with information about the types of urban land use desired and specifications of the requirements for each of these should be given (Branagan, 1972). They should, in their turn, provide the planners and engineers with factual data about the environmental situation, natural hazards, etc. The planners can, to an important degree, contribute to the optimum use of urban space and to the mitigation of hazards. This can be achieved by enforcing strict building codes and ordinances for risk zoning, etc. (Olson and Wallace, 1970). The municipal, provincial or national authorities may also give subvention for certain measures of site improvement, for example, slope stabilisation, the cost of which would otherwise be prohibitive (O'Riordan, 1973). The engineers will have to respond by finding ways and means of adapting building design and construction materials to the requirements of the building site at hand (Degenkolb, 1970).

In fact, in urban areas, comparatively expensive measures for the safety of the population and prevention or minimization of damage to property are often justified because of the enormous losses which may otherwise occur. It has been calculated that in one single year in Alameda County (California), USA, damage amounting to US dollars 5,000,000 occurred which averages US dollar 400 per acre on slopes steeper than 15%. Alfore et al. (1973) estimate that US dollars 55 billion worth of damage will be caused in California in the period 1970-2000 as a result of processes such as landsliding, flooding, erosion, swelling of soils and fault displacement. If no proper measures are taken, insurance firms may cease to cover the damage and earth scientists will become reluctant to be involved in matters of siting for urban construction (Anon., 1974; Yelverton, 1973).

Numerous studies on the physical environment of urban areas have been executed in recent years. This has resulted in the publication of several textbooks and substantial articles on the matter. Specifically mentioned are those by: Daneky and Harding (1969), Fallat and Tetherow (1973), Kaye (1968), Leggett (1973), Price (1971), Radbruch (1968), Rockaway (1972). Interesting case studies also exist, such as those by Grube and Germany (1972) on Hamburg and by Schmol and Dobrovolin (1968) on Anchorage (see also Hoffa, 1968; Hogberg, 1971). Diversified factors are considered in these studies, such as urban extension and renewal, control of erosion, transportation systems, underground utility networks, waste disposal and aspects of open space and recreation (Guy, 1975; Hilpman, 1970). The studies carried out can be incorporated into one of the two following categories:

1. a geomorphological break-down of the urban and sub-urban zone into approximately homogeneous units with indication of the inherent characteristics and resources of each unit, and
2. geomorphological studies concerning one or more active geomorphological processes affecting the urban area, parts of it or specific sites.

The mapping scales for each of these two categories of geomorphological surveys concord with those previously indicated for class 1 and 3 respectively. The two categories are further elaborated upon in the following sub-sections.

7.3.2 Geomorphological terrain classification of urban areas

Geomorphology as a basis for terrain classification of urban and sub-urban areas has found application in many countries. This is explained by the fact that geomorphological processes and other environmental parameters which affect the inherent properties and engineering problems of the land and thus its potential use, are associated in one way or another to geomorphological terrain classes. The geomorphological approach to the assessment of the physical characteristics of these terrain forms and their evaluation for various types of urban land use is a comparatively cheap and rapid one, and is thus increasingly appreciated by urban planners and engineers.

The studies made usually include mapping at scales of 1:25,000 to 1:10,000 but the method has also been used for more detailed investigations. Apart from general geomorphological mapping with emphasis on landforms and actual processes, and the assessment of the resources of every geomorphological terrain unit, special maps have been made which concentrate on specific aspects of the terrain. Among these, slope maps indicating slope angle classes and intensity of dissection, rank high. Other kinds of special maps relate to matters such as slope stability, bearing strength and groundwater supply. Maps indicating modes of land improvement as required at various localities and maps of hazard/risk zoning are also often prepared. A common method of presenting the information in a condensed and ready-for-use form is by way of tables in which all geomorphological terrain units are listed together with the inherent characteristics and properties of each of them (Knight, 1971). Table 7.1 gives an example.

Among the numerous studies of this kind, those reported on during the Bucharest Symposium of the IGU Commission on Applied Geomorphology in 1967 deserve mention, particularly as regards the European context. Mihailescu and Martiniuc (1969) give the general outline of the surveys for urban development carried out in Romania; Klimaszewski (1969) elaborates upon the work done in Poland, and several Romanian case studies (Morariu, 1967; Martiniuc and Bacauanu, 1967; Iancu and Velcea 1967) can be found in the Proceedings. Good examples are also at hand from other countries such as the geomorphological study on the planned urban extension of Suez by Cooke et al., (1979), by Clootz-Hirsch and Maire (1972) on Strasbourg and by Font and Williamson (1970) on Waco (USA).

Fig. 7.9 gives an example of geomorphological terrain classification and relates to the town of Quetta, Pakistan. The town is situated at the foot of steep and barren limestone mountains and is built partly on gravel fans where sheetflood phenomena occur and partly on the fine-textured valley fill. The town centre in particular is built on these alluvial deposits where groundwater occurs at a shallow depth. The double line (10) in the map indicates the approximate boundary between the gravel fans on the right and the much finer-textured (silt-loam) deposits of the adjacent alluvial plain situated near the left edge of the map. Since earthquake tremors transmit better in the fine-grained and often water-saturated alluvial deposits than in the dry gravels of the fans, earthquake risk is considerably greater in the alluvial plain. This became evident during the ill-famed Quetta earthquake of 1935,whereby damage in the town centre was considerably more than in the cantonment area. Local building codes now prohibit the construction of buildings of more than two storeys in those parts of the town situated in the alluvial plain and the use of earthquake-resistant, reinforced concrete struc-

Major regions	Landforms / Agencies	Resources: Sand & gravel	Resources: Housing development	Resources: Recreation	Resources: Recreation	Drainage: Soil erosion: water	Drainage: Sedimentation	Drainage: Flash flooding	Drainage: Debris flows	Drainage: Machine erosion	Drainage: Soil erosion: wind	Surface stability, Landsliding: Bedrock	Surface stability, Landsliding: Overburden	Surface stability, Landsliding: Snow avalanche	Surface stability, Subsidence: Hydro-compaction	Surface stability, Subsidence: Water extraction	Surface stability, Subsidence: Expansive soil
REGIONS & LANDFORMS																	
ALLUVIAL FANS	Fan surface	●	●			o		◎	◎	◎	ø						
	Deeply incised flood channels					◎	◎	⊙									
MOUNTAINS (ROCK OUTCROP)	Very steep slopes											ø		◎			
	Terrace remnants		o			ø								◎			
VALLEY FLOORS	Old alluvial floodplain	o	●								⊙					◎	◎
	Stabilised dunes		o							◎	●/						
	Active X-cut channels					◎	◎	⊙			●/						
MAIN STEM EPHEMERAL STREAM FLOOD PLAINS	Channel	o	o				◎	◎			ø						◎
	Terrace	●	●														
	Artificial swamp						◎	◎		◎							◎
PRINCIPAL MANAGEMENT AGENCIES																	
PRIVATE	Property owners	●	●					●	●	●	●			●			●
	Insurance companies		O					⊙								O	O
	Engineering and geol. consultants	O						⊙	O		O					O	⊙
	Citizens groups							O		O	O						
CITY	Planning		O					O									O
	Building & safety		O					O								O	O
	Engineering	O						⊙		O						O	O
COUNTY	Flood control							⊙									
	Sanitation		O				O	⊙								⊙	
	Planning	O	O			O	O	O									O
	Engineering					O	O	⊙		O	O					O	⊙
	Building & safety		O					O	O							O	O
STATE	Division of Mines and Geology	O						O								O	O
	Department of Transportation							⊙									
	Environmental Review Board							O		O							
FEDERAL	U.S.D.A. Soil Conservancy Serv.	O				O	O	O									O
	Bureau of Land Management					O	O	⊙						O			O
	U.S. Geological Survey							O								O	
	Bureau of Reclamation							⊙									
	Corps of Engineers							⊙									O
	Dept. of Interior Fed. Housing Assc.							O								O	
	Department of Transportation							⊙									⊙

● Widespread

o Localised

O Severe

/ Slight

O Survey, planning, assessment

● Exploitation, repair, maintenance

Table 7.1 Relations between landforms, geomorphological resources, problems and hazards, and management agencies in Las Vegas.

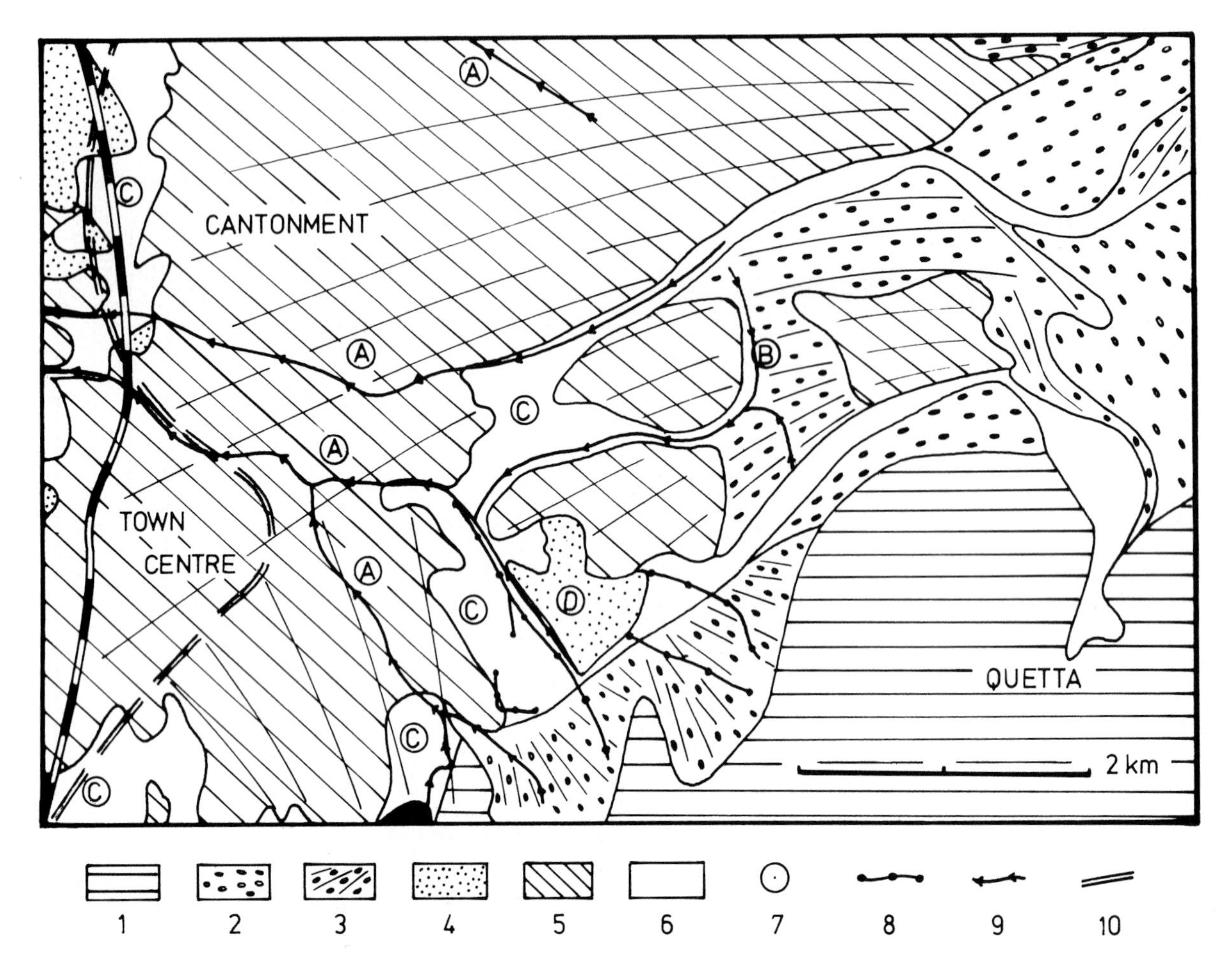

Fig. 7.9 Geomorphological situation of part of Quetta, Pakistan.
Key: 1. mountains, 2. old gravelly deposits, 3. active gravel fans, 4. rain-dependent fields (sheet flood), 5. built-up area, 6. waste land with some grazing, 7. locations (A-D) mentioned in text, 8. karezes (chain wells), 9. open irrigation canal, 10. boundary between fans and alluvial plain forming limit of earthquake hazard zones.

tures has been imposed.

Another geomorphological phenomenon of interest to urban planners is the (imperfect) sheetflood in the suburban fringe on the middle parts of the gravel fans. Sizeable acreages of water-dependent wheat fields are watered by difuse run-off on the fans. It is obvious, however, that where urban extension occurs on the fans, sheetflood is undesirable and urban and rural uses of the land may thus clash unless appropriate provisions are made in time. The map of fig. 7.9 pictures the situation in the eastern part of Quetta and its surroundings. It shows several subhorizontal drainage tunnels or chain wells, locally called karezes (A), originating in the upper, gravelly portions of the fans. They traverse the town farther downslope as open canals, carrying off sizeable quantities of urban waste. A canal and a ditch (B) are constructed parallel to the contour lines to prevent flooding of the quarters situated to the left. As a result,the rural land farther downslope (C) is deprived of irrigation water from sheetflood. Only where water can be obtained from smaller fans to the south, do rain-dependent wheat fields still occur (D).

The studies required may also relate to specific aspects of urbanisation, such as the location of an air field, the municipal water supply, the anthropogenic relief in the area, spoil heaps, etc. In any of such cases the specifications should be clearly established prior to the joint survey of planners and engineers.

For an air field, for example, a fairly extensive area of very good drainage and minimum grading should be available for the construction of long runways in one or more directions. There should be no topographic obstacles, either natural or man-made, and the approach and take-off routes in particular should be free of them. Other critical factors are flood hazard and high frequency of fog which may occur, for example, near a lake or other humid areas. Dune sand and dust may be factors in arid and semi-arid areas, for example, at Bahrain (Doornkamp et al.,1979), but protective measures may be suggested in such cases.

Water supply is another important area of study in the

urban context, particularly where cities rapidly grow and the average water use per capita rises. Where groundwater resources are available a detailed geomorphological study of origin and texture of the unconsolidated materials such as river terraces, buried river systems, etc. may contribute to the solution of the problem.

Numerous borings are usually required and the possibility of water resources in hard rock should be investigated by the geologist. In those parts of northern Europe and North America where the Pleistocene continental ice sheet has substantially altered the drainage pattern and pre-existing valley systems or proglacial valleys formed at the ice front are filled up with glacial sand and gravel deposits, the water supply for municipal and industrial uses is often derived from goundwater available in these valley systems and their study is thus of considerable economic importance.

Where surface water is to be used the study may encompass the siting of a suitable reservoir and the location of the conduct line. In settlements in semi-arid and arid areas or in more humid areas where water is scarce for other reasons, the storage of rainwater in reservoirs, so-called 'tanks', at topographically suitable low sites with impervious soil or in cisterns, is often of crucial importance to provide sufficient drinking water, particularly in periods of drought. The situations where cisterns can be constructed are diversified and only a few examples can be given here. Fig. 7.10 shows a traditional cistern as is found near almost every village in Rajasthan, India. The rainwater flows towards the cistern over the almost barren soil and enters the rectangular concrete cistern by way of a silt trap which is cleared at regular intervals. The water can be obtained through a square hole in the concrete roof of the cistern. Figure. 7.11 shows a cistern on the tiny volcanic island of Saba (Neth. Antilles). The catchment is in concrete and the collected water enters the cistern by way of a hole in the lower end of the catchment. Figures 7.12 and 7.13 show the conduct canal leading to the reservoir which until a decade ago was used for the water supply of La Muela, Spain. Fig. 7.14 shows a small cistern built in an ephemeral riverbed in Spain, where river water collected during the rainy winter season can be used in the summer. Fig. 7.15 shows a line of small cisterns that have been built in the past along a now abandoned road. The barren surface of the road acted at the time as the catchment. This kind of 'intermediate technology' is of great importance for providing drinking water for humans and animals in Sahelian villages, particularly in times of drought.

Fig. 7.10 Cistern in Rajasthan, India, with silt trap and catchment.

The urban and in particular the sub-urban zones of many settlements are anthropogenic forms resulting from cut and fill during construction, the extraction of construction materials, the disposal of urban waste, etc. The resulting quarries, spoil heaps or mine tips, are the subject of another kind of special studies. These may aim at landscaping and further rehabilitation of these often delapitated sites, or may serve to investigate possible hazards that might result from their presence. The rehabilitation aspect may englobe matters such as drainage, wind and water erosion control, vegetation, etc. It is particularly in industrialized areas that these problems sometimes are very outspoken (Wrona, 1973). Some of the quarries or spoilheaps may be put to good use, for example, for recreation. Stauffer (1964) discusses various uses and solutions and also gives good bibliographic information. The safety aspect has become a focus of interest particularly after the slide-disaster of the Aberfan coal tip in the U.K. (Bischop et al., 1969; Nash, 1969; Rosswell Moore, 1969; Woodland, 1969). The problem has since been investigated in many areas (Bischop 1973; Holubec, 1976; McKechnie, Thomson and Rodin, 1972).

Geomorphological terrain classification and related matters can, of course, be optimally applied where entire new cities are built, such as Milton Keynes in the U.K. (Cratchley and Denness, 1972; Lukey, 1974). Sequential aerial photography is a very useful tool for the geomorphologist to monitor the effects of urban growth on the environmental conditions. He will thus be better prepared to recommend certain protective measures during the progress of construction and also after the new town has been fully developed. An interesting example in this respect is

Fig. 7.11 Cistern on steep volcanic slope. Saba (Neth. Ant.).

Fig. 7.12 Catchment of water reservoir pictured in fig. 7.13, La Muela, Spain.

Fig. 7.13 Conduct canal and part of reservoir until recently providing drinking water for La Muela, Spain (cf.7.12).

Fig. 7.14 Small cistern in ephemeral riverbed, Northern Spain.

Ciudad Guyana (Yanez, 1979), which has grown rapidly in the last 15 years as a consequence of the development of mining and related heavy industries. The rapid urbanisation caused many problems to the planners, such as environmental degradation, a great demand for building materials and large amounts of urban waste. Careful planning was thus required and a geomorphological terrain classification was carried out in this context. Yanez could make use of a pre-urbanisation coverage of aerial photographs dating from 1958 and of four subsequent coverages, the most recent dating from 1974. The scales range from 1:15,000 to 1:25,000. Geologically the area forms part of the Precambrian basement and is composed of highly metamorphosed sedimentary rocks up to the granulite facies. It is unconformably covered by Plio-Pleistocene, sub-hor-

Fig. 7.15 Line of cisterns near La Muela, Spain, built along a now abandoned road which served as a catchment. The inlet of the water can be seen at the foot of the structure which prevents evaporation. Shallow, perched groundwater adds to the yield.

izontal, alluvial, lagoonal and deltaic sediments, mainly gravel, sands, silts and clays. It proved indispensible for the geotechnical purpose of the survey, to base the breakdown of the urban area under development on the types of the surface materials because all foundation and drainage problems relate to the properties of these materials. It was found that below the altitude of 85 metres mostly clays and silty clays occur,where springs and soil-creep phenomena can be observed at many places. The sediments above 85 metres are mostly sands and silty sands and consequently a tendency to the formation of ravines and badlands exist. Among the observations made, three are of particular interest: 1) a gradual lowering of the watertable was observed due to the increased surface run-off and decreased infiltration; 2) badlands developed where previously none existed as a result of bad drainage design and several improvements could be indicated; 3) the deforestation of the area has led to surprisingly rapid erosion in the unconsolidated sediments. The aerial photographs of figs. 7.16 and 7.17, dating from 1958 and 1974 respectively, illustrate the environmental changes provoked by urbanisation.

7.3.3. Geomorphological processes and urban sites

In areas of violent morpho-dynamics such as alpine regions and some semi-arid areas, the magnitude, force and velocity of certain geomorphological processes of fluvial, aeolian or other origin and in particular slope processes may become so overwhelming that they dominate all other environmental factors. They then turn into natural hazards that may endanger parts of the urban zone, or, in case of smaller rural settlements, the entire site of the town. If remedial measures are too costly or not feasible, the location may have to be abandoned. If so, a less endangered site should be searched for.

A classic example of landslide danger in a large settlement in a mountainous terrain is La Paz, Bolivia (Salgueiro, 1965), where sliding has developed and is seriously endangering parts of the town. A major landslide occurred in March 1958 and the changes in the terrain configuration could be studied and quantified by detailed photogrammetric mapping,using 1:12,000 scale airphotos taken in 1956 prior to the event and 1:6,000 airphotos of 1958, taken immediately afterwards. The subsequent further growth of the slide was studied from airphotos at 1:14,000 dating from 1963. The direction or axis of the movement, the situation of semi-circular slump planes and the total volume of the material involved in the slide (122,800 m^3) could be precisely calculated. Fig. 7.18 pictures the situation of 1958.

The effect of slope instability and related processes on smaller settlements is well illustrated in a special volume of the Bulletin of the Geological Commission of Peru which contains case studies of several Andean settlements in Peru. Ballon (1966) studied the situation of the town of Chumuch where the Quaternary overburden is subject to sliding and fracturing. The situation is considered hazardous, particularly when an earthquake occurs. He suggested relocation and recommended improvement of the irrigation canals. Cossio (1966) likewise recommended relocation of Conchucos because of landslide hazard in detrital material. He also studied Pilipampa and some other settlements which are endangered by cracks produced in steep slopes due to tremors. Relocation of Pilipampa is recommended whereas the situation near the other two towns could be improved by the construction of a drainage canal on the slope of the towns and a defence wall on the bank of the river. Garcia (1966) recommended reafforestation and the construction of protective embankments along the Ichu River to protect the village of Acoria. Other interesting papers by Mendivil (1966), Morales (1966) and Wilson (1967) discuss comparable problems and solutions for a number of other settlements in the Peruvian Andes.

The extended stereopair of fig. 7.19 pictures part of Lima, Peru where considerable problems exist with the

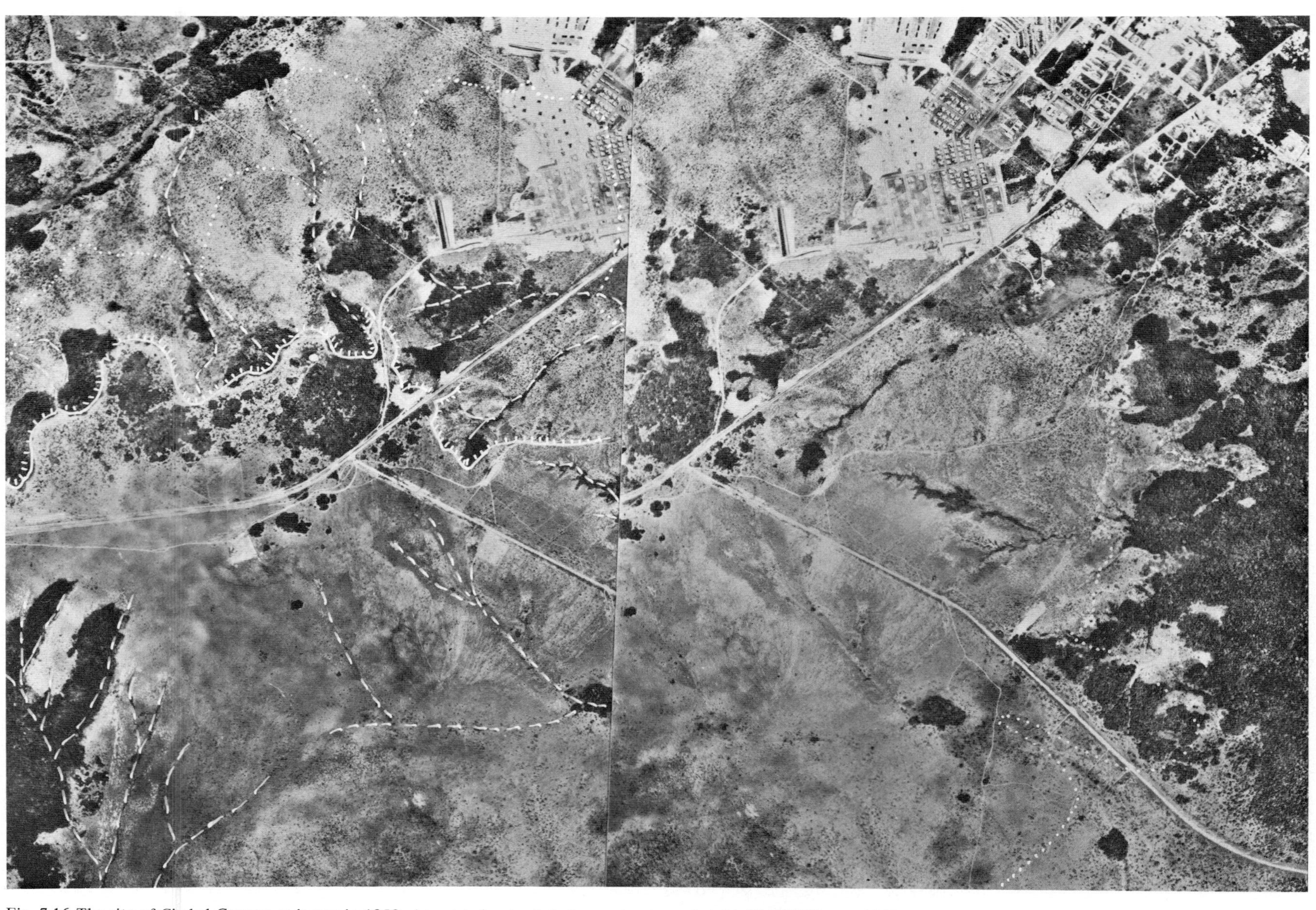

Fig. 7.16 The site of Ciudad Guyana as it was in 1958. Annotated extended stereopair at a scale of 1:25,000 (Yanez, 1979). Compare with 7.17.

Fig. 7.17 The same area as pictured in fig. 7.16 as it was after urban development in 1974. Annotated extended stereopair at a scale of 1:25,000 (Yanez, 1979).

Fig. 7.18 Extended stereopair showing a major landslide that occurred in La Paz, Bolivia, in 1958. Scale 1:14,000.

collapse of terrace edges and other escarpments introduced by lateral sapping of the rivers and ravines. The aerial view clearly shows that the bedload of the rivers is high as a result of the accelerated erosion in their upstream parts. The river regime is highly torrential. Sliding is activated where water-saturated slopes are deprived of their support after the collapse of the scarps at their lower end. Sliding affected the suburb of Villa San Antonio, for example, situated slightly above the centre of the stereopair. The mechanism of the slope process is illustrated by the ground-view of fig. 7.20. The large scale aerial photograph of fig. 7.21 demonstrates that similar phenomena also affect certain parts of the town centre.

Similar problems have been reported from many other urban areas situated in zones of very active morphodynamics (Kehle 1970; Kellaway and Taylor, 1968; Ludgren and Rapp, 1974; Nossin, 1972; Sauer, 1975; Varnes, 1969). Some examples from Spain and southern Italy are given below. Avci (1972, 1978) describes the situation of two small settlements in the Spanish Pyrenees, Tendruy and Puigcergos. The initial site of the village of Tendruy was on top of a gently inclined promontory of Palaeocene red marls covered by unconsolidated gravel, which is surrounded at three sides by the meander of a small river with a 50-metre deep incision. A gentle depression separates it at the fourth and southern side from geologically comparable areas of greater extent. The rivulet vertically cuts and saps the steep sides of its narrow valley, particularly in the meander bends just below the old site of Tendruy. The contact between the marls and the covering gravels has an inclination of 7^{0} towards the valley and acts as a slipface when water infiltrates in the gravel during the winter rains. The hazardous situation is clearly depicted on aerial photographs of 1957 and in 1969 the danger became so imminent that it was decided to relocate the village farther south and at a safe distance from the rivulet. Fig. 7.22 gives the situation in plan and in cross-section. The old village still exists,but its houses have become ruins.

A comparable, though slightly different morpho-dynamical complex characterizes the situation of the village of Puigcergos, which has been similarly relocated. The old site was on a hilltop rising approximately 100 metres above the small Espona River. It was built on Eocene limestone and marls which are underlain by the Palaeocene red marls. At the contact plane spring water emerged in the winter. The beds of both formations gently dipped to the southwest where the Espona Rivulet laterally sapped the red marls, thereby increasing the slope instability.

Cracks appeared in the substratum of the village, whereafter it was rapidly abandoned before the disaster ultimately occurred. It was rebuilt at a safer place farther to the northwest. Figure 7.23 gives the situation both in plan and in cross-section.

Sometimes, in case of urban extension, the slope instability is released shortly after the terrain has been stripped of vegetation, but before building. This is a comparatively favourable situation because appropriate measures can then be immediately taken. An example is the extension of Oriolo, Calabria, Italy, where a major slide occurred under such conditions in March 1974. The tongue of the flow reached the valley bottom just at the junction of two tributaries where a small lake was ponded up. A bridge situated about 100 metres from the slide was destroyed due to the deformation of the subsoil caused

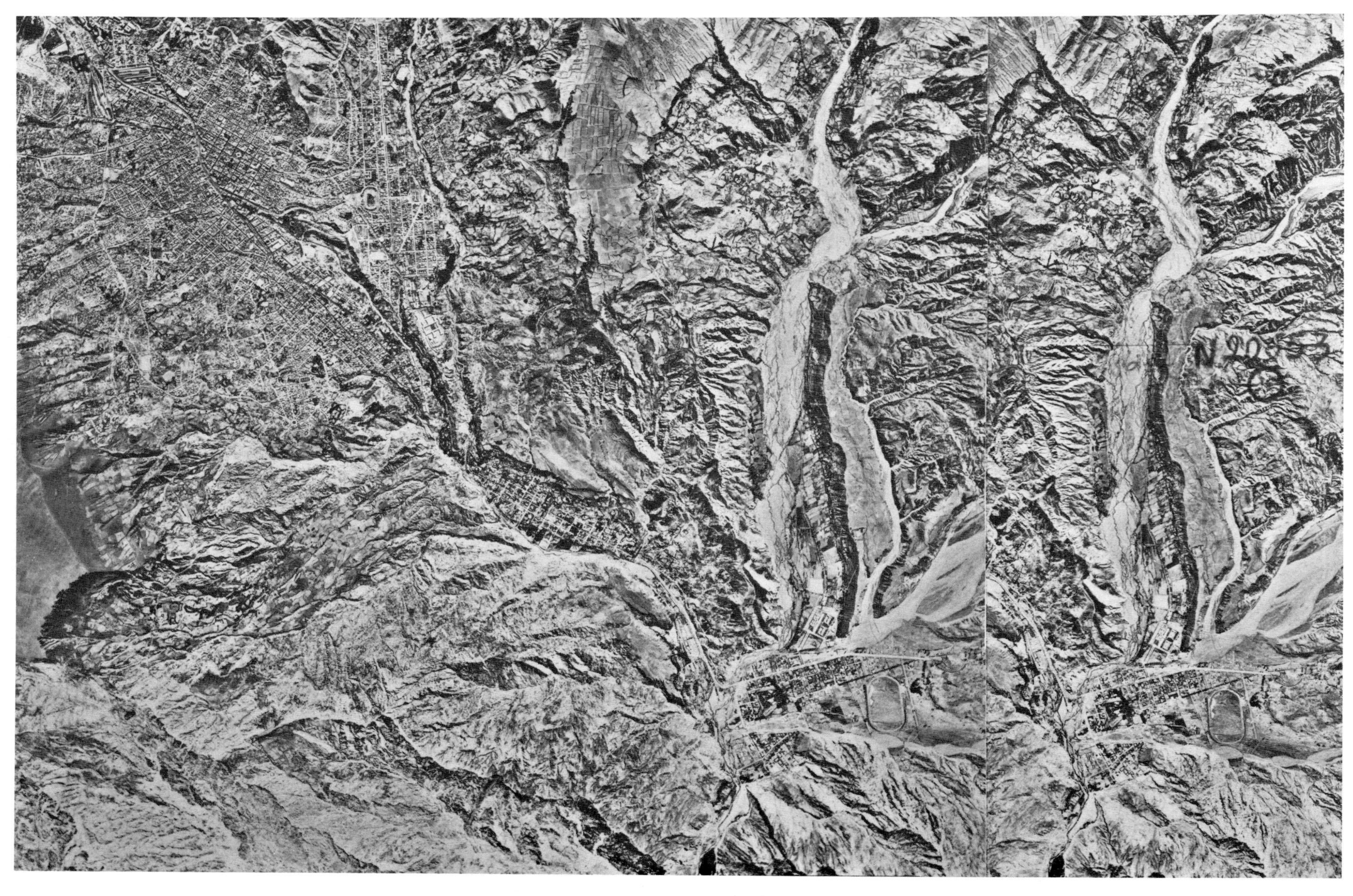

Fig. 7.19 Extended stereopair showing gullying and collapse of escarpments in Lima, Peru, at a scale of 1:50,000.

Fig. 7.20 Ground view of scarp collapse and related phenomena in Villa San Antonio, Lima, Peru, resulting from lateral sapping of torrent.

by the pressure exerted by the material of the slide. Improvements included the construction of drainage canals and tubes within and adjacent to the slide to stabilize this zone and the erection of concrete embankments in the valley bottom to ensure the proper discharge of the tributaries and the drainage of the lake. Fig. 7.24 pictures the situation including the improvement works.

Many examples of medieval settlements threatened by badland formation can also be found in southern Italy, particularly where the ravines grow more rapidly than elsewhere, along faults or fractures. The stereopair of fig. 7.25 shows how the village of Aliano (Basilicata) was gradually invaded from two sides by the upper reaches of deep ravines and at places was reduced to a narrow crestline where only a single row of houses remained. The airphoto (scale 1:17,000) clearly shows the effect of faults stretching across the village as well as the checkdam built in the ravine at the most endangered site. Badlands are also intruding in the nearby larger ancient town of St. Arcangelo. Two faults are situated near the town and although not endangering the site of the settlement itself, the progressing erosion renders its future access increasingly difficult. (Cotecchia et al., 1980). Aliano is situated in fine-textured Plio-Pleistocene Aliano sands where active badland formation is widespread, particularly on sun-exposed slopes where the face-slopes of the beds face south and west and are strongly exposed to the drying effect of the sun (Rabelo, 1980). St. Arcangelo is located on comparable sandy deposits but is surrounded by older marls and clays where the same processes occur, though in slightly deviating forms.

7.4 Two Case Studies

7.4.1 Torrent control in Matrei in East Tirol, Austria

Coarse-grained alluvial fans are often formed where in mountainous terrain torrents enter the main stem of the river. The torrent may occasionally shift its position(s) on the fan, particularly during very high discharge. Active deposition of normally coarse material may occur on the entire fan in some cases, but more commonly, an older, slightly higher and vegetated part of the fan can be distinguished from the active barren part. Increased deposition on the fan may result from deforestation of the torrent's drainage basin and associated deterioration of the watershed. An exceptional discharge may trigger the reactivation of the processes involved.

For various reasons, alluvial fans have been selected for the site of human habitats in many mountainous parts of the world. One reason is that the fans often have less agricultural value than the surrounding land. Furthermore, the external and internal drainage of most fans is satisfactory and water supply for an average village is not a problem. In arid and semi-arid areas the selection of fans for human settlement is usually based on the fact that the irrigation systems originate near their apex.

It is often forgotten by the settlers that the fan they live on is a dynamic feature where deposition continues and where the torrent may change its course(s) overnight. Flooding and serious damage to life and property may then occur. The more rapid the sedimentation is (or becomes, due to accelerated erosion of the basin), the more frequent such changes will be and the more dangerous the site of the settlement. Tricart (1962) describes the effects of sudden, high-discharge and violent deposition in the village of Ristolas, situated in the Gardon d'Arles Valley, French Alps, during the catastrophic flood of 1958. Similar situations have occurred in the nearby Guil Valley.

It is evident that under these conditions new building sites should be carefully scrutinized with respect to flood susceptibility. Inactive parts of the fans, which are somewhat higher than their surroundings, should be preferred. The construction of houses on the sloping fan surface should be on top of a wedge-shaped fill rather than cut

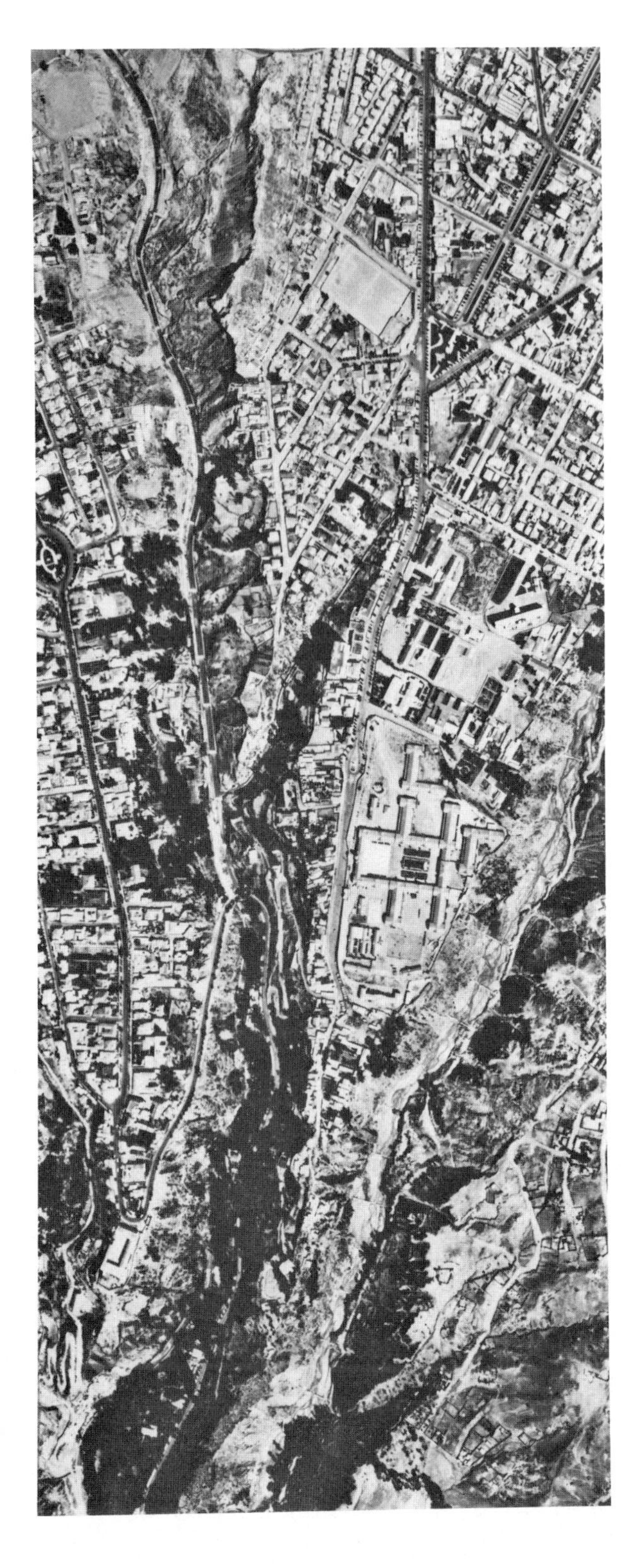

Fig. 7.21 Large scale aerial view of part of Lima, Peru,showing how geomorphological processes pose numerous problems to urban planners and engineers. Compare fig. 7.19.

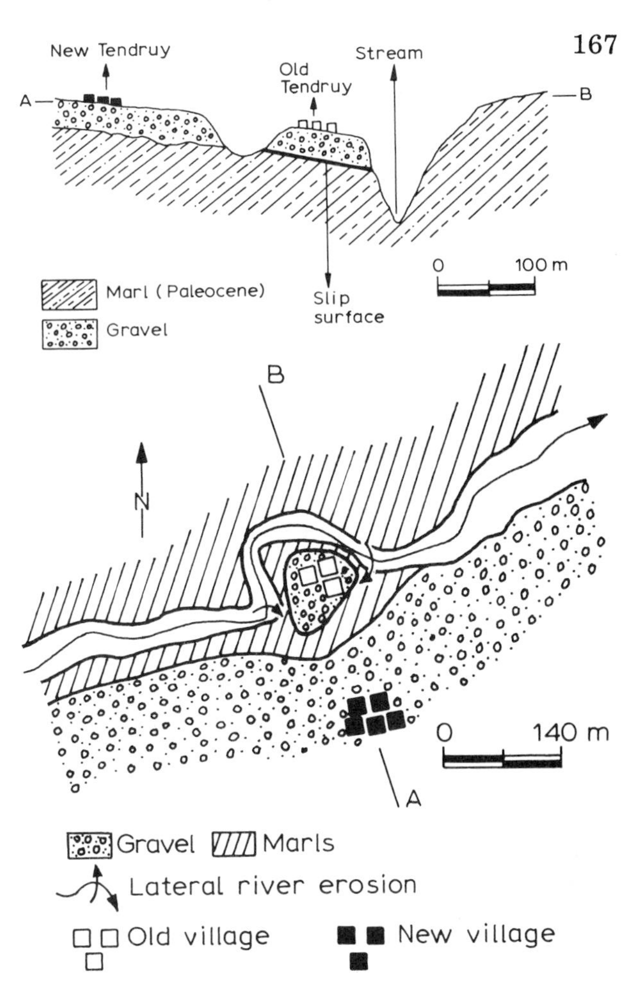

Fig. 7.22 Plan view and cross-section of Tendruy, northern Spain (Avci, 1978).

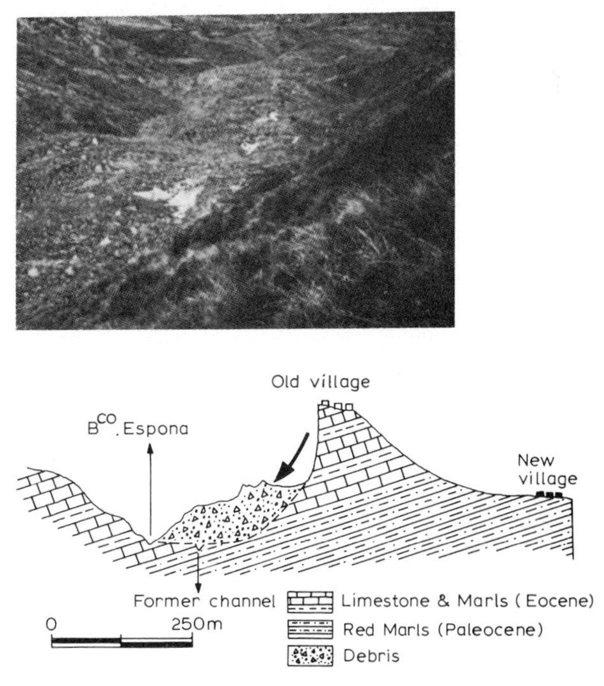

Fig. 7.23 Plan view and cross-section of Puigcergos, northern Spain (Avci, 1978).

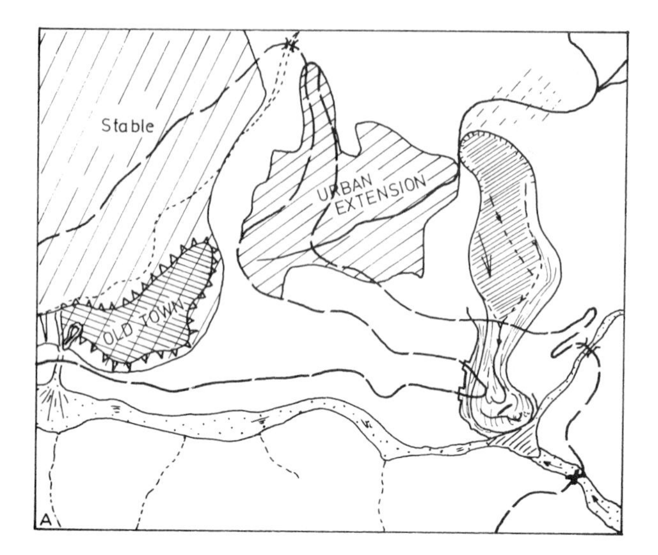

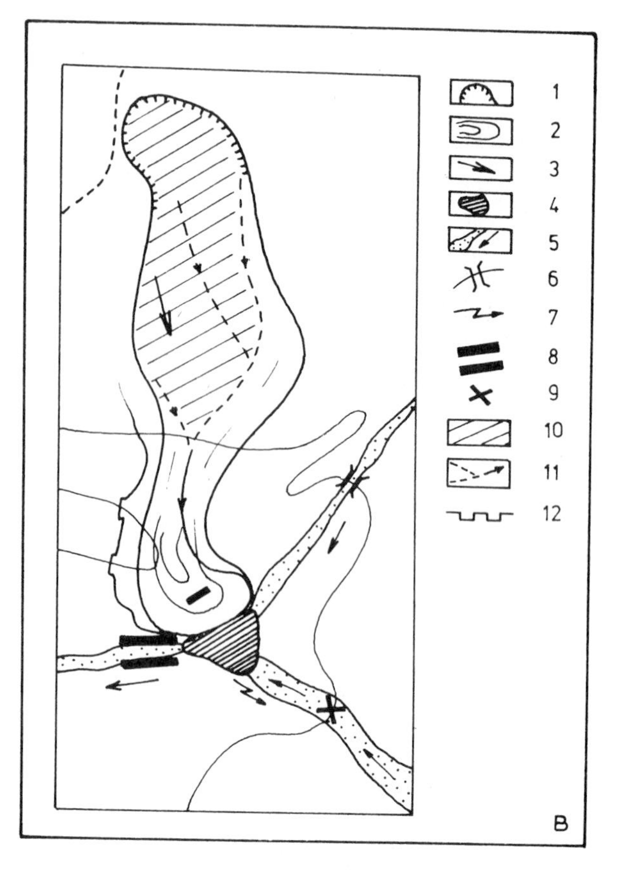

Fig. 7.24 The old part of the small town of Oriolo, Calabria, Italy is located on stable rock, but the urban extension planned in the early nineteen seventies is on unstable flysh slopes. A major slide occurred when the land was cleared for building and is pictured in more detail to the right.

Key: 1. scarp of slide, 2. tongue of slide, 3. surface flow, 4. lake ponded up by slide, 5. gravelly riverbeds, 6. bridge, 7. push induced by slide that destroyed bridge, 8. concrete revetment, 9. destroyed bridge, 10. levelled barren terrain, 11. corrugated iron drainage lines on slide, 12. drain at side of slide, with culverts.

Scale approximately 1:50,000 and 1:15,000 respectively.

Fig. 7.25 Extended stereopair at the scale of 1:17,000 showing the small town of Aliano, Basilicata, Italy, situated at the crossing of several small faults which govern the growth of ravines endangering the site of the town.

into the fan. Concrete walls and other structures may be erected to protect existing settlements. Such works on the fan should be complemented by harnessing of the torrent upstream by groynes/revetments.

An example of successful harnessing of a torrent for the protection of a settlement situated on its fan is the work carried out in the Bretterwand Bach for the protection of Matrei in East Tirol, Austria. Fig. 7.26 shows the situation. The Bretterwand Bach fan is formed just upstream of the junction of the Tauern Bach, coming from the north, with the Isel River. The erosion in the torrent's basin has, in the past, caused flooding and deposition in the village. It has also pushed the Tauern Bach towards the western side of the main valley. To prevent further damage to the village, high retaining walls have been constructed at either side of the torrent, along its bed from upstream of the village to the lower end of the fan. Fig. 7.27 shows a section of this wall, as seen from the torrent bed and with part of the village in the background. The new Felbertauern motorway passes high above the torrent bed and the walls to the east of the village. The main street of the village, however, crosses the torrent by way of a rather low wooden bridge; there the retaining walls had to be interrupted, though in case of emergency, closure by way of thick wooden planks is possible, see fig. 7.28.

Numerous solid gabions now stop vertical incision and drastically diminish the transportation of sediment further upstream in the Brettenbach. Fig. 7.29 show some of these structures. It is worth noting that the Bretterwand Bach now enters the Tauern Bach inversely at a rather acute angle. If there was to be a substantial transportation of coarse debris on the fan, this situation could impede the drainage in case of high discharges. Since, however,

Fig. 7.27 Detail of the retaining wall of the Bretterwand torrent in Matrei in East Tirol.

Fig. 7.28 Interruption of the retaining wall where the main street passes over the torrent. Provisions are made to close it by plankings in case of emergency.

the sedimentation has been strongly reduced, this danger does not appear serious. The decrease in gradient of the lower reaches of the Brettenwand Bach, provoked by this inverse course, is even favourable in this case, since it counteracts the tendency of incision in the fan which could develop as a result of the decreased sedimentation and the lower ruggedness of the bed. The fact that the torrent is now confined to a rather narrow sector of the fan, tends to increase the water velocity in case of flood and is thus favourable for the evacuation of waste. Obviously, the substantial costs of engineering works carried out have resulted in an effective protection of Matrei and its inhabitants against the caprices of the Bretterwand Bach.

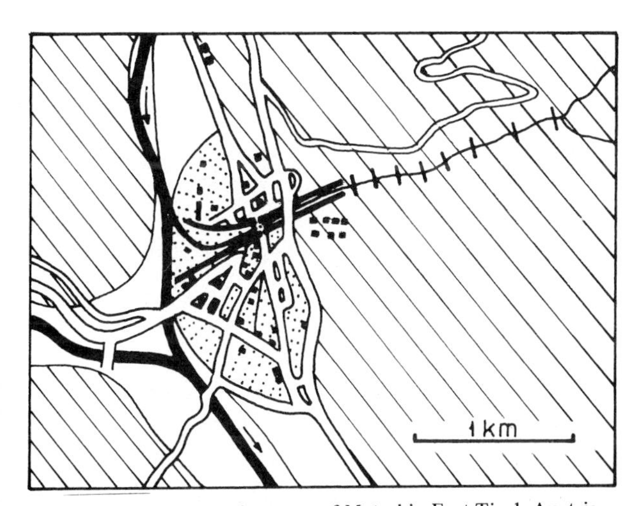

Fig. 7.26 Plan view of the town of Matrei in East Tirol, Austria, situated on the fan (dotted) of the Bretterwand torrent which originates in the mountains (hachured) to the east. It is now controlled by a series of check dams and confined between stone revetments in and near the settlement.

Fig. 7.29 The Bretterwand torrent in the mountains upstream of Matrei in East Tirol with some of the gabions indicated in fig. 7.26.

7.4.2 A study of the mudflow hazard at Craco, Basilicata, Italy using orthophotographs

Orthophotographs are increasingly being used in surveys of urban areas; this is mainly due to their good geometric properties. Realizing this, it is important to stress that interpretation of orthophotos need not always be required in thematic and resource mapping, but can be effected with normal aerial photographs, after which the annotations made are transferred to the orthophotos. This can be done with great precision owing to the wealth of detail on the photographic image. In this case, orthophotos are mainly used as a base for thematic information. Stereoscopy can also be obtained with orthophotos. If the orthophotos are produced on an enlarged scale, as is often the case, an additional advantage is that the information to be mapped, can be plotted in more detail and with even greater precision. This is especially important for detailed mapping, for example, that related to certain urban engineering works.

Orthophotographs have been successfully applied by the author (Verstappen, 1977) in the study of a disastrous mudflow which destroyed a large part of the old village of Craco (Basilicata), Italy, and of an incipient one threatening new extensions of this settlement. This demonstrates the importance of geomorphology for urban development and planning. The aerial photography used for the interpretation was panchromatic, infrared and true colour at the scale of 1:15,000 dating from 9th August 1976. The orthophoto of fig. 7.30 is at a scale of approximately 1:10,000 and was prepared from the panchromatic photography using a Santoni Simplex 2 C. Generally, the panchromatic emulsion suited the purpose of the mudflow study, though the infrared readily showed moist zones and the true colour photos gave improved interpretation of relevant vegetation elements, such as the zoning of grass and scrub on the pressure ridges of the mudflows.

Craco is situated on a narrow strip of subhorizontal Middle to Lower Pleistocene Tempa del Muto conglomerates. The formation is only a few tens of metres thick and the Craco outcrop is a relic of a formerly more extensive deposit, preserved because of its position on the divide between the Cavone River basin in the north and the Agri River basin in the south. Many other settlements in the area are similarly situated on these (and other) conglomerates, since in other parts landsliding and badland formation are a widespread threat to building and urban development. Water supply is another reason for this preference; a perched watertable occurs at the contact of the conglomerates with the underlying rocks of low permeability. In the past, the elevated hilltop position was also favoured for reasons of defence.

Surrounding Craco and underlying the conglomerates mentioned, Cretaceous (possibly lower Miocene) clays occur. These are very susceptible to flowage when water is absorbed. The slopes south of Craco are the most unstable ones and are depicted in the orthophoto of fig. 7.30. Near Craco the formation is subhorizontal. Numerous small springs, moist zones and wells mark the contact between the conglomerates and the clays. The road Pisticci-Stigliano skirts the southern edges of Craco, around this contact area. The road has mostly a crestline location, to avoid cross-drainage and slope instability problems where possible.

The Cretaceous clays are limited to the south by a fault scarp downslope of which greyish-blue Gravina clays of Calabrian-Pliocene age outcrop. They dip gently southwest towards the Bruscata Rivulet which flows eastward at the foot of the slope. This formation is subject to the formation of ravines rather than to mudflows. Parts of the slope are intensely dissected, although some rather well-preserved, non-dissected areas, where creep dominates, can also be observed. The faultscarp substantially adds to the

Fig. 7.30 Annotated orthophoto of the town of Craco, Basilicata, Italy, at a scale of 1:10,000. The major mudflow that destroyed the old town in 1964 and the incipient flow that endangers the houses of the new town can be clearly seen.

instability of the slope below Craco,because it decreases the strength and increases the general gradient of the slope. Even more important is the groundwater that emerges in the vicinity of the fault scarp causing continuous mudflow activity where ravines cut across the scarp. It is uncertain whether tectonic movements along the fault scarp still occur.

A major mudflow extends from the lower part of the old nucleus of Craco in a south-west direction, almost to the Bruscata Rivulet. Its total length is approximately one kilometre and its development has been gradual, possibly with violent phases in the remote past. The long tongue of the flow downslope of the fault scarp is under vegetation and its surface shows flow structures and pressure ridges. The active parts of the flow are near the fault scarp and farther upslope where water is injected at the contact of the conglomerates with the Cretaceous clays. By far the most active portion of the mudflow is at and near the fault scarp. It is probable that the mudflow originated there,as is the case with two much smaller mudflows situated farther to the east. Up to the present these have not reached farther upslope into the Cretaceous zone and are, therefore, not hazardous to Craco.

However, the major mudflow has developed headward to such an extent that the whole Cretaceous formation has been affected and the contact with the conglomerates has been reached. Eventually, after a period of heavy rains, disaster overcame Craco in January, 1964. Houses began to crack and move downslope until they finally collapsed. Since the movement was gradual, the area could be evacuated in time and no casualities occurred. When the ground came to rest, it appeared that the settlement was heavily damaged, leaving only a narrow zone in the north and west untouched. The terrestrial stereophoto of fig. 7.31 shows the situation. Also the Pisticci-Stigliano road was destroyed over a length of about 150 metres, as can be seen in the terrestrial stereophoto of fig. 7.32. The new alignment of the road constructed after the disaster can be seen in the orthophoto of fig. 7.30, on which the old road is indicated by a dashed line.

Craco was not rebuilt on the same spot, but slightly farther to the north-west, again on the conglomerates and alongside the road. The street pattern of the new settlement is clearly distinguished from that of the old. A comparatively small but active mudflow can be seen in in the north-west part of the orthophotograph just downslope of the new buildings. Its activity centres at the fault-scarp and the mudflow tongue extends in the Gravina clays immediately downslope. It is clear, however, that the mudflow is developing headward with several branches and incipient branches which stretch halfway up to the conglomerates. Beyond the contact more water will be fed into the system. There can be little doubt of the fate which will sooner or later befall the new parts of Craco if no adequate protective measures are taken. Drainage of the slope, controlled land use and careful handling of urban disposal water rank first among the remedial measures to be contemplated. The ravine occurring between the major mudflow and the incipient one described above also merit extreme care although its upper parts have been controlled to some extent and no recent mudflow activity can be observed.

Fig. 7.31 Stereoscopic ground view of the destroyed old town of Craco, Basilicata, Italy.

Fig. 7.32 Stereoscopic ground view of the destroyed section of the Pisticci-Stigliano road destroyed by the Craco slide.

References

Abreu, E., and Verhasselt, Y., 1981. Quelques aspects géographiques du développement de Caracas. Cahiers d'Outre Mer, 34(134): 180–186.

Alfore, J.T., et al., 1973. Urban geology master plan for California. The nature, magnitude, and costs of geologic hazards in California and recommendations for their mitigation. Final report, Cal. Div. Mines and Geol., Sacramento, Bull. 198, 111 pp.

Anon., 1969. Geographical zonation of the plain skirting Bucharest and the micro-regions of a special interest. Inst. Geol. Geogr. Acad. Romania: 114–139.

Anon., 1974. South Santa Clara County study. Cal. Geol., 27(3): 64–67.

Avci, M., 1972. Landclassification for engineering purposes. M.Sc. Thesis ITC, 112 pp.

Avci, M., 1978. Airphoto interpretation of mass movements with special reference to the Spanish Pyrenees, Tremp., N. Spain. Proc. 3rd I.A.E.G. Conf.: IV 2: 72–78.

Ballon, L., 1966. Fracturamentos del suelo en el pueblo de Chumuch (Prov. de Celendin, Dept. de Cajamarca). Bol. Com. Carta Geol. Nac. Peru, 13: 133–141.

Becker, C., and Mulhern, J., 1975. Sediment yield and source prediction for urbanizing areas. In: Present and prospective technology for predicting sediment yields and sources. Proc. Sediment-yield Workshop, USDA Sedimentation Lab. Oxford (Miss.), Nov. 1972, USDA Agric. Res. Serv. ARS-S-40: 83–88.

Berlin, G., et al., 1975. Seismology. CRC Critical Reviews in Environmental Control, 5(3): 275–396.

Berry, B.J.I., and Neils, E., 1969. Location, size and shape of cities as influenced by environmental factors: the urban environment. In: H.S. Perloff (Editor), The quality of the urban environment, Baltimore.

Bishop, A.W., et al., 1969. Geotechnical investigation into the causes and circumstances of the disaster of 21st October, 1966. In: A selection of technical reports submitted to the Aberfan tribunal. HMSO, 80 pp.

Bishop, A.W., 1973. The stability of tips and spoil heaps. Quart. J. Eng. Geol., 6(3–4): 335–376.

Bolt, A., et al., 1969. The safety of fills. In: Geologic and engineering aspects of San Francisco Bay fill. Cal. Div. Mines and Geol., Spec. Report 97: 119–130.

Bouma, J., 1973. Application of soil surveys to selection of sites for on-site disposal of liquid household wastes. Geoderma, 10(1–2): 113–122.

Branagan, D.F., 1972. Geological data for the city engineer: a comparison of five Australian cities. Proc. 24th Int. Geol. Congr. Montreal, Section 13, Eng. Geol: 3–12.

Brandt, H., et al., 1971. An economic analysis of erosion and sediment control methods for watersheds undergoing urbanization, 191 pp.

Burton, I., et al., 1969. The shores of Megalopolis: coastal occupance and adjustment to flood hazard. Final Rep. O. N.R. Contract no. 388–073, Elmer (New Jersey): 1–603.

Blanc, P., and Cleveland, B., 1971. Natural slope stability as related to geology. San Clemente Area, Orange and San Diego Counties, California. Cal. Div. Mines and Geol., Special Report, 98: 1–19.

Calembert, L., 1958. Le sous-sol. Étude de l'influence des facteurs géologiques et miniers sur les déformations du sol de la région liégeoise. Plan d'aménagement de la région liégeoise, 1, l'enquete: 57-76.
Chandler, T.J., 1976. Physical problems of the urban environment. Geogr. J. 142(1): 57-80.
Clootz-Hirsch, A.R., and Maire, G., 1972. Map of the socio-economic sectors of Strasbourg. The physical constraints which influence urbanization. Cartographie Géom.: 205-210.
Cluff, L.S., 1968. Urban development within the San Andreas fault system. Proc. Conf. geol. problems of San Andreas fault system, Stanford (Cal.), 1967, 11: 55-66.
Cluff, L.S., 1969. Risk from earthquakes in the modern urban environment, with special emphasis on the San Francisco Bay area. Urban environmental geology in the San Francisco Bay region: 25-64.
Cooke, R.U., et al., 1979. Assessment of geomorphological problems in urban areas of dry lands. U.N.U. Publ., 2 Vols., 490 pp.
Cooke, R.U., et al., 1982. Urban geomorphology in dry lands. Oxford Univ. Press, Oxford,
Cossio, A., 1966. Landslides and stability of lakes in the area of Conchucos. Bol. Com. Carta Geol. Nac. Peru, 13: 7-73.
Cossio. A., 1966. Deslizamientos de tierras y condiciones de seguridad de las lagunas en el área de Conchucos (Prov. de Pallasca, Dept. de Ancash.). Bol. Com. Carta Geol. Nac. Peru, 13: 57-73
Cossio, A., 1966. Fracturamientos de tierras en las áreas de Pillipampa, Santa Rosa, y Miraflores (Prov. de Corongo y Pallasca, Dept. de Ancash). Bol. Com. Carta Geol. Nac. Peru, 13: 95-110.
Cotecchio, V., et al., 1980. Studio sulle possibilitá di stabilizzazione del rilievo collinare di S. Arcangelo, in Basilicata. Geol. Appl. e idrogeol., 15: 305-327.
Cratchley, C.R., and Denness, B., 1972. Engineering geology in urban planning with an example from the new city of Milton Keynes. Proc. 24th Int. Geol. Congr. Montreal, Section 13, Eng. Geol: 13-22.
Cypra, K., 1969. Technical services for the urban floodplain property manager: Organization of the design problem, 1969.
Danehy, A., and Harding, C., 1969. Urban environmental geology in the San Francisco Bay region, 162 pp.
Davis, D., and Meyer, P.A., 1972. Sand and gravel versus urban development. Environmental geology of the Wasatch Front: Q1-Q9.
Degenbold, H.J., 1970. An engineer's perspective on geologic hazards. In: R.A. Olson and M.M. Wallace (Editors), Geologic hazards and public problems. U.S. Office Emergency Preparedness, Region 7, Proc. Conf. San Francisco, Calif. U.S. Govt. Printing Off.: 183-195.
Demirev, A., et al., 1974. Influence of volumetric changeability of clays on the cracking of buildings in the town of Pernik (in bulgarian). Godisnik na Vissija Minnogeol. Inst. Sofia, 18(4): 193-201.
Doornkamp, J.C., et al., 1979. Geology, geomorphology and pedology of Bahrein. Geo Books, Norwich, U.K.
Fallat, C., and Tetherow, T., 1973. A discussion of methodology used in the analysis of geomorphic phenomena encountered in the environmental assessment of a proposed personal rapid transit system in Denver (Col.). Great Plains – Rocky Mts. Geogr., 2: 12-17.
Folk, R.L., 1975. Geological Hindplanning: An example from a Hellenistic-Byzantine city: Stobis, Yugoslavia, Macedonia. Environmental Geology, 1(1): 5-22.
Font, R.G., and Williamson, E.F., 1970. Geologic factors affecting construction in Waco. Urban geology of greater Waco-Pt. 4, Engineering: 1-34.
Gambolati, G., et al., 1974. Predictive simulation of the subsidence of Venice. Science, 183(4127): 849-851.
Garcia, W., 1966. Derrumbes de tierras en el cerro Puca-Puca (Dist. Acoria, Prov. y Dpto. de Huancavelica). Bol. Com. Carta Geol. Nac., Peru, 13: 153-158.
Golany, G. (Editor), 1978. Urban planning in arid zones. Wiley, New York
Goldman, H.B., 1969. Geology of San Francisco Bay. Geologic and engineering aspects of San Francisco Bay fill: 11-29.
Grube, F., and Germany, F.R., 1972. Urban and environmental geology of Hamburg (Germany). Proc. 24th Int. Geol. Congr., Montreal, Section 13, Eng. Geol: 30-36.
Grumazescu, H.C., 1967. L'étude des conditions géomorphologiques des localités du delta du Danube en vue de leur systématisation. Symp. Int. Géom. Appl., Bucharest: 223-232.
Guy, H.P., 1975. Urban sediment problems: a statement on scope, research, legislation and education. Hydraul. Div. ASCE, 101(HY4), Proc. Paper, 11256: 329-340.
Hacket, J.E., 1968. Quaternary studies in urban and regional development. The Quaternary of Illinois: 176-179.
Hamilton, J.L., and Owens, W.G., 1972. Geologic aspects, soils and related foundation problems, Denver Metropolitan area (Col.), Colorado Geol. Survey, Environmental Geology, 1: 1-20.
Harjoaba, I., and Donisa, I., 1967. Le critérium géomorphologique dans la systématisation du territore et du réseau des localités. Symp. Int. Géom. Appl., Bucharest: 233-236.
Hart, S., 1974. Potentially swelling soil and rock in the Front Range urban corridor, Colorado. Colorado Geol. Survey, Environmental Geology, 7
Hewitt, W.P., 1970. Construction hazards in Utah. In: Governor's conference on geologic hazards in Utah, December 14, 1967. Utah Geol. and Mineral Survey., Spec. Studies, 32: 19-20.
Hilpman, L., 1970. Urban growth and environmental geology. Governor's conference on environmental geology: 16-19.
Hoffa, M., 1968. Some physical and geographical features of the Kolobrzeg region and their significance to the harbour and health resort. Poznaskie Towarzuatwo Przyjaciol. Prace Komisji Geogr.-Geol., 9(3): 1-115.
Hogberg, R.K., 1971. Environmental geology of the Twin Cities Metropolitan Area. Minnesota Geol. Survey Educ. Ser., 5: 1-63.
Holubec, I., Geotechnical aspects of coal waste embankments. Can. Geotech. J., 13(1): 27-39.
Horton, A., et al., 1974. The geology of Peterborough. Explanation of 1:25,000 special geological sheet including parts of TF 00, 10, 20, TL 09, 19, 29. U.K. Inst. Geol., Sci. Report, 73/12: 1-86 pp.
Houghton, P.D., 1975. Soil erosion within the urban environment. Part I. Recognition of urban erosion. Soil Conser vation Service New South Wales, 31(3): 172-178.
Hunter, J.H., and Schuster, R.L., 1968. Stability of simple cuttings in normally consolidated clays. Géotechnique, 18(3): 372-278.
Hutchinson, J.N., 1959. The landslides of February, 1959 at Vibstad in Namdalen. Norges Geotekniske Inst. Publ., 61.
Iancu, M., 1967. Les prognoses géomorphologiques et l'extension de certains centres urbains. Symp. Int. Géom. Appl., Bucharest: 41-48.
Johnson, A.I., 1970. Suggested method for geologic reconnaissance of construction sites. Special procedures for testing soil and rock engineering purposes (5th ed.): 43-55.
Kates, R.W., 1962. Hazard and choice perception in floodplain management. Univ. Chicago, Dept. of Geogr. Res. Paper, 78.
Kates, R.W., 1965. Industrial flood losses: damage estimation

in the Lehigh Valley. Univ. Chicago, Dept. of Geogr. Res. Paper, 98, 25 pp.

Kaye, A., 1968. Geology and our cities. Trans. New York Acad. Sci., 30: 1045–1051.

Kehle, R.O., 1970. Earth movements–an increasing problem in the cities. Environmental geology–AGI short course lecture notes, Milwaukee (Wis.),: 1–78.

Kellaway, G.A., and Taylor, J.H., 1968. The influence of landslipping on the development of the City of Bath, England. Proc. 23rd Int. Geol. Congr. Prague, Section 12: 65–76.

Ker, D.S., 1970. Renewed movement on a slump at Utiky. New Zealand J. Geol. & Geoph., 13(4): 996–1017.

Kiersch, G.A., 1969. Pfeiffer versus General Insurance Corporation–landslide damage to insured dwelling, Orinda, California, and relevant cases. In: Legal aspects of geology in engineering practice. Geol. Soc. Amer. Eng. Geol. Case Histories, 7: 81–93.

Klimaszewski, M., 1967. Geomorphological research for town planning purposes. Symp. Int. Géom. Appl., Bucharest: 17–22.

Knight, F., 1971. Geologic problems of urban growth in limestone terrains of Pennsylvania. Bull. Ass. Eng. Geol., 8(1): 91–101.

Kollmorgen, W.M., 1953. Settlement control beats flood control. Econ. Geogr.,29: 208–215.

Kolomensky, N.V., 1959. Buts principaux des recherches géomorphologiques lors de l'évaluation géotechnique d'une localité. Prospection et protection, Moscow, 7:42–48.

Kotlov, F.V., 1972. Modifications in the geological environment resulting from growth of cities. Proc. 24th Int. Geol. Congr. Montreal, 1972, Symp. 1, Earth Sciences and the Quality of Life: 36–40.

Kugler, H., 1978. Einheitliche Gestaltungsprinzipien und Generalisierungswege bei der Schaffung geomorphologischer Karten verschiedener Massstäbe. Peterm. Geogr. Mitt. Erg. H. 271, 1968: 259–279 (see also: Guide to medium-scale geomorphological mapping, Demek-Embleton (Editors), Brno, 348 pp.

Leggett, F., 1973. Cities and geology, 624 pp.

Leighton, F., 1972.Origin and control of landslides in the urban environment of California. Proc. 24th. Int. Geol. Congr. Montreal, Section 13, Eng. Geol.: 89–96.

Leser, M., 1976. Medium-scale geomorphological mapping in connection with settlement and recreation. Guide medium scale geom. mapping, Demek-Embleton (Editors), Brno: 176–181.

Lobdell, L., 1970. Settlement of buildings constructed in Hackensach meadows. Amer. Soc. Civil Eng. Proc. 96, paper 7398, J. Soil Mech. and Found. Div. SM4: 1235–1248.

Loughry, F.G., 1973. The use of soil science in sanitary landfill selection and management. Geoderma, 10(1–2): 131–140.

Lukey, M.E., 1974. Milton Keynes new city–a site survey challenge. Ground Eng., 7(1): 34–37.

Lundgren, A., and Rapp, A., 1974. A complex landslide with destructive effects on the water supply for Morogoro town, Tanzania. Geogr. Ann.,A, 56(3–4): 251–260.

Martiniuc, C., and Băcauănu, V., 1969. The town of Jassy. Problems of applied geomorphology in town planning. Symp. Int. Géom. Appl., Bucharest: 33–39.

Matsuda, I., 1974. Distribution of the recent deposits and buried landforms in the Kanto Lowland, Central Japan. Geogr. Reports Tokyo Metropolitan Univ., 9: 1–36.

McCulloch, D.S., et al., 1969. San Francisco peninsula trip for planners and public officials, Field Trip 5.,Field Trips–Ass. Eng. Geol. Annual Meeting, San Francisco (Cal.): E1–E10.

McKechnie Thomson, G., and Rodin, S., 1972. Colliery spoil tips–after Aberfan. The Inst. Civil Eng., London: 1–59.

Mendivil, S., 1969. Deslizamiento de tierras en el cerro Martin Capasha de Huariaca (Prov. de Cerro de Pasco, Dept. de Pasco). Bol. Com. Carta Geol. Nac., Peru, 12: 45–56.

Mendivil, S., 1966. Remoción de tierras en el anexo de Anascapa (Prov. General Sánchez Cerro, Depto. de Moquegua). Bol. Com. Carta Geol. Nac., Peru, 13: 191–204.

Merril, C.L., et al., 1960. The new Aklavik, search for the site. Eng. J., 43: 52–57.

Mihailescu, V., and Martiniuc, C., 1967. Géomorphologie et urbanisme en Rumanie. Symp. Géom. Appl., Bucharest:11-15.

Mittelstedt, H., 1976. Terraces protect Glen Ullin. Soil Conservation, 41(7): 9–10.

Morales, L., 1966. A landslide in the Yuncanpata area, province and dept. of Pasco. Bol. Com. Carta Geol. Nac., Peru, 13: 35–43.

Morariu, T., and Mac, I., 1967. Influence of relief on the planning and development of the town of Cluj. Symp. Int. Géom. Appl., Bucharest: 23–31.

Nakano, T., 1970. Lands below sealevel due to land subsidence in the urban areas of Japan. Ass. Japanese Geogr. Spec. Publ. 2: 237–244.

Nash, J.K.T.L., 1969. Report on the stability of Aberfan Tip No. 7. In: A selection of technical reports submitted to the Aberfan tribunal. HMSO: 83–88.

Nimigeanu, G., 1967. Les conditions géomorphologiques et l'organisation du réseau de localités rurales au environs de la ville de Dragasani. Symp. Géom. Appl., Bucharest: 237-242.

Nossin, J.J., 1972. Landsliding in the Crati Basin, Calabria, Italy. Geol. en Mijnbouw, 51(6): 591–607.

Olson, R.A., and Wallace, M.M., 1970. Geological hazards and public problems. U.S. Office of Emergency Preparedness, Region 7, Proc. Conf. San Francisco, Calif., 1969, U.S. Govt. Printing Off., 335 pp.

Oriordan, T., 1973. The Tanunanui landslip of August 1970: an investigation of landslip protection policy in New Zealand. New Zealand Geographer, 29(1): 16–30.

Pavoni, J.L., and Hagerty, D.J., 1973. Geologic aspects of landfill refuse disposal. Eng. Geol., 7(3): 219–230.

Pestrong, R., 1968. The role of the urban geologist in city planning. California Div. Mines and Geol.,Mineral Inf. Serv., 21(10): 151–152.

Powell, M.D., et al., 1970. Urban soil erosion and sediment control. Nat. Ass. Counties Res. Found., Washington D.C., Water pollution control research series, W71–02276, FWQA-15030–DTL–05/70, 113 pp.

Price, D.G., 1971. Engineering geology in the urban environment. Quart. J. Eng. Geol., 4(3): 191–208.

Rabelo, R.-M., 1980. Active geomorphological processes in Basilicata, Italy. ITC M.Sc. thesis, Enschede, 102 pp.

Radbruch, D.H., 1968. Engineering geology in urban planning and construction in the United States. Proc. 23rd. Int. Geol. Congr., Prague, Sect. 12, Eng. geol. in country planning: 105–111.

Rockaway, J.D., 1972. Evaluation of geologic factors for urban planning. Proc. 24th Int. Geol. Congr., Montreal, Section 13, Eng. Geol.: 64–69.

Roswell Moore, L., 1969. Geological report on the tipping site and its environs at Merthyr Vale and Aberfan. In: A selection of technical reports submitted to the Aberfan tribunal. HMSO: 147–185.

Rowan, C.L., 1973. It's not nice . . . Soil Conservation, 38 (9): 195–197.

Salguiero, P.R., 1965. Landslide investigation by means of photogrammetry. Photogrammetria, 20: 107–114.

Sandy, I.M., 1977. Atlas land use Indonesia. Jakarta

Sauer, E.K., 1975. Urban fringe development and slope instability in southern Saskatchewan. Can. Geotechn. J., 12(1): 106–118.
Schmidt, G., 1969. Contribution of photointerpretation to the reconstruction of the geographic-topographic situation of the ancient ports in Italy. Proc. 10th I.S.P. Conference, Lisbon, 32 pp.
Schmol, H.R., and Dobrovoin, E., 1968. Geology as applied to urban planning—an example from the Greater Anchorage Area Borough, Alaska. Proc. 23rd Int. Geol. Congr., Prague, 1968, Sect. 12, Eng. Geol. in country planning: 39–56.
Stauffer, T., 1974. Occupance and use of underground mined-out space in urban areas: an annotated bibliography. Council of Planning Librarians, Exchange Bibl., 602: 1–50.
Taylor, R.K., 1971. The functions of the engineering geologist in urban development. Quart. J. Eng. Geol., 4(3): 221–240.
Tricart, J., 1962. L'épiderme de la terre, Esquisse d'une géomorphologie appliquée. Masson et cie., Paris: 54–75.
Varnes, D.J., 1969. Stability of the west slope of Government Hill, port area of Anchorage, Alaska. U.S.G.S. 1258–D, 61 pp.
Verstappen, H.Th., 1977. Orthophotos in applied geomorphology: the mudlfow hazard at Cráco, Italy. ITC Journal, 4: 717–722.
Wayne, J., 1969. Urban geology—a need and a challenge. Indiana Acad. Sci. Proc. 78: 49–64.
White, G.F., et al., 1968. Change in Urban Occupance of floodplains in the United States. Univ. Chicago, Dept. of Geogr. Res. Paper, 86: 1–235.
Wigginton, B., 1969. Bay mud—Why does it have such a bad reputation? Urban environmental geology in the San Francisco Bay region: 75–81.
Williams, G.M.J., 1969. Inquiry into the Aberfan disaster. In: A selection of technical reports submitted to the Aberfan tribunal. HMSO: 91–115.
Wilson, J.J., 1966. A landslide at Mushuquino mountain District and Province of Pallasca, Dept. of Ancash. Bol. Com. Carta Geol. Nac., Peru, 13: 25–34.
Wilson, G., and Grace, H., 1942. The settlement of London due to underdrainage of the London Clay. J. Inst. Civ. Eng., 19–20: 100–127.
Wolman, M., and Schick, A.P., 1967. Effects of construction on fluvial sediment in urban and suburban areas of Maryland. Water Resources Research, 3(2): 451–464.
Woodland, A.W., 1969. Geological report on the Aberfan tip disaster of 21st October 1966. In: A selection of technical reports submitted to the Aberfan tribunal. HMSO: 119–145.
Wrona, A., 1973. Influence of industrialization in Upper Silesian Industrial District upon changes in land surface contours. Przeglad Geografiszny, 45(3): 557–572.
Yanez, G.A., 1979. Geomorphological applications using aerial photographs, two case studies in Venezuela, ITC Journal, 1: 85–98.
Yelverton, Ch.A., 1973. Land stability insurance company, origin, goals, procedures and guidelines. Landslide, 1(2): 53–56.
Young, E., 1968. Urban planning for sand and gravel needs. California Div. Mines and Geol. Mineral Inf. Serv., 21(10): 147–150.

GEOMORPHOLOGY AND ENGINEERING

8.1 Introduction

In order to be successful, engineering works have to be adequately implanted in the natural environment. They almost invariably require a considerable modification of the environment for which a thorough knowledge of its characteristics is indispensable. Pre-existing equilibria are disturbed in many cases and new, reactivated or accelerated processes are then initiated and may lead to drastic re-adjustments, such as slope failure of new road cuts, during or shortly after the construction work. However, slow, almost imperceptible deterioration of the environment also may result. This determines to an important degree whether or not the structure will have a lasting beneficial effect. An almost classical example is the gradual salination and subsequent abandonment of land where irrigation schemes have been devised which lack adequate drainage conditions.

History has proved that the survival or decline of civilizations has been strongly affected by the environmental impact of their engineering works (Camp, 1963). Some ancient engineers have achieved a degree of environmental adaptation in the design of their structures that is unequaled by modern technology. This fact is partly explained by the smaller scale and lower pace of ancient engineering, but also the fact that in those days the engineer was usually less remote from nature and from people who, by tilling the land generation after generation, accumulated a profound understanding of the environmental situation. When, in the seventeenth century, the industrial growth gained momentum in Europe and led to rapidly increasing engineering activities, civil engineers through their work were suddenly faced with matters such as slope instability, river work, coastal recession, etc. (Price, 1977). Their practical experience contributed not only to our knowledge of geomorphological processes and environmental dynamics, but also to the development of fundamental concepts in earth sciences, as mentioned in section 1. The fact that thereafter earth sciences and engineering developed rather independently is a regrettable turn of events for both parties and has left its traces in numerous engineering failures. Substantial emphasis on environmental issues is a *conditio sine qua non* for modern engineering and geomorphology has its distinct role in this context. (Geologie en Mijnbouw, 1979; Kugler, 1963; Largiller, 1970; Pecsi, 1970).

It is of utmost importance for successful cooperation between an earth scientist such as the geomorphologist and the engineer, that the participation of the former is sought from the beginning and not as an afterthought once problems have arisen and the engineer is in trouble. It is only in this way that disappointments, misjudgement of natural hazards, etc.,can be avoided and that failure of a structure, unexpected extra costs and loss of time can be prevented. Too often the engineer seeks advice in a late stage, when during construction a land slide or river work threatens the successful implementation of the structure. Characteristically, the engineer will insist on factual, if possible, quantitative information solely related to matters of relevance to engineering and adequately presented in a concise and understandable way (Coates, 1981; Matula, 1979). The detail of the information required will vary with the phase of the work, but if the criteria mentioned above are adhered to, the geomorphologist will be capable of making a valid contribution,not only during the planning, but also in the design, construction and maintenance phases of the work.

Although it is undeniable that geomorphology can contribute considerably to the successful planning and implementation of engineering works (Woods, 1967), the type of contribution varies in character during the subsequent phases of the work even though these phases often tend to overlap or merge. During the reconnaissance phase of the work,when comparatively large areas have to be investigated, emphasis will be laid on major terrain characteristics and a broad terrain classification will have to be made in order to arrive at a first breakdown of the land into major units. Aerial photographs,and in an increasing degree also other types of aerospace imagery,play an important role in providing the required, relevant information within acceptable time limits (Voûte, 1964). Landform analysis will also assist in revealing major geological factors and data on engineering soils. It will also contribute to the efficient planning and execution of field investigations at representative sites of easy access. Last but not least, the search for building materials can be optimized because the availability and type of construction materials is normally associated with certain geomorphological types, the occurrence and distributional pattern of which may offer important clues. This matter is elaborated upon in subsection 8.4.

Feasibility studies are a second phase in engineering works and likewise relate to fairly large areas. This phase differs from the reconnaissance phase in that more detailed investigations form a part of the work, normally including a fair amount of fieldwork, the digging of test pits and field and laboratory investigations on soil mechanics. The role of geomorphology in this phase is comparable to that during the reconnaissance phase, although more detailed studies and the use of larger scale aerial photographs will be required. In the subsequent 'avant-project' phase,the studies are mainly restricted to one or more sites considered to be the most favorable. For the geomorphologist this means a complete and detailed geomorphological survey of these limited areas. The emphasis of the survey work will gradually move from landforms to geomorphological processes. Attention should be given to both active and dormant processes which may be activated or re-activated as a result of the engineering activities. The specific problems of each of the sites should be studied in detail and clearly formulated. Detailed investigations, for example, the effect of palaeoclimates on weathering and slope stability are relevant in this phase, in which large-scale aerial photographs, thermographs, elaborate field studies and laboratory investigations should be efficiently incorporated. It is on the basis of the results obtained that adequate advice can be given on problems cropping up during the execution of the engineering work.

The geomorphologist should be capable of interpreting the geomorphological situation of landforms and processes in terms of environmental conditions and recognizing potential engineering problems such as land slides, hard rock cuts or river cutting. He is also expected to communicate his insight on the engineering properties and the active processes inherent to the natural environment in which the structure is to be erected to the engineer. Only then can a dialogue with the engineer be generated, which is of crucial importance in making proper decisions. The contribution of the engineer to this dialogue is defining the requirements and specifications of modes of engineering technology and design which might apply to the problem at hand.

The potential contribution of geomorphology to engineering also varies with the nature of the engineering problem. Basically, the impact of engineering works is restricted in area though more violent than, for example, the usually rather diffuse effect of rural land use. It may range from a 'point' impact where the construction of a building or a dam is concerned to a 'line' impact in the case of road, railway or pipeline construction. Sometimes, however, one may rightly speak of an 'area' impact where major engineering activities also affect the surrounding terrain, either directly - as is the case with roadside erosion due to drainage problems - or indirectly when the engineering activities spark other activities such as settlement, thereby causing a sprawl in the environment at large. An example of this often neglected effect of engineering works is given in the vertical aerial photograph of fig. 8.1 which depicts the damage to the primeval forest along a mountain road in northern Sumatra, Indonesia. Concentration of burning, agricultural activities, and settlements along a new road, accompanied by depopulation of more remote areas,is a common situation in developing areas. It will be clear that different geomorphological approaches are required for matters such as road location, airfield construction, selection of dam sites and reservoir lakes, earthquake engineering, location of planned settlements and industry, river engineering, harbour works, coastal engineering, etc. River and coastal engineering in particular are completely different subjects requiring an input from the geomorphologist which is mainly geared to a qualitative and quantitative analysis of the littoral and fluvial processes involved; an assessment of their effect on a particular engineering site and to the chances of effectively harnassing a coastline or training a river by appropriate protective measures and structures. The environmental relations between river and coastal engineering,on the one hand,and other fields of civil (for example, highway) engineering on the other, are mostly limited to matters such as the design and location of bridges, etc.

Summarizing, it can be said that geomorphological terrain classification and assessment during the planning and reconnaissance phases of the work and,subsequently, detailed engineering geomorphological studies of the proposed site(s) for identifying potential engineering problems and environmental impacts are major fields of application of geomorphology in engineering. In the field of engineering the study of relief forms and of active or dormant processes together with their assessment in terms of environment is evidently the main input from the side of the geomorphologist. The information gathered should be scrutinized for engineering relevance and grouped in the following four headings:

1. Relief and genetic landform types.
2. Processes and natural hazards.
3. Soil and sub-soil conditions (which have to be evaluated on the basis of data on soil and rock mechanics and the geomorphological situation).
4. Construction materials.

Any terrain classification for engineering should take these four aspects into consideration. It should be borne in mind, however, that the ultimate location and design of a structure is not based on matters of physical environment alone, but also on factors related to the human environment, for example, property severance and technological level. To ensure a successful implementation of the project, the geomorphologist should cooperate with fellow earth scientists, engineers and also with social scientists and planners (Telford, Ed., 1979).

8.2 Relief as a Factor in Engineering

The fact that relief is a complicating and cost increasing factor for building activities and for establishing adequate urban communication systems has been discussed in section 7 dealing with urban area development. The relief factor also greatly affects the grade and alignment of roads, railways, pipelines, etc. particularly in hilly and

Fig. 8.1 Aerial photograph (1:20,000) showing deforestation following the construction of a road in northern Sumatra, Indonesia.

mountainous terrain. Relief is a particularly dominant factor in densely dissected areas of high relief. Many cuts, fills and also bridges or culverts are then required. It is important in this context, when mapping landforms to distinguish from the onset between hard rock sections and parts of the road where unconsolidated materials or weathering products occur; it makes a great difference whether road cuts have to be made in hard rock or in unconsolidated materials. Type and engineering properties of the material are of importance because they will have an effect on the angle of slope of the cut and on its shape, for example, straight or step-wise. It is thus evident that relief is seldom a factor in itself but is interlinked with other aspects of the physical environment in which the engineering work is to be carried out. The specifications set by the engineer with regard to structure also have an important effect on his response to the relief problems, as will be evident from the examples given below.

The profile of fig. 8.2 demonstrates the importance of the relief factor for highway location in mountainous terrain. The two-level section of the motorway between Salerno and Reggio Calabria pictured, is a succession of tunnels or deep road cuts and viaducts. The ultimate location has been determined mainly by the altitudinal range covered, the maximum permissible gradient of the road and the cost per metre of tunneling as compared to the

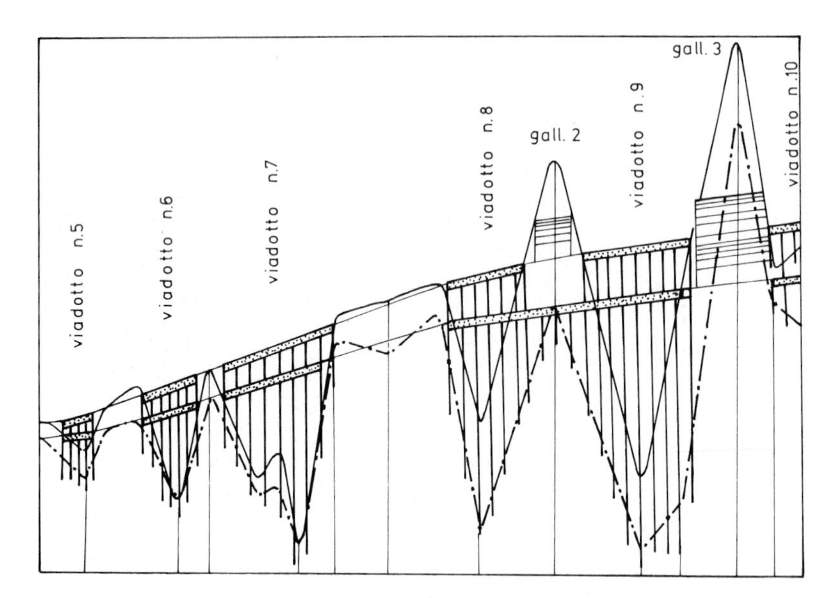

Fig. 8.2 Section of part of the motorway Salerno-Reggio Calabria, Italy, showing alternation of viaducts and tunnels and two-level design as to cope with relief problems.

cost of viaducts of a certain height. The geological and geomorphological situation of the proposed route, if taken into consideration, may change the ultimate location selected for construction of a motorway. Traffic requirements are that curves be kept to a minimum. Secondary roads normally have more curves, less expensive bridges and less deep road cuts. Forest roads are the other extreme when compared to motorways: they should have maximum contact with the ground and follow the terrain precisely because cuts and bridges would interfere with the loading of logs on the trucks (Remeyn, 1978). Therefore, the type of road and the specifications as to maximum gradient, minimum radius of the curves, etc. must be determined prior to the highway location survey. Fig. 8.3 illustrates the effect of relief on the construction of an irrigation intake. Depending on the height above the river of the land to be irrigated, a direct intake will be possible (A), a slight rise of the river (B) or pumping (C,D) may be needed. The siting of siltation basin, flush, sluices, pumping station, etc. depends on the terrain configuration. Examples are from the Serayu River basin, Java, Indonesia.

In some cases a seemingly obvious location on the strength of relief considerations may, for other reasons, be less appropriate. The map of fig. 8.4 gives an example of a road and a railroad in Central Sumatra, Indonesia, situated exactly on an active fault. The straight and comparatively flat fault zone in the otherwise mountainous terrain has lured engineers to opt for this location, without being conscious of the inherent earthquake risk. It should be understood that under such conditions proper weighing of the earthquake risk against other factors, engineering problems and natural hazards of alternate routes may ultimately lead to construction in the earthquake risk zone. Obviously, it is difficult to clearly separate the relief factor in all cases from other relevant factors involved.

The relief factor is minimum in almost flat terrain; there emphasis is placed on the bearing strength and drainage conditions of the land. An important consideration is whether the site is workable year round without flood or other hazards. A terrace-site as compared to a location in the floodplain may therefore be preferred. This is important for the speedy implementation and to keep the cost of construction as low as possible. The roads in flat terrain usually form a straight line, deviations being caused mainly by matters of property severence and by hydrographic features such as rivers, creeks, lakes and swamps. Lakes and swamps should be by-passed whenever possible. If they are unavoidable, the shortest route should be chosen. Rock fill is then preferred, provided it is not too deep; otherwise bridges will be necessary. Steep-sided swamps are often deep, whereas shallowness is likely where gentle shores occur. This requires further proof in the field and vegetation indications may also be of use in ascertaining the terrain conditions in the swamps. Some examples from Canada have been given in this respect when discussing the role of geomorphology in vegetation survey (section 5).

Various possibilities concerning relief exist when constructing a road in rolling or hilly terrain. If the road is basically parallel to the ridges, the main choice falls between a valley line and a crest line. The valley line results in flat grades, numerous curves, many culverts and bridges and in an excess in the amount of fill as compared to the amount of cut. The engineering characteristics of the crest line are quite different. The drainage problems are usually simpler because of the crest being the divide. Bridges and culverts are thus only required where the road deviates from the divide line. The retrogressive erosion of ravines draining the slopes does not normally reach the road.

The choice between these two alternatives depends on

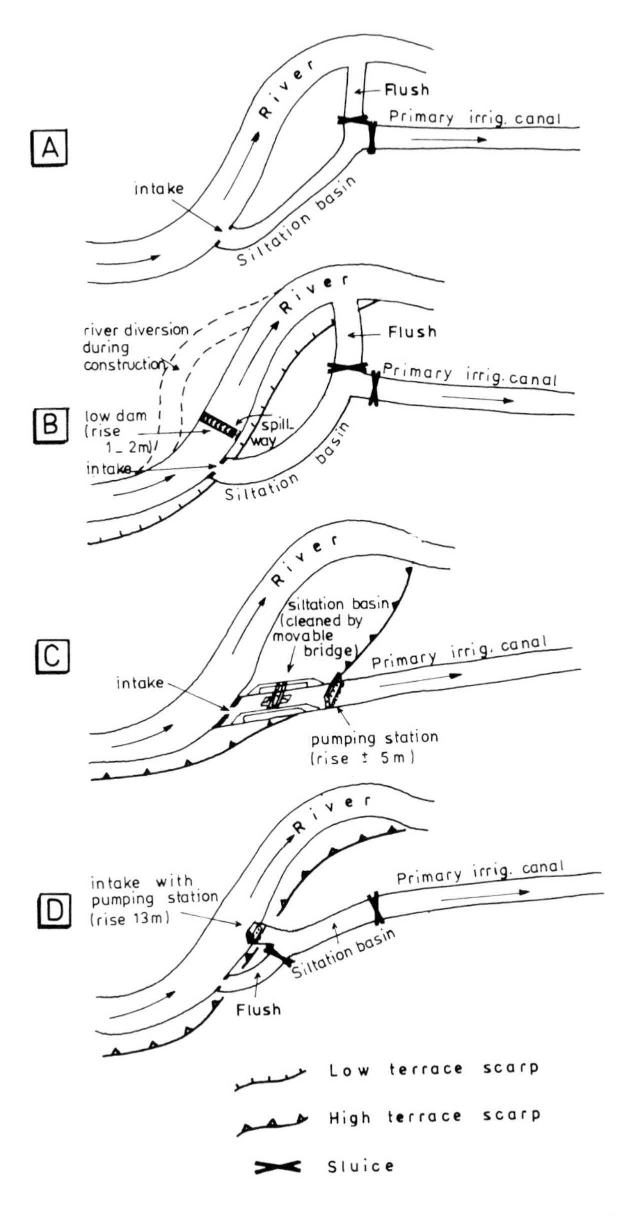

Fig. 8.3 Types of irrigation intakes in the Serayu Basin, Central Java, Indonesia as affected by the terrain configuration. A: direct intake to irrigate alluvial plain; B: with low dam to rise the river level; C: with pumping station to irrigate river terrace; D: as C, but without space for siltation basin in alluvial plain.

the smoothness of the crest, the purpose of the road and the needs of its users. Differences in length, cost of expropriation, construction and maintenance, together with many other (also non-technical) reasons affect the ultimate location of the road. The stereo triplet of fig. 8.5 , scale 1:30,000 shows some roads in Uruguay which are predominantly constructed in crest line positions. Only where the crest became too sinuous or too irregular in relief has the divide line been abandoned for the construction of minor bridges and culverts.

When the general direction of the road makes a more or less acute angle with the ridges, a side-hill line of the road is common. This is most effective where the drainage density is rather low and the slopes long and comparatively smooth. Under such conditions the road is characterized by a rather low and regular grade, either up or downhill. It will then exactly follow the shapes of the hills with a well-balanced grading. Bridges or culverts will be required at places and cross-drainage problems may arise in certain critical sites. Another disadvantage, at least in cool climates, is that in the spring, snow takes more time to melt in places that are protected from wind and/or sun than is the case on a hill crest. An advantage of 'riding side-hill' is that in most cases construction materials can be found near the road,which reduces the cost of construction.

If the road is more or less perpendicular to the ridges, relief may cause serious construction problems. Curving parts of steep gradient usually alternate with comparatively straight, less steep stretches. The maximum permissible gradient for the road becomes an important factor. Heavy duty roads require less steep gradients than roads intended to accommodate lighter traffic. Under such difficult conditions,it is advisable to locate the most suitable sites for the bridges needed for crossing the major rivers first and thereafter to locate the road accordingly. Inadequate sites for bridges often result in high maintenance costs or in collapsed structures. In rugged, mountainous terrain the relief problems for road and railway location are maximum and the development of such communication lines by way of switch backs, loops or the construction of tunnels is often unavoidable. Cuts in steep rock slopes should be avoided because they are expensive and side-hill lines may become problematic where very steep slopes are concerned.

The ruggedness factor (HD), being the product of relief amplitude (H) and drainage density (D), is a useful parameter for the general assessment of relief problems in the planning and reconnaissance phase of road location. It has been successfully applied in a study on the location of a projected Trans Sumatra Highway, Indonesia (Verstappen, 1977). Quantitative morphometrical information was considered an essential prerequisite for this study, for two reasons:

a. the relief characterizing the terrain has a pronounced effect on the location of the road,
b. relief factors also have a great influence on the peak flow conditions of the streams to be crossed by culverts or bridges.

The terrain was not only classified according to ruggedness number (HD) by Meijerink (ITC, 1967), but the D and H values were also plotted separately on the map, to indicate the influence of each of these two values on the ruggedness number. A low HD number generally indicates favourable topographic conditions for road location, particularly when both D and H are low. If D is high, greater steepness of slope will occur and the number

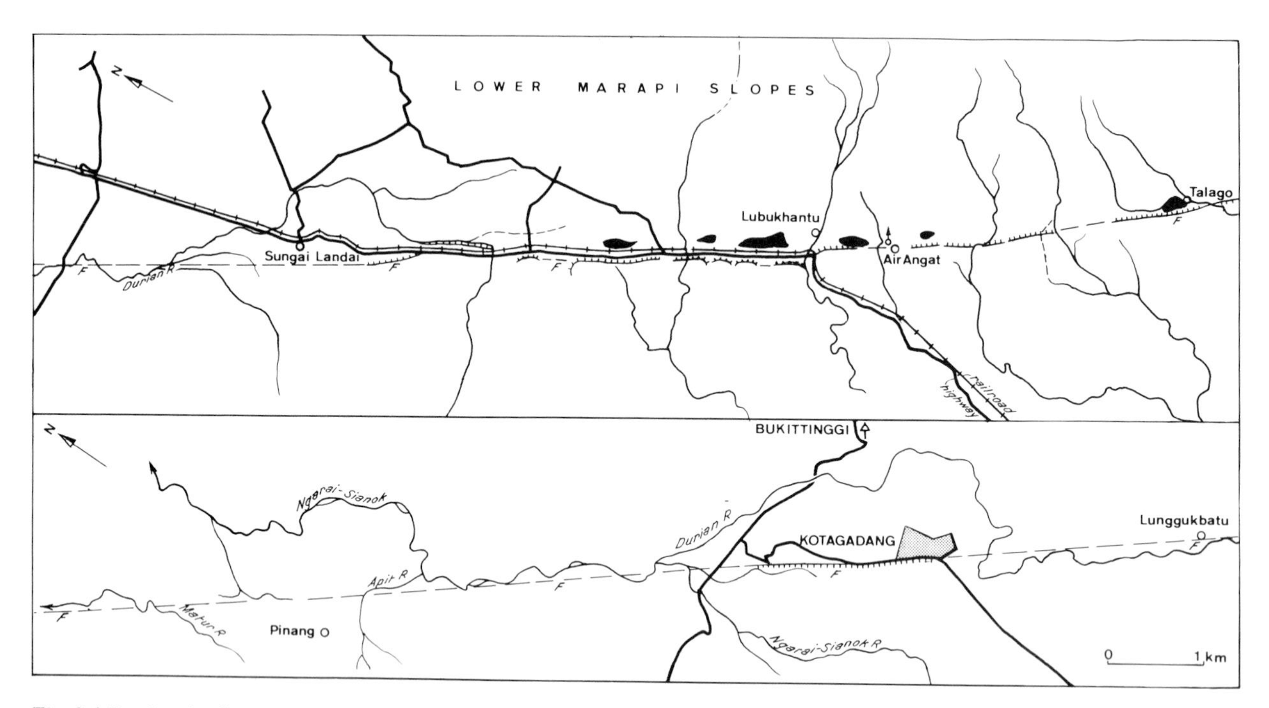

Fig. 8.4 Road and railway constructed along active fault in Central Sumatra, Indonesia. Note stream off-sets, hotsprings and tectonic lakes.

of channel crossings will be great. Low D-values, even if combined with high H-values, indicate greater slope lengths and are favourable for the construction of roads which are not too sinuous. It was considered appropriate to classify the channels in various major groups of peak run-off rates under wet monsoon conditions to estimate the capacity required of culverts, small bridges, etc. This matter was dealt with in section 4.

Another aspect of the relief factor is that the excavation of sizeable quantities of material may produce a substantial environmental impact at the site of the cuts produced as well as at the location where the material is deposited. The material excavated in road cuts,which has to be dealt with by highway engineering in mountainous terrain, may only be partly successfully used in the fills, provided that it suits the purpose. Since long hauling distances may cause a marked rise in construction cost, the tendency often exists for engineers to dump the material from cuts in the immediate surroundings of the excavation, without much thought of the effect this may have on the morphodynamics. The problems met in waste disposal are diversified and include slope instability induced by the overweight caused by the dumped material, linear erosion in the barren slopes, impeded river flow, etc. The photo of fig. 8.6 gives an example and shows the situation on a steep slope in the Venezuelan Andes after the improvements of a road zigzagging upwards from the bottom of the Chama Valley to the village of Villareal. The following examples from S. Italy illustrate the problems that may arise from dumps for which no suitable site was found at an acceptable distance from the excavation site. The dual highway Salerno-Reggio Calabria,situated in the upper Crati and in particular in the Savuto Valley, illustrates this. Some kilometres to the south of Cosenza the highway enters the steep-sided narrow valley of the Iassa River where a two-level design of the highway was inevitable due to the rugged relief. The palaeozoic igneous and metamorphic rocks are strongly fractured and deeply weathered. This made it difficult to find a solid foundation for the pillars on which the road was to be constructed,even though it is normally about 30 metres deep. The permanent danger of major landslides also had to be faced. Normal solutions,such as improved drainage of the slope, retaining walls at the hill-side of the road, wire netting of slopes with imminent rock fall danger, protective vegetation, low hedges of twigs woven in mesh, and even artificial tunnels at places where the road was most endangered by falling debris, could only solve a small portion of the problem.

It was decided that in certain localities the only solution was to strip the slopes completely of the overburden. In total 10^6 cbm. of unconsolidated material was excavated and subsequently disposed of in dumps on the steep valley sides below the road. The photo of fig. 8.7 shows a part of the stripped slope and a major dump can be seen on the photo of fig. 8.8. The river is pushed against the opposite side of the valley by the dumped material,which causes minor slides. More important is the huge amount of debris deposited in the riverbed which is evacuated, with great difficulty by the river during the winter rains.

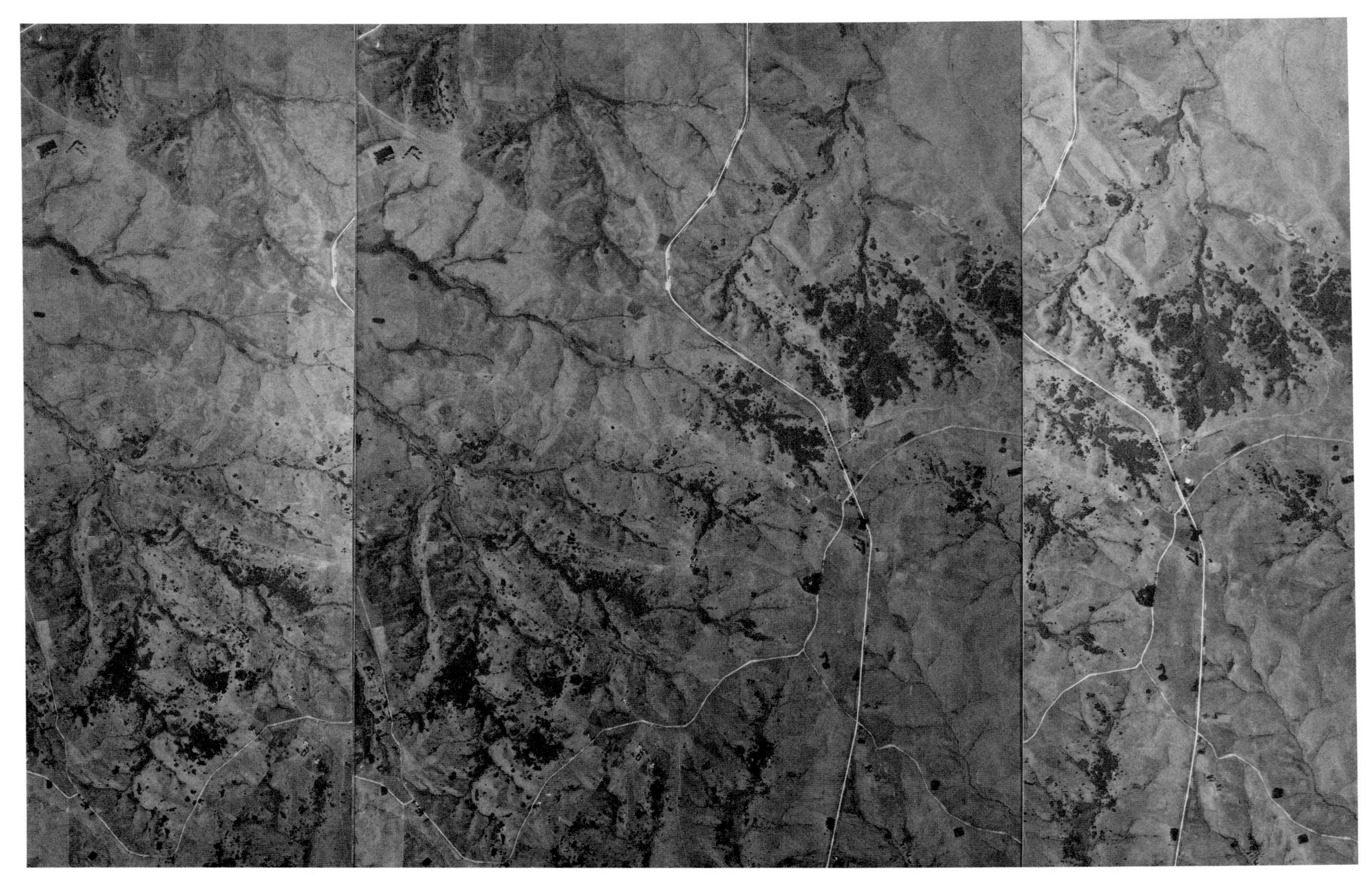

Fig. 8.5 Stereotriplet at a scale of 1:30,000 showing crestline location of a road in Uruguay. Cross drainage problems are practically absent.

Fig. 8.6 Barren slope of the Chama Valley, Venezuelan Andes, resulting from widening and improvement of hairpins of the road connecting the village of Villareal with the motorway in the valley. An example of destructive roadside dumping of excavated materials.

Fig. 8.7 Earth work for the construction of a dual highway south of Cosenza, southern Italy.

Fig. 8.8 Dumped material resulting from the excavations shown in fig. 8.7. The river touches the toe of the dump and lateral sapping causes sliding of debris and excessive bed load downstream.

In the Savuto Valley so much debris resulting from highway engineering works was transported by the river that rapid accretion of its delta resulted in the formation of a new beach several kilometres long and 50 or more metres wide.

Another interesting case of a dump produced by excavation for the construction of this dual highway is in the lower Savuto Valley where approximately 350,000 cbm. of debris was dumped in the riverbed, narrowing it considerably immediately downstream of the Giurio, a torrent which used to produce large quantities of bedload material because of intensive erosion of its basin. The map of fig. 8.9 depicts the situation and figs. 8.10 and 8.11 show the dump and the harnassed torrent.

The Giurio torrent drains an area of green schist, serpentine and phyllitic schist. The latter rock type occurs particularly in its upstream parts and is strongly eroded. The slopes are very unstable and form a continuum of slides and rock falls. The eroded material was deposited where the torrent entered the Savuto Valley. The gravel fan developed there slowly pushed the Savuto to the opposite side of the valley (dashed line in fig. 8.9). The torrent had to be harnessed in connection with the construction of the highway,as it was of course not permissible for the road to be covered by floodwaters and debris. Two long concrete checkdams now stretch along both sides of the torrent and the highway passes over it by way of a small bridge. Furthermore, a whole series of solid gabions was built perpendicular to the torrent channel, ranging in height from 5-6 m in the upstream parts to 2-3 m downstream. Vertical incision was thus checked and the gradient of the torrent became step-wise and reduced.

The highway construction resulted in considerable earthwork with excavation for road cuts and tunnels exceeding the amount of fill. A dump of this material (approx. 350.000 cbm) is situated in the Savuto floodplain downstream of the Ciurio fan and adjacent to the Tribito tunnel,as can be seen from fig. 8.10. In fact, the dump protects the Tribito tunnel from lateral sapping by the Savuto

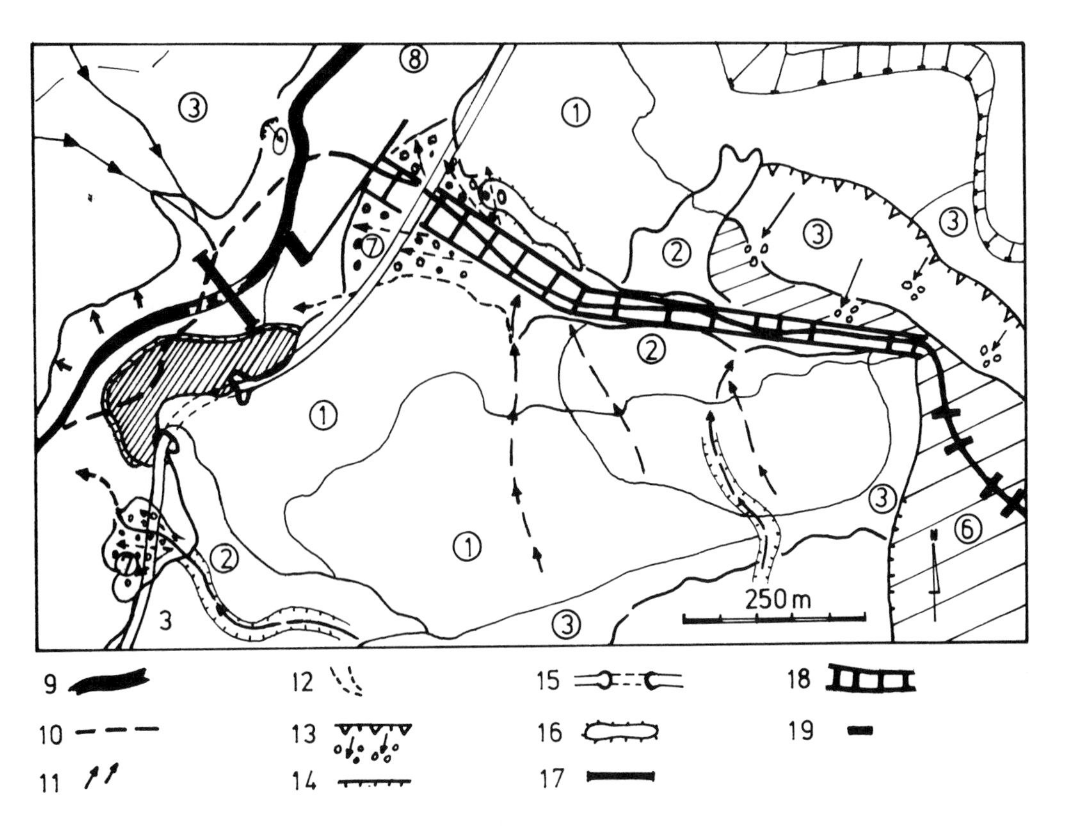

Fig. 8.9 Map showing a road side dump (hachured) situated in the flood plain and protecting a tunnel in the Savuto Valley, S. Italy.

Legend: 1-8 various geomorphological units.
1. denudational hills in green schist; 2. denudational hills in serpentine; 3. denudational hills in phyllitic schist;
4. high denudational hills in phyllitic schist; 5. fluvial accum. terrace; 6. major erosion scar with detritus;
7. gravel fans; 8. alluvial plain.
9-19 minor forms (line symbols)
9. minor bed of Savuto River; 10. minor bed of Savuto River prior to dump; 11. lateral river sapping; 12. ravines; 13. major rock fall; 14. scarps; 15. highway with tunnel; 16. dump; 17. check-dam of Savuto River; 18. revetments; 19. single groynes.

Fig. 8.10 The tunnel and the dump of fig. 8.9 looking downstream.

Fig. 8.11 Harnassed torrent upstream of the dump pictured in figs. 8.9 and 8.10.

River which, by way of one checkdam and two revetments is directed around the dump. Downstream of the dump a small, protected part of the floodplain is put to agricultural use. Lateral sapping, which will possibly induce slides in the future, now affects the opposite valley side. Since the floodplain is reduced to half its width near the dump, floodwaters may rise higher than previously was the case.

A fatally situated dump of excavated material can be found along the 'transversal' highway Cosenza-Paola in Calabria and is described further in the next sub-section. In the same area,though slightly farther to the east and near the eastern end of the major tunnel which cuts through the highest parts of the coastal range, a large dump composed of material excavated from the tunnel was formed on potentially unstable ground which could not bear the overweight caused by this dump. Considerable protective measures including improved drainage were required to stabilize the situation.

8.3 Geomorphological Processes and Natural Hazards as a Factor in Engineering

The study of geomorphological processes is among the major contributions of geomorphology to engineering. Slope processes of various kinds rank high among these but fluvial and littoral processes are of special importance for river and coastal engineering, discussed in sub-sections 8.8 and 8.9. In fact, a diversity of processes has to be taken into consideration such as: avalanches, fan building, swelling of the ground under wet conditions, sliding and slumping, lateral sapping by rivers and torrent cutting. (Bell et al., 1969; Malkin and Wood, 1972; Surumi, 1973; Wilson and Hilts, 1972)

The general philosophy of the study should be to indicate the most favourable areas to the engineer and to avoid potential trouble spots whenever possible (Chowdury, 1978). If a difficult area such as one of unstable slopes (Derbyshire-Miller, 1981; Sarma, 1981) cannot be bypassed, it is the geomorphologist's task to try and find the least troublesome sites. For example, if the foot of an instable slope is undercut by a river and water infiltrates in the higher parts through more pervious overlying rocks (such as limestone), the middle portion of a slope may prove to be the most stable part and construction there is thus to be preferred. In such difficult cases, the engineer may have to adapt the design of the structures to be erected to the conditions of the site. A more sinuous route, requiring less deep cuts may be selected, considering the consequent speed limitation for traffic.

It is thus evident that in addition to the mapping of geomorphological forms and features and the breakdown of the land into geomorphological units, the indication of geomorphological processes at the appropriate localities on the map is essential to the engineer. Mapping of active processes is usually comparatively easy because fresh traces can be observed in the field or on aerial photographs. The slow and sometimes hardly perceptible diffuse processes such as surface wash and sheet erosion can be detected by normal survey methods and the often more violent, concentrated, linear processes such as gullying can be immediately detected. Some examples of geomorphological processes affecting road construction are given below.

Fig. 8.12 shows an aerial photograph at the scale of 1:20,000 of the headward erosion of a flat-bottomed valley with paddy fields in western Java, Indonesia,and its effect on a surfaced, secondary road. How headward erosion (in this case generated by mudflow activity farther downslope) endangers lasting road connections in mountainous terrain is shown by photo 8.13 from the Pyrenees, Spain. The photograph of fig. 8.14 shows three small bridges in southern Italy which have been successively used for crossing the ravine in the foreground. Headward erosion in combination with some other processes is the cause of the destruction of the two older bridges. Grave difficulties may be encountered in badland areas such as the one depicted in fig. 8.15 from the Basilicata, Italy. Piping often aggravates the problems connected with checking the cutting and headward erosion of the ravines. The photo of fig. 8.16 shows the quite different problems met in the same area on the slopes of Mt. Sirino with the effects of melting and sliding of snow in the spring. Voight, (1979) discussed special engineering problems related to rockslides and avalanches.

Older, presently inactive processes may be more difficult to map,as their traces are often partly or completely obliterated. The location of past morphodynamics and the proper assessment of its dormant or episodic character is of considerable importance. It has been observed

Fig. 8.12 Vertical aerial photograph (1:20,000) showing headward erosion with related processes of a flat-bottomed minor valley affecting a road in western Java, Indonesia.

Fig. 8.13 Headward erosion and related processes affecting a mountain road in the Pyrenees, Spain.

Fig. 8.15 Badlands cause serious road engineering problems. An example from the Basilicata, Italy.

Fig. 8.14 Rapid headward erosion in the Basilicata, Italy leading to the destruction of two bridges which were successively in use and to the construction of a third one.

Fig. 8.16 Slope processes affecting a mountain road on Mt. Sirino, Basilicata, Italy; debris falls induced by the melting and avalanching of snow.

Fig. 8.17 Damage caused by inadequate drain of surface run-off from asphalt roads in Calabria, Italy.

in many cases that the initiation of landsliding following road construction occurs in areas of former instability (Chandler et al., 1973; Ingold, 1975). Voûte (1969) gives an interesting example of this phenomenon from the Zagros Mountains in Iraq. Large slides have occurred there during Pluvial periods of the Quaternary,whereas at present gullying and deep river cutting are the dominant processes. The tedious task of the highway engineers in this area is to avoid re-activation of the old, Pleistocene slides.

Engineering works can easily disrupt subtle morphodynamical equilibria,for example, by introducing slope instability (Parizek, 1971). The ways in which this happens are diversified but in the end all cases can be attributed to ignorance about or lack of adaptation to the natural environment. Highway engineering works may induce slope instability by: deep road cuts, the weight of the road and its sub-grade, the weight of dumps of excavated material, denuded natural slopes or disturbance of the groundwater situation (Hall, 1968). The highway engineer may also interfere with river work. Bridge pillars may impede the peak discharge of rivers and the same happens when culverts having a low capacity are used. Inadequate engineering constructions may lead to an irreversible morphodynamical sequence of events causing lasting damage to the environment. They may also lead to the failure and ultimate destruction of the engineering work which provoked them.

Serious problems are frequently encountered when evacuating rainwater from highways (Leaf, 1974; Meyer and Schönberger, 1975). Sizeable quantities of water may run in the road side ditches and cause their erosion particularly on sloping terrain (De Belle, 1971). The transfer of this water to the natural surface drainage system is often a delicate matter. The photo of fig. 8.17 shows an inadequate solution which has led to substantial damage of the road. Critical situations may also arise where the water drains over recently excavated material which is deposited downslope of the road. Fig. 8.18 shows an example from S. Italy where erosion is imminent despite careful grassing and the application of a protective meshwork of twigs.

Quarrying for obtaining construction materials is another potential engineering impact on the environment. In areas where sand and gravel is taken from riverbeds, floodplains or beaches, the fluvial and/or littoral dynamics are affected and cutting or lateral sapping by the rivers or abrasion of the beach near their outfalls may result. The narrow coastal plain bordering the Tyrrhenian Sea near Longobardi in Calabria, Italy, exemplifies such an interaction of geomorphology and man. The plain is only 2-3 metres above sea level and is accompanied by gravel fans where torrents emerge from the steep coastal range rising landward. The fans presumably date from the historical past and may be caused by deforestation during the days of ancient Graecia Magna. Higher, Pleistocene sea levels are marked by a staircase of marine terraces with related old fans and other deposits.

The fans vary in size and development due to differences in the erosion phases of the various basins. Some of the minor basins have not developed a fan at all. After their formation, the fans were substantially modified by man by: agricultural utilization of the fans and nearby parts of

Fig. 8.18 Attempts at stabilization of a minor road-side dump along a road in Calabria, Italy.

Fig. 8.19 Coastal highway passing under a torrent bed by way of a tunnel and a deep cut in the gravel fan. Calabria, Italy.

the coastal plain, and building activities such as the construction of check dams alongside the torrents for protection against the threat of floodwaters and debris. As a result, torrent beds extended several metres above the level of the fans. Parts of the fans were excavated and the gravel was used for construction material. Quarrying activity has strongly increased in recent years as a result of the economic revival of the area and some fans may disappear altogether in due course.

Large stretches of coast are presently subject to abrasion and at places the railway which was constructed at a short distance from the beach (to avoid tunneling through the fans) had to be relocated farther inland. A primary road was constructed landward of the railway in the late Sixties.

Torrents which have not developed major fans are led under the bridges or pass over the road by way of a viaduct. Sometimes deep road cuts are made in the fan and the torrent is then led over the road, but most fans are tunneled through as can be seen on the photo of fig. 8.19. Evidently, in such cases the torrent bed had to be lined with concrete over its full extent and confining concrete dams also had to be built. The map of fig. 8.20 shows a variety of solutions found to solve the encountered engineering problems. The railway (blocked line) is constructed in front of or on the lowest parts of the fans where no cuts are needed and small bridges suffice to cross the torrents which are harnessed by stepped concrete beds. The new coastal highway passes under the torrents and deep cuts in the fans have been made (see fig. 8.19). The winding old road (thick black line) is situated near the apex of the fans at the foot of or just above the lower marine terrace (horizontal hachures). It crosses the torrents by way of small bridges. An older marine terrace (oblique hachures) occurs farther inland. The railway may at places be affected by coastal abrasion. The harnassing of torrents, including the revetments built, decreased the amount of fluvial debris that ultimately reaches their mouth and thereafter contributes to feeding the beach. The abrasion problem of the coast was particularly aggravated by the large amounts of gravel taken from the beach and also from the torrent beds for the booming construction activities (roads, hotels, etc.). From this point of view the gravel fans are a better source of construction material (Verstappen, 1977).

Engineering experiences of past decades have resulted in a wide range of techniques, methods and designs to cope with morphodynamically difficult situations. Monitoring of processes, in particular of those affecting stability of soil and rock slopes,has become increasingly important. The methods of geodetic control by repeated measurements of the position of pegs in land slide zones are well-known,but other simple as well as sophisticated techniques were developed. Zaruba and Mencl (1969) devote ample attention to this matter and special techniques are described by Boyd et al. (1973); Franklin and Denton (1976); Ross-Brown (1974); Savage (1973); and Wilson (1972). Troeh (1975) discusses methods for the measurement of soil creep.

The kinds of stabilization work and the remedial measures at the disposal of the engineer are diversified and a careful selection of the most appropriate techniques or

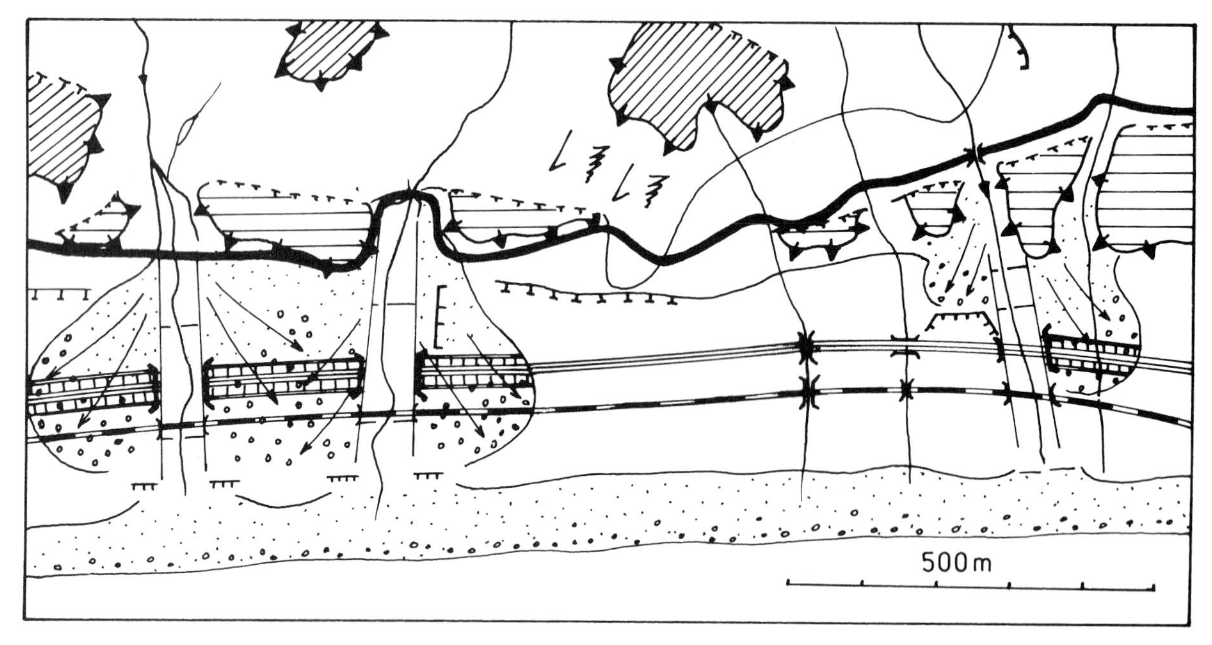

Fig. 8.20 Map showing different road engineering solutions for traversing torrents and their gravel fans along the Tyrrhenian coast of Calabria, Italy.

combination of techniques is required for every particular situation (Hayes, 1970; Kochenderfer, 1970;Komarnitskii, 1968; Patton, 1970; Pocock, 1970; Royster, 1975; Kezdi, 1979). Stabilization of fresh and barren artificial slopes by traditional ways of regrassing is sometimes ineffective due to the agressive impact of the rain and the rapid development of rills and gullies. Low barriers of twigs may assist in retarding rill development but spraying with bitumen containing selected grass seeds is sometimes required. The bitumen then gives stability until the grass cover is firmly established. This method is often applied in creating so-called grassed waterways (see fig. 8.26). The addition of small quantities of hydrated lime or portland cement to the uppermost few centimetres of the unconsolidated slope material reduces rain impact to almost zero and does not interfere with the germination and growth of grass (Macha, 1974).

Drainage improvement of instable slopes is an almost indispensable step towards stabilization (Post and Robinson, 1967; Jones and Larsen, 1970; Popescu, 1967; Posea, 1967). The measures to be taken may incorporate deflection of the surface water upslope of the critical area, controlled surface drainage in the critical area and the instalment of subhorizontal drainage pipes which may be several tens of metres long. Other measures to be contemplated in cases of potential sliding or slumping of slopes include matters such as coating the higher part of the slope by spraying polyester or 'shooting' cement and the support of the foot of the instable slope by concrete structures. The latter may comprise of a concrete fill of an overhang or deep-seated, arch-shaped supporting devices. Even the strength, expansion and permeability of certain swelling (for example, montmorillonitic) clays may be altered by certain chemicals (Arora and Scott, 1974).

In areas where rockfall problems are faced, various methods of stabilization can be applied. The controlled removal of loose rock, under a nylon net or by excavation can be contemplated. The strength of the rock face can be increased by grouting of cement. Further reinforcement is possible using techniques such as buttressing with outside concrete structures, applying rock bolts, various kinds of anchors and/or iron bars across zones of weakness, or by fixing loose blocks in place. If rockfall cannot be effectively stopped, defense works are required either close to the rock wall, for example, iron nets, or close to the structure to be protected. The latter may be flexible iron nets to withstand the impact of the falling rocks, V-shaped concrete walls upslope of high tension poles, pillars or viaducts (Fookes and Sweeney, 1976).

Not all structures made and methods used are properly adapted to the morphodynamic situation in which they are applied. Simple concrete walls almost never significantly contribute to the reinforcement of an unstable slope. The photograph of fig. 8.21 shows the collapse of a sizeable concrete wall in Calabria, Italy following a landslide. In case of rock slopes concrete walls can be anchored to the rocks,as can be seen in fig. 8.22 from the surroundings of Caracas, Venezuela. Good drainage (fig. 8.23) is essential in all cases.

The following photographs give some examples of engineering solutions for morphodynamical problems. Fig. 8.24 shows a concrete 'staircase' constructed on a gravel fan in the Armento Valley, Basilicata, Italy, preventing

Fig. 8.21 Concrete supporting wall of a road cut in Calabria, Italy, collapsed due to landsliding.

Fig. 8.22 Anchored concrete supporting wall in a road cut near Caracas, Venezuela.

Fig. 8.23 Drainage of a road cut near Caracas, Venezuela.

Fig. 8.24 Torrent confined to a concrete 'staircase' and passing under the road in the foreground with the aim of preventing flooding and deposition of debris where the road traverses the fan.

the torrent in spate from using the entire fan and thus covering a long stretch of the road at its foot with debris. The structure adequately confines the torrent and renders only a minor bridge necessary. The velocity of the torrent waters is expected to be sufficient to evacuate the debris from the 'staircase' and to transport it to the nearby river. Occasionally the need for cleaning the structure may arise. The figs. 8.25, 8.26 and 8.27 show the attempts at stabilization of a steep ravine in a badland area to protect the bridge in the highway which connects the town of Matera with the Basento Valley, Basilicata, Italy. A concrete channel has been built upstream of the bridge with two steps and artificial ruggedness to minimize the flow velocity (fig. 8.25). Downstream of the bridge overlapping corrugated iron segments have been placed as earth movements in this section can be expected,which could cause cracking of concrete structures. A grassed waterway has been established by spraying with bitumen containing seeds (fig. 8.26). Further terracing of nearby slopes has been carried out as a preparation for replanting (fig. 8.27).

Notwithstanding all these precautions the situation could not be adequately stabilized and as a consequence the bridge collapsed some years ago.

Two examples of major engineering problems related to geomorphological processes encountered during highway construction in southern Italy are given below. The first relates to a major slump which occurred in Calabria in 1968 during the construction of a section of the Autostrada del Sol south of the 'Ospedale' tunnel on the east side of a minor valley forming part of the Crati River basin. Fig. 8.28 is a location map at the scale of 1:100,000 showing the tunnel and the slide together with another, older, major slide which occurred on the opposite side of the valley during the construction of a railway line in 1904. The latter slide caused four casualities among the railway workers and forced the engineers to adopt another location for the railway line some kilometres farther on. The memory of this slide is still vivid among the local population,and as the geomorphological situation at both sides of the valley is identical its occurrence should have

Fig. 8.25 Concrete bed of high ruggedness so as to reduce current velocity where the steep ravine passes under the highway connecting Matera with the Basento Valley, Basilicata, Italy. The sides of the waterway are grassed.

Fig. 8.26 Grassed waterway with channel made in corrugated iron downstream of the bridge pictured in fig. 8.27 and forming the continuation of the concrete bed of fig. 8.25.

Fig. 8.27 Terracing of the slopes near the ravine pictured in the figs. 8.25 and 8.26 as preparation for reafforestation.

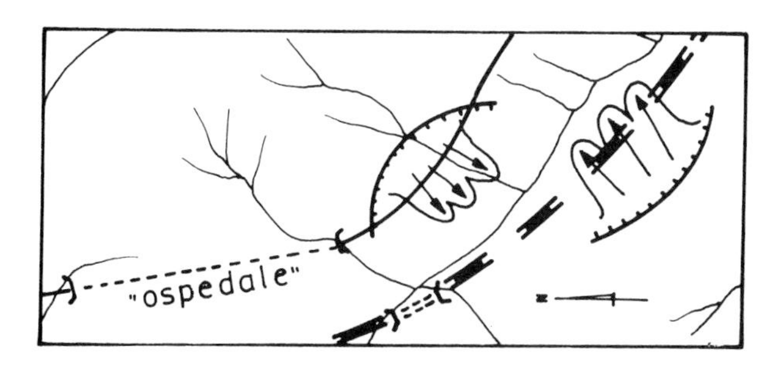

Fig. 8.28 General situation of the slumps that stopped railway construction in 1904 (right) and that delayed highway construction (centre) near the entrance of the Ospedale tunnel in Calabria, Italy,(1:100,000).

Fig. 8.29 The slump that covered the railway track of fig. 8.28. The foot of the Ospedale slump can be seen in the foreground.

been a warning to the road engineers in the late Nineteen-Sixties. When the 'Ospedale' slide occurred after heavy winter rains, probably provoked by the extra weight of the road under construction, it came unexpectedly and caused a delay of more than two years in the opening of this section of the highway. The photograph of fig. 8.29 shows the railway slide of 1904 with the railway track which was never completed at its foot and in the foreground the foot of the 'Ospedale' slide. The map of fig. 8.30 gives a detailed picture of the latter slide and of the remedial works required to control the situation. The major scarps of the slide are indicated as is the line of horizontal drainage tubes (about 30 m deep) which drain off the excess water from the otherwise frequently saturated silty-clays occurring. The excess water is drained off to a major ravine by way of two collecting tunnels. A thirty metre deep concrete wall in the form of a series of arches was constructed to support the foot of the slide. Repeated measurement of eventual displacement of a number of of fixed points on the slide is effectuated to ensure that no further movements occur. Thorough studies prior to the construction of this road section could have saved substantial sums of money.

The second example relates to the highway connecting the town of Cosenza, Calabria,with the Tyrrhenian coast. The diversified problems met during the construction of the road in a stretch of about one kilometre are pictured in the map of fig. 8.31. The major slide which occurred at the southern end of the Cupo viaduct pictured on the photo of fig. 8.32 is just outside the lower left corner of the map. More serious was the huge slide at the site of the Bocca Lepre II viaduct. The basic cause of this slide was the hydrological situation caused by the impervious phyllite underlying the schist of the higher parts of the slope. The contact between these two formations is indicated on the map. The photograph of fig. 8.33,taken from the opposite side of the valley, shows the elaborate system of drainage canals required to drain off both the surface (rain) water and the excess water collected by the sub-horizontal drainage pipes that had to be installed. Ultimately, all the water is collected in a ravine. A series of gabions was built in the major ravines to prevent their vertical cutting which would increase the slope instability.

Another problem resulting from the construction of the tunnel is visible at the right-hand side of the map of fig. 8.31 and the photograph of fig. 8.33. The excavated material from the tunnel was dumped on the steep slope immediately outside the tunnel entrance to minimize the cost of haulage. The disastrous results are pictured on the photo of fig. 8.34. The previously largely grass-covered slope, where rills and ravines were nearly completely absent, was transformed into a barren slope of loose, dumped materials in which the rain water rapidly formed

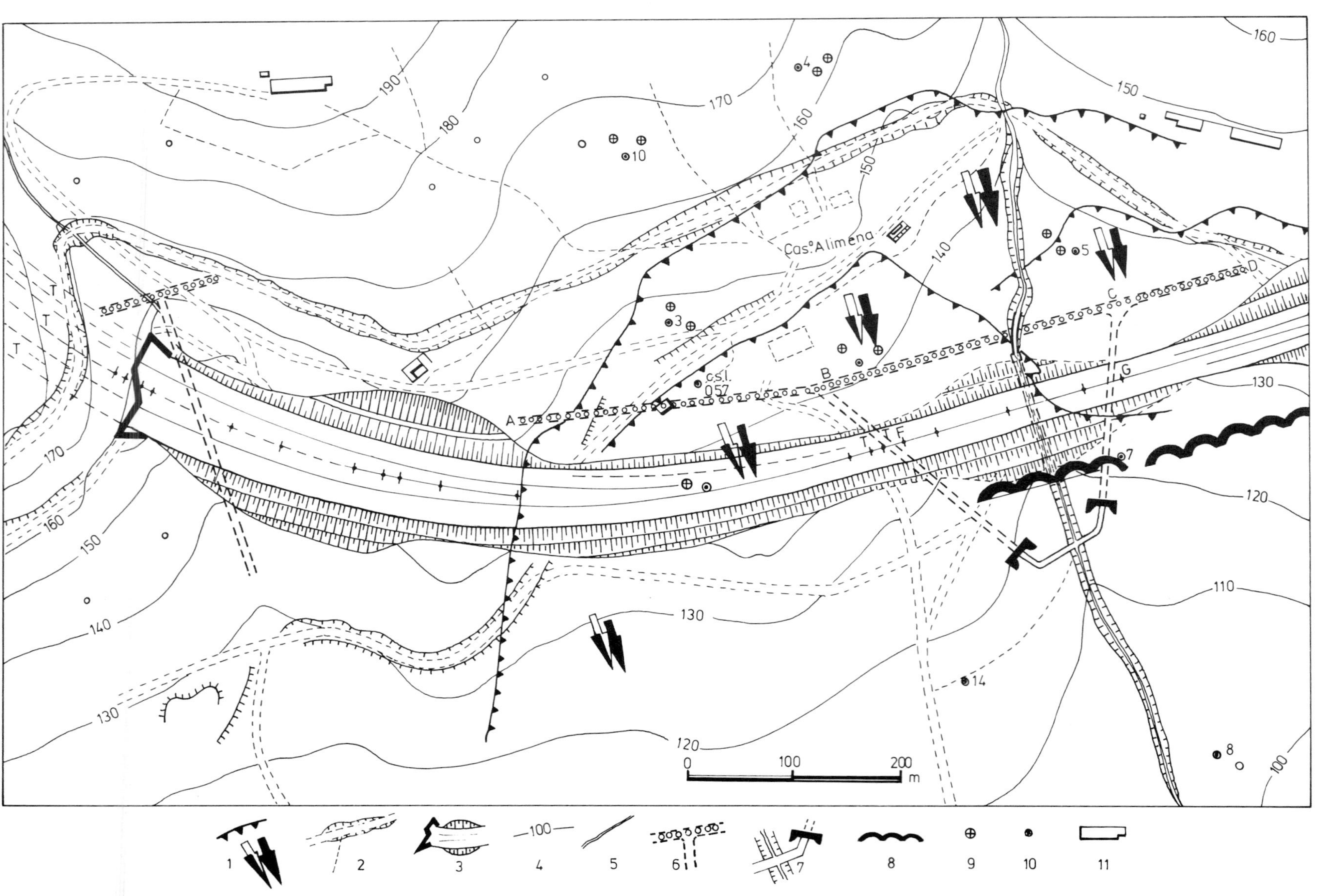

Fig. 8.30 Detailed map of the Ospedale slump in the Crati basin, Calabria, Italy. Legend: 1. main scarps with direction of movement; 2. existing road with cut; 3. highway with tunnel entrance and deep cuts; 4. contour lines; 5. ravine; 6. aligned drainage pipes with drainage tube leading to ravine; 7. open drainage line forming continuation of 6; 8. deep concrete support for toe of slump; 9. soil moisture check; 10. geodetic control; 11. tool shed

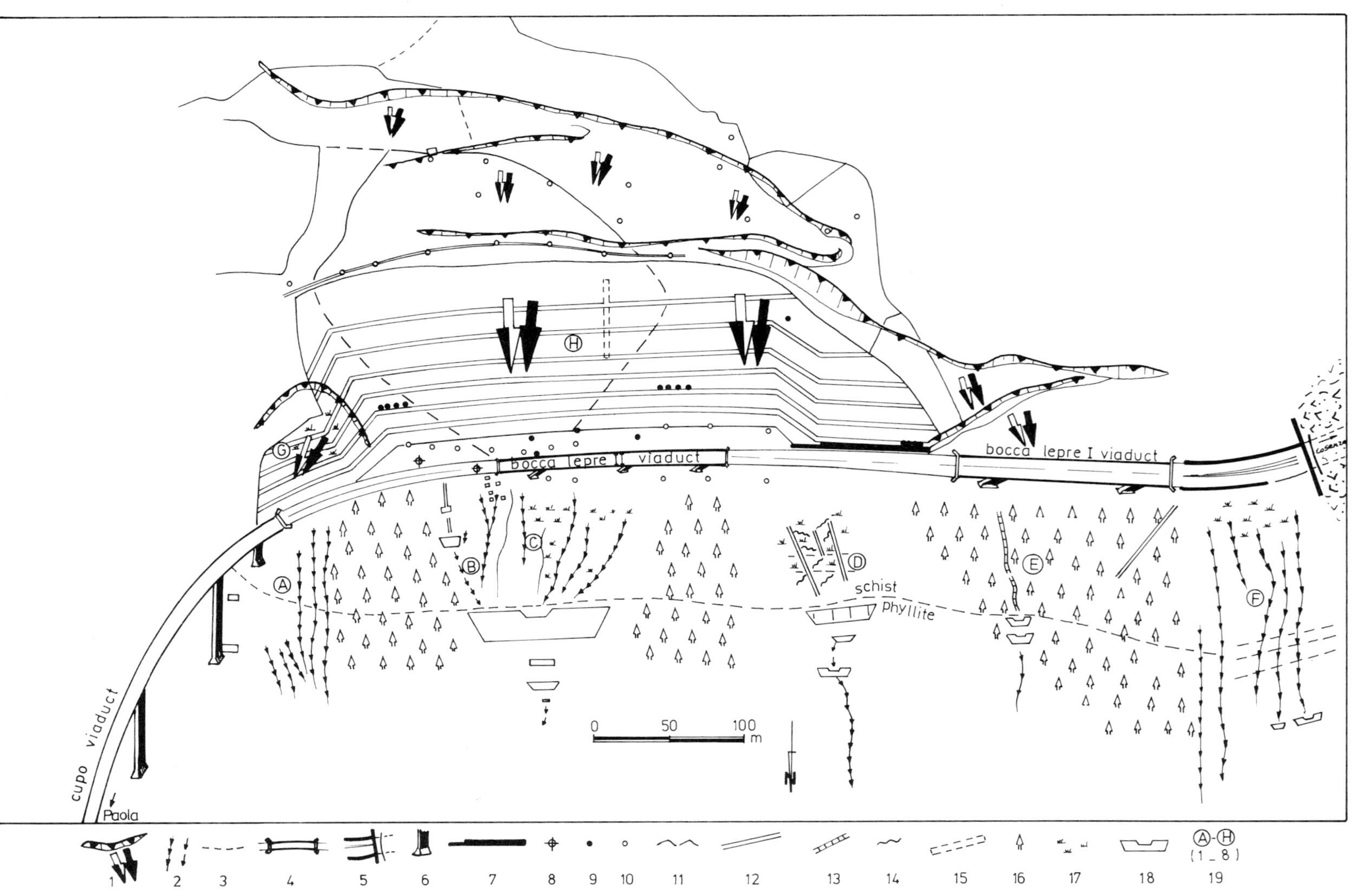

Fig. 8.31 Major slump and related phenomena along the highway Cosenza-Paola, Calabria, Italy. The tunnel mentioned in the text occurs in the right and the Cupo viaduct in the lower left. Legend: 1. main scarps with direction of movement; 2. ravines; 3. boundary schists-phyllites; 4. viaduct; 5. tunnel entrance; 6. bridge pillar; 7. concrete wall in road cut; 8. boring; 9. drainage pipe; 10. geodetic control; 11. tunnel roof; 12. open drainage line; 13. stepped drainage line; 14. grassed waterway; 15. cross drainage; 16. trees/scrub; 17. grass cover; 18. concrete gabion. Characteristic parts of affected slope: A. severe gullying beneath bridge pillars; no protection. B. drainage pipes and gabions have largely stopped gullying. C. higher parts: active gullying attacking grass cover and stepped drain levels. Lower parts: Gabion stopped active gullying. D. higher parts: erosion checked by grassed waterways, drainage pipes, contour bunds, wooden meshworks and gabions. Lower parts: active erosion. E. Gullying controlled by stepped drain levels and gabions. F. Active gullying particularly beneath tunnel entrance destroying remnants of grass cover. Some contour bunds made. Gabions only in lower part of slope. G. Landslide scarp partly stabilized by contour bunding and grassing. Upper layers deeply weathered. H. System of drainage pipes, contour bunds and grass cover stabilize slope. (From ITC fieldwork report H. Ghanem, 1972).

Fig. 8.32 Part of the Cupo viaduct with the slide at its northern end as seen from below.

Fig. 8.34 Severe gullying beneath the tunnel entrance of fig. 8.31.

Fig. 8.33 Ground view of the slope pictured in fig. 8.31.

rills and increasingly deep ravines cutting across the buried grass cover which had previously protected the slope. Violent processes which are difficult to control and which may ultimately lead to the collapse of parts of the tunnel have thus been provoked. The selection of a somewhat more remote and less critical site for dumping the excavated material could have saved serious problems and cost which is a multiple of the small extra cost of hauling required.

8.4 Soil and Sub-soil Conditions

Soil and sub-soil conditions are also influenced to a considerable degree by geomorphological conditions. (Broms, 1973). The field is diversified and includes matters such as the shearing strength of the sub-soil, potential sliding and plastic deformation caused by the pressure of the overlying beds or planned structures (Scott and Schoustra, 1970; Stewart, 1971; Struble and Mintzer, 1967). A number of geological and pedological factors become of particular importance here. The strike and dip of beds and the direction of shear and tension joints are matters to be thoroughly studied, especially where heavy structures such as storage dams are concerned. Shear joints are the most dangerous and may cause collapse of structures particularly when pressure is exerted under angles of 45°. Tension joints, usually perpendicular to the bedding planes,will need grouting to avoid water losses near the dam. Detailed geomorphological survey and aerial photograph interpretation may contribute to the solution of these problems,provided that a thorough follow-up in the field and the laboratory is carried through.

The soil and sub-soil conditions influence engineering works such as road location in various ways (Legget, 1974; Myslivec and Kysela, 1978; Wong, 1975; Wu, 1975). They may influence the site of the structure or have an important effect on the type of sub-grade to be used, particularly for heavy duty roads. This means that they will influence construction cost and also cost of maintenance. Poor drainage may easily result in 'pumping' in the case of a concrete road, which is very destructive. Data on some highways in the USA demonstrate the importance of sub-soil conditions for minimizing the cost of maintenance. For example, in parts of Highway 31, Indiana, which are located in aeolian sands, only 9,7 cracks per 1000 ft. were observed after a number of years, whereas in glacial till 23,8 - 27,5 cracks per 1000 ft. occurred. The maintenance in glacial till is, therefore, considerably more expensive. Drainage conditions are always important in this respect. In glacial till seepage often occurs along slopes. In gravel terraces and in uplifted coral reefs drainage is good and the cost of maintenance low. On the contrary, in alluvial plains drainage is often poor and the bearing strength of the soil low.

The determination of the engineering properties of soil and sub-soil material is of the utmost importance for the location, design and ultimate success of engineering structures (Terzaghi and Peck, 1967; Wilun and Starzewski, 1972; Gidigash, 1975; Bazaut, 1979; West and Dumbleton, 1975). It is, however, for various reasons often fraught with difficulties. There are numerous factors to be considered in soil mechanics,such as the physical characteristics and types of soils, the water present and percolating in the soils, the consolidation and settlement of the material as well as its shearing strength (Olson, 1974) and its stability when forming slopes (Capper and Cassie, 1969). Evidently, the in situ properties (compressibility, swelling, etc.) of the material have to be determined and evaluated for foundation engineering (Castro, 1975),whereas for the evaluation of construction materials (sub-section 8.5) the characteristics of disturbed material have to be considered (Handy, 1968). A second difficulty is that the physical characteristics of certain soils such as quick clays may change completely when exposed to extra loading due to the weight of engineering structures. Crawford (1968) studied quick clays in eastern Canada which, though in many respects very similar to other glacial/marine clays, give rise to important changes in volume and dramatic loss of strength when loaded. A third and particularly important problem is that the reliability of the results obtained by standard testing methods are sometimes questionable as the natural dynamics have not been considered. Examples of this are numerous. A clayey soil of low plasticity limit may be suitable for construction on horizontal terrain but is dangerous on slopes because of the consequences of excessive rainfall. The mineralogical composition of the clays is a matter of prime importance where slides are concerned. It has, for example, been found that the engineering properties of some residual soils in Kenya were inconsistent with index properties and standard control tests show marked variations. Sijmons (1976) discusses difficulties in the assessment of shearing strength and mentions the reduction in shear strength where micro fractures have caused strong but localised leaching. Trow and Morton (1969) mention the difficulty of interpreting the results of standard laboratory tests such as plasticity, compaction and grain size of ferruginous laterites developed by tropical weathering on ultramafic peridotite bedrock in the Dominican Republic. The combination of high clay content, low compressibility and high shear strength was quite unexpected.

The bedrock occurring has a large effect on soil and sub-soil conditions (Dearman and Fookes, 1974; Jennings, 1966; Wilson, 1972; Winert, 1975). Many schists, for example, are unsuitable for high unit loading and sometimes rather permeable in the direction of the schistosity. The strength of igneous rocks is usually good,but much depends on the texture and degree of weathering. Greatest strength can be expected in the case of interlocking crystal grains. Many lavas function as well as intrusives but volcanic breccias and soriaceous material may be unreliable. Pyroclastics and tuffs have a generally low compression and shearing strength and swelling may occur when saturated with water. Coarse clastic sediments are almost incompressible, whereas shales and clays are highly so. Compaction shales in particular are unreliable for construction. All shales need special precaution because of possible alteration when exposed to air. Many clastic sediments lose their coherence to a significant degree when exposed to moisture or to alternating freezing and thawing conditions. Collapse may occur in limestone areas where fissures and cavities weaken the conditions. Organic soils are poor foundations whereas thin soil covers over solid rock are excellent foundations. A very detailed study of the very diversified foundation characteristics of the wide range of materials is therefore required.

Among the various soil materials, clays often pose problems for engineering, therefore, a wide range of soil stabilization techniques has developed (Gillot, 1968). Swelling and expansion of clays under wet conditions can be a particularly disturbing factor for which correction is required (Chen, 1975). Stabilization may, for example, be reached by chemical means, pre-saturation, or by compaction. A special design of the foundation may also offer a solution. Teng and Clisby (1975) describe the use of sand drains, lime treatment and bituminous membranes to combate active, swelling clays. Expansion is a main problem in arid and semi-arid areas, but shrinkage during dry periods may pose serious problems in areas where soils are usually comparatively humid (Jones and Holtz, 1973). Montmorillonitic clays are particularly susceptible to expansion and often create serious problems (see Kerr and Drew, 1969) but salt may also give rise to soil heave in (semi) arid areas under the influence of great daily temperature changes,provided that ample soil moisture is available. Chemical treatment and blending with other materials may be contemplated (Blaser and Scherer, 1969).

Soil and sub-soil problems are extremely diversified. The photo of fig. 8.35 shows a part of a badly deformed surfaced mountain road, inadequately constructed on swelling clays in the Basilicata, south Italy. Snow melt in spring causes particularly devastating effects. Improved drainage and a better sub-grade ultimately kept the road motorable, but maintenance remains expensive. Repeated freezing and thawing is a widespread engineering problem in the moderate climates (Philippe, 1972; Richard, 1972). In (semi) arid areas the salts precipitated at and near the surface and carried in solution by the (sub) surface water are often very agressive towards cement and are therefore damaging to engineering structures unless special precautions are taken. Sulphidic or potentially acid sulphatic sub-soil layers ('cat clay') may be similarly detrimental to the strength of concrete foundation piles (Holst and Westerveld, 1975). Subsidence and the sudden formation of sinkholes may cause serious problems in limestones and dolomites (Newland et al., 1973). The surface configuration of the karst phenomena then becomes an important indication on which the detailed drilling programmes and seismic or electrical resistivity methods in critical areas are based (Vineyard and Williams, 1967). On the other hand sinkholes may be used to drain off rainwater (see fig. 8.36).

Many regional studies on the subject have appeared in recent years but only a few can be mentioned here to give some examples. Anderson and Trigg (1970) discussing the geotechnical factors of valleys in South Wales, emphasize the soil slip hazard on the Coal Measures formation. the solution caves formed in Carboniferous limetones, the unstability of the boulder clays, the foundation problems related to soft marine clays and the importance of mapping buried valleys having an abnormally thick cover of superficial materials. Wesley (1973) has made an excellent study on the engineering properties of latosols and andosols, two widespread soil groups in the volcanic terrain of Java and Sumatra. The engineering characteristics of glacial deposits and particularly glacial tills have been the subject of numerous studies, such as by Wakelin Norman (1969); McGown (1971); White (1964, 1972); Horton, (1975); Olmstead (1969) and Tushinskiy (1971).

Soil and sub-soil conditions are a particularly critical factor in the far north where permanently frozen ground is of widespread occurrence (Croney and Jacobs, 1967). These permafrost conditions are distributed over one-fifth of the land area of the world (Ives, 1974). The base of the permafrost may reach 300-400 metres below sea level, such as in certain parts of Alaska (Thomas, 1969) and only one or two metres at the top will thaw during the summer season.

This permafrost terrain is extremely sensitive to disturbance, for example by engineering works, as the equilibrium between frozen and unfrozen conditions is very subtle. A rise in temperature to 0.2°C above freezing point is enough to initiate thawing and to set a whole chain of almost completely uncontrollable events in motion (Halmos, 1968). The susceptibility to degradation varies with the ice content of the frozen soil and inversely with the insolation properties of the surface layer. It is greatest in the northernmost areas where the organic layer is thin and the ice content high and generally less - though often difficult to precisely assess - near the southern fringe of the permafrost zone where the ice is

Fig. 8.35 The effect of snow melt on a mountain road with inadequate sub-grade in swelling clays. Basilicata, Italy.

Fig. 8.36 Sink hole in the Central Range of Irian Jaya (New Guinea), Indonesia,used for draining an air strip.

discontinuous and the organic layer thicker (Haugen and Brown, 1971).

The engineering properties of frozen soils also depend to an important degree on temperature in the range of 0-10°C and especially so when considerable amounts of unfrozen water occur, for example, in clays (Hoekstra 1969). Freezing of the active layer results in frost heaving which becomes particularly disturbing when water from outside is attracted to the freezing zones. Subsidence and the formation of thermokarst depressions associated with thawing of permafrost are equally disturbing to the engineer. Any change in surface heat exchange, for example, by surface disturbance caused by off-road vehicle movements, forest fires, etc.,should be avoided as much as possible. Vehicle transport increases soil density, reduces moisture content and results in accelerated and deeper thaw. Furthermore, organic surficial horizons can be physically altered,particularly in relatively wet areas (Brown, 1971, 1976; Gersper and Challinor, 1975). Vodolazkin (1966) gives examples of ice-saturated soils showing sharp drops in bearing strength when changing from a frozen to a non-frozen state following engineering works. Clay particles become separated, the moisture content becomes less and thawed soils are compacted though the rate of settlement decreases rapidly with time until finally a stabilized state is reached. Speer et al. (1973) relate thaw settlement to frozen bulk density in a graph. In many cases it has been proved to be good practice to leave the natural insolation of organic soils intact, for example by applying gravel pads on the surface with cold air ventilation or by construction on piles which are frozen in or seated on the hard substratum (Brooker and Hayley, 1970).

It is evident that the key for good construction is an understanding of the nature of permafrost (Dingman, 1975). Mackay (1972) states that a knowledge of the distribution and abundance of permafrost is an essential pre-requisite for development of northern territories. The source of water before freezing and the processes causing the transfer of water to the freezing plane are considered important criteria for the classification of underground ice in various types. He explains the excess water found in the ice as water is expelled from coarse-textured sediments by the downward growth of the permafrost. Relic permafrost under shallow Arctic coastal waters, dating from colder periods of the past is also an interesting aspect. Prince (1967) observed that permafrost is generally considerably thinner under deglaciated areas as compared to non-glaciated areas. The age and origin of permafrost varies from one region to another. The surveying and mapping of permafrost conditions is solidly rooted in geomorphology. It essentially amounts to a breakdown of the land in geomorphological units and the mapping of individual geomorphological features using photo-interpretation (Leighty, 1962; Mollard and Pihlainen, 1966; Owen, 1972; Richardson and Sauer, 1972) and other remote sensing techniques as well as data collection (augering, etc.) in the field. Geophysical investigations (seismic, electrical, etc.) contribute invaluable information about the thickness and condition of the underground ice (Bernes, 1966; Ferrians and Hobson, 1973). In the USSR the so-called landscape approach has been applied to field and aerial photographic permafrost studies (Soloyev, 1971). Shalikov (1974) used this approach in his studies on the depth of frost penetration in various types of lowland terrain near Smolensk and established the relationship between the depth of freezing, the sum of negative air temperatures and the thickness of the snow cover.

The engineering response to the permafrost problem has grown rapidly with the development of northern latitudes in recent decades. (Jessop, 1970; Leschaele and Kachadoorian, 1975; Linell, 1960). Considerable experience has been gained in Canada, the USSR and the USA (Alaska),and numerous publications on the subject exist. Initially mainly wooden houses were erected which led to subsidence particularly of the warmer, central parts. Heavier concrete buildings are now common and although cracking of walls may occur, their performance has been satisfactory, provided that adequate precautions were taken. Among these, proper settlement of the soil prior to and during construction is essential, as has been mentioned above. The role of grain size and chemical composition of the soil is also understood and even chemical treatment of the soils is investigated (Corte, 1969). Surficial salt treatment for the attenuation of frost heave (Young et al., 1973) has been successfully applied and improved construction methods and design techniques have been introduced in highway engineering (Esch, 1973; Wallace and Williams, 1974). The construction and maintenance costs have been considerably reduced by applying a range of methods, from simple methods such as painting the pavement surface white as to achieve maximum heat reflection (Berg and Aitken, 1973) to the use of frozen-in piles. A surcharge of gravel, squeezing out the excess moisture from muskeg and gravel embankments and providing insolation against thawing of permafrost has also proved most useful in the construction of the Anchorage-Fairbanks highway in Alaska (Becker, 1972). Avoiding excavation by careful grading of the road is another important rule as is careful clearing of organic debris by hand methods. When road cuts exposing ice are unavoidable,Smith and Berg (1973) recommend: (i) avoiding slopes facing north, (ii) making vertical back slopes with a wide ditch at the base, (iii) clearing vegetation from the top of the slope and (iv) not revegetating the cut until the slope is re-stabilized. The design of pile construction is commented upon by Nixon and McRoberts (1976) and on bridge foundation by Crory (1975). A good review of general engineering problems can be found in Linell and Johnston (1973). Special problems are met in the case of (gravel) placer mining (Pettibone, 1973; Swinzow, 1969).

Particularly severe engineering problems have been met when constructing oil and gas pipelines in permafrost areas (Palmer, 1972; Peyton, 1971). Whereas a gas pipeline may be chilled well below freezing point in critical sections (Hardy and Morrison, 1972), this is not feasible for oil because of viscosity problems. In a pipeline carrying warm oil, permafrost thawing with the resulting subsidence and solifluction can be expected, particularly where the ice content of the ground is high (Kachadoorian and Ferrians, 1973; Lachenbruch, 1970). Pipelines in the Arctic are normally constructed in a special zig-zag layout to achieve maximum flexibility, with fixed points, for example every 500 metres. A general review of the problems is given by Williams (1979). More detailed information on permafrost engineering can be found in various modern handbooks on civil engineering.

8.5 Construction Materials Survey

The search for considerable quantities of appropriate building materials required for purposes of construction is a matter of great concern for the civil engineer. In many cases the outcome of such surveys has a profound effect on the execution of engineering works and specifically on the design and location of roads, buildings and other structures. Therefore, these surveys should always be effectuated in the early planning and reconnaissance phases of the work. The availability or non-availability of sufficiently large quantities of certain materials such as clay, and scree may, for example, in the case of dam construction, dictate the choice between an earthen or a rockfill dam whereas, if no great quantities of such construction materials are found nearby, a concrete dam may be contemplated. The type of material available and its inherent characteristics determine the suitability for certain engineering purposes, for example, sub-grading in highway construction. It may lead to the necessity of specific procedures or designs and involve extra drainage requirements. The cost factor is always an important element in such matters and exerts an influence in three different directions. First, there is the cost of its excavation. Where hard rocks are concerned it is important that the spacing of faults and joints should be sufficiently close to lessen blasting requirements but not too numerous to impair the stability of the excavation slope. Secondly, there is the cost to prepare the material for use. If chemicals have to be added to make the material suitable, precautions are required and this will add to the expense. Most important is, however, the cost of hauling. This rapidly increases when construction materials have to be hauled over great distances or have to be derived from areas of difficult access.

The survey concerning material location should be carried out by an experienced geomorphologist capable of identifying landforms generally associated with building materials, such as sandy and gravelly river terraces, eskers and scree fans. Therefore, a landform classification geared to this purpose is required. It is customary to start the work by using aerial photography; this is logical as otherwise useful occurrences might easily be overlooked. Furthermore, it is the easiest way to precisely outline the areas where construction materials are likely to occur. One cannot always obtain detailed information from the aerial photographs about the precise nature of the materials occurring and about their suitability for purposes of construction. However, interesting indications can be found in many instances and it can be established with fair accurracy whether the deposits of a river terrace are gravelly/sandy, silty or clayey. The density of drainage lines is one clue and the cross section of the ravines is a second. Sandy/gravelly materials tend to develop inclined straight slopes; vertical slope angles are common in silty materials and broad convex-concave vales point to clayey deposits. More precise information has to be gathered through field and laboratory methods.

A common procedure in material location surveys and particularly when granular materials (gravel, sand) are concerned, is as follows. On the basis of existing geological and soil maps, literature and other available information, a preliminary map is made on which existing quarries and known outcrops are plotted and formations of possible interest are outlined. Aerial photo-interpretation is started in this phase with emphasis on the photographic characteristics such as drainage patterns and gully profiles of the various landforms.

A first field check concentrating on an examination of existing quarries and the study of relevant geological and geomorphological features is executed and some preliminary sampling may also be done. The results of the survey are condensed, preferably in tabular form and the prospect sites are located and classified mainly on the basis of three criteria:

1. haulage distance and accessibility
2. suitability of the material
3. quantity of material or estimated size of the deposit

Credit points are given for each of these three items: good (1), medium (2), poor (3), and subsequently added. During a second field check the most promising prospect sites are investigated; sampling and augering is done and appropriate geophysical surveys (electrical resistivity, seismic) are executed. In the final report field findings and laboratory results are correlated. The various deposits are described and the criteria used to identify and locate them are given. Details of laboratory results, quantity estimates, haulage distance, etc. are discussed and aerial photographs of the sites and a map of their location are added to the report.

It is evident from the aforesaid that the essence of material location surveying is in the identification of landforms and landform associations where certain kinds of construction materials can be expected (SCLSERP, 1973). An outspoken example of this is the mapping of eskers in areas of Pleistocene continental glaciation in relation to granular materials required for engineering. In Sweden more than 80% of such material is derived from eskers. Other sources are for example alluvial fans which may be sites of angular debris of variable size, floodplains and river terraces where granual materials (Hester, 1970; Christiansen, 1970) may occur although a high silt content and heterogeneity of the material can be expected, and talus cones which may be sources of a limited amount of angular and often too coarse material. Residual soils are poorly sorted and the pebbles are sometimes too strongly weathered. Fluvial sands may be a good aggragate for concrete, whereas aeolian sands are normally less suitable for this purpose because of the roundness of the grains. Crushed igneous rocks may also be good concrete aggregates, provided that they are not too strongly weathered (Simpson and Horrobin, 1970). The suitability of bedrock for construction materials varies with lithology, weathering and fracturing and in all cases thorough investigations are required.

Publications on the methods of search for construction materials are numerous (Molyn, 1961; Bergström, 1960; Mintzer, 1959; Joliffe, 1980). Several publications of the British Transport and Road Research Laboratory (TRRL) deserve special mention in this context (Schofield, 1962; Dowling, 1963; Clare and Beaven, 1965; Beaven, 1966), as they relate to tropical conditions such as the use of coral limestone which in this respect have been studied much less than, for example, areas of Pleistocene glacial deposits. Grant and Aitchinson (1970) have pointed to the utility of silcretes and ferricretes in (semi) arid areas and Simon et al. (1973) have emphasized the use of some lateritic soils after cement stabilization.

All these authors utilize aerial photographs for the search for building materials applying a geomorphological approach. Bergström, in searching for concrete aggregate and impervious till for the Swedish State Power Board, bases his interpretation on dimensions and morphology, situation and pattern, texture and grey tone. Sand occurrences are associated with phenomena such as eskers, outwash plains and dunes. Mintzer (1959) links construction sources of gravel with beach ridges terraces and fluvioglacial plains. Drainage characteristics and vegetation types are mentioned as major criteria by him. Apart from the type of deposit, the distance of transportation by water is important. If transportation distance is short, gravelly deposits are more likely to occur than fine-grained materials. This is important since gravel is often more in demand than sand. The material of kames is often poorly sorted and sometimes there are extensive transitions to till.

Schofield (1962) mapped landforms inNyasaland with the

aim of locating new sources of gravel. In only a few of these landform types suitable material for sub-grades and pavements of roads was found. He maintains that the study of the conditions under which workable deposits were formed made deductions possible as to the presence of useful material at certain places. Such deductions were, for example, the occurrence of thick deposits of grey sands in the heads of valleys at the bottom of concave slopes and laterite 'reefs' and gravel from truncated profiles where the clayey topsoil was removed by erosion.

8.6 Various Kinds of Engineering Works and Geomorphology

The mode of application of geomorphology in engineering depends largely on the type of project involved and the phase of its realization,as has already been indicated in the beginning of sub-section 8.1. In some cases landform classification with certain clearly defined objectives may be most important whereas in other cases more stress must be put on morphodynamical factors. When a geomorphologist and an engineer work together on a project,a dialogue should be aimed at with the geomorphologist communicating the potentialities, limitations and hazards of the environment to the engineer and the latter specifying the requirements for the structure and carefully selecting the design best suited to the environmental conditions. A few examples may clarify this.

Geomorphological terrain classification is of importance, for example in earthquake engineering and will lead to earthquake hazard zoning. Experience from numerous earthquakes has shown that, depending on the distance from the epicenter,the major damage usually is concentrated in well-defined parts of the affected areas. Structures built across faultlines in seismic areas are particularly vulnerable,whereas those situated at even comparatively small distance from these faults are much less seriously damaged. Some geomorphological environments such as alluvial plains saturated with water are particularly prone to earthquake damage. The quake under such conditions is optimally transmitted, therefore causing serious damage. The drier parts of a valley are considerably safer, particularly when gravel occurs, as the shock is less effectively transmitted.

This explains why during the severe earthquake of 1935 the low-lying old parts of the town of Quetta were almost completely destroyed, whereas the cantonment area located on gravel fans was much less affected. Fig. 7.7 (see chapter 7) shows the urban area of Quetta with the limit between fan and alluvial plain. The disastrous earthquake that struck Chile in 1960 gave evidence that the edges of river terraces are potentially dangerous zones because these scarps in unconsolidated materials may easily collapse as a result of a shock and cause destruction of nearby buildings. Slopes covered by clayey glacial deposits from the upper Pleistocene also proved to be danger zones under wet conditions: the shocks may easily initiate major mass movements. Artificial fills are notorious danger areas in the San Francisco Bay area where special building codes have been developed. Mapping lithology is important in bedrock as the transmission of seismic tremors varies with the type of rock and particularly with hardness.

Kurth (1971) gives another example of the use of geomorphological terrain classification for the entirely different purpose of forestry. When cable ways have to be constructed for the evacuation of logs in logging operations, the siting of such transportation devices is of great importance. Their spacing depends on the productivity of the forest and the terrain configuration may either reduce or enlarge the area that can be served by one single cable way. Therefore, the terrain configuration has an important effect on the economics of logging. Kurth rightly claims that proper siting is greatly facilitated by the use of aerial photographs.

The importance of proper specification of requirements is illustrated by the example of airfield location. The main requirements are (i) the possibility for the construction of a long runway in at least one but usually in various directions, (ii) very good drainage, no danger of flooding and low fog frequency and (iii) minimum grading. Evidently, these requirements can only be met in a limited number of geomorphological situations. The usual situation is that of a trade-off of various pros and cons with the ultimate choice also being strongly affected by non-geomorphological, for example economic factors.

A floodplain may offer a suitable site in some cases, provided it is sufficiently wide. Grading here will be minimum but drainage may be a problem. Flood hazard may exist and necessitate the construction of dikes around the field; fog may also occur. Because of these disadvantages it may be preferred to locate the airfield on a river terrace which should be wide, flat and not dissected by ravines. Depending on the composition of the terrace material, drainage may be better than in the floodplain and flood hazard will be absent. Long runways may be possible with little grading on a lake plain, but drainage is usually poor. There will be no flood hazard and fog usually occurs near the lake only. In areas of Pleistocene glaciation, sandy outwash plains offer excellent sites. Till plains may also be considered though they are not always perfectly flat and drainage may be poor. Tiling and a suitable subgrade are therefore, required.

A karst plain may offer a suitable site in limestone regions. Its internal drainage does not normally offer major problems and rainwater may be drained to sinkholes. Photo 8.36 gives an example of a sinkhole in the Central Range of Irian Jaya, Indonesia, used to drain a minor strip. Regular cleaning of these sinkholes is essental to prevent their clogging, which could impede the drainage of the field. Sinkholes situated in the planned runway have to be filled by non-compressible material. Drilling and geophysical methods should be applied to trace possible cavities which could reduce the bearing strength of the runway.

Raised coral reefs frequently offer a suitable site in humid tròpical oceanic areas such as the Pacific. The construction is normally very simple, drainage is excellent and the cost of maintenance very low. Numerous airfields have been constructed on raised coral reefs during World War II in the Pacific area. The most appropriate parts of the reef should be sought as the properties of the uplifted coral reef are very varied. The reef edges are composed of hard and porous coral colonies in their original position; the old moats and sandy cays are composed of coral debris, sand, vegetal remains and peat. The geomorphological study of raised coral reefs is greatly facilitated by the interpretation of aerial photographs. Cavities in raised reefs are a common occurrence and drilling and geophysical research are needed for their location, particularly as the present weight of the aircraft is much greater than during World War II.

The role of geomorphology in engineering may change with time, either because the requirements evolve or due to changes in the environment. During World War I, military engineers were mainly interested in geology due to stabilized trench warfare which made the type of rock occurring of prime importance for digging trenches, for mining and counter mining, and also for water supply and the search for building materials. However, in World War II, the situation was completely different and the armies had to cover large distances and conquer extensive territories. Geomorphology became important then as a basis for terrain analysis and evaluation. This was fundamental for proper planning of military campaigns. Emphasis was on matters such as trafficability, movement and cover, water supply and construction materials. Changes of the environment affecting the geomorphological applications to engineering are mainly man-induced. Flood peaks, for example, may increase as a result of smoothing the alluvial plain by agricultural practices resulting in a reduced depression storage. Urbanisation of formerly natural terrain, leading to extensive surfacing with asphalt, may have a comparable effect. Growing towns may, in the course of time, become true bottle necks of rivers in spate, the bridges being too narrow and the houses too close to the river.

In surveying for highway engineering, two major phases can be distinguished: the 'location for area' and the 'location for line'. During the location for area, emphasis is on a geomorphological breakdown of the land leading to a terrain classification for engineering. A fairly broad zone or region is investigated and all feasible locations are considered. Medium scale aerial photographs and photomosaics have an important place in this phase of the work. The various alternative locations resulting from this location for area are subsequently compared and ultimately a choice will be made; it is only thereafter that the location for line can start. Important factors in this choice are often suitable river crossings and year-round workability of the site. It may be preferred to locate the road on a well-drained river terrace instead of on the valley bottom where occasional river floods or heavy rains may interfere with the continuity of the work. During the location for line, all relevant information is given per kilometer.

Particularly in the second phase, i.e. 'location for line', emphasis is on morphodynamics and particularly slope and fluvial processes. The photograph of fig. 8.37 shows how, when constructing the road connecting Briançon, France, with Turino, Italy, the highway engineer coped with the problem of traversing a major scree slope on the eastern slope of the Crete de la Portiala, South of the Mt. Chaberton in the Alps,by constructing a circular tunnel. Several examples of the effect of mass movements on highway engineering have been given in sub-section 8.3.

Mass movements have always been of particular concern to engineers and diversified investigations have been carried out in this context. These include aerial photographic and thermographic studies, morphometric analysis (Blong, 1973; Brunsden et al., 1975) and a range of mineralogical and geotechnical methods, such as grain size and clay mineral analysis, pH and water content determination, Atterberg limits measurements, triaxial and direct shear testing, etc. On the basis of the data obtained, engineers have developed approaches to predict the behaviour of slopes under certain conditions. Since various assumptions have to be made to carry out the calculations, the outcome always has to be utilized with care. No universal system for the classification of landslides has as yet been developed (Zaruba and Mencl, 1969; Garland, 1975).

In case of slopes formed in unconsolidated materials, grain size, clay minerals and liquid and plastic Atterberg limits are of particular importance. The latter indicate the amount of water needed for the material to assume liquid and plastic properties. The strength of the material, being the combination of cohesion and internal friction, is normally expressed in terms of drained and/or undrained shear strength and displayed using a Coulomb Mohr diagram. Bearing capacity, being the load per unit area which

Fig. 8.37 Road traversing a major scree slope in the Alps.

can be safely supported is another useful parameter for arriving at an engineering classification and evaluation of unconsolidated materials or engineering soils (Terzaghi and Peck, 1967). The stability of slopes in these materials can be approximated assuming, for example, isotropy of the material or situation of the water table at the surface, the slip surface being a plane or an arc passing through the toe of the slope. The matter is elaborated upon by Carson and Kirkby (1973).

In case of rock slopes, not only the mineralogical characteristics of the rock material have to be considered but also the discontinuities occurring, such as joints, faults, bedding and foliation planes. An important parameter for the engineer is the unit weight of the rock material, usually expressed in g/cm^3. The strength of the rock slope is determined by cohesion and internal friction, as is the case with slopes in unconsolidated material. The main parameters in expressing rock strength are unconfined compressive strength, point load strength and shear strength. The elastic properties of the rock also are an important factor. Because of faults and joints, a large body of rock is rather a discontinuum of complex mechanics; therefore, the detailed study of all discontinuities is always required. Obviously, stability of rock slopes is a complex matter. Determination of the so-called critical height (maximum height without failure) and of safety factors are major items in rock slope analysis for engineering (Jaeger, 1971; Kutter, 1974).

The proper siting of river crossings is of prime importance in highway engineering and may be the decisive factor in the selection of the most appropriate highway location. Oya (1977) gives an example of the siting of a major bridge across the Jamuna River in Bangladesh. Since it proved impossible to permanently fix the riverbed because of the strong bank erosion and constantly shifting channels characterizing this huge river, the problem was to find a naturally stable point along the river course where the bridge could be built with an acceptable life expectancy. Landsat imagery and aerial photographs were used to map the numerous former courses of the river. A site was finally selected outside the alluvial fan area (where the river continuously divagates) and within a stretch where a significant natural levee has developed. Slight tectonic upheaval also contributes to the stability of the site where villages have been present for at least 180 years, unaffected by lateral sapping of the river, which therefore must be slight.

Design of bridges is also a matter of careful consideration. If the pillars are too thick and the passage for the water too narrow when the river is in spate, this construction may easily be transformed into a temporary dam with logs and debris remaining at the bridge. Ponding up of the flood waters will then result until the bridge collapses under the excessive pressure and serious flood damage may than occur downstream. Tricart (1959) mentions such a case from the Hautes Alpes, France, where a serious flood occurred in 1957. The RN 202 in the Guil Valley was largely destroyed on that occasion mainly due to the fact that in the Guil-gorge old bridges were present,with narrow arches which did not allow the flood discharge to pass, particularly because the river carried logs and boulders which blocked the arches and transformed the bridges into dams. As a result, sedimentation of boulders upstream of the bridges up to 5-10 m occurred. Several bridges finally collapsed and caused calamities downstream. The road is now reconstructed in more or less the same position, but well-designed light bridges have been built. Reconstruction of the road higher on the slopes was initially considered by the engineers. However, the geomorphologists pointed out that the rather clayey glacial deposits occurring there presented danger of sliding and should be avoided.

The bridge design should always be such that the flood waters can be easily coped with to avoid the problems outlined above. The photo of fig. 8.38 shows a bridge in the Ebro Basin, Spain, where-although vegetal debris has accumulated behind one of the pillars - the flow of the flood waters has not been unduly impeded. The photo of fig. 8.39 shows a small bridge across a wadi in Tunesia; flood waters found their way to either side where no concrete could check their force.

Bridge pillars reduce the space available for flood waters and consequently tend to increase their flow velocity under the bridge. Vertical incision of the river is introduced and may ultimately cause the instability of certain slopes, for example scree slopes which under normal conditions are only just in equilibrium. The river may then be temporarily ponded up. Another possible consequence of incision is the destruction of the bridge pillars themselves. When the current velocity downstream of the bridge decreases, deposition of coarse debris tends to occur which often leads to the development of a shoal in the riverbed, introducing lateral sapping. To prevent cutting under the bridge, a concrete flooring is constructed which forms a fixed point in the long profile, an artificial local baselevel of erosion. The photo of fig. 8.40 gives an example of the Ebro Basin in Spain. The result of such construction may, however, be a tendency to lateral sapping upstream of the bridge which may lead to a loss of agricultural land or to damage of structures bordering that part of the river. Downstream of the bridge vertical incision of the river often leads to partial exposure of the concrete flooring as can be clearly seen in the photograph. Where bridges cross deep, steep-sided, active ravines, further cutting of these ravines may endanger the foundation of the bridge pillars if they are in exposed position. The photo of fig. 8.41 shows a highway bridge under construction in Calabria, Italy one of the pillars of which is midslope of the ravine. Although the foundation is 30 metres deep, it may ultimately be a victim of the violent linear processes in this ravine unless proper management of its basin upstream checks erosion.

It is difficult to lay out general rules about the most appropriate bridge design. One must carefully study every individual case and adjust the structure to the river (or

Fig. 8.38 Bridge pillar impeding the transportation of trees, shrub and other debris. Ebro Basin, Spain.

other natural features) involved and not inversely force the river into an inadequate structure (Oya, 1977). If possible, natural processes (river, surf, longshore current, etc.) should be left intact and in any case the natural dynamics should not be opposed. It is not always easy to convince those concerned of the usefulness of a proposed design, because it may be regarded as unfavourable under normal conditions (for example, less intense agricultural possibilities), even though it is very wise when seen in the long run and under exceptional conditions such as flood. As an example the bridge crossing the Citarum River near Batujajar, Java, Indonesia, and built on horizontally bedded lake sediments forming a rather unrealiable foundation for the abutments of the bridge can be mentioned. Wet-rice cultivation near the bridge had to be prohibited to safeguard the structure as the irrigation caused almost constant saturation of the sediments, decreasing the bearing strength and causing danger of slumping. Improved drainage of the site also contributed to the solution of the problem.

Railway engineers face problems comparable to those mentioned above concerning highway engineers. The railway bridge across the Hadjel River, slightly upstream of its junction with the Zeroud River in Central Tunisia, is an interesting example of different engineering solutions to master environmental factors; in this case a wadi occasionally subject to violent floods. Initially a technically advanced design was made and the railway traversed the river in a faint curve by way of a steel bridge on high pillars. The bridge was thoroughly destroyed when the river was in spate in 1934. Several parts of the bridge were carried downstream and many pillars even completely disappeared (Ghanem, 1973). It is clear that the pillars of a bridge of this type need a very solid and deep foundation to hold out against the force of the flood waters and of the moving bedload. The bridge was reconstructed using

Fig. 8.39 Bridge in a minor wadi in Central Tunesia, incapable of coping with the flood waters.

Fig. 8.40 Foundation of a bridge in the Ebro Basin, Spain, acting as a local base level of erosion.

a more simple technology: a low bridge was designed instead. For this purpose the railway was lowered into the major riverbed by a newly constructed sharp loop and traverses the river a few metres above the water over a bridge built from river boulders kept in place by steel mesh.

The photo of fig. 8.42 shows the Hadjel River viewed upstream with the initial bridge in the distance and the simple low bridge now in operation in front of it. One of the steel segments of the old bridge, carried off downstream during the flood, can be seen in the left. The photo of fig. 8.43 shows a detail of the new bridge. The water can pass not only between the pillars, but to a certain extent also between the boulders upon which the pillars

Fig. 8.41 Bridge pillar constructed on the steep slope of a ravine, Calabria, Italy.

Fig. 8.42 Hadjel River, Central Tunesia, looking upstream with the destroyed railway bridge in the distance and the low new bridge in front of it.

Fig. 8.43 Detail of the low new bridge showing the construction made up of boulders and steel meshwork.

are built. When the river is in spate, it will overflow the structure completely. On such rare occasions the railway is out of operation and parts of the bridge may be destroyed. However, reconstruction is comparatively easy, and can be realized with material and labour locally available. The map of fig. 8.44 shows the situation at a scale of approximately 1:20,000. Note that three check dams have been constructed in the direction of the minor riverbed both up and downstream of the low bridge in order to lead the river water under the bridge perpendicular to the structure. Since the whole width of the valley is filled

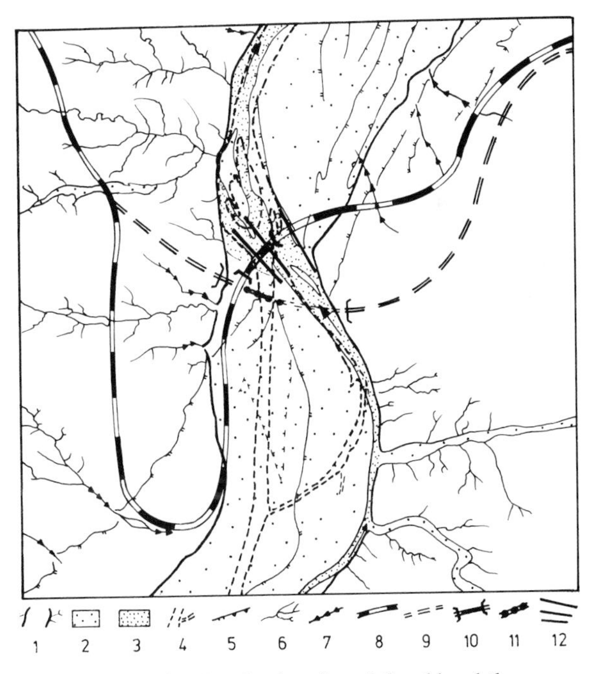

Fig. 8.44 Map showing the situation of the old and the new bridge in the Hadjel River valley pictured in figs. 8.42 and 8.43. Legend: 1. riverbed; 2. sand shoals; 3. minor riverbed before the 1969 flood; 4. same, after the 1969 flood; 5. escarpments; 6. ravines; 7. active gullying; 8. new railway location; 9. old railway location; 10. new (low) railway bridge; 11. destroyed old (high) railway bridge with pillars; 12. check dams.

with water during a flood, one has to face the fact that on those occasions parts of the water will hit the bridge at a slightly oblique angle. The main current will, however, meet the bridge at a right angle,because it is situated at the eastern valley side as a result of the curve of the river occurring just upstream of the bridge.

During the ill-famed 1969 flood, about 40 cm of sand and gravel was deposited in this part of the Hadjel River and also covered the railway stretch in the valley bottom. It only remained in the left-hand part of the riverbed; elsewhere erosion of the newly deposited material followed almost immediately during the later stages of the flood. The minor riverbed is more clear at the right bank where marked bank erosion resulted from the lateral sapping by the flood waters. The ravines entering the Hadjel actively eroded their beds, damaging and threatening parts of the railway line and some roads. At their mouths 0.5 - 1.5 metres of deposits formed due to the back water effect of the Hadjel during the flood.

8.7 River Engineering

Where river or coastal engineering is concerned, emphasis is particularly on the natural processes responsible for the fluvial and coastal dynamics. A careful analysis and evaluation of these processes is essential. Landforms, soil and subsoil conditions and construction materials should not be neglected, especially when matters like dam site location are at stake, as will be elaborated upon in this sub-section. It is possible in many cases to modify the natural processes to advantage, or at least to change the localities where they are operating (Nixon, 1963). One can also try to stop these processes altogether by harnessing the river by way of dikes and/or groines and stabilizing the coastline by way of concrete walls. It should be understood, however, that these are usually costly and often vain efforts, particularly when larger rivers and coasts exposed to heavy surf are concerned. It should always be attempted to allow natural processes to work favourably and in unison with the intended structures.

It is essential to know that any artificial structure or modification tends to modify the fluvial or coastal dynamics to some extent, if not fundamentally. (Blench, 1969; Lagasse and Simons, 1976; Lane, 1955). Such effects should be evaluated properly in advance and the structure adapted to the provoked changes such as cutting of a river directly downstream of a dam. For example, the straightening of a river course, effectuated by the artificial cut-off of meanders or of other bends of the river, will result in an increased gradient of the riverbed. This will affect the velocity and transporting capacity of the river and may lead to a tendency towards vertical incision. Reinforced erosion of the ravines debouching in that stretch of the river may subsequently be introduced.

If such a development is considered undesirable, an artificial device may be built in the river, the height of which covers the extra-gradient caused by the shortening of the river course and which leaves the gradient of the new course in the same order of magnitude as that previously existing in the river. Fig. 8.45 illustrates the principle of the method. The concrete step is indicated by hachures. In case of a navigable river, provisions for shipping will have to be made.

Ryckborst (1980) elaborates on the response to straightening rivers by way of artificial meander cut-off and gives examples from the Rhine, the Mississippi and the West Prairie River (Canada). He concludes that a meandering river only seems to be an inefficient system. Meander cutting results in excess energy which is spent on erosion of riverbeds and banks, while the eroded material is deposited farther downstream. Man-made interference with such rivers is, therefore, a costly operation requiring constant maintenance and improvement. He mentions the 14% artificial shortening of the Rhine channel achieved between Basel and Strasbourg in 1800 by replacing the natural meanders by a limited number of bends of much greater radius. Notwithstanding some reinforcement of certain parts of the banks, the Rhine started to form new meanders of the type previously existing. Subsequently, the whole Rhine channel was fully harnessed by the engineers and successfully kept in place. Since then, the riverbed is continuously shifting between sand banks, making this stretch of the river so treacherous that pilot services are required. Fig. 8.46 illustrates the sequence of events.

Stabilisation of river channels in mountainous areas is fraught with special difficulties because of the torrential run-off characteristics and the often high bedload. (Krofellner-Kraus, 1981). The stereotriplet of fig. 8.47 gives an example from the Spanish Pyrenees and pictures part of the Ribagorzana Valley at the scale of 1:20,000. The approximately southward flowing river merges with the Rio de Llauset downstream of the village of Senet, situated on a large alluvial fan near the upper limit of the triplet and runs into the Rio Noguera. The upstream part of the valley (lower and central portions of the image) is fairly wide and carved in Late-Carboniferous biotite-granodiorites. Farther downstream, near Senet, partly metamorphosed sedimentaries of the Lower-Devonian Ruela Formation outcrop. This formation is composed of graywacke, sandy shales (Gelada formation) and slates with some limestone beds (Aneto formation). The beds strike E W and the dip is almost vertical. The village of Aneto can be seen in the upper right corner of the photo and the village of Senet is situated within the Aneto formation, the rocks of which are considerably weaker than

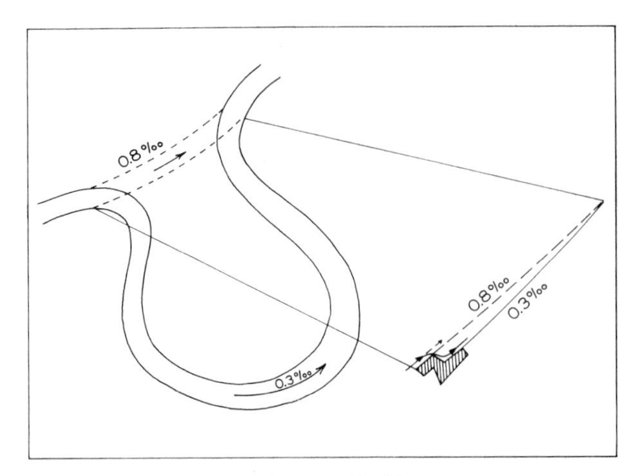

Fig. 8.45. Artificial meander cut-off with concrete step to prevent headward cutting due to increased gradient.

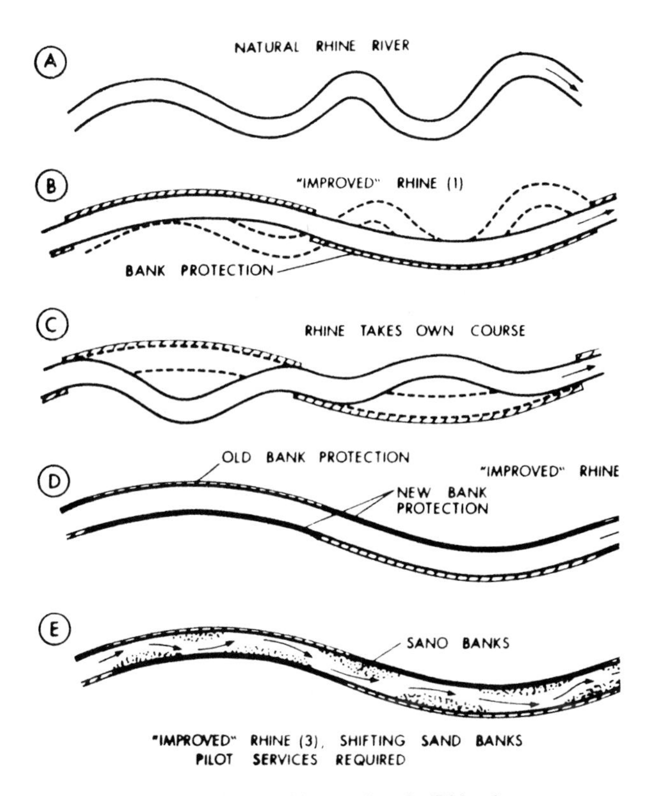

Fig. 8.46. Various phases of harnessing the Rhine between Basel and Strasbourg and the response of the river (Rijckborst, 1980, Geol. Mijnbouw, 59:122).

those of the Gelada formation outcropping near the upper photo edge where the valley becomes substantially narrower.

The main engineering problem in the area is associated with the steep torrent that originates on the slopes of the Pico Munido (2650 m),debouching in the main river directly north of Senet. It is located in the easily erodable rocks of the Aneto formation and it has formed a sizeable fan in the past which is presently under cultivation. The slopes of the torrent basin are severely affected by accelerated erosion and the great amounts of debris produced have formed a new, active fan (light tones on the photograph) which pushes the main river towards the opposite valleyside and impedes its flow when in spate. Excessive sedimentation and lateral sapping resulting in a widening of the floodplain, can be clearly seen upstream of the bottleneck. The bridge situated here is in a precarious position. Undercutting of both the old and the new active fan is obvious from the scarps that have been formed.

A series of checkdams has been constructed across the torrent bed to reduce the throughflow of debris as illustrated in the aerial photograph of fig. 8.47. The success is, at best,partial. One might consider constructing a new channel for the torrent in the downstream part of the (old) alluvial fan where traces of a previously existing course of the torrent can be seen. This would create a hydrologically more favorable acute angle of confluence with the main river but brings the torrent across or near the village,which does not seem desirable. Thorough conservation of the torrent basin will be required to get the situation ultimately under control. The Barranco de Gelada enters the main river directly downstream of Senet and can be seen near the upper photo edge. It also formed a fan which though not large, clearly presents a second bottleneck for the main river when in spate. torrent has a most unfavorable, inverse angle of confluence and its artificial downstream deflection on the fan, parallel to the forest limit on the airphoto, is recommendable. Since this would increase its gradient to some extent, a stepwise concrete structure may be required to prevent renewed cutting and increased production of debris. In the parts of the main river valley depicted in the lower part of the aerial photograph, no major problems are present. Note the safe position of the bridge in the middle of the image.

The photographs of figs. 8.48 through 8.51 show some other problems, related to excessive bedload of mountain rivers. Fig. 8.48 shows excessive sedimentation of mainly medium-sized sands in a tributary of the Crati River, Calabria, Italy,near its junction with the main river. Fig. 8.49 shows the remedial measures in a comparable adjacent tributary of the Crati River. A narrow, concrete bed has been constructed, which will result in a higher velocity of flow and hopefully enable the river to carry through the bedload to the main river, eliminating further troubles. An entirely different case is the torrent in the Himalayan foothills near Dehra Dun, India,pictured in fig. 8.50 where a torrent bed, following a major landslide, was transformed in a large debris flow, the material

Fig. 8.47. Road engineering and fluvial dynamics; an example from the Ribagorzana Valley, Spain. Scale 1:20,000. For explanation see text.

Fig. 8.48. Excessive sedimentation by a tributary of the Crati river, Calabria, Italy,resulting from strongly eroded upstream parts of the basin.

Fig. 8.50. Debris flow in a small valley near Dehra Dun, India.

Fig. 8.49. Regulation of another tributary of the Crati river, Calabria, Italy. By confining the river to a rather narrow concrete bed increased velocity and transportation of debris has been achieved.

Fig. 8.51. Bridge situated downstream of the debris flow of fig. 8.50, the pillars of which have almost been completely buried by the debris.

of which almost completely buried the pillars of the bridge of fig. 8.51 which originally was several tens of metres above the valley bottom. Because the accumulation of debris was a gradual process, the bridge was unaffected; the torrent water now percolates through the gravel deposits.

In areas where a settlement is located in a narrow valley, it may occur that the main street is immediately adjacent to the riverbed. In (semi)arid areas main street and river channel may even coincide. An example is given in the photo of fig. 8.52 showing the main street (named Calle Baranco) of the village of Aguilon in the Cordillera Iberica, Spain. In such villages this situation may be acceptable, though it certainly becomes a hindrance when larger settlements are concerned and when the floods of the torrent are of destructive magnitude. An example is the old fortified town of Daroca, Spain where a French engineer, Pierre Bedel constructed a tunnel in the sixteenth century to deflect the waters of the ephemeral river which previously traversed and occasionally inundated the town.

River works can be carried out for numerous reasons: navigation, the use of water for irrigation (Oliver, 1972), generating electricity, flood control, etc. (Tricart, 1959). The structures required are therefore, diversified. Most frequently the attempt at harnassing rivers are partly or wholly related to flood control. This is of particular

Fig. 8.52. Riverbed forming mainstreet of the village of Aguilon, Spain.

importance along the lower reaches of the rivers and in deltaic areas. The need for flood control arises early in the development if rivers characterized by torrential, swift current and flash floods are concerned. In the SE Asian theatre the Nobi River (Japan), the Cho Shui (Taiwan) and the Red River (Vietnam) are examples. Where the rise of the river water is gentle and the flood volume is spread over several months, modest flood control measures per unit area will suffice. Some major SE Asian rivers fall into this category: the Irrawaddy, the Chao Phya and the Mekong rivers. The latter benefits from the excellent detention basin formed by Lake Tonle Sap. The embankments in the upper part of the Irrawaddy Delta, though only 1 metre above the 100 yr. flood level, are never overtopped,but serious problems are created by the three earlier mentioned, much smaller rivers (ECAFE, 1963; UNESCO, 1966). The degree of flood control that can be achieved is not only affected by the characteristics of the rivers concerned, but also to an important degree by the economic feasibility, which depends on the stage of development reached, the intensity of use and density of population. This is reflected in matters such as height of embankments, etc.

The simplest response to flooding is to live with it by establishing agricultural systems based on floods as elaborated upon in section 6 and by building settlements on adjacent higher ground or on artificial mounds. This is how it has always been possible to raise a crop along the Mekong River, notwithstanding the almost-absence of flood control prior to the execution of the Mekon River works. In Thailand, crop failure is imminent if the annual flood fails or is below average. Local and usually minor corrections of the river are a first step towards improvement. Clearing of obstructions in the channel of small rivers and straightening the river course by cutting off meanders will lead to increased gradient and higher velocity and may lead to deepening of the channel and to greater channel storage. These measures are a futile effort, however, if the river is rapidly aggrading. In such cases works in the headwaters are required if feasible. Flood detention in natural channels is nevertheless important in many areas and widening, deepening and rectification of existing drainage channels are widespread measures.

Embanking is a more rigorous method of flood control which affects the flood stages and the position of the bed and therefore, the river regimen. The water tends to build up when the river is in spate,because it no longer has free access to the adjacent land. This introduces the danger of rupture of the dikes,especially when the riverbed is topographically several metres higher than the backswamps, as is the case in many lowland areas. Rupture of the dikes may also result from lateral shifting of a river which is too big to be effectively harnassed or when seepage through the dike body reduces its strength (Sluys and Ovaa, 1973). In any case the dikes should be constructed at some distance from the river to provide adequate storage for the floodwaters. Further reduction of flood crests can be obtained through retarding basins, to which the flood water has access through low sections of the embankment. Decrease of the natural storage capacity may result from artificially raising the land to make it suitable for agricultural or other activities. It is particularly common in backswamps and on tidal flats. The construction of sea dikes in the lower zone of deltaic areas also decreases the storage and tends to increase the storm surges. However, during short duration storms the maximum levels will not be reached because of the time needed for filling the flooded areas. In some major deltas such as the Irrawadi and the Ganges deltas in SE Asia, sea dikes are absent and protection from river floods is obtained by way of horseshoe-shaped embankments on the natural levees.

In many cases it is clear that flood control cannot be mastered by measures in the affected lowland areas alone but that flood and erosion control of the whole river basin and the construction of storage reservoirs upstream is required. A flood control controversy may then arise between those advocating the establishment of large dams in the main stem of the river and others favoring a larger number of small retention dams in the headwaters. This is often a complex matter and much depends on the (usually multi)-purpose of the dam (Leopold and Maddock, 1954).

Various geomorphological factors have a bearing on the selection of a dam site and a careful study of the fluvial dynamics is useful for evaluating the effects of the dam on the river behaviour and on the environment as a whole. The effect of a dam on these matters is usually profound (Wahlstrom, 1974; Walters, 1971). Among the major requirements for a dam site, the following can be listed:

- a water-tight basin of adequate size
- a narrow outlet of the basin (for economical construction of the dam)
- an opportunity to build a spillway for surplus water
- the availability of construction materials (especially for earthen dams)
- siltation which is not overly excessive (in future reservoir).

Special attention should be paid to the study of matters such as the occurrence of faults and joints, the evidence of former floods, the thickness of alluvium and in the case of an arch dam, joints and weak zones in the abutments. Since various terrain factors such as width of the outlet, bearing strength of rocks at damsite and construction materials influence the type of dam (gravity dam, buttress dam, arch dam, or earthen dams of various types), close co-operation with the engineer is required from the onset of the survey work (Stephenson, 1979). Earthen dams run the risk of being destroyed by overtopping or by seepage if not concentrated at the foot of the structure. Seepage in the basin may be a problem, particularly if the water table is above the groundwater table.

The studies should encompass the whole upstream drainage basin in order to estimate the rate of silting up of the future reservoir.

The lower courses of the tributaries entering in the reservoir will be drowned and form indentations where sedimentation will start. With larger reservoirs the promontories will be eroded by the waves, and spits and bars may be formed. More common also with smaller reservoir lakes, is erosion of the slopes at the water level. Since the lake level normally fluctuates with the season, a concavity will be formed which may introduce slope instability. Another factor which may introduce slope instability is the higher groundwater level around the lake. All these matters require careful consideration prior to construction of the dam (Bojoi, 1969; Slazik, 1978; Szupryczynski, 1974; Stoilov, 1970).

Downstream of the dam the floodplain will be narrowed and the bed stabilized. The decrease in load may lead to cutting and to retrogressive erosion towards the dam (Komura and Simmons, 1967). Since the main river is often no longer capable of evacuating the debris produced by its tributaries, they may build up gravel fans, as is illustrated in fig. 8.53 from the Agri Valley, Italy (Maruyama, 1976).

The waters passing over a dam or through a spillway may have various effects on the channel configuration as is illustrated by two examples from Spain and southern Italy. The first example of how a dam affects the erosion sedimentation, flow and further behaviour of a river relates to the Embalse de Pina, which was build some decades ago in the Ebro River downstream of Zaragoza, Spain. It is a very low dam, almost continuously overtopped over its full length by the river, its purpose being to raise the water level in order to create an adequate gradient in the canals of the irrigation system, the intake of which is immediately upstream of the dam. The situation at and near the dam is depicted in the aerial photograph of fig. 8.54 which has an approximate scale of 1:35,000 and was taken on August 15, 1956 when the discharge of the Ebro River was low. The interpretation of the photograph shows the floodplain (white), the alluvial plain with its agricultural fields (oblique hachures) and an accumulation terrace (vertical hachures), which are the main geomorphological elements.

The Ebro River is indicated with dots as it was at the time of the air photography and its position as appearing on a map of 1938 is added for information. It is obvious that the river meanders changed position both up and downstream of the dam. This natural process caused some loss and gain of land. The alluvial plain was attacked by lateral riverwork at several localities. The only fixed point in the river is where it overflows the dam: it is the task of the engineer to prevent any excursion of the river at this site which might result in its bypassing the dam, if unchecked.

A striking change in riverbed morphology can be seen

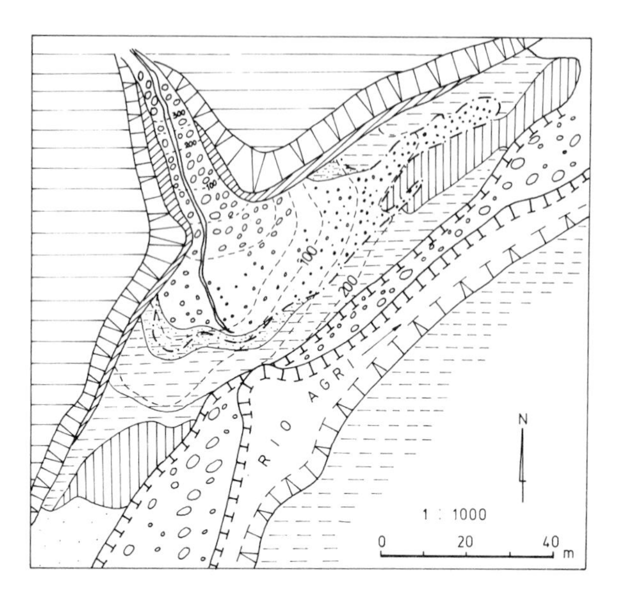

Fig. 8.53. Growth of a gravel fan where a minor tributary joins the Agri River, Basilicata, Italy. After the construction of a major reservoir lake upstream, the Agri River is unable to transport this material (Maruyama, 1976).

downstream of the dam over a distance of about one kilometre from the comparison of the 1938 and 1956 situation. Since the construction of the dam this section of the river has widened considerable: it is now braided and most likely more shallow than previously. Active lateral erosion downstream of the dam is evident and was still continuing when the author visited the damsite on several occasions in the late nineteen-sixties and early nineteen-seventies. Fig. 8.55 shows the sapping of the south bank (where the spillway is situated) which has not yet been effectively brought to a standstill by a groyne and concrete blocks.

The dam has actually modified the river flow in two different ways. In the first place the water now overflows the dam over its full length,whereas it formerly concentrated in the main channel. This assists in preventing vertical incision of the river downstream of the dam,which could endanger the structure. The V-shaped ground plan of the dam is a second aid in the dispersion of the river water. It splits the current into two channels situated near the north and south banks, separated by a gravel island.

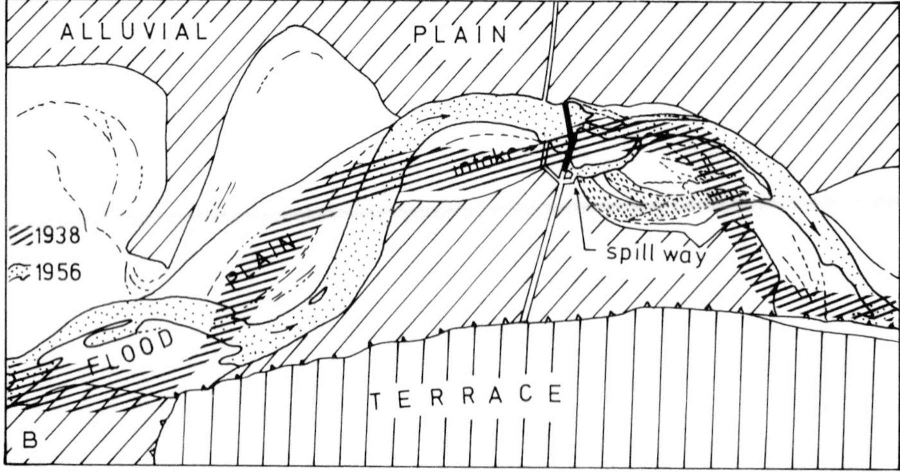

Fig. 8.54. Aerial photograph and interpretation showing the modification of the riverbed and the introduction of lateral river work following the construction of a low dam in the Ebro River, Spain.

Fig. 8.55. Concrete blocks supposed to stop lateral sapping downstream of the dam of fig. 8.54.

It is evident that the V-shape of the dam is the main cause of the lateral sapping immediately downstream. After about one kilometre the river resumes its normal, meandering pattern.

The second example (fig. 8.56) is from a movable dam in the Crati River, Calabria, Italy. The water of the reservoir needed only to bridge a shortage of irrigation water during a few months in summer. Since the bedload of the main river and that of a ravine entering into the reservoir is high, the lake is repeatedly flushed in the wet winter season to evacuate the debris. Large movable steel doors have been built for this purpose. The spasmodic flow of the river water causes deposition of debris at some distance downstream of the dam which deflects the riverwater sideways, causing sapping of the riverbed. Directly below the dam the debris is washed away and cutting has occurred. To control this a checkdam has been constructed across the river. The fact that major storage lakes have an effect on the climatic aspects of the environment and on health conditions is only briefly mentioned here.

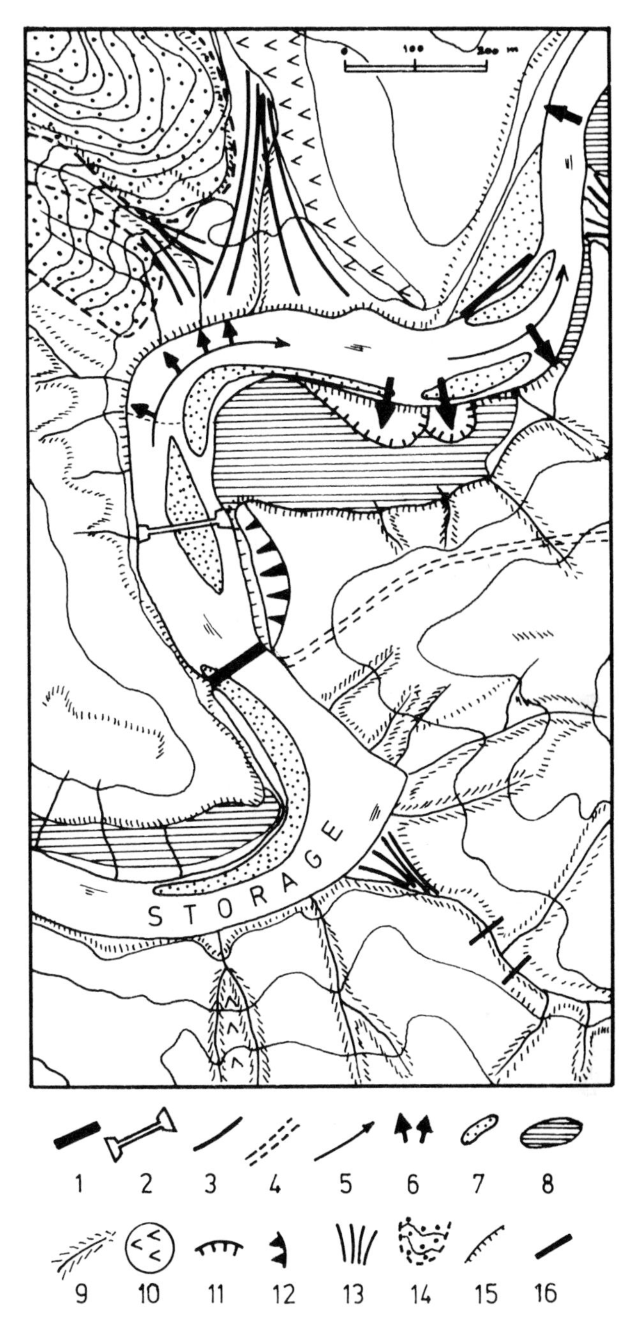

Fig. 8.56. Movable dam in the Crati River and its geomorphological context. Legend: 1. movable dam, 2. low check dam to stop vertical cutting, 3. revetment, 4. irrigation tunnel, 5. direction of flow, 6. lateral sapping, 7. sand/gravel shoals in riverbed, 8. river terrace, 9. ravines, 10. badlands, 11. low scarp, 12. high scarp, 13. fan, 14. re-afforestation, 15. floodplain scarp, 16. gabions.

8.8 Coastal Engineering

In coastal engineering the study of morphodynamics is beyond doubt the most important contribution to be made by the geomorphologist. (Allen and Spooner, 1968; Mitchell, 1968; Bruun, 1970, 1978; Charlier, 1968; Chuzhmir, 1968; Dulan, 1973; Komar, 1970, 1974; Prior and Ho, 1972). The study of currents, tides, waves and swell rooted in measurements off-shore can be greatly assisted by the interpretation of aerial photographs and other types of remote sensing imagery (Koopmans, 1971). The same applies to studies of littoral drifting of sand and the deposition of fine detritus carried in suspension. (Erchinger, 1970; Fedorovskii, 1968; Sellert, 1967; Godfrey and Godfrey, 1973; Hagyard and Gilmour, 1969; Luck, 1970). The imagery gives an overview of the situation which cannot be otherwise obtained. This is demonstrated by the example given in fig. 8.57 which is a vertical aerial photograph at the scale of 1:6,000 showing incoming sea waves refracted by the two moles of

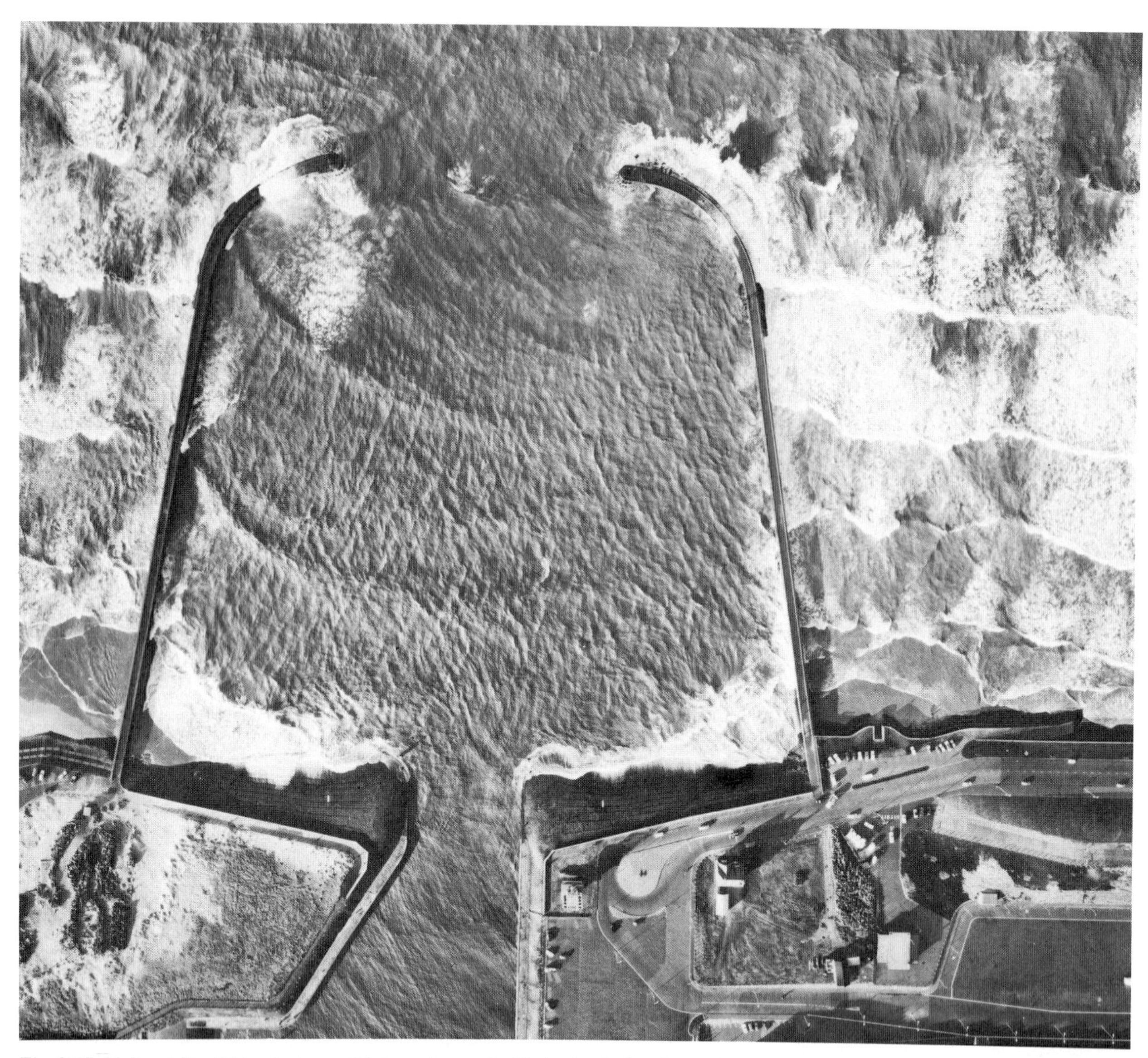

Fig. 8.57. Moles of the fishing harbour of Scheveningen, the Netherlands showing reflected waves and rough water caused by 'clapotis' in the middle of the harbour entrance.

the fishing harbour of Scheveningen, The Netherlands. They give rise to extremely rough water in the middle of the harbour entrance, causing problems for incoming and outgoing vessels under bad weather conditions. A different configuration of the moles, recently effectuated, precludes these difficulties. Photography and particularly terrestrial photogrammetry can also be applied to mapping exact position and shapes of the waves by means of contours. As roughness of the water surface has a profound effect on radar reflection, SLAR imagery has been successfully applied to the study of the state of the sea over large areas. The bottom configuration of shallow coastal waters and matters such as the drifting of sand along beaches can be studied from refraction patterns of swell and waves and from the angle of these with the coastline (Verstappen, 1977). The effect of engineering works and other objects on the coastal configuration and modification (Zabawa et al., 1981) are often clearly visible on remote sensing imagery. The aerial photograph of fig. 8.58,at a scale of 1:2,000, shows a part of the Dutch coast, the so-called Hondsbossche Sea dike, protected by groynes. In front of one of these a ship has sunk and caused a marked deviation of the normal sedimentation pattern along the coast.

Even a single aerial photographic coverage, though picturing only a momentary situation, may significantly reveal the coastal dynamics as is demonstrated by the 1:40,000 aerial photograph of fig. 8.59 showing the coastal development near Negombo along the west coast of Sri Lanka. Abrasion is characteristic for the coast stretching from Negombo northward. This has resulted in a recession of the coast which started during the early Holocene,as is evident from the occurrence of old beach ridges and swales cut off by more recent ones which in turn are cut off by the present coastline. To the north of Negombo the coast was formerly considerably more sea-

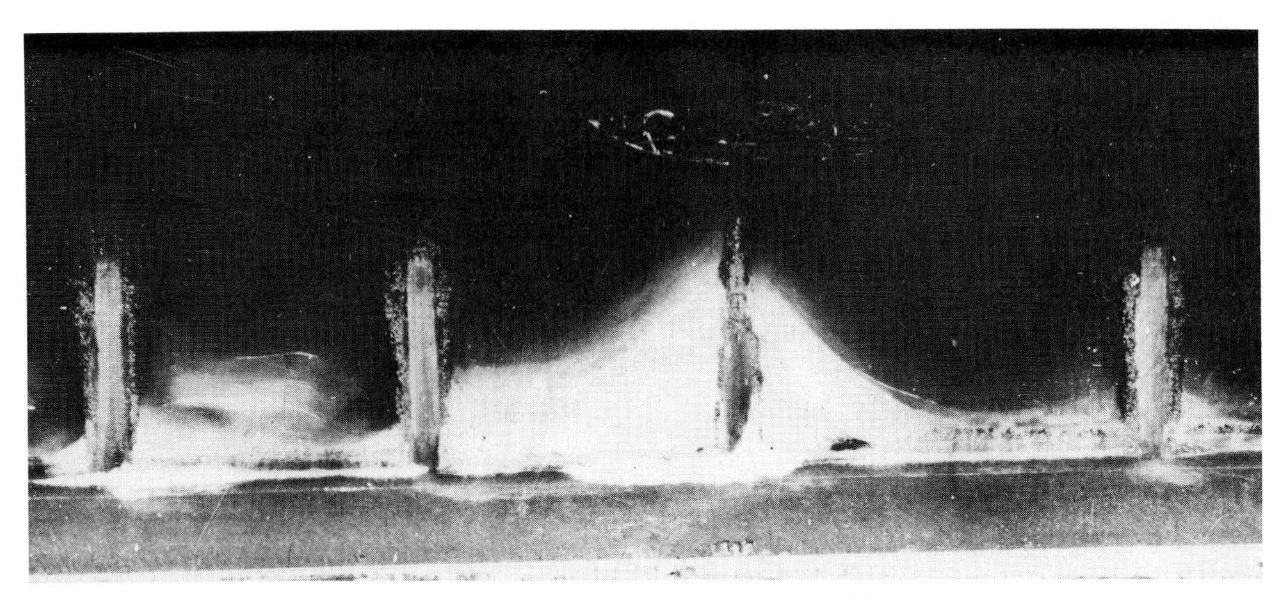

Fig. 8.58. Changes in sedimentation pattern near a groyne along the Dutch coast due to a shipwreck.

ward than at present. To the south of Negombo the situation is different. An off-shore bar and a broad lagoon are found there and the coast is either advancing or stable; in any case not receding. Airborne geophysical observations indicate that a fault zone occurs here, running more or less perpendicular to the coast. This is evidently an old feature, as most of the faults of Sri Lanka, but the coastal configuration points to the fact that in all likelihood movements also occurred during the Holocene. Tectonic subsidence is probably the cause underlying the abrasion at Negombo and this cannot be changed. It is, however, possible to attempt to diminish the harmful results of this subsidence by changing the coastal configuration. To this aim the barrier bar to the west of Negombo has been extended northward by constructing a concrete jetty. This does not withstand the strong surf, however. One can also try to break the force of the surf at Negombo by keeping the depth of the immediate foreshore at a minimum. It appears that material is transported out of this critical area, is sucked into the lagoon under the influence of the tidal currents and forms a tidal delta in the lagoon. The material deposited here could be very useful for the protection of the coast at Negombo. Furthermore, it appears from the airphoto that a southward transport of sand occurs along the coast to the north of Negombo. No groynes perpendicular to the coast should therefore be permitted along this northern part of the coast; the sand should reach the Negombo off-shore area unhampered to provide maximum protection there. Slight abrasion of less valuable land farther to the north should remain unchecked for this purpose (Herath, 1962).

Much more information can be gathered if sequential imagery is available as is evident from a study by Schmitt-Taverna (1975) of the Europoort area in The Netherlands. Fig. 8.60 pictures the coastal configuration before (dashed) and after the engineering works. Artificial beach feeding contributed substantially to the growth of the recurved spit at the SW end of the Westplaat shoal. Guilcher (1974) gives several examples of the application of dynamic geomorphology to coastal engineering. He mentions the role in alternating abrasion and accretion of oblique funnels and ridges and the problems in delta growth and erosion. (Berg and Duane, 1968). Coastal spits may also cause particular problems and force the closing-off of harbours from the sea such as that of Lobito in Angola.

Hite and Stopp (1971) and Joliffe (1968) have placed coastal protection in the context of coastal resource management. Silvester (1979) elaborates on matters of coastal defense such as the shoreline changes provoked by groynes and headland control. Specific emphasis on on the shape of breakwaters which refract the wave energy in such a way that sediment transport is facilitated.

Vertical walls or revetments meant for coastal protection have in many cases proved to cause erosion due to the refraction of the surf (Steers, 1971). Walls or breakwaters stretching perpendicular to the coast, e.g., near harbour entrances, may also cause coastal recession, at least at the side which is in the lee of the littoral drifting. This is exemplified by fig. 8.61 showing the ancient Sunda Kelapa harbour (now only a fishing harbour) of Jakarta with the changes in the position of the coast in the period 1625-1977 and the effect of the long vertical walls bordering the harbour entrance (Bird and Ongkosongo 1980).

In many cases not only actual coastal processes have to be studied (Quat. J. Eng. Geol., 1979) but also the earlier geomorphological development, as demonstrated

Fig. 8.59. Coastal configuration near Negombo, Sri Lanka. For explanation see text.

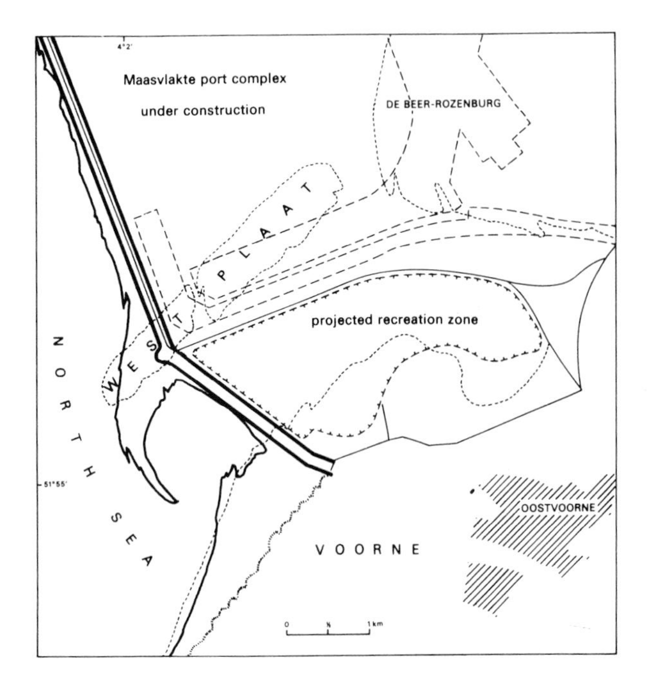

Fig. 8.60. Coastal changes resulting from the interaction of coastal engineering works and natural processes near the Westplaat, The Netherlands.

by the extended harbour of Jakarta at Tanjung Priok implemented in the nineteen-fifties. The genesis of this cape largely influenced the engineering works carried out at that locality. This muddy cape cannot be explained as a delta, as no present or former river course occurs there. A row of coral islands stretching approximately perpendicular to the coast, reaches the shore exactly at this locality. To a geomorphologist it is,therefore,evident that a buried coral reef with associated tombolo accounts for the presence of the cape. This is indeed the case,and the location of the reef is indicated in fig. 8.62. This invisible reef necessitated a considerable change original harbour extension scheme. Its presence had a great influence on the dredging operations and on the final location of the new parts of the harbour. Many other examples can be given to illustrate the importance of a genetical understanding of the landforms and processes occurring in a coastal area. (Thelb, 1969; Valentin, 1952; Verger, 1959; White, 1970).

8.9 Surveying Methods and Aerospace Imagery for Engineering

The orientation of the land survey work required for engineering purposes varies with the phase and scope of the project involved (Dumbleton, 1973). When starting the planning of linear structures such as highways, railroads and pipelines, two phases have to be distinguished from the onset: the location for area and the location for line. During the reconnaissance for area, all feasible locations within the region are considered, from which a final selection is later made. During the location for area the interpretation of aerial photographs and of other types of aerospace imagery is of particular importance. The location for line amounts specifically to a comparison of the various alternate routes and their pros and cons. Once the final choice is made, further work is restricted to a strip along the selected line. Medium to large scale aerial photographs may be used in this context for purposes of terrain appreciation, but emphasis will be on the photogrammetric uses of imagery in defining cut and fill quantities, gradient of the road, etc. Fieldwork and related laboratory investigations become more important although airphotos can be used for matters such as staking the line in the field. Aerial photographs are of widespread use among engineers,as they save time, money and effort; increased accuracy of the work and efficient planning can thus be achieved. Where non-linear, localized engineering structures such as major buildings, airfields or dams are concerned, the same approach applies with the location for area being followed by the location for site. In well-developed areas the phase of location for area can sometimes be reduced or omitted.

The complete sequence of the land survey operations required can be summarized as follows: once the general concept of the structure and its requirements are agreed upon, a field reconnaissance of the terrain is carried out of the entire area (not of one line only). Based on this reconnaissance a preliminary survey of one or more locations is carried out, followed by office studies on projecting a tentative alignment and profile and estimating the quantities of cut and fill, cost, etc. A location survey is then required for staking out the projected location in the field with only minor adjustments of grade and/or alignment. Further office work will ultimately lead to the construction survey phase. Experience has shown that the most serious errors in location usually originate in the reconnaissance phase of the work. Thorough location for area is of the utmost importance (Lawrence, 1972). During the location for line,the selection of the route is greatly facilitated by marking the location on the aerial photographs using transparant straight edges and stencils. The feasibility of the various alternate routes discussed in the preliminary report is preferably carried through per kilometre,while treating the various factors or parameters separately, for example, in tabular form with columns for each of them. Matters such as relief, bridges and culverts, earthwork in rock, earthwork in unconsolidated material, vegetation, curves and required slope stability, have to be included. Staking out the line in the field is usually done on the basis of annotated

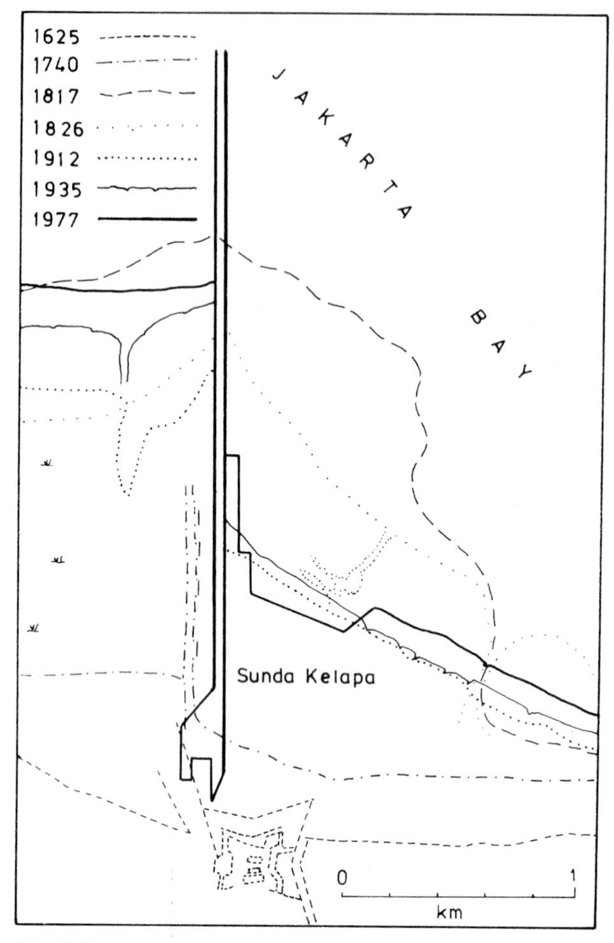

Fig. 8.61 Map showing the effect of the ancient harbour of Jakarta at Sunda Kelapa on the coastal development (Bird and Ongkosongo, 1980).

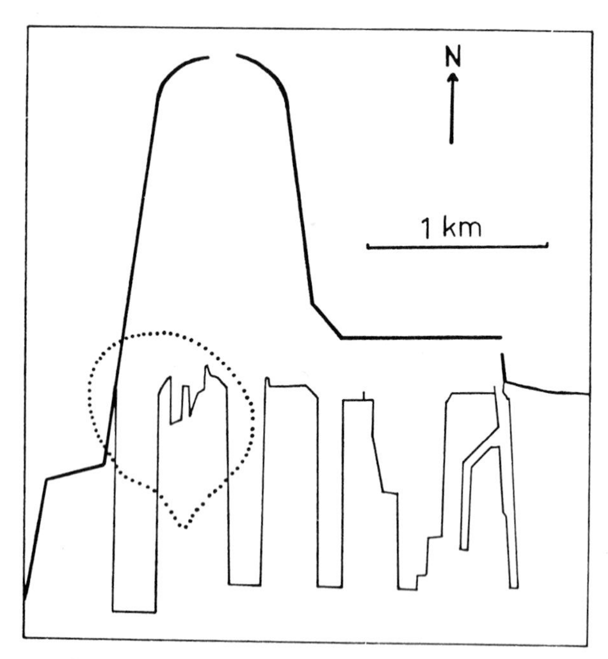

Fig. 8.62. The position of a buried coral reef within the confines of the harbour of Tanjung Priok, near Jakarta, which affected the harbour extension works considerably.

aerial photographs. Only minor alterations will be needed, assuming good photo reading capability of the field party chief.

It is obvious that the uses made of aerial photographs by the civil engineer only partly relate to geomorphology. Photogrammetric uses are equally important and other applications such as planning the field survey logistics should be mentioned. The aerial photographs may directly serve as a basemap or form the documents from which study plans in scales ranging from 1:5,000 - 1:15,000 are derived photogrammetrically. Detailed plans in scales of approximately 1:2,500 may be required in built up areas. Pryor (1960) mentions even larger scales in the order of 1:1,000 for rural areas and 1:500 or 1:600 for urban areas. These plans may serve as reference maps for topographic detail or ground survey data. Computation of cut and fill quantities is another photogrammetric application of aerial photographs in engineering (Holder, 1962). Among the main advantages of the aerial photographic approach are to be listed:

. the best possible alignment with least cut and fill can easily be found
. engineering problems and trouble spots can be located
. no field stake out of the preliminary line is required
. the property severance and land value survey is quicker, cheaper, and more precise.

The photo-interpretation carried out in the context of engineering studies mainly relates to terrain classification rooted in geomorphology and using vertical aerial photographs or photomosaics. (Atkinson and Brown, 1962; Blesch and Ta Liang, 1962; Burton, 1969, 1970; Dolezal, 1962, 1966; Dumbleton and West, 1978, 1972; Koopmans, 1971; Kurth, 1971).

Uncontrolled mosaics may be useful in the reconnaissance phase when geometrical distortions are not critical. Data about drainage network characteristics, linear erosion features and land sliding can be derived from them. Controlled mosaics with contours are more applicable for more detailed work. Spagna (1974, 1979) compared the suitability of photogrammetric maps and photomosaics in highway engineering studies. Ortho-photographs are a powerful tool in the detailed phases of the work. Soeters and Rengers (1979) elaborate on the use of aerial photographs for engineering geological mapping at the scales of 1:5,000 - 1:15,000. Photoscales of 1:15,000 - 1:20,000 are considered suitable,especially if blow-ups at the scales of 1:3,000 - 1:5,000 are stereoscopically studied using the ITC stereo-scan. Preliminary photo-interpretation, mapping and final reporting are the main phases of work distinguished by these authors. Tanguay et al. (1969) mention the use of multispectral imagery for engineering soil mapping and the uses of other remote sensing techniques are elaborated upon by Adams et al. (1970),

Parker (1968) and Rib (1975). Mollard (1962. 1972) discusses the special problems met in cold areas.

Notwithstanding the widespread use made of aerial photograph interpretation as a basis for terrain evaluation by civil engineers, one cannot escape the impression that relief factors and metrical aspects have often been overemphasized compared with the effects of active processes. The study of relief can be made more effective by applying morphometrical terrain analysis. This approach may be an aid in estimating the size of bridges and culverts required. Geomorphological photo-interpretation may assist the engineers in three major fields:

1. Quantitative terrain analysis. This includes both purely photogrammetric aspects such as the computation of cut and fill quantities and morphometrical methods such as the compilation of slope distribution maps and quantitative analysis of drainage basins. (Hobson, 1972).
2. The study of geomorphological processes such as colluviation, solifluction, landsliding, fluvial erosion and deposition and beach processes.
3. Terrain classification on the basis of landform, soil, geology and vegetation. To achieve this aim a thorough study of the two previously mentioned fields is required. Emphasis should be on the engineering properties of the geomorphological units distinguished as affected by soil, rock and other characteristics.

Among the pioneering work in this field the publications by Dowling (1963, 1968) on engineering soil conditions in part of northern Nigeria and by Schofield (1962) on road construction in Malawi should be mentioned. Dowling found a distinct relation between airphoto patterns and engineering soil conditions and stressed the importance of unraveling the erosional history of the area concerned. Schofield stressed the importance of geomorphology and particularly of colluvial soil movements for the search of pavement material, etc. in the area of study. The terrain classification and evaluation for engineering purposes (Miles, 1962) is often realized by applying a multi-disciplinary approach of land surveying (NIIR, 1971) developed by the CSIRO in Australia, the MEXE in U.K. (Beckett, 1972) and other organisations (Brink, 1968, 1982; Beaven, 1972; Grabau 1968). The so-called PUCE system is probably the most common among engineers (Aitkinson, 1967, 1971; Grant 1978, 1974; Grant and Aitkinson, 1970; Avci, 1972). The varous methods are elaborated upon in section 12 on Synthetic Land Surveys.

References

Adams, J.A.S., 1970. Development of remote methods for obtaining soil information and location of construction materials using gamma ray signatures. Rice Univ., Houston, Tex., Dept. Geol. Report 718519: 1–275.

Aitchison, G.D., and Grant, K., 1967. The P.U.C.E. programme of terrain description, evaluation and interpretation for engineering purposes. Publ. Soil Mech. Section CSIRO, Australia

Aitchison, G.D., et al., 1971. Terrain evaluation for engineering. Proc. 4th Reg. Conf. Soil Mech. and Found. Eng., 1: 1–8.

Allan, R.H., and Spooner, E.L., 1968. Annotated Bibliography. Coastal Eng. Res. Center, M–P 1–68: 1–141.

Anderson, J.G.C., and Trigg, C.F., 1970. Geotechnical factors in the redevelopment of South Wales valleys. In: Civil Engineering problems of the South Wales valleys. Inst. Civil Eng., London: 13–22.

Arora, H.S., and Scott, J.B., 1974. Chemical stabilization of landslides by ion exchange, California Geol., 27(5): 99–107.

Atkinson, J.R., and Brown, N.B., 1962. Airphoto Interpretation Study for the Division of Roads and Road Traffic of the S. Rhod. Gvt., Proc. 1st Symp. Comm. VII, I.S.P., Delft: 487–491.

Avci, M., 1972. Mühendislik maksatlari icin arazi siniflandirmasi, Tremp, Ispanya. (Land classification for engineering purposes, Tremp, Spain). Jeomorfoloji Dergisi, 4: 121–132.

Barnes, D.F., 1966. Geophysical methods of delineating permafrost. Proc. Intern. Conf. Permafrost Nat. Acad. of Sci.-N.R.C. Publ. 1287: 349–355.

Bažant, A., 1979. Methods of foundation engineering. Elsevier Publ. Comp., Amsterdam, 616 pp.

Beaven, P.J., 1966. Road making materials in Basutoland. Classification of soils. Report 47, Br. Transp. and Road Res. Lab.: 1–24.

Beaven, P.J., et al., 1971. A study of terrain evaluation in West Malaysia for road location and design. Proc. 4th Asian Reg. Conf. Soil Mech. and Found. Eng., 1: 411–416.

Becker, J.C., 1972. Alaska builds highway over muskeg and permafrost. Civil Eng. - ASCE, 42(7): 75–77.

Beckett, P.H., et al., 1972. Terrain evaluation by means of a data bank. Geogr. J., 138(4): 430–456.

Bell, J.R., et al., 1969. Lessons from an embankment failure analysis utilizing vane shear strength data. Proc. 7th Ann. Eng. Geol. and Soil Eng. Symp., Moscow (Idaho): 199–210.

Berg, D.W., and Duane, D.B., 1968. Effect of particle size and distribution on stability of artificially filled beach, Presque Island Peninsula, Pennsylvania. Proc. 11th Conf. Great Lakes Res., Milwaukee (Wisc.): 161–178.

Berg, R.L., and Aitken, G.W., 1973. Some passive methods of controlling geocryological conditions in roadway construction. Permafrost: the N. Amer. contrib. to the Second Int. Conf., Irkutsk, U.S.S.R., U.S. Nat. Acad. Sci. Publ.: 581–586.

Bergström, E., 1960. Some experiences of mapping surficial deposits in Northern Sweden by means of air photo interpretation. Int.Arch.Photogramm.: 1–25.

Bird, E.C.F., and Ongkosongo, O.S.R., 1980. Environmental changes on the coasts of Indonesia. U.N.U. Publ., Tokyo, 109 pp

Blanck, J.P., 1972. Investigación geomorfológica aplicada a proyectos de aprovechamiento hidroagrícola del valle medio del rio Niger (Républica de Mali). Revista Geografica (Mérida), 10(22–23), 1969: 5–30.

Blaser, H.D., and Scherer, O.J., 1969. Expansion of soils containing sodium sulfate caused by drop in ambient temperatures. In: Proc. Int. Conf. Effects temperatures and heat on engineering behaviour of soils, Washington D.C.: 150–160.

Blench, T., 1969. Mobile-bed fluviology. A regime theory treatment of canals and rivers for engineers and hydrologists. Univ. Alberta Press, Edmonton, 2nd Ed., 168 pp.

Blesch, R.R., and Ta Liang, 1962. Application of Photo Inter-

pretation in Route Location. Proc. 1st Symp. Comm. VII, I.S.P., Delft: 477–486.
Blong, R.J., 1973. Relationship between morphometric attributes of landslides. Ztschr. f. Geom., Suppl 18: 66–77.
Bojoi, I., 1969. Processes géomorphologiques actuels dans la zone du lac de barrage Izvorul Muntelui sur la Bistrita (Carpates Orientales). Proc. Symp. Int. Géom. Appl., Bucharest: 131–138.
Boyd, J.M., et al., 1973. The simple devices for monitoring movements in rock slopes. Quart. J. Eng. Geol, 6(3–4): 295–302.
Brink, A.B.A., 1962. Airphoto interpretation applied to soil engineering mapping in South Africa. Proc. 1st. Symp. Comm. VII, I.S.P., Delft: 498–506.
Brink, A.B.A., et al., 1968. Land classification and data storage for the engineering usage of natural materials. Proc. 4th Conf. Austral. Road Res. Board: 1624–1647.
Brink, A.B.A., et al., 1982. Soil survey for engineering. Oxford Univ. Press,
Broms, B.B., 1973. The geotechnical aspects of moraine. Bull. Geol. Inst. Univ. Uppsala, 5: 51–60.
Brooker, E.W., and Hayley, D.W., 1970. A new look at the permafrost problem. Oilweek, 21: 47–52.
Brown, R.H., 1976. Permafrost engineering (Citations from engineering index). Report for 1970–75, U.S. Nat. Tech. Inf. Service, Springfield, Va.: 1–76.
Brown, R.J.E., 1971. Proc. Seminar on the permafrost active layer, May 1971. Nat. Res. Council, Can. Ass. Comm. Geotechn. Res., Techn. Mem., 103, 69 pp.
Brunsden, D., et al., 1975. Geomorphological mapping techniques in highway engineering. J. Inst. Highway Engineers, 22(1): 35–41.
Bruun, P., 1970. Use of tracers in harbor, coastal and ocean engineering. Eng. Geol., 4(1): 73–88.
Bruun, P., 1978. Stability of tidal inlets. Elsevier Publ. Comp., Amsterdam: 510 pp.
Burton, A.N., 1969. Air photograph interpretation in site investigation for roads. Roads & Road Constr., 47(555): 72–76.
Burton, A.N., 1970. The influence of tectonics on the geotechnical properties of Calabrian rocks and the mapping of slope instability using aerial photographs. Quart. J. Eng. Geol., 2: 237–254.
Camp, L.S., de, 1963. The Ancient Engineers. Garden City, New York, 409 pp.
Capper, P., and Cassie, W., 1969. The mechanics of engineering soils. 5th edition. E.P.F.N. Spon Ltd., London, 309 pp.
Carson, M.A., and Kirkby, M.J., 1973. Hillslope form and process, Cambridge Geogr. Studies, 3, Cambridge Univ. Press,
Castro, G., 1975. Liquefaction and cyclic mobility of saturated sands. J. Geotechn. Eng. Div., ASCE, 101(GT6): 551–569.
Chandler, R.J., et al., 1973. Four long-term failures on embankments founded on areas of landslip. Quat. J. Eng. Geol., 6(3–4): 405–422.
Charlier, R.H., 1968. North Sea beach erosion in Belgium. Proc. 23rd Int. Geol. Congr., Prague: 167–171.
Chen, F.H., 1975. Foundations on expansive soils. Elsevier Publ. Comp., Amsterdam, 280 pp.
Chowdury, R.N., 1978. Slope analysis. Elsevier Publ. Comp., Amsterdam, 422 pp.
Christiansen, E.A., 1970. Sand and gravel resources. In: Physical environment of Saskatoon, Canada. N.R.C. Canada Publ., 11378, 56 pp.
Chuzhmir, A.A., 1968. Theory and practice of multi-directional reinforcement of shores in the Crimea and at Odessa (in russian). Inzh. zashchita beregov Chernogo morya i rats ispol'z priberezin territorii (Engineering defence of the shores of the Black Sea and rational use of coastal territories): 16–19.
Clare, K.E., and Beaven, P.J., 1965. Roadmaking materials in northern Borneo. Transp. and Road Res. Lab., Techn. Paper 68: 1–20.
Coates, D.R. (Editor), 1981. Geomorphology and engineering. G. Allan and Unwin, London, 384 pp.
Corte, A.E., 1969. Geocryology and engineering. Rev. Eng. Geol., 2: 119–185.
Crawford, C.B., 1968. Recent quick-clay studies, 6, Quick clays of eastern Canada. Eng. Geol., 2(4): 239–265.
Croney, D., and Jacobs, J.C., 1967. The frost susceptibility of soils and road materials. U.K. Min. Transport, Road Res. Lab., Report L.R. 90: 1–68.
Crory, F.E., 1975. Bridge foundations in permafrost areas, Moose and Spinach Creeks, Fairbanks, Alaska. U.S. Cold Reg. Eng. Lab., Hanover NH, Techn. Rep., TR–266, 37 pp.
Dearman, W.R., and Fookes, P.K., 1974. Engineering geological mapping for civil engineering practice in the U.K. Quat. J. Eng. Geol., 7(3): 227–256.
Derbyshire, E., and Miller, V., 1981. Highway beneath the Ghulkin. Karakorum glacier menace. Geogr. Mag., 53(10): 626–635.
Dingman, L.S., 1975. Hydrologic effects of frozen ground: literature review and synthesis. U.S. Cold Reg. Res. Eng. Lab., Spec. Rep. SR–218: 1–60.
Dolan, R., 1973. Barrier Islands: natural and controlled. In: D.R. Coates (Editor), Coastal Geomorphology (Publ. in Geom., State Univ. New York, Binghamton): 263–278.
Dolezal, R., 1962. Das Luftbild im Dienste der Landerhaltung und Landgewinnung. Proc. 1st Symp. Comm. VII, I.S.P., Delft: 322–326.
Dolezal, R., 1966. Das Luftbild im Dienste wasserwirtschaftlicher Massnahmen, 1966. Deutsche Gewässerk. Mitt., 10 (6): 174–182.
Dowling, J.W.F., 1963. The use of aerial photography in evaluating engineering soil conditions in a selected area of northern Nigeria. Publ. TRRL, 12 pp.
Dowling, J.F.W., 1968. The classification of terrain for road engineering purposes. Proc. C.E.P.O. Conf., session 5, 6 pp.
Dowling, J.W.F., 1968. Land evaluation for engineering purposes in northern Nigeria. In: G.A. Stewart (Editor), Land evaluation, Melbourne: 146–159.
Dumbleton, M.J., and West, G., 1970. Air-photograph interpretation for road engineers in Britain. U.K., Min., Transport, Road Res. Lab.,RRL Report LR 369: 50 pp.
Dumbleton, M.J., and West, G., 1972. Preliminary sources of site information for roads in Britian. Quart. J. Eng. Geol.: 15–18.
Dumbleton, M.J., 1973. Available information for route planning and site investigation. Report on a TRRL Symposium U.K., Transport and Road Res. Lab., TRRL Report, LR 591, 25 pp.
ECAFE, 1963. Proceedings of the regional symposium on flood control, reclamation, utilization and development of deltaic areas, Bangkok 1963. Water Resources Series, 25, New York, 224 pp.
Ellis, C.I., 1973. Arabian salt-bearing soil (Sabkha) as an engineering material. U.K. Transport and Road Res. Lab., TRRL Report, LR 523, 19 pp.
Erchinger, H.F., 1970. Küstensschutz durch Vorlandgewinnung Deichbau und Deicherhaltung in Ostfriesland. Die Küste, 19: 125–185.
Esch, D.C., 1973. Control of permafrost degradation beneath a roadway by subgrade insulation. In: Permafrost, the North American contribution to the 2nd Int. Conf., 13–28 July 1973, Irkutsk, U.S.S.R.: U.S. Nat Acad. Sci.

Publ.: 608–622.
Fedorovskii, N.N., 1968. Characteristics of shore-protection measures at the health resort of Sochi (in russian). Inzh. zashchita beregov Charnogo morya i rats ispol'z probrezhn territorii. In: Engineering protection of the shores of the Black Sea and rational use of coastal territories: 22–26.
Ferrians, Jnr. O.J., and Hobson, G.D., 1973. Mapping and predicting permafrost in North America: a review, 1963–1973. Permafrost: The N. Amer. contrib. to the Second International Conference, Irkutsk, U.S.S.R., U.S. Nat Acad. Sci: 479–498.
Fookes, P.G., and Sweeney, M., 1976. Stabilization and control of local rock falls and degrading rock slopes. Quart. J. Eng. Geol., 9(1): 37–56.
Franklin, J.A., and Denton, P.E., 1973. The monitoring of rock slopes. Quart. J. Eng. Geol., 6(3–4): 259–286.
Garland, G., 1975. Engineering geomorphology–A review illustrated by case studies in Cochem (W. Germany) and Isarco valley (Italy), M.Sc. thesis, ITC, Enschede, 101 pp.
Gellert, J.G., 1967. Geomorphologische Studien an den durch Kunstbauten veränderten Küsten der D.D.R. C.R. Symp. Int. de Géom. Appl., Bucharest: 61–68.
Geologie en Mijnbouw, 1979. Engineering in Delta areas, 58, (4): 385–480.
Gersper, P.L., and Challinor, J.L., 1975. Vehicle perturbation effects upon a tundra soil-plant system: 1. Effects on morphological and physical environmental properties of the soils. Soil Sci. Soc. Amer. Proc., 39(4): 737–744.
Ghanem, H., 1973. A study on the morphological changes in part of the Zeroud River basin (Central Tunisia), induced by the flood of 1969. M.Sc. thesis, ITC, Enschede, 103 pp.
Gidigasu, M.D., 1975. Laterite soil engineering. Elsevier Publ. Comp., Amsterdam, 554 pp.
Gillott, J.E., 1968 Clay in engineering geology. Elsevier Publ. Comp, Amsterdam, 296 pp.
Glazik, R., 1978. The effect of a reservoir on the Vistula near Wocawek on the changes of aquatic conditions in the valley. (in polish, with english summary). Dok. Geogr. Inst. Geogr. Spat. Org., P.A.N., Warszawa, 6(2, 3): 1–119.
Godfrey, P.J., and Godfrey, M.M., 1973. Comparison of ecological and geomorphic interactions between altered and unaltered barrier island systems in N. Carolina. In: D.R. Coates (Editor), Coastal Geomorphology. Publ. in Geom. State Univ. New York, Binghamton : 239–258.
Gottschalk, L.C., 1964. Reservoir sedimentation. In: Ven te Chow (Editor), Handbook of Applied Hydrology. McGraw-Hill, New York, Section 17, 34 pp.
Grabau, W.E., 1968. An integrated system for exploiting quantitative terrain data for engineering purposes. In: G.A. Stewart (Editor), Land evaluation. Melbourne: 211–220.
Grant, D.R., 1974. Granular resources inventory of New Foundland. Can. Geol. Survey, Open File Report, 194.
Grant, K., and Aitchison, G.D., 1970. The engineering significance of silcretes and ferricretes in Australia. Engineering Geology (Elsevier), 4(2): 93–120.
Grant, K., 1973. Terrain classification for engineering purposes: Queenscliff area, Victoria. Div. Appl. Geomech., CSIRO, Austral., Techn. Paper, 12, 217 pp.
Grant, K., 1974. The P.U.C.E. programme for terrain evaluation for engineering purposes, II, Procedures for terrain classification. Div. Appl. Geomech., CSIRO, Austral., Techn. Paper, 19, 82 pp.
Guilcher, A., 1959. Travaux de Géomorphologie appliquée dans le domaine littoral et estuarien. Revue Géom. dyn., X: 145–147.
Hagyard, T., and Gilmour, I.A., 1969. A proposal to remove sand bars by fluidisation. New Zealand J. of Sci., 12(4): 851–864.
Hall, C.E., 1968. Utilization of a landslide area for a structure location. Proc. 6th Ann. eng. geol. and soils eng. symp., Boise, Idaho: 312–327.
Halmos, E.E., 1969. Permafrost will be a key factor in Artic pipelining. Pipelining Eng., 41(13): 44–46.
Handy, R.L., 1968. The Pleistocene of Iowa–an engineering appraisal. Proc. Iowa Acad. Sci., 75: 210–224.
Hardy, R.M., and Morrison, H.L., 1972. Slope stability and drainage considerations for Artic pipelines. Proc. Can. North Pipeline Res. Conf. Nat. Res. Council Can. Techn. Mem. 104: 249–266.
Haugen, R.K., and Brown, J., 1971. Natural and man-induced disturbances of permafrost terrain. In: D.R. Coates (Editor). Environmental Geomorphology: 139–149.
Hayes, C.J., 1970. Landslides and related phenomena pertaining to highway construction in Oklahoma. Proc. Symp. Environmental aspects of geology and engineering in Oklahoma: Stillwater (Oklahoma).
Herath, L., 1962. Shoreline development and protection of Negombo Beach, Ceylon. An aerial photographic approach. Proc. 1st Symp. Comm. VII, I.S.P., Delft: 453–460.
Hester, N.C., 1970. Sand and gravel resources of Sangmon County, Illinois. Illinois Geol. Survey Circ., 452, 20 pp.
Hite, J.C., and Stepp, J.M., 1971. Coastal Zone Resource Management. Praeger, New York, 169 pp.
Hobson, R.D., 1972. Surface roughness in topography: quantitative approach. In: R.J. Chorley (Editor), Spatial analysis in geomorphology, London: 221–245.
Hoekstra, P., 1969. The physics and chemistry of frozen soils. Proc. Int. Conf. on effects of temperature and heat on engineering behaviour of soils, Washington D.C., NRS–NAS –NAE Publ. 1641: 78–90.
Holst, A.F. van, and Westerveld, G.J.W., 1975. Aantasting van betonpalen in potentiële katteklei. Boor en Spade, 19: 152–162.
Horder, R.I., 1962. Road designing from the air. New Scientist, 7: 589–591.
Horton, A., 1975. The engineering geology of the Pleistocene deposits of the Birmingham district, U.K. Inst. Geol. Sci., Report, 20 pp.
Ijima, T., et al., 1951/56. Measurements of ocean waves. Month. Rept. Trans. Techn. Res. Inst., Tokyo, Vol. 1–7,
Ingold, T.S., 1975. The stability of highways in landslipped area. Highway Engineer, 22: 14–22.
ITC, 1967. Engineering–geological appraisal of the Trans-Sumatra Highway. Appendix II, Explanatory notes to the hydro-morphological map. Unpublished. ITC Rep.,
Ives, J.D., 1974. Permafrost. In: Ives and Barry (Editors). Artic and alpine environments: 159–194.
Jaeger, J.C., 1971. Friction of rocks and stability of rock slopes. 11th Rankine Lecture. Géotechnique, 21(2):
Jennings, J.E., 1966. Building on dolomites in the Transvaal. Civ. Eng. in S. Africa: 41–62.
Jessop, A.M., 1970. How to beat permafrost problems. Oilweek, 20(47): 22–25.
Joliffe, A.P., 1968. Planning and research problems in the exploitation of coastal areas. Proc. 23rd Int. Geol. Congr., Prague: 173–185.
Joliffe, A.P., and McLellan, A.G., 1980. Prospectors for sand and gravel. Geogr. Mag., 52(9): 615–617.
Jones Jr., D.E., and Holtz, W.G., 1973. Expansive soils–the hidden disaster. Civil Eng. ASCE, 43(8): 49–51.
Jones, W.V., and Larsen, R., 1970. Investigation and correction of a highway landslide (special emphasis on horizontal drains). Proc. Symp. Eng. geol. and soils eng. symposium: 123–144.
Kachadoorian, R., and Ferrians Jr., O.J., 1973. Permafrost

related engineering geology problems posed by the Trans-Alaska pipeline. Permafrost: the N. Amer. contrib. to the Second Int. Conf., Irkutsk, U.S.S.R., U.S. Nat Acad. Sci. Publ.: 684–687.

Kerr, P.F., and Drew, I.M., 1972. Clay mobility in ridge route landslides, Castaic, California. Amer. Ass. Petrol. Geol. Bull., 56(11): 2168–2184.

Kézdi, A., 1979. Stabilized earth roads. Elsevier Publ. Comp., Amsterdam, 328 pp.

Kochenderfer, J.N., 1970. Erosion control on logging roads in the Appalachians. U.S. Dept. Agric., Forest Service, Parsons, W. Va., Timber and Watershed Lab., Res. Paper NE–158, 28 pp.

Koh, S.L. (Editor), 1980. Mechanics of landslides and slope stability. Eng. Geol. Special Issue, 16(1–2), 194 pp.

Komar, P.D., and Inman, D.L., 1970. Longshore sand transport on beaches. J. Geophys. Res.,75: 914–927.

Komar, P.D., 1974. The mechanics of sand transport on beaches. J. Geophys. Res., 79: 713–718.

Komarnitskii, N.I., 1968. Zones and planes of weakness in rocks and slope stability. (Transl. from russian). Consultants Bureau, New York, 108 pp.

Komura, S., and Simons, D.B., 1967. Riverbed degradation below dams. Proc. Amer. Soc. Civ. Eng., 9, J. Hydrualics Div., H73, 533 pp.

Koopmans, B.N., 1971. Interpretatión de fotografías aéreas en morfología costera relacionada con proyectos de ingeneria. Publ. CIAF, Bogotá, 23 pp.

Krofellner-Kraus, G. (Editor), 1981. Beiträge zur Wildbacherosion und Lawinen forschung. Publ. Forstl. Bundesversuchsanstalt, Vienna, 182 pp.

Kugler, H., 1963. Zur Erfassung und Klassifikation geomorphologischer Erscheinungen bei der ingenieurgeologischen Spezialkartierung. Ztschr. f. angew. Geologie, 11, Berlin: 591–598.

Kurth, A., 1971. Luftbild und Plannung von Erschliessungsanlagen. Ztschr. f. Forstwesen: 149–153.

Kutter, H.K., 1974. Analytical methods of rock slope analysis. In: L. Müller (Editor), Rock mechanics.

Lachbruch, A.H., 1970. Some estimates of the thermal effects of a heated pipeline in permafrost. USGS Circular, 632, 23 pp.

Lagasse, P.F., and Simons, D.B., 1976. Impact of dredging on river system morphology. In: Rivers 76. Vol. 1. Proc. Symp. Inland waterways f. Navigation, Flood control and Water diversions. ASCE, New York: 434–458.

Lane, E.W., 1955. The importance of fluvial morphology in hydraulic engineering. Proc. Amer. Soc. Civ. Eng., 81, 17 pp.

Largiller, J.F., 1970. Utilisation de la géomorphologie dans l'étude d'un tracé d'autoroute. Proc. 1st Congr. Int. Ass. Eng. Geol., Vol. 2

Lawrance, C.J., 1972. Terrain evaluation in West Malaysia, Part I, Terrain classification and survey methods. U.K. Transport and Road Res. Lab. Report, LR 506, 19 pp.

Leaf, C.F., 1974. A model for predicting erosion and sediment yield from secondary forest road construction. U.S. Dept. Agric., Forest Serv. Res. Nore, RM 274, 4 pp.

Legget, R.F., 1974. Glacial landforms and civil engineering. In: D.R. Coates (Editor), Glacial geomorphology. Fifth Ann. Geom. Symp., State Univ. New York, Binghamton: 351–374.

Leighty, R.D., 1962. Engineering information of arctic areas obtained from aerial imagery. Proc. 1st Symp. Comm. VII, I.S.P., Delft: 383–389.

Leopold, L.B., and Maddock Jr., T., 1954. The flood control controversy. Ronal Press, New York.

Leschack, L., and Kachadoorian, R., 1975. How the Soviets build in permafrost. Civil Eng., ASCE: 57–59.

Linell, K.A., 1960. Frost action and permafrost. In: K.B. Woods (Editor), Highway Engineering Handbook. McGraw-Hill, New York, section 13.

Linell, K.A., and Johnston, G.H., 1973. Engineering design and construction in permafrost regions: a review. Permafrost: the North American contribution to the Second Int. Conf., Irkutsk, U.S.S.R., U.S. Acad. Sci. Publ.: 553–575.

Luck, G., 1970. Die Forschungsstelle für Insel- und Küstenschutz auf Norderney. Die Küste, 19: 1–28.

Macha, G., 1975. Stabilization of soils for erosion control on construction sites. Interim report, 1974. Purdue Univ., Lafayette, Ind., Joint Highway Res. Project, JHRP–75–5, 125 pp.

Mackay, R.J., 1972. The world of underground ice. Ann. Ass. Amer. Geogr., 62(1): 1–22.

Malkin, A.B., and Wood, J.C., 1972. Subsidence problems in route design and construction. Quart. J. Eng. Geol., 5(1-2): 179–194.

Maruyama, Y., 1976. Hydro-geomorphological study of part of the Agri River basin, S. Italy; with emphasis on the application of Landsat imagery. M.Sc. thesis, ITC, Enschede, 76 pp.

Matula, M., 1979. Regional engineering geological evaluation for planning purposes. Proc. Symp. Eng. Geol. Mapping, Newcastle-upon-Tyne. Bull. I.A.E.G., 19: 87–92.

McGown, A., 1971. The classification for engineering purposes of tills from moraines and associated landforms. Quart. J. Eng. Geol., 4(2): 115–130.

Meyer, G.J., and Schoenberger, P.J., 1975. Sediment yields from roadsides: an application of the universal soil loss equation. J. Soil and Water Conserv., 30(6): 289–292.

Miles, R.D., 1962. A concept of landforms, parent materials and soils, in airphoto interpretation studies for engineering purposes. Proc. 1st Symp. Comm. VII, I.S.P., Delft: 462–476.

Mintzer, O.W., 1959. Using airphotos to identify construction sources of gravel. The Ohio State Univ. Res. Paper, 18 pp.

Mitchell, J.K., 1968. A selected bibliography of coastal erosion, protection, and related human activity in North America and the British Isles. Natural Hazard Research,Working Paper 4, 66 pp.

Mollard, D., 1962. Photo interpretation in prospecting for granular construction materials. Proc. 1st Symp. Comm. VII, I.S.P., Delft: 514–523.

Mollard, J.D., 1972. Airphoto terrain classification and mapping for northern feasibility studies. Proc. Can. Northern Pipeline Res. Conf.: 105–128.

Molyn, J.C.M. de, 1961. Inleiding tot de opsporing van zanden grindvoorkomens via geomorphologische luchtfoto-interpretatie. De Ingenieur, 73: B 113–120; B 123–126.

Myslivec, A., and Kysela, Z., 1978. The bearing capacity of building foundations. Elsevier Publ. Comp., Amsterdam, 238 pp.

Newland, J.G., et al., 1973. Sinkhole problem along proposed route of Insterstate Highway 459, near Greenwood, Alabama. Alabama Geol. Surv. Univ. Circular, 83, 63 pp.

N.I.I.R., 1971. The production of soil engineering maps for roads and the storage of materials data. Nat. Inst. Road Res., Pretoria, South Afr., Techn. Recomm. f. Highways, 2.

Nixon, M., 1963. Flood regulation and river training in England and Wales. Inst. Civ. Eng., Symp. Conserv. Water Res., Sessions III, IV: 35–48.

Nixon, J.F., and McRoberts, E.C., 1976. A design approach for pile foundations in permafrost. Can. Geotechn. J., 13(1: 40–57.

Oliver, H., 1972. Irrigation and Water Resources Engineering. Arnold, London, 190 pp.

Olmsted, L., 1969. Geological aspects and engineering properties of glacial till in the Puget Lowland, Washington. Proc. 7th Annual Eng. Geol. and Soils Eng. Symp., Moscow, Idaho: 223–233.
Olson, R.E., 1974. Shearing strength of kaolinite, illite and montmorillonite. J. Geotechn. Eng. Div., ASCE, 100 (GT11): 1215–1229.
Owen, L.H., 1972. Surficial geology and land classification, Mackenzie valley transportation corridor. Proc. Can. Northern Pipeline Res. Conf.: 17–24.
Oya, M., 1977. Applied geomorphological study on the selection of the proposed bridge site along the Jamuna River, Bangladesh. Nat. Geographer, 12(2): 101–113.
Palmer, A.C., 1972. Settlement of pipeline on thawing permafrost. Transportation Eng. J., ASCE, 98(TES): 477–491.
Parizek, R.P., 1971. Impact of highways on the hydrogeologic environment. In: D.R. Coates (Editor): Environmental Geomorphology: 151–190.
Parker, D.C., 1968. Developments in remote sensing applicable to airborne engineering surveys of soils and rocks. Materials Res. & Standards, 8(2): 22–30.
Parker, G.G., and Jenne, E.A., 1967. Structural failure of western highways caused by piping. Proc. Symp. subsurface drainage. Highway Res. Rec., 203: 57–76.
Patton, F.D., 1970. Significant geologic factors in rock slope stability. In: P.W.J. van Rensburg (Editor): Planning Open Pit Mines, S. Afr. Inst. Min. Metall.
Pecsi, M., 1970. A mérnöki geomorfológie problematikâia (Problem of geomorphology for engineers). Földrajzi Ertesito, 19(4): 369–380.
Pettibone, H.C., 1973. Stability of an underground room in frozen gravel. Permafrost: the N. Amer. contrib. to the Second Int. Conf., Irkutsk, U.S.S.R., U.S. Nat Acad. Sci. Publ.: 699–706.
Peyton, H.R., 1971. Arctic pipelines. Proc. 8th World Petrol. Congr. (Moscow), Paper PD 25(5),
Philippe, A., et al., 1972. Etude en simulation des effets du gel sur les structure routières et leurs sols supports. Etude des phénomènes périglaciaires en laboratoire. Col. Int. Géom., Liège - Caen, CNRS Bull., 13–15: 77–99.
Pocock, R.G., 1970. The use of cement-stabilized chalk in road construction. U.K. Transport and Road Res. Lab., Report LR 328, 32 pp.
Popescu, N., 1967. Aménagements routiers sur la vallée de Cerna et les problèmes de la stabilité des versants. Proc. Symp. Int. de Géom. Appl. Bucharest, Mai: 145–150.
Posea, G., 1967. Glissements, méandres et voies de communication dans la vallée du Buzau. Proc. Symp. Int. de Géom. Appl. Bucharest: 139–144.
Post, J.D., and Robinson, C.S., 1967. Geological, geophysical and engineering investigations of the Loveland basin landslide, Clear Creek County, (Col.), 1963–65. USGS Prof. Paper 673–D: 21–26.
Price, D.G., 1977. Engineering geology–past, present, future. Geol. en Mijnbouw, 56(2): 161–167.
Prince, P.A., 1967. Environmental effects of permafrost. Pennsylvania Geogr., 5(4): 4–10.
Prior, D.B., and Ho, L., 1972. Coastal and mountain slope instability on the islands of St. Lucia and Barbados. Eng. Geol., 6(1)
Pryor, W.T., 1964. Evaluation of aerial photography and mapping in highway engineering. Photogr. Eng., 30: 111–123.
Quart. J. Eng. Geol., 1979. Proc. Meeting Coastal Engineering Geology, Southampton 1978, 12(4)
Remeijn, J.M., 1978. Forest road planning from aerial photographs. ITC Journal, 3: 429–444.
Rib, M.T., 1975. Engineering regional inventories, corridor surveys and site investigations. In: Manual of Remote Sensing (Amer. Soc. Ph.), Vol. 2: 1881–1933.
Richard, W., et al., 1972. The performance of a frost-tube for the determination of soil freezing and thawing depths. Soil Science, 113(2): 149–154.
Richardson, N.W., and Sauer, E.K., 1975. Terrain evaluation of the Dempster Highway across the Eagle Plain and along the Richardson Mountains, Yukon Territory. Can. Geotechn. J., 12(3): 296–319.
Ross-Brown, D.M., 1974. A simple device for monitoring large pre-failure movements on a slope. Quart. J. Eng. Geol., 7(3): 315–316.
Royster, D.L., 1975. Tackling major highway landslides in Tennessee mountains. Civil Eng., ASCE, 45(9): 1485–1487.
Ryckborst, H., 1980. Geomorphological changes after river-meander surgery. Geol. en Mijnbouw, 59(2): 121–128.
Sarma, S.K., 1981. Stability analysis of embankments and slopes. J. Geotechn. Eng. Div., 107.
Saunders, M.K., and Fookes, P.G., 1970. A review of the relationship of rock weathering and climate; its significance to foundation engineering. Eng. Geol.: 189–325.
Savage, R.J., 1973. Soil and rock slope instrumentation. Quart. J. Eng. Geol., 6(3–4): 287–294.
Schmitt-Taverna, M.R., 1975. The Westplaat (SW-Netherlands). A study of coastal dynamics from sequential aerial photographs. ITC Journal, 2: 173–185.
Schofield, A.N., 1962. The use of aerial photographs in road construction in Nyasaland. British Road Res. Lab. Publ. 4, 21 pp.
Sclserp, 1973. Gravel for Western London. Final report of the working party for the Western and Maidenhead service areas to SCLSERP. Cement, Lime & Gravel, 48(8): 166–170.
Scott, R.F., and Schoustra, J.J., 1970. Soil Mechanics and Engineering, McGraw-Hill, New York, 314 pp.
Sen Mathur, B., and Gartner, J.F., 1964. Principles of photo interpretation in highway engineering practice. Ontario Dept. Highways, Toronto, 237 pp.
Shalikov, V.A., 1974. An attempt of a landscape approach in the study of seasonal thawing and freezing soils. Vestnik Moskowsk. Univ., Ser. Geogr., 5: 55–60.
Silvester, R., 1979. Coastal engineering. 2 Vols. Elsevier Publ. Comp., Amsterdam, 459 pp; 338 pp.
Simon, A.B., et al., 1973. Use of lateritic soils for road construction in north Dahomey. Eng. Geol., 7(3): 197–218.
Simpson, J.W., and Horrobin, P.J. (Editors), 1970. The weathering and performance of building materials (M.T.P., Aylesbury), 286 pp.
Sissons, J.B., 1970. Geomorphology and foundation conditions around Grangemouth. Quart. J. Eng. Geol., 3(3): 183–191.
Sluys, P. van der, and Ovaa, I., 1973. Dike breaches and soil conditions. Geoderma, 10(1–2): 141–150.
Smith, N., and Berg, R., 1973. Encountering massive ground ice during road construction in Central Alaska. Permafrost: the N. Amer. contrib. to the Second Int. Conf. Irkutsk, U.S.S.R., U.S. Nat. Acad. Sci. Publ.: 730–736.
Soeters, R., and Rengers, N., 1981. An engineering geological map from large-scale aerial photography. ITC Journal, 2: 140–152.
Soloyev, P.A., 1966. Zonality of the strength of the seasonally thawing layer and its mapping in western and southern Yakutiya. Draft translation of Seasonal Thawing and Freezing of the Ground in the Northeast Territory of the U.S.S.R., Cold. Reg. Res. Eng. Lab., Techn. Rep., TL 283: 14-20.
Spagna, V.,1978. Fotointerpretazione in geomorfologia applicata alla progettazione stradale. Atti XVII Conc. Naz. Stradale Venezia, 3–7 giugno
Spagna, V., 1979. The use of aerial photographs in engineering

geomorphological mapping for road planning and maintenance in the Alps Area. ITC Journal, 2: 99-106.
Speer, T.L., et al., 1973. Effects of ground-ice variability and resulting thaw settlements on buried warm-oil pipelines. Permafrost: the N. Amer. contrib. to the Second Int. Conf., Irkutsk, U.S.S.R., U.S. Nat. Acad. Sci. Publ.: 746-752.
Steers, J.A. (Editor), 1971. Applied Coastal Geomorphology. Macmillan, London, 227 pp.
Stephenson, D., 1979. Rock fill in hydraulic engineering. Elsevier Publ. Comp., Amsterdam, 216 pp.
Stewart, M., 1971. The use of seismic refraction in a route feasibility study in St. Lucia. U.K. Transport and Road Res. Lab., Report LR 424, 23 pp.
Stoilov, D., 1970. Geomorphological consideration of the relief with reference to the construction of the dam-lake "Mihajlovgrad", (in bulgarian). Izv. Balgarsk. Geogr. Druz., 10(20): 51-73.
Struble, R.A., and Mintzer, O.W., 1967. Combined investigation techniques for procuring highway design data. Proc. 18th Ann. Highway geol. symp. Purdue Univ. Eng. Bull., 51(4): 27-43.
Swinzow, G.K., 1969. Certain aspects of engineering geology in permafrost. Eng. Geol., 3(3): 177-215.
Symons, I.F., 1976. Assessment and control of stability for road embankments constructed on soft subsoils. U.K. Transport and Road Res. Lab., TRRL Report 711, 33 pp.
Szupryczynski, J., 1974. Veränderungen des geographischen Milieus im Weichseltal als Folge des Erbauung des Wasserstaubeckens in Wkockawek. Kül. Földr. Ert., Budapest, 23(2): 175-180.
Tanguay, M.C., et al., 1969. Multispectral imagery and automatic classification of spectral response for detailed engineering soils mapping. Proc. 6th Int. Symp. R.S. Environm., Ann Arbor (Mich.): 33-63.
Telford, T. (Editor), 1979. Design parameters in geotechnical engineering. Proc. 7th Europ. Conf. Soil Mechanics and Foundation Engineering,
Teng, C.T.P., and Clisby, M.B., 1975. Experimental work for active clays in Mississippi. Transp. Eng. J., ASCE, 101 (TE1): 77-95.
Terzaghi, K., and Peck, R.B., 1967. Soil Mechanics in Engineering Practice, 2nd Ed., Wiley & Sons, New York, 629 pp.
Thomas, A.N., 1969. Permafrost, the major challenge. Oilweek, 20(37): 40-46.
Tricart, J., 1959. Géomorphologie d'aménagements hydrauliques, l'Universitaire, 1: 31-45.
Tricart, J., 1959. L'épiderme de la Terre, esquisse d'une géomorphologie appliquée, Masson et Cie., Paris, 167 pp.
Troeh, F.R., 1975. Measuring soil creep. Proc. Soil Sci. Sco. Amer., 39(4): 707-709.
Trow, W.A., and Morton, J.D., 1969. Laterite soils at Guardarraya, La Republica Dominicana, their development, composition and engineering properties. Proc. Speciality Session Eng. Properties lateritic Soils, 7th Int. Conf. Soil Mech. Found. Eng., Mexico: 75-84.
Tsurumi, E., 1973. Estimation of the probability of slope disasters along national highways. Bull. Geogr. Surv. Inst., Tokyo, 19(1): 81-88.
Tushinskiy, G.K., et al., 1971. Engineering-glaciological regionalization of the Soviet Union. Soviet Hydrology Selected Papers 1: 91-94.
UNESCO, 1966. Scientific problems of the humid tropical zone deltas and their implications. Proc. Dacca Symposium, 1964, Paris, 422 pp.
Valentin, H., 1952. Der Landverlust in Holderness, Ostenglad von 1820 bis 1952. Die Erde, 6: 296-315.
Verger, F., 1959. Colmatage et endiguements sur les rivages de la baie de Bourgneuf. Bull. du C.O.E.C., IX: 179-188.
Verstappen, H.Th., 1977. A geomorphological survey of the SW Cosenza Province, Calabria, Italy. ITC Journal, 4: 578-594.
Verstappen, H.Th., 1977. Remote Sensing in Geomorphology. Elsevier Publ. Comp., Amsterdam, 214 pp.
Vineyard, J.D., and Williams, J.H., 1967. A foundation problem in cavernous-dolomite terrain. Proc. 18th Annual Highway geol. symp. Purdue Univ. Eng. Bull., 51(4): 49-59.
Vodolazkin, V.M., 1966. Use of thawed soils of the Vorkuta region as a foundation for buildings and other structures. Problems of the North, 10: 1313-1341.
Voight, B. (Editor), 1979. Rock slides and avalanches. Elsevier Publ. Comp., Amsterdam. Part 1 Natural phenomena, 834 pp; Part 2 Engineering sites, 850 pp.
Voûte, C., 1960. Climate and landscape in the Zagros Mountains, Iraq. Proc. 21st Inst. Geol. Conf.: 81-87.
Wahlstrom, E.E., 1974. Dams, dam foundations and reservoir sites. Elsevier Publ. Comp., Amsterdam, 278 pp.
Wakelin Norman, J., 1969. Photo-interpretation of boulder clay areas as an aid to engineering geological studies. Quart. J. Eng. Geol., 2(2): 149-157.
Wallace, A.E., and Williams, P.J., Problems of building roads in the north. Can. Geogr. J., 89(1, 2): 40-47.
Walters, R.C., 1971. Dam Geology, 2nd Ed. Butterworths, London, 470 pp.
Wesley, L.D., 1973. Some basic engineering properties of halloysite and allophane clays in Java, Indonesia. Géotechnique, 23(4): 471-494.
West, G., and Dumbleton, M.J., 1975. An assessment of geophysics in site investigation for roads in Britain. U.K. Transport and Road Res. Lab., Report 680, 24 pp.
White, G.F., 1964. Choice of adjustment to floods, Univ. Chicago, Dept. Geogr., Res. Paper 93, 164 pp.
White, G.F. , 1972. Engineering implications of stratigraphy of glacial deposits. Proc. 24th Int. Geol. Congr., Montreal, 1972, Section 13 Eng. Geol., Geol. Surv. Canada, Ottawa: 76-82.
White, H.H., 1970. Die Schutzarbeiten auf den Ostfrisischen Inseln. Die Küste, 19: 68-124.
Wickens, E.H., and Barton, N.R., 1971. The application of photogrammetry to the stability of excavated rock slopes. Photogrammetric Record, 7(37): 46-54.
Williams, P.J., 1979. Pipelines and permafrost. Topics in Applied Geography. Longman, London, New York, 98 pp.
Wilson, S.D., and Hilts, D.E., 1972. Instrumentation applied to slope stability problems. Transportation Eng. J., ASCE, 98 (TE3), Proc. Paper 9101: 562-576.
Wilson, H.E., 1972. The geological map and the civil engineer. 24th Int. Geol. Congr., Montreal, Section 13 Eng. Geol.: 83-86.
Wilun, Z., and Starzewski, K., 1972. Soil mechanics in foundation engineering. Intertext Books, London, 2 Vols, 252 pp; 222 pp.
Winert, H.H., 1975. A climatical index of weathering and its application in road construction. Géotechnique, 24(4): 475-488.
Wong, R.T., et al., 1975. Cyclic loading liquefaction of gravelly soils. J. Geotechn. Eng. Div., ASCE, 101 (GT6): 571-583.
Woods, K.B., 1967. Some highway problems of the United States correlated with physiographic provinces. Proc. 18th Ann. Highway geol. symp., Purdue Univ. Eng. Bull., 51(4), 16 pp.
Wu, T.H., et al., 1975. Stability of embankment on clay. J. Geotechn. Eng. Div., ASCE, 101 (GT9). Proc. Paper 11584: 913-932.
Young, R.N., et al., 1973. Some aspects of surficial salt treatment for attenuation of frost heaving. Permafrost: the N.

Amer. contrib. to the Second Int. Conf., Irkutsk, U.S.S.R., U.S. Nat. Acad. Sci. Publ.: 426–431.
Zabawa, C.F., et al., 1981. Effects of erosion control structures along a portion of the northern Chesapeake Bay shoreline. Envir. Geol., 3(4),
Zaruba, Q., and Mencl, V., 1969. Landslides and their control. Academia, Prague, and Elsevier, Amsterdam, 205 pp.

GEOMORPHOLOGY AND MINERAL EXPLORATION/ RESEARCH

9.1 Main Types of Mineral Deposits

As type and degree of applicability of geomorphological approaches in the search for exploitable mineral resources vary with the mineral deposit and its mode of formation, the brief review of the main types of mineral deposits given in this sub-section is an essential introduction to the discussion of the uses of geomorphology in the field of mining included in the following sub-sections (Bashenina et al., 1974; Chervanev and Prokhodskii, 1968; Institution, 1973/1979; Kostryukov, 1969; Seyhan, 1972; Smirnov, 1977).

A major distinction between primary and secondary mineral deposits is to be noted. Primary mineral deposits have maintained their chemical composition and spatial relationship with the surrounding rock material since their formation. They may have been formed simultaneously with their igneous or sedimentary surroundings (syngenetic mineral deposits) or at some later stage by filling voids or replacing other minerals (epigenetic mineral deposits). In case of secondary mineral deposits, diversified erosional processes have led to a modification of the chemical composition, physical characteristics, and the spatial position of the minerals and have caused a sufficiently high degree of chemical or mechanical concentration, making them economically exploitable. (Sidorenko, 1971).

Among syngenetic primary mineral deposits, magmatic ores rank high. Magmatic separations occurring in the early phases of the process of cooling and gradual solidification in the lower portion of sills composed of mafic magma, include: magnetite, ilmenite, chromite and platinum. Rare minerals and gems are often associated with later phases of magmatic separation in pegmatite dykes situated in the outer parts of a batholith. Syngenetic primary mineral deposits formed in a sedimentary environment include, for example, phosphate deposits and beds of hematitic iron ore.

Epigenetic primary mineral deposits are essentially hydrothermal in nature. Not considering volcanic sublimate and hotspring deposits (sulphur) and tabular deposits such as magnetite, tungsten, etc., formed by contact metamorphism around an intrusive body, epigenetic primary mineral deposits are mostly linked to veins. In addition to the chemical composition of the magma and the pressure prevailing, the minerals formed largely depend on the temperature during their formation which relates to their distance to the intrusive body. E.g., casserite (SnO_2), magnetite (Fe_3O_4) and molybdenite (MoS_2) deposits require high temperatures for their formation; smaltite ($CoAs_2$) and niccolite (NiAs), medium temperature and cinnabar (HgS) and silver (Ag) low temperatures. Not all types of hydrothermal mineral deposits are conditioned by temperatures and galena (PbS) can develop under medium to low temperatures.

Secondary mineral deposits are diversified and can be grouped under the following main headings:
Sedimentary deposits resulting from diversified sedimentation processes. Placer deposits, mainly fluvial and marine placers but also aeolian, glacial, and other types of placers, are the most important in this group and include magnetite, casserite, gold, platinum and diamond deposits. (Gardner, 1955; Guilcher, 1959; Helgren, 1979; Ludwig and Vollbrecht, 1957; Sigov, 1971). Evaporite deposits are another type of sedimentary mineral occurrences. They are either associated with marine evaporation or playa environments. (Johnson, 1968; Reimnitz, 1970; Watson and Ansino, 1969). A variety of evaporite deposits exists including sodium, potassium and magnesium salts, anhydrate, gypsum and borate.

Weathering deposits resulting from chemical alteration and disintegration of certain types of rock under the influence of moisture and oxygen having percolated for long periods of time in the weathered regolith of old land surfaces. These processes may lead to laterization or bauxitization and result in the formation of iron ore, bauxite, nickel, manganese and other ores. (Faniran, 1971; Fratschner, 1960; Dulemba, 1969; Molengraaf, 1959; Verstappen, 1959; Thomas, 1966). Another type of occurrence is the weathered outcrop of ore bodies, the so-called gossan, where oxidized and hydrated products may be found such as limonitic or hematitic iron and copper, lead and zinc compounds.

Groundwater deposits which result from concentration of minerals by percolating groundwater but without the contribution of weathering mentioned above. Leaching and re-deposition may lead to further concentration of mineral deposits such as gossan. Deposition may also occur in gash veins, sinkholes and bedding planes. The concentration of well soluble minerals such as uranium compounds often is governed by repeated groundwater action.

9.2 Terrain Configuration and Mineral Deposits

In a limited number of cases mineral deposits are directly reflected in the terrain configuration and can be located on the strength of their relief characteristics alone. This is the case, for example, where gossan form a resistant capping of the underlying rock material and consequently rise above their surroundings. It is also common where veins are concerned with which hydrothermal mineral deposits are frequently associated. Dykes and sills, sources of various magmatic mineral deposits, also have distinct relief characteristics in many cases. Both veins and dykes may differ from the adjacent rock in colour and/or vegetation; this fact facilitates their identification, particularly when aerospace survey techniques are applied (Thornbury, 1969; Trushin, 1974).

Evaporites are confined to certain characteristic geomorphological situations and the same applies to many placer deposits. The economic concentration of weathering (e.g., iron) deposits requires long periods of weathering and their occurrence is therefore associated with old planation surfaces mostly of Tertiary age and with some other landforms which have been subjected to prolonged periods of chemical weathering. Some iron ores of different origin also have a distinct relief expression. Examples are the high grade iron ores of Liberia, Venezuela, and the Lake Superior region, U.S.A.

Sometimes it is not so much the overall form of the deposit which contributes to its recognition but rather certain details of slope profiles which are characteristic for a particular deposit. In this context attention should be drawn to the importance of mineralisation slumping and the (partial) collapse of the capping of a mineral deposit as a result of the shrinkage of the ore body due to oxidation and weathering.

It is much more common, however, for a mineral deposit to not be directly reflected in the terrain configuration. A thorough study on the geology and geomorphology of the region concerned geared toward the purpose of mining is required in order to locate areas of possible economical importance or, inversely, to point out areas where further search for minerals is not recommendable in view of their low economic potentialities. Further prospecting can then be concentrated on the most promising areas such as the contact metamorphic zone around an intrusive body with associated dikes and veins. Well-directed and more efficient search in areas of more limited extent is thus possible. The stereopair of fig. 9.1. illustrates this and shows a concentration of mining activities in the Bolivian Andes, associated with a granitic intrusion.

It should be stressed here that the relation between mineralization and relief is considerably more complex than may appear from the lines above. Matters of palaeo-relief are often involved and it is especially in those cases that morphogenetical research has proved its value. In fact mineral deposits may be associated with:

a. Actual relief forms, resulting from present geomorphological processes/developments.
b. Relic relief forms, for example Tertiary planation surfaces, which have been formed in the past but still form part of the earth's surface.
c. Buried relief forms, covered by more recent deposits, the distributional pattern of which they largely determine
d. Fossil relief forms, so deeply buried that they in no way affect the outcrop pattern of younger formations.
e. Exhumed relief forms, from which the covering layers have been removed by erosion, bringing them to the surface again.

The relations of the mineral deposits with the above-mentioned types of relief forms vary with their inherent or genetic characteristics (Gellert, 1974; Piotrovski et al., 1972). Basically, the relations fall under the following four headings:

1. The mineral deposit is directly related to the relief development.
2. The mineral deposit results from factors among which relief plays a part.
3. The mineral deposit is not genetically related to the relief but is nevertheless reflected in the terrain configuration due to geomorphological processes and developments which have no bearing on the formation of the mineral deposit.
4. The mineral deposit is not genetically related to the relief and is not reflected in the terrain configuration.

Further subdivision of these four headings is possible on the basis of the processes involved. Mixed types also occur, mainly because the enrichment of a mineral deposit and its development into an economically exploitable concentration is a gradual and often complex process. Minerals formed by magmatic processes may pass into eluvial ore deposits and these may in time become the source of placer deposits of various kinds. The geomorphological approaches to mineral exploration vary accordingly. Orlov and Piotrovski (Piotrovski et al., 1972) have listed the objects and methods of geomorphological surveys used in prospecting for various mineral deposits in tabular form (Table 9.1). It is evident that most mineral deposits are in one way or another associated with matters of relief and morphogenesis (Patyk-kara nad Iloginova, 1968; Piotrovski and Lukashov, 1976). Many mineral deposits are even directly related to the actual or palaeo relief forms and are affected by specific endogenous and exogenous processes. Geomorphological studies often emphasizing palaeogeomorphological aspects (McKee, 1963) are commonly carried out in the framework of mineral exploration (Anan'yev, 1969, 1972; Helwig, 1967; Krook, 1969; Paraschiv, 1976; Trescases, 1975).

Among the mineral deposits directly related to the relief (group 1), those associated with actual relief forms (a), are comparatively rare. Most lignites, some placers and evaporites belong to this class. Sulphur deposits (fig. 9.2) are also often linked to present (volcanic) landforms. More commonly they are associated with relic relief forms (b) or buried relief forms (c). Most placer deposits, weathering ores, young coal deposits, and ceramic clays

Fig. 9.1. Stereopair at a scale of 1:40,000 showing mining activities linked to a granitic intrusion in strongly folded Tertiary sediments, Llallagna area, Bolivia.

are in this group, whereas high grade (non lateritic) iron, manganese, and coal deposits (Piotrovski, 1976), are associated with fossil relief forms (d). Mineral deposits associated with buried or fossil relief forms may also be associated with exhumed relief forms (d) where the geomorphological development has caused them.

A good example of an ore body associated with a palaeo-relief is found in the copper belt of Zambia (Garlick, 1972). Voet and Freeman (1972) showed that in the Chingola open pit mine, a close spatial relationship exists between palaeo ridges on one hand and metasediments and copper concentrations on the other. Contours of the buried top of the Basement Complex reveal that the basement was deeply eroded and the ridges were formed at the onset of the deposition of the Lower Roan beds. Detritus from nearby ridges subsequently filled the valleys formed with Katanga basal sediments (basal conglomerate and arkose), where the copper is mined. Both the metasediments and the copper are believed to originate from the palaeo-ridges. The copper derives from nearby lodes in the gneisses and schists of the basement which are still partly traceable under the cupriferous sediments. Reconcentration in the sediments was brought about by detrital and/or chemical processes. The sections of fig. 9.3 show the present situation as well as two earlier inferred sedimentation phases.

Among the mineral deposits in the formation of which the relief is only a part of the causitive factors, oil and natural gas rank high (Leontev and Lukyanova, 1976; Moore, 1969; Shumilov, 1969). These result primarily from the accumulation of organic material in favorable palaeo-relief conditions. When in some later stage migration occurs, they may be trapped in certain structures which in turn may be part of a fossil relief such as coral reefs, delta or tidal flat deposits. Paraschiv (1976) mentions the manner in which fossil planation surfaces, fossil valleys, karst phenomena and other palaeo-relief forms have contributed to the hydrocarbon deposits of Rumania (fig. 9.4).

OBJECTS AND METHODS OF PROSPECTING AND MAPPING / TYPE OF MINERAL DEPOSITS	Deposits in genetic correlation to the Quaternary relief and Quaternary sediments							Deposits in genetic correlation to the pre-Quaternary relief, partly preserved and inherited, and corresponding to related deposits												
II described method of prospecting is indispensable; I described method of prospecting is important; symbol is not specified - described method of prospecting can be used according to the type of territory and the latter's characteristic features.	Continental placers, mostly fluvial	Littoral marine and lacustrine placers	Peat bogs	Clays and sandy loams	Gravel-sand deposits	Salt-lakes - minerals muds	Groundwaters of Quaternary deposits	Continental placers, mostly fluvial	Littoral marine and lacustrine placers	Brown coal deposits, limnic	Brown coal deposits, paralic	Bauxites and refractory clays of Mesozoic and Tertiary age	Nonmetallic industrial minerals of poorly diagenetically consolidated deposits (glass and moulding sands, cretaceous marly limestones, etc.)	Deposits of weathered crusts (nickel-cobalt, iron, magnesite), residual	Weathering deposits (sulphur, uranium, gypsum), infiltrated	Deposits of translocated weathered crusts (cobalt, iron, etc.)	Salt deposits (potassium, sodium salts, gypsum, etc.)	Sedimentary, ferro-manganese deposits	Phosphorites	Waters of water-bearing layers in buried valleys
	1.	2.	3.	4.	5.	6.	7.	8.	9.	10.	11.	12.	13.	14.	15.	16.	17.	18.	19.	20.
1. Regional morphostructural ",base" (on the basis of the analysis of dome - block morphotectonics).	II	I				I		II	I	II	II	II	I	II	II	II	I	I	I	II
2. Relief of folded crystalline fundament, buried under sedimentary mantle, old structures of sedimentary mantle, dependent on movement of blocks of the fundament.	I	I						I		II	II	II		I	I	I				II
3. Morphostructural plan of region of detailed mapping.	II	I	I	I	I	I	I	II	I	II	II	II	I	II	II	II	I	I	I	II
4. Expression of deep intrusions and their structures in the relief (mantle relief, petrographic differentiation and later dislocations).	I							I				II		II		I				
5. Expressions of hypoabyssal and subvolcanic intrusions in the relief: volcanogenic morphostructures.	II	I						II	I					I		I				
6. Circular and radially circular morphostructures: magmatogenic and conditioned movements of buried blocks of the fundament.	II	I						II	I					I		I				I
7. Modification of structures of stratified sedimentary and metamorphosed rocks.	I	I						I	I	I	I	I	I	I	I	I	I	I	I	I
8. Modification of ore bodies and zone of contact and adjoining changes.	II	I						II	I					I						
9. Morphostructures dependent on free tectonics and tectonies associated with mud volcanoes.						I	I													I
10. Correlation between groundplan of valley system and valley pattern and tectonics. - a) folded	I	I						I	I	I	I	I		I	I	I				
- b) dome	I	I						I	I	I	I	I		I	I	I	I	I	I	I
- c) faulted - bock-type	II	I						II	I	I	I	II		II	II	II	I	I	I	I
11. Old valley system. Reconstruction of system conditioned by - a) tectonics and volcanism	II	I	I	I	I		I	II	I	I	I					I				II
- b) exogenous factors: glaciation, lacustrine - marine fluvial and colluvial sedimentation and subsequent epigenesis (sometimes influenced by tectonics).	II	I	I	I	I		I	II	I											II
12. Longitudinal valley profiles	II							II								I				
13. Valley sides, slopes of ridges etc. and colluvial deposits	II			II	I		I	II				I		I			I	I	I	I
14. River terraces and	II		II	II	II	II		II						II						
15. Flood-plains their	II		II	II	II	II		II						I						
16. River-beds and their shapes sediments	II				II	I		II												
17. Surfaces of planation	II	I	I	II	I	I	I	II	II	I	I	II		II	II	II	I	I		II
18. Eluvium and weathered crusts	II	I		II	I	I	I	II	II	I	I	II		II	II	II	I	I	I	I
19. Present-day karst	II	I	I	I		I		II				II		II	II	I				
20. Fossil karst	II		I	I		I		II				II		II	II	II	II	II		II
21. Lake basins and lacustrine forms, lacustrine-river and marshy accumulations. - a) in humid regions			I	I	I	I	I			I	I			I		I	II			II
- b) in arid and semi-arid regions			I	I	I	I	I			I	I						II			
22. Marine and lacustrine littoral forms (primarily of wave and accumulation types).		II	I	I	II	I	II		II	I	I		I			I				
23. Glacial (exaration and accumulation) and fluvioglacial forms of mountain glaciation.	I			I	II	II		I						II						II
24. Glacial (exaration and accumulation) and fluvioglacial forms of continental glaciation.	I		II	II	II			I						II						II
25. Cryogenic forms - a) present-day	I		II	II	I									II						II
- b) relict	I		II	II	I			I						II						II
26. Relief buried by unconsolidated and poorly consolidated and young volcanogenic rocks.	II				II			II	II					II	II	II				II
27. Complete diagram of denudation chronology (stages of dissection, levelling, accumulation) tectonically and climatically conditioned.	II	II	I	I	I	I	I	II	II	II	II	II	I	II	II	II	II	II	I	II
28. Complete diagram and location of regions of denudation, lines of transport and regions of accumulation of clastic material. Characteristics of processes and laws governing the changes in the composition of the material according to relief elements.	II	II	I	I	I	I	I	II	II	I	I	I	I	II	II	II	I	I	I	II
29. Complete diagram and location of regions of denudation, transport and regions of accumulation of sediments in solutions and colloids according to relief elements and geomorphologically conditioned ways of groundwater movement.						I	I					II		II	II	II	II	II	I	I

Table 9.1 Prospecting and mapping for mineral resources. I.V. Orlov (1962), revised by M.V. Piotrovski.

TYPE OF MINERAL DEPOSITS / OBJECTS AND METHODS OF PROSPECTING AND MAPPING	Deposits in genetic correlation to the fossil relief and its forms and geological structures										Deposits connected with relief development indirectly, in co-action with tectogenesis									
II described method of prospecting is indispensable **I** described method of prospecting is important symbol is not specified - described method of prospecting can be used according to the type of territory and the latter's characteristic features.	Continental fossil placers	Littoral marine fossil placers	Black coal deposits, limnic	Black coal deposits, paralic	Bauxites of Paleozoic age	Sedimentary ferro-manganese deposits	Transitory and polymetallic sedimentary deposits	Salt deposits (potassium, sodium salts, etc.)	Waters of water-bearing seams and joints	Oil and natural gas deposits	Magmatic ore deposits of metals, diamonds, etc.	Albitic-greisen deposits of rare metals and earths	Pegmatitic deposits of muscovite, quartzy-bituminous raw materials, optical raw materials, etc.	Carbonate deposits of rare minerals	Skarn deposits of iron, cobalt, copper, gold etc.	Hydrothermal plutogenic deposits of non-ferrous, rare, and noble metals, optical raw materials, etc.	Hydrothermal volcanogenic deposits	Hydrothermal deposits of light-colour metals	Metamorphic deposits of iron, manganese, etc.	Fissure waters
	21.	22.	23.	24.	25.	26.	27.	28.	29.	30.	31.	32.	33.	34.	35.	36.	37.	38.	39.	40.
1. Regional morphostructural ",base" (on the basis of the analysis of dome - block morphotectonics).	II	II	I	I	I	I	I	I	I	II	II	I	I	I	I	II	II	II	I	II
2. Relief of folded crystalline fundament, buried under sedimentary mantle, old structures of sedimentary mantle, dependent on movement of blocks of the fundament.	I	I	II	II					I	II	I	I				I	I	I	I	II
3. Morphostructural plan of region of detailed mapping.	II	II	II	II	I	I	I	I	I	II	II	I	I	I		II	II	II	I	II
4. Expression of deep intrusions and their structures in the relief (mantle relief, petrographic differentiation and later dislocations).	I	I			I	I					II	II	II	I	I	I	I	I	II	I
5. Expressions of hypoabyssal and subvolcanic intrusions in the relief: volcanogenic morphostructures.	II	II					I		I	II	II	II	I	II	II	II	II	II	II	II
6. Circular and radially circular morphostructures: magmatogenic and conditioned movements of buried blocks of the fundament.	II	II	I	I	I				I	II	II	I	I	I	II	II	II	I	I	I
7. Modification of structures of stratified sedimentary and metamorphosed rocks.	I	I	I	I	I	I	I	I	II	II	I	I	I	I	I	I	I	I	I	I
8. Modification of ore bodies and zone of contact and adjoining changes.	II	II						II	II	II	II	II	II	II	II	II	II	II	I	I
9. Morphostructures dependent on free tectonics and tectonies associated with mud volcanoes.			II	II	I	I	I	I	II	II	I	I	I	I	I	I	I	I	I	I
10. Correlation between groundplan of valley system and valley pattern and tectonics. - a) folded	I	I	II	II	I	I	I	I	I	II	I	I	I	I	I	I	I	I	I	I
- b) dome	I	I	I	I	I	I	I	I	I	II	I	I	I	I	I	I	I	I	I	I
- c) faulted - bock-type	II	I	II	II	I	I	I	I	II	II	II	II	II	II	II	II	II	II	II	II
11 Old valley system. Reconstruction of system conditioned by - a) tectonics and volcanism	II	II	I	I					II	II						II				II
- b) exogenous factors: glaciation, lacustrine - marine fluvial and colluvial sedimentation and subsequent epigenesis (sometimes influenced by tectonics).	II	II	I	I					II	II						II				II
12 Longitudinal valley profiles										II										
13 Valley sides, slopes of ridges etc. and colluvial deposits	I	I	I	I	I	I	I	I	I	I	I	I	I	I	I	I	I	I	I	I
14 River terraces and their sediments	II	II								II										
15 Flood-plains and their sediments	II	II								I										
16. River-beds and their shapes and their sediments	II	II																		
17. Surfaces of planation	I	I	I	I																
18. Eluvium and weathered crusts	I	I			II	I	I	I	I	II	I		I	I	I	I	I	I	I	I
19. Present-day karst	I	I			I			II	II	I	I									II
20 Fossil karst					II			II	II	II										II
21 Lake basins and lacustrine forms, lacustrine-river and marshy accumulations. - a) in humid regions	I	I				I		II	II	I										II
- b) in arid and semi-arid regions.					I			II	II	I										II
22 Marine and lacustrine littoral forms (primarily of wave and accumulation types).								I												
23 Glacial (exaration and accumulation) and fluvioglacial forms of mountain glaciation.										I										
24. Glacial (exaration and accumulation) and fluvioglacial forms of continental glaciation.																				I
25. Cryogenic forms - a) present-day									I	II										II
- b) relict.									I	II										II
26. Relief buried by unconsolidated and poorly consolidated and young volcanogenic rocks.																	II			
27. Complete diagram of denudation chronology (stages of dissection, levelling, accumulation) tectonically and climatically conditioned.	II	II	II	II	II	II	II	II	II	II	I	I	I	I	I	I	I	I	I	I
28. Complete diagram and location of regions of denudation, lines of transport and regions of accumulation of clastic material. Characteristics of processes and laws governing the changes in the composition of the material according to relief elements.	II	II	I	I	I	I	I	I	I	I	I	I	I	I	I	I	I	I	I	I
29. Complete diagram and location of regions of denudation, transport and regions of accumulation of sediments in solutions and colloids according to relief elements and geomorphologically conditioned ways of groundwater movement.					II	II	II	II	II	II	I	I	I	I	I	I	I	I	I	I

Fig. 9.2. Oblique aerial photograph of the exploitation of sulphur deposits in a crater of the Patuha Volcano, Java, Indonesia.

Helwig (1967), Martin (1966, 1967), Aristarkhova and Polkanova (1975), and Aristarkhova et al. (1976) have also pointed out the importance of palaeogeomorphological studies in oil and gas exploration.

Zvonkova (1972) advocated the importance of structural geomorphology for the search of oil in areas where inherited surface forms prevail. Where active crustal movements occur, a geomorphological study of the neotectonics may prove useful since the recent trends in many cases correspond to major tectonic lines of the past and thus to possible oil traps (Aristarkhova, 1968; Lastochkin, 1970).

Other types of mineral deposits having only an indirect relation with the relief (group 2) are mineralised dykes associated with deep-seated fault scarps which have left a trace in the terrain configuration.

Among mineral deposits not genetically related to relief but standing out due to other factors (group 3), magmatic mineral deposits and in particular veins and gossan should be mentioned. Many other magmatic mineral deposits have no reflection in relief and thus belong to group 4.

9.3. Geomorphological Processes and the Search for Minerals

It is evident from the preceding sub-section that the link between geomorphology and mineral deposits is not restricted to matters of terrain configuration alone. Geomorphological processes of various types play a special role where groundwater, placer and weathering deposits are concerned. In the case of the latter and also for many other types of mineral deposits the regional geomorphological development with emphasis on morphogenesis and morphochronology is of great practical importance. Since mineralization and the further reworking of the minerals into economically important concentrations and quantities is effectuated under the influence of endogenic and/or exogenic processes, both should be equally stressed. Geological factors are of importance in indicating the source areas of useful minerals, but also endogenic processes such as faulting and other kinds of crustal movements affect the distributional pattern of mineralization.

Geomorphology in many instances also assists in

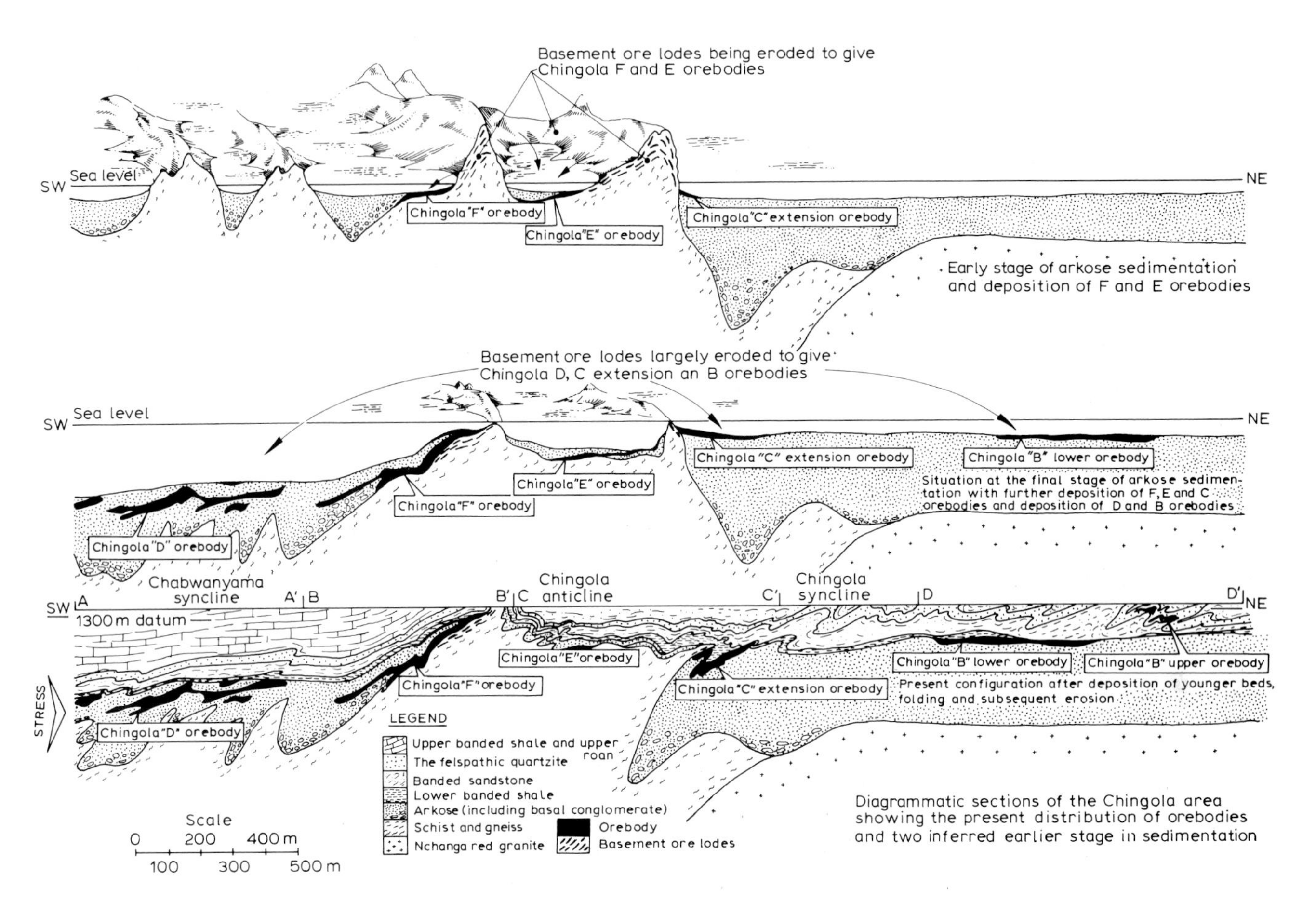

Fig. 9.3. Diagrammatic sections of the Chingola area, Zambia, showing the present distribution of ore bodies and two inferred earlier stages in sedimentation (Voet and Freeman, 1972, Geol. Mijnbouw, 51: 307).

the study of these endogenic processes because of their reflection in the regional geomorphology. Old structural features may be unveiled this way, particularly where recent crustal movements correspond with earlier tectonic trends (Berlyant, 1969). The position of spits, bars, beach ridges and other coastal features is often greatly influenced by structural elements. Position and type of living or upheaved coral reefs also reveal pre-existing structural elements. Inland, the distributional pattern of lakes, swamps and other drainage characteristics may yield similar clues. In areas of denudational relief certain structural trends can often be recognized on the strength of anomalies in the drainage pattern or the occurrence of fault scarps.

It should be understood that not only the endogenic processes but also various exogenic, geomorphological processes may have changed in type and intensity in the course of time. This is of particular importance in the search for minerals such as diamonds which are very resistant to erosion and thus long-lived (Bobrievitch and Bondarenko, 1959). The search for soluble minerals such as uranium compounds can be largely concentrated on the transportation in solution under the groundwater conditions of the present and the near past. Clark et al. (1967) detected three periods of supergene enrichment of copper deposits in the Athacama Desert, Chile, and associated each of these with a period of pediment dissection and not to 'mature Tertiary land' as previously believed.

Geochemical investigations substantially benefit from a knowledge of the geomorphological context in which the groundwater circulates. Where easily disintegrating minerals such as cassiterite are concerned, the present weathering conditions and slope processes are of prime importance and mineral deposits occur near the source area in the form of eluvial or colluvial placers. However, in the case of hard minerals such as diamonds, one process may follow an other and the ultimate deposit is the result of a complex series of processes covering long geological periods. For example, the diamond placers found in the recent alluvium and in the Quaternary and Tertiary deposits of certain rivers in the Urals are derived from littoral placers of Palaeozoic age and their source has not yet been ascertained.

As an example of geochemical investigations the study of Meyer and Evans (1973) on the use of mercury as an aid in zinc/lead/cadmium prospecting in a glacial drift area in Ireland can be mentioned. Complex combinations of periglacial weathering, mechanical disintegration under glacial conditions and soil forming processes are involved in this case (see also Maurice and Meyer, 1975).

A succession of alternating processes can be expected even if the transporting and concentrating factors have only been in operation since a comparatively short period, for example the Quaternary, due to pronounced climatic changes in many parts of the world in that period. Changes in sea level, the formation of temporary lakes and the intermittent aeolian activities connected with these climatic changes also have to be taken into consideration. Consequently, several diamond placers such as those in southern Africa went through repeated fluctuations in type and force of littoral processes, sheetwash, river work, and aeolian activity. In many cases fluvial placers, for example gold placers (fig. 9.5), are affected by the changes in the river regimen brought about by Quaternary climatic fluctuations. Three matters are of basic importance in the search for fluvial placer deposits: the location of the source area, the mapping of river alluvium and terraces where placers can be expected and the identification of 'traps' including the study of the trapping mechanism which in the end leads to the deposition of exploitable gold placers. When traces of gold are found in certain places, efforts can be gradually limited in areal extent (Postolenko et al., 1969). In the case of littoral placers, source area, transporting/concentrating processes and possible mineral 'traps' are foci of interest. Potential source areas are cliffs abraded by the surf as well as rivers entering the sea. If the minerals of littoral placers are brought by rivers these in turn may have collected them from denudational areas or volcanic terrain. The processes have to be carefully studied in both space and time.

A well-known example of the linkage of geomorphology and placer deposits are the tin placers in Malaysia and the Indonesian tin islands Banka, Belitung and Sinkep. These placers are associated with existing and former river valleys and can be traced to the off-shore extensions of these valleys which were used during the cold periods of the Pleistocene when the Sunda shelf ran dry. Echo sounding, sonic surveys and other sophisticated techniques have been employed to detect the submarine valley system which is filled with younger sediments (Adam, 1960).

Fig. 9.6. shows part of NW Belitung with the 'Granite I' (hachures), the sedimentaries (white), and the eluvial (so-called 'kaksa') cassiterite placers. The valley deposits have a clear affinity for the contact zone between the intrusive 'Granite I' and the sedimentary rocks (Krol, 1960; Overeem, 1960). The primary source of the tin is in numerous veins where it is also exploited. The extensive base-levelling occurring throughout the island left isolated hills rising above a planation surface which is dissected by the valleys where the eroded cassiterite from the veins is deposited in the form of 'koelit' and 'kaksa' placers.

Michel (1960) pointed out the effect of various geomorphological factors on the formation and distributional pattern of bauxites in Senegal and Gambia, W. Africa. His study is of general importance as similar situations have led to bauxite formation in several other parts of the world. It appears that the bauxites of economic importance are restricted to two high planation surfaces, probably dating from the lower Cretaceous and the Eocene respectively. They are of considerable extent and covered by a thick lateritic cover. The favourable, humid tropical climatic conditions prevailing during long periods of time on these planation surfaces account for this cover. Within these surfaces, bauxite occurrences are only found where schists or dolerites are the parent rock. Two other factors affect the situation of bauxite occurrences. First, the position with respect to the drainage net: only larger relics of the old planation surfaces which are far away from the main rivers have escaped from the erosion that followed after their uplift. Good drainage resulting in the removal of iron compounds is a second essential factor in bauxite formation. Optimal conditions exist where tributary rivers have dissected the planation surfaces into narrow strips.

In the western part of Suriname, South America, a similar situation is found in the Nassau Mountains (564 m), depicted in the stereopair of fig. 9.7. A flat-topped relic of an old planation surface is covered by a thick weathering cap and high grade bauxite. It rises above a lower lying planation surface from which it is separated by a rather steep slope actively dissected and attacked by headward erosion by ravines. The hydrological conditions are not ideal for supporting a tropical forest and irregular patches of low vegetation are clearly visible. Nevertheless, the laterite is fairly pervious and the infiltrated rain emerges in springs along the edges of the planation surface, carrying off iron compounds and activating the incision of the ravines. The bauxite exploitation of the area is in progress (Aleva, 1979).

It is relevant in this context to mention the lowland bauxites of Suriname which are of an entirely different origin and did not originate as weathering deposits of the basal complex. They were formed by way of desilication of clayey sediments, rich in kaolinite and deposited near an old coast (Bakker et al., 1953). This makes a great difference in the distributional pattern of these bauxite

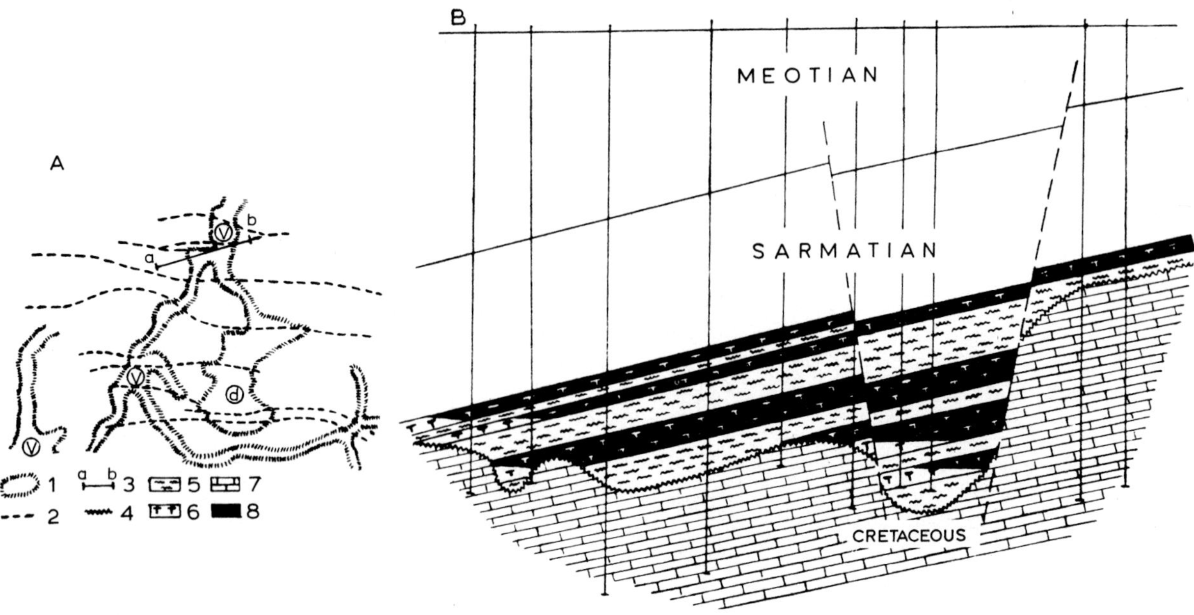

Fig. 9.4. Palaeo-relief west of Bucarest, Rumania (Paraschiv, 1976).

A. Palaeo Arges System: 1. fossil palaeo relief (vz. fossil palaeo valleys, d = fossil palaeo dolines),
2. fault
3. direction of cross section (B).

B. Geological cross section of Palaeo Arges Valley indicating oil distribution depending on fossil palaeo-relief:
4. stratigraphic lacuna (palaeo relief),
5. marl
6. sandstone
7. limestone
8. oil

occurrences. The map of fig. 9.8, after Montagne (1964) demonstrates the situation. Subsidence of the coastal zone, maximum along a SSW-NNE stretching axis in the west, and subsequent sedimentation periods are leading factors in the geomorphological development. Bauxitization is thought to have occurred in the Middle Eocene to Late Oligocene.

Krook (1969) maintains that simultaneously kaolinization occurred in the underlying sediments and mentions the relatively high feldspar content in these beds, found in one of the bore holes,as a possible key to the solution of the problem concerning the bauxites.

Yanez (1979) reported on the geomorphological context of various bauxites and aluminum laterites in Venezuela. The weathering deposits, approximately 20 m thick, are derived from diabase which outcrops on the backslope of a cuesta and overlies sandstones of the Roraima formation. A ferruginous crust, 0 - 5 m thick, forms a covering layer and is in its extent controlled by joints. He stressed that geomorphological studies, emphasising internal and external drainage conditions together with sampling and laboratory testing of pH and Eh, etc. conditions, optimize the search and economic development of such ore deposits. The stereotriplet of fig. 9.9 and the section of fig. 9.10 explains the situation.

The thick laterites found on top of the ultrabasic rocks occurring along the north coast of the island of New Guinea and on New Caledonia are of economical interest due to their nickel and cobalt content. The subdued forms of these hills and plane surfaces together with their scanty vegetation due to poor hydrological conditions render these weathering deposits clearly visible, also from the air. The vertical airphoto of fig. 9.11 shows these laterites overlying peridotites to the north of Fofak Bay on the island of Waigeo, Indonesia. The more dissected parts to the NE of the bay entrance are less characteristically developed and are less promising from a mining point of view. Colour aerial photographs facilitate the identification of promising sites (Verstappen, 1959).

Recently, Trescases (1975) has made a quantitative study of the peridotite massif of S. New Caledonia, using the geomorphological analysis of the area as a framework for mineralogical, geochemical and other investigations. It is evident from the aforesaid that the application of geomorphology in the field of mineral exploration is a diversified matter. Special attention should be paid not only to matters of actual relief, but also to palaeo-relief Processes of the past and present, long-term morphological development and finally neotectonics and geological aspects including lithology should receive ample attention.

Fig. 9.5. Vertical aerial photograph at a scale of 1:20,000 showing placer gold mining activities in northern Sumatra, Indonesia.

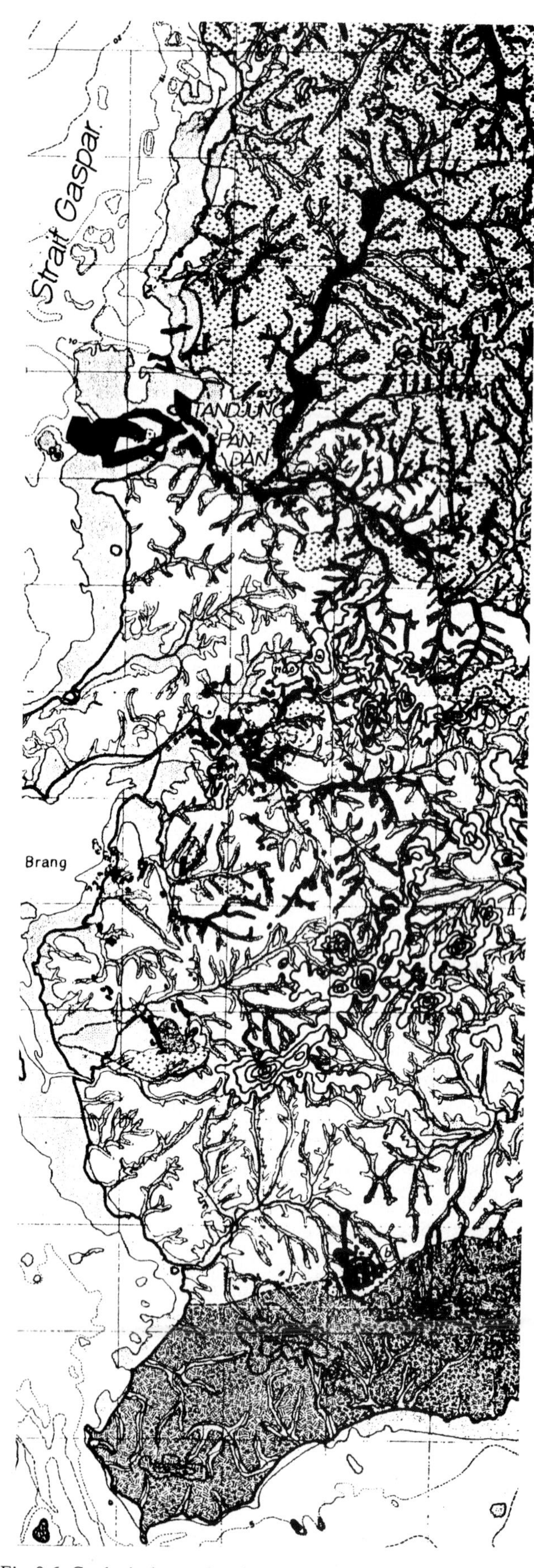

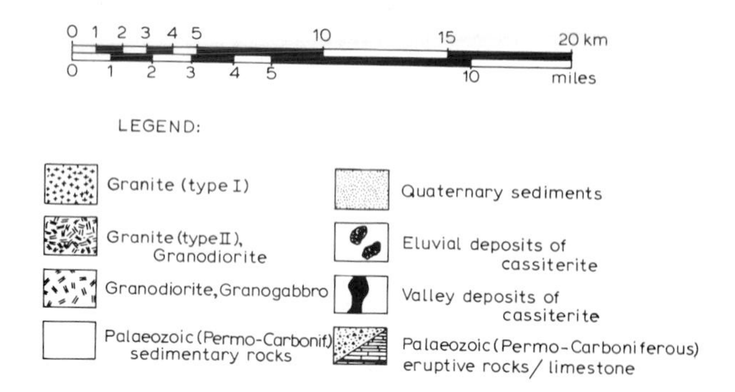

Fig. 9.6. Geological map showing part of NW Belitung, Indonesia with eluvial and alluvial cassiterite placers (Overeem, 1960, Geol. Mijnbouw, 39: 448).

Fig. 9.7 Extended stereopair showing an old planation surface with thick bauxitic weathering cap in the Nassau Mountains, Surinam (1:40,000).

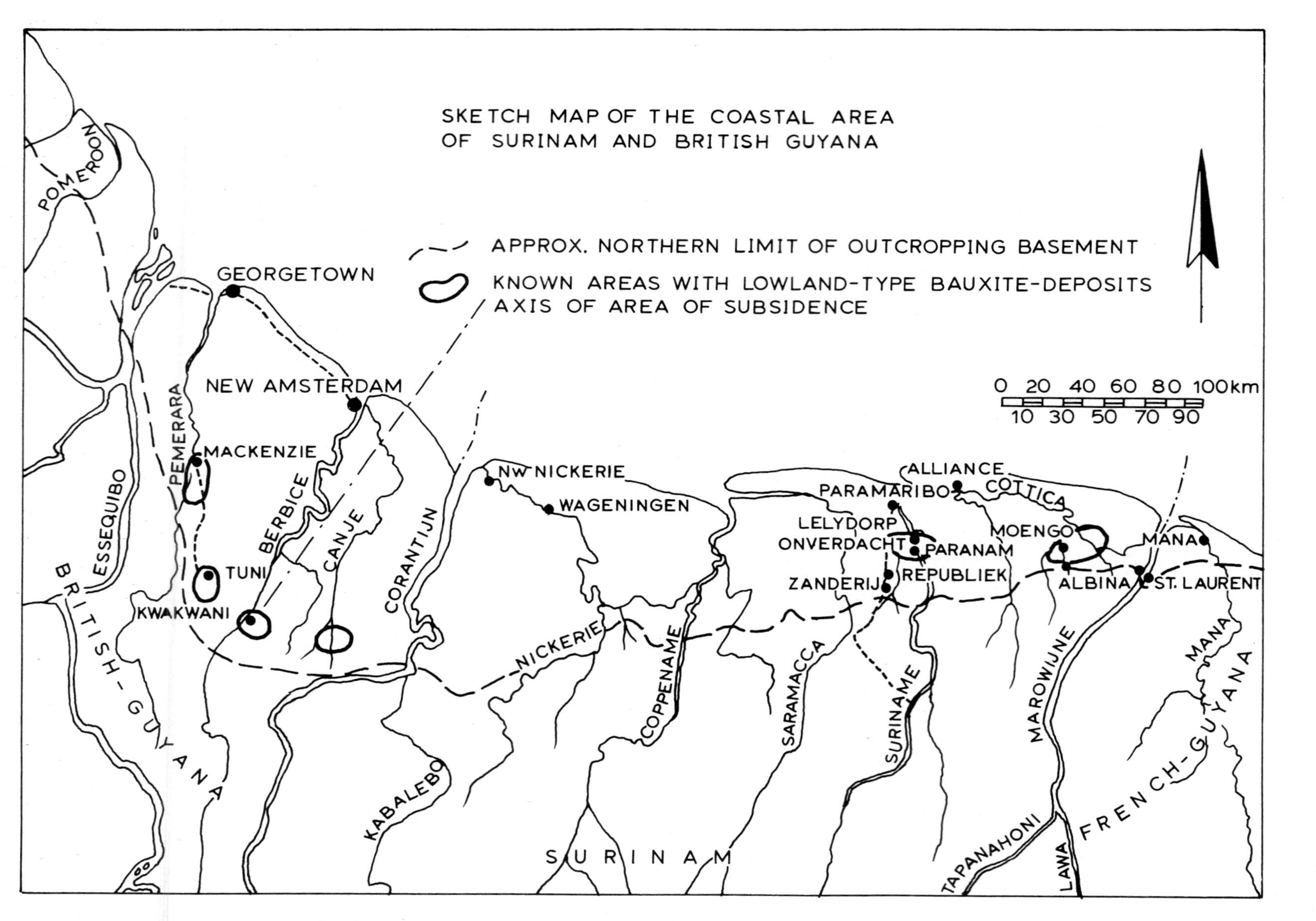

Fig. 9.8. Geomorphological situation of the lowland bauxites of Surinam. The distribution of the bauxite is associated with an old coastline (Montagne, 1964).

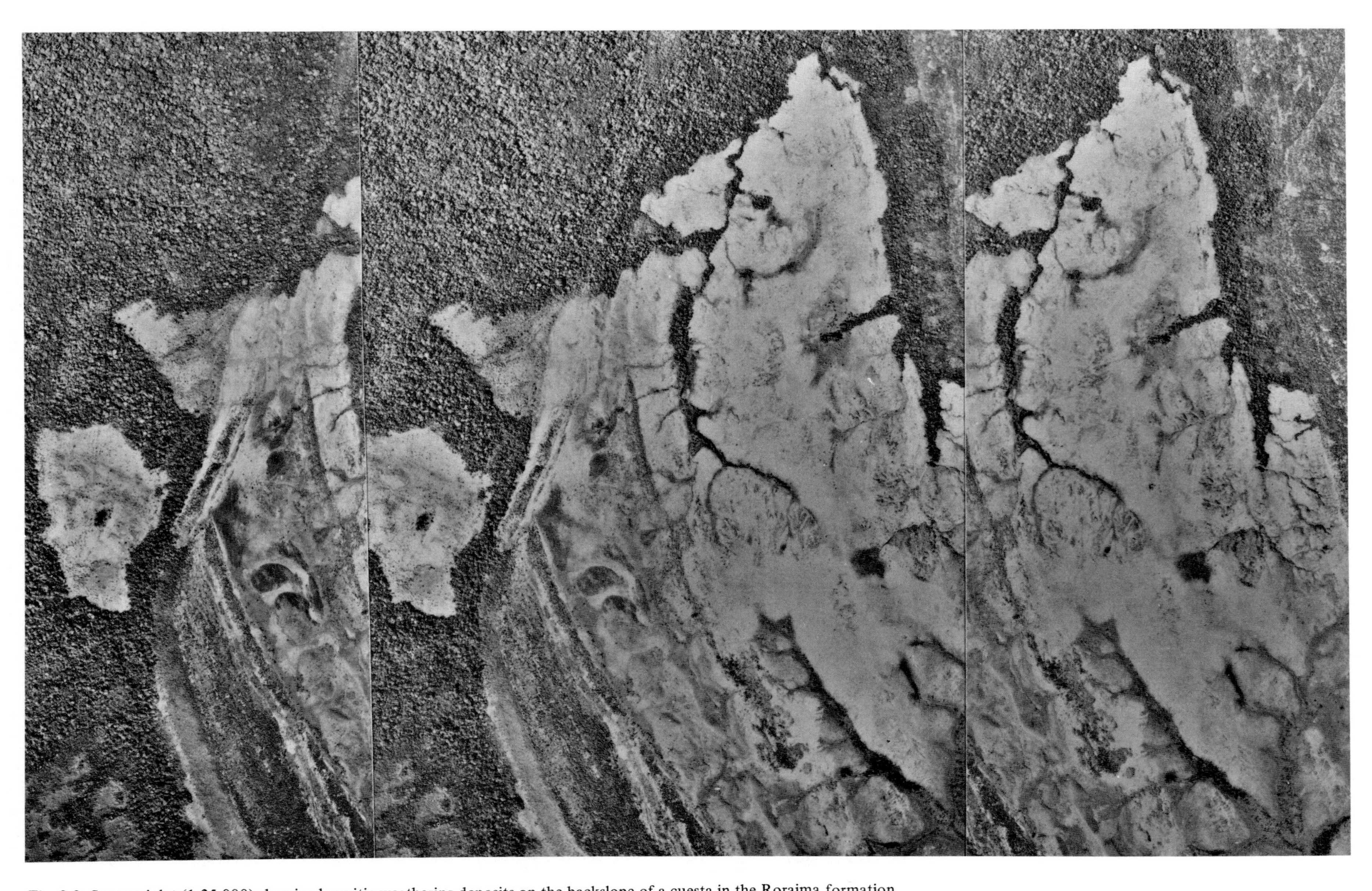

Fig. 9.9. Stereotriplet (1:25,000) showing bauxitic weathering deposits on the backslope of a cuesta in the Roraima formation Venezuela (Yanez, 1979).

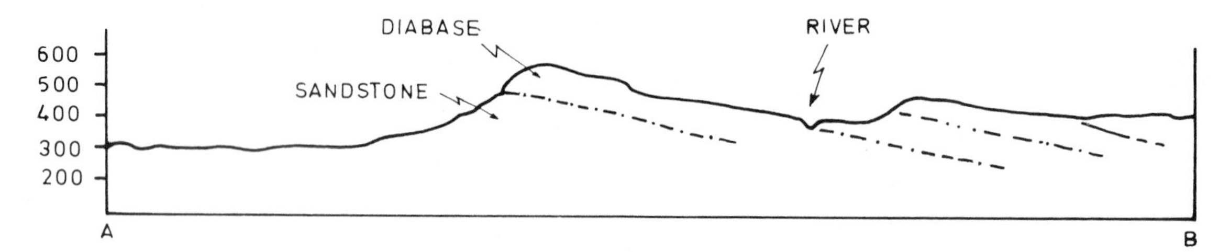

Fig. 9.10 Schematic cross section of the cuesta pictured in fig. 9.9

Fig. 9.11 Vertical aerial photograph (1:40,000) of the poorly vegetated lateritic nickel deposits of Waigeo, Indonesia.

References

Adam, J.W.H., 1960. The geology of the tin lodes in the sedimentary formation of Billiton. Geol. en Mijnbouw, 39 (10): 405–436.

Aleva, G.J.J., 1979. Bauxitic and other duricrusts in Suriname: a review. Geol. en Mijnbouw, 58: 321–340.

Anan'yev, G.S., 1969. Problems of geomorphology in searches for placer deposits of mineral resources. Geomorfologiya. Vyp. 3. (Coll. Geomorphology, 3), Moscow: 19–20.

Anan'yev, G.S., 1969. Geomorphic mapping in prospecting for mineral placer deposits. Soviet Geography: review and translation, 13(6): 369–374.
Translated from Vestnik Moskovskogo Univ., geografiya, (2): 25–30.

Aristarkhova, L.B., 1968. A rational complex of methods of detailed structural-geomorphological studies in the salt-dome region of the Sub-Urals plateau (in russian); Sbornik: Geomorfologisch. analiz pri geologich. issled. v Prikaspijks. vpadine: 5–21.

Aristarkhova, L.B., and Polkanova, L.P., 1975. The objectives of geomorphological research in oil and gas prospecting (in Russian). Vestnik Moskovskogo Univ., Seriya Geogr., 2: 80–85.

Aristarkhova, L.B., et al., 1976. Geomorphological mapping in prospecting for oil and gas. In: Demek and Embleton (Editors): Guide to medium-scale geomorphological mapping, Brno: 213–218.

Bakker, J.P., et al., 1953. Bauxite and sedimentation phases in the northern part of Surinam (Neth. Guyana). Geol. en Mijnbouw, 6(15: 215–226.

Bashenina, N.V., et al., 1974. Legend for 1:200,000 and 1:500,000 scale geomorphological maps compiled with the aims of searching for mineral resources. Studia Geogr. Brno, 41: 115–150.

Berlyant, A.M., 1969. Regularities in the connection between the most recent and the oldest tectonic structures in the N. Perchskaya Depression (in russian). Sovetskaya geol., 1: 115–120.

Bobrievitsch, A.P., and Bondarenko, M.N., 1959. Les gisements de diamants de Yakoutie. Gosgeoltekhizdat,Moscow, 514 pp.

Chervanev, I.G., and Prokhodskii, S.I., 1968. The problem of the structural-geomorphological and neotectonic investigation of plain regions of the Ukraine (in russian). Materialy Kharkovskii otdel Geografichesko Obshchestvo Ukrainy, 6: 35–39.

Clark, A.H., et al., 1967. Relationships between supergene mineral alteration and geomorphology, Southern Atacama Desert, Chile.Trans. Inst. Mining and Metallurgy, B76: 89–96.

Dulemba, J.L., 1969. Sédimentation ferrugineuse actuelle et quaternaire: technique d'étude et remarques pratiques. Noticia Geomorfológica, 9(17): 67–73.

Faniran, A., 1971. Implications of deep weathering on the location of natural resources. Nigerian Geogr. J., 14(1): 59–69.

Fratschner, W.Th., 1960. Die Laterite des südöstlichen Boé (Port. Guinea). Geol. en Mijnbouw, 39(10): 500–511.

Gardner, D.E., 1955. Beach sand heavy mineral deposits of Eastern Australia. Comm. of Australia, Bur. Min. Res., Geol. and Geoph., Bull., 28, 10 pp.

Garlick, W.G., 1972. Sedimentary environment of Zambian copper deposition. Geol. en Mijnbouw, 51(3): 277–298.

Gellert, J.F., 1974. Geomorphologischen Methoden und Karten bei der Lagerstättensuche und Erkundung in der Sowjetunion. Ztschr.f. angewandte Geol., Berlin, 20(1): 40–43.

Guilcher, A., 1959. Travaux de géomorphologie appliquée dans le domaine littoral et estuarien. Rev. Géom. dyn., 10: 145–147.

Helgren, D.M., 1979. River of diamonds; an alluvial history of the Lower Vaal Basin, S. Africa. Univ. Chicago, Dept. Geogr., Res. Paper 185, 389 pp.

Helwig, J., 1967. Paleogeomorphology and its applications to exploration for oil and gas. Bull. Amer. Ass. Petrol. Geol., 51(12), 2468 pp.

Institution Mining and Metallurgy, 1973, 1975, 1977, 1979. Prospecting in areas of glaciated terrain, Proc. 1st–4th symp. I.M.M., London.

Johnson, J.H., 1968. The geology and mineral resources of the Hudson estuary. Hudson River Ecology Symp. 1966, Tuxedo, New York: 8–40.

Kostryukov, M.S., 1969. An attempt to compile medium-scale geomorphological prediction maps for lands with placer potentialities (in russian). Geomorfologiya. Vyp. 3. (Coll. Geomorphology 3): 22–24.

Krol, G.L., 1960. Theories on the genesis of "Kaksa". Geol. en Mijnbouw, 39(10): 437–443.

Krook, L., 1969. Investigations on the mineralogical composition of the Tertiary and Quaternary sands in N. Surinam. Verh. Kon. Ned. Geol. Mijnbouwk. Gen., 27: 89–100.

Leontev, O.K., and Lukyanova, S.A., 1976. Geomorphological mapping of continental shelves in prospecting for gas and oil. In: Demek and Embleton (Editors): Guide to medium-scale geomorphological mapping, Brno: 218–231.

Ludwig, G., and Vollbrecht, K., 1957. Die allgemeinen Bildungsbedingungen litoraler Schwermineralkonzentrate und ihre Bedeutung für die Auffindung sedimentärer Lagerstätten. Geologie, VI: 233–277.

Martin, R., 1966. Paleogeomorphology and its application to exploration for oil and gas (with examples from Western Canada). Bull. Ass. Amer. Petr. Geol., 50(10): 2277–2311.

Martin, R., 1967. Reply to discussion by J. Helwig of "Paleogeomorphology and its application to exploration for oil and gas", 1966. Bull. Amer. Ass. Petrol. Geol. 51(12): 2469–2470.

Maurice, Y.T., and Meyer, W.T., 1975. Influence of preglacial dispersion on geochemical exploration in County Offaly, central Ireland. J. Geochem. Explor., 4(3): 315–330.

McKee, M., 1963. Paleogeomorphology, a practical exploration technique. O. Geol. J., 61(42), 12 pp.

Meyer, W.T., and Evens, D.S., 1973. Dispersion of mercury and associated elements in a glacial drift environment at Keel, Eire. In: Jones, K.M. (Editor), Prospecting in areas of glacial terrain. Inst. Mining and Metallurgy, London: 127–138.

Michel, P., 1960. L'évolution géomorphologique des bassins du Sénégal et de la Haute-Gambie, ses rapports avec la prospection minière. Rev. Géom. Dyn., 10: 117–143.

Molengraaf, G.J.H., 1959. Economisch-Geologisch Rapport over West-Waigeo, Report S 22507/ZO Min. Z.O., The Hague, 34 pp.

Montagne, D.G., 1964. New facts on the geology of the "young" unconsolidated sediments of northern Surinam. Geol. en Mijnbouw, (12) 43: 499–515.

Moore, G.T., 1969. Interaction of rivers and oceans–Pleistocene petroleum potential. Bull. Amer. Soc. Petrol. Geol., 53(12): 2421–2430.

Overeem, A.J.A., 1960. The geology of the cassiterite placers of Billiton (Indonesia). Geol. en Mijnbouw, 39(10): 444–526.

Paraschiv, D., 1976. The contribution of the paleorelief to the hydrocarbon deposit formation in Romania. Rev. Roum. Géol., Géophys. et Géogr., 20: 81–88.

Patyk-kara, N.G., and Iloginova, I.E., 1968. Gravitational differentiation of heavy minerals in eluvial-slope deposits as

an index of intensity of denudation processes (in russian). Geomorphological methods in searches for endogenous mineralisation, Chita: 63–67.
Piotrovski, M.V., et al., 1972. Detailed geomorphological mapping in mineral prospecting. In: Demek (Editor), Manual of detailed geom. mapping,Brno: 267–278.
Piotrovski, M.V., and Lukashov, A.A., 1976. Geomorphological mapping in searches for minerals. In: Demek and Embleton (Editors), Guide to medium-scale geomorphological mapping, Brno: 185–210.
Piotrovski, M.V., 1976. Geomorphological mapping and evaluation of coal and lignite deposits. In: Demek and Embleton (Editors), Guide to medium-scale geomorphological mapping, Brno: 210–213.
Postolenki, G.A., et al., 1969. Principles of, and an attempt at making a special geomorphological map of a gold-bearing region (in russian). Geomorfologiya. Vyp. 3 (Coll. Geomorphology, 3): 20–22.
Reimnitz, E., et al., 1970. Detrital gold and sediments in Nuka Bay. USGS Prof. Paper 700-C: C35–C42.
Seyhan, I., 1972. Maden aramalarinda jeomorfolojinin önemi. (Geomorphology and the search for ores). Jeomorfoloji Dergisi, 4: 47–56.
Shumilov, Y.V., 1969. Some signs of neotectonic movements in the basin of the Malyi Anyui (in russian): Kolyma, 1: 36–39.
Sidorenko, A.W., 1971. Geomorphologie und Volkswirstschaft. Fragen der praktischen Geomorphologie. Ztschr.f. angewandte Geol., 7: 257–263.
Sigov, A.P., et al., 1971. Placers of the Urals, their formation, distribution and elements of geomorphic prediction. Soviet Geography, 13(6): 375–387. Transl. from: Geomorfologiya (1): 28–38.
Smirnov, V.I. (Editor), 1977. Ore deposits of the U.S.S.R. Pitman Publ., London, S. Francisco, Melbourne, 3 Vols, 1268 pp.
Thomas, M.F., 1966. Some geomorphological implications of deep weathering patterns in crystalline rocks in Nigeria. Trans. Inst. Brit. Geogr., 40: 173–193.
Thornbury, W.D., 1969. Principles of Geomorphology. John Wiley & Sons, New York, 594 pp.
Trescases, J.J., 1975. L'évolution géochimique supergène des roches ultrabasiques en zone tropicale. Formation des gisements nickélifères de Nouvelle-Calédonie. Mém. ORSTOM, 78, 259 pp.
Trushin, A.V., 1974. Some problems of relief analysis when searching for near-surface deposits: exemplified by the ore fields in north-east U.S.S.R. (in russian). Vestnik Moskovskogo Univ., Geografiya, 3: 90–92.
Verstappen, H.Th., 1959. Foto-geologische resultaten van de Tamrau Expeditie, 1958. Rapport GONGG., 9, 19 pp.
Voet, H.W., and P.V. Freeman, 1972. Copper orebodies in the basal lower Roan meta-sediments of the Chingola open pit area, Zambian copperbelt. Geol. en Mijnbouw, 51(3): 299–308.
Watson, A., and Angino, E., 1969. Iron-rich layers in sediments from the Gulf of Mexico. J. Sedim. Petrol., 39(4): 1412–1419.
Yanez, G.A., 1979. Geomorphological trends in Venezuela, with two case studies. ITC Journal, 1: 85–98.
Zvonkova, T.V., 1970. Geomorphic methods in oil and gas prospecting. Soviet geography 13(6): 353–363 (translated from: Prikladnaya geomorfologiya: 99–114).

GEOMORPHOLOGY AND DEVELOPMENT PLANNING

10.1. Geomorphological Inputs

It is evident from the previous sections (6-9) that geomorphology plays an important part in the diversified man/environment relationships. On one hand the environment provides the setting for the utilization of the land by man. It may be favourable for certain types of uses or cause adverse conditions for others. On the other hand the environment is also modified by man, either for the sake of or as a result of certain economic or other activities. This applies to both rural and urban areas and also where engineering works or mining activities are concerned. However, geomorphology can only play its part if placed in an environmental context (Leser, 1976). Important inputs from other disciplines have to be integrated.

The position of geomorphology among earth and environmental sciences has been touched upon in Part I of this book: sections 3,4 and 5 which deal with a-biotic factors, land (geology, soils), water (hydrology) and with biotic factors (vegetation). In order to integrate geomorphological studies in the environmental context effective for purposes of development, they should be pursued using man-centered approaches (part II). The survey methods outlined in part III of this book (sections 11-17) are based on this principle. In the past man has through generations built up an 'ethnoscientific' knowledge of his land by trial and error and discovered the types of land utilization suiting the environment and his own needs best. The floating rice practices, e.g. in the Mekong River basin and the flood retreat cultivation, common in several of the larger flood plains in Africa, are good examples, elaborated upon in section 6.

The unraveling of these systems assists in assessing the potentialities of the land and in guiding 'technoscientific' impacts required for raising the productivity (Farvar, 1972; McJunkin, 1975; White, 1978; WHO, 1972).

In a study of coastal lagoons, Lasserre (1979) illustrates how these subtle ecosystems have long been exploited by man under various conditions of climate, salinity, etc. Sophisticated fish breeding and fish farming systems such as the Indonesian fish ponds or 'tambaks', and ingenious fishing and fish-farming systems such as the 'Acadjas' in West Africa (based on submerged heaps of branches where the fish seek refuge and multiply) compete in efficiency with profitable shellfish industries, salt production and reclamation.

Traditional societies have not always managed to live in balance with their environment, particularly when their population increased, when unduly large herds were kept or when monocultivation for cash crops was introduced. However, the impacts of modern technology are generally much more drastic and may easily and often in a short time, lead to lasting, irrevocable damage to affected ecosystems. For this reason alone, careful planning for development and proper implementation of investigations and surveys (for the benefit of the planner) are essential. Another good reason for the proper planning is the urgency of the development issue and the limited funds and manpower available for it (Falque et al., 1979; Farmer, 1971; Vink, 1974).

Three phases can be distinguished in a rational planning programme:

Pre-planning phase, during which the need for action and planning is recognized, the planning objectives are provisionally outlined and the terms of reference are defined.

Actual planning phase, during which the planning objectives are translated into weighted operational goals, the necessary data are collected, various alternative models are evaluated, and the optimal one identified.

Post-planning phase, during which the plan is approved, the implementation is effectuated, and a feed-back exists with the planning authority. It is evident that terrain evaluation and geomorphology in particular should be considered in all three above-mentioned phases. In pre-planning it is essential that the importance of physical environment and terrain factors is realized by the planners, which unfortunately is not always the case. In the actual planning phase, data collection should include geomorphological surveying and terrain analysis. The maps and reports produced should be at the service of the planners on time, and in a form that is useful to them. In the post-planning phase the further advice of the geomorphologist is required in implementing the plan, e.g. during construction when certain specific problems arise. Reference is made to the examples and further literature mentioned in section I of this book.

Development planning of a comprehensive nature and concerning smaller or larger regions should encompass four integrated categories of operational ends: social, economic, administrative and physical, and spatial. The contribution of geomorphology is mostly evident in the fourth, the spatial category; it relates to spatial distributional patterns and location. Fig. 10.1 gives an example of coastal abrasion in Sri Lanka detrimental to the coastal ecosystem and to the development of tourism.

Aerial photographs and other remote sensing images serve as an essential source of information in this connection. The survey work has to be carried out in good cooperation with planners and decision makers as to adequately serve all four operational ends of the development planning (Olofin, 1974; Panos, 1973; Raeside and Deuning, 1974;

Fig. 10.1. The west coast of Sri Lanka near Negombo, showing inadequate coastal protection by rock debris: the abrasion to the landward is not effectively prevented, damage to property is imminent in the distance and touristic development of the beach is hardly feasible.

Rozyeha, 1962).

It is equally important to seriously consider the environmental perception of the population which is to benefit from the development planning. Its 'ethnoscientific' knowledge, already mentioned, may give very useful clues. Also, its attitudes and views may substantially differ from those of outsiders. Whyte (1977) illustrates the value of inside views using the example of the Mixtec farmers of southern Mexico. For centuries they successfully used gully erosion for transporting soil material from the eroding hills to the valley floors which gradually widened as a result of accelerated deposition, thereby gradually changing their poor hill-side fields into more productive fields on fertile alluvial soils. Their perception of fertile and easily erodable deposits differs considerably from that of modern conservation experts who might be tempted to introduce erosion control measures that would upset the whole Mixtec agricultural system. Whyte lists five goals of perception research in envirnomental survey and assessment:

1. contributing to the more rational use of biosphere resources by harmonizing local knowledge and outside expertise,
2. increasing the understanding of all sides of the rational basis for different perceptions of the environment,
3. encouraging local involvement in development and planning as the basis for more effective implementation of more appropriate change,
4. helping to preserve or record the rich environmental perceptions and systems of knowledge that are rapidly being lost in many rural areas, and
5. acting as an educational tool and change agent as well as providing a training opportunity for those involved in the research.

The role of perceptional aspects in the assessment of environmental resources and their economic utilization is particularly evident, also in the evaluation of the visual and scenic qualities of the land (Anon. 1979; Craik, 1972; Kane, 1981; Linton, 1968; Lowenthal and Prince, 1965). Although these matters are difficult to quantify, systematic studies (De Veer, 1977, 1978; Weddel, 1969) and even numerical approaches (Melhorn et al., 1975) have been carried through.

The routine procedure of development planning consists of three parts: inventory of resources by way of surveying, followed by appraisal of resources taking technological means of improvement and matters of agricultural and other optimization into consideration, and finally, development through physical planning, taking the four main operational ends already mentioned into consideration and ultimately leading to the implementation of the plan so devised (Allen, 1959; Grumazescu, 1969; Kregh, 1962; Stärkel, 1972; Verstappen et al., 1972).

Dawson and Doornkamp (1973) mention two alternative approaches. The first of these is the economic approach in which the economic feasibilities are initially considered whereafter the study of environmental suitability is limited to the economically most important areas and subjects. A difficulty in this approach is for the economist to come to conclusions prior to having adequate information about the environment and resource situation. The second alternative is that which Dawson and Doornkamp describe as one of contemporary functional relationships leaning heavily on the study of the existing ecological landscape situations and the related man/environment interactions. It comes close to the ethnoscientific and perceptional concepts given above. Moss (1968, 1969), who did important research in this respect in SW Nigeria, states that ecology should replace geomorphology as a basis for resource assessment. This author disagrees with this viewpoint,but feels that Moss' criticism rather concerns the inadequacy of the geomorphological approaches that have been applied by many for purposes of land assessment (Leser, 1976; Larsen, 1974; Legge, 1974; Leighton, 1970; Olson and Wallace, 1970).

The methods of geomorphological surveying for purposes of development are further specified in Part III of this book (sections 11-17). The geomorphologist should not only provide relevant information pertaining to the geomorphological situation in sensu-stricto by way of the analytical geomorphological surveys outlined in section 11, but coupled with multidisciplinary data so that synthetic surveys of the land of the types described in section 12 can be carried out, placing them in proper environmental context. The two approaches should go hand in hand. Apart from this, a range of differently devised survey methods exists geared to specific applications. The most important ones are described in sections 13-17. They are intended to be used for general physical planning of spatially defined human activities and/or for rare, extreme situations when natural hazards may create disasters. This special aspect is outlined in the following sub-section.

10.2 Physical Planning and Surveying for Disaster Mitigation

Disaster prevention and mitigation is a specific aspect of physical planning which is of great importance in vulnerable areas. It does not suffice to properly plan for the optimal future utilization of the environment under 'normal' conditions as rare events can be expected to leave deep and long-lasting traces. Certain types of land utilization may even be precluded by them or the potentialities of parts of the land may be severely limited, particularly if no appropriate countermeasures are taken. Obviously the aim of the responsible authorities should be to keep the losses of human life and property at a minimum and to maximize the productivity of the land and the quality of life for its inhabitants. Both planners and a diversity of scientists have given these complex problems ample thought, especially in recent years (White, 1945, 1961; Burton et al., 1968, 1978; O'Keefe and Westgate, 1977; Sowell and Fostern, 1976; Tag Eldeen, 1980; UNDRO, 1977).

In order to mitigate the damage if and when the disaster strikes, the hazard situation should be studied and assessed in advance. A hazard or risk zoning should be established and the state of preparedness for disaster should be optimized. The methodology involved includes matters such as:

. monitoring the hazard (e.g. gauging of rivers, surveillance of volcanoes)
. surveying spatial patterns of hazard susceptibility
. zoning of hazard/risk
. setting up an early warning system
. formulating building codes
. land use planning
. planning emergency and relief operations (evacuation, fire squads, etc.)
. post-disaster planning

Since disaster mitigation is a society-centered issue, socio-economic, perceptional and behavioural, technical and managerial aspects are involved to which the geomorphologist cannot contribute anything of importance (Dacy and Kunreuther, 1969). His task is related to monitoring, surveying and mapping of hazards and to susceptibility, assessment and zoning of the potentially affected land (Quicly, 1975). In this context two main types of maps can be prepared which are of use to the planner, namely:

1. Hazard or risk zone maps of various types and complexities which indicate the spatial distribution of the type and magnitude of the hazard.
2. Potential damage or vulnerability maps on which the number of affected houses, settlements,other structures and social activities are indicated, in case of a disaster of a certain magnitude.

From these two types of maps a clear picture can be obtained concerning the extent to which the community will be disrupted and uprooted if no countermeasures are taken. On this basis a study of possible remedial measures can be made. This may result in a modification of the hazard zoning by land improvement works of various kinds, adaptation of the land occupancy, technical recommendations, early warning systems and in a disaster preparedness plan.

As far as their causative factors are concerned, natural hazards can be grouped into three main categories, namely:
. natural hazards of exogenous origin (flooding, drought, landsliding, etc.)
. natural hazards of endogenous origin (volcanic and earthquake hazard)
. natural hazards of anthropogenic origin (subsidence, etc.)

It should be emphasized that hazards of endogneous and anthropogenic origin are experienced through exogenous processes and phenomena. One should also be aware of the fact that hazards of exogenous origin may be sparked or aggravated by anthropogenic causes. This holds true for certain (storage lake induced earthquakes) natural hazards of endogenous origin. This classification is of limited value only, though it serves to systematize the matter.

The major hazards of exogenous origin are discussed in sections 13 on flood susceptibility surveys, 14, on drought susceptibility surveys and 15 on slope stability and erosion surveys. The avalanche hazard, significant in many alpine zones, is discussed in section 16. Therefore, the hazards of exogenous origin are only briefly mentioned here since information about natural hazards of endogenous origin and related surveys is given in section 17 (volcanic hazards and earthquake hazards), only the hazards of anthropogenic origin are discussed in the context of this section on planning (sub-section 10.3).

It should be emphasized that these natural hazards and the potentially consequent disasters are at present a matter of major concern for government authorities and for international organizations (UNDRO, 1977).

Their effects on society have been seriously studied in critical areas. It has already been pointed out that disaster mitigation and prevention is a complex matter of interdisciplinary nature, encompassing not only the physical aspects of the phenomena but also socio-economic, psychological, technical and managerial aspects. The perception and response of society towards the natural hazard concerned and human behaviour under stress when disaster strikes, are distinct foci of interest in these society-centered studies. The literature on the subject has grown rapidly in recent years (Bolt et al., 1975; Burton et al., 1978). The monitoring of the hazard and the assessment and zoning of natural hazards have received little attention until recently except in certain frequently affected countries. The contribution of the geomorphologist to disaster studies, particularly in these fields, is great, as exemplified in the following sections on specific natural hazards.

10.3 Natural Hazards of Anthropogenic Origin

The natural hazards of this category are in most cases not of a separate kind, but exogenous (flooding, landsliding) or endogenous (earthquakes) hazards which are provoked or aggravated by human action. Since the natural and anthropogenic causes of such calamities are not always easily separable, it may be preferred to list the latter simply under the causative factors in disaster studies. But in some fields the anthropogenic origin of the hazard is so evident that mention under a separate heading is justified.

SUBSIDENCE ranks high among natural hazards of anthropogenic origin and it is a major cause of increased susceptibility to flooding in areas previously unaffected by flood waters. Subsidence can be accurately measured, e.g., by precise levelling, as exemplified by fig. 10.2, concerning the Tokyo lowland. Of course it may result from natural causes, but in such cases it is usually localized. Sinks in limestone areas and ignimbrite plateaus are examples. However, generalized subsidence in densely populated lowlands is in most cases of anthropogenic origin (Allen, 1969; Foose, 1968). Endogenous causes and sea level rise may also occur,but are usually of secondary importance.

Subsidence can be measured by precise levelings as exemplified by fig. 10.2 but it can also be predicted, provided that cause, location and further parameters are known. Geertsema and Opstal (1973) have calculated the amount and distributional pattern of the subsidence that can be expected over the large Slochteren gas fields and adjacent areas in the northeastern part of the Netherlands in 1990, 2005, 2050 and 2150. The subsidence caused is maximum over the approximately east-west stretching, depressed older parts of the Holocene alluvial plain,which are situated between the Drente Plateau in the south and the slightly higher, younger parts in the north. The latter have been formed when the sea level was relatively higher. Fig. 10.3 pictures the situation.

Obviously the outlook is for increased drainage problems and inundation hazard in the depressed zone in the future. Inundations have occurred in several parts of the depressed zone,such as near the village of Hoogkerk, west of Groningen, following rapid snowmelt on the Drente Plateau in early 1979. The extent of these has been surveyed by the author using overboard photos taken from low-flying small aircraft and mapped on a 1:25,000 scale. The location of these floodings is plotted in fig. 10.3.

If no precautions are taken, a similar event occurring in the future may result in more extensive inundations, possibly centering slightly more westward.

The subsidence that has occurred in the past due to tectonic and other causes has been studied by Edelman (1954) on the basis of two precise levelings with a time interval of approximately 80 years. This subsidence is maximum in the north and extrapolation in time of these data also affects the future distributional pattern of flood hazard in the area.

Major anthropogenic causes of subsidence are:

1. The withdrawal of fluids or natural gas.
   This relates mostly to groundwater, crude oil or natural gas. The amount of subsidence and the surface areas affected may be considerable. Bull and Poland (1975) mention that in the San Joaquin Valley in California, USA, a zone of approximately 3200 sq. km. has subsided more than 30 cm as a result of groundwater withdrawal. The maximum subsidence measured in this case amounts to almost eight metres.

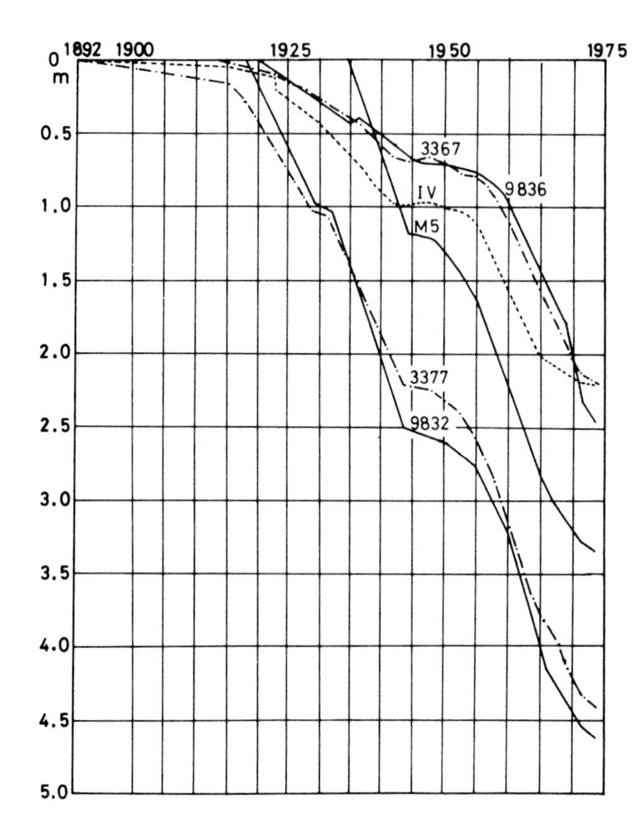

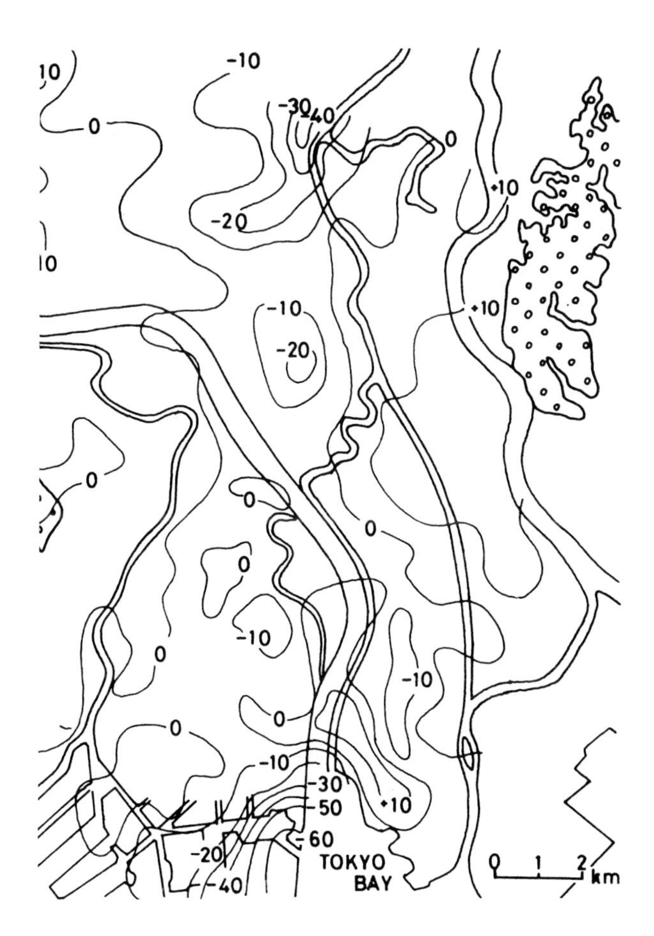

Fig. 10.2 Accumulated land subsidence in the Tokyo lowland since 1892 (top) and the subsidence measured in 1974 (bottom). After Nakano and Matsuda (1976).

Lofgren and Klausing (1968, 1969), Gabrysch (1975), Dunham and Gray (1972), Mitsui (1980), Sato (1973) and Nakano and Matsuda (1976) point out other examples, and UNESCO (1969) gives a full account of the problems involved. A particularly striking example is the subsidence of Venice due to groundwater extraction for industrial purposes, endangering the cultural monuments in this old city (section 7). The effects of the extraction of natural gas has similar effects as the withdrawal of groundwater as shown in the case of the large gas field in NE Groningen, the Netherlands (NAM, 1973), as already mentioned.

The damage caused by subsidence may be considerable. Jones and Larson (1975) estimate the annual damage due to groundwater withdrawal in a 1,700 sq. km. area along the Texas Gulf coast at US dollars 37,8 million and give specifications as to the types of damage including matters such as temporary flooding, permanent inudation, landfilling and structural damage. The problem is notinsolvable, however, once the withdrawal is stopped or strongly reduced, the subsidence will come to an end (Poland, 1973, 1075) and the land may even rise again, although normally not entirely up to its initial height. The injection of saline water following oil production may have a similar effect (Pierce, 1970).

2. Improved drainage of peat layers.
   This normally results in an irreversible compaction of the peat due to drying. The amount of subsidence depends on matters such as the old and new drainage conditions and on the clay and organic matter content. It can in most cases be predicted (Stephens and Speir, 1969; v.d. Molen, 1975: Glopper, 1973).
3. Hydrocompaction.
   This is a widespread cause of land subsidence particulary in moisture deficient alluvial deposits and in loose formations. The process is also known to occur following irrigation schemes in semi-arid areas (Lofgren, 1969; Prokopovich, 1972).
4. Mining activities.
   Extensive underground mining activities may give rise to subsidence although this is usually of a more localized nature. Built-up areas may be seriously affected by the structural damage caused in this way. Heavy mine tips may also be affected by it (HMSO, 1929, 1949; Murr, 1973; Wardell and Piggot, 1969; Shadbolt and Mabe, 1970).

EARTHQUAKES triggered by human action are another type of natural hazard of anthropogenic origin. Observations made in recent years have revealed that earthquakes may involuntarily be caused by human action. Although most of these are minor quakes, magnitudes of over 6 are also on record. These shocks may follow after injection of water or other fluid wastes (Bardwell, 1970; Evans, 1970; Schleicher, 1975; USGS, 1973) or they may occur some years after the con-

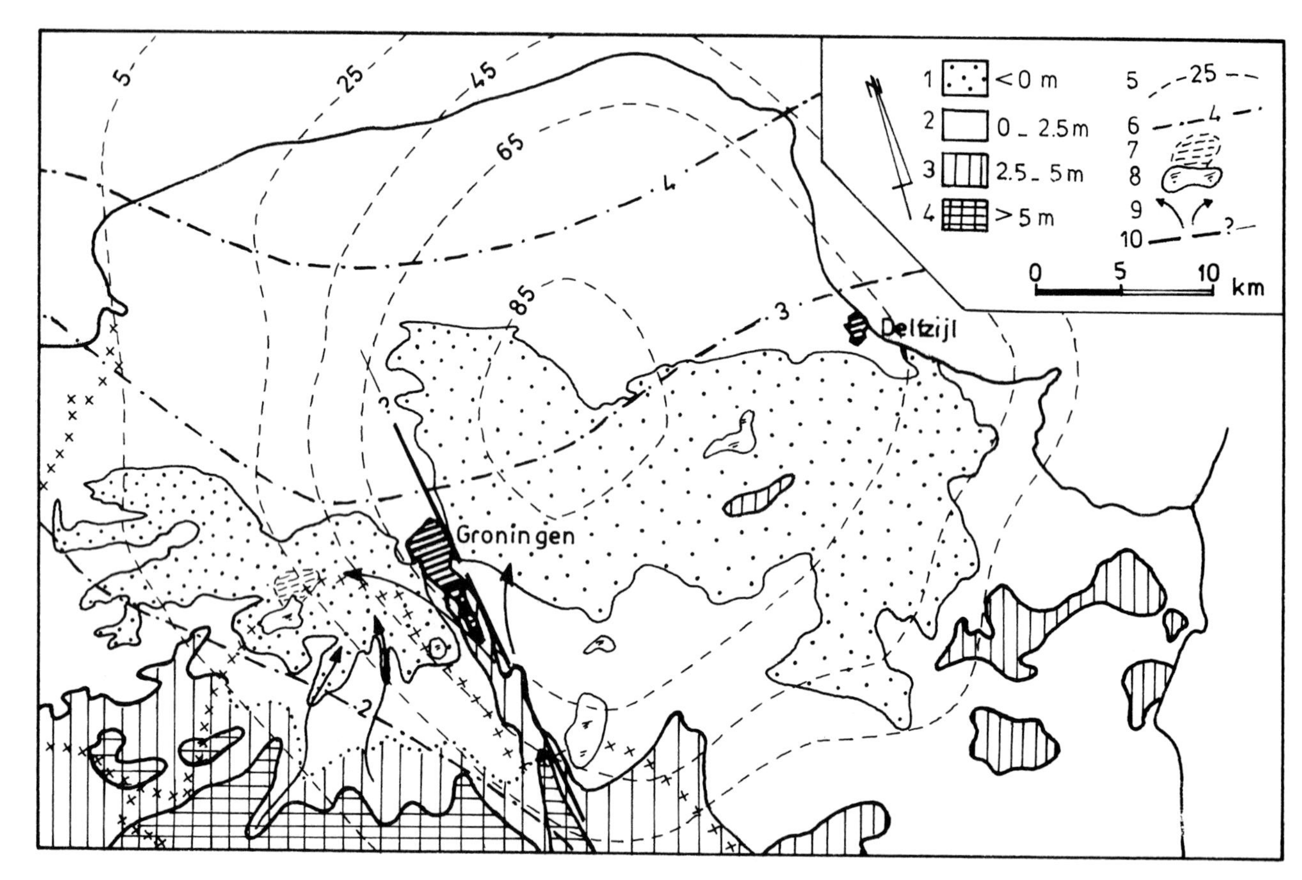

Fig. 10.3. Contour map showing the NE part of the Netherlands with the depressed zone between the low Plateau of Drente in the south and the slightly higher and younger parts of the alluvial plain in the north which was formed with the rising sea level. Contour interval is 5 m. The relative subsidence as calculated from the two precise levelings (Edelman, 1954) is stronger in the north (dashed lines),but the subsidence expected towards 2005 due to the extraction of natural gas (Geertsema and Opstal, 1973) is stronger over the already depressed zone (full lines). This may cause drainage problems in the future. The extent of the 1979 inundations in the depressed zone are indicated in black.

struction of a major dam (Long, 1974; Lomnitz, 1974; Rothe, 1970). Lubrication of faults or slipping planes is the obvious cause when quakes follow the injection of fluids, whereas the extra stress produced by impounding a lake is an important factor in dam-related quakes, although lubrication by groundwater may also contribute under such circumstances. Rothe (1970) maintains that the depth of the storage lake is more important than its volume in this context. Contrary to the trend in many natural earthquakes, the major shock is usually preceded, not followed, by minor ones. Examples of earthquakes induced by reservoir lakes are numerous, e.g. Lake Mead, USA (Carder, 1970), the Mangla Reservoir, Pakistan (Brown, 1974), the Koyna Reservoir, India (Gupta et al., 1974; Gupta and Rastogi, 1974, 1975), Lake Kariba, East Africa (Lane, 1974); Lake Nourek, USSR (Nikolaev, 1974); Lake Benman, New Zealand (Adam, 1977).

References

Geomorphology and development planning

Allen, S.W., 1959. Conserving natural resources. Principles and practice in a democracy. McGraw-Hill, New York, 369 pp.

Anon., 1979. Land capability classification for outdoor recreation. Report 6, Dept.Nat.Cons.Recr., Ottawa.

Bolt, B.A., et al., 1975. Geological hazards. Springer Verlag, Stuttgart, 328 pp.

Burton, I., et al., 1968. The human ecology of extreme geophysical events. Nat. Hazard Res. Working Paper, Univ. Toronto, 33 pp.

Burton, I., et al., 1978. The environment as a hazard. Oxford Univ. Press, New York, 240 pp.

Craik, K.H., 1972. Psychological factors in landscape appraisal. Environment and Behaviour, I: 255–266.

Dacy, D.C., and Kunreuther, H., 1969. The economics of natural disasters. Free Press, New York, 270 pp.

Dasmann, R.F., 1976. Environmental conservation, 4th Ed., Wiley & Sons, 529 pp.

Dawson, J.A., and Doornkamp, J.C. (Editors), 1973. Evaluating the Human Environment, essays in applied geography. Edward Arnold Ltd., London, 288 pp.

Falque, M., et al., 1974. Note on geological methodology used in ecological planning studies. Revue Géogr. Phys. et Géol. Dyn., 16(5): 459–464.
Farmer, B.H., 1971. The environmental sciences and economic development. J. Dev. Studies,7: 257–269.
Farvar, M.T., and Milton, J.P. (Editors), 1972. The careless technology: ecology and international development. Nat. His. Press, New York.
Grumazescu, H., and Grumazescu, C., 1969. A geomorphological study of the Danube Delta localities for planning purposes. Int. Symp. géom. appliquée, Bucharest: 223–232.
Kane, Ph.S., 1981. Assessing landscape attractiveness: a comparative test of two new methods. Appl. Geogr., 1(2): 77–96.
Kragh,H., 1962. Die Hilfe des Luftbildes bei der Landschaftsplanung. First Symp. Comm. VII, I.S.P. (Delft): 333–337.
Larsen, J.I., 1974. Geology for planning in Lake County (Ill.) Illinois State Geol. Survey, Circular 481, 43 pp.
Lasserre, P., 1979. Coastal lagoons. Sanctuary ecosystems, cradles of culture, targets for economic growth. Nature and Resources, 15(4): 2–21.
Legge, M., 1974. Geology tries to protect envinronment. Can. Geogr. J., 88(4): 30–37.
Leighton, F., 1970. Landslides. In: R.A. Olson and M.N. Wallace (Editors), Geologic hazards and public problems. U.S. Office Emergency Preparedness, Region 7, Proc. Conf. San Francisco, Cal., 1969. U.S. Govt. Printing Off., Washington: 97–132.
Leser, H., 1976. Landschaftsökologie. Eugen Ulmar Verlag, Stuttgart, 432 pp.
Linton, D.L., 1968. The assessment of scenery as a natural resource. Scottish Geogr. Mg., 84: 218–238.
Lowenthal, D., and Prince, H.C., 1965. English landscape tastes. Geogr. Rev., 54: 309–346.
Lowenthal, D., and Prince, H.C., 1972. English landscape tastes. Geogr. Rev. 55, 1965: 186–222. Reprinted in: P.W. English and R.C. Mayfield (Editors), Man, Space and Environment. Oxford Univ. Press, New York: 81–114.
McJunkin, F.E., 1975. Water, engineers, development and disease in the tropics: Schistosomiasis engineering applied to planning, design, construction and operation of irrigation, hydroelectric and other water development schemes. USAID publ., Washington.
Melhorn, W.N., et al., 1975. Landscape aesthetics numerically defined (Land system): application to fluvial environments. Studies in fluvial geomorphology, 1, Purdue Univ., Lafayette (Ind.), Water Resources Res. Center, Techn. Rep., TR-37, 169 pp.
Moss, R.P., 1968. Land use, vegetation and soil factors in SW Nigeria: a new approach. Pacific Viewpoints, 9: 107–127.
Moss, R.P., 1969. The appraisal of land resources in tropical Africa: a critique of some concepts. Pacific Viewpoint, 10: 18–27.
O'Keefe, P., and Westgate, K., 1977. Preventive planning for disasters. Long Range Planning, 10/3: 25–29.
Olofin, E.A., 1974. Classification of slope angles for land planning purposes. J. Trop. Geogr., 39: 72–77.
Olson, R.A., and Wallace, M.M., 1970. Geological hazards and public problems. U.S. Office Emergency Preparedness, Region 7, Proc. Conf. San Francisco,(Cal.), 1969, U.S. Govt. Printing Off., Washington, 335 pp.
Panoš, V., 1973. Project of rehabilitation of devastated part of the South Karst Plain in Western Cuba. Acta Univ. Palackianae Olomucensis,Fac. R.N., 42, Geogr. Geol., 13: 81–108.
Quicley, R.M., 1975. Comments on hazard land zoning. Geoscience Canada, 2/2: 111–112.
Raeside, J.D., and Rennie, W.F., 1974. Soils of Christchurch region, New Zealand: the soil factor in regional planning. New Zealand Soil Survey,Report, 16, 74 pp.
Rozycha, W., 1962. Physiographic research in town and country planning. Problems of applied geography, II. Geogr. Polonica, 3, Proc. Anglo-Polish Seminar, Keele: 251–262.
Sewell, W.R.D., and Fosterm, H.D., 1976. Environmental risk: Management strategies in the developing world. Environm. Management, 1/1: 49–59.
Stärkel, L., 1972. An outline of the relief of the Polish Carpathians and its importance for human management. Probl. Zagospodarowania Zien Gorskich, 10: 75–150.
Tag Eldeem, M., 1980. Pre-disaster physical planning: integration of disaster risk analysis into physical planning—a case study in Tunisia. Disasters, 4/2: 211–222.
UNDRO, 1977. Disaster prevention and mitigation: A compendium of current knowledge, 5, land use aspects, New York.
UNDRO, 1977. Review of the priority subject areas natural disasters, 3, measures for mitigation of natural disasters, New York.
UNESCO, 1969. Annual summary of information on natural disasters. Earthquakes, tsunamis, volcanic eruptions, landslides, avalanches.
Veer, A.A., de, et al., 1977. Vergelijking van Nederlandse methoden van landschapskartering en hun toepassingsmogelijkheden. STIBOKA/PUDOC, Wageningen, 65 pp.
Veer, A.A., de, and Burrough, P.A., 1978. Physiognomic landscape mapping in the Netherlands. Landscape planning, 5(1): 45–62.
Verstappen, H.Th., et al., 1972. Studie over het ontwikkelings potentieel van de bovenwindse eilanden Saba, St. Eustatius en St. Maarten (N.A.). Deel 1: Terreingesteldheid en landgebruik, 65 pp.
Vink, A.P.A., 1974. Physical geography for land use planning. Univ. Reading, Dept. Geogr., Geogr. Papers, 29, 2 pp.
Weddle, A.E., 1969. Techniques in landscape planning. J. Town Planning Inst., 55: 387–398.
White, G.F., 1945. Human adjustment to floods. Univ. Chicago, Dept. Geogr., Res. Paper, 29, 236 pp.
White, G.F. (Editor), 1961. Papers on flood problems. Univ. Chicago, Dept. Geogr., Res. Paper,70, 26 pp.
White, G.F. (Editor), 1978. Environmental effects of arid land irrigation in developing countries. UNESCO, M.A.B. Techn. Notes, 8 .
Whyte, A., 1977. The role of environmental perception research. Nature and Resources, 13(4): 19–21.
W.H.O., 1972. Health hazards of the human environment. Geneva, World Health Org., 387 pp.

Disasters of anthropogenic origin

Adams, R.D., 1974. Statistical studies of earthquakes associated with Lake Benmore, New Zealand. Eng. Geol., 8(1–2): 40–45.
Allen, A.S., 1969. Geologic settings of subsidence. In: D.J. Varnes and G. Kiersch (Editors), Reviews in Engineering Geology, 2: 305–342.
Bardwell, G.E., 1970. Some statistical features of the relationship between Rocky Mountain arsenal waste disposal and frequency of earthquakes. Geol. Soc. Amer. Eng. Geol., Case Histories, 8: 33–37.
Brown, R.L., 1974. Seismic activity following impounding of Mangla reservoir. Eng. Geol., 8(1–2): 79–93.
Bull., W.B., and Poland, J.F., 1975. Land subsidence due to groundwater withdrawal in the Los Banos-Kettleman City area, California, Part 3. Interrelations of water level change, change in aquifer-system thickness, and subsidence. USGS, Prof. Paper, 437-G, 60 pp.

Carder, D.S., 1970. Reservoir loading and local earthquakes. Geol. Soc. Amer. Eng. Geol. Case Histories, 8: 51–61.

Dunham, K.C., and Gray, D.A., 1972. A discussion on problems associated with the subsidence of southeastern England. Phil. Trans. Roy. Soc., London, A 272(1221): 81–274.

Edelman, T., 1954. Tectonic movements as resulting from the comparison of two precision levelings. Geol. en Mijnbouw, 16, 1–4.

Evans, D.M., 1970. The Denver area earthquakes and the Rocky Mountain arsenal disposal well. Geol. Soc. Amer. Eng. Geol. Case Histories, 8: 25–32.

Foose, R.M., 1968. Surface subsidence and collapse caused by groundwater withdrawal in carbonate rock areas. Proc. 23rd Int. Geol. Congress, Prague: 155–166.

Gabrysch, R.K., 1975. Landsurface subsidence in the Houston-Galveston Region, Texas, R.K. Texas Water Dev. Board, Austin, Report 188, 19 pp.

Geertsema, J., and Opstal, G. van, 1973. A numerical technique for predicting subsidence above compacting reservoirs, based on the nucleus concept. Verh. Kon. Ned. Geol. Mijnbouwk. Gen., 28

Glopper, R.J. de, 1973. Van Zee tot Land. Subsidence after drainage of the deposits in the former Zuyder Zee and in the brackish and marine forelands in the Netherlands. Staatsuitgeverij, 's-Gravenhage, 50, 205 pp.

Guha, S.K., et al., 1974. Case histories of some artificial crustal disturbances. Eng. Geol., 8(1–2): 59–77.

Gupta, H.K., and Rastogi, B.K., 1974. Investigations of the behaviour of reservoir associated earthquakes. Eng. Geol., 8(1–2): 29–38.

Gupta, H., and Rastogi, B.K., 1974. Will another damaging earthquake occur in Koyna . Nature, 248 (5445)b: 215–216.

Gupta, H., and Rastogi, B.K., 1975. Dams and earthquakes. Elsevier Publ. Comp., Amsterdam, 230 pp.

H.M.S.O., 1929. Royal Commission on Mining Subsidence, Final Report, Cmd. 2899

H.M.S.O., 1949. Inter-Departmental Committee Report on Mining Subsidence, Report of Committee on Mining Subsidence, Cmd. 7637

Jones, L.L., and Larson, J., 1975. Economic effects of land subsidence due to excessive groundwater withdrawal in the Texas Gulf Coast area. Texas A & M Univ., College Station, Water Res. Inst., Techn. Report, TR-67; 44 pp.

Lane, R.G.T., 1974. Investigation of seismicity at dam/reservoir sites. Eng. Geol., 8(1–2): 95–98.

Lofgren, B.E., 1968. Analysis of stresses causing land subsidence. USGS, Prof. Paper, 600–B: 219–225.

Lofgren, B.E., and Klausing, R.L., 1969. Land subsidence due to groundwater withdrawal, Tulare-Wasco area (Cal.), USGS, Prof. Paper, 437–B, 101 pp.

Lofgren, B.E., 1969. Land subsidence due to the application of water. In: D.J. Varnes and G Kiersch (Editors), Re Reviews in engineering geology, 2: 271–303.

Long, R.E., 1974. Seismicity investigations at dam sites. Eng. Geol., 8(1–2): 199–212.

Lomnitz, G., 1974. Earthquakes and reservoir impounding: state of the art. Eng. Geol., 8(1–2): 191–198.

Mitsui, K., 1980. Changes in inundation patterns by complicated environment. Proc. IGU Conf. Japan.

Molen, W.H. van der, 1975. Subsidence of peat soils after drainage. In: Hydrology of marsh-ridden areas. Proc. Minsk Symposium, June 1972, Paris/IAHS: Studies and reports in hydrology 19: 183–186.

Murr, K., 1973. Rohstofferschlieszung in der USA verursacht Geländesenkungen. Ztschr. f. Wirstschaftsgeogr., 17(6), 184 pp.

N.A.M., 1973. The analysis of surface subsidence resulting from gas production in the Groningen area, the Netherlands. Verh. Kon. Ned. Geol. Mijnbouwk. Gen., 28, 109 pp.

Nakano, T., and Matsuda, I., 1976. A note on land subsidence in Japan. In: Der Staat und sein Territorium. Steiner Verlag, Wiesbaden: 111–125.

Nikolaev, N.I., 1974. The first case of induced earthquakes during construction of a hydro-electric power station in the U.S.S.R. Eng. Geol., 8(1–2): 107–108.

Nikolaev, N.I., 1974. Tectonic conditions favourable for causing earthquakes occurring in connection with reservoir filling. Eng. Geol., 8(1–2): 171–189.

Pierce, R.L., 1970. Reducing land subsidence in the Wilmington oil field by use of saline waters. Water Res. Research, 6(5): 1505–1514.

Poland, J.F., 1973. Land subsidence in western United States. In: R.A. Olson and M.M. Wallace (Editors), Geologic Hazards and Public Problems. U.S. Govt. Printing Off.: 77–96.

Poland, J.F., et al., 1975. Land subsidence in the San Joaquin Valley, California, as of 1972. USGS.,Prof. Paper, 437–H, 78 pp.

Prokopovich, N.P., 1972. Land subsidence and population growth. Proc. 24th Int. Geol. Congr., Montreal, Section 13: 44–54.

Rothe, J.P., 1970. Séismes artificiels. In: T. Rikitake (Editor) Earthquake mechanics, symposium, Madrid, 1969 (Tectonophysics, 9: 2–3)): 215–238.

Sato, H., 1973. The survey of ground subsidence. Bull. Geogr. Survey. Inst., 19(1): 30–31.

Schleicher, D., 1975. A model for earthquakes near Palisades reservoir, SE Idaho. J. Res. USGS, 3(4): 393–400.

Shadbolt, C.H., and Mabe, W.J., 1970. Subsidence aspects of mining development in some northern coalfields. In: Geological aspects of development and planning in Northern England: 108–123.

Stephens, J.C., and Speir, W.H., 1969. Subsidence of organic soils in the USA. Pub. Int. Ass. Sci. Hydrol., 89: 523–534.

UNESCO, 1969, Land Subsidence. Publ. Inst. Sci. Hydrol., 88(2 Vols).

USGS, 1973. Man-made earthquakes at Denver and Rangely, Colorado. Earthquake Information Bull. 5(4): 4–9.

Wardell, K., and Piggott, R.J., 1969. Report on mining subsidence. In: A selection of technical reports submitted to the Aberfan Tribunal, HMSO: 187–204.

PART C. GEOMORPHOLOGY AND SURVEYING FOR PLANNED DEVELOPMENT AND HAZARD ZONING

ANALYTICAL GEOMORPHOLOGICAL SURVEY AS A TOOL

11.1 The Concept and its Evolution

Geomorphological survey and mapping is presently considered an essential pre-requisite for applied geomorphological research. This concept was not initially accepted; in fact, the methodology was largely developed from the nineteen-fifties onward and the importance of the techniques involved for surveys and development are not, even to date, universally understood. The precise mapping of landforms, exogenous processes and all other relevant geomorphological information at appropriate scales is only feasible if all aspects of the relief, including genesis and ecological landscape context, have been thoroughly studied and analysed. Since the spatial distribution and classification of landforms and related phenomena is of considerable importance for the study of renewable and non-renewable natural resources and for the rational utilization of these resources by man, the results of geomorphological investigations as represented in the form of a map will serve not only for the advancement of science but can also be applied to numerous aspects of environmental management (Barsch, 1981; Brunsden, 1981; Mensching, 1979; Tricart, 1965).

When viewed from this angle, it is surprising that geomorphological mapping began more than a century after systematic geological mapping was introduced in the beginning of the nineteenth century. A first attempt is from Passarge (1914) who emphasized morphological and morphometrical elements in particular. His work failed to arouse the interest of geomorphologists at that time and they continued to (verbally) describe landforms and to produce rather sketchy maps, mostly at small scales indicating either major structural elements only or giving no more than a pictorial view of the land (Raisz, 1941, 1956). The few more detailed maps related to one or more selected groups of features, such as river terraces and glacial forms and left large areas unmapped.

The mapping of morphological processes was a particularly neglected field; this prevented the timely acceptance of geomorphology by soil and other scientists as a useful scientific discipline when incorporating the effect of terrain factors in their survey work. As a consequence of their limited appreciation of geomorphology, they restricted themselves to matters of relief, dividing the terrain into areas of comparable relief amplitude and form. They applied a straightforward toposequence concept, dividing slopes into high, central and low parts. It is evident that neither of these methods is very satisfactory from a geomorphological point of view since genetic, chronological and other aspects are not considered. This situation resulted in insufficient emphasis on the role of dynamic geomorphology and lasted for many years. From these early interdisciplinary contacts, however, methods evolved which gave an impetus to landscape ecological concepts and the position of geomorphology therein. These 'synthetic' surveys of the land are the counterpart of the 'analytical' geomorphological surveys now under discussion and are treated in the following section (12). In the author's opinion, both types of surveys are mutually complementary.

The modern concept of analytical geomorphological survey and mapping only rather recently took shape. Helbing (1952) produced a 1:25,000 geomorphological map (of the Sern Valley) in Switzerland; Annaheim (1956) was also among the early advocates here. In France, Cholley (1956) and his co-workers published a geomorphological map of the Paris Basin and Tricart (1959) and his group of the CGA at Strasbourg surveyed the Senegal Delta at the scale of 1:50,000 for purposes of development. A particularly important impetus came from Poland where Klimaszewski (1956, 1968) and his school in Krakow and Galon (1962) in Torun launched a country-wide, systematic, geomorphological survey resulting in a map series at the scale of 1:50,000. Other countries in eastern and western Europe developed similar maps each with a slightly different emphasis and varying in cartographic elaboration. Most of these maps are detailed at scales of 1:50,000 or larger but several medium and small scale maps soon were produced. The need for some degree of standardisation of geomorphological maps arose and matters of generalisation in case of scale reduction had to be considered. The Commission on Geomorphological Survey and Mapping of the International Geographical Union played an essential part in this process.

The increased interest among geomorphologists for the systematic survey of landforms and geomorphological processes finds its explanation in several factors, among which the more scientific attitudes and research methods rank high. Other factors are the greatly improved technology for survey and mapping brought about by the use of aerial photographs and other kinds of remotely sensed imagery such as side-looking radar (SLAR). The advent of orbiting satellites for earth observation sparked entirely new developments including digital image enhancement, computer-assisted image interpretation and monitoring of dynamic processes. Furthermore, the growing awareness of the importance of the geomorphological configuration of the earth's crust for the study and proper utilization of the environment as a whole has stimulated geomorphological survey and mapping techniques, particularly because of the urgent need for resource development and environmental management, now existing in almost every part of the world.

The specific aim of geomorphological mapping is the representation of terrain configuration, landforms being the main subject matter (Christian et al., 1956; Linton, 1951). The cartographical elaboration of the map should allow for the identification and exact description of the landforms and landform complexes and should indicate their position and arrangement as well as their genesis. The spatial relationships with adjacent forms and the temporal or chronological relationship with features of other, e.g. fossil, form generations should be emphasized. To fulfil this aim, an analytical approach was devised distinguishing between four different types of information about landforms:

1. Morphographical information
   The forms should be identified from a geomorphological point of view, for example, river terrace and erosion surface, which is not the case on a topographic map. Since the identification of forms is normally based on their mode of formation, the morphographical aspect is closely related to the morphogenesis mentioned under heading 2.
2. Morphogenetic information
   The forms should be represented in such a way that their origin and development is immediately clear. In the legend accompanying the map, descriptive indications such as 'sandy plain' should be avoided; instead, genetic descriptions ('sandy alluvial plain', 'sandy pro-glacial plain') should be used. Since form is the outcome of the combined effect of exogenous geomorphological processes of the past and present, as well as endogenous neo-tectonic processes, it should be properly mapped. Lithology and geological structure, forming the morphostructural framework of landform development, merit full attention also. These factors have a profound effect on forms and processes occurring to date.
3. Morphometrical information
   Topographical (contour) maps give much essential information in this respect therefore, the geomorphological map can best be printed on an orohydrographical base, but certain geomorphologically important data, such as height of trough shoulders or river terraces, must be added. Slope steepness is another important element contributing to an understanding of the geomorphological situation.
4. Morphochronological information
   Since every form is characterized by the period of its formation and further development, it is essential to make a distinction between forms of different ages, in particular between recent forms and those inherited from earlier periods when different climatic conditions prevailed. It is essential to keep age indications flexible since it is this part of the map that is most apt to need revision with the advance of geomorphological knowledge.

It is evident from the aforesaid that, for the proper compilation and successful application of geomorphological maps, the following factors should be considered: genesis, age sequence and morphometrical characteristics of the landforms; also, the processes forming them and those which have formed them in the past and the material in which they have been formed. To a large extent these factors (landform, morphodynamic process and surface material/parent rock) govern the inherent properties of the various parts of the terrain and thus deserve specific attention in applied geomorphological mapping. The landform aspect may relate specifically to morphometric characteristics (Richter, 1962; Ridd, 1963), such as slope steepness and ruggedness; slope processes effectuating accelerated erosion and slope stability rank high in the morphodynamic aspect; parent rock and the nature of the overburden are important especially because of their effect on hydrological and other conditions that influence the process. Depending on the purpose of the study, the accent of the geomorphological analysis may be on either of these aspects or on a combination of them. The legend required for the work varies accordingly.

Standardisation of legends for applied geomorphological maps is only feasible up to a certain degree; in the first place, survey and cartographic elaboration should be optimal for the aim of the map and satisfactory for its user. The great heterogeneity of the configuration of the earth's surface which encompasses alpine regions as well as plain lands, areas of accumulation as well as areas of denudation, and various quite distinct climatic zones, renders it difficult to achieve a reasonable degree of standardisation when geomorphological maps are made for reasons of academic research. The diversified features outnumber by far the amount of suitable distinguishable colours and handy line symbols. Any attempt at establishing a detailed unified key for universal use is therefore bound to be impractical (Bashenina et al., 1968; Demek et al., Ed., 1972). In fact complete standardisation is an absolute necessity only when a map series comprised of several sheets has to be prepared.

Evidently, standardisation with its distinct advantages should be aimed at as much as possible, though it should not result in the introduction of impractical or inconvenient cartographic solutions. The techniques of geomorphological survey and mapping and the legends applied are gradually evolving under the influence of new scientific developments, increasing technological capabilities and the requirements of the users. (Ashwell, 1964; Ganeshin, 1975; Grimm et al., 1964; Hagedorn, 1967; Journaux, 1973; Leser, 1972;

Melander, 1975; Poser and Hagedorn, 1966; Troppmair, 1975; Waters, 1958).

11.2 The Cartographic Elaboration

The basic problem of the cartographic elaboration of geomorphological maps is the great variety in types of information that one could include. Restraint in this respect is an absolute necessity; otherwise the maps produced will be complex, costly in printing and difficult for the user to read. The limited range of cartographic means of expression restricts the tendencies to undue perfectionism, although great care is always required to not overload the maps. Overloading has other disadvantages: it tends to turn users away from geomorphological maps and reduces their appreciation. The application of geomorphology may easily be adversely affected. Too high printing cost reduces the divulgation of geomorphological maps and has a comparable, adverse effect. Simplicity by emphasizing the most essential information and generalizing or omitting the less important information should be a guiding principle. Part of the information required may be included in the description of the geomorphological phenomena given in the legend and not in the map itself. Another possibility is to produce several types of geomorphological maps, the content of each of them depending on the purpose. If all information is considered indispensable, the data may be divided in an appropriate way over two maps covering the same area. Optimum results can only be obtained by a geomorphologist setting his priorities and cooperating with a skilled cartographer who knows how to get the most out of the available cartographic means. Proper selection of drainage lines is an additional means of clarifying the geomorphological situation. Fig. 11.1 gives an example from Central Java where the drainage network clearly reflects the occurrence of two major volcanic bodies and an explosion crater.

The means of cartographic expression can be summarized as follows:

1. coloured area symbols
2. degree of saturation of these colours
3. screens (hachures, cross hachures, etc.)
4. line symbols
5. indices
6. numbering/lettering systems

The expressiveness of these six types of symbols decreases in the order of their listing. The coloured area symbols are by far the most expressive and should preferably be used for those aspects of the geomorphological situation which are considered the most essential. They represent the highest level in the hierarchic classification of landforms defined in the legend. The sub-division of these coloured areas using different degrees of saturation can be used for the mapping of the second-highest level of hierarchic classification but this means can be used only if adequate printing facilities and funds are available. Screens, hachures, etc., represent the following lower hierarchic level. These first three means of cartographic expression are suitable for indicating phenomena of a mappable areal extent and can be used for major landforms or landform complexes, further referred to as geomorphological units. The symbols listed under headings 4 to 6 are used for individual geomorphological features of a distinctly smaller magnitude. Line symbols are applicable to minor forms of mappable extent (e.g. landslides) ranking low in the hierarchy of landforms. Their discontinuous occurrence would adversely affect the legibiltiy of the map if they were represented by coloured area symbols. Line symbols may also be used for mapping processes. Indices fall in an even lower level in the hierarchy and are often used for features which are not extensive enough for the use of area symbols. The indices may result in an oversized representation of a feature or may be used for the simplified representation for a group of features of the same nature. Letters and numbers may serve the same purpose but should be used with care since otherwise overcrowding and poor legibility of the map may result. They may also be used for indicating the code numbering of the phenomena mapped and listed in the legend and may assist in discriminating between the colours and saturations of area symbols, which are difficult to separate visually. Alternatively, they may replace (part of) the colours or saturations when simpler and less expensive printing procedures are applied.

The morphographical, morphogenetic, morphochronological and morphometrical information to be mapped and discussed in sub-section 11.1 can be grouped for cartographic purposes under the following headings:

1. Morphographical/morphogenetic information
2. Morphochronological information
3. Lithological data of bedrock and unconsolidated materials
4. Morphostructural data
5. Slope categories (steepness, length)
6. Other morphometrical data (relief amplitude, altimetric situation)

In a taxonomic classification of landforms, each of these six categories might be selected to represent the highest hierarchy of the legend for which the coloured area symbols (the most expressive means of cartographic expression) should be applied. It is now agreed by most geomorphologists that the morphographical/morphogenetic information merits recognition as the highest hierarchical level since this is the most essential aspect of geo-

Fig. 11.1. Vertical aerial photograph (1:40,000) of a volcanic terrain in Central Java showing the importance of the drainage network for visualizing the morphological situation: the rivers running diagonally across the picture mark the limit between two major volcanic bodies in the lower left and the upper right. The Menjer Lake, situated in an explosion crater is another noteworthy feature.

morphology. However, in special cases one may deviate from this rule. If,for example,a geomorphological map is made for the specific purpose of agricultural development, the importance of slope steepness may be so overwhelming that this factor is given priority and thus is indicated by coloured area symbols. Lithological information may be allotted the highest hierarchic level in certain engineering geomorphological studies. The earlier existance or simultaneous preparation of other,geological or soil maps of the surveyed area at the same scale may affect the legend ultimately selected for the map.

The categories 2 to 6 are not normally selected for the highest level of the classification and the legend because they are of less dominant geomorphological significance and are less easily established. Chronology, for example, is insufficiently known in many areas and not easily established. Using coloured area symbols for lithology has the disadvantage that it is excluded in areas of accumulation which may encompass entire countries (Bangladesh, The Netherlands, etc.). It is considered preferable,therefore, to include essential lithological data in the morphographical/morphogenetic information which makes it possible to apply coloured area symbols.

The distinction between areas of degradation and areas of accumulation has been used as a major distinction, indicated by differences in coloured area symbol. Although this distinction is indeed important, this cartographic solution has the disadvantage that morphogenetically very different low land areas are put under the same heading. Another complication is that in areas of degradation an overburden may occur which is thick and important enough to merit separate cartographic treatment. An additional map may even be required to give full details of the nature of the unconsolidated materials. Morphostructures can be partly incorporated into the morphological/morphogenetic information of the highest hierarchy and can also be indicated by line symbols. The first solution is most practical in case of small scale mapping whereas the second is usually more practical for detailed geomorphological maps.

11.3 Diversified Approaches

A variety of attempts to match the cartographic means with the geomorphological needs has been tried in the nineteen-fifties and nineteen-sixties as is demonstrated by the examples given below.

Most modern analytical geomorphological maps justly put ample emphasis on the morphological/morphogenetic aspects of geomorphology. This is already the case with the geomorphological map of the Sern Valley, Switzerland, produced by Helbing (1952). However, chronological indications fail on this map, which renders it difficult to grasp the interrelation between the various phenomena and the geomorphological evolution of the area surveyed.

In the geomorphological maps of the hills and mountains of southern Poland prepared by Klimaszewski (1956) and his school in Krakow at the scales of 1:25,000 and 1:50,000, colours are used in a subtle way to indicate genesis and chronology combined. Orange, for example, is used for forms resulting from Pleistocene fluvial erosion and denudation. One may also contemplate the use of various degrees of colour saturation to express chronology (Klimaszweski, 1968; Demek, 1976). The advantage of this method is that genesis and chronology are equally stressed, although lithology and structure are not emphasized (Klimaszewski, 1978). Since structure becomes increasingly important for geomorphological mapping with decreasing scale, substantial changes of the legend will be required in case of medium- and small scale mapping. Galon (1962) used the same principle for the mapping of the lowlands of northern Poland. Due to the clear-cut sequence of Pleistocene glacial and fluvio-glacial forms, and Late Glacial and Holocene aeolian, fluvial and denudational forms, maps of a remarkable simplicity and expressiveness could be produced. Fig. 11.2 gives an example.

In the DDR geomorphological maps have been compiled on which the morphogenetical units and forms are coloured according to chronology (Gellert et al., 1960; Brunner and Franz, 1962). These maps concur well with geological maps. Almost one and a half centuries ago, geologists had to take decisions about the contents of geological maps; lithology was considered to be of secondary importance, at least in the case of sedimentary rocks, and it was decided to use the coloured area symbols to indicate the age of the geological formations. This certainly was a sound decision for geological maps and the choice was

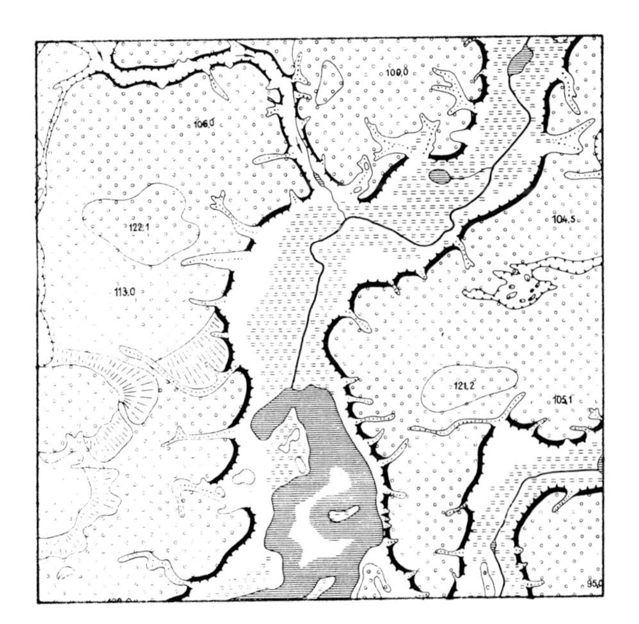

Fig. 11.2. Simple, detailed geomorphological map of an outwash landscape of the Poznan Stage of the Weichsel glaciation in NW Poland. Scale 1:50,000. (Galon, 1954).

also justified in the lowlands of the DDR where the age of most of the forms is well-established. It is usually difficult to establish the age of forms and in many yet insufficiently studied areas, age remains obscure. Therefore, it does not seem to be good policy to reserve the coloured area symbols for chronology which is certainly not the most important aspect of geomorphology. In the FRG, detailed geomorphological mapping has recently been furthered by the German Research Council and a number of map sheets have been produced according to a legend devised by the working group for the geomorphological map of the FRG (Leser, 1974).

At the Centre for Applied Geography (CGA) in Strasbourg, France, great attention was paid to lithology (for which the coloured symbols were used) and to the successive landform generations (Bourdiec et al., 1963). Tricart (1955, 1959, 1972) produced many maps of this type, placing the accent on granulometry and either chemical nature or mechanical characteristics of the basement rocks, indicating chronology by the colour of the line symbols. A slightly different approach was adhered to by others, e.g. Joly (1963), in Paris. Cooperative research on geomorphological mapping was recently organized and sponsored by the Centre National de la Recherche Scientifique (CNRS). A survey of France at 1:50,000 with sample areas at 1:20,000 or 1:25,000 was undertaken (Tricart, 1970). The legend depicts the surficial deposits with their thickness and granulometry. The colours are used for the different morphogenetic systems differing in hue for the successive generations of landforms. For the taxonomical classification of the landforms, a decimal system was developed (CNRS, 1970).

In the USSR the legend is also essentially morphogenetic and the coloured area symbols are used for major landform types and complexes, called mesoforms. Emphasis is also on chronology, which is indicated by the density of the colours, though lithology is rather neglected. Morphometric indications are scarce and morphostructural matters are particularly emphasized in cases of medium and small scale mapping. Where surficial deposits exceeding 10 metres in thickness occur, a map of the Quaternary deposits is then made, in addition to the geomorphological map. The distinction made between major and minor form groups is useful if the legend is to be applied for mapping at various scales (Bashenina, 1960; 1964; 1972; 1974). In Hungary, Pecsi (1962, 1964) used the coloured area symbols for the major morphogenetic landform types but he also stressed lithological influences. Processes are mapped by hachuring and chronology is indicated by ciphers.

In a number of modern Belgian analytical geomorphological maps, initially prepared under the auspices of the National Centre for Geomorphological Research, morphometric data rank high next to the morphogenetic information given. A detailed classification of slopes based on gradient and also breaks of slopes, valley forms, etc. is mapped in detail (Macar et al., 1960/61; Fourneau, 1963). Savigear (1965) has emphasised morphometric aspects in geomorphological mapping in the U.K.

The legend of the geomorphological map of the Netherlands at the scale of 1:50,000 (Maarleveld et al., 1974) is structured in such a way that the eight relief classes defined by slope gradient and length are indicated by coloured area symbols. They are subdivided into 18 relief types according to the relief amplitude. The genetic terrain forms and the past and present processes which cause them are listed in the legend where they can be determined on the basis of a hachuring system.

A lettering system is used for the coding of geomorphological terrain types. The map, therefore, has a strong morphometrical/morphographical trend in its cartographic representation though the genesis and the processes are indicated by code numbers in the map and by hachures in the legend. The map is complementary to the soils map and the geological map of the Netherlands at the same scale, which justifies the structure mentioned above of the legend which was used.

The mapping system devised at the International Institute for Aerial Survey and Earth Sciences (ITC) in Enschede, the Netherlands (Verstappen, 1970; Verstappen and Van Zuidam, 1975),and designed for both purely scientific and applied mapping in different scales in a variety of geomorphological situations, will be dealt with in a separate sub-section (11.6).

A comparison of the various mapping systems and legends has been attempted by some geomorphologists (Gilewski, 1967; Van Dorsser and Salome, 1973, 1974). Considerable efforts have been made to unify the legends used in various countries or at least to make them more readily comparable. From 1960 onward, the sub-commission on Geomorphological Mapping of the IGU Commission on Applied Geomorphology (since 1968 the IGU Commission of Geomorphological Survey and Mapping) has been particularly stimulating and coordinating in the field of geomorphological mapping. (Gellert, 1964). It even made an attempt at compiling a unified key for worldwide use (Gellert, 1971; Gellert and Scholz, 1964, 1974). Although a universal key elaborated to the smallest detail is unlikely to be practical, the fact that general principles have been agreed upon is most valuable: coloured area symbols, for example, are now generally used for large morphogenetic (groups of) landforms (St. Onge, 1968).

11.4 Surveying and Mapping of Major Geomorphological Units and Minor Forms

The representation of geomorphological phenomena varies inevitably with their areal extent and with the hierarchical importance attached to them. Important phenomena of mappable areal extent become geomorphological (mapping) units and depending on their hierarchy, may be indicated by coloured area symbols, hues, various degrees of saturation or by hachures of different types, as explained in sub-section 11.2.

Phenomena which are mappable but represent a lower hierarchic level may be indicated by true-to-scale line symbols within a geomorphological unit. An example is the presentation of geomorphological features such as a sizeable landslide by a line symbol within a denudational slope, considered a geomorphological unit and indicated by a coloured area symbol.

Smaller geomorphological phenomena can be indicated by oversized symbols only, unless they can be conveniently grouped and then represented by group symbols, (e.g., sinkhole area). Among other things, line symbols may be used for adding morphometrical information, e.g. breaks of slope, or geomorphological processes (see sub-section 11.5). Indices and lettering systems may be applied, e.g., for plotting chronological information. Line symbols, indices and lettering may have various colours, provided that sufficient funds and printing facilities are available, though in many cases it is a wise policy to use only one colour, preferably black, in order to simplify the map reproduction. One may consider, however,whether it is useful to highlight certain, e.g., active, processes by using a different colour, e.g., red.

Since morphogenetic classifications have proved their validity and are now commonly used, one may apply the following classification at the highest hierarchic level, preferably using the indicated colours for the coloured area symbols:

1. Forms of structural origin	purple
2. Forms of volcanic origin	red
3. Forms of denudational origin	brown
4. Forms of fluvial origin	dark-blue
5. Forms of lacustrine/marine origin	green
6. Forms of glacial/peri-glacial origin	light-blue
7. Forms of aeolian origin	yellow
8. Forms of solutional (karst) origin	orange
9. Forms of biological origin	black
10. Forms of anthropogenic origin	grey

Although this broad classification, also used in the ITC System of Geomorphological Survey (sub-section 11.7) provides a sound basis for arriving at a morphogenetical break-down of the land, in actual practice many problems are still left unsolved and therefore, much is left to the discretion of the geomorphologist, the author of the map. For example, it may not be easy to decide whether a certain faulted area should be incorporated into the forms of structural origin or whether it is more justifiable to map it as a form of denudational origin with a number of additional line symbols for the various faults and major fractures. Likewise, it may be questionable to classify a strongly eroded volcanic area under the forms of volcanic origin. The destruction of the volcanic body(namely) may have progressed to such an extent that the characteristic volcanic forms are completely obliterated, in which case it would be more appropriate to map the area as a form or unit of denudational origin, possibly adding: 'carved in volcanic rocks'.

Difficulties arise in areas of complex morphogenetic development. In areas where an old planation surface has been block-faulted or otherwise deformed and subsequently covered by aeolian, glacial, volcanic or other deposits of considerable thickness (10 metres or more), one may wish to optimally indicate the information on the morphostructural characteristics of the substratum, the nature of the major landforms and also the type and thickness of the covering surficial materials. This example explains why in the morphological maps of some countries emphasis is on morphostructures, especially in case of small scale mapping, notwithstanding the unavoidable greater complexity of the legend and the risk of overloading the map to the detriment of its legibility. The example also shows that it may be good practice, for scientific as well as for certain applied maps, to produce a separate map for surficial deposits having a thickness of more than, e.g., 10 metres.

Lithology is often considered the second highest hierarchy and it may, for example, be indicated by hachures or by various screens in a subdued colour so as to render it visible without emphasising it unduly. However, in certain cases one may prefer to raise the lithological information hierarchically and to include it in the breakdown of the land into major geomorphological units. Coloured area symbols can then be used for indicating lithology. This approach is particulary useful where lithological conditions are of special geomorphological importance, e.g., if they give rise to specific landforms or processes. Where very extensive morphogenetic units occur, e.g., of denudational origin, their subdivision according to lithology may add to the quality of the map.

In such cases, it may be preferable to arrive at a more detailed breakdown of the land into morphogenetic units on the strength of relief characteristics. This may be expressed verbally, using terms such as 'low-rolling denudational hills' and 'intensely dissected, steep-sided, high mountains', or it may be based on certain well-defined morphometric parameters. Specifically, the height of the terrain above sea level and slope angle have been used by some as criteria for these purposes; different saturations of the colours used for the geomorphological units concerned have been applied to this aim. The height above sea level, however, is of relatively minor geomorphological consequence. Sub-division of geomorphological units of various kinds according to this principle, although convenient in areas where contour maps exist, serves no real purpose except that of the plasticity of the map and a better visualisation of the terrain configuration, particularly on small scale maps. One disadvantage is that the extra lines and coloured areas may lead to overcrowding of the map. Slope angle appears to be a geomorphologically more relevant criterion if sub-division of geomorphological units is desired. It is best suited in cases of detailed mapping whereas much generalisation is required on small scale maps on which only average slope angle groups can be given.

In addition to the possibility of sub-dividing geomorphological units,it may also be considered appropriate to group them and to indicate where structural elements are of prime importance, where the lithological conditions play a leading part in the terrain configuration and where the occurrence of active processes is a major element. The ultimate aim should always be to give a clear and concise picture of the geomorphological situation, emphasising the data which are considered essential to the map and omitting or suppressing the information which is irrelevant in this context. The legend should optimize the cartographic expression of the relevant geomorphological information. A standard legend is only essential when a coherent map series has to be produced. It should never become a harness resulting in cumbersome maps.

Fig. 11.3 gives an example of a simple break-down of the the terrain into geomorphological units in an area of southern Italy. Major geomorphological units of fluvial, marine and denudational origin are depicted and represent the mesoforms occurring in the area. Minor forms are added by applying black line symbols. Processes and chronological data have been omitted so that the attention of the reader will not be diverted.

11.5 Surveying and Cartographic Presentation of Geomorphological Processes

The qualitative and quantitative study of geomorphological processes, both of the present and of the past, is of considerable significance in surveys for development. It is therefore important that proper attention is paid to all types of morphodynamic phenomena in geomorpho-

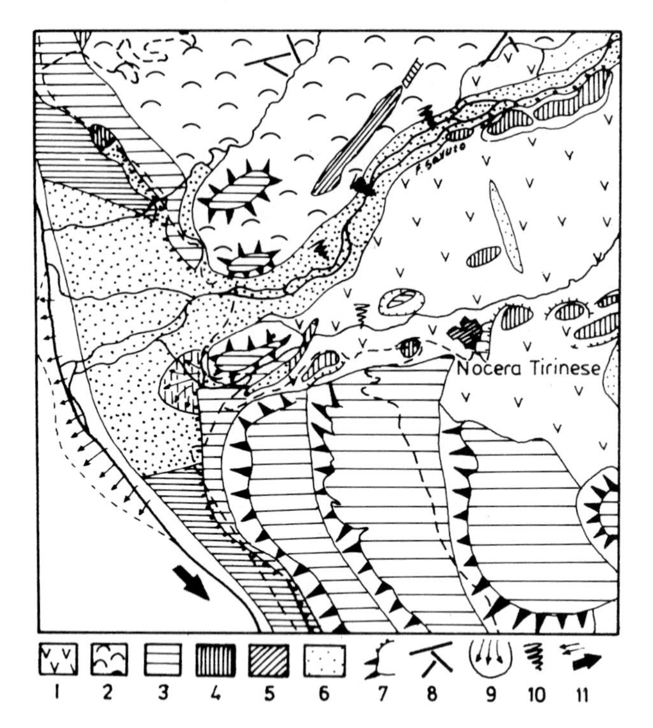

Fig. 11.3 Black and white geomorphological map of the lower Savuto Valley and adjacent areas of Calabria, Italy at a scale of 1:100,000.
Key:
1. denudational mountains in soft metamorphic/igneous rocks
2. denudational hills in unconsolidated Quaternary sandstones/conglomerates
3. marine terraces
4. fluvial terraces
5. erosion glacis
6. floodplain and delta
7. major and minor scarps
8. faults
9. fan
10. sheet erosion
11. coastal accretion and beach drifing

logical surveying and mapping for such purposes. The relevance of gathering data on geomorphological processes is evident considering that the exogenous processes.

1. govern the evolution of the landforms and are essential to arrive at a full understanding of the geomorphological situation and optimization of the contribution of geomorphology to applied research
2. influence matters such as the soil profile development and and hydrological conditions and therefore accentuate the position of geomorphology in the context of natural environment,
3. have a pronounced effect in everyday life and provoke marked changes within the span of the human time scale.

It is therefore rather surprising that the representation of morphological processes and of morphodynamics as a whole is rather neglected in many geomorphological maps.

The best possibilities for mapping processes exist in detailed geomorphological mapping up to the scale of 1:50,000. It should be understood, however, that very often it is the forms created by the processes rather than the processes themselves which are mapped. Major and medium-sized phenomena such as large ravines and land slides can then be accurately mapped in their proper size and form using various line symbols. Care should be taken to select these symbols to match the actual configuration of these phenomena, observed on vertical aerial photographs. Minor linear features, such as rills, and diffuse processes, such as creep and surface wash, will have to be indicated by using generalised symbols, even in case of detailed mapping, because of their limited or diffuse areal extent. It has been explained in sub-section 10.2 that the emphasis on medium and small scale geomorphological maps is primarily on major landforms, landform complexes, and structural influences. The processes in these scales are usually too limited and the forms created by them too small to be individually mapped true to scale. They can at best be generally mapped or mentioned in the legend in the description of the geomorphological units. In small scale mapping, even this may not be feasible and the processes should be properly described in the explanatory notes. Only in case of special purpose maps, e.g., those related to accelerated erosion, processes may become a matter of higher order in the hierarchic classification of the land.

The following examples may serve to illustrate the geomorphological processes and the forms caused by them as observed in the field and subsequently mapped either exact of generalized. The photograph of fig. 11.4 shows a wheat field in Calabria, south Italy, which is severely affected by rill erosion. The slope is under agricultural use due to the scarcity of level land. The fields are barren in the rainy winter season of the Mediterranean and are exposed to plowing in the slope direction in order to avoid overturning the heavy machinery. After plowing the slope will be largely smoothed and only the main rills may persist. Gullies or ravines may develop at the edge of the fields where the farmer does not plow but turns to and from the field with his machinery (fig. 11.5). The degree of erosion may be easily underestimated in the growing season unless the soil profile is carefully studied. Through the centuries, these processes have resulted in the development of extensive badlands, locally known as Calanchi.

Another process also occurring in the wheat fields of the area is shallow landsliding which only affects the plowed topsoil. Plowing has aerated the topsoil and has facilitated its saturation with rain water. This layer may easily slide downslope on top of the less pervious and more compact soil which remained untouched when tilling the land. The barren patches formed (fig. 11.6) may then be further degraded by rill formation. These patches are almost invisible

Fig. 11.4 Wheat field in Calabria, south Italy, severely affected by rill erosion.

the cracks in the soil, forming tunnels; the roof of these tunnels finally collapses and creates headward extension of the ravine. Fig. 11.8 shows the phenomenon which poses serious problems in conservation projects as the protected head of a ravine is often quite unexpectedly by-passed. Conservation bunds and terraces may become ineffective by this suffosion or piping process.

Mass movements are also very diversified and the precise mapping of their type, intensity, location and areal extent is of equal importance as that of linear and sheet erosion. The occurrence of creep can be easily established in the field on the strength of features such as the overhanging, humus-rich root zones over road cuts or natural escarpments (fig. 11.9) and the classical downslope bending of trees (fig. 11.10).

Fig. 11.5 Field boundary gully between two wheat fields in Calabria, south Italy. Rill erosion can be noted in the fields.

the next season when the same phenomenon may affect other parts of the cultivated slope, rendering its topography slightly more irregular than previously was the case. The process gradually decreases the yield of the fields and the affected parts of the slope clearly show up in the growing season; the crop turns yellow and fails because of the low water retention capacity of the B-horizon forming the soil surface. The photo of fig. 11.7 shows an example of this development in a wheat field (with olives) on a partially affected slope. The wheat grows satisfactorily only where the A-horizon is still present.

The headward erosion of ravines in silty materials in which cracks develop in dry periods often precedes a process referred to as suffosion. The surface run-off enters

Fig. 11.6 Fresh shallow slide of the plow-layer in a wheat field in Calabria, south Italy.

Fig. 11.7 The effect of sheet erosion on wheat growth in Calabria, south Italy. The light toned patches mark areas devoid of the A-horizon where the wheat does not survive the hot summer.

Fig. 11.8 Example of piping in silty clays, Calabria, south Italy.

Fig. 11.9 Overhang of grass and topsoil over a road cut due to soil creep.

In areas where surface wash occurs and results in diffuse erosion, the terrain at the upslope side of the bent tree is distinctly higher than at the downslope side, while the slope is uninterrupted at either side of the tree. Shallow slides and slumps give rise to irregularities in slope profile, tilted (telephone) poles and trees as illustrated by fig. 11.11. Major slides are usually deeper seated and affect a larger surface area. Fig. 11.12 gives an example from Calabria, south Italy, with an attempt at reconstruction of a road situated mid-slope and wiped out by the slide.

Fig. 11.10 Downslope bend in tree trunks: another indication of soil creep.

Fig. 11.11 Slope in Basilicata, Italy, affected by shallow sliding as evidenced by irregular topography and tilted telephone poles and trees.

Fig. 11.12 Major slide in Calabria that partly destroyed the village at the top and wiped out a secondary road which is being reconstructed.

Fig. 11.13 shows a major mudflow, the size of which permits it being mapped true-to-scale on detailed (1:25,000 to 1:50,000) maps. The upper portion shows a number of cracks in the break-off zone from where the material drops on (and slides down with) the tongue-shaped mudflow. The distribution of these phenomena is governed by factors such as composition of the affected material and of the underlying beds, dip of the bedding planes and the hydrological situation. Only careful study of the situation and of the causative factors can give a clue concerning the stabilization of such slopes. The same applies to rock falls which may be induced by lateral sapping of a river, such as is the case with the rockfall pictured on fig. 11.14 along the Ebro River. For active endogenous processes, such as recent faulting or generalized subsidence, geomorphological evidence can often be found. Fig. 3.3 shows a recent faultscarp in a young Pleistocene gravel fan in the Magdalena Valley, Colombia. The mapping of such features is an important issue in earthquake hazard mitigation studies. The role of geomorphology in such neotectonic studies has been discussed in section 3.

Fig. 11.13 Tongue of a major mudflow in the Basilicata, Italy. Mappable detail depends on mapping scale.

11.6 Mapping at Different Scales and the Problems of Generalization

Apart from the early small scale geomorphological maps which were not based on adequate field observations but had a sketchy, pictorial or structural character (Raisz, 1941, 1956), geomorphological maps were initially essentially detailed analytical maps. Recently, medium and small scale analytical geomorphological maps also received attention. It soon became clear that the contents of geomorphological maps varies to some extent with the mapping scale. Several authors have elaborated on this matter (Demek, 1967; Kugler, 1965, 1968, 1974; Scholz, 1973, 1974) which is also touched upon in the 'Manual of medium scale geomorphological mapping' (Demek and Embleton, [Eds.] 1978).

It has been attempted to classify geomorphological maps by scale into geomorphological plans, basic geomorphological maps, detailed geomorphological maps, medium scale geomorphological maps and small scale maps. In addition, maps of entire countries, continents and of the world have been separated into categories. This author considers such a classification irrelevant; even if these types actually exist, the transition of one type into another is so gradual that it is not possible to indicate exactly where the scale limits of the so-called 'types' should be drawn. The terms 'plans' and 'basic maps' are also confusing. It is clear, however, that on very detailed maps (e.g., 1:10,000) miniature landforms and landform elements must be shown. Since these are usually of a rather uniform morphogenesis, differentiation within the map sheet will be mainly based on sedimentological and lithological criteria. Morphometric and morphodynamic information can also be indicated. In detailed maps (e.g.,1:25.000 - 1:50,000), the morphogenetic diversity is a key element in the classification of landforms. Information on landforms is further specified by data on morphodynamics, morphometry, morphochronology and lithology. The same applies to medium scale geomorphological maps (e.g.,1:100,000 to 1:250,000) although major landforms and landform complexes or associations are gradually emphasized, whereas morphodynamics, and to a certain extent also morphometry, is generally less convenient to map (Marchesini and Pistolesi, 1964). On small scale maps (1:500,000 and smaller) large groups of landforms and morphostructures become dominant features (Neugebauer, 1971).

It is thouht by some that lithology becomes increasingly important with decreasing scale however, this depends on the climate. Joly (1963) devised a legend for geomorphological mapping in arid and semi-arid areas at a scale of 1:1,000,000. He successfully used coloured symbols for lithology and put much emphasis on structural elements. A genetic classification of landforms is used and minor differences in hues discriminate between Pleistocene and Holocene features. In the humid tropics lithology is usually less dominant as a factor in landform development due to the strong chemical weathering. When preparing a 1:2,000,000 geomorphological map of Sumatra, Indonesia, Verstappen (1973) used coloured area symbols for major landform associations, distinguishing between forms of

Fig. 11.14 Rock fall caused by lateral sapping of the Ebro River, west of Zaragoza, Spain.

depositional, structural and volcanic origin. Lithology was indicated by symbols and chronology by indices. In the USSR the breakdown of the land into units in small scale mapping is based mainly on the morphostructural position and morphogenesis. Auxiliary data on matters such as the extent of Pleistocene glaciations, the limit of permafrost and marine regressions may be given (Rostovceva, 1960).

The mode of compilation of geomorphological maps is probably the most reliable approach to the assessment of their degree of detail and validity. Detailed maps (up to 1:50,000 or 1:100,000) are fully field-checked and presently compiled using aerial photographs. Maps of the same scale that are field checked in all essential areas, but in the compilation of which some extrapolation has been applied, fall under the heading of semi-detailed geomorphological maps. Medium scale maps are often made by production from detailed maps; generalisation and also extrapolation may have occurred in their compilation. Small scale maps are usually field checked in key areas only; some traverses are usually made in addition. Orbital imagery and SLAR imagery have brought direct compilation of medium and small scale maps within reach and the lengthy traditional procedure of scale reduction and generalisation of detailed maps can now often be avoided (Verstappen, 1977). The matter will be discussed in subsection 11.7. Field checking in key areas and traverses are minimum requirements of ground truth gathering in this case. In a later stage of scientific research, such maps may be superseded by 'second generation' maps at the same scale compiled in the traditional way whereby the whole area is mapped in detail in the field with the use of aerial photographs and other kinds of remote sensing imagery.

A particularly interesting paper on the problems concerning the generalisation of geomorphological maps is that of Boyer (1981). She clearly states that it is not only a matter of generalisation of the cartographical and geometrical aspects of the map but that a conceptual generalisation is also involved. The latter is mainly the task of the geomorphologist carrying out the survey and the former is the competency of the cartographer. Both map and legend are affected by these two types of generalisation.

Conceptual generalisation includes the selection of features to be mapped and their classification according to importance. Features or units may be combined or emphasized in order to visualize the concept. The second phase of (graphical and geometric) generalization is effectuated during the cartographic elaboration of the concept. It includes matters such as exaggeration, simplification, combination or even displacement and omission of minor features. The aim of the exercise is to optimize legibility and plasticity of the map. Boyer deals with the matter in a systematic way, discussing all map elements: geomorphological units and individual features, lithology, chronology and topography and also the map legend. Because of the limited acuity of the human eye and of perception limitations and also because of limits of cartographic techniques and printing capabilities, the minimum size of coloured area symbols when viewing the map at approximately 30 cm distance should not be less than 2,5 mm^2 and their separation not less than 0,1 mm. Line symbols should not be less than 0,3 mm thick. Their separation and length also should be within acceptable limits. Fig. 11.15 gives an example of generalization of geomorphological units and fig. 11.16 of generalization of geomorphological line symbols (Boyer, 1981).

11.7 Applications of Geomorphological Maps and the ITC System of Geomorphological Survey

In many countries, analytical geomorphological survey and mapping are considered instrumental in strengthening pure and systematic geomorphological research and in representing the results in the form of a map. It has been understood, however, from the early beginning of geomorphological survey that the wealth of thematic information contained in analytical geomorphological maps is of considerable interest for a variety of purposes. Marchesini et al. (1963) and Tricart (1969), for example, have stressed the role of these maps in Third World development. But several other fields of application have also been mentioned, such as land evaluation (Crofts, 1974), land utilisation (Le Coz, 1972, Kugler, 1976), erosion and conservation studies (Rao, 1975; Tricart, 1972), geology/mining (Rubina, 1974; Tricart, 1972) and engineering (Kugler, 1963;

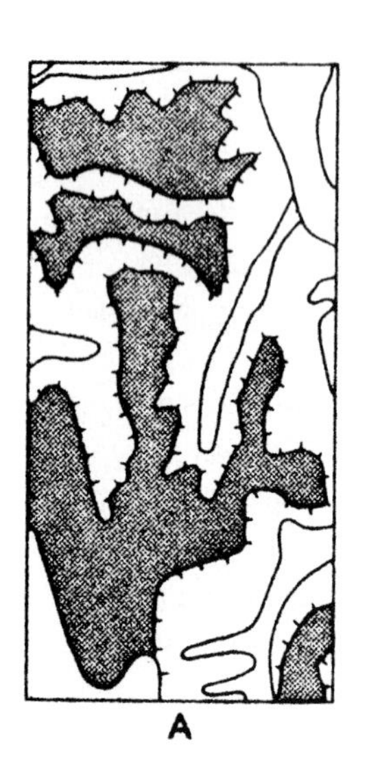

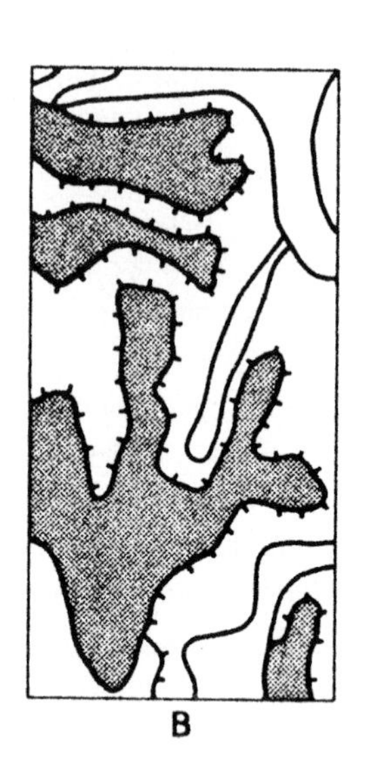

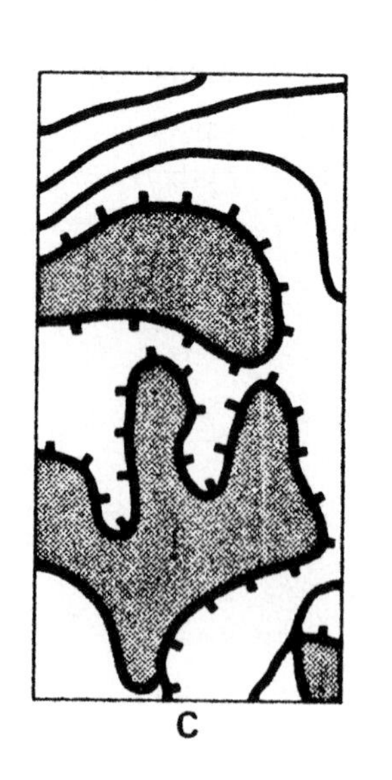

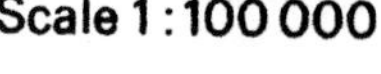

Fig. 11.15 Generalization of geomorphological units in an area in northern Spain, after Boyer (1981), by combination of areas and simplification of boundaries. A: scale 1:50,000, B: scale 1:100,000 (enlarged 2x), C: scale 1:200,000 (enlarged 4x). To the right: the same map fragments but in the true scale.

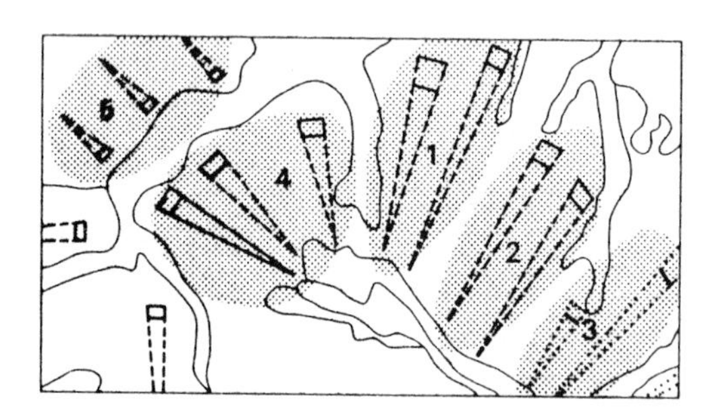

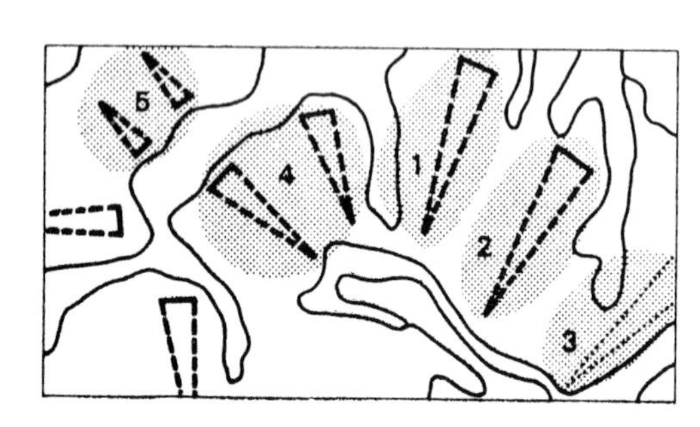

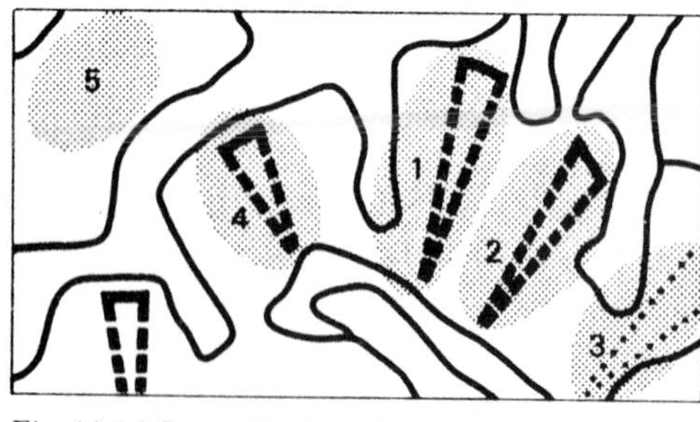

Fig. 11.16 Generalisation of geomorphological details (glacis line symbols) in an area in northern Spain, after Boyer (1981). A: scale 1:50,000; B: scale 1:100,000 (enlarged 2x); C: scale 1:200,000 (enlarged 4x).

Garland, 1975; Spagna, 1979). More general papers on the applicability of geomorphological maps are from Antonin (1971), Fourneau (1973), Guldali (1973), Kop, (1972), Kugler (1975), and Mazur and Mazurova (1965).

The usual situation is that analytical geomorphological maps contain so much diversified information that they are not easily utilized by non-geomorphologists. Therefore, a 'derived' map has to be presented to the user (engineer, agriculturist or planner) on which only the relevant information for a specific application is emphasized In some instances geomorphological information which is not normally incorporated in analytical geomorphological maps has to be added. This applies to slope gradient and other morphometric information which does not rank high in the pure geomorphological hierarchy but which is essential for some applications such as in the field of agriculture. The map contents of 'bonity maps', as they are sometimes called, varies with the aim of the survey, and the scope for standardisation of legends is thus limited (Scholz, 1974; Spiridonov, 1952, 1975).

Applied geomorphological maps have always been emphasized at the International Institute for Aerial Survey and Earth Sciences (ITC), The Netherlands. When it became clear that most of the existing systems were not quite optimal in the sense of being universably applicable and were often cartographically too complex to meet the requirements of speedy production inherent to most development projects, methods for surveying and mapping were devised not only related to pure geomorphological maps maps but also covering a variety of map types, tailor made for specific applications. A true mapping system gradually developed which is known as the 'ITC System of Geomorphological Survey' (Verstappen, 1970; Verstappen and Van Zuidam, 1975).

Several ideas and cartographical solutions of pre-existing geomorphological legends were incorporated in the ITC System of Geomorphological Survey, and it became an original mapping system. Coloured area symbols are used for the major morphographic/morphogenetic units. Lithology is given in grey zipatones (screens) on the same plate as the topography; this is done so that the lithological factor is not overstressed. Whenever lithology is of particular importance, it should be raised hierarchically and entered under the heading of morphographic/morphogenetic units. Minor forms and processes are indicated by black line symbols. However, it can be contemplated to use red symbols for active processes to differentiate between inactive, dormant and active processes. This use of colours is usually reserved for certain types of applied maps related to erosion/conservation and engineering. Chronology is indicated by a lettering system which gives maximum flexibility: the detail of the lettering system increases with the geomorphological knowledge. In insufficiently studied areas the chronology can simply be omitted. Morphometric information is limited to line symbols (e.g., for breaks of slope) and contour lines or spot heights.

Within this framework,much is left to the discretion of the map authors using the system. Since completely uniform legends are only required when a consistent, e.g., nation-wide systematic geomorphological survey is concerned, covering a smaller or larger number of map sheets, it is preferable to adapt the legend to the purpose of the survey. This is particularly clear when dynamic-geomorphological maps are made in which the geomorphological changes are emphasized which have occurred in the past or can be expected in the future. These maps are normally based on sequential aerial photographic or orbital imagery and pose specific and often difficult cartographic problems (Ashwell, 1964; Schmitt-Taverna, 1975; Verger, 1968).

The guiding principles when devising the system were the following:

1. The approach of the survey should be analytical and include morphometrical, morphographical, morphogenetic and morphochronological data as well as information on the lithology of the bedrock and the nature of unconsolidated deposits. This approach is adhered to because it is strongly felt that the genetic and chronological characteristics of landforms are of great scientific and practical value. Engineers, agriculturists and town planners have little use for knowledge of the study of the relief alone but require a proper evaluation of the morphogenetic processes that modify the landforms, affect the distribution and development of soil, have a bearing on the hydrological conditions, and even may result in natural hazards of several kinds. It is in this field that the most essential contribution of geomorphology to surveys for development occurs.
2. The analytical geomorphological surveys are complementary to synthetic surveys of terrain (see section 12). Therefore, the coloured area symbols are not reserved for lithology or chronology, but for major genetic landform units (see sub-section 11.2), since they tend to coincide with synthetic terrain units such as land systems/units. Apart from these landform units, a diversity of additional information is included in the analytical geomorphological maps.
3. The system should be flexible in its elaboration to accomodate mapping in the most diversified types of terrain formed in the various climatic zones of the globe. So, instead of giving an exhaustive list of symbols, much should be left to the discretion of the map author. Otherwise it might happen that the most suitable symbols cannot be used because they have been reserved for phenomena that do not occur in the survey area.
4. The maps produced should be as simple as possible to avoid insurmountable cartographic problems, low production rates and high printing cost. Much will depend on the art of the geomorphologist to select the essential information in such a way to present a complete picture without adding undue details that would conflict with the principles outlined above. For purposes of speedy production and inexpensive reproduction, it may be required to produce black and white maps, in which then the essential information is portrayed. The geomorphological map of the island of St. Eustatius, Netherlands Antilles, given in fig. 11.17 is an example of this. This map was prepared by the author in the framework of a land evaluation survey for purposes of rural development (Verstappen, 1977) and served as a base map for the terrain classification map of the island discussed in section 12 (Synthetic mapping of terrain).
5. The legend should be applicable for mapping at all scales. Evidently, the content of the map changes with scale. On small scale maps, symbols for small sized features have to be omitted or replaced by, e.g., group symbols. At the same time accent is placed on major landforms, landform complexes and morphostructures. The rules on conceptual and cartographical generalisation, outlined in sub-section 11.6,are applicable.
6. The system should comprise standard or 'general purpose' geomorphological maps on which full geo-

morphological information is indicated, as well as various derived or 'special purpose' maps, the contents of which are geared to specific fields of application. If both kinds of maps are to be made of the same area they should, for reasons of efficiency, be compiled using the same topographical and lithological basis.

The so-called morpho-conservation map is the first type of special purpose maps referred to above. The purpose of its compilation is to present a document on which type, distributional pattern and intensity of all erosional processes occurring are precisely mapped. Clear distinction should be made between inactive and active process, e.g.,by using red colours for the latter. All factors influencing the occurrence and intensity of these processes, such as slope form and gradient, nature of surficial materials, and cover type should also be indicated. Coloured area symbols, colour intensities or hachures may be used for slope angle classes. The image interpretation and ground truth gathering is basically process oriented. Attention should not be focussed on cultivated fields or grazing areas only but on the whole area surveyed. The intensity of processes can be quantified following detailed field observations covering a sufficiently long, representative period, but since this kind of mapping is particularly valuable in the semi-detailed to reconnaissance phases of the work, it normally suffices to give qualitative data on process intensity only, e.g., in terms of high, moderate, and low. Further, more precise evaluation is possible on the strength of an estimation of the combined effect of the various factors affecting the erosion processes involved. Data which are not adequately obtainable on the strength of image interpretation alone, should be collected in the field. The observations may include (i) the type and erodibility of soil, (ii) the frequency and amount of surface run-off and deposition to evaluate the erosivity of the water, (iii) the conservation measures taken in the area, (iv) the age and growth rate of gullies badlands and landslides. The various parts of the terrain are subsequently classified according to the dominant type and intensity of process. As an example the following classification is given:

Dominant processes:	Sub-differentiation according to relative intensity:
River erosion	
Lateral erosion of banks. Vertical erosion of beds. Deposition on floodplains.	Two or three intensity classes, if feasible.
Gully erosion	
Differentiation according to morphometry (density average dimensions, average slope steepness) state of activation and type (discontinuous, piping, gullies).	Two or three classes of rate of erosion or rate of growth of the gullies.
Sheet and rill erosion	
Differentiation of areas affected by sheet and rill erosion and areas subject to deposition from sheetwash.	Three or more intensity classes, e.g, severe, moderate and slight erosion; erosion alone is not significant.
Mass wasting	
Differentiation according to type, such as soil flows, slumps & falls.	If possible, the relative frequency of the movements or estimate of volume involved may be given.
Man-made sources	
Mining dumps, eroding drainage lines, road fills and road cuts.	Differentiation of relative amount of debris produced by erosion.

The hydro-morphological map is the second main type of special purpose map. The aim is to map the characteristics of the terrain features affecting the water resources of the area concerned and the hydrological events, such as peak discharge characteristics, occurring there. Hydrological assessment of the land, based on image interpretation and related field data gathering, is therefore the essence of hydro-morphological survey. The maps produced give inferred data rather than the measured hydrological data given in pure hydrological maps. Therefore, they are particularly useful in areas or river basins where hydrological observations are scarce, which is often the case in development studies. A preliminary assessment of the potential retention, storage and uses of surface run-off and of the exploitability of groundwater is often of considerable interest.

Hydro-morphological maps may emphasize either surface water resources or ground water resources. In the first case, attention should be given to (i) run-off intensities of interfluves based on the assessment of infiltration, retention and depression storage, (ii) estimated peak discharge of small streams based on catchment characteristics as assessed in the imagery and channel characteristics observed in the field, (iii) areas with stagnant water or those susceptible to flooding.

Hydrological maps emphasizing groundwater resources should stress: (i) assessment of groundwater recharge by infiltration, (ii) indication of shallow and deep groundwater bodies, (iii) assessment of base flow characteristics (duration, volume) of the rivers as this reflects the groundwater outflow. Obviously, lithological and relief data as well as information on cover type are of interest in hydro-morphological surveys. The precise contents of the maps may vary with the purpose but it is clear that they are a valuable tool for purposes of water conservation and for assessing the capacity of culverts in case of road construction.

Among the other types of special purpose maps, engineering geomorphological maps should be mentioned. The purpose of these maps is to give adequate data on terrain configuration, lithology of bedrock, thickness and type of unconsolidated surficial materials and on the geomorphological processes and potential morphodynamic instabilities, presented in a form suitable for use by civil engineers (Brunsden, 1975; Garland, 1975; Spagna, 1979). The data on rock type and surficial materials are relevant in combination with the processes for assessing the prevailing soil and subsoil conditions, and for the construction materials location survey which should be effectuated in the reconnaissance phase of the engineering project. Coloured area symbols may be advantageously used in indicating lithology and/or relief characteristics, using different degrees of colour saturation for discriminating between these two groups of factors. Thickness of superficial materials may be indicated by hachures or screens; however, where matters such as slope instability are of prime importance, it may be preferred to use the coloured area symbols. A good solution is to outline critical areas by thick (red) lines. The matter is elaborated upon in section 8.

The essential issue is to define the specific needs of the user(s) of the maps required, to assess the geomorphological situation accordingly and to find the most satisfactory cartographic representation for the purpose. In cases of mapping for agricultural development, the following should be stressed: (i) relief factors and processes affecting the use of land and water and thus the productivity; (ii) the factors presently limiting the land use and the potential of the land for improvement. A hierarchy of relevant factors should be established and applied accordingly.

11.8 The Use of Aerospace Imagery

The use of aerial photographs and other types of aerospace imagery such as Landsat and SLAR imagery has been advocated by several authors in the past (Nakano, 1955; Webster, 1962; Kesik, 1972; Verstappen, 1977; Cochrane and Browne, 1981). Aerial photographs have always played an essential part in the ITC system of geomorphological survey. With the advent of modern remote sensing techniques in the nineteen-sixties and seventies, visual and digital interpretation of Landsat imagery and the use of SLAR have also become common practice.

Aerial photographs are the logical starting point for compiling any detailed or semi-detailed geomorphological map. The interpretation of these images, together with the study of existing literature on the area, is the basis for the preliminary map which should be produced before going to the field. It serves to give a first idea about terrain configuration, type of geomorphological phenomena and problems likely to appear and is an aid in planning the field survey. The classification of geomorphological units and features may, in this stage, be rather descriptive as it is not uncommon that the morphogenesis of the forms becomes clear only after field investigations. Aerial photographs are also very useful in giving information on inaccessible terrain, as demonstrated by fig. 11.18 showing mega-ripples on the sea bottom several kilometres off the Dutch coast.

The geomorphological (general and/or special purpose) maps are made after the fieldwork is terminated, after the samples taken are investigated in the laboratory and a second thorough interpretation of the images has been carried out. The topographic base prepared for the preliminary map is either abstracted from existing topographic maps or obtained from aerial photographs and can also be used for the final map. The same usually applies for the lithological base if reliable geological maps of the area already exist or the lithological interpretation of the imagery has been proved correct. However, for practical purposes, it may be decided after fieldwork to apply a different grouping of lithological types.

Adequate topographical and hydrographical detail and also the plotting of principal points of all aerial photographs (with run and photo number) are essential for proper location of the field observations in the map. The difference between a detailed and semi-detailed map is only partly a matter of scale. Initially it indicates whether a map has been fully field-checked or that some extrapolation has been used to speed up the survey.

Medium- and small scale geomorphological maps should be obtained through reduction and generalization of detailed geomorphological maps of the same type, covering the whole area. In the absence of such map coverage, some authors are tempted to produce sketch maps on these scales, based on insufficient survey; however, this is not to be recommended. Because the degree of detail and precision expected from maps in these scales is less than from detailed maps, complete detailed coverage is not required for producing adequate medium and small scale maps. These should be labelled 'reconnaissance maps' to indicate that extrapolation has been used.

Formerly, in medium or small scale reconnaissance mapping, photointerpretation had to be carried out by way of photographic mosaics or, in their absence, by a quick scanning of photo pairs, strips or blocks. Reduction of the interpretation results and, if desired, of the photographic material to the mapping scale was realised before going to the field. The aerial photographs were also taken

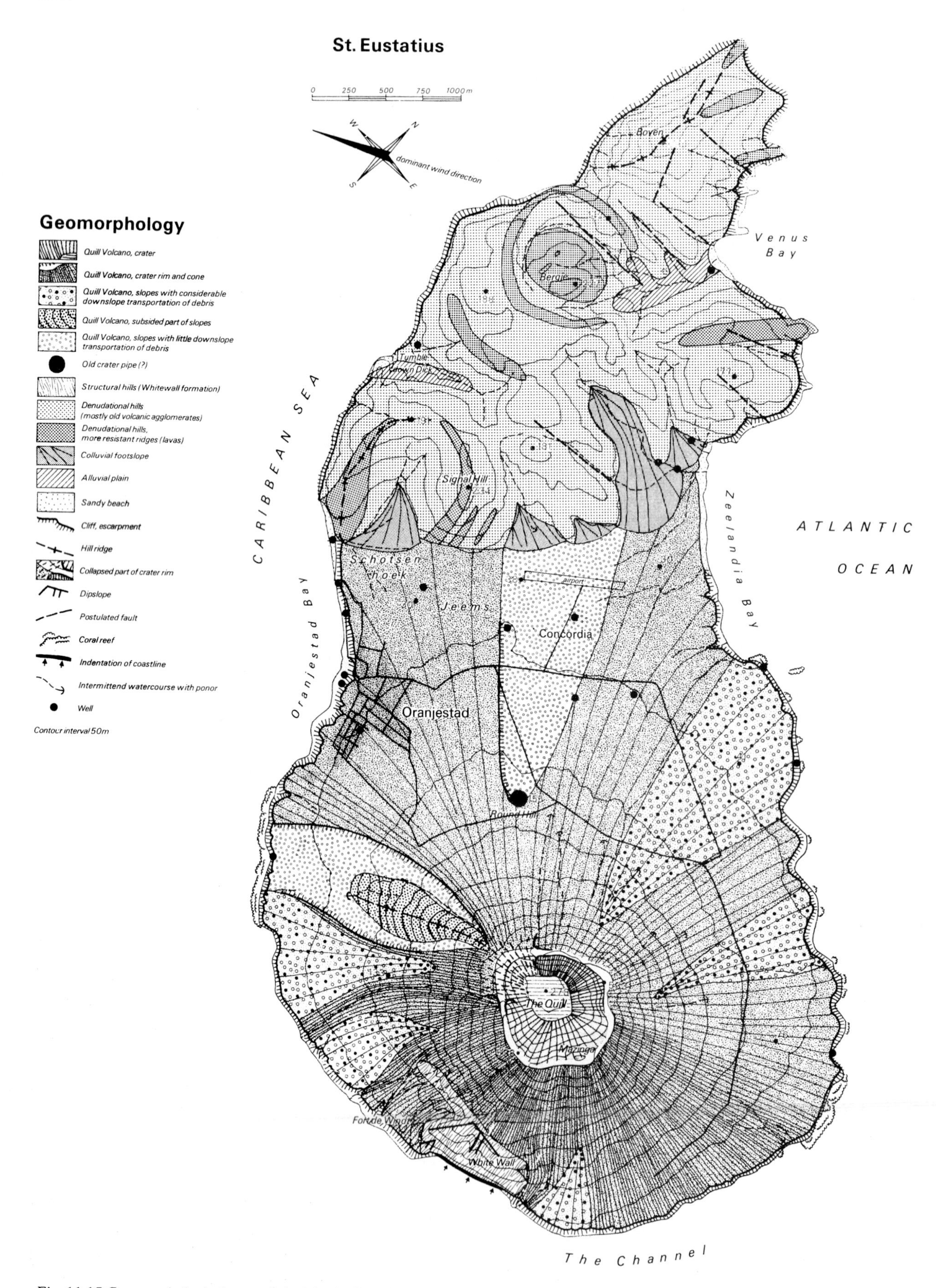

Fig. 11.17 Geomorphological map of the island of St. Eustatius, Netherlands Antilles,showing that considerable detail can be given on black and white maps.

Fig. 11.18 Vertical aerial photograph showing mega ripples on part of the sea bottom off the Dutch coast, scale 1:10,000.

to the field for quick reference. The preliminary map obtained was subject to partial field checking, whereafter a final medium scale map was produced.

Presently, reconnaissance medium and small scale geomorphological maps can be prepared directly, using Landsat and/or SLAR imagery without the traditional procedure of scale reduction and generalization.

This is an advantage as terrain configuration is conspicuous on these images and enables the geomorphologist to recognize familiar geomorphological patterns and complexes. The preliminary map produced is field checked along well-chosen traverses (in field vehicles and/or small aircraft) and by systematic site investigations in representative localities within each of the major geomorphological units mapped. A rough scanning of aerial photographs or photo-mosaics existing in the area (or part of it) may be a useful intermediate phase linking the interpretation of Landsat or SLAR imagery with the field survey. In this way it is often possible to produce small and medium scale maps simultaneously. The figs. 14.7 and 14.9 give an example of northern Botswana. Fig. 14.7 shows a Landsat image with distinct geomorphological features at the scale of 1:1,000,000 and fig. 14.8 depicts a medium scale map of the same area prepared on the basis of both Landsat and aerial photo interpretation. For many purposes this 'multiphase' approach or 'zooming-in' principle suffices in gaining an insight into the geomorphological situation and its inherent characteristics adequate for most development surveys. Detailed survey work can then be restricted to limited zones of special interest. The methodology is elaborated upon in section 14 dealing with surveying for drought and desertification.

References

Annaheim, A., 1956. Zur Frage der geomorphologischen Kartierung. Peterm. Geogr. Mitt., 80: 315–319.

Antonin, I., 1971. Applied geomorphological map of the Pisárcy Basin. Studia Geogr. Brno, 21: 33–49.

Ashwell, I.Y., 1964. Morphological mapping in western Iceland. Geogr., J., 130: 440–441.

Barsch, D. (Editor), 1979. The geomorphological approach to environment. Geojournal 3(4): 329–416.

Bashenina, N.V., 1972. Geomorphologische Kartierung des Gebirgsrelief im Massstab 1:200,000 auf Grund einer Morphostrukturanalyse. Ztschr.f.Geom., NF. 16(2): 125–128.

Bashenina, N.V., 1974. Legend of medium-scale maps as basis for geomorphological mapping. Studia Geogr. Brno, 41:105–113.

Bashenina, N.V., et al., 1960. Legendy geomorfologiceškoj karty Sovetskogo Sujuza masštab 1:50,000–1:25,000. Mosk. Gos. Univ., Geogr. Fak., Moscow.

Bashenina, N.V., et al., 1964. Legendy geomorfologices'kich kart dja kompleksnych atlasov (with english summary). In: Sov. probl. geogr., Contrib. 20th IGU Congress (London) 1964, Moscow: 399–406.

Bashenina, N.V., et al., 1968. The unified key to the detailed geomorphological map of the world 1:25,000–1:50,000. Folia Geogr., 2, ser. geogr. fys., 7, 40 pp.

Boerdiec, F., et al., 1963. Légende des cartes géomorphologiques détaillées. Univ. Strasbourg, CGS

Boyer, L., 1981. Generalization in semi-detailed geomorphological mapping: a practical example using an ITC student project from Spain. ITC Journal 1; 98–123.

Brunner, H., and Franz, H.J., 1962. Morphogenetische Karte der DDR 1:200,000 Blatt, Potsdam

Brunsden, D., 1975. Geomorphological mapping techniques in highway engineering. Highway Eng., 22: 35–41.

Brunsden, D., 1981. Geomorphology in practice. Geogr. Mag., 53 (8): 531–533.

Cholley, A., et al., 1956. Carte géomorphologique du Bassin de Paris

Christian, C.S., et al., 1956. Geomorphology. Ch. 5 Guidebook to Research Data for Arid Land Development. UNESCO, Paris: 51–56.

C.N.R.S., 1970. Légende pour la carte géomorphologique de la France au 1:50,000. CNRS-RCP, 77, 78 pp.

Cochrane, G.R., and Brown, G.H., 1981. Geomorphic mapping from Landsat-3 Return Beam Vidicon (RBV) Imag. Ph. Eng. and R.S., 47

Coz, J. le, 1972. Carte géomorphologique et géographie agraire: Aspects de l'occupation humain du Massif de la Clape. Cartographie géomorph., Serv. Doc. et Cartogr. Géogr., CNRS, 12

Crofts, R., 1974. Detailed geomorphological mapping and land evaluation in Highland Scotland. Progress in Geomorphology (papers in honour of D.L. Linton): 231–252.

Demek, J., 1967. Generalization of geomorphological maps. In: Progress made in geomorphological mapping. Geografický ústav ČSAV, Brno, 9: 36–72.

Demek, J. (Editor), 1972. Manual of detailed geomorphological mapping. IGU Comm. geom. survey and mapping. Academia, Prague, 344 pp.

Demek, J., 1976. Geomorphological mapping: progress and problems. Geografický Casopis, 28(2): 112–121.

Demek, J., and Embleton, C. (Editors), 1978. Guide to medium-scale geomorphological survey and mapping, IGU Comm. geom. survey and mapping. Schweizerbart, Stuttgart, 348 pp.

Dorsser, H.J. van, and Salomé, A.I., 1973. Different methods of detailed geomorphological mapping. Geogr. Tijdschr., 7(1): 71–74.

Dorsser, H.J. van, and Salomé, A.I., 1974. Two methods of geomorphological mapping. Geogr. Tijdschr., 8(5): 467–468.
Fourneau, R., 1963. Essai de cartographie géomorphologique. Rev. Belge Géogr., 87(3)
Fourneau, R., 1973. Un exemple d'application de la carte géomorphologique de Belgique. Ann. Soc. Géol. Belgique, 96(2): 361–362.
Galon, R., 1962. Instruction to the detailed geomorphological map of the Polish Lowland. Geogr. Geom. Dept., Polish Acad. Sci. Torun.
Ganeshin, G.S., 1975. Problemy geomorfologicheskego kartirovanya, Leningrad, 134 pp.
Garland, G.C., 1975. Engineering geomorphology–A review illustrated by case studies in Cochem (W. Germany) and Isaro Valley (S. Tirol), Italy. M.Sc. thesis, ITC Enschede, 112 pp.
Gellert, J.F., et al., 1960. Konzeption und Methodik einer morphogenetischen Karte der DDR. Geogr. Berichte, 14, 19 pp.
Gellert, J.F., and Scholz, E., 1964. Katalog des Inhaltes von geomorphologischen Detailkarten aus verschiedenen europäischen Länder. Potsdam.
Gellert, J.F., 1968. Das System der Komplex-geomorphologischen Karten. Peterm. Geogr. Mitt., 112(3): 185–190.
Gellert, J.F., 1969. The system of the morphogenesis and the morphogenetic classification of the earth surface forms as basis of the concept of geomorphological maps. Studia Geogr. Brno, 4: 32–36.
Gellert, J.F., 1971. Internationale Konzeption und unifizierte Legenden für geomorphologische Karten verschiedener Massstäbe. Wiss. Ztschr. Pädag. Hochschule Postdam: 423–429.
Gellert, J.F., and Scholz, E., 1974. Bemerkungen zur international vereinheitlichten Legende für mittelmassstäbliche Uebersichtskarten von 1:200,000–1:500,000. Studia Geogr. Brno, 41: 31–102.
Gilewska, S., 1967. Different methods of showing the relief on the detailed geomorphological maps. Zeitschr. f. Geom., Neue Folge 11(4): 481–490.
Grimm, F., et al., 1964. Empfehlung für den Inhalt und die Bearbeitung einer geomorphologischen Grundkarte im Massstab 1:10,000. Peterm. Geogr. Mitt., 108(1/2): 150–157.
Guldali, N., 1973. Tatbiki jeomorfoloju ve Türkiye'nin byük ölçekli jeomorfoloji haritsasini alma sorunu (applied geomorphology and the question of the large-scale geomorphological map of Turkey). Jeomorfologi Dergisi, 5: 77–81.
Hagedorn, J., 1967. Uber die Konzeption neuer geomorphologische Karten kleinen Massstabes. Wiss. Redaktion, Bibil. Inst. Mannheim, 4(1): 65–80.
Helbing, E., 1952. Morphologie des Sernftales. Thesis Univ. Bern.
Joly, F., 1963. Recherche d'une méthode de cartographie géomorphologique pour une carte des pays arides et semi-arides du monde à l'échelle du 1:1,000,000. B.S. Hellénique, Athènes, 4: 82–99.
Journaux, A., 1973. Cartes des formations superficielles et cartes géomorphologiques de Basse-Normandie au 1:50,000 (feuille de Bayeux-Courseulles), CNRS Centre Géom. Bull. Caen., 17.
Kesik, A., 1972. Supplementary reconnaissance airphotography as an aid for detailed geomorphological mapping. Proc. 1st Can. Symp. Remote Sensing, Ottawa, Vol. 2: 353–365.
Klimaszewski, M., 1956. The principles of the geomorphological survey of Poland. Przeglad Geograficzny, 28 suppl.: 32–40.
Klimaszewski, M., 1968. Problems of the detailed geomorphological map. Folia Geogr., Ser. Geogr. Phys., Polska Akad. Nauk. Krákow, 2, 40 pp.
Klimaszewski, M., 1978. A detailed geomorphological map. Folia Geographica. Ser. Geogr. Phys., 11: 9–25.
Kop, P.M.J., 1972. Het buitendijks deltagebied – een geomorfologische kartering. Rapport Milieu Dienst RWS, Middelburg, 45 pp.
Kugler, M., 1963. Zur Erfassung und Klassifikation geomorphologischer Erscheinungen bei der ingenieur-geologischen Spezialkartierung. Ztschr. f. angewandte Geol.: 591–598.
Kugler, M., 1965. Aufgaben, Grundsätze und methodische Wege für grossmassstäbliches geomorphologisches Kartieren. Peterm. Geogr. Mitt., 109(4): 241–257.
Kugler, M., 1968. Einheitliche Gestaltungsprinzipien und Generalisierungswege bei der Schaffung geomorphologischer Karten verschiedener Massstäbe. Peterm. Geogr. Mitt., Ergänzungsheft Neef-Festschrift, 271: 259–279.
Kugler, H., 1974. Das Georelief und seine kartographische Modellierung. Diss. B, Martin Luther Univ., Halle-Wittenberg
Kugler, H., 1976. Geomorphologische Erkundung und Landnutzung. Geogr. Ber. VEB H.Haack, Gotha/Leipzig, 21(3): 190–204.
Kugler, H., 1975. Grundlagen und Regeln der kartographischen Formulierung geographischer Aussägen in ihrer Anwendung auf geomorphologische Karten. Peterm. Geogr. Mitt., 119 (2): 145–149.
Leser, H., 1972. Inhalt und Form als Problem gross- und kleinmassstäbliger geomorphologische Karten. Kart. Nachr. 22: 156–165.
Leser, H., 1974. Geomorphologische Karten im Gebiete der BRD nach 1945. Bericht über die Aktivität des Arbeitskreises "Geomorphologische Karte der BRD. Catena, 1: 297–326.
Linton, D.L., 1951. Delimitation of morphological regions. Essays in Geography. Longmans Green, London.
Maarleveld, C.G., et al., 1974. Die geomorphologische Karte der Niederlande. Zeitschr. f. Geom., NF 18(4): 484–494.
Macar, P., et al., 1960/61. Travaux préparatoires à l'élaboration d'une carte géomorphologique détaillée de Belgique. Ann. Soc. Géol. Belge, 84: 179–198.
Marchesini, E., et al., 1963. Reconnaissance mapping for developing countries. CNR, Rome.
Marchesini, E., and Pistolesi, A., 1964. Landform maps of intermediate scale. Ass. Ital. Cartogr., 22.
Mazur, E., and Mazurova, V., 1965. Maps relatívnej vyškovej ǒlenitosti Slovenska a možnosti jej použitia pre geografickǔ rajonizáciu (Relief energy map of Slovakia and its use in geographical regionalization) (with german summary). Geogr. Casopis, 17(1): 3–18.
Melander, O., 1975. Geomorfologiska Kartbladet 29 i Kebnekaise. Statens Naturvardsverk, Stockholm, 7–37/74, 75 pp.
Mensching, H.,1979. Angewandte Geomorphologie, Beispiele aus den Subtropen und Tropen. Abh. 42 D. Geogr. tag Göttingen: 25–34.
Nakano, T., 1955. The use of aerial photographs in landform classification in Japan. Bull. Geogr. Survey Inst., 4(2), 21 pp.
Nakano, T., and Shiki, M., 1957. Landform classification survey in Japan. Proc. Reg. IGU Conf., Tokyo
Neugebauer, G., 1971. Entwurf einer geomorphologischen Uebersichtskarte des westlichen Mitteleuropa 1:1,000,000. Mitt. Inst. Angewandte Geod., Frankfurt, 123
Onge, D.A. St., 1968. Geomorphological maps. In: R.W. Fairbridge (Editor), Encycl. of Geom., Reinhold, New York: 388–403.
Passarge, S., 1914. Morphologischer Atlas Erläuterungen zu Lief. I, Morphologie des Messtischblattes Stadtremda (1:25,000). Mitt. Geogr. Ges. Hamburg, 28.

Pécsi, M., 1964. Geomorphological mapping in Hungary in the service of theory and practice. Appl. Geogr. in Hungary. Akad. kiadó, Budapest.
Pécsi, M., et al., 1962. Zeichenschlüssel zu der genetischen geomorphologische Uebersichtskarte Ungarns. Hung. Acad. sci. Budapest.
Poser, H., and Hagedorn, J., 1966. Problems of geomorphological mapping. Geogr. Studies, Warszawa, 46, 78 pp.
Rao, D.P., 1975. Applied geomorphological mapping for erosion surveys: an example of the Oliva Basin, Calabria. ITC Journal, 3: 341–351.
Raisz, E.J., 1941. The physiographic method of representing sceneries on maps. Geogr. Review, 31: 297–304.
Raisz, E.J., 1956. Landform maps. Peterm. Geogr. Mitt., 100 (2): 171–172.
Richter, H., 1962. Eine neue Methode der grossmassstäbigen Kartierung des Reliefs. Peterm. Geogr. Mitt., 104(4): 309–312.
Ridd, M.K., 1963. The proportional relief landform map. Ann. Ass. Amer. Geogr., 53(4): 569–576.
Rostovcera, E.P., 1960. Experience with the compilation of a geomorphological map of the U.S.S.R., Geodezija i Cartografija, 8: 50–56.
Rubina, E.A., 1974. Legend for 1:200,000–1:500,000 scale geomorphological maps compiled with the aim of searching for mineral resources. Studia Geogr., 41: 115–150.
Savigear, R.A.G., 1965. Technique of morphological mapping. Ann. Ass. Amer. Geogr., 55(3): 514–538.
Schmitt-Taverna, M.R., 1975. The Westplaat (SW Netherlands). A study of coastal dynamics from sequential aerial photographs. ITC Journal, 3: 173–185.
Scholz, E., 1973. Geomorphologische Karten und Legenden ausgewählter Massstabsgruppen. Studia Geogr., 32, 119 pp.
Scholz, E., 1974. Zur Klassifikation geomorphologischer Karten nach Massstab und Inhalt. Wiss. Zeitschr. Päd. Hochschule Potsdam, 18(3): 393–402.
Scholz, E., 1974. Zur Klassifikation geomorphologischer Karten. In: Problems of medium-scale geomorphological mapping. Studia Geographica, 41: 15–30.
Spagna, V., 1979. The use of aerial photographs in engineering geomorphological mapping for road planning and maintenance. ITC Journal, 1: 99–106.
Spiridonov, A., 1961. Geomorfologicheskoe kartografinovanie. Gosud. izd. geogr. lit., 1952: 1–185. (German translation: Geomorphologische Kartografie. VEB Deutsch. Verlag d. Wiss., Berlin, 60 pp.)
Spiridonov, A., 1975. Geomorfologischeskye kartografovanya, Nedra, Moscow, 183 pp.
Tricart, J., 1954. Un complément des cartes géologiques: les cartes géomorphologiques. Bull. Soc. Géol. de France, 6e sér., 4: 739–750.
Tricart, J., 1955. Un nouvel instrument au service de l'agronomie. African Soils, 4(1): 1–12.
Tricart, J., 1959. Présentation d'une feuille de la carte géomorphologique du delta du Sénégal au 1:50,000. Revue Géom. dyn., 10: 106–116.
Tricart, J., 1965. Principes et Méthodes de la géomorphologie. Masson et Cie., Paris, 496 pp.
Tricart, J., 1969. Cartographic aspects of geomorphological surveys in relation to development programmes. World Cartography UN/ECOSOC, 9: 75–83.
Tricart, J., 1970. Normes pour l'établissement de la carte géomorphologique détaillée de la France. Mém. et Doc. CNRS, Paris, 12
Tricart, J., 1972. Cartographie géomorphologique. Mém. et Doc., Service Doc. et Cartogr. Géogr., CNRS, 12, 267 pp.
Tricart, J., 1972. Cartographie géomorphologique et classement des terres pour la conservation. Cartogr. géomorphologique: 215–222.
Troppmair, H., 1970. Estudio comparativo de mapeamentos geomorfologicos. Noticia Geomorfológica, 10(2): 3–11.
Verger, M.F., 1968. Statique, cinématique et dynamique en cartographie géomorphologique. Bull. de l'Ass. Géogr. français, 359/360: 3–4.
Verstappen, H.Th., 1970. Introduction to the ITC-system of geomorphological survey. Geogr. Tijdschr., 4(1): 85–91.
Verstappen, H.Th., 1973. A geomorphological reconnaissance of Sumatra and adjacent islands. Wolters, Groningen, 182 pp.
Verstappen, H.Th., and Zuidam, R.A. van, 1975. ITC-system of geomorphological survey. ITC-Textbook VII.2, (3rd Ed.), 52 pp.
Verstappen, H.Th., Methodische Probleme der geomorphologischen Kartierung mittels Fernerkundung. 2nd Basler Geometh. Coll.: 19–43.
Verstappen, H.Th., 1977. Geomorphology and Terrain analysis of Saba and St. Eustatius (Neth. Antilles). ITC Journal, 4: 675–682.
Waters, R.S., 1958. Morphological mapping. Geography, 43: 10–17.
Webster, R., 1962. The use of basic physiographic units in airphoto-interpretation. Trans. 1st Symp. Photo-interpretation Comm. VII, I.S.P., Delft: 143–148.

SYNTHETIC SURVEY OF TERRAIN

12.1 Introduction

The analytical geomorphological surveys outlined in the previous section give full information on the geomorphology of the area of study, including processes and genesis. They do not, however, give the data which may be required about other environmental parameters in the fields of geology, soils, hydrology, vegetation, etc. Another, complementary approach is therefore needed to provide this diversified information which is essential for placing the geomorphological situation into an environmental context, making it operational for planned resource development and proper environmental management. This synthetic approach has the advantage that many aspects of the land are considered, but the genesis of the land and the processes operating, are usually neglected. Since the break-down of the land into units in synthetic or 'regional' land surveys is in most cases rooted in geomorphology and especially in landforms and topographic terrain patterns as appearing in aerial photographs and orbital images, the integration of the monodisciplinary analytical approach and the multidisciplinary synthetic approach, to combine the advantages of both, does not pose major problems. In fact, from a geomorphological view point, the second approach can be considered a logical extension of the first.

Initially analytical geomorphological surveying was applied mostly for large-scale detailed surveys, with synthetic, regional surveys mostly for reconnaissance, small-scale mapping. These differences are gradually disappearing, however, since more detailed regional surveys have been introduced and medium to small scale analytical geomorphological mapping has been developed. The advantages of each of these methods can be easily combined as follows: the analytical, geomorphological survey is carried out as usual and the distinguished geomorphological (landform) units are subsequently used as land systems and land units, after slight adaptation if necessary. Data on soils, geology, hydrology, vegetation, land utilization, etc. are then gathered for each unit, using site analysis techniques and with the aid of check lists and mapping codes, to ensure a uniform description of the terrain. The total information thus gathered, then serves to classify the terrain and to evaluate it for various purposes.

It should be understood, however, that there are types of terrain classification for specific purposes for which collecting detailed information on every aspect of the land may be superfluous. This is particularly the case for certain problem-oriented surveys, e.g., related to natural hazards, discussed in the following sections of this book. The analytical geomorphological survey can then be followed immediately by the classification or hazard/risk zoning. In some cases the morphogenetic analysis can also be reduced in extent or in depth. A pragmatic approach to the selection of phenomena to be mapped and criteria to be used may in fact be more efficient. It is not always an easy task, however, for the geomorphologist to judge the precise requirements of the survey and to translate these in terms of reference and an adequate legend. The examples given in the following sections may serve as a guide. Careful weighing of the merits of the three approaches, analytical, synthetic and pragmatic, is essential before embarking on any geomorphological survey for purposes of resource development and environmental management.

The aim of the surveys should be to provide all relevant information about the terrain in an organized manner and in such a way that the data required can be efficiently retrieved by a user, either for scientific or for practical purposes. Therefore, a user-friendly system must be devised in each case which meets three basic requirements (Mitchell, 1973):

1. It should be able to deal with requests for information from intended users.
2. It should have a capacity for acquiring, analysing and storing data about terrain and its actual or potential uses.
3. It should comprise a method of data retrieval from storage and translation into a form suitable for the user.

Extrapolation from known to unknown parts of the terrain is often an essential part of the work. Proper definition of terrain characteristics at known localities, in combination with the proper recognition of comparable situations in unknown localities is a pre-requisite for the successful application of this principle. Apart from data gathering, the mode of data processing and data presentation also have to be considered to arrive at an efficient and timely availability to the user of the information required.

Mitchell (1973) distinguished five subsequent phases in this procedure, namely the acquisition, coding, storage decoding and issue of data. Data acquisition can be realised by a variety of means; coding involves classification of the data (if possible into a numerical form); storage may be in written form, or as maps or photos and also on cards, magnetic tape, etc., but in any case in a way that permits quick retrieval through the phases of decoding and issue.

12.2 Some Basic Concepts

Terrain analysis and classification are usually based on the ecological qualities of the land (Chorley, 1969; Coates,

1971, 1972/74; Flawn, 1970; Fränzle, 1971), and the physical characteristics of the land are then systematically related to the ecological situation and the requirements of the user. One can also, however, emphasize the visual qualities of the land (Duffiend, 1975; Dunn, 1974; Jacobs and Way, 1969; Lintuus, 1960), including scenery, etc., for purposes of landscape architecture, recreation, management of national parks and related subjects. In both cases a descriptive and usually qualitative survey can be carried out or attempts can be made at some degree of quantifications.

Parametric analysis of the land is a method that has been successfully applied for purposes of quantification, e.g., in military trafficability studies. It amounts to rating a number of specific parametric qualities or characteristics of the terrain, known as parameters, attributes or elements (Buringh, 1960; Mabbutt, 1968) and their mathematical combination - by addition, substraction, multiplication or otherwise - to arrive at an index of suitability for a defined purpose. The success of the method depends on the proper choice of the attributes and on the limiting values between the various classes (Speight, 1968; Cadigan, 1972; King, 1970; Teaci, 1972; Turner and Miles, 1968). The parametric methods sometimes lack comprehensiveness and the recognition of the specific problems of the various parts of the terrain may be less easy. Advantages are, e.g., that it is less dependent on subjective interpretation of landform, statistically more reliable for measuring variances, formulating rational sample policy and expressing probability limits of findings. Flexibility in the selection of parameters, easy data handling and versatility, particularly when computerized processing is used and the unbiased, quantitative output are other assets (Verstappen, 1977).

The most appropriate scale(s) of the survey is another matter of interest. In the early part of this century emphasis was on small scale mapping resulting in a breakdown of the land in major physiographic provinces. The boundaries of these very large terrain units were often rather poorly defined and the units themselves insufficiently homogeneous for association with specific land utilization purposes. The small-scale maps so produced had little practical value and mainly served for educational purposes. Since the morphogenetic principles underlying the surveys thus conceived are long since obsolete, it seems inappropriate to apply the term 'genetic approach' (Mabbutt, 1968). Confusion may arise from this terminology since modern morphogenetic mapping is often executed at large scales. Small scale mapping of terrain has recently received new impetus following the availability of small scale imagery from orbiting satellites or by way of side-looking airborn radar (SLAR). If properly integrated in a multi-phase surveying system, mapping at this scale now is a valid starting point also for 'zooming in' leading to downward regionalisation (Verstappen, 1970). Groundtruth gathering by way of traversing (road or small aircraft), detailed site observations in key areas and subsequent extrapolation procedures have to be thoroughly planned (see section 14 on drought susceptibility surveys).

Land system mapping and related types of synthetic surveys in one way or another are based on landscape ecological principles and follow the so-called 'landscape approach' (Mabbutt, 1968). Usually these surveys are carried out in medium scales and characteristically only the land systems are mapped and not the land units of which they are composed. In recent years the 'landscape approach' also has been used for detailled mapping. The land units are then determined first and the land systems or land complexes which they build thereafter (Wright, 1972). 'Upward regionalisation' (Verstappen, 1970) has become more systematically applied.

The geomorphological input in such surveys (Thomas, 1969; Townshend, 1981; Tricart, 1973) should be based on morphogenetic considerations using the analytical survey method already outlined in the previous section. At all scales, full integration of the analytical and synthetic surveying methods is essential for obtaining optimum results. The parametric approach (Mabbutt, 1968) just mentioned is not to be considered a separate method on equal footing with the analytical and synthetic surveys but rather a means of data gathering and processing often with emphasis on quantification.

The phases of synthetic surveys of terrain encompass analysis, classification and evaluation. The analysis is basically geared to terrain factors only; the classification is often at least also influenced by the potential user requirements and the boundary between classification and evaluation therefore is not always a very sharp one, as will be evident from the forthcoming sub-sections. In essence, however, the evaluation aspects and particularly the human factors then involved are only briefly touched upon. Emphasis is on the terrain factors which are, of course, more directly linked to geomorphology. For further details on evaluation aspects the reader is referred to the numerous excellent recent textbooks in this field (UNESCO, 1968; Nossin, 1977; Vink, 1960, 1980).

12.3 Major Types of Synthetic Surveys

It has already been mentioned in the previous subsection that synthetic surveys - with the exception of those emphasizing the visual, scenic qualities of the land and, in part, those related to military assessment of terrain - are rooted in landscape ecological concepts (van

Dyne, 1969). These fundamental concepts of the complex interrelations between biotic and abiotic factors characterizing units of areal extent have first been clearly defined by Troll (1939) but their origins can be traced back in physiographical regionalization and landscape science even as far as the latter part of the nineteenth century. This field of science has traditionally been strongly developed in countries such as Germany and the USSR (Haase, 1964; Klink, 1966; Neef, 1964, 1967; Schneider, 1969; Troll, 1939, 1966; Isachenko, 1973; Leser, 1973, 1974). A good review of the modern ideas on this matter is given by Leser (1976), who also provides numerous elucidating examples.

These theoretical concepts have a very pragmatic counterpart and thus received new impetus with the advent of aerial photographic interpretation, which greatly facilitated the recognition of more or less homogeneous land units and their arrangement in systems or complexes. Verstappen (1977) elaborates on this matter and especially draws attention to the role of geomorphology in the study of airphoto patterns. Efficient methods for mapping land(scapes) or terrain, particularly on reconnaissance and semi-detailed scales, thus rapidly developed (Zonneveld, 1979). Sizeable portions of many countries have been the subject of these types of surveys which use a two-level classification; namely from the climatological, geomorphological, pedological, hydrological, etc. view point more or less homogeneous usually small land surfaces, known as sites, facets, ecotopes, land units, etc. and their supposedly recurrent patterns within a larger surface area, designated as region, land system, etc. (Bawden, 1967; Stewart, 1968; Mitchell, 1971, 1973). The term 'land unit' is applied by some as a separate level of division of the terrain between the site and the land system.

It is logical that a great distance exists between fundamental landscape ecology, not directly suited for surveying and mapping, and these pragmatic synthetic surveys which sometimes give the impression of the tabular summation of parametric information without revealing much of interdependencies. One may question whether landscape /terrain patterns are indeed as recurrent as claimed to be (Terzaghi, 1965). One may also wonder whether a two- or three-level classification is not an overgeneralisation for complex matters such as landscape ecological situations. One may even doubt whether a 'smallest homogeneous unit' does in effect exist at all. The fact that the units of the lower category (land units) are not mapped but only the assemblages (land systems) are outlined, although largely due to scale limitations, may in part reflect this dilemma. Notwithstanding the reservations just made, the study of terrain units, and their grouping into assemblages, has proved its practical and fundamental importance on many occasions (Blakes and Paymans, 1973). On the other hand, it has also gained considerably from being placed in a larger physiographic or landscape ecological context. Geomorphology serves as an integrating science in explaining the assemblages and their genetic internal and external relations (Verstappen, 1966).

The most common of these methods is the land system survey as developed by the Land Research and Regional Survey Division of the CSIRO (Scientific and Industrial Research Organisation) in Australia. The land systems, recurring patterns of topography, soil and vegetation (Christian and Stewart, 1953, 1968; Christian, 1958; Christian et al., 1957), are mapped, usually at a scale of approximately 1:250,000, and diagrammetric cross sections, tables, etc., provide further information about details within the land systems. When, more recently, the same approach was also applied to more detailed mapping in scales of 1:100,000 or larger, land units were individually mapped. They are considered to be composed of homogeneous 'sites' (Bourne, 1931; Christian and Stewart, 1968) which are, in fact, a third level of break-down of terrain into units. For mapping on reconnaissance scales of 1:500,000 - 1:1,000,000, rather heterogeneous 'complex' or 'compound' land systems have been used as basic mapping units.

Land system surveys (CSIRO, 1953, 1974) have evolved considerably within a few decades (Haantjens, 1975; Scott and Austin, 1971) and after having been tried out in the almost uninhabited Australian dry lands have been subsequently used in the humid tropical conditions of Papua New Guinea (Loffler, 1974) and have even been introduced in areas that have witnessed centuries of intensive land utilization. The surveys mostly served the purpose of general inventory of land resources, initially with emphasis on the extensive pastoral land use of the Australian interior. When attempts were made to assess the potentialities of the land for other more intense types of land utilization, engineering, etc., modifications of the approach became necessary, as will be discussed farther on in this sub-section.

Aerial photograph interpretation has always been an essential part of the survey work and serves to identify the general lay of the land, the patterns developed, the morphometric characteristics, etc. (Reiner, 1970). Field surveying consists of quick and continuous observations by a team of experts along traverse lines which are generally several kilometres apart. In cultivated areas their position is usually determined by the existing road network. Detailed observations on landform, soils, vegetation, hydrology, etc., are made along these traverse lines every few kilometres at sites considered characteristic for dis-

Table 12.1 Principle types of L.R.D. surveys (D.O.S.) with information shown at various scales.

Survey type and Survey area covered (mi^2)	Objective	Typical map scale	Landscape analysis	Landscape components				Economic analysis
				Geomorpho-logy	Soil	Vegetation	Land use	
1. Reconnaissance. More than 2,000. May be as large as 100,000.	National or regional inventory	1:250,000 and 1:500,000	Land systems	Major relief	Order, sub-order, great soil group or associations	Climatic and edaphic formation types	Agroecological groups	National or regional economy
2. Extensive. 1,000-10,000	Detailed inventory; broad assessment of agricultural potential	1:100,000 to 1:250,000	Land systems	Relief units or major landforms	Great soil groups, or associations of series	Climatic and edaphic formation types and plant associations	Land use systems and cultivation density	Regional economy and/or sector analysis
3. Intensive. 500 - 5,000	Location and definition of development projects	1:25,000 to 1:100,000	Land systems and facets	Detailed landforms	Series or associations of series	Plant associations	Land use and farming systems, plus specific parameters	Sector analysis including market prospects and cost-benefit analysis of development projects
4. Development study. Usually more than 50	Resource analysis and development planning	1:10,000 to 1:25,000	Land facets and elements	Landform elements or slope units	Phases of series and/or selected parameters	Plant associations and species distribution	Specific parameters, eg. crop distribution, field patterns	Detailed examination of development projects, cost benefots analysis and commodity studies

tinct land units. The systematic execution of the surveys and unbiased data gathering by various surveyor teams is furthered by the use of check lists and standardized mapping codes for all items to be investigated at the sites. Mapping cards finally give all information gathered for the various land systems with indication of the subdivision in units by way of block diagrams, profiles, etc.

The Land Resource Division of the DOS (Directorate of Overseas Surveys) in U.K., applies approximately the same principles and also heavily relies on aerial photograph interpretation and airphoto patterns (Brunt, 1967). Land systems are being mapped in surveys at the scales of 1:100,000 and smaller. In an earlier stage, however, more detailed surveys have been carried out, particularly in east Africa, there terrain units of lower order, called land facets of even land elements, are mapped. Table 12.1 summarizes the type of information provided at various mapping scales.

Landform plays an important role in the types of synthetic surveys mentioned above. Christian (1958) states that the order of a land unit depends on the nature of the landform accepted as a unit of study; he thus clearly ly emphasizes the important position of landform characteristics also in land system surveys. Solntsev (1962) states that the geological - geomorphological foundation is always the principal factor in segregating landscapes and he thus insists that a genetic system of landscape classification must be based on it. This line of thought is also accepted by Nakano (1962) who introduced a method of landform classification in Japan in which landform types, - series, - associations, - sections and - provinces are outlined on the strength of geomorphological genesis and evolution. Combined with geological and soils information these maps are used for purposes of planned development. A similar approach was also followed by Verstappen (1967) who, in an integrated survey of the Porali Plain (Pakistan), outlined the main landscape units on the basis of a preliminary study of landforms. Using this map as a starting point, more precise maps were subsequently produced on the geomorphology, soils, hydrology and vegetation/land use of the area (Pakistan Meteorological Service, 1965).

The figs. 12.1, 12.2 and 12.3, as well as table 12.2 give examples of the tabular and graphic modes of data presentation used in the kinds of synthetic surveys of terrain outlined above. The tables 12.1 and 12.3 give examples of mapping cards used during the survey work.

The methods of synthetic mapping of terrain applied in the USSR are strongly rooted in landscape science and bear resemblance to the structure of the CSIRO approach outlined above, with 'facies', 'urochishcha' and 'mestnosti' representing sites, land units and land systems, respectively (Vinogradov, 1968; Isachenco, 1973). Above these three categories there is a taxonomic level of higher order: the landscape. Solntsev (1962) distinguishes even more hierarchical levels and subdivides the urochischcha into sub-urochishcha. Apart from this, the Soviet landscape approach also distinguishes itself by the emphasis put on three complementary and interrelated fields of study, namely (i) the study of the distributional patterns of the various land(scape) elements, (ii) the relation between the landscape elements which are visible in aerial photographic images and those which are not directly visible, and (iii) the study of landscape dynamics based on genetic evolution. The surveys are effectuated by studies in key areas using cross-sections (see fig. 12.4) and by the analysis of spatial complexes (Nechayev and Fedorin, 1973; Nikolayev, 1974; Tagunova, 1968). Some similarities with the so-called physiographic approach developed in the Netherlands for purposes of soil survey exist (see section 3).

Botanical elements may also serve as a key factor in landscape classification,which explains the emphasis put on vegetation, e.g., in the Canadian type of land surveys. Even more apparent is the place of the botanical factor in the vegetation surveys developed by Gaussen and his school in Montpellier, France. Vegetation is then considered to be an adequate and reliable reflection of all environmental factors combined. Consequently, a study of the vegetation from this point of view will result in detailed information on pedological, hydrological and other factors which not only effect the natural vegetation but also the growth of agricultural crops and have a bearing on the overal potential for development of the area concerned. The botanical approach is also strongly rooted in the interpretation of aerial photographs and has recently received new impetus with the advent of other remote sensing techniques. In areas where vegetation occurs, it is a major factor effecting the reflection received or, more generally speaking, of the signals so recorded, especially in areas of low relief.

A difficulty encountered in the landscape ecological evaluation of the vegetative patterns mapped is the discrepancy existing in almost every part of the world nowadays between the vegetation that theoretically should be present in view of the ecological situation and the vegetation actually occurring because of human interference of varying intensity and duration. This discrepancy certainly is a major reason why the botanical approach in synthetic land surveys so far has not yet gained general acceptance, contrary to the geomorphological approaches mentioned earlier. The work by Thalen (1978) in Iraq shows, however, that the two approaches are compatible.

It should be emphasized once more that in the geo-

SURFACE PATTERNS

FACET DESCRIPTION

VEGETATION AND LAND USE OF FACET	Cult. this year	Bush fallow	Village bush	Non-agric. land
	Form^n			
	Assoc^n			
	% facet			
	Crops			

DRAINAGE IRRIGATION RELIEF MODS.	

ACCELERATED EROSION	Sheetwash	Gullies

STONINESS ROCKINESS					
	Size				
	% surface				
	American class				
	% facet area				

SOIL DEPTH VARIATION	

WATER TABLE PROFILE	Wet season	Dry season

MICROCLIMATIC FEATURES	Aspect	Surface	Shelter

PROFILE PIT

LOCATION

Moisture condition

Recent weather

Pedological drainage

Site drainage

Infiltration

Permeability

Groundwater

Parent material

PROFILE PATTERNS

0 -

1 -

2 -

3 -

4 -

5 -

6 -

7 -

8 -

9 -

10 -

LAND CAPABILITY ASSESSMENT

ROOT ROOM	EROSION SUSCEPTIBILITY
DROUGHT/FLOOD	MECHANISATION LIMITATION
CHEMICAL LIMITATIONS	OVERALL CLASS

Table 12.2 Mapping card used for landscape facet description in the DOS type of synthetic surveys of terrain.

FACETS OF MASAKA LAND SYSTEM

	Form	Soils, materials and hydrology	Land cover
1	Plateau crest, flat usually several hundred metres across. Abrupt short steeply sloping margins.	Thick massive laterite over weathered pre-Cambrian metamorphic rock, mainly schist. Above laterite is dark brown sandy soil, humus stained and frequently containing murram to surface. Above groundwater influence.	*Themeda - Cymbopogon* grass savanna
2	Quartzite ridge, steep sided, with rounded crest. Usually occur on interfluve summits but may also occur on mid slopes.	Quartzite, bare rock or with shallow stony soil. Above groundwater influence.	*Themeda - Cymbopogon* grass savanna
3	Convex interfluve and slope steepening to ca 7° lower and midslopes mainly straight. Where adjacent to laterite margins (land facet 1) uppermost part of slope steeper. May possess steeper portion on lower slope immediately above valley floor.	Two variants: a) reddish yellow loam, 1 m or more deep, over stone line, over weathered schist; little or no concretionary iron material. b) red clay loam or clay with little quartzite over murram, partially indurated laterite or both over weathered schist. Drainage free above groundwater influence.	Cultivated, or *Pennisetum* fallow. Moist deciduous forest on steeper sites.
	Small valley. Narrow drainage lines with rounded bottoms, occasionally broadening to 100 m with flat bottoms.	As 3 above but grading to mixed alluvial deposits in flatter occurrences within the influence of groundwater.	Moist deciduous forest.
5	Main valley floor. Wide 100-500 m across, and flat.	Sand with occasional layers of gravel or mottled clay. Humus stained but little peat. High ground water table.	*Papyrus* and *Miscanthidium*

Table 12.3 Facets of Masaka Land system

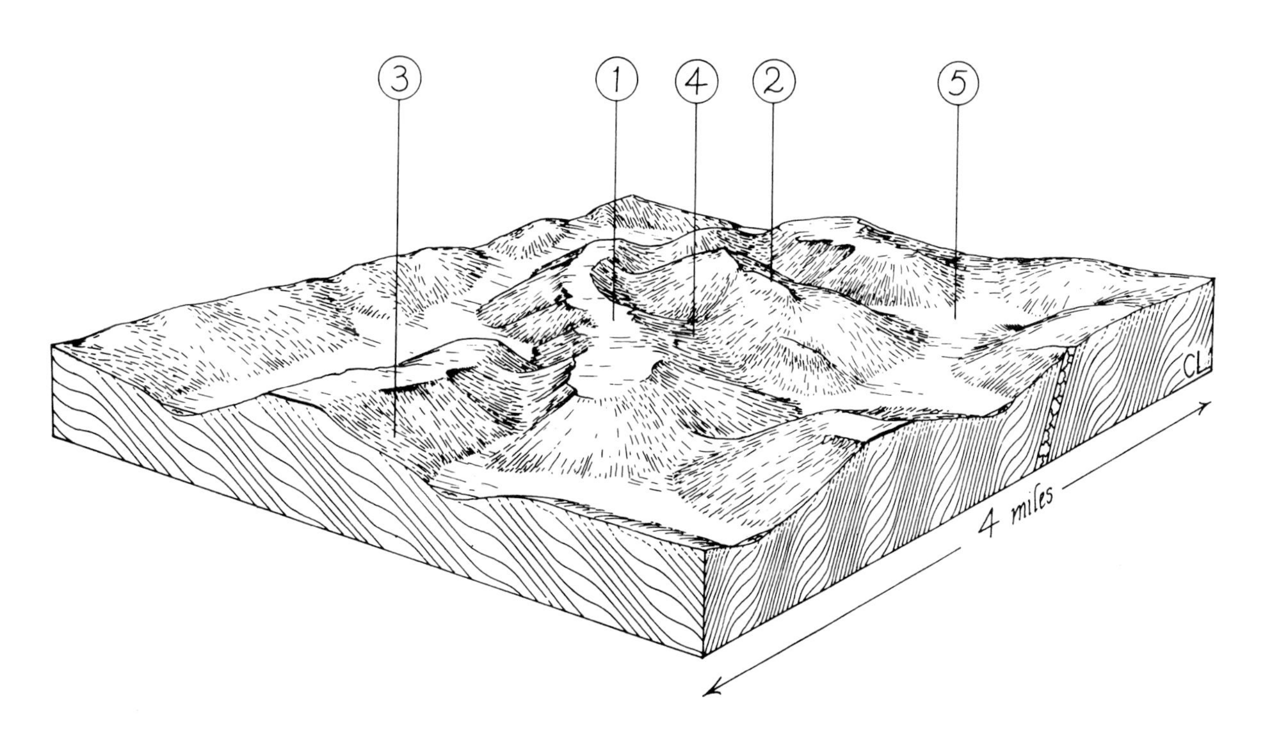

Fig. 12.1 Block diagram of the Masaka land system, Uganda, illustraging a DOS resource survey. A general description of the land system is added and the five facets into which it is subdivided are described subsequently (table 12.2).

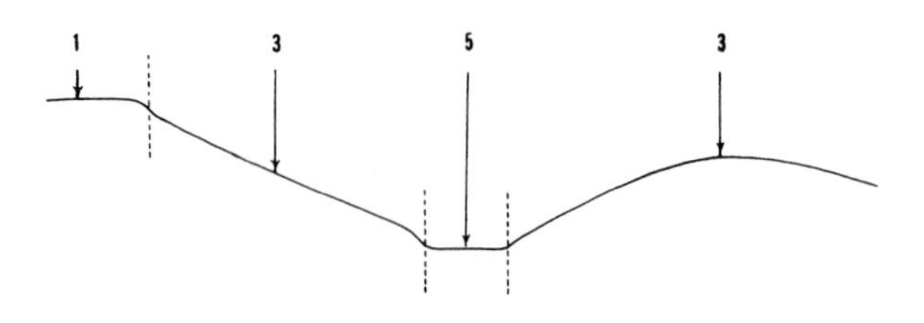

Representative Section

Fig. 12.2 Representative section across the Masaka land system of fig. 12.1 with indication of the situation of three of the facets occurring.

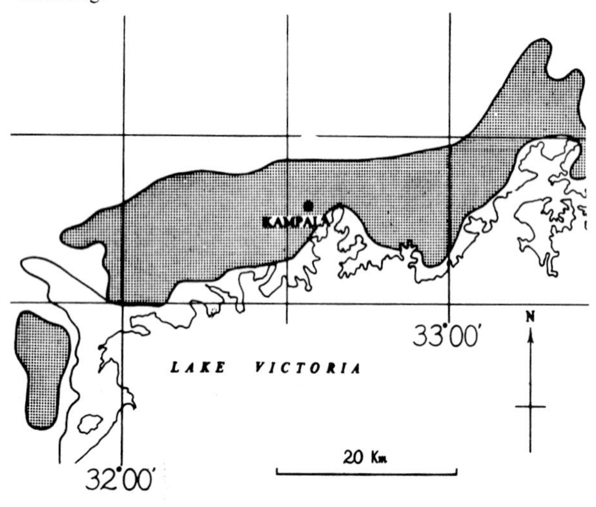

Fig. 12.3 Location and extent (stippled) of the Masaka land system of fig. 12.1.

General description of Masaka Land system

Climate:	1000-1300 mm. rainfall, bimodal; mild dry season
Rock:	Pre-Cambrian basement complex, mainly schists and gneisses mainly deeply weathered and lateritised.
Morpho-genesis:	Dissected old land surface in which massive laterite is preserved as level caps to major interfluves. Below these are long hill slopes leading to wide aggraded and frequently swampy valleys.
Soils:	A variety of red loam,lateritic (ferrallitic) type. (Buganda catena Kifu and Kaku series)
Vegetation:	Forest/savanna mosaic with forest dominant along valleys
Altitude:	1300 m approx.
Relief:	120-150 m.

morphological approach a mere descriptive treatment of the land components is insufficient: the more profound the study is (and thus allows for discovering the causes of the differentiation) the more useful the survey tends to be. Also the biocenological approach to terrain classification for agriculture advocated by Moss (1969) is worth mentioning in this context.

The acceptance of the importance of these interrelationships automatically clarifies the role of landscape genesis. If one agrees with the landscape approach as a physiographic classification (Becket and Webster, 1962, 1969, 1971; Conklin, 1959), a geomorphological view point also taken by Mabbutt and Stewart (1965), it becomes obvious that the modern analytical approach in geomorphological mapping may be advantageously integrated in the survey work. This concept of a 'genetic' approach differs essentially from the approach with the same name discarded by Mabbutt (1968). It stems from the conviction that it is the very insight into the genesis of the land which clarifies its physical (and engineering) properties and its economic potentiality.

12.4 Adaptation of the Concepts in Cultural Landscapes

Synthetic surveys of the land are a useful inventory of natural resources. They may provide essential basic information for purposes of development (Marsz, 1974). However, their immediate application is limited if the technological means are not studied that could improve the land for human use. Some tracts of land may be suitable for specific purposes without further improvement, whereas others may require technological 'upgrading', sometimes by sophisticated, costly means. Technological 'intrusions' will fail to sort the effects desired, however, if implanted in an area without taking the human factors into consideration. Thus, three interacting spheres have to be taken into consideration, during the survey, viz. the natural environment with its ecosystems and its inherent (climate, soil) and produced (mineral/agricultural products) resources; the human factor with the associated economic conditions and the technological possibilities to render the natural environment better suited to human needs (Hunter, 1980; Lewin, 1975), by way of land improvement and engineering works.

Managerial aspects related to the formulation and execution of integrated surveys covering these three spheres, being preparatory to planning and development, also merit full attention. It is preferred by some to first arrive at an assessment of the physical characteristics of the land and thereafter give a socio-economic evaluation. Others, however, consider it more appropriate to study the physical environment and the socio-economic factors simultaneously, thus profiting from a continuous exchange of ideas. The choice between these two approaches will be influenced by various matters and intermediate working methods also may be applied, see fig. 12.5.

The intricate relations existing between the three spheres will be clear from the following considerations. A natural resource is culturally defined; thus it varies with the human factor both in space and in time. Products once highly valued have lost their importance and vice versa. Also a natural resource depends on technology since it only becomes a true resource if it can be econ-

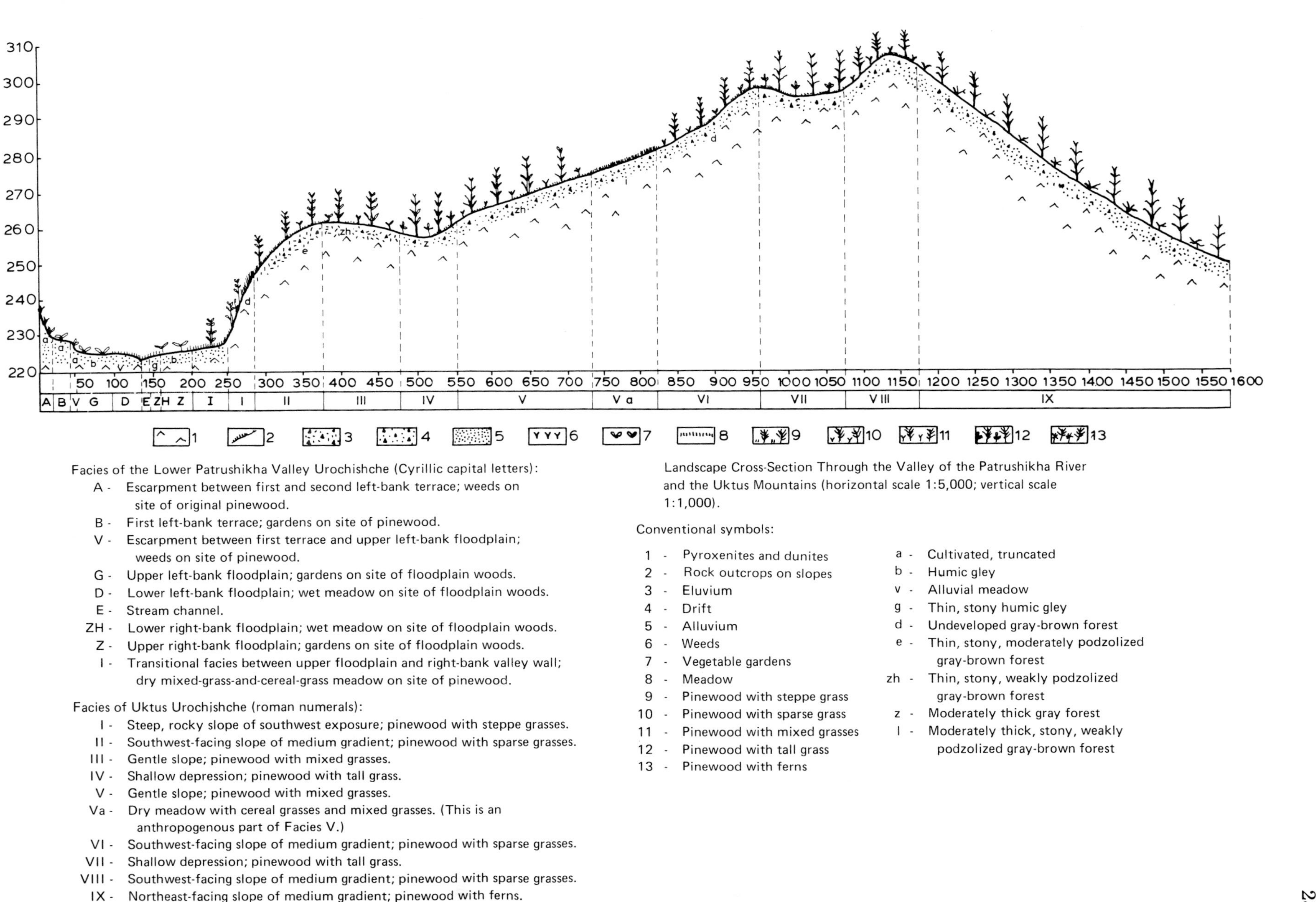

Fig. 12.4 Landscape cross section with facies description as used in synthetic mapping of terrain in the USSR.

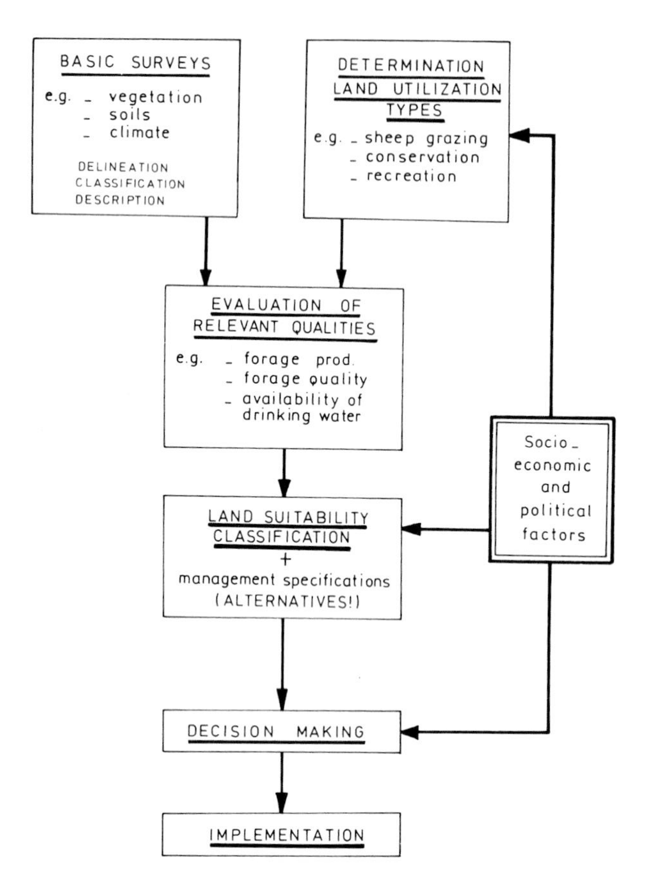

Fig. 12.5 Schematic diagram showing the input of basic resource surveys and of land utilization types for the assessment of the land suitability. Socio economic and political factors affect the ultimate decision making and implementation process (After Thalen, 1978).

omically exploited (Brookfield, 1969).

In order to render the various methods of synthetic survey mentioned above applicable to inhabited environments, they should be geared to specific land use(s); modifications of the survey methods are therefore required (Schreier, 1972). Because in such cases emphasis is usually on more intense utilization of already occupied land rather than opening up of virgin areas, they tend to be of a more detailed nature than those executed for purposes of resource inventory only. As has already been explained in section 5, mankind has,in the course of generations of agricultural activities, location of settlements, etc.,adapted his utilization of the land by trial and error to the properties of the natural environment in such a way so as to make optimum use of it, given his cultural heritage and the technological level attained. It is thus evident that a thorough study of the uses and misuses of the land and of the ecological changes provoked this way may, in many cases, offer a valuable clue to environmental factors that influence and/or limit the utilization of the land. Inversely, the study of environmental conditions may clarify the type and distribution of land utilization which otherwise could remain unexplained. It is thus evident that,as soon as inhabited areas are concerned, the traditional utilization of the land in the broadest sense should be incorporated in the survey work.

A new, more evolved land utilization type, possibly less rigidly adapted to the environmental conditions, that to some extent may have to be artificially changed for the purpose of increasing the potentiality of the area, may result from such survey as a consequence of improved technological levels and modifications in the socio/economic conditions.

An interesting example of such incorporation of land occupancy is a study made by the Soil Conservation Authority of Victoria, Australia (see: Gibbons and Downes, 1964). The survey is based on the land system concept but includes a chapter on the present land use. The information contained therein is included in tabular form in the diagrams given for each land system, distinguishing between present and potential land use and indicating, under separate headings, the actual erosion, the hazard of erosion and the specific problems associated with cultivation. Land use is also considered in integrated surveys of various kinds carried out in several other countries, such as those carried out in Pakistan by the Arid Zone Section of the Geophysical Institute, Quetta, and the Central Arid Zone Research Institute at Jodhpur, India. Land system surveys carried out in Pakistan revealed the importance of the usually complex relationships existing between geomorphology, ground- and surface water conditions and agricultural land use that are especially pronounced in arid and semi-arid areas. In a survey of the Porali Plain, Las Bela, Pakistan (Verstappen, 1967), information on mostly agricultural land use was indicated together with the data on the natural vegetation in a map which, in combination with other maps on (i) geomorphology, (ii) soil conditions and (iii) hydrology,gave a full account of the elements of the natural and cultural landscape occurring in the area that were of interest to the purpose of the survey. Although the survey was basically executed along the lines of the land system concept, no separate land system map was prepared, but the land system boundaries were indicated by thick lines on the four thematic maps listed above. The philosophy behind this procedure was that a more convenient visual correlation between the land systems and the four landscape elements studied could be achieved.

It is evident from the aforesaid that when studying the human factors in the landscape one should not restrict oneself to crop survey or to the purely agricultural aspects of land utilization. Much valuable information can be gained from the distribution of accelerated erosion and other geomorphological phenomena in combination with the study of field pattern, parcellation, settlement types, etc.,particularly when seen in the perspective of the historical evolution of the cultural landscape (Butzer, 1971; Demek, 1971; Detwyter, 1971).

Physio-geographical regions or natural landscapes (Kugler, 1974; Pecsi, 1974) in intensely cultivated areas may differ substantially from the actual, cultural landscapes. There exists extensive literature, particularly by German authors,about the differences between 'Naturräume' and 'Reallandschaften'. Terms such as cultural ecotope, cultural landscape unit, socio(eco)tope, anthropotope, etc., are in use for indicating the smallest units (Uhlig, 1967). Although developments in this field are promising and also urgently needed in the framework of regional planning and development, substantial systematic research in the field of cultural landscape studies still is required before the same degree of precision and integration can be achieved as has been reached in surveys related to natural landscapes.

The study of cultural landscapes becomes particularly appealing when also the actual functions of the cultural elements of the landscape are considered. The functions, however, are frequently not visualized in the landscape. It would be of interest to investigate to which extent the criteria used in the various types of regionalization are related to observable elements in the cultural landscape. Conclusions as to non-observable elements are only within reach on the strength of correlation with observable elements by way of deductive reasoning. As many of the interrelations in the field of socio-economics are usually not of a simple causal type and only relate partly to environmental factors, deduction and extrapolation are fraught with difficulties.

In actual practice,adaptation of the synthetic survey concept to purposes of intended land use in inhabited areas or in areas to be put under use is effectuated by using a double classification system, namely: (i) classification of the land on the basis of the prevailing environmental conditions, and (ii) evaluation or assessment of the potential utilization of the various parts of the terrain. In the latter, emphasis is on potentialities and on limiting factors affecting present and future uses of the land. Table 12.4 gives an example of this approach from a FAO survey in Iran (Bordbar and v.d. Weg, 1969). Terrain analysis and classification thus are clearly separated from land evaluation aspects. It may be useful, however, to structure the classification and to select the class limits between the various classes already with an eye on future uses. In such cases the terrain classification and the evaluation cannot be fully separated (Sheng, 1972).

12.5 Some Specific Applications

12.5.1. Engineering

Civil and military engineers in several countries have tried out the applicability of land system and other, comparable, synthetic approaches for their particular purposes. The Soil Mechanics Section of CSIRO, Australia, found that the complex land systems mapped by the Land Research and Regional Survey Section of the same organisation were not of much use for matters such as highway location, construction materials survey, etc. More detailed surveys resulting in smaller sub-divisions of the land and emphasis on land properties directly relevant for engineers with avoidance of any abstraction appeared to be required (Aitchison and Grant, 1967, 1968; Grant, 1968, 1971; Haantjens, 1968; Muller, 1972). A system of terrain classification thus was devised, dividing the land into terrain patterns, terrain units and terrain components. The latter is the smallest unit in this so-called P.U.C.E. (Pattern, Unit, Component Evaluation) programme. It is characterized by great homogeneity in slope form, soil association (within one class of the Unified Soil Classification, USC), and vegetation. It is normally too small to be mapped using (medium-scale) aerial photographs alone. Three phases are distinguished in this type of terrain classification, namely establishment, quantification and interpretation/application. The first phase is, in fact the actual mapping, resulting in a break-down of the terrain into units of the taxonomic levels just mentioned. The quantification phase amounts to the assessment of engineering properties/qualities to the mapping units and the interpretation/application phase concerns the data handling in general.

The British Transport and Road Research Laboratory (TRRL) has applied the land system concept to engineering assessment of terrain, particularly for reconnaissance purposes in tropical areas (Dowling, 1967, 1968). The land systems and facets mapped with the use of aerial photographs are matched with field and laboratory data about the engineering properties of the occurring materials to arrive at a reliable assessment. Similar approaches have been used by the National Institute of Road Research (NIRR) in the Republic of South Africa (Brink et al., 1966). Aerial photos were used to arrive at a terrain classification which subsequently serves to assess the (soil) engineering properties of the mapping units with a limited amount of ground truth gathering. The work done by the Canada Geological Survey (1973) also deserves mention (see also Hemstock and MacFarlane, 1972; Hughes, 1978). Garland (1975, 1976) has emphasized the importance of morpho-genetic concepts for these types of surveys.

MAPPING CARD

1	2 3	4	5 6 7 8
OSTAN	AREA NAME	SHEET	PHOTO NUMBER

21	22
F	N R
4	0

A								
B								
C								
D								

9 10 15 16 17 18 19 20 23 24 25 26 27 28 29 30 31 32 36 37 38 39 40 41 42 43 44 45 46 47 48 49 50 51 52 53 54 55 56 57 58 59 60 63 64 65 66 67 68 69 70 71 72 73 74 75 76

TRACT NO	MAPPING UNIT	SERIES NAME	LAND SYSTEM			LAND UNIT					Micro-relief		STONINESS				ROCKINESS		WIND ERO		WATER ERO		DRAINAGE	GROUND WATER				IRRIG WATER		LAND USE	PARENT MAT	SOIL DEPTH	INCLUSIONS	LAND CLASS FORMULA	LAND CLASS	HA
			NAME	K	S	K	Ag	S	SLOPE Sh	SLOPE %	K	I	K	St	Cb	gM	K	I	K	I	K	I		DEPTH dcm	N	quality C	quality S	C	S							
01																																				
02																																				
03																																				
04																																				
05																																				
06																																				
07																																				
08																																				
09																																				
10																																				
11																																				

Table 12.4 Another example of a mapping card that has been used in Libanon.

The Military Engineering Experimental Establishment (MEXE), U.K.,has also applied synthetic surveys for purposes of terrain classification. This so-called MEXE-Oxford method is rather similar to the CSIRO land system approach. Although seven taxonomic levels are distinguished, only three of these are,in fact, in use, namely the land system, the land facet and the land element. The

The land system, initially called 'recurrent landscape patterns', is the highest category of these. It is considered suitable for mapping at scales of 1:250,000 to 1:1,000,000 in particular. The land facet is the basic terrain unit and can be mapped at scales of 1:10,000 to 1:50,000. The land element is a further sub-division of the facet which, though recognizable on aerial photographs, is not normally mapped (Webster and Beckett, 1970; Beckett and Webster, 1968, 1971; Beckett, 1975; Brink et al., 1966). The method initially aimed particularly at military applications such as the extrapolation of terrain conditions in known areas to similar unknown terrain. Some morphogenetic principles were applied for this purpose. Civil applications in the field of resource assessment soon became a dominant aim, however, and as such the method is not essentially different from those already mentioned. At a symposium held in Oxford (Brink et al., 1965), a certain degree of standardization was reached among the MEXE-Oxford, the TRRL, the CSIRO-SM and the NIRR groups that all adhere to the landform concept. The lowest category of the classification, the land element, is generally homogeneous as far as slope angle, soil type, drainage and lithology are concerned. It is usually visible on aerial photographs of scales in the order of 1:10,000 to 1:50,000, but not mappable at these scales. The next-higher category, the land facet, is defined as a part of the landscape, usually with simple form, on a particular rock or superficial deposit, with soil and water regimes that are either uniform over the whole facet or, if not, vary in a simple and consistent way (Webster and Beckett, 1970). Recurrent facets together form land systems.

The U.S. Army Engineers Waterways Experimental Station (USAEWES) in Vicksburg (Miss.) developed a different, mainly parametric classification method for

LAND RESOURCES MAPPING LEGEND

TYPE	MAPPING SYMBOL	LAND UNIT	TOPOGRAPHY	CLIMATE	SOILS	NATURAL VEGETATION	DRAINAGE	SOIL SALINITY	PRESENT USE	OTHER FEATURES and AREA (HA)
1 MOUNTAINS	11 D-GREEN	FORESTED ELBURZ MOUNTAINS FRONT 400/2700m. ALTITUDE	SLOPE: 25-60% ROUNDED CRESTS UP TO 1200m. HIGH.	P=1000 mm.? COLD WINTERS WITH SNOW	MOD. DEEP (50 cm.) CONTINUOUS COVER, BROWN FOREST & PODZOLIC OVER LIMESTONE & SCHISTS	DENSE DECIDUOUS FOREST	MODERATE RUNOFF. MOD. SLOW INTERNAL DRAINAGE	NO	FOREST - WITH LOCAL GRAZING AND WHEAT CULTIVATION	378.000
	12 D-GREEN	LOESS COVERED MOUNTAINS WITH SPARSE FOREST 400/900m. ALT.	ROUNDED RIDGES SLOPE: 15-40% WITH VERY DEEP GULLIES.	P=500 mm.? COOL WINTERS OCCASIONAL SNOWS.	DEEP (>1m.) CONTINUOUS SOIL COVER-LIGHT BROWN FOREST OVER LOESS	SPARSE DEGRADED DECIDUOUS FOREST	RAPID RUNOFF AND INTERNAL DRAINAGE	NO	FOREST-GRAZING AND DRY FARMED WHEAT	107.000
2 HILLS	21 L-GREEN	LOESS FOOTHILLS AND TOE SLOPES PARTLY DEFORESTED 150/400m. ALT.	ROUNDED RIDGES 10-60% SLOPE.	P=6-700 mm.? COOL WINTERS OCCASIONAL SNOW.	DEEP (>1m.) CONTINUOUS COVER, BROWN FOREST SOILS ON LOESS	DENSE DECIDUOUS FOREST WITH LOCAL CLEARINGS	MODERATE RUNOFF-RAPID INTERNAL DRAINAGE	NO	FOREST-DRY FARMED WHEAT-GRAZING	31.000
	22 D-BROWN	LOESS HILLS-BARE AND STRONGLY DISSECTED 100/300m. ALT.	"BADLANDS" DENSELY GULLIED 20-40% SLOPE.	P=300 mm? COOL WINTERS 4-5 DRY MONTHS.	DEEP (>1m.) CONTINUOUS REGOSOLS	VERY SPARSE-STEPPIC	RAPID RUNOFF AND INTERNAL DRAINAGE	NO	SEASONAL GRAZING	143.000
	23 D-BROWN	LOESS RIDGES-BARE AND STRONGLY DISSECTED. ALT. <50m.	ROUNDED TOPS AND GULLIED SLOPES UP TO 15%	P=200-300mm? MILD WINTERS 4-5 DRY MONTHS.	DEEP (>1m.) CONTINUOUS REGOSOLS	VERY SPARSE-GRASS	RAPID RUNOFF & INTERNAL DRAINAGE	NO	OCCASIONAL GRAZING	LOCAL SALINE DEPRESSIONS & SHALLOW LAKES 12.500
3 PLATEAUS	31 L-BROWN	UNDULATING UPPER TERRACES OF ATRAK. ALT. <150m.	NARROW VALLEYS AND FLATS. SLOPE <15%	P=200-300mm? MILD WINTERS 4-5 DRY MONTHS.	DEEP (>1m.) CONTINUOUS REGOSOLS & SIEROZEMS ON LOESS & SANDS	VERY SPARSE-STEPPIC	RAPID RUNOFF & INTERNAL DRAINAGE	MODERATE	SEASONAL GRAZING	NARROW MARSHY VALLEY FLOORS 74.000
	32 L-BROWN	DISSECTED PIEDMONT PLATEAU ON LOESS AND MARLS +100/200m. ALT.	FLAT TOPPED RIDGES AND DEEP GULLIES SLOPE <30%	P=300 mm? COOL WINTERS 5-6 DRY MONTHS	DEEP (>1m.) CONTINUOUS REGOSOLS & SIEROZEMS ON LOESS-SANDS-MARLS	SPARSE-STEPPIC	RAPID RUNOFF & INTERNAL DRAINAGE-	LOCAL SLIGHT	SEASONAL GRAZING	22.000
4 PIEDMONT PLAINS	41 ORANGE	KORDKUY PIEDMONT PLAIN -20/+50m. ALT.	REGULAR, VERY GENTLE SLOPE 1-2%	P=600 mm. MILD WINTERS 3 DRY MONTHS	DEEP CLAYS AND CLAY LOAMS-BROWN FOREST SOILS-(CULTIVATED)	LOCAL REMNANTS OF OAK FORESTS	MODERATE RUNOFF, SLOW INTERNAL DRAINAGE G.W. 1.5-6m.- CARBONATIC	SLIGHT, INCREASING TO THE WEST	IRRIGATED COTTON-DRY FARMED WHEAT	G.W. SALINITY-FROM C2 TO C4 TO N.& W. 18.500
	42 ORANGE	GORGAN PIEDMONT PLAIN WITH LARGE ALLUVIAL FANS. +60/150m.	REGULAR, GENTLE SLOPE 1-4% WITH WIDE UNDULATIONS	P=500-600mm. MILD WINTERS 3 DRY MONTHS	DEEP CLAY LOAMS BROWN FOREST SOILS-(CULTIVATED)	LOCAL REMNANTS OF OAK FORESTS	RAPID RUNOFF-MODERATE INT. DRAINAGE- G.W. 6-30m.- C3 S1/CARBONATIC	NO	IRRIGATED COTTON-DRY FARMED-IRRIGATED WHEAT	FREQUENT FLOOD TRANSIT 58.500
	43 ORANGE	TURENG TAPPEH PIEDMONT PLAIN (FANS PERIPHERY) -10/+60m.ALT.	NEARLY LEVEL SLOPE <1%	P=400-500mm. MILD WINTERS 3 DRY MONTHS	DEEP HEAVY CLAYS & LOAMS-BROWN FOREST-ALLUVIAL & HYDROMORPHIC SOILS	FULLY CULTIVATED	SLOW RUNOFF & INT. DRAINAGE- G.W. 1.5-4m.- SULPHATIC- C3 S1	SLIGHT	IRRIGATED WHEAT & COTTON	MODERATE FLOODING HAZARD 38.000
	44 ORANGE	DALAND PIEDMONT PLAIN WITH SMALL FLAT RIVER FANS. +30/120m. ALT.	REGULAR, NEARLY LEVEL SLOPE ~1%	P=500mm.? COOL WINTERS 3-4 DRY MONTHS	DEEP CLAY LOAMS BROWN FOREST TO CHESNUT SOILS.	LOCAL OAK FOREST REMNANTS	SLOW RUNOFF-MODERATE INT. DRAINAGE- G.W. 4.5m.- CARBONATIC/C1S1	NO	PARTLY IRRIGATED WHEAT & SOME COTTON.	22.000
	45 ORANGE	MINU DASHT PIEDMONT PLAIN WITH NARROW VALLEY FLOORS. 60/200m. ALT.	GENTLE SLOPE 1-3%. UNDULATING	P=400-500mm. COOL WINTERS 4 DRY MONTHS	DEEP CLAY LOAMS CHESNUT & ALLUVIAL SOILS	FULLY CULTIVATED	MODERATE RUNOFF & INT. DRAINAGE- G.W. 7-20m.-	NO	DRY FARMED WHEAT-IRRIGATED IN THE VALLEYS.	39.500
5 ALLUVIAL PLAINS	51 L-YELLOW	PAHLAVIDEJ LOWER REACHES GORGAN RUD RIVER BANKS. -12/-25m. ALT.	SLOPE <1%, REGULAR	P=400mm. MILD WINTERS 3 DRY MONTHS	DEEP LOAMS & SILT LOAMS-SALINE ALLUVIAL SOILS.	FULLY CULTIVATED	SLOW RUNOFF-MODERATE INT. DRAINAGE- G.W. 1-4m.- C4-CHLORIC	MODERATE TO HIGH	WHEAT & SOME IRRIGATED COTTON (BY RIVER WATERS)	37.000
	52 L-YELLOW	SANGAR SAVAR MIDDLE REACHES GORGAN RUD RIVER BANKS. 0/+40m. ALT.	SLOPE <2%. AWAY FROM RIVER, IRREGULAR	P=300 mm. MILD WINTERS 4 DRY MONTHS	DEEP SILTY CLAY LOAMS-ALLUVIAL & LIGHT BROWN SOILS.	STEPPIC/50% CULTIVATED	MODERATE RUNOFF & INT. DRAINAGE G.W. 2-20m.- C4-CHLORIC	MODERATE IN SUB SOIL	IRRIGATED WHEAT AND COTTON (BY RIVER WATERS)-GRAZING	FORMER RIVER COURSES 35.000
	53 L-YELLOW	GERI MIDDLE REACHES GORGAN RUD PLAIN 0/20m. ALT.	LEVEL WITH LOCAL DEPRESSIONS	P=300-350 mm. MILD WINTERS 4 DRY MONTHS	DEEP CLAY LOAMS & CLAYS-ALLUVIAL & SALINE LOW HUMIC GLEY SOILS	CULTIVATED	VERY SLOW RUNOFF-SLOW INT. DRAINAGE- G.W. 0-4m.- C4-SULPHATIC	MODERATE TO HIGH	DRY FARMED WHEAT-IRRIGATED COTTON (BY WELLS)	MARSHES & SHALLOW LAKES IN PLACES 61.000
	54 L-YELLOW	GONBAD QABUS MIDDLE REACHES GORGAN RUD PLAIN +30/80m. ALT.	LEVEL, MOSTLY REGULAR	P=300-350 mm. MILD WINTERS 4-5 DRY MONTHS	DEEP CLAY LOAMS-ALLUVIAL SOILS	CULTIVATED	SLOW RUNOFF-MODERATE INT. DRAINAGE- G.W. 4-10m.	SLIGHT OR NONE	DRY FARMED WHEAT & IRRIGATED COTTON.	FEW INCISIONS BY WATER COURSES 25.500
	55 L-YELLOW	UNDULATING ALLUVIAL PLAIN N.N.W. OF GONBAD +10/60m. ALT.	SLOPE 1-2% WITH ELONGATED WIDE DEPRESSIONS	P=300mm.? MILD WINTERS 4-5 DRY MONTHS	DEEP LOAMS-SALINE ALLUVIAL & BROWN SOILS	STEPPE-LOCALLY CULTIVATED	MODERATE RUNOFF & INT. DRAINAGE- G.W. VERY DEEP?	MODERATE, LOCALLY HIGH	GRAZING-SOME DRY FARMED/IRR. WHEAT.	MANY WIDE INCISIONS BY FLOOD CHANNELS 32.000
	56 L-YELLOW	SUFEYAN PLAIN INCISED BY UPPER GORGAN RUD +100/40m. ALT.	SLOPE 1-3%, REGULAR WITH DEEP WIDE GULLIES	P=400-600 mm. COOL TO COLD WINTERS 3-5 DRY MONTHS	DEEP CLAY LOAMS-LIGHT CHESTNUT SOILS & ALLUVIAL SOILS	CULTIVATED/GRASS COVERED GULLIES	RAPID RUNOFF & INT. DRAINAGE- G.W. VERY DEEP?	NONE TO SLIGHT	DRY FARMED WHEAT.	INCREASINGLY DISSECTED TO THE EAST. 55.000
	57 L-YELLOW	CHAT PLAIN INCISED BY ATRAK RIVER ALT. <100m.	SLOPE 1-2% WITH UNDULATIONS & DEEP GULLIES	P=200mm.? MILD WINTERS 5-6 DRY MONTHS	DEEP SILT LOAMS-SANDY LOAMS-SALINE ALLUVIAL SOILS.	DEGRADED/STEPPIC	RAPID RUNOFF & INT. DRAINAGE-G.W. ?	HIGH TO VERY HIGH	GRAZING	11.000
6 LOWLANDS	61 L-BLUE	COASTAL SAND BARS AND MARSHES -15/-25m. ALT.	LEVEL WITH SLIGHT UNDULATIONS	P=400-500 mm. MILD WINTERS 3 DRY MONTHS	DEEP SANDY LOAMS-SALINE ALLUVIAL SOILS/SOLONCHAKS	STEPPIC & SALINE MARSH VEGETATION	VERY SLOW RUNOFF AND INT. DRAINAGE G.W. 0-1.5m.- EXTREMELY SALINE	VERY HIGH	GRAZING-WASTE LAND-SOME WHEAT	29.000
	62 L-BLUE	QARAH SU DEPRESSION -10/-25m. ALT.	LEVEL SLIGHTLY CONCAVE	P=400-500mm. MILD WINTERS 3 DRY MONTHS	DEEP CLAYS LOW HUMIC CLAY & ALLUVIAL SOILS	MARSHY VEGETATION-70% CULTIVATED	VERY SLOW RUNOFF & INT. DRAINAGE-PONDING-SULPHATIC G.W. 0-2m.- C4	MODERATE	IRR./DRY FARMED WHEAT AND COTTON WASTE LAND	MARSHES AND SHALLOW PONDS IN PLACES 25.000
	63 L-BLUE	LOWLANDS OF THE ATRAK GORGAN RIVER INTERFLUVE WITHOUT BASINS -10/-20m. ALT.	NEARLY LEVEL, REGULAR	P=300 mm. MILD WINTERS 3-4 DRY MONTHS	DEEP FINE SANDY LOAMS-SILTY CLAY LOAMS, SOLONCHAKS AND SALINE ALLUVIAL SOILS	SALINE STEPPIC 15% CULTIVATED	VERY SLOW RUNOFF-SLOW INT. DRAINAGE G.W. 1-3m.- EXTREMELY SALINE	HIGH	GRAZING-WASTE LAND-SOME WHEAT	46.000
	64 L-BLUE	FORMER GORGAN RUD MEANDERING FLOOD PLAIN -15/-20m. ALT.	NEARLY LEVEL-IRREGULAR	P=300 mm. MILD WINTERS 3 DRY MONTHS	DEEP FINE SANDY LOAMS-SILTY CLAY LOAMS SALINE ALLUVIAL SOILS & SOLONCHAKS	SALINE STEPPIC 15% CULTIVATED	VERY SLOW RUNOFF & INTERNAL DRAINAGE-SOME PONDING-G.W. 0-3m.- EXTREMELY SALINE	HIGH	WASTE LAND-GRAZING-SOME WHEAT	MANY TRACES OF FORMER RIVER COURSES-A FEW ELONGATED HILLS 42.000
	65 L-BLUE	LOWLANDS AND BASINS OF THE ATRAK-GORGAN RIVER INTERFLUVE -20/+60m.	NEARLY LEVEL-REGULAR	P=250-300 mm. MILD WINTERS 4 DRY MONTHS	DEEP FINE SANDY LOAMS-SILTY CLAY LOAMS/SOLONCHAKS	SALICORNIA STEPPE	MODERATE RUNOFF-SLOW INTERNAL DRAINAGE G.W.3-7m. EXTREMELY SALINE	HIGH	WASTE LAND-GRAZING	MANY SMALL BASINS AND FLOOD CHANNELS MANY HILLOCKS 164.000
									TOTAL AREA →	1.503.500 ha

EVALUATION of LAND POTENTIALITIES (1968)

LAND UNIT	IRRIGATED ANNUAL CROPS	IRRIGATED ORCHARDS	DRY FARMING	RANGE	FOREST	RECREATION	URBAN and INDUSTRIAL AREA
11	6/-	6/-	6/-	(4)/(Ao)	1/Ao	(1)/B+	6/-
12	6/-	6/-	(2)/(Bo)	(1)/(Ao)	2/Ao	-/-	6/-
21	6/-	6/-	(2)/Co	(1)/Ao	1/Ao	(1)/B+	(6)/-
22	6/-	6/-	6/-	2/Bo	(3)/C+	-/-	-/-
23	6/-	6/-	6/-	3/Bo	6/-	6/-	(4)/(C+)
31	5/-	6/-	6/-	2/Ao	6/-	6/-	6/-
32	6/-	6/-	6/-	2/Bo	5/-	6/-	6/-
41	1/C+	(2)/A+	1/B+	-/-	(4)/(Ao)	-/-	2/B+
42	1/(Bo)	1/(Bo)	1/(Bo)	-/-	(1)/(Ao)	(1)/(Ao)	1/(Bo)
43	2/Bo	2/Bo	2/Bo	-/-	(4)/-	6/-	2/(Bo)
44	1/(Ao)	1/(Ao)	1/(Ao)	-/-	(1)/(Ao)	(1)/-	1/Ao
45	1/(Ao)	2/(Ao)	1/Ao	1/Ao	(1)/(Ao)	-/-	1/(Ao)
51	2/C+	5/-	5/-	2/Ao	6/-	6/-	3/Bo
52	2/B+	2/B+	5/-	3/Ao	6/-	6/-	2/Bo
53	2/(B+)	(2)/(B+)	3/Ao	(2)/(Ao)	(4)/(Ao)	6/-	(2)/Ao
54	1/A+	2/A+	3/Ao	-/-	6/-	6/-	1/Ao
55	(2)/(A+)	(2)/(A+)	3/Bo	1/Ao	6/-	6/-	(2)/Ao
56	2/(Co)	2/(Co)	1/(Co)	1/(Co)	(3)/(Bo)	6/-	(1)/(Ao)
57	2/C+	5/-	6/-	3/Bo	6/-	6/-	1/Ao
61	6/-	6/-	6/-	2/Ao	6/-	6/-	(3)/(Bo)
62	2/C+	3/C+	1/B+	1/A+	(4)/A+	6/-	(3)/(B+)
63	5/C+	5/C+	5/B+	1/Ao	6/-	6/-	2/B+
64	5/C+	5/C+	5/C+	1/Ao	6/-	6/-	(2)/B+
65	5/D+	5/D+	6/-	1/Bo	6/-	6/-	(3)/B+

EXPLANATION of LAND EVALUATION SYMBOLS

LAND SUITABILITY CLASS (FOR A GIVEN TYPE OF LAND USE AND AFTER INPUT) ← 1/Ao → INITIAL INPUT REQUIREMENT LEVEL (LAND CONSERVATION AND IMPROVEMENT WORKS)

LAND SUITABILITY CLASS		INITIAL INPUT REQUIREMENT LEVEL	
1	HIGH SUITABILITY	A	LOW
2	MODERATE SUITABILITY	B	MODERATE
3	MARGINAL SUITABILITY	C	HIGH
4	RESTRICTED SUITABILITY (UNDER SPECIAL CONDITIONS)	D	VERY HIGH
5	UNDETERMINED SUITABILITY	o	LAND CONSERVATION WORKS
6	UNSUITABLE	+	LAND IMPROVEMENT WORKS

WHEN THE EVALUATION SYMBOL APPLIES ONLY PARTLY WITHIN THE LAND UNIT, THE SYMBOL IS PLACED BETWEEN PARANTHESIS.

REGIONAL MAP of LAND RESOURCES and POTENTIALITIES

GORGAN REGION-EAST MAZANDERAN

FIRST EDITION 1968*

BY: R.F. VAN DE WEG F.A.O. ASS. EXPERT.
P.J. MAHLER F.A.O. EXPERT.
M. BORDBAR ENGINEER SOIL INSTITUTE
M. ASKARI ENGINEER SOIL INSTITUTE

Table 12.5 Two separate legends used in a FAO survey of the regional land resources in the Gorgan Region, Iran. Left: the land resources mapping legend giving the physical characteristics of all terrain units mapped. Right: the assessment of the land potentialities for specific types of land utilization (6 classes are distinguished) (Bodbar and v.d. Weg, 1969).

purposes of assessing the terrain for such military purposes as trafficability/passability, sensor emplacement and performance, concealment, etc. Terrain analogs are an essential part of the concept and serve to predict cross-country vehicle performance and various terrain conditions relevant for military needs in remote, unknown terrain on the basis of conditions found in test areas. Difficulties of this approach are, of course, numerous and relate, e.g., to the inadequacy of aerial photographs to detect essential microrelief features and the seasonal changes in terrain conditions (summer/winter; wet/dry season) Terrain in this MEGA (Military Evaluation and Geographic Areas) programme was analysed and quantified factor-wise. These factors were grouped in the following three factor families:

. general factors (physiography, surface conditions, etc.)
. form factors (slope, relief, etc.)
. ground and vegetation factors

Benn and Grabau (1968) elaborate on cross-country vehicle performance and distinguish four main factor families, namely: surface geometry, surface composition, vegetation and hydrologic geometry.

Factor maps are combined to factor family maps and these subsequently to terrain complex maps (Anstey, 1974). The terrain ultimately was assessed on the basis of the gathered, processed and presented terrain data and on specific military requirements, vehicle performance, etc. (Parry et al., 1968; Parry and Bleswick, 1973). The terrain classification and the trafficability assessment are used for several other purposes too. Another considerably less elaborate method for military terrain classification has been developed in Canada (Parry et al., 1968). It is evident from the aforesaid that many civil engineers in Anglo-Saxon countries have opted for a synthetic land system type of survey, whereas the military engineers have preference for a parametric approach.

12.5.2. Agriculture

Agriculturalists are another large group of specialists with a long-standing interest in terrain classification and land evaluation (Vink, 1975). Since optimization of the agricultural land utilization is a major aim,logically marked emphasis is on matters related to soil conditions. Initially, in fact, soil survey alone was considered sufficient for the purpose but gradually the importance of other terrain factors such as slope angle, slope processes, hydrological conditions, etc. was more appreciated and the pragmatic methods now in use for land evaluation thus evolved.

The method of land capability mapping developed in the nineteen-thirties by the Soil Conservation Service of the US Department of Agriculture is probably the most widely used land evluation system. In this USDA system, the land is divided into eight capability classes on the basis of inherent, permanent characteristics that limit its use or bring about deteoriation by erosion, flooding, etc. Class I has none or only minor limitations and the risks or limitations gradually increase to extremely severe (Class VII) and absolute (Class VIII). Only the classes I-IV are suitable for cultivation. Class VIII is considered suitable solely for watershed protection, wildlife and recreation. The areal extent of the various capability classes in fact amounts to grouping of soil mapping units and it is thus evident that soil characteristics such as texture, structure, drainage are important criteria. Other partly geomorphological criteria are also considered, however, and include matters such as slope angle, erosion susceptibility, liability to flooding, rooting depth and climate. Fig. 12.6 gives an example. Further sub-division in capability sub-classes and units also is pursued, and here relief and drainage factors also play a role. It should be emphasized, however, that the classification is not devised as a productivity rating but primarily concerns the risks of erosion. Therefore, cultivation of land in the classes II through IV requires in this order increasingly substantial conservation measures. On the other hand, this concept also implies that some highly productive land (e.g., vineyards) may be found in the classes V, VI and VII,which are considered suitable for a limited number of uses only.

Another, more recent method of land evaluation has been developed by the Bureau of Reclamation of the US Department of the Interior (USBR). This land classification system is meant for irrigated agriculture (USBR, 1953; Klingebiel and Montgomery, 1961) and, contrary to to the USDA system, is oriented in the first place towards the productivity of the land. Another difference is that the classification is not solely based on physical factors but also on economic factors. These two groups of factors determine the so-called payment capacity which can be defined as the actual money gained by the farmer after all costs (including cost of living) have been substracted (Beek, 1980). Six classes are distinguished in the system. Class 1 is most suitable for irrigation and consequently has the highest payment capacity. Class 3 has the lowest potential in this respect. Class 4 is used for land that is suitable for specific uses only or that has severe limitations but nevertheless might be put under irrigation. Land mapped as class 5 requires further study before ultimate assessment can be made and all land unsuitable for irrigation is grouped in class 6. Among the physical factors that affect the classification soil conditions rank high. Soil texture, depth, moisture retention, permeability of least permeable layer, salinity, sodicity and the percentage cobble/gravel/stoniness are parameters that, for every class, should be within strict limits specified in the system. Other physical parameters relate to relief, tree cover and

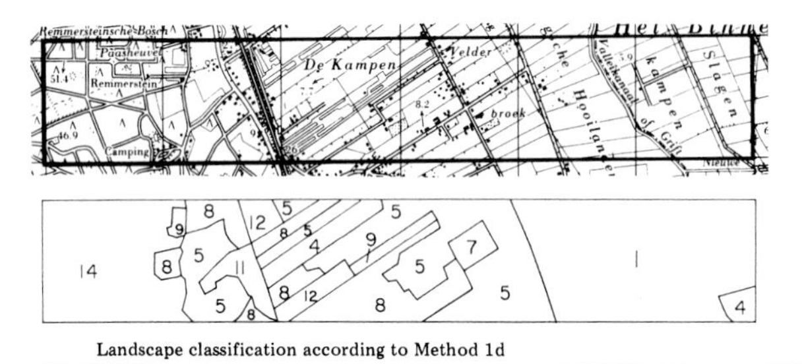

Landscape classification according to Method 1d

Spaces edge transparency	Size of space		
	Large (>100 ha)	Medium (25–100 ha)	Small (<25 ha)
Open-edge >75% transparent	1	4	7
Half-open-edge 25–75% transparent	2*	5	8
Enclosed-edge <25% transparent	3*	6*	9

Associations of space-mass		Mass	
Mass: transparent	10*	Transparent	13*
Mass: transparent and non-transparent	11	Transparent and non-transparent	14
Mass: non-transparent	12	Non-transparent	15*

*Does not occur in this area.

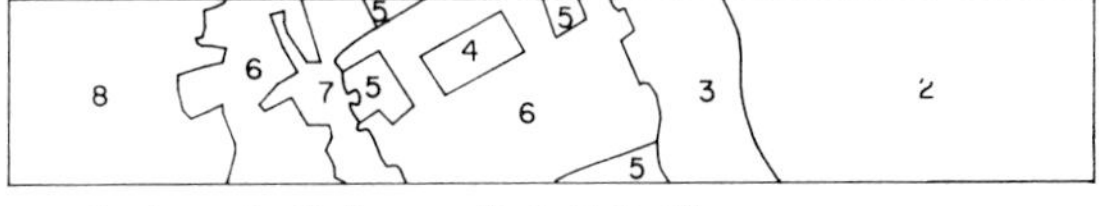

Landscape classification according to Method 2

Space

Map code	Classification of observation points	Summed angles of objects in viewing-circle (degrees)		
		1200 m Extraocular	500–1200 m Ocular	500 m Intraocular
1*	Very wide view	>180	<180	<180
2	Wide view	>60 SB^f or 100–180	<180	<180
3	View bounded, but with far outlooks	5–60 SB^f 5–100	>240	<120
4	Bounded view	<5	>300	<60
5	Closely bounded view but with far outlooks	5–60 SB^f 5–100	<120	>240
6	Closely bounded view	<5	<60	>300

Other landscape types: 7 = periphery area; 8 = forest.

^{f}SB = angle of view within a single arc. *Does not occur in this area.

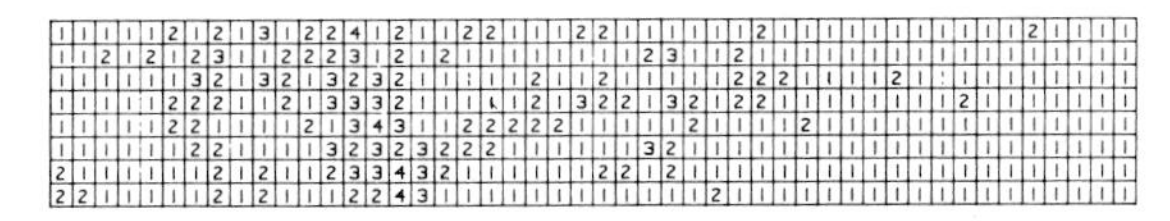

Landscape classification according to Method 3 Visual complexity is ranked from 1 (small) to 5 (large). (Type 5 does not occur in this area.)

Fig. 12.6 Topographical map of an area near Rhenen, the Netherlands, 1:50,000, with a visual landscape classification according to a compartment (1), breadth of view (2), and a gridcell (3) approach as described by De Veer and Burrough, 1978.

drainage conditions. The geomorphological contribution to surveys related to this classification is not in the first place the analysis of relief conditions, as one might expect, but rather in the prediction and explanation of the soil factors just mentioned and in particular matters such as cobble/gravel occurrences, the distribution of saline and sodic conditions, etc. A disadvantage of the USBR system is that it can only be directly applied if it is clear from the onset that irrigated agriculture is to be aimed at. The situation becomes more complex when a choice has to be made from several alternative uses.

The matter of multidisciplinary terrain classification for agriculture has been given ample thought in recent years, following an experts' consultation meeting organised by FAO in 1972 (Brinkman and Smyth, 1973). This resulted in a FAO framework for land evaluation (FAO, 1976). Other important contributions to this field were made by Van der Kevie (1976), Beek (1978), and others. Essential concepts in the methodology developed by Beek are land utilization types (LUT), land (use) requirements (LR), with their determinants (LRD), and land qualities (LQ). Land utilization types are defined much more specifically by Beek when compared with section 12.4 of this book. Ultimately, a land (mapping) unit (LU) is combined with a land utilization type (LUT) to form a land use system (LUS). The basic concept of the FAO framework is now to assess the land in terms of land use systems which are composed of a mappable physical constituent and a land use type. The aim is to arrive at a comparison of the performance of use in the various mapping units. The inherent present or alternative types of land use and land properties or characteristics of the land, for these purposes have been combined to translate them into land qualities for specific land use types. These combinations and their limiting values change with the land use type concerned. The geomorphologist has a distinct part in this multidisciplinary approach to the problem of agricultural assessment of the land. Close cooperation with other earth scientists, particularly soil scientists and agriculturalists, economists, etc., is required to obtain optimum results.

12.5.3 Landscape architecture and environmental management

Needs for terrain classification and evaluation have also arisen in several circles outside those of engineers and agriculturalists. The importance of assessing the visual scenic qualities of the land has already been mentioned in section 10 on the application of geomorphology in planning. The scenic qualities may relate to the qualities of a certain site and/or to the surroundings visible from that site. Cooke and Doornkamp (1974) give examples from each of these approaches in the U.K. De Veer and Burrough (1978) describe various types of so-called physiognomic landscape mapping developped in the Netherlands, where the problem is basically different because of the absence of any sizeable relief. They distinguish three major approaches, which they indicate as the compartment, breadth of view and grid cell system respectively. Much in this system also depends on the view obtainable and on the open or closed character of the land. The maps and legends of fig. 12.6 give an example of each of these three approaches (De Veer and Burrough, 1978).

The visual assessment of terrain is a subject of its own. Evidently, these methods relate to geometrical, psychological and other aspects of the land but have only a remote relation within synthetic surveys based on landscape ecology. When mapping the environment, and in particular its deterioration due to pollution, erosion, building activities, quarrying, etc.,is concerned, still other aspects have to be considered which in part involve landscape ecology (Chapman, 1969; Craik, 1970; Dasmann, 1968; Hedberg, 1976; Lowenthal, 1967; Masters, 1974; Tank, 1976). Journaux (1975) has developed a legend for this purpose which is composed of two parts, namely a description of the various aspects of the environment and a specification of the environmental dynamics. The first group includes matters such as relief, water features, climate, built up areas, agricultural fields, and 'green' areas. The specification of the environmental dynamics includes the land degradation due to various natural and anthropogenic causes and the effects of water and air pollution. Protective measures and structures are also indicated in the maps.

12.6 The Role of Geomorphology, a Summary

The geomorphological contribution to synthetic survey of terrain and to land evaluation merits some further discussion. Although the importance of landform is generally recognized and also the role of other geomorphological factors, such as processes, is increasingly understood, the proper position and the mode of integration of geomorphology in the context of terrain classification and land evaluation are still subject to much confusion. Genetic aspects are often insufficiently emphasized and the integrating function of geomorphology neglected. As a result,such kinds of synthetic surveys risk becoming tabular summations of physical terrain data to the detriment of the applicability of the survey results. Several authors have in recent years made substantial contributions to the optimization of geomorphological approaches to terrain classification,such as Way (1978) and Van Zuidam and Van Zuidam-Cancelado (1979). The latter devised an 'ITC System of Terrain Classification' complementing the existing 'ITC System of Geomorphological Survey', outlined in the previous section. The parametres used in the ITC system to arrive at the terrain classification are grouped under the following headings: relief, geomorphological processes, rock type, soil characteristics, surface and groundwater and natural vegetation/land use.

The class limits used for the individual parameters are in part borrowed from other systems of terrain classification and land evaluation to concur with existing international trends. The hierarchic levels of the ITC terrain classification system are given in table 12.6 with their approximate equivalent in some major classification systems mentioned in this section.

According to the author, ideally the following procedures should be applied:

1. Analytical geomorphological surveys, taking into consideration: landform and relief characteristics, materials (hard rock and unconsolidated materials), morphogenesis (including processes, chronology, etc.)
2. Delimitation of geomorphological units on the basis of 1.
3. Translation of the geomorphological units into terrain units, for which some adaptation and re-arrangement may be required to achieve practical mapping units.
4. Synthetic surveys of the terrain in conjunction with 3 by integrating the geomorphological data with the pedological, hydrological etc.,information on the area in cooperation with other experts.
5. Storage of the terrain data obtained in a handy form for easy access. The inventory of the terrain is now completed, but no specific application is possible until the following steps are completed.
6. Selection of relevant parameters and the determination of critical class limits for specific purposes.
7. Classification of the terrain for specific purposes on the basis of the data selection and processing mentioned under 6.

The contribution of the geomorphologist to synthetic surveys (Young, 1969, 1970, 1976) requires integration on the level of physical aspects of the land; a higher level of integration is required to incorporate economic and other aspects touched upon in the previous sub-section.

It is evident from the variety of synthetic surveys outlined in this section that the aspects or parameters to be used,and thus the legends and the maps produced,depend first and foremost on the purpose of the survey. In many cases a complete analytical geomorphological survey may not be required; similarly, a fully integrated synthetic survey for many purposes may not be needed. Therefore, it is advisable to take a practical attitude to the matter of surveying for development and to adapt the scope of the survey and the aspects to be incorporated to the aim of the project. In other words, it is appropriate in many cases to proceed directly to pragmatic, problem-oriented surveys aiming at terrain classification or hazard zoning for special purposes and including a selection of analytical and synthetic information relevant for a specific aim and in a given situation. Several types of those special purpose surveys are described in the following sections of this book, together with legends and examples. They relate to matters such as flooding, drought, avalanche hazard, earthquake risk zoning, etc.

Level	ITC–Geomorphological classification system			Physiographic classification	ITC Terrain classification	Oxford–MEXE	CSIRO Geomech. Div. (Grant)	CSIRO Land Research and Reg. Surv. (Christian)	DOS	Soviet system
	Level	Main characteristics	Scales							
1	Geomorphological province	Highly generalised. Genesis and lithology are most important. Displays a small range of surface form and properties expressive of a lithological unit or a close lithological association of comparable geomorphic evolution	⩽1:250,000	Physiographic province	Terrain province	Land region (and Land system)	Terrain province	Complex land system (and land system)	Land region/province (as a part of the Land system)	Landscape (and Mestnosti)
2	Main geomorphological unit	Moderately generalised. Relief, lithology and genesis are the main criteria for the classification. Displays a recurrent pattern of genetically linked terrain components	⩾1:250,000	Main physiographic unit	Terrain system (pattern)	Land system	Terrain pattern	Land system	Land system	Mestnosti (and Urochischa)
3	Geomorphological unit	No or minor generalisations in areal classes. Details may be generalised. Relief, lithology and genesis are the main classification criteria. Reasonably homogeneous and fairly distinct from surrounding terrain	⩾1:50,000	Detailed physiographic unit	Terrain unit	Land facet	Terrain unit	Land unit	Land facet	Urochischa
4	Geomorphological detail	No generalisation in areal classes. No or minor generalisation in details. Relief is most important classification criterion. Basically uniform in landform lithology, soil, vegetation and processes.	⩾1:10,000	Physiographic element	Terrain component	Land element	Terrain component	Site	—	Facies

Table 12.6 Table showing the four main hierarchic classification levels of the ITC terrain classification system as compared to some other approaches to synthetic surveys of terrain (van Zuidam and van Zuidam-Cancelado, 1979).

References

Aitchison, G.D., and Grant, K., 1967. The P.U.C.E. programme of terrain description, evaluation and interpretation for engineering purposes.

Aitchison, G.D., and Grant, K., 1968. Terrain evaluation for engineering. Proc. 4th Reg. Conf. Soil Mech. and Found. Eng., 1: 125–146.

Anstey, R.L., 1974. The Natick landform classification system. Final report. U.S. Army Material Systems, Analysis Activity, Aberdeen Proving Ground (Md.) Techn. Rep. TR-100, 45 pp.

Bawden, M.G., 1967. Applications of aerial photography in land system mapping. Photogrammetric Record 5: 461–464.

Beckett, P.H.T., and Webster, R., 1962. The storage and collation of information on terrain. MEXE, Interim report, Christchurch, U.K., 39 pp.

Beckett, P.H.T., and Webster, R., 1965. A classification System for Terrain. Methods and Principles. MEXE Report 872, 247 pp.

Beckett, P.H.T., 1965. Field trials of a terrain classification system-statistical procedure, MEXE report 873.

Beckett, P.H.T., 1968. Method and scale of land resource surveys in relation to precision and cost. In: Stewart, G. A. (Editor), Land evaluation, Proc. of a CSIRO Symp.: 53–63.

Beckett, P.H.T., and Webster, R., 1969. A review of studies on terrain evaluation by the Oxford MEXE–Cambridge group 1960–1969. MEXE Report 1123, Christchurch, U.K.

Beckett, P.H.T., and Webster, R., 1971. The development of a system for terrain evaluation over large areas. Royal Engineers Journal, 85: 243–258.

Beckett, P.H.T., et al., 1972. Terrain evaluation by means of a data bank. Geogr. J., 139: 430–456.

Beckett, P.H.T., 1975. The cost-effectiveness of terrain evaluation. 2 Vols. Review of fieldwork 1971–74. Final techn. Report. (Oxford Univ., Dept. of Agric. Sci.), 217 pp.

Beek, K.J., 1978. Land evaluation for Agricultural Development. ILRI publ., Wageningen, 23, 333 pp.

Benn, B.O., and Grabau, W.E., 1968. Terrain evaluation as a function of users requirements. In: Stewart, G.A. (Editor), Land evaluation. Proc. of a CSIRO Symp.: 64–77.

Blakes, D.H., and Paymans, K., 1973. Reconnaissance mapping of land resources in Papua New Guinea. Austral. Geogr. Stud., 11(2): 201–210.

Bordbar, M., and Weg, R. v.d., 1964. Presentation of regional land resources and potentialities maps–methods and working procedures. Cycl. Report Land evaluation seminar, Teheran, 6 pp.

Bourne, R., 1931. Regional survey and its relation to stock taking of the agricultural resources of the British Empire. Oxford Forestry Memoirs, 13: 16–18.

Brink, A.B., et al., 1966. Report of the working group on land classification and data storage. MEXE Report 940.

Brinkman, R., and Smyth, A.J., 1973. Land evaluation for rural purposes. Summary of an expert consultation, Wageningen, the Netherlands, 6–12 Oct. 1972. (International In-

stitute for Land Reclamation and Improvement, Wageningen). ILRI Publ. 17, Wageningen, 116 pp.
Brookfield, H.C., 1969. On the environment as perceived. Progress in Geography, I, London: 51–80.
Brunt, M., 1967. The methods employed by the Directorate of Overseas Surveys in the assessment of land resources. Etudes de Synthèse, 6: 3–10.
Buringh, P., 1960. Photo-interpretation in the ITC soil section. ITC Publ. B 2, 16 pp.
Butzer, K.W., 1971. Environment and archaeology: an ecological approach to prehistory, 2nd edition, 703 pp.
Cadigan, R.A., 1972. Terrain classification–a multivariate approach. U.S.Geol. Survey, Washington D.C., USGS–GD–72–024 , 65 pp.
Canada Geological Survey, 1973. Terrain classification and sensitivity maps. Open File Reports, 144, 145, 151 and 157
Canada Geological Survey, 1973. Terrain maps–Mackenzie Valley. Open File Reports, 96, 97, 108, 117, 119, 120, 121, 125, 155, 158, 167, 189, 191
Chapman, J.D., 1969. Interactions between man and his resources. In: Committee on resources and man, San Francisco
Chorley, R.J. (Editor), 1969. Water, Earth and Man–Synthesis of hydrology, geomorphology and socio-economic geography. Methuen & Co., Ltd., London, 588 pp.
Christian, C.S., and Stewart, G.A., 1953. General report on survey of the Katherine-Darwin region, 1946. CSIRO Austral., Land Res. Ser., 1
Christian, C.S., et al., 1957. Geomorphology. Ch. 5. In: Guidebook to research data for arid zone management, UNESCO, Paris: 51–65.
Christian, C.S., 1958. The concept of land units and land systems. Proc. 9th Pacific Sci. Congr., 20: 75–81.
Christian, C.S., and Stewart, G.A., 1968. Methodology of integrated surveys. In: Proc. UNESCO Conf. Aerial Surveys and Integrated Studies, Toulouse, 1964. UNESCO, Paris: 233–288.
Coates, D.R., 1971. Legal and environmental case studies in applied geomorphology. Environmental Geomorphology, Binghamton, New York: 223–242.
Coates, D.R., 1971. Environmental geomorphology. State Univ. New York, Binghamton, 262 pp.
Coates, D.R. (Editor), 1972/74. Environmental geomorphology and landscape conservation. 3 Vols., Benchmark papers in geology, 485 pp.
Conklin, H.E., 1959. The Cornell system of land classification. J. Farm Econ., 41: 548–557.
Cooke, R.U., and Doornkamp, J.C., 1974. Geomorphology in environmental management: an introduction, Clarendon Press, Oxford, 413 pp.
Craik, K.H., 1970. Environmental psychology. New Direction in Psychology, 4, 21 pp.
C.S.I.R.O., 1953–74. Land Research Series, Melbourne
Dasmann, R.F., 1968. Environmental Conservation Wiley, New York , 375 pp.
Demek, J., 1973. Quaternary relief development and man. Geoforum, 15: 68–71.
Detwyler, T.R., 1971. Man's impact on environment. McGraw-Hill, New York.
Dowling, J.W.F., 1967. The classification of landforms in northern Nigeria and their use with aerial photographs in engineering soil survey. Publ. British Transport and Road Res. Lab.
Dowling, J.W.F., 1968. Land evaluation for engineering purposes in northern Nigeria. In: Stewart, G.A. (Editor), Land evaluation. Proc. of a CSIRO Symp.: 147–159.
Dowling, J.W.F., 1968. The classification of terrain for road engineering purposes. Proc. Conf. Civ. Eng. Problems Overseas. Inst. Civ. Eng., London: 285–310.
Duffield, D.B., and Coppock, J.I., 1975. The delineation of recreational landscapes: the role of a computer-based information system. Trans. Inst. Br. Geogr.: 141–148.
Dunn, M.C., 1974. Landscape evaluation techniques–an appraisal and review of the literature. Centre for urban and regional studies, Birmingham, Working paper 4, 68 pp.
Dyne, G. van, 1969. The ecosystem concept in natural resource management. Acad. Press, New York, London
F.A.O., 1974. Approaches to land classification. F.A.O. Soil Bull., 22, 120 pp.
F.A.O., 1976. A Framework for Land evaluation. ILRI Publ., 22, Wageningen. F.A.O. Soils Bull., Rome 32, 72 pp.
Flawn, P.T., 1970. Environmental geology: conservation, land use planning and resource management. Harper Ltd., N.Y., 313 pp.
Fränzle, O., 1971. Physische Geographie als quantitative Landschaftsforschung. Schriften Geogr. Inst. Univ. Kiel: 297–312.
Garland, G.G., 1975. Engineering geomorphology–a review with two case studies in Cochem (B.R.D.)and Isarco Valley (Italy). ITC M.Sc. thesis, Enschede, 102 pp.
Garland, G.G., 1976. A note on two different methods of engineering geomorphological mapping. S. Afr. Geogr. J., 58(2): 134–140.
Gavrilyuk, F.Ya., and Val'kov, V.F., 1972. Land capability criteria (in russian). Pochvovedeniye, 2: 14–21.
Gibbons, F.R., and Downes, R.G., 1964. A study of the land in southwestern Victoria, 289 pp.
Gimbarzevsky, Ph., 1972. Terrain analysis from small-scale aerial photographs.Resource satellites and remote airborne sensing for Canada. Proc. 1st Can. Symp. Remote Sensing, Ottawa, 2: 367–378.
Grant, K., 1968. A terrain evaluation system for engineering. CSIRO Austral. Div. Soilmech. Techn. Paper, 27 pp.
Grant, K., 1971. Terrain classification for engineering purposes. Photo interpretation 1(3): 15–23.
Grant, K., 1972. Terrain classification for engineering purposes of the Melbourne area, Victoria. CSIRO, Div. of Appl. Geomech. Techn. paper 11, 216 pp.
Grant, K., 1974. The P.U.C.E. programme for terrain evaluation for engineering purposes. Procedures for terrain classification. CSIRO Div. Appl. Geomech. Techn. Paper, 82 pp.
Haantjens, H.A., 1968. The relevance for engineering of principles, limitations and developments in land system surveys in New Guinea. Proc. 4th Conf. Austral. Road Res. Board, 4(1) 593–1: 612.
Haantjens, H.A., 1975. Procedures for computer storage of soil and landscape data from Papua New Guinea. Geoderma 13(2): 105–140.
Haase, G., 1961. Hanggestaltung und ökologische Differenzierung nach dem Catena Prinzip. Peterm. Geogr. Mitt., 105: 1–8.
Hedberg, O. (Editor), 1976. Ecological studies in the Mtera Basin, Tanzania, SWECO Publ., Stockholm, 173 pp.
Hemstock, R.A., and Macfarlane, I.C., 1972. Problems of northern terrain classification. Proc. Can. Northern Pipeline Res. Conf., NRC Techn. Mem. 104: 267–276.
Herz, K., 1973. Beitrag zur Theorie der landschaftsanalytischen Massstabsbereiche. Peterm. Geogr. Mitt., 117(2): 91–96.
Hughes, O.L., 1972. Surficial geology and land classification Mackenzie Valley transportation corridor. Can. Northern Pipeline Res. Conf., NRC Techn. Mem. 104: 17–24.
Hunter, J.M., 1980. River blindness in Nangodi, Northern Ghana: a hypothesis of cyclinal advance and retreat. In: N.D. McGlashan (Editor), Medical geography techniques and field studies, London.
Isachenko., A.G., 1973. Principles of landscape science and physical geographic regionalisation . (J.S. Massey, Editor Melbourne Univ. Press, 311 pp.

Jacobs, P. and Way, D., 1969. Visual analysis of landscape development. Harvard Univ., Dept. Landscape Architecture, Cambridge Mass., 53 pp.

Journaux, A., 1975. Légende pour une carte de l'environnement et de sa dynamique. Publ. Fac. Lettres et Sciences Humaines. Univ. de Caen, 15 pp.

Kenzie, G.D., Mc., 1975. Investigations in environmental geoscience. Burgess Publ. Co., Mineapolis, 173 pp.

King, R.B., 1970. A parametric approach to land system classification, Geoderma, 4: 37–46.

Klingebiel, A.A., and Montgomery, P.H., 1961. Land-capability classification. Soil Conservation, Agric. Handb. 210, Washington D.C.

Klink, H.J., 1966. Die naturräumliche Gliederung als Forschungsgegenstand der Landeskunde . Ber.z.D. Landesk., 36: 223–246.

Kugler, H. (Editor), 1979. Relief und Naturraumkomplex. Wiss. Beitr. Martin Luther Univ., Halle-Wittenberg, 101 pp.

Lafarque, B.C., 1977. Review of concepts of land classification. In: Stewart, G.A. (Editor), Land evaluation, Proc. of a CSIRO Symp.: 11–28.

Leser, H., 1973. Zum Konzept einer angewandten Physchischen Geographie. Geogr. Ztschr., 61(1): 36–46.

Leser, H., 1974. Angewandte physiche Geographie und Landschaftsökologie als regionale Geographie. Geogr. Ztschr., 62(3): 161–178.

Leser, H., 1976. Landschaftsökologie. Eugen Ulmer, Stuttgart, 432 pp.

Lewin, J., 1975. Geomorphology and environmental impact statements. Area, 7(2): 127–129.

Lintuns, D.L., 1960. The assessment of scenery as a natural resource. Scott. Geogr. Mag., 84: 219–235.

Löffler, E., 1974. Land resources surveys in Ostneuguinea. Ein Beispiel aus der angewandten Geomorphologie (Land resources survey in East New Guinea–an example of applied geomorphology). Geographische Rundschau, 2: 60–63.

Lowenthal, D. (Editor), 1967. Environmental Perception and Behaviour. Univ. Chicago, Dept. Geogr., Res. Paper 109, 88 pp.

Mabbutt, J.A., and Stewart, G.A., 1965. Application of geomorphology in integrated resources surveys in Australia. Revue Géom. dyn., 6 : 1–13.

Mabbutt, J.A., 1968. Review of concepts of land classification. In: Stewart, G.A. (Editor), Land evaluation, Proc. of a CSIRO Symp.: 11–28.

Maletic, J.T., 1970. Land classification principles. Soil Sci. Training Inst. U.S. Bur. Reclam., Denver (Col.) (cyclostyled)

Marsz, A., 1974. A new method of physiographic regionalization. Questiones Geographicae, 1: 97–107.

Masters, G.M., 1974. Introduction to environmental science and technology. John Wiley & Sons., New York, London, 404 pp.

McKenzie, G.D., and Utgard, R.O. (Editor), 1975. Man and his physical environment. Readings in environmental geology. 2nd Ed., Burgess Publ., Minneapolis, 388 pp.

Mitchell, C.W., 1971. An appraisal of a hierarchy of desert land units. Geoforum, 7: 69–79.

Mitchell, C.W., 1973. Terrain evaluation. An introductory handbook to the history, principles and methods of practical assessment. Longman Group Ltd., London, 221 pp.

Möller, S.G., 1972. A system of describing and classifying information concerning landforms. (National Swedish Inst. for Building Research, Stockholm, Building Research, D8), 81 pp.

Moss, R.P., 1969. The appraisal of land resources in tropical Africa. Pacific Viewpoint, 10: 18–27.

Nakano, T., et al., 1962. Geographical surveys by means of aerial photo interpretation in Japan. Trans. 1st Symp. Comm. VII, I.S.P., Delft: 307–310.

Nechayev, L.A., and Fedorin, Yu. V., 1973. Method for classifying land capability in southeastern Kazakhstan. Pochvovedeniye, 12: 121–126.

Neef, E., 1964. Zur grossmassstäbigen landschaftsökologischen Forschung. Peterm. Geogr. Mitt., 108; 1–7.

Neef, E., 1967. Die theoretischen Grundlagen der Landschaftslehre. Haack Verlag, Leipzig, 167 pp.

Nikolayev, V.A., 1974. Principles of landscape classification. Soviet Geography: Review & Translation, 15(10): 654–661.

Nossin, J.J. (Editor), 1977. Proc. ITC Symp. Surveys for Development. Elsevier Sci. Publ. Comp., Amsterdam, 182 pp.

Pakistan Meteorological Service, 1965. An Integrated Survey of the Porali Plain 1964, Karachi, 120 pp.

Parry, J.T., et al., 1968. Terrain evaluation in mobility studies for military vehicles. In: Stewart, G.A. (Editor), Land evaluation. Proc. of a CSIRO Symp.: 160–171.

Parry, J.T., and Bleswick, A.A., 1973. The application of two morphometric terrain classification systems using air-photo interpretation methods. Photogrammetria, 29: 153–186.

Pečsi, M., 1974. Complex environmental studies. Geographical questions. Man and environment: 59–65.

Reiner, E., 1970. Das Land-System–Konzept und das Muster ('Pattern') in der Luftbild-Interpretation. Bildmessung u. Luftbildwesen, Karlsruhe, 38: 303–306.

Sheng, T.C., 1972. A treatment-oriented land capability classification scheme for hilly marginal lands in the humid tropics. J. Sci. Res. Council of Jamaica, 3(2): 93–112.

Schneider, S., 1969. Methoden der Raumgliederung mit Hilfe des Luftbildes. Ber.z.D.Landesk., 42: 147–156.

Schreier, H.P., 1972. Land evaluation of the Savuto River basin in Calabria, Italy. ITC Fieldwork report.

Scott, R.M., and Austin, M.P., 1971. Numerical classification of land systems using geomorphological attributes. Austral. Geographical Studies, 9(1): 33–40.

Solntsev, N.A., 1962. Basic problems in Soviet Landscape Science. Soviet Geography, 3: 3–15.

Speight, G., 1968. Parametric description of landforms. In: Stewart, G.A. (Editor), Land evaluation. Proc. of a CSIRO Symp.: 153–186.

Stewart, G.A.,(Editor), 1968. Land evaluation, Papers of a CSIRO Symp. McMillan of Australia, Melbourne, 392 pp.

Tagunova, L.N., 1968. Landscape indicators of engineering geological conditions in the W. Siberian forest tundra (in russian). Biogeography Phenology (Moscow), 2,: 14–16.

Tank, R., 1976. Focus on environmental geology. A collection of case histories and readings from original sources. 7th ed. Oxford Univ. Press, NY, 538 pp.

Teaci, D., 1972. Mechanographic and computer methods used in soil and ecological investigations to substantiate land capability evaluation methods. Pochvovedeniye, 1: 128–139.

Terzaghi, R.D., 1965. Sources of error in joint surveys. Géotechnique,15: 287–304.

Thalen, D.C.P., 1978. Complex mapping units, geosyntaxa and the evaluation of grazing areas. In: Ber. Int. Symp. Ver. Vegetationskunde: Assoziationskomplexe und Ihre Anwendung. (R. Tüxen, Editor). Cramer Verlag, Vaduz: 491–514.

Thomas, M.F., 1969. Geomorphology and land classification in tropical Africa. In: Thomas, M.F., and Whittington, S. W., (Editors), Environment and Land Use in Africa. Methuen, London: 103–145.
Townshend, J.R.G., 1981. Terrain analysis and remote sensing. George Allan & Unwin Ltd., London, 232 pp.
Tricart, J., 1973. La géomorphologie dans les études intégrées d'aménagement du mileu naturel. Ann. Géogr., 82(452): 421–453.
Troll, C., 1939. Luftbildplan und ökologische Bodenforschung. Ztschr. Gesellsch.f. Erdkunde, Berlin, 53: 241–298.
Troll, C., 1966. Landscape Ecology. ITC Publ. S 4 Delft, 16 pp.
Turner, A.K., and Miles, R.D., 1968. Terrain analysis by computer. Proc. Indiana Acad. Sci., 77: 256–270.
Uhlig, H., 1967. Methodische Begriffe der Geographie, besonders der Landschaftskunde. Westermanns Lexikon der Geographie, Braunschweig, 18 pp.
U.K. Min. Agric. Fisheries & Food, 1974. Land Capability Classification. Techn. Bull., 30, 141 pp.
UNESCO, 1968. Proc. Conf. Aerial Surveys and Integrated Studies. Toulouse 1964, UNESCO, Paris, 575 pp.
USAEWES, 1967. The unified soil classification system. Techn. Mem. 3-357, 2nd Ed., Vicksburg, USA , 121 pp.
USBR, 1953. Irrigated land use, Part 2 Land classification, Denver, Colorado. Manual U.S. Bureau of Reclamation, 5
Veer, A.A. de, and Burrough, P.A., 1978. Physiognomic landscape mapping in the Netherlands. Landscape Planning, 5(1): 45–62.
Verstappen, H.Th., 1964. Geomorfologia y conservación de recursos naturales. Revista Geografica Merida, Venezuela, 4(5): 69–82.
Verstappen, H.Th., 1965. Geomorphology and the conservation of natural resources. ITC Publ. B 33: 24–35.
Verstappen, H.Th., 1966. The role of landform classification in integrated surveys. Proc. 2nd Symp. Comm. VII, I.S.P., VI: 35–39.
Verstappen, H.Th., 1967. Landform classification as a basis for regionalization in Pakistan. Proc. Meeting IGU Comm. Interpr. Aerial Photogr., Ottawa, 7 pp.
Verstappen, H.Th., 1970. Aerial imagery and regionalization. Proc. 3rd Symp. Comm. VII, I.S.P., Dresden: 25–45.
Verstappen, H.Th., 1977. Remote Sensing in geomorphology. Elseviers Publ. Comp., Amsterdam, 214 pp.
Vink, A.P.A., 1960. Quantitative aspects of land classification. Trans. 7th Int. Congr. Soil Sci., 5: 371–378.
Vink, A.P.A., 1975. Land use in advancing agriculture. Adv. Series Agric. Sci., Springer Verlag, Heidelberg, 1, 394 pp.
Vink, A.P.A., 1980. De rol van fysische geografie en bodemkunde in de ontwikkelingssamenwerking. Geogr. Tijdschr., 14(5): 354–357.
Vinogradov, B.V., 1968. Airphoto methods of geographical research in the U.S.S.R. Photogrammetria 23: 77–94.
Way, D., 1978. Terrain analysis. A guide to site selection using aerial photographic interpretation. Dowden, Hutchingson and Ross, Stroudburg, 2nd Ed., 392 pp.
Webster, R., et al., 1969. A review of studies on terrain evaluation. MEXE Test Report, 1123, 23 pp.
Webster, R., and Beckett, P.H.T., 1970. Terrain classification and evaluation using airphotography, a review of recent work at Oxford. Photogrammetria, 26(2/3): 51–75.
Weir, M.J.C., 1976. Landscape evaluation in Great Britain–a short review. Meded. Werkgemeenschap Landschapsecologisch onderzoek, 3: 15–20.
Whyte, R.C., 1976. Land and land appraisal. W. Junk. Publ., The Hague, 280 pp.
Wright, R.L., 1972. Principles in a geomorphological approach to land classification. Ztschr.f.Geom., 16: 351–373.
Young, A., 1969. The appraisal of land resources in tropical Africa. A critique of some concepts. Pacific Viewpoint 10: 18–27.
Young, K., 1970. Man in the geobiocoenose. Environmental geology – AGI short course lecture notes, Milwaukee (Wis.), 23 pp.
Young, A., 1976. Tropical soil and soil survey. Cambridge Univ. Studies, 9, Cambridge Univ. Press.: 382–424.
Zonneveld, I.S., 1979. Land evaluation in land(scape) science. ITC Textbook VII.4, 2nd ed., ITC Enschede, 106 pp.
Zuidam, R.A. van, and Zuidam-Cancelado, F.I. van, 1979. Terrain analysis and classification using aerial photographs. ITC Textbook VII. 6, 348 pp.

FLOOD SUSCEPTIBILITY SURVEYS

13.1 Introduction

In many parts of the world, floods that invade river plains and coastal lowlands are very serious natural hazards. Flood-prone low areas are often densely populated and form the economic mainstay of numerous countries. The food and health situation may be adversely affected by floods, and loss of life and property becomes even more severe where urban and industrialized areas are concerned. The annual losses caused by floods in the USA, e.g., amount to approximately US dollars $1.600.10^6$ (data US Water Resources Council) and for Japan a figure of US dollars 450.10^6 has been quoted by Tada and Oya (1971). It is therefore justified to spend substantial funds for surveying flood susceptibility with the aim of improving the situation whenever possible and to avoid unsuitable land use and the erection of inadequate buildings in the most endangered zones (Howells, 1977; Penning Rousell, 1977). In fact, in the USA many surveys of this kind are undertaken for purposes of insurance.

In Bangladesh, a country largely occupied by deltaic and flood plain environments, flood protection is a major issue in the planning of national development. The maps of figs. 13.1 and 13.2, (hydrological map 1:500,000 of Bangladesh (1971), based on EPWAPDA Annual Reports) illustrate the magnitude of the problem in that country.

Fig. 13.1 shows the large tracts of land which were seasonally flooded more than once in the six years under consideration; some parts were flooded every year during part of the wet season. Fig. 13.2 reviews the flooding conditions in the country in connection with the potential drainage. Improvements are possible by the construction of high or low embankments and by gravity drainage, using tidal sluices, pumps, or tidal regulators. Only a relatively small percentage of the surface area has no major drainage or flood problems. The problem with embankments along the lower Ganges River is that they cannot be protected from lateral sapping by this huge river. Once an embankment is attacked the only solution is a quick adaptation to the new situation.

A particular problem in Bangladesh is the surges caused by tropical cyclones which frequently hit the low deltaic coast from the WSW or S,as indicated in fig. 13.3. A problem here is that the low coast is virtually unprotected from the sea only in the so-called Sundarbans does a mangrove belt provide (an inadequate) protection. The lower reaches are somewhat more protected from river floods, as high, well-developed natural levees have been formed by the river due to the effect of the tides. Natural, horseshoe-shaped levees between the various distributaries have been formed, now crowned by embankments around areas which drain towards the sea by gravity. Strong and costly embankments along the seaward side would be required to protect against the invasion of the sometimes suddenly upsurging sea water.

The causes of floods are diversified and vary with the river basin or the region (Gronet, 1979; Lambert and Vigneau, 1981; Pirazelli, 1973; Schick, 1971; Stuckmann, 1974). Local rains falling in the flood susceptible areas and their immediate surroundings are a first factor. More common, however, are heavy rains and/or snow melt in the upper catchment. Incursions of sea water (fig. 13.4) may be a major cause along exposed coasts and particularly where an important tidal range exists and strong onshore winds occasionally occur (Schroeder Lanz, 1962). Such surges may be caused by cyclones or hurricanes as mentioned for Bangladesh, but also by sea waves generated by earthquakes, landslides or volcanic eruptions, which may give rise to serious or even catastrophic floods, also along far-away shores. Tsunami waves in mid-ocean are low (approx. 1 metre) but they travel fast (up to 700 km per hour) and may have a wavelength of more than 100 km. They build up importantly along some types of affected coasts.

Many factors influence the type and degree of flooding basically brought about by one or more of the three major causes listed above (Benson, 1962; Parker, 1981). First, there are the climatological characteristics of the area and, second, the hydrological and environmental conditions of the drainage basin. Together these determine the hydrological regime (discharge characteristics, sediment load, geomorphological dynamics) of the main river and its tributaries (Various, 1980; Viereck, 1973; Wisner, 1979). The physical environment of the basin is affected by relief, geology (particularly lithology), soils and vegetation. Infiltration where permeable rocks occur, retention by vegetation and depression storage are among the matters to be emphasized in this context. The drainage basin has to be studied in its totality for purposes of river basin development, otherwise it will be difficult to 'harness' the river adequately. The effect of the works of man also has to be investigated as the river regime may be substantially altered by deforestation or the construction of embankments (Santema, 1966).

For relief from the more immediate flood danger, a flood susceptibility survey of the affected lowland areas is very useful. The important matters to be studied are: micro relief, the geomorphological units of the plain and their inherent properties with respect to flooding, the deposition of sediments, bank erosion and other characteristics of the channel bed (Monnet, 1959; Hugues, 1980; Newson, 1980). Neotectonics and specifically subsidence of land may have a pronounced effect on the distributional

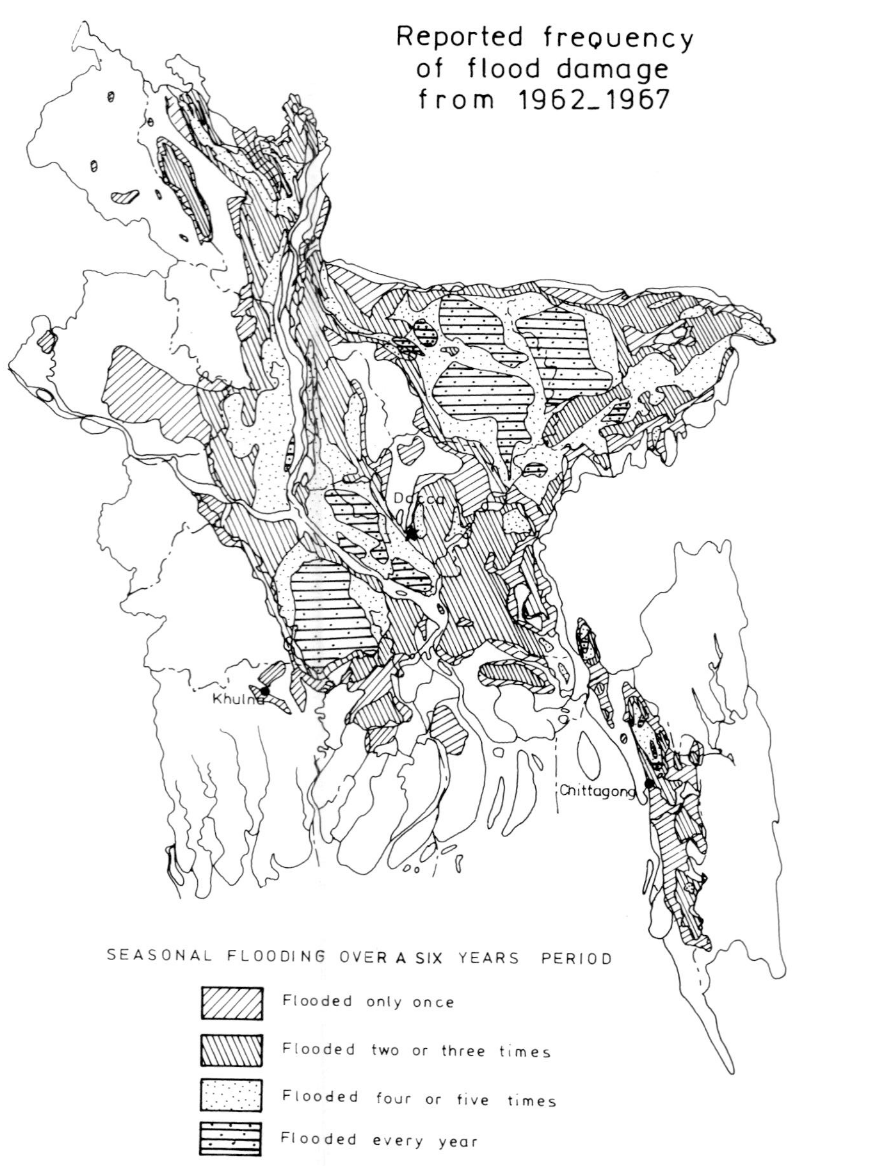

Fig. 13.1 Source: Hydro geological map of Bangladesh 1:500,000 of 1971 based on EP - WAPDA Annual Reports.

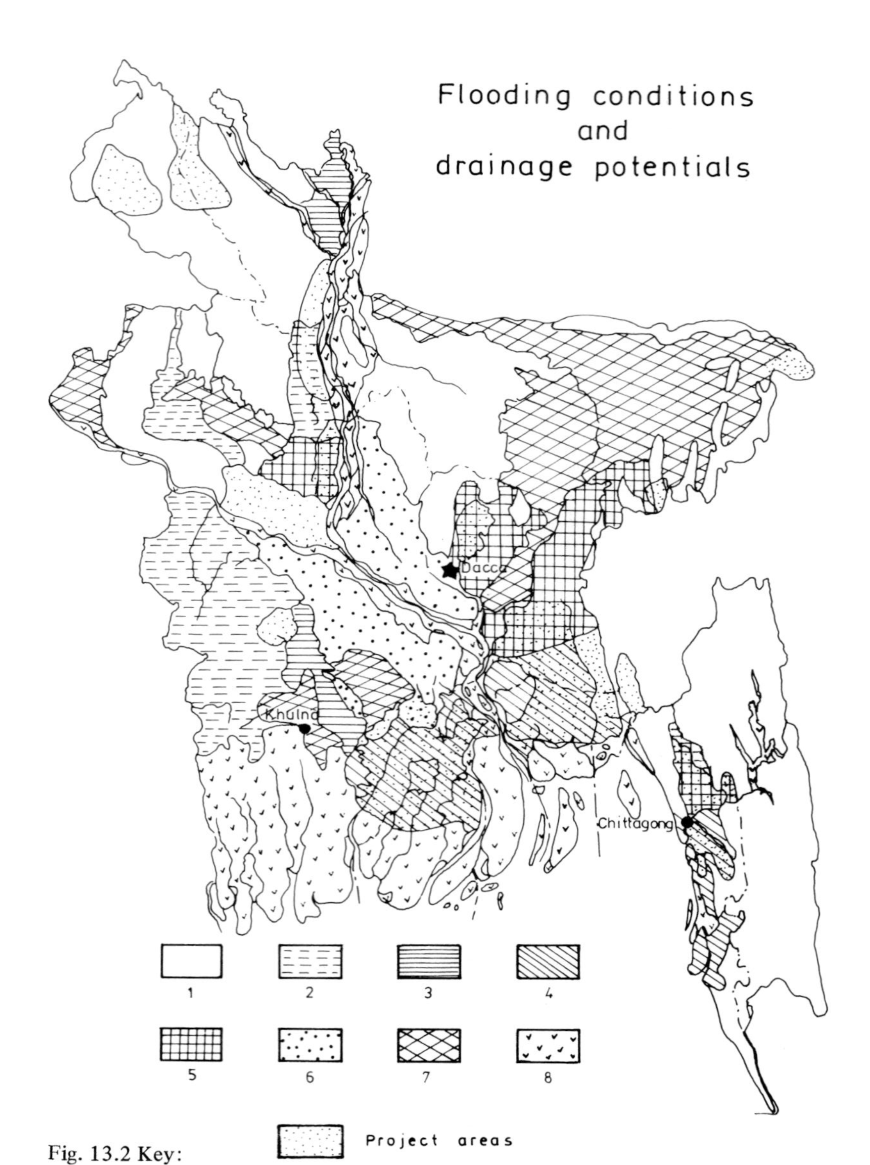

Fig. 13.2 Key:

1. no major drainage or flood problems 2. improvement by gravity drainage
3. improvement by gravity drainage and low embankments
4. improvement by gravity drainage, low embankments and tidal sluices
5. improvement by medium embankments, pump drainage 6. improvement by major embankments, pumping and tidal regulators 7. severely flooded, no major improvement envisioned 8. areas subject to active fluvial or tidal actions in all seasons

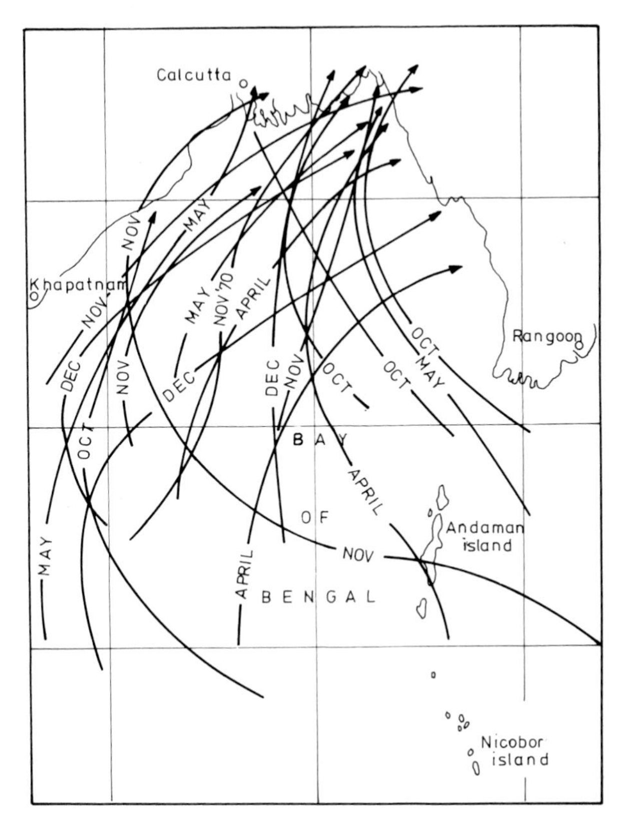

Fig. 13.3 Cyclonic storm tracks in the Bay of Bengal causing storm surges and flooding in the Ganges Delta.

pattern of the inundations. Examples of this are given in section 3 on the relation between geomorphology and geology. The width of a coastal plain may, in tectonically unstable areas be not at all correlated to the size of the river and the amount of sediment delivered because subsidence and changes of sea level during the Quaternary are more dominant determining factors.

It should be understood that the flood plain and deltaic areas are natural sites for fluvial (and/or marine) sedimentation and have in fact been built and fertilized by deposits of these rivers during floods. The Nile Valley is a classic example of this, but the same applies to the plains of the Euphrate and Tigris and for almost any other lowland plain. Occasional shifts of river channel position in deltas and alluvial plains and on fans are natural phenomena. When prevented from flooding adjacent zones and from lateral shifting by the construction of embankments, the water level will tend to build up vertically and the sedimentation will be localized along the bed. This will increase the height difference between rivers and adjacent low-lying land. Unless adequate measures are taken to accommodate for such situations, they can become increasingly dangerous in case of an exceptionally high discharge.

The population of the flood-prone areas usually adapts its way of life in one way or another to the flood situation (Burton et al., 1969; Ericksen, 1971; Kates, 1962; Kates and White, 1964; Smith and Tobin, 1982; White, 1949, 1961, 1964). Settlements are preferably built on natural levees or other elevated parts and are rare in swampy zones, though they may be located there for reasons of defense or or fishery. If no adequate natural rise exists in a certain locality, houses or entire settlements may be built on artificial mounds as demonstrated in fig. 13.5 of the Netherlands and fig. 13.6 of Bangladesh. The latter example is interesting especially because the dug-out is transformed into a fish pond and thus put to efficient use. Embankments along the river are a technologically more advanced solution. Often a low embankment is constructed along the minor river bed and a higher dike is built at a more adequate distance from the river to check the flood waters. The area between these two embankments is subject to seasonal or occasional flooding in order to give the water lateral space and prevent it from rising to dangerous heights. Fig. 13.7 gives an example of some flooded orchards in such zones along the lower Danube, Romania. Further control of floodwater can be achieved by creating overflow possibilities: selected backswamps or other low areas suitable for depression storage whereby the rise of the riverwater farther downstream can be reduced. The construction of storage reservoirs at suitable upstream locations within the drainage basin is another, technologically advanced solution.

Drainage of the areas encircled by embankments may be effected by gravity, making use of the tides or done artificially using windmills (fig. 13.8) or pumps.

Not only the location of the settlements and the roads, but also the agricultural system, is often strongly influenced by the flood situation as explained in section 6. The floating rise in the backswamps of the Chao Phyo Plain, Thailand, and along the Mekong River is a classic example, as is the flood retreat cultivation along rivers such as the Niger and the Senegal. The flood control measures and land management practices in turn affect the flooding patterns (Bird, 1980; Burby and French, 1981; Sheaffer, 1960). Once the hydrological regime of the river is changed by man for reasons of integral river basin development, flood control, navigation, etc., the whole agricultural way of life has to be changed to adapt to the new situation.

In the forthcoming subsections, the use of aerospace technology in flood susceptibility surveys will be elaborated upon and the characteristics of flooding situations caused by excessive water upstream, the sea or local rains will be discussed. Special applications of such surveys are also described.

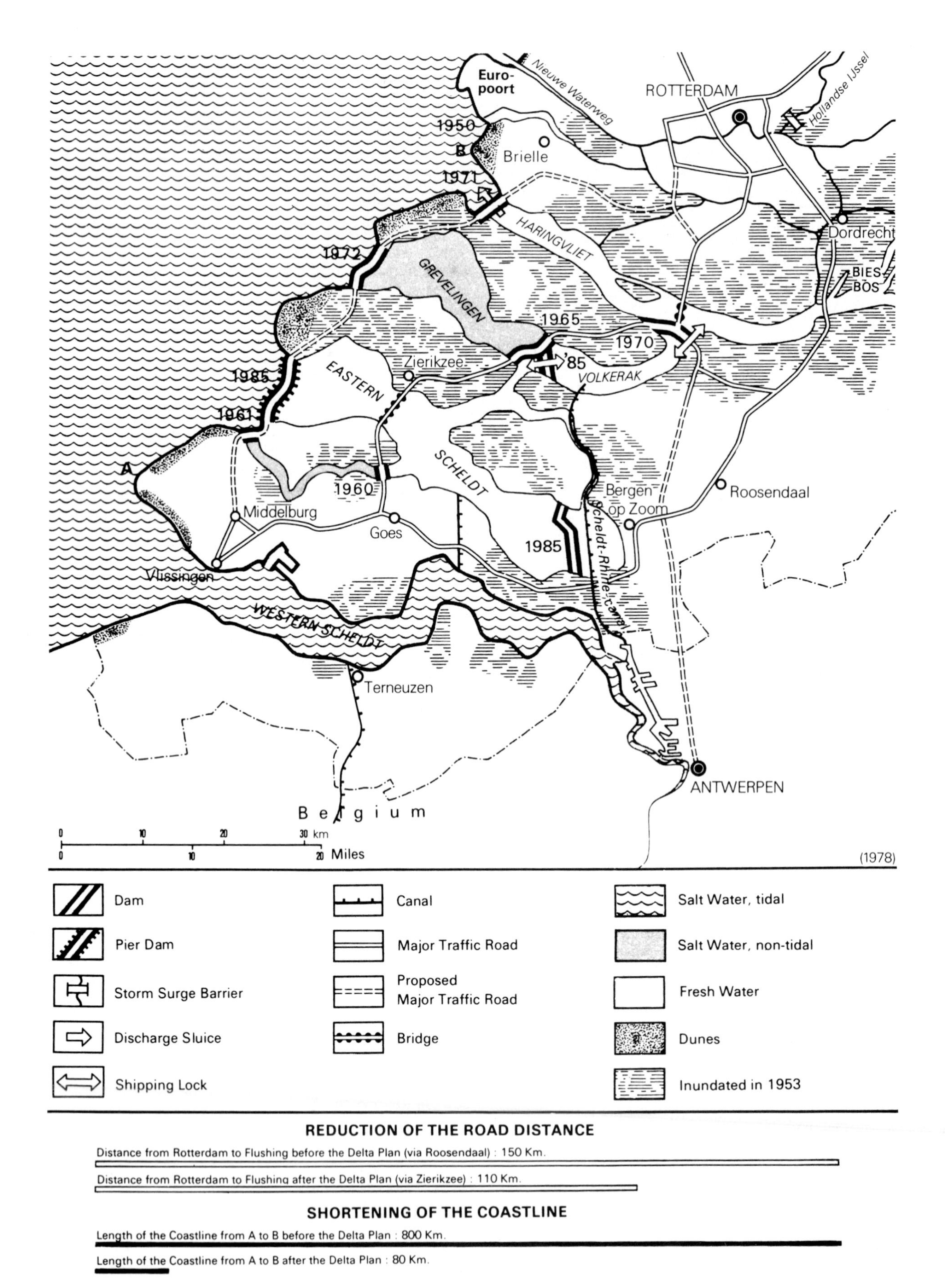

Fig. 13.4 The extent of the 1953 floods in the estuarine areas of the south western parts of the Netherlands and the engineering works executed to prevent such disasters in the future. Source: Geografisch Documentatie Centrum, Utrecht, 1978.

Fig. 13.5 Farm house built on a low artificial mound in the low parts of the Netherlands.

Fig. 13.6 Farmhouse in Bangladesh built on a small artificial mound in the vicinity of a major Ganges branch. The soil material required was excavated nearby where a fish pond resulted.

Fig. 13.7 Inundated orchards in the zone between the low summer quay and the winter dike (left) during a flood of the Danube in Romania.

Fig. 13.8 Water management in the low western part of the Netherlands: windmills pumping water from the polder on the right into the canal in the foreground where a small lock regulates boat traffic.

13.2 Survey Methods and the Use of Aerospace Imagery

In addition to hydrological data on river discharge, sediment production and frequency and areal extent of floods, a detailed knowledge of the climatological and meteorological conditions prevailing in the drainage basin is important information for the geomorphologist engaged in flood susceptibility surveying. To arrive at a better prediction of hazards and future changes, his survey work should make use of this information, emphasize the relief and microrelief of the flood-prone area, the individual processes and the complex morphodynamics and also touch upon past developments of the plain.

Accurate topographic maps with detailed contours and small contour interval (0.1 - 0.5m for detailed surveys; 1-2m for semi-detailed mapping) are an essential prerequisite for the work. Absolute data on terrain elevation is not a stringent requirement for reconnaissance surveys at small scales. The general lay of the land may serve as a first indication of relative height and of flood susceptibility, though the vertical accuracy required generally surpasses the necessary horizontal geometry. The latter may be obtained from aerial photographs, the former may have to be obtained from detailed levelling, e.g., along traverses whereby the position is precisely plotted on the aerial photographs. In this way it is possible to acquire or approximate the necessary vertical precision, at least for semi-detailed work (1:25,000 - 1:50,000 scales), with a minimum of time, effort and cost.

It is obvious that aerial photographs and in particular, overlapping verticals which can be stereoscopically studied are another essential tool for the survey. They permit the rapid delineation of geomorphological units and smaller features. If they are taken during or immediately after the flood,they may be used for mapping the extent of inundated areas by a flood of known magnitude (discharge, water level), as is illustrated by the example from Java, given in subsection 13.6. In addition to outlining geomorphological units, they may serve to evaluate fluvial and other processes and to map their effects on the inundations of the land.

This is demonstrated by the monoscopic vertical aerial photograph (1:20,000) of fig. 13.9. The area depicted is an alluvial lowland in St. Landry County, Louisiana, bordering the river which is visible in the upper left and right. The river is actively laterally sapping which results in bank erosion by way of slumping. In due time the dike constructed parallel to the river will be endangered by this process.

The larger part of the photograph is occupied by the natural levee built by the river. The gently sloping back slope of the levee is under agricultural use and light toned in the photographic image. The rectangular field patterns form a remarkable contrast with the irregular sinuous pattern of tiny, abandoned, sinuous channels (dark toned), separated by lighter toned lenses of slightly coarser textured soil. Some of these former channels are more or less parallel and extend perpendicular to the river, such as that in the upper right. They indicate general overtopping of the levee when the river was in spate. Other channels, however, converge to certain places where the levee was breached and a true crevasse was formed. Scouring caused a crevasse lake in one of these localities. The coarse crevasse deposits near the river were laid down on the back slope of the levee where vertical accretion occurred and the flood waters were drained off to the back swamp which appears in the lower right. The back swamp, being too low for agriculture, is under forest and dark toned.

The vertical aerial view of fig. 6.7, showing a part of the north coast of Western Java,is another example of a process visualized in the imagery. Abrasion of the clayey lowland coast has occurred in the recent past following nearby coastal engineering works. The dark-toned mangrove belt has gradually been reduced in width by the invasion of the sea until some of the low bunds surrounding the fish ponds directly inland were ultimately destroyed. Subsequently, new ones have been constructed farther inland. The creek system which drained the area prior to the construction of the fish ponds can be clearly seen in the photograph. Many other examples are given in this section and illustrate the importance of photo-interpretation, not only for delineating geomorphological units and smaller 'static' features but also for the study of geomorphological processes related to flooding.

New developments in the field of remote sensing (Verstappen, 1977) have resulted in the successful application of other types of imagery and have added digital approaches to the previously evolved visual methods of analysis.

Fig. 13.10 exemplifies the applicability of visual interpretation of Landsat imagery for flood susceptibility surveying on the reconnaissance scale. The map results from the study of a 1:500,000 colour composite and relates to a flood of the recent past and shows a part of the Peoples' Republic of China, south of Beijing,where the San-Kan Ho River has flooded large sectors of its alluvial fan; this event is probably of a recurrent nature. The extent of the flood deposits was mapped using an image acquired on 12 March 1975. The interpretation was done at a scale of 1:500,000 and subsequently the data were reduced and generalized, as shown in the figure. Certain details of the patterns traced indicate that the whitish, barren areas are subject to aeolian activity during the dry winter season when northerly winds prevail. Similar but vaguer patterns in the agricultural fields on the darker parts of the fan point to effective wind work in other periods of the recent past. Storage of flood waters is realised in a back swamp zone slightly south and just outside the figure, canals have been dug for the purpose. The fact that the flood waters cause a substantial amount of fluvial material to be deposited on the fan when the river is in spate affects agricultural productivity adversely. Storage basins and possibly other upstream measures will be required to control the situation. It is clear that the information obtained is an approximation of the problem only and that much more detailed information will be required to arrive at conclusions of practical importance. The interpretation will have to be based on aerial photographs, field observations and hydrological data. A general appreciation of the environmental situation in which the floods occur and of their broad distributional pattern can be easily obtained using Landsat information.

Data obtained from orbiting satellites, e.g., Landsat, may also assist in the rapid and accurate mapping of the extent

Fig. 13.9 Vertical aerial photograph at a scale of 1:20,000 showing a natural levee in Lousiana (USA) crowned by a dike and with signs of a breach (top) and of overtopping (right).

of inundations in alluvial plains. Limitations are not set by height of the satellites and scale of the imagery, but by the spatial resolution. Digital analysis may result in print outs of various types at scales of 1:25,000. The problems at the present state of the art are twofold: the interval between two subsequent passes of the satellite being 18 days, no complete monitoring of the various stages of the oncoming and abating flood can as yet be obtained. More frequent passes are required for this. Secondly, it can be expected that during and shortly after the floods substantial cloud cover will occur over the inundated areas which cannot be penetrated by the sensors of Landsat. Radar is required for this, however,the power requirements for this sensor are high and recording is only possible for limited zones on request unless on-board nuclear power is utilized.

In spite of these limitations, some interesting results have been obtained. Fig. 13.11 gives an example related to inundations along the Tulangbawang River, Sumatra, Indonesia, recorded by Landsat-2 in 1973. Band 7 gives the clearest delineation of the flooded areas as expected within this infra-red part of the EM spectrum. The flooded areas are pictured in black. Band 6 shows the inundated zones, though less clearly, and Band 5 (red) shows the river which is more easily traced here than on the other bands. The distinction in photographic density ('grey tone') between inundated and non-inundated zones is faint. This is mainly due to the fact that the inundated parts are vegetated (flooded forests and fields) and thus their spectral characteristics do not substantially differ from the non-flooded vegetated areas, at least in this wavelength band. Cloud penetration is less in Band 4 compared to Band 7; this makes coastal delineation considerably easier when Band 7 is used.

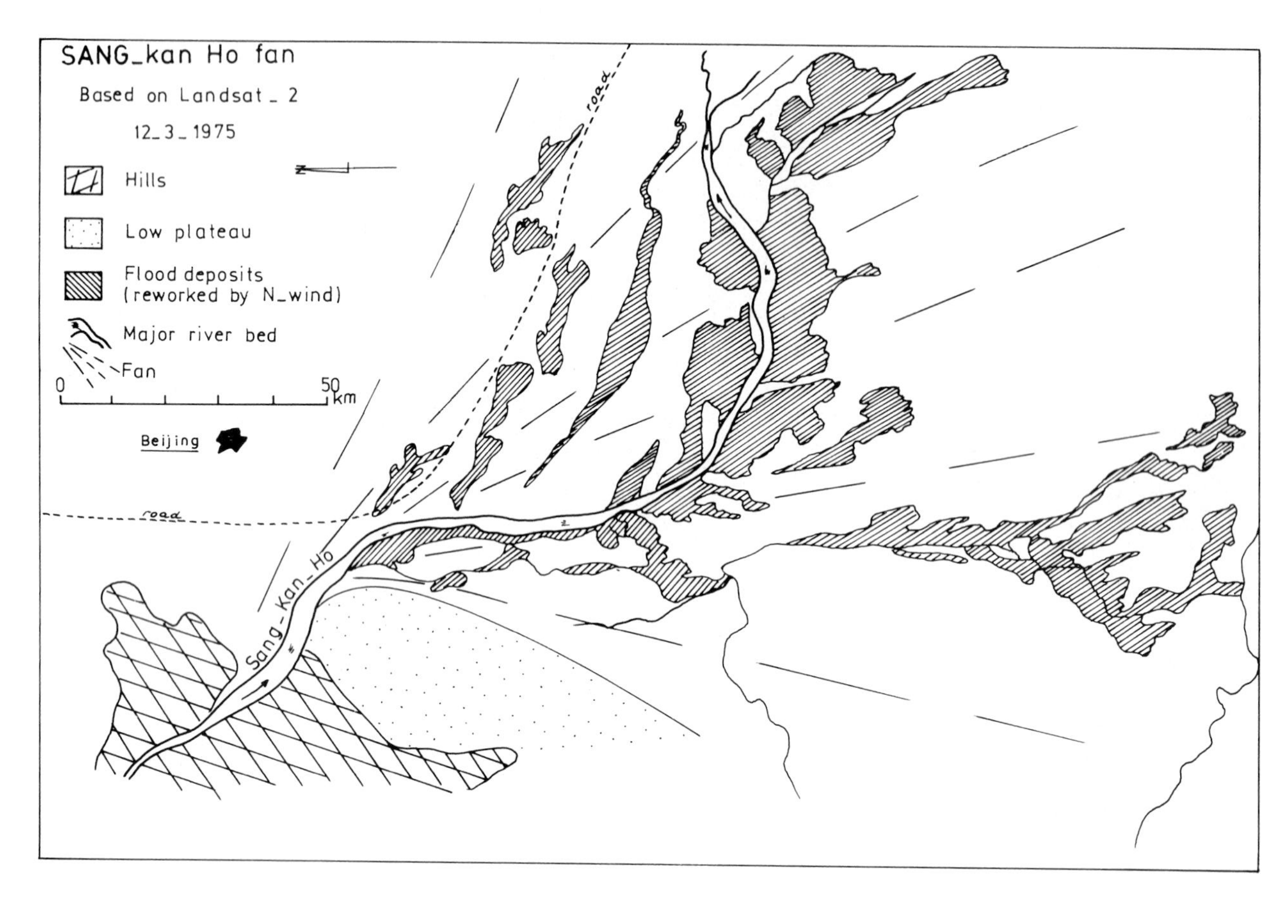

Fig. 13.10 Floods of the past recorded by Landsat: an example from China.

It is worth mentioning that no flooding occurs in the plains of the Seputih and Mesuji rivers, which are situated to the north and south,respectively. The heavy rains that have caused the inundations in the Tulangbawang plain have been strongly localized.

Multi-spectral aircraft data may also be used; more detail can be obtained this way, but the satellite data will be superior to elucidate the general situation. Sollers et al. (1978) have compared aircraft and satellite borne MSS data in a study of the floodplain of the Susquehanna River, USA, applying the digital approach. It proved to be less difficult to delineate flood-prone areas under agricultural use of limited development than in forest covered zones. The data obtained were found suitable for providing preliminary planning information and for studying previous floods and actual inundations. The monitoring capacity was found to be of particular importance. The MSS approach has the obvious disadvantage that no stereoscopic impression of (micro) relief can be obtained; stereoscopy is one of the major advantages of the application of overlapping vertical aerial photographs for this purpose, together with the good metrical properties of these images.

Monitoring is a major asset of orbiting satellites,such as the satellites of the Landsat series which pass over the larger part of the world at regular intervals. Comparison of the situation recorded in flood susceptible lowland areas during subsequent passes may give important information on the changes caused by gradual processes of river work, or those that represent sudden changes due to floods. Longer time intervals may be required for such studies,in which case comparison of older aerial photographic coverage with Landsat imagery may be useful. Fig. 13.12 gives an example of the possibilities of this approach and depicts a part of the Ganges Delta in Bangladesh. It shows the changes in the position of various Ganges branches in the period between the aerial photograph coverage of 1952,on which the topographical map of the area is based, and a Landsat pass of 1973. The major changes are due to the flood of 1972. The inset shows changes of considerably smaller magnitude that occurred in and near the riverbed in 1973/1974 and 1974/1975 as recorded by sequent Landsat passes. The map results from a project carried out at ITC by Mr. J. Stein, Australia, in the framework of a M.Sc. course in

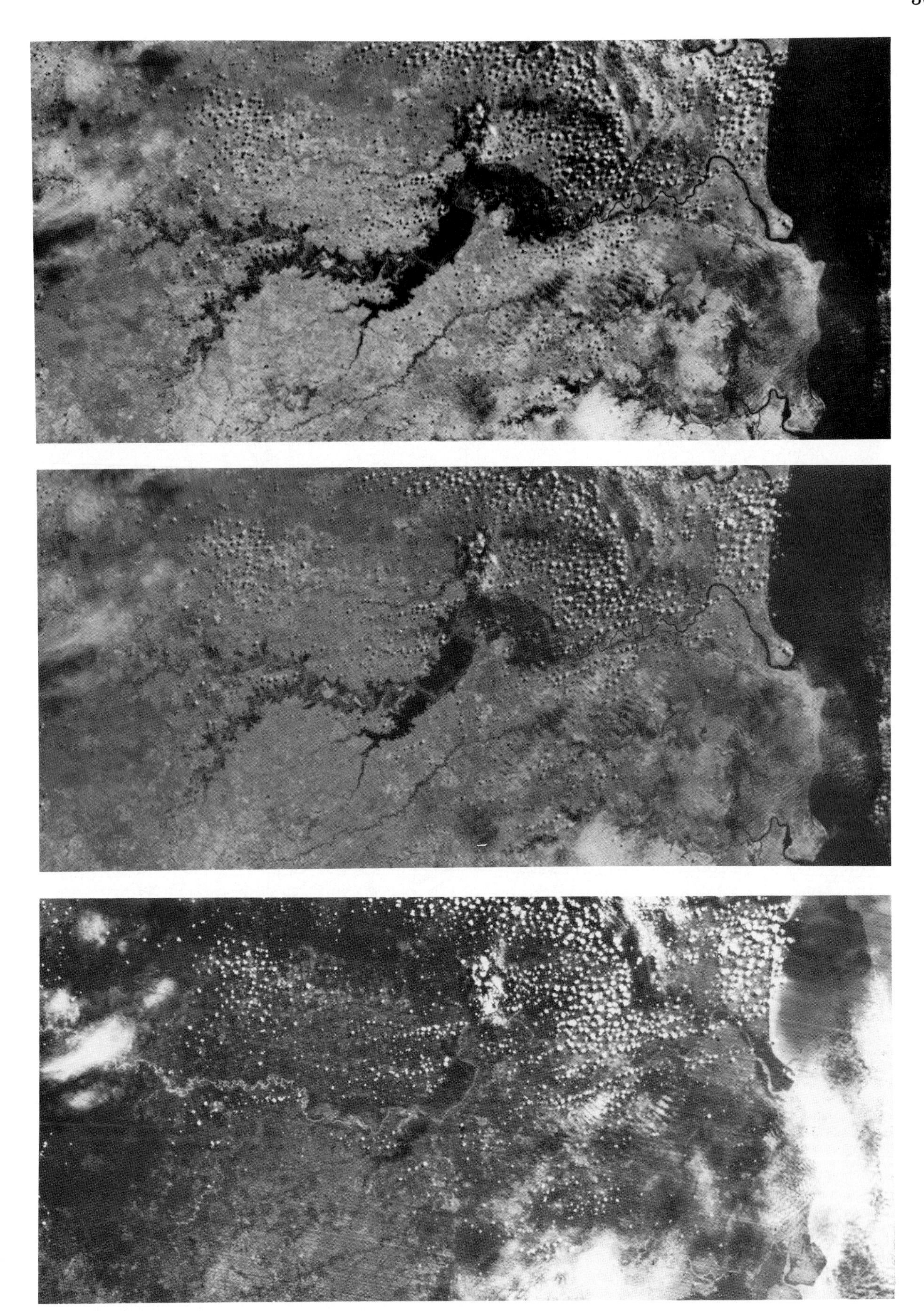

Fig. 13.11 A flood in the lower Tulangbawang River basin in south Sumatra, Indonesia,pictured on a Landsat frame of 1973. Delineation of the inundated areas is optimal on band 7 (top), moderate on band 6 (centre) and poor on band 4 (bottom).

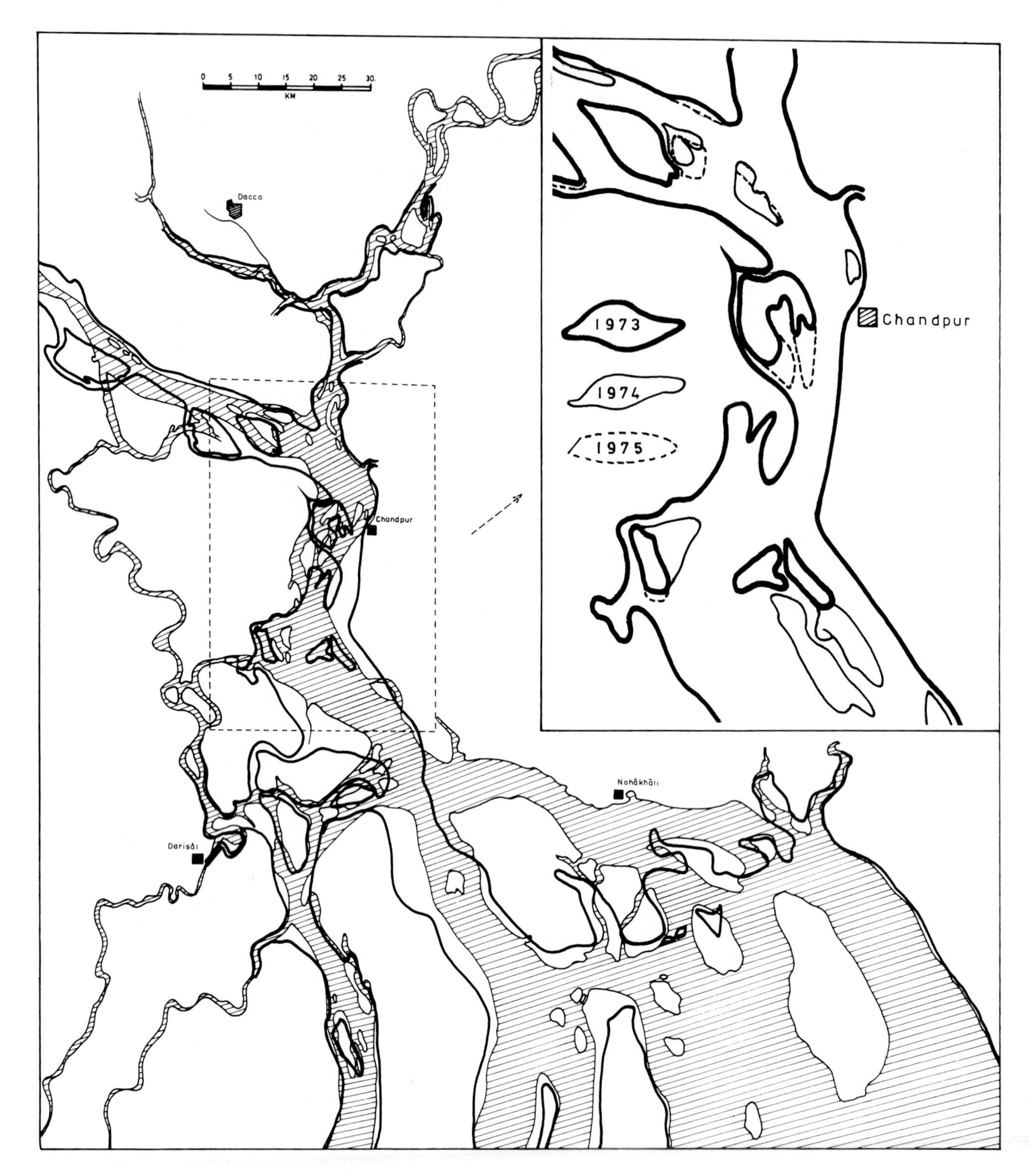

Fig. 13.12 Aerospace technology as a tool in river survey: the Ganges distributaries and the Mechra confluence to the south of Dacca as pictured on a photogrammetric map based on aerial photographs dating from 1952 (hachured) compared with the pattern obtained from a Landsat image dating from 1973 (thick lines). Inset: detail of an area near Chandpur giving the situations of the years 1973, 1974 and 1975 as recorded by Landsat.

cartography and clearly shows the importance of the approach for map revision and for recording the effects of fluvial dynamics and floods.

A considerable amount of data on flood plain geomorphology and the areal extent of major geomorphological units can also be obtained by a thorough visual and/or digital geomorphological analysis of a single Landsat image, followed by the interpretation of aerial photographs in key areas and with the support of existing or gathered field data. Fig. 13.13, picturing part of a geomorphological map with the corresponding flood susceptibility classification, compiled by Mr. Shaheed ul Islam of of the Water and Power Development Authority (WAPDA), Bangladesh, in the framework of an ITC geomorphology course in 1978, gives an example, picturing a part of NE Bangladesh in the Brahmaputra plain.

The main geomorphological terrain units distinguished in the deltaic plain comprise:

1. Palaeozoic hills of Shillong Plateau
2. Pleistocene outcrops of Madhapur (dissected and bordered by old Brahmaputra sediments)
3. Piedmont alluvium
   fans, coalescent fans
   depressions within fan or between fans
4. Surma alluvium (levee-basin complex bordering large low basin)
   forming flood susceptible 'Haor' region with often submerged point bar/levee system
5. Older deposits of the Old Brahmaputra (point bar system)
   oldest deposits, predominantly of Old Brahmaputra
   oldest deposits, of Old Brahmaputra with piedmont alluvium
   oldest deposits, of Old Brahmaputra with Surma alluvium
6. Active flood plain of the old Brahmaputra
   youngest flood plain (levee system)
   young flood plain (levee/levee-point bar system)

Features such as major and minor river courses, old channels, low-lying marshes, ponds, lakes, point bars, submerged point bars,levees and young and older Meghra deposits could be mapped at the scale of 1:500,000. The area was classified according to the relative height and susceptibility to flooding, following the classes commonly distinguished in Bangladesh by WAPDA, as follows:

Flood classes	Topographic situation	Flood susceptibility
1.	hilly zones and other high areas	above highest flood level
2.	areas of medium height	subject to exceptional flooding; in part infrequent shallow flooding
3.	medium height to low areas	infrequent flooding, in part seasonal shallow flooding
4.	low lying areas	subject to annual flooding. In part frequently and deeply flooded.
5.	local depressions	deep flooding, often moist.
6.	local depressions lowest parts	frequently very deeply flooded and in part waterlogged throughout the year.

Research on the use of aerospace imagery and the related field techniques is furthered in many countries in view of its great potentialities (Carter et al., 1979; Deutsch et al., 1974; v. Es, 1976; Eyre, 1980; Hallberg et al., 1973; Harker, 1974; Rango and Anderson, 1974; Reckendorf, 1973; Sollers et al., 1978; Williamson, 1974). The use of soil data as an indicator has been stressed by Cain and Beatty (1968); Lee et al., (1972); Yanggen et al., (1966) and many others; the use of vegetation has been studied by Everitt (1968); Sigafoos (1964, 1974) and geomorphological aspects are dealt with by Burgess (1967); Reckenfor (1963) and Wolman (1971).

Increasingly effective uses can be expected in the future with the development of remote sensing technology and the further exploration of the modes of its optimal application and its integration in other approaches to flood surveying.

13.3 Surveying for River Floods

The purpose of flood susceptibility maps is not to map the areal extent and further characteristics of floods that have occurred in the past, but rather to predict the effect of future floods of various magnitudes and to assess the damage that they may provoke (Cochrane, 1981). In its simplest form it amounts to mapping the limits of the flood plains and the extent of some recent major floods. This approach is common, particularly in the USA and Canada where it serves, among others, purposes of insurance (Environment Canada, 1978; Hanke, 1972; Kates, 1965; Kesik, 1981; Platt, 1976).

Considerable work has been done in Japan to develop the concept of flood susceptibility. Emphasis is on: the distributional pattern of the areas subject to flooding, the duration and depth of inundation, the direction of the current, possible changes in the river course and erosion and deposition due to flooding. It is the forecasting potential which renders these geomorphological maps of great value, particularly in the densely populated lowlands of Japan and other countries of the Far East.

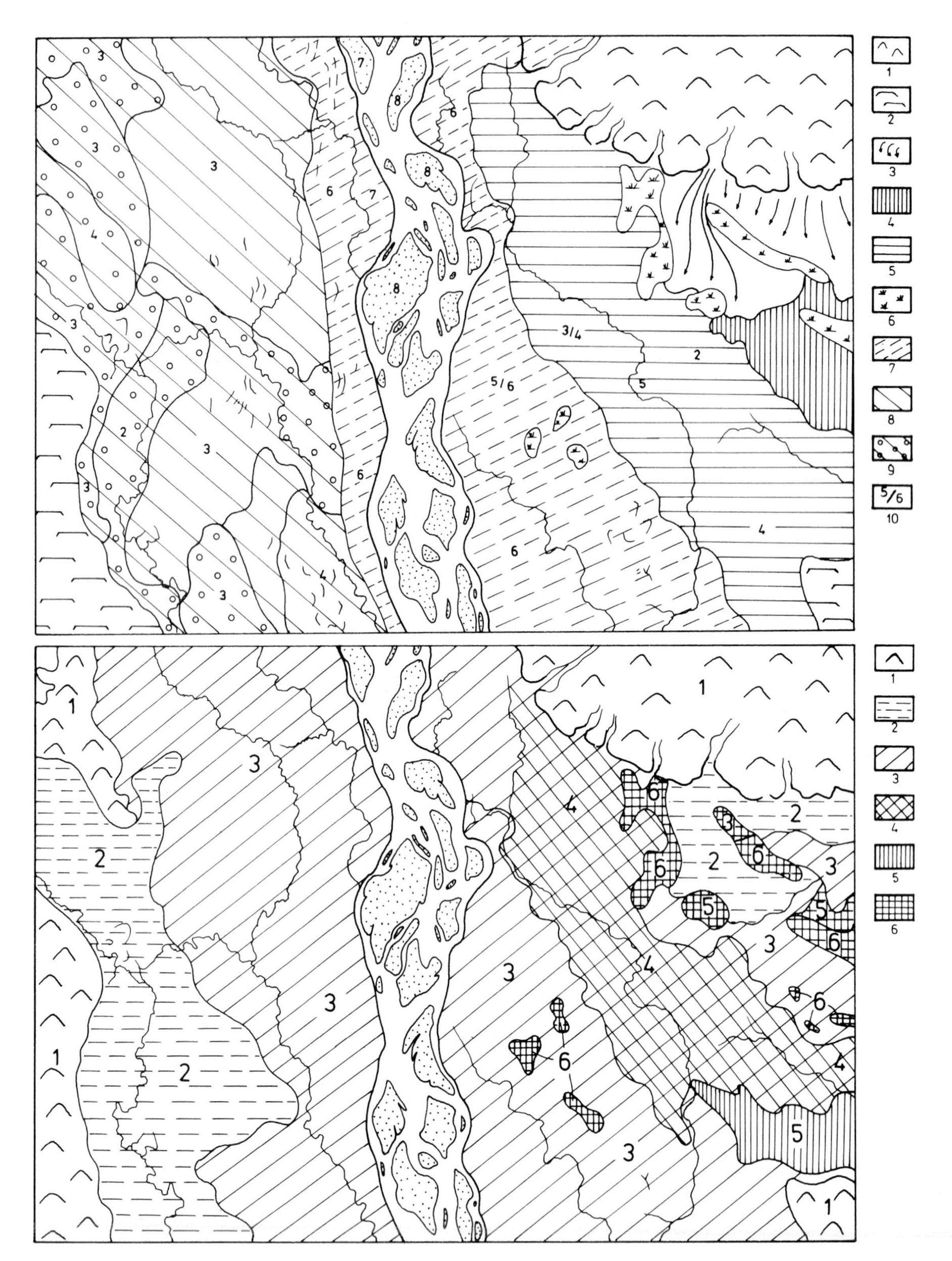

Fig. 13.13 Geomorphological (top) and flood classification (bottom) map of part of the Gangetic Plain/Delta in Bangladesh based on Landsat imagery; generalised after S. Islam, (ITC, 1978).

Key:
1. hills of Shillong Plateau - (Palaeozoic)
2. old Brahmaputra plain; Barind and Madhapur tracts (Pleistocene)
3. alluvial fans bordering on Shillong plateau (Piedmont alluvium)
4. alluvial plain bordering on Shillong plateau/fans
5. older Brahmaputra river system (point bar complex)
6. local depression, swamp
7. younger Brahmaputra river system
8. younger Teesta river system (point bar complex)
9. younger Teesta river system (levee-back swamp/point bar system)
10. deposition phases

Key: Flood classification: 1-6 = WAPDA flood susceptibility classes (see text)

(Oya, 1955, 1963, 1964, 1967, 1968, 1971, 1973, 1976, 1977; Oya and Akagiri, 1968).

This forecasting capacity is basically rooted in two facts. Firstly, it is the (micro) relief of the plain land which governs the flow of the floodwater, its accumulation in depressions, etc.; thus it is logical that the study and detailed mapping of the relief substantially contributes to the forecasting of future flood conditions. Secondly, and more important than the relief situation, is the fact that the terrain configuration of the lowland plains and the distributional pattern of gravel, sand, silt and clay deposition result from fluvial deposition of the past by the same river(s) responsible for the future floods. Their study contributes to an understanding of the hydrological regime of the river,and therefore, to the nature of future floods.

The reliability of the predictions made based on these maps has been proven on several occasions. For example, when the southern Nobi Plain was flooded by the sea and storm surges of the Isefan typhoon in 1959, the characteristics of the inundations,such as areal extent, depth and duration and the flow direction of the flood waters, were precisely as predicted by the geomorphological flood susceptibility map prepared in 1955.

The compilation of these maps, most of which are at the scale of 1:50,000, normally starts with the interpretation of aerial photographs at a scale of 1:40,000. The main geomorphological units of the plain, such as flood plains, alluvial plains, natural levees, back swamps, oxbow lakes, former river courses, terraces, fans, deltas and beach ridges,can be easily mapped. The data obtained are subsequently checked in the field and hydrological and sedimentological data which cannot be derived from the aerial photographic images are added.

Information about flooding obtained from the Department of Public Works,other official sources or the local population,is gathered and included in the map when appropriate. Ultimately, the maps are valid in the context of establishing an early warning system (fig. 13.14).

An essential element of the maps is that for every geomorphological unit indicated, the type and degree of flood susceptibility is evaluated (Tada and Oya, 1968). This is done qualitatively by giving descriptions such as 'never submerged', 'submerged, but water drains off well', 'shallowly submerged' and 'subject to deep and long-lasting inundation'. Further specification is feasible: remarks such as 'seasonal flooding', 'inundation by rainfall', and 'flooded by storm tides', are common in the legends of these maps. Reliability and the precise quantification of flood prediction can be improved if observations can be made in the field during or shortly after a flood situation. Reconnaissance flights, overboard photographs, or full coverage of vertical photographs of these conditions are also extremely useful in this respect. Tada et al., (1977) were in a position to survey the Tsugara Plain, Japan, shortly after the severe flood of August, 1975. The flood current on certain active fans was so fast that houses and bridges were washed away. Former river courses channelled the invading flood waters and flat deltaic plains were easily submerged though less deeply than backswamp areas.

Fig. 13.14 An efficient early warning system: continuous digital recording of the level of the river water in the centre of a Japanese town (Nobi plain).

Oya (1967) observed a flood of the Mekong River near Vientiane and surroundings, and plotted his data on a 1:50,000 map. His legend is as follows:

1. inundated areas (blue hachures)
2. direction of flood current (blue arrows)
3. direction of drainage (blue dashed arrows)
4. date of beginning of inundation (red lettering)
5. period of inundation in days (red italics)
6. depth of inundation in cm (blue ciphers)
7. thickness of deposits in cm (red ciphers)
8. areas of sand deposition (red dots)
9. areas where erosion occurred (green hachures)
10. temporary dikes (red lines).

If the data of a recent flood have to be added to the flood susceptibilty map, this can be done by using an overlay. Tada and Oya (1968) applied this method in their map of the Kuzuruyu River when indicating the effects of the flood of September 1965. Torrential rains caused landslides and substantial amounts of sand and gravel reached the valley plain. Erosion, in particular bank erosion, was widespread and dikes were destroyed over considerable distances. A fortunate condition in this basin is that a large depression south of the city of Ono acts as a retarding basin, preventing serious flooding farther downstream. The legend of the 1965 flood overlay includes: landslide, debris avalanche, eroded land, areas of both thick and thin sand/gravel deposition, areas of driftwood deposits, areas with both erosion and deposition, bank erosion, damaged or destroyed dikes, direction of flood current, direction of drainage, depth of inundation, thickness of deposits, period of inundation, time of maximum inundation and inundated areas. Landform units and general flood susceptibility are incorporated in the flood susceptibility map.

The same overlay method was applied earlier by Oya (1966) when surveying the Kano River basin with an indication of the effects of the flood of September 1958.

In cultivated and/or densely populated coastal areas it may be important to precisely map where, when and how man has altered the natural conditions of the lowlands. Tada and Oya (1963), on their flood susceptibility map of the Yoshino River basin, mapped the artificial land fills, dikes and the gradual occupation of the lowlands by paddy fields since about 1500.

This type of information is important for two reasons: Firstly, the flood flow and drainage conditions may be modified to an important degree by man-made structures and, secondly, flooding becomes devastating in terms of loss of life and property only when densely inhabited zones with costly structures are affected.

When mapping the Neyagawa River basin near Osaka, Japan, Oya (1971) followed an approach slightly different from that mentioned above concerning the Tsugara Plain. The main geomorphological units (steep mountain and hill slopes, terrace, terrace or dissected fan, shallow valley in terrace, alluvial fan, natural levee, valley plain, wave cut bench, sand bar, lower natural levee, delta, lower delta/marsh, dry river bed, former river channel) were indicated by coloured area symbols on a topographic map. The flood susceptibility was not given for each of these units but independently, by hachures, specifying whether the flooding was due to the Yodo River, the Yamato River, local rains or the sea. Furthermore, contour lines, river embankments, likely dike breaches, direction of flood current, inland limit of tides and reclaimed land, were indicated. Fig. 13.15 shows a part of the map at a reduced scale.

13.4 River Flood Surveys and Damage Prediction

In 1974, Kim, c.s., of the Research and Development Department of Asia Aero Survey Co., Ltd. carried out a flood susceptibility survey of part of the Han Valley near Deogso, Korea (scale 1:25,000). From their work another concept of flood susceptibility mapping has evolved, with emphasis on the actual damage that can be expected. Apart from a geomorphological flood map, they also made a map indicating possible flood damage and an overlay showing sequential inundation stages by floods of different magnitudes (fig. 13.16). The overlay can be placed over the geomorphological flood map to classify the terrain into flood susceptibility classes.

A number of geomorphological units and features are indicated on the geomorphological flood map as listed in the legend accompanying fig. 13.14. Furthermore, the flood limit of 1972 and the areas of maximum inundation during an exceptional flood are plotted. On the possible flood damage map, houses, paddy fields, crop land and industrial land susceptible to flooding are mapped, and structures such as bridges and roads endangered by flood waters are also indicated. Potential sites of secondary geomorphological phenomena, such as landslides and bank erosion which may be caused by the flood waters, are also located in this map. On the overlay of sequential inundation stages, four inundation stages bank full stage, dangerous flood, emergency flood and exceptional flood are distinguished and the areal extent of land affected by direct flooding and/or ponded-up water of tributaries is given.

For compilation of data for the sequential inundation stages, photomorphological methods of classifying and mapping the inundation stages were used, based on geomorphological units and elements. Also given was information about the 1972 flood as well as absolute and relative relief. The gauging station at the foot bridge served for establishing the frequency of flood water levels and river discharges.

	Discharge m^3/sec	Water level (m)	Frequency in 100 yrs
Warning flood (1)	up to 14,700	up to 8.70	50-100
Dangerous flood (2)	up to 23,200	up to 10.3	10-20
Emergent flood (3)	up to 37,300	up to 11.69	2-4
Exceptional flood (4)	up to 37,600	up to 12.75	0.5-1

The effect of improvement of the existing embankments was studied and may result in a full-class decrease of flood hazard (dangerous flood conditions having warning flood consequences, etc.)

The information given in this type of flood susceptibility

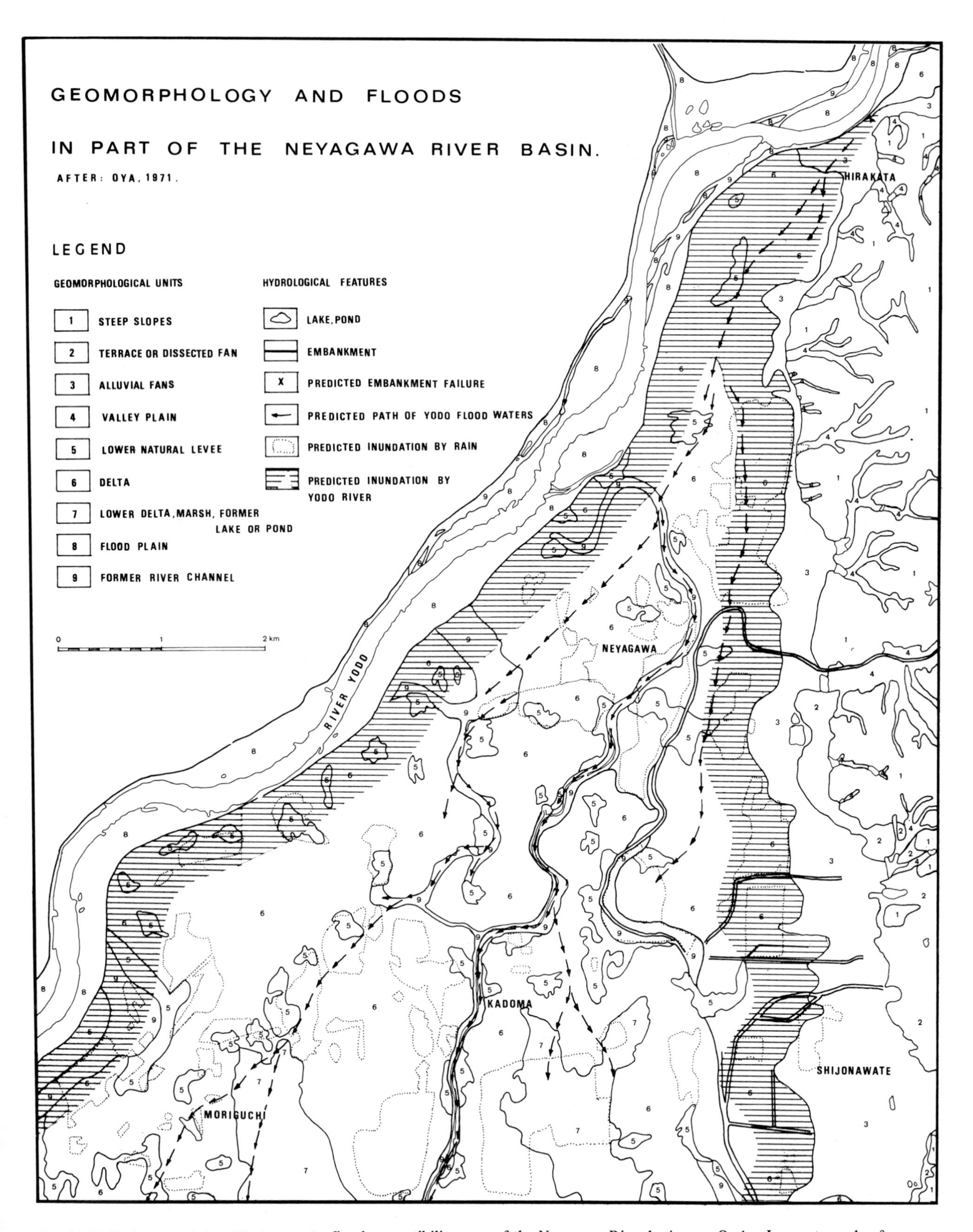

Fig. 13.15 Reduced and simplified part of a flood susceptibility map of the Neyagawa River basin near Osaka, Japan at a scale of 1:50,000 (Oya, 1971).

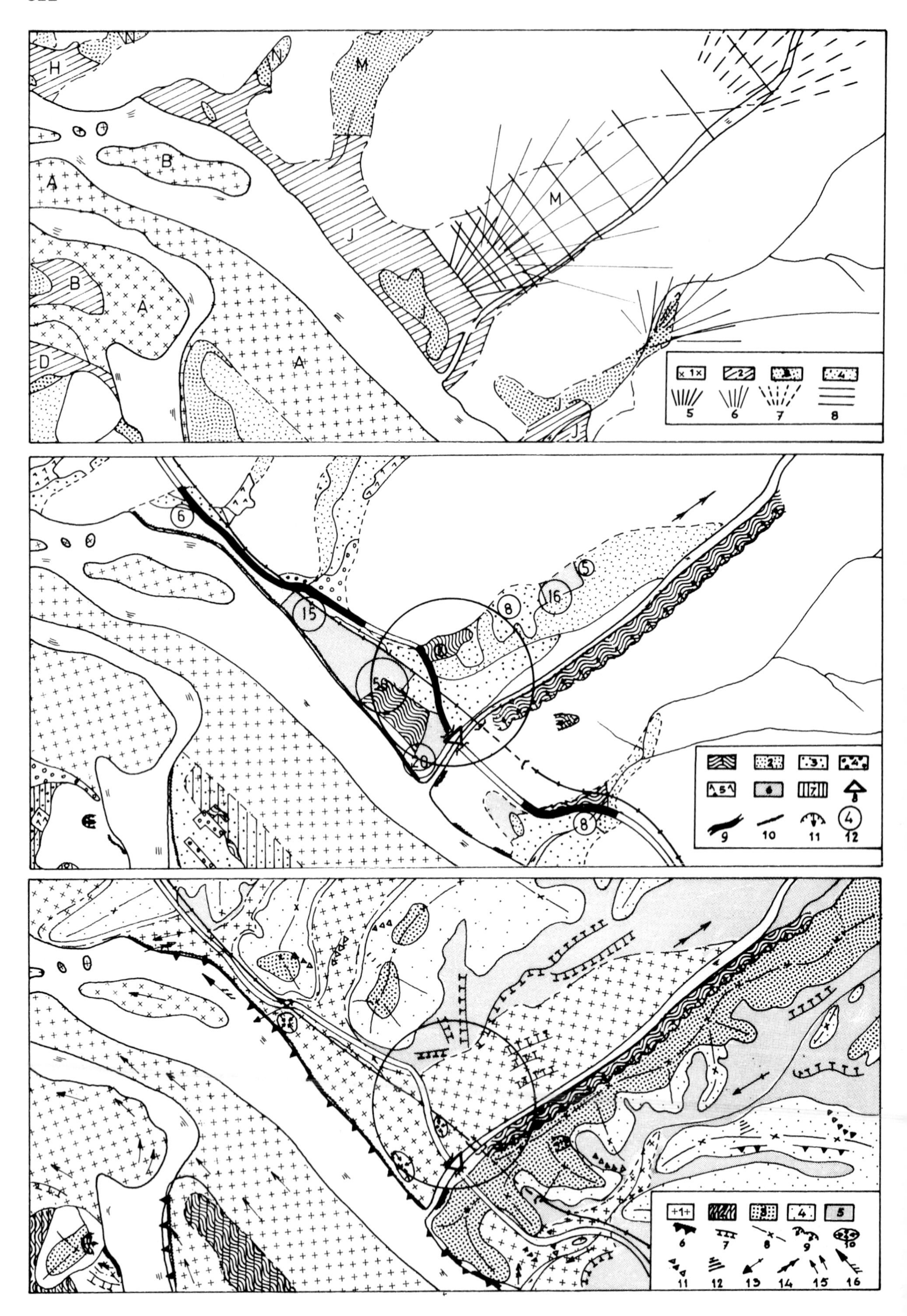
H
N
M
B
A
J
D
1
2
3
4
5
6
7
8
6
15
16
8
5
20
9
10
11
12
13
14
15
16

Fig. 13.16 Part of a flood susceptibility map series at a scale of 1:25,000 of part of the Han river valley, near Deogso, South Korea. The map series comprises of three specific discharge volumes (top), the potential loss of life and property (centre) and the geomorphological classification of the flood plain (Kim et al., R & D Dept., Asia Aero Survey Co., Ltd.).

Legend of fig. 13.16. Flood susceptibility mapping in Han river plain near Deogso, South Korea 1:25,000.

OVERLAY SEQUENT INUNDATION STAGES

1. first inundation stage (channel full stage)
2. second inundation stage (dangerous flood)
3. third inundation stage (emergent flood, 1972)
4. fourth inundation stage (exceptional flood)
5. areal extent influenced by backwaters of dangerous flood
6. same, of emergent flood
7. same, of exceptional flood
8. same, influenced by both backwater of mainstream and tributary runoff.

A: dry riverbed
B: sandbar
D: third river terrace
H: backswamp
J: lower alluvial plain
M: higher alluvial plain
N: alluvial slope

MAP OF POSSIBLE FLOOD DAMAGE

1. inundation of houses and structures
2. inundation of paddy fields
3. inundation of crop land
4. inundation of perennial crop land
5. inundation of forest
6. inundation of industrial area
7. area of maximum inundation during exceptional
8. bridge susceptible to destruction
9. inundation of road
10. erosion and collapse of river bank
11. landslide
12. number of houses and structures of possible inundation

GEOMORPHOLOGICAL FLOOD MAP

1. flood limit of 1972
2. areas of maximum inundation during exceptional flood
3. steep slope on hill
4. gentle slope on hill
5. alluvial slope or plain
6. escarpment
7. natural levee
8. divide line
9. landslide
10. depression without drain
11. gully erosion
12. sheet erosion
13. concentrated linear erosion
14. gully start
15. direction of flood current
16. rapids

survey is of great concern for decision makers on various governmental levels. The detailed data on potential damage is not always easily obtainable, particularly when rivers characterized by frequent flash floods are concerned or where typhoons or tsunami flood waves, caused by distant earthquakes,are involved. However,the approach is a practical and useful one in many cases.

13.5 Surveying for Inundations in Coastal Lowlands

An interesting study on the flood hazard originating from Tsunami waves is that by Nakano (1961, 1962). He investigated the effects of the Tsunami caused by the ill-famed Chilean earthquakes of May 24, 1960. Several interesting geomorphological maps accompany the report. It was found that a distinct relationship exists between the magnitude and force of the Tsunami waves, the configuration of the coast and the off-shore sea floor. As long as they move forward in oceanic waters of sufficient depth, the height of these sea waves is uniform, but when they approach a coast, appreciable differences develop in their height and velocity. In the case investigated, the height varied from 1-8 metres and velocity from approximately 5-15 km/hr. Coastal exposure and the distance over which the sea waves have travelled are other essential factors in hazard evaluation. It appears that sea waves from far sources affect large coastal areas, that the surface of the water remains relatively calm, and that their height is usually larger than that of locally generated Tsunami.

Generally speaking, Tsunami from close sources produce violent waves and surges that attack promontories but die out rapidly when entering bays and estuaries (Roy. Soc. N.Z., 1976). On the contrary, long-distance Tsunami have a tendency to build up to an important degree when moving inland over gradually shallower water. The progress and force of the sea waves is strongly influenced by the submarine topography. Deep waters in front of the coast increase the hazard, whereas broad shallow stretches of sea tend to diminish the force of Tsunami, provided no funneling topographic situation exists, such as estuaries. Submarine valleys and deep bay entrances provide easy access for sea waves. Once they invade the land,the forward thrust is directed and furthered by sea walls, river courses, canals, roads and rail ways and penetration is facilitated. However, the general roughness of the land rapidly decreases the force of the waves; they may end near the landward side of deltas and not invade the alluvial plain farther inland.

In mapping for protection against such hazards, much attention should be paid to coastal configuration and classification, sub-marine topography of the near-shore zone, and to natural and cultural phenomena which favour the invasion or 'attract' the sea waves. Nakano classified the coasts into 7 types with respect to their Tsunami susceptibility. Type of flood, velocity of water, depth and period of inundation and erosion and deposition depend largely on the geomorphological situation of the affected land areas. In the area of study, only 6 types were distinguished, namely:

valley plain type
wave cut bench type
delta type
sand spit type
sand dune type
filled-in land type

Coasts safe from Tsunami waves may not be so often waves caused by typhoons and vice versa. Shallow, straight coasts may be relatively safe from earthquake-caused Tsunami waves though they could be affected by sea waves produced by typhoons.

An example of the flooding caused in an urban area by a typhoon is given by Oya (1961, 1970) who studied the inundations in and near the town of Nagoya, Japan,following the Isewan typhoon of 1959. The water level in the harbour rose to 3.89 m above mean tide T.P. (Tokyo Port), consequently, extensive areas in and around the town where inundated, causing 5,300 casualities and a material loss of approximately 1,5 billion US dollars. The distributional pattern of the devastation was strongly influenced by the geomorphological situation. Areas of the plain situated farther inland and characterized by a natural levee and backswamp configuration were not flooded because of their distance from the sea, nor were the terraces flooded, due to their height. The flooded areas were affected in three different ways: a) the delta plain was slowly invaded by seawater through breaches in the dikes, b) the recently reclaimed paddy fields directly behind the coastline were severely damaged by in and out going tidal currents of high velocity. The depth of inundation reached 5.7 metres, c) the artificially filled urban land was directly affected by high velocity tidal currents but the flood waters drained off well.

Based on the experience and investigations gained by the 1959 flood, the municipal authorities divided the urban and sub-urban areas of the town into five zones (fig. 13.17). In each zone strict rules were established concerning permission to build, construction materials to be used, minimum height of the houses and installation of refuge rooms (fig. 13.18). The vertical aerial photographs of figs. 13.19 and 13.20 show the effects of the 1959 inundation in a rural area and an urban (port) area,respectively, at a scale of 1:16,500. Fig. 13.21 shows paddy fields immediately bordering the coast where high velocity currents flow through breaches in the dikes.

Another investigation by Oya et al. (1963) on floods

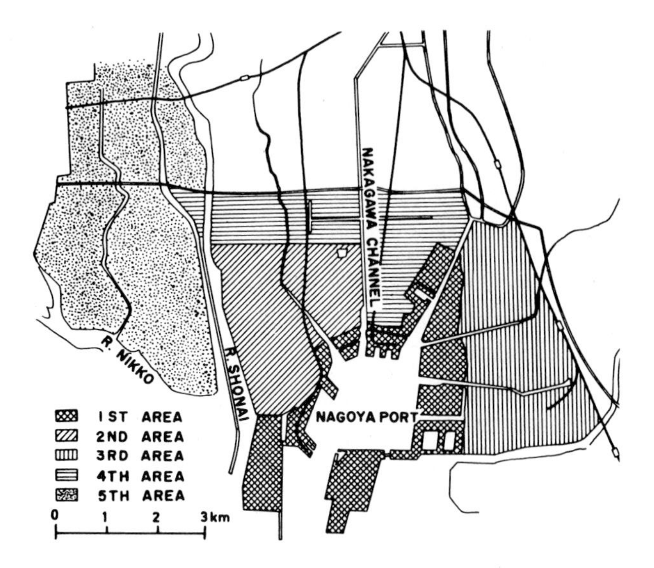

Fig. 13.17 Five flood hazard zones established in the Nagoya port area on the basis of a geomorphological flood susceptibility survey (Oya, 1970).

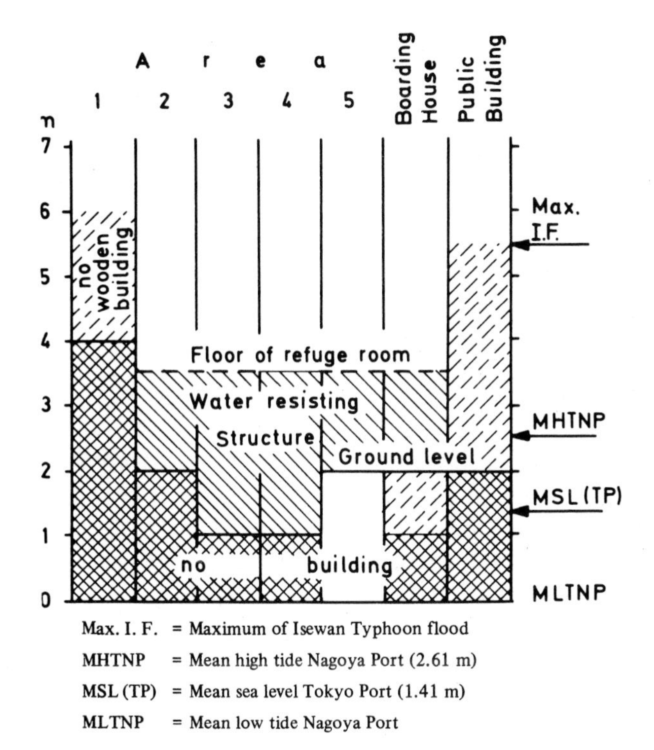

Fig. 13.18 Building codes for the five zones of fig. 13.17.

in coastal zones is in the Ariake-Kai area (Kyushu) where they prepared a geomorphological flood map indicating detailed contour lines, types of inundations and risk zones.

Fig. 13.19 Vertical infra-red aerial photograph at a scale of 1:16,500 of a flooded rural area near Nagoya following the Isewan typhoon of 26 September, 1959.

13.6 Surveying for Floods by Local Rains and Impeded Drainage

An example of a flood susceptibility survey where local or nearby rainfall is the main source of water leading to inundations, is given by Verstappen (1975). His study relates to a lowland area along the south coast of Central Java, Indonesia, near the town of Kroya, a part of which is subject to flooding every year. Although the periods of inundation are usually comparatively short (7-10 days), they suffice to exclude these areas from technologically improved irrigation of the paddy fields.

The area is traversed by the Serayu River, the natural levees of which are never overtopped in its lower course. The tides have a slight effect in some minor areas only.

Fig. 13.20 Vertical infrared aerial photograph at a scale of 1:16,500 of a flooded part of the Nagoya port area following the Isewan typhoon of 26 September, 1959.

The tidal range is 0.5 - 1.5 metres and saline water penetrates a few kilometres upstream in the Serayu and other rivers without causing much harm. Although groundwater is at shallow depth in the plain, the distributional pattern of the inundation areas and the short time between rains and flooding indicate that rainwater and the impeded drainage of surface water due to certain aspects of the terrain configuration are responsible for the situation and not the rising groundwater.

The general situation is shown in the map of fig. 13.22 on which major landforms and flood-classification zones are indicated. The map of fig. 13.23 gives a detailed picture of one of the most seriously affected localities. A set of 1:20,000 black and white infrared aerial photographs of the area dating from 22 December 1972 were used for the study and served to map the various geomorpho logical features and also the flood stricken areas, since inundation was actually taking place at the time of air photography.

The plain is bordered on the landward side by the South Serayu mountain range, which is almost connected to the Karangbolong limestone hills marking the eastern end of the plain and
it in the NW. There is no high ground to define the western limit of the plain where it gives way to swamps and surface water of the Segara Anakan north of the Nusa Kambangan limestone peninsula. A broad sandy beach ridge occurs along the coast.

Numerous older beach ridges, approximately parallel to the present coastline,can be traced inland and record the effect of waves in the past. They are absent or nearly absent in the Segara Anakan area to the west, which was always protected from waves by the hills of Nusa Kambangan and are best developed along those parts of the (former) coastline most exposed to the SE winds. Immediately east of the Serayu River, beach ridges are conspicuous only in the near-coastal zone. Farther inland, former natural levees and point bars of the Serayu are dominant relief elements.

The area is traversed by the Saporegel and Donan Creeks near Cilacap, as well as by the small Kedungbaya River which joins the Serayu from the west near its mouth, and by the Bengawan and Ijo Rivers, the drainage basins of which occupy the greater part of the plain.

The plain can be divided into the following flood classification zones:

1. Hills bordering the plain
2. A zone bordering the hills to the north with sufficient gradient to preclude inundation. Large parts are irrigated; a large non-irrigated area stretches far to the south, to the NNW of Cilacap and marks the east rim of the Segara Anakan depression.
3. A more or less wedge-shaped area of Serayu deposits having its apex where this river enters the plain and its base at the south coast from a point some kilometers west of the present river mouth to the present Bengawan river mouth (G Selok). The west strip of this unit is formed by the present Serayu natural levee and point bars, whereas the eastern part is made up of older deposits of the same type referred to by the author as older Serayu deposits. Since the Serayu River almost never overtops its levee, inundations in this zone are very rare.
4. Seaward of 2, in the west, swamps and marshes occur. These low, wet areas drain mostly to the Segara Anakan which acts as a flood storage. No beach ridges impede the drainage in this waste land area which is connected to the sea by the tidal creeks of the Saporegel River.

Fig. 13.21 Vertical infrared aerial photograph at a scale of 1:16,500 of flooded paddy fields and breached sea dikes following the Isewan typhoon of 26 September 1959. Note the strong tidal currents visualized by the turbidity patterns.

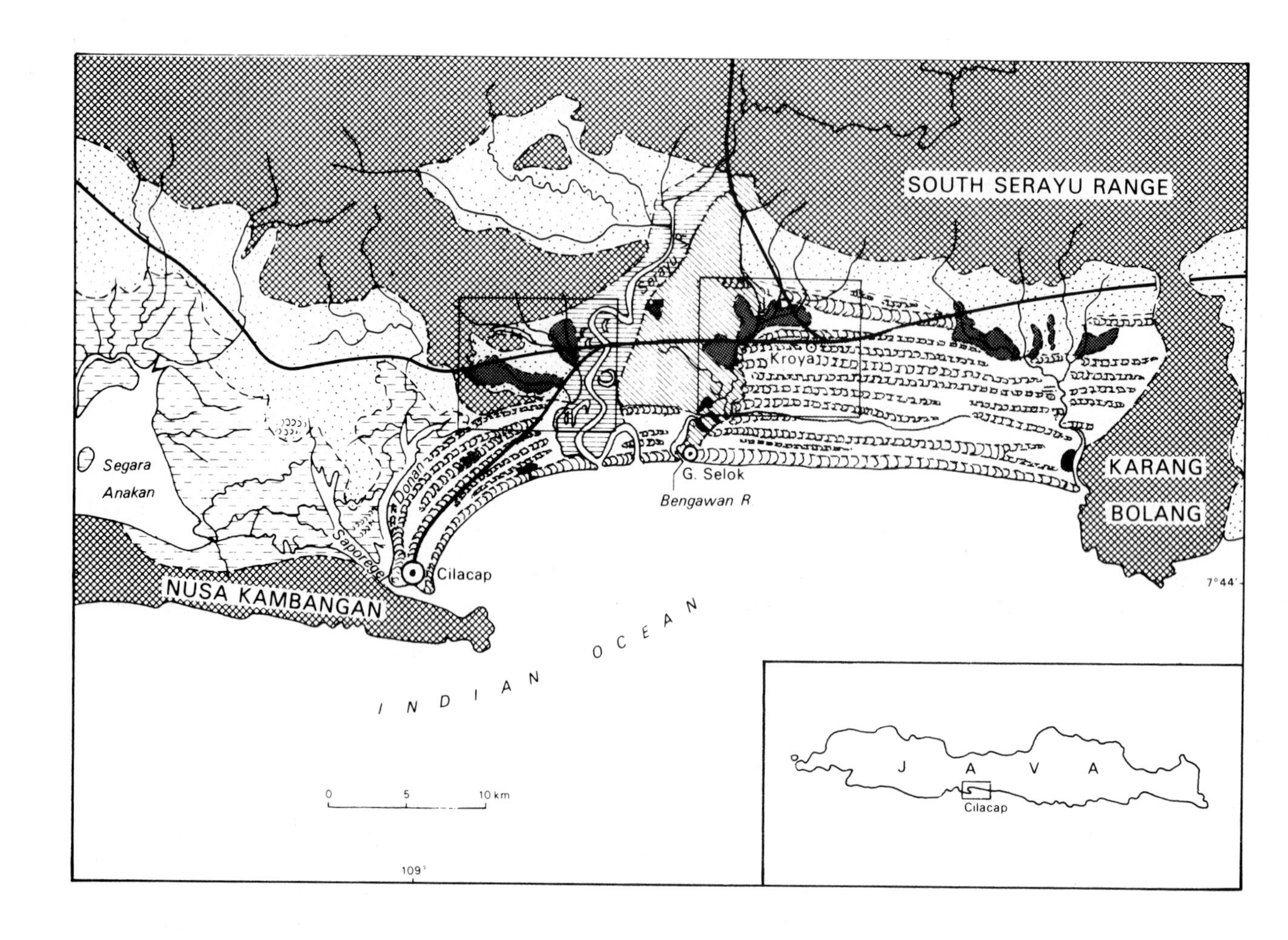

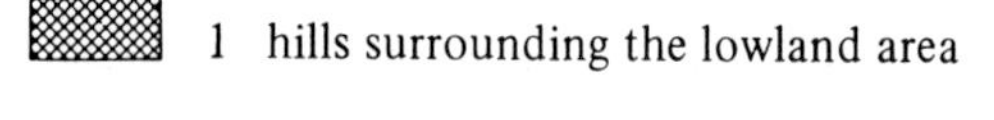

1 hills surrounding the lowland area

2 higher (inland) parts of the plain - - - normally not inundated

3 Serayu deposits (recent deposits in the west; Older Serayu deposits in the E)
- - - normally not inundated

4 seaward of area 2 in the W: swamps and marshes draining to the
Segara Anakan inland sea which acts as a flood storage; mostly semi-permanent flooding

5 seaward of area 2 in the E: impeded drainage caused by low ridges,
Serayu deposits and/or beach ridges; - - - *area of annual inundations;* duration up to
one week

6 beach ridges and alternating swales situated seaward of area 5; some swales may
be inundated but conditions generally better than in area 5

7 near-coastal areas affected by tides

Fig. 13.22 Generalised geomorphological map of the Kroya plain, Central Java, with the main flood hazard zones. Scale 1:250,000.

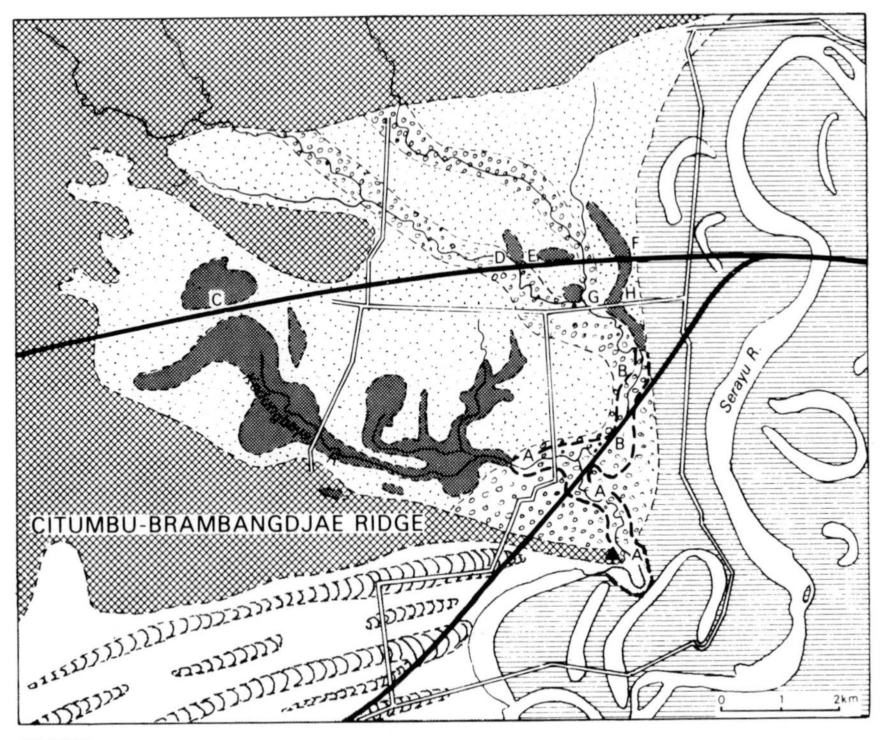

hills surrounding the lowland area

natural levee, point bars etc of the Serayu River

deposits of minor rivers draining the area

areas most seriously affected by inundation

present and former position of Serayu riverbed

beach ridges and intervening swales

primary cause of inundations; point where the eastern end of the Citumbu-Brambangdjae Ridge meets with the Serayu deposits, thus giving rise to impeded drainage of the small Kedungbaya River

riverbed to be enlarged

A-H impeded drainage

Fig. 13.23 Main inundation zones to the west of the Serayu River, Central Java, Indonesia, scale 1:100,000. For location see fig. 13.22. It is evident that the inundations are in the first place due to the impeded drainage in the SE where a low ridge meets with the natural levee, point bars, etc. of the Serayu River; upstream of this locality the minor rivers draining the plain forcedly deposited their load (A,B). The railway (C, D, E. F) and the road system (G, H) locally add to the problem.

The easternmost parts of this unit are situated landward of the beach ridges near Cilacap and are drained by the Donan tidal creek.

5. Seaward of 2, in the east, a zone of impeded drainage occurs where the annual inundations are concentrated. The latter occur, from west to east, in three areas:
   a) along the small Kedungbaya River, upstream of the point where this river has to force its way between the EW stretching Citumbu-Brambangjae Ridge and the actual natural levee of the Serayu;
   b) where the upper reaches of the Bengawan River, draining part of the South Serayu Range, join and break through a zone where numerous parallel beach ridges meet with the Older Serayu deposits and with the more recent deposits of the Bengawan River which overly parts of the former;
   c) where the upper reaches of the Ijo River, draining part of the South Serayu Range, join and break through the beach ridges of b) near their eastern ends, where sedimentation by the Ijo River has also occurred. Tectonic subsidence may have contributed to the concentration of inundations in zone 5.
6. The area seaward is composed of beach ridges with alternating swales. The swales in this zone may be inundated after heavy rains, but are mostly inundated by water collected locally. Flooding is thus much less serious, if it occurs at all, than in the zones mentioned at 5. The seaward parts are generally the least susceptible to flooding.

The most favourable situation exists in the broad swale drained by the Secang River which connects the Bengawan and the Ijo.

7. Near-coastal areas affected by the tides, comprising: 1) the economically unimportant areas to the west of Cilacap, 2) a smaller area near the former Serayu River mouth about 13 km NE of this town. and 3) a few small areas near the mouth of the Bengawan River and possibly also near the mouths of the other rivers draining the area.

The map of fig. 13.23 giving a more detailed picture of area 5a to the west of the Serayu River, clearly shows where the Citumbu-Brambangjae Ridge and the natural levee of the Serayu River meet, thus ponding up the tiny Kedungbaya River and causing annual inundations in the surrounding low areas. The inundated zones are indicated in black on the basis of the aerial photographs (1:20,000) of December 1972. Although tectonic subsidence may be a secondary factor in the distributional pattern of the inundations, it is evident that widening and deepening of the drainage lines in the SE near A could improve the situation considerably. Further improvements can also be expected from similar works near B. Culverts of adequate capacity underneath the railway at C, D, E and F, and under the road near G and H also are important.

It could not be directly ascertained whether the inundation was at its peak when the aerial photographs were taken on 22 December 1972. Data on inundations in the area of the years 1969-1972, obtained from the Department of Public Works (PUTL) in Purwokerto, revealed that to the east of the Serayu River, and thus in areas 5b and 5c, up to 3,438 ha are subject to annual flooding. The areas flooded on 22 December 1972 amount to 920 ha in area 5a, 1640 ha in area 5b and 1290 ha in area 5c. Comparison of the total acreage flooded on 22 December 1972 in areas 5b and 5c (2930 ha) with the PUTL data on these areas indicates that the time of air photography coincided with a situation of serious flooding, the total extent of which is only 15% less than the maximum extent of inundations over the period 1969-1972.

13.7 Flood Susceptibility Mapping for Specific Development Purposes

It is evident from the previous sections that geomorphological mapping for purposes of flood susceptibility classification has found many applications. It served for providing decision makers and planners with information for minimizing the loss of life and property and for arriving at an optimal land use planning of the often densely populated affected areas. It could also be that a virgin or sparsely populated area susceptible to flooding or a swamp zone has to be opened or developed for more intensive settlement and land use. Deciding upon the most appropriate location of the future road network, the settlements, etc. is then a major issue.

As an example, the swamp classification of the coastal zone of the West Paseman, Sumatra, Indonesia, is given (Verstappen, 1975) which was carried out in the framework of a regional development plan. These swamps are situated between the sandy beach ridges bordering the coast and the hills and volcanic footslopes of the interior. They are narrow where the vast, gentle lower slopes of the Malintang and Talamau volcanoes closely approach the coast and they reach their maximum width between these two volcanoes and to the north and south of them. The local rainfall is not exceptionally high though there is an ample supply of surface water, both channelled and diffused, coming from the steep mountain ranges occurring inland and particularly from the extensive fluvio-volcanic lower slopes of the two volcanic cones mentioned above. The radial drainage of these cones promotes the rather diffuse supply of water to the edges of the swamps, the effect of which is strengthened by emerging groundwater around the lower part of the fluvio-volcanic slopes.

At places deforestation, due to shifting cultivation, has led to increased surface runoff and deterioration of the swamps.

Detailed aerial photo analysis and the study of Landsat imagery shows that the coastal swamps and marshes are not homogeneous, but can be subdivided into eight classes according to differences in landform, soil conditions, flood susceptibility and vegetation. The distributional pattern of these classes is indicated in the map of fig. 13.24.

It should be understood that the patterns in the swamps are complex and transitions between classes are gradual. The general pattern is made up of the natural river levees and old and young coastal beach ridges which, together with the fluvio-volcanic slopes, enclose a number of partly interconnected compartments where the water from the higher surroundings is collected. Flooding is almost continuous and organic deposits are deep. Drainage in these lowest central parts of the swamps is impeded; flooding is less frequent on the slightly higher zones surrounding these hard-cores of the swamps. The varying degrees of flood susceptibility basically indicate the potentialities of these sites.

Class 1: Brackish Tidal Swamps

This area type occurs only in the north along the lower course of the Tamak River. Vegetation is mangrove, mostly Rhizophora, which shows up dark and fine-textured on the

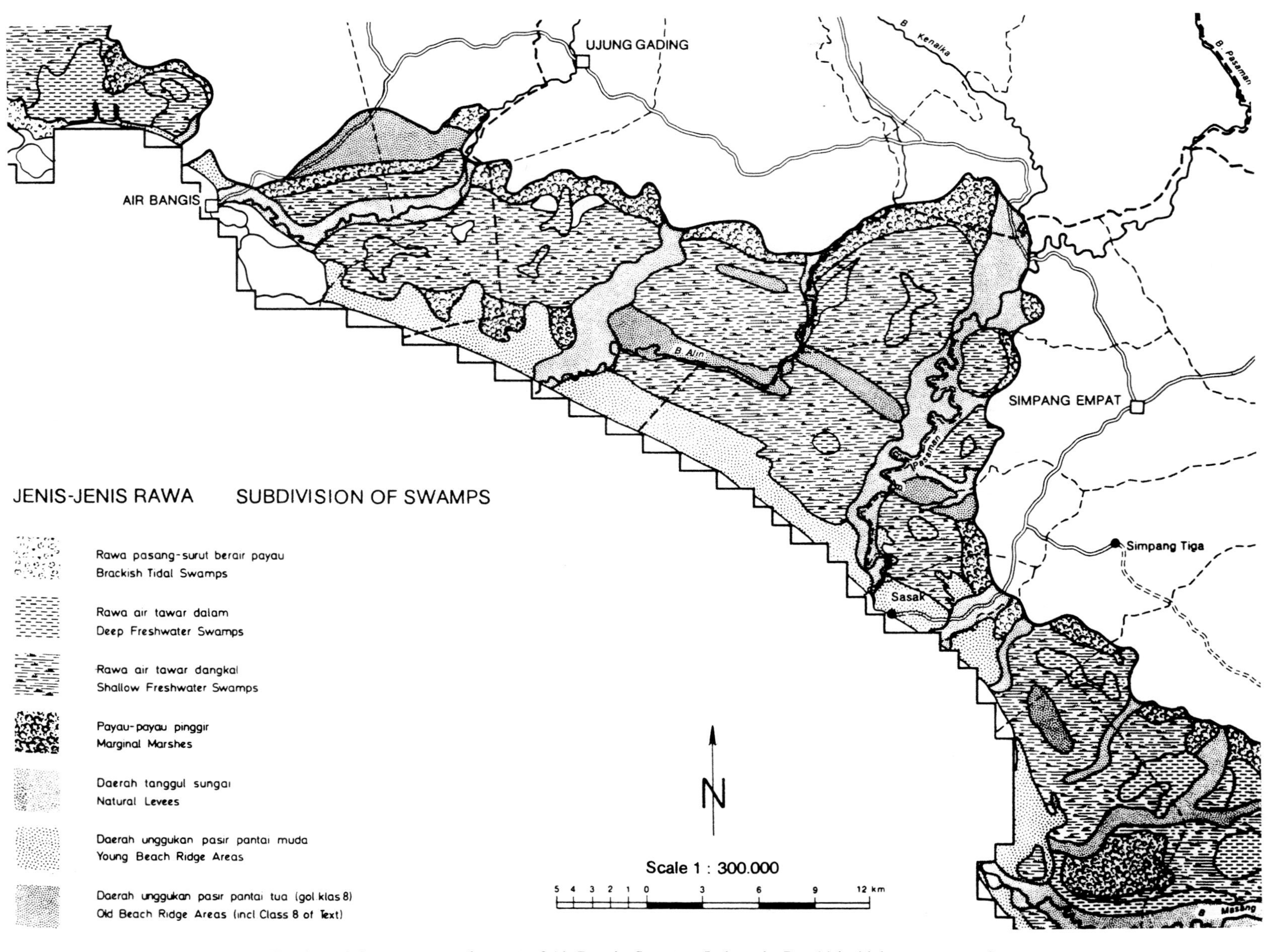

Fig. 13.24 Geomorphological classification of the swamps to the west of Air Bangis, Sumatra, Indonesia. Brackish tidal swamps are of limited extent; deep freshwater swamps, shallow freshwater swamps and marginal marshes are main classes. Natural levees and beach-ridges are relatively dry zones.

aerial photographs. A narrow whitish strip, almost devoid of vegetation occurs at places at the transition with the fresh water swamps of class 2.

Class 2: Deep Fresh Water Swamps
This area type is covered by low vegetation and occurs in the lowest parts, the true back swamps. It shows up as fine-textured light-toned areas.

Class 3: Shallow Fresh Water Swamps
These are covered by swamp forest. They surround the deep fresh-water swamps of class 2 and are bounded by natural levees, beach ridges of fluvio-volcanic slopes (classes 4-8) at the other drier side. On the aerial photos they border zones of the back swamps, are rather dark-toned and show a somewhat coarse and irregular texture.

Class 4: Marginal Marshes
These marshes are characteristic for the zones bordering the fluvio-volcanic slopes. Natural vegetation is rain forest rather than true swamp forest. On the air photos the colour tone is dark and the texture coarse.

Class 5: Natural Levees
These silty-loamy ridges of varying height and width accompany the smaller and larger rivers that traverse the swamps and marshes. They are covered by intermediate forest and the highest parts have some agricultural use. The susceptibility to flooding varies according to the height and can be roughly estimated by the width of the natural levees indicated.

Class 6 : Young Beach Ridges and Intervening Swales
These sandy ridges are fairly high and border the sea. They occur in sets approximately parallel to the coast and the individual ridges are separated from each other by narrow, low and swampy swales. On the aerial photographs the more or less barren sandy ridges are light-toned, whereas the vegetation of swales appears much darker. A distinct, striped texture thereby results, which facilitates the recognition of these sites.

Rivers breaking through these beach ridges are sometimes diverted parallel to the coast in a swale until they ultimately debouch into the sea. Natural vegetation consists of Casuarina and only small areas are used for coconut cultivation.

Class 7 : Old Beach Ridges and Intervening Swales
These zones are situated farther inland and are surrounded by swamps. They mark former positions of the coastline. These ridges have become lower and more subdued in the course of time. The soils are more mature and are not pure sand but contain a finer-textured material. The striped texture mentioned for the young beach ridges can also be recognized here, but is less distinct. The colour tone is generally darker. Susceptibility to flooding is greater than for class 6, and agriculture is almost completely absent.

Class 8: Isolated Old Beach Ridges
At places a rather faint striping of the swamp vegetation indicates the occurrence of low beach ridges. Moderate flooding periods occur; soils have a much higher sand content than in the adjacent parts of the swamp.

It is obvious that the classification presented here and the distributional pattern of the eight classes will facilitate road location once the decision has been made to further develop these swamp areas. Comparable applications of geomorphological surveying for a more efficient utilization of swamp zones by regulating levels, partial or complete drainage and/or by using the water elsewhere, can be given from different parts of the world also. An interesting example is the geomorphological study of Lake Faguibine and nearby swamps in Mali for purposes of land development reported by Blanck (1960) and Tricart (1960). The aim of the project was to regulate the intake of water by the Goundam swamp and thus to lower the level of Lake Fabuibine by natural evaporation. Subsequently, sufficient supply of water to the lake must be guaranteed in years of below-average rainfall to optimize the agricultural production around it.

The geomorphological survey served to assist the engineers in the proper location of the various structures required, to understand the dynamics of the swamps and to evaluate the danger of reactivating aeolian processes in the ancient dune areas by lowering the lake (and groundwater) level.

Another example is the geomorphological survey of the Entrerriano Delta, Argentina, by Pasotti et al. (1976). The map of fig. 13.25 shows a geomorphological terrain classification on which inundation conditions and flow direction of the water are indicated.

13.8 Excessive Sedimentation and Erosion Caused by Floods

In many cases of flood susceptibility surveys, not only the immediate effects of the flood waters have to be studied, but also the (more lasting) adverse consequences of the sediments carried and deposited by them. This is particularly the case when flash floods are concerned in

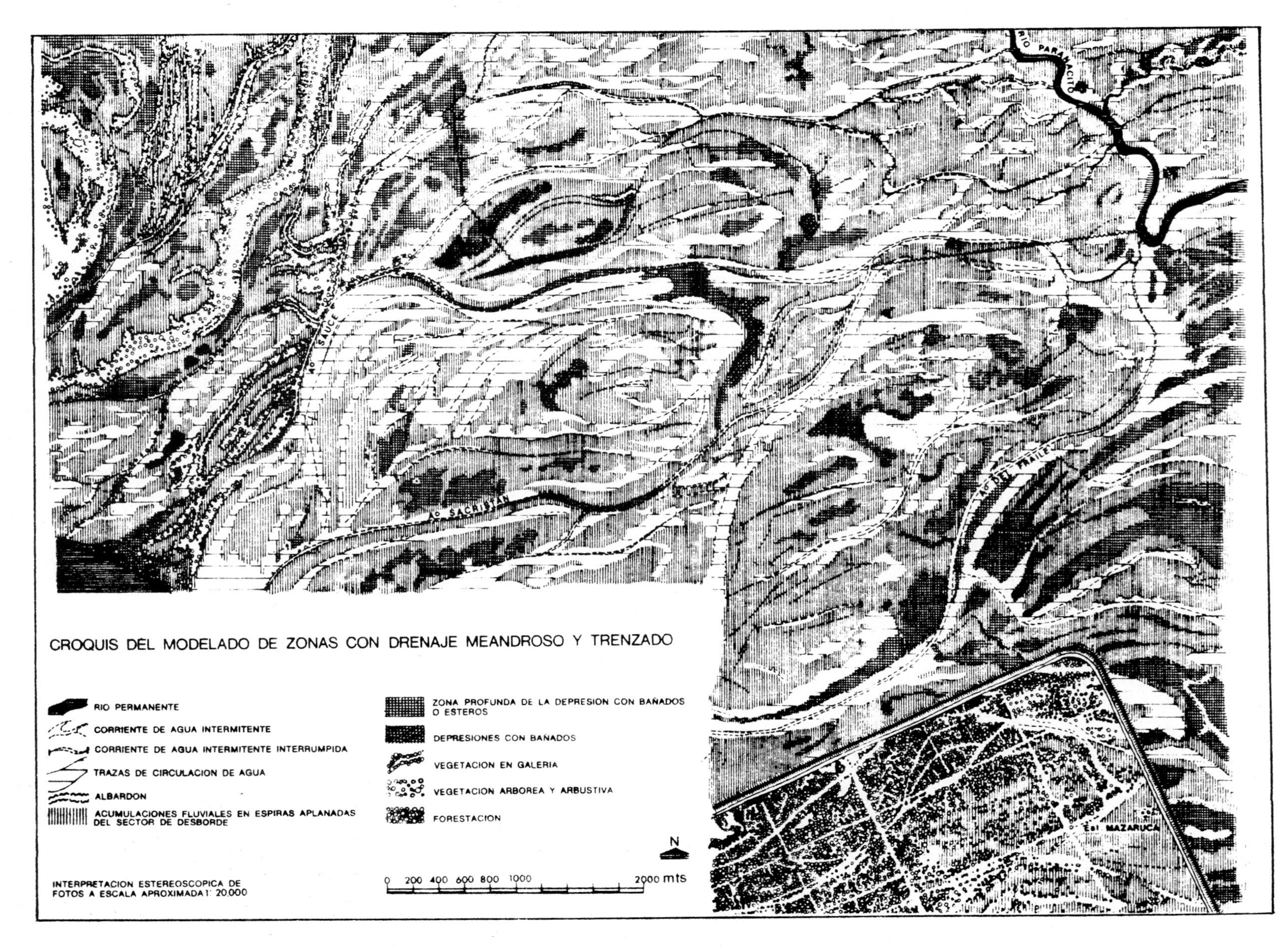

Fig. 13.25 Detailed geomorphological classification of a part of the Entrerriano Delta, Argentina, with indication of inundations and flow direction in the channels (Pasotti et al., 1976).

poorly vegetated or deforested watersheds where the vegetative cover insufficiently contributes to the retention of surface runoff on the interfluves. Sudden decreases in river gradient where the mountains give way to plains also add to this problem. These conditions prevail particularly, though not exclusively, in numerous river basins in (semi)arid countries.

The sediments deposited are often of considerable thickness and may cover valuable agricultural land, invade settlements, disorganise irrigation systems, etc. Since they tend to be sandy or even gravelly and are devoid of humus and other essential elements for plant growth, their fertility is usually low. They dry out easily and then act as a source of sand for further wind work from which additional loss of land may result. The fine-textured sediments carried in suspension may reach the sea but are often deposited in closed depressions leading to the traditional problem of salinity after subsequent evaporation has taken place.

An important part of the high sediment yield of such floods is derived from bank erosion caused when large amounts of water are forced into a narrow riverbed, causing a sharp rise. Sand and gravel from terraces or fans contribute to the bed load and where clayey materials drop into the turbulent water, clay balls covered with gravel are formed, which are characteristic of flood deposits. The rest of the fines are carried off in suspension.

Another source of flood sediments is found on the interfluves where raindrops introduce splash erosion and large quantities of sediment are carried downslope by surface wash, partly channelled runoff and (imperfect) sheet flood. The material ultimately reaches the main river where it is deposited in the form of fans at the mouth of tributary streams or ravines. The material of these rapidly formed fans is transported further downstream until it comes to rest when the flood abates. A series of low, horizontal agricultural bunds following the contours may be laid out on those slopes affected by unconcentrated overland flow and sheet wash. The water can thus be collected and given a chance to infiltrate and evaporate, thereby increasing the surface retention.

The distance between these bunds depends on various factors such as slope angle, roughness and permeability of the soil. It should be calculated in such a way that the danger for overtopping is minimum. Once a breach is formed in one of the bunds, the water will form a gully and add to the water retained by the lower bund which as a consequence will also be breached. Proper lay-out and good maintenance of the bund systems is therefore essential for their effectiveness.

Characteristically, the flood deposits, fans, etc.,are immediately dissected again in the later part of the flood period when the sediment load of the river-in-spate decreases. A rapid succession of alternating sedimentation and erosion and important overnight changes of the riverbed and the alluvial plain are common features in such cases, as demonstrated by the examples given.

An example of the problems arising from such floods with high sediment production is the catastrophic flood of the Zeroud and Merguelil rivers in Central Tunisia in September/October 1969 (Mensching, 1970). Floods of this type are common in these river basins. In April 1973, when the author studied the effects of the 1969 flood using sequential aerial photography from before and immediately after this flood as a basis, he found himself caught in a flood which enabled him to study the various phenomena directly (fig. 13.26). The 1969 flood route of the Zeroud River is indicated in fig. 13.27 and shows that the river overflowed its bank at two places in particular, namely near Sidi Saad and downstream of the Dj. Hallouf gorge. At the last mentioned locality the flood waters passed to the north of the village Menzel el Mhiri and finally reached the depression of the Sebkha of Cherita and Sidi el Hani. The Merguellil River passed south of the town of Kairouan and reached the sea via the Sebkha Kelbia far away from its usual outfall. The deposits were sands, gravels, boulders and clay balls. The thickness of the deposits was 0.5 m on the average, but locally considerably more. Occasional huge boulders, up to 4.5 x 2.5 x 1 metre,were also deposited.

Fig. 13.26 Ground view of the flood waters of the Zeroud River, near Hadjeb el Aioun, Central Tunisia, in May 1972.

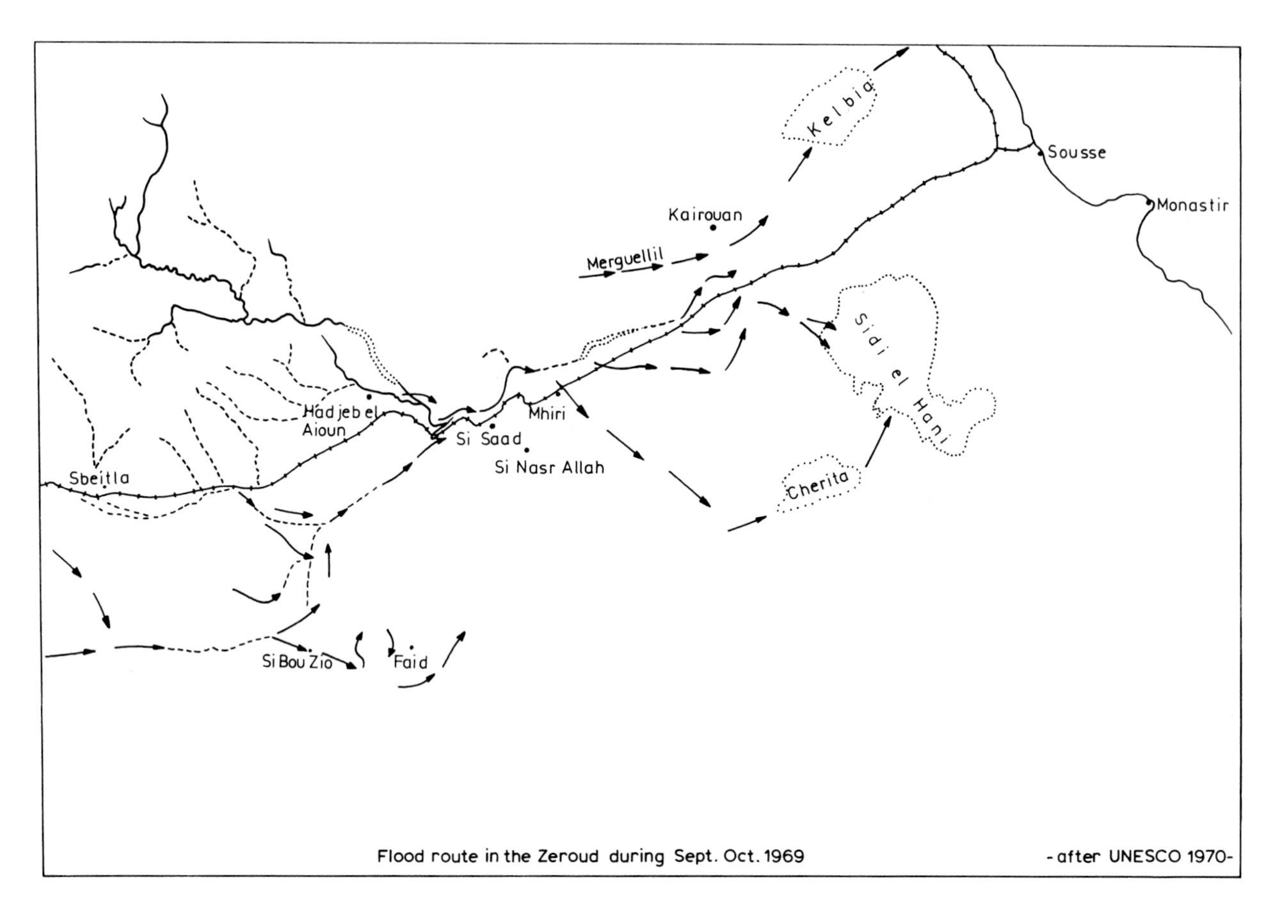

Fig. 13.27 General pattern of the floods in the Zeroud River basin, Central Tunisia, in September/November 1969. Depressions were filled with water and the river flowed in channels that were not normally used.

The stereotriplet of fig. 13.28 (1:25,000) shows the immediate surroundings of the town of Kairouan and how they were affected by the flood. Houses, gardens and agricultural land were invaded and destroyed by the flood waters and subsequently covered or surrounded by the coarse and infertile flood deposits. The terrestrial photograph of fig. 13.29 shows the railway disrupted by flood waters and sediments. The stereotriplet of fig. 13.30 (1:25,000) shows how the flood waters of relatively small tributaries coming down from the mountains with rather steep gradients spread out laterally when leaving the mountains and invaded the fields on the more gentle slopes and plains before reaching the main river. Rapid sedimentation immediately followed by erosion and the cutting of a new channel in the newly laid sediments are obvious. The terrestrial photograph of fig. 13.31 shows a flood-produced sandy fan formed where a tributary river enters the Zeroud. Also here, the incision following the deposition is evident. The extended stereogram of fig. 13.32 (1:25,000) illustrates the effect of overland flow and the flood-laid deposits in the wheat fields and orchards. The slopes used for wheat cultivation are barren in the rainy winter season and are provided with the low earthen agricultural bunds mentioned earlier which are locally known as 'tabia dams'. Obviously, many of them could not accommodate the consequences of conditions that occurred during the flood of September/October 1969. A full description of the geomorphological characteristics and consequences of this flood is given by Ghanem (1973) who studied the fluvial dynamics using sequential aerial photography (Tag Eldeen, 1980).

Also outside the (semi)arid zone such catastrophic floods with high sediment production may occur, particularly where alpine mountains occur. Numerous violent slope processes may be triggered by severe rains and subsequent peak runoff conditions.

Major landslides, debris slides, snow and debris avalanches, etc.,may cause disasters of various kinds, including the temporary damming of larger rivers. Fig. 13.33 depicts a part of a geomorphological map made on such an occasion. It is evident from this and other examples that adequate flood protection can only be obtained when the flooding conditions are mapped not only in the lowlands but when a survey is carried out of the drainage basin (Castiglioni et al., 1974). The river basin surveys mentioned in section 4 and the erosion/conservation surveys described in section 15, together with flood susceptibility surveys in the lowlands dealt with in this section, provide the information required to improve the environmental conditions in the watershed and thus to arrive at long-term benefit and safety.

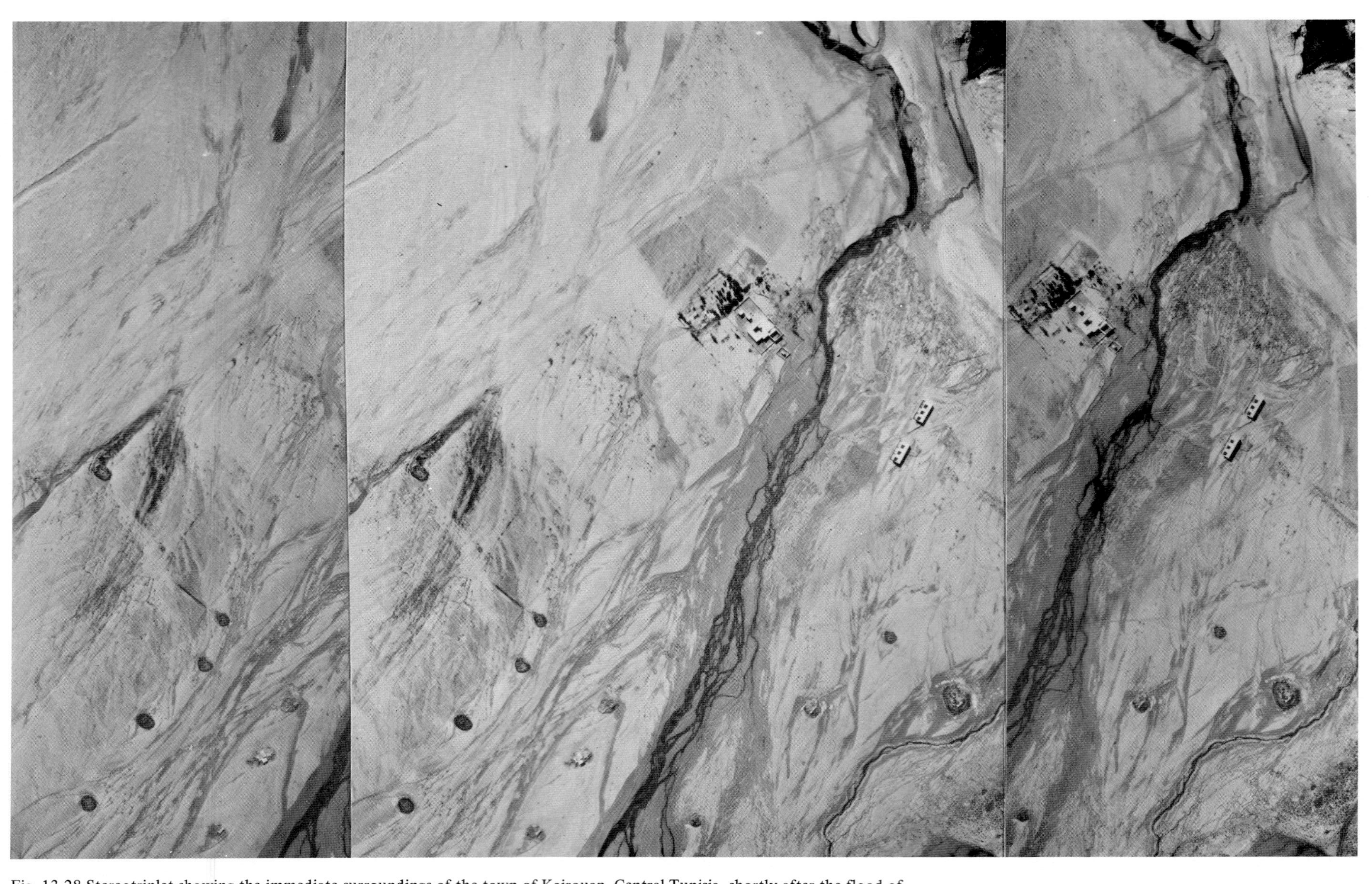

Fig. 13.28 Stereotriplet showing the immediate surroundings of the town of Kairouan, Central Tunisia, shortly after the flood of 1969: sand covers agricultural fields and houses are destroyed. Scale 1:25,000.

Fig. 13.29 The railway line near Kairouan after the 1969 flood: the track is disrupted and sand covers the area.

For fig. 13.30 see next page.

Fig. 13.31 Ground view of the 1969 flood deposits in the Zeroud River plain near Hadjeb el Aioun. The fan of a minor tributary was formed during the peak of the flood and immediately dissected when the flood abated.

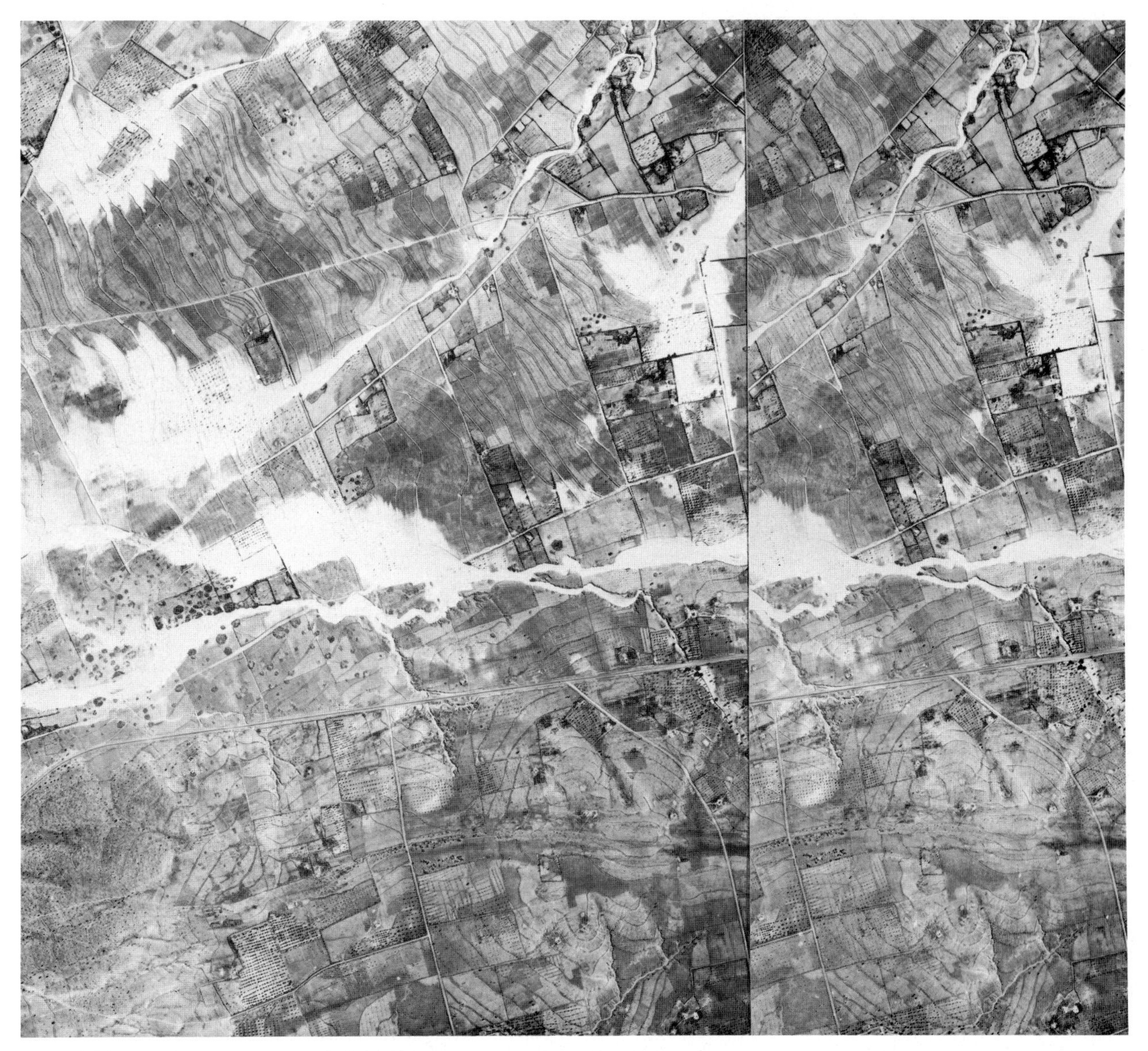

Fig. 13.32 Extended stereogram, scale 1:25,000 showing the effect of overland flow and flood-laid deposits dating from 1969 on wheat fields, orchards and flood protection bunds ('tabia').

Fig. 13.30 Stereotriplet showing how the flood waters and the coarse sediments of even relatively small tributaries devastated agricultural fields during the floods in Central Tunisia in 1969. Scale 1:25,000.

Fig. 13.33 Devastation in upstream and downstream stretches of rivers in part of the Italian Alps near Longarone caused before and during the 1966 floods (Castiglioni et al., 1974).

Key:
1. minor landslides (less than 10 ha.), pre-1966
2. same, formed in 1966
3. major slides (more than 10 ha.), pre-1966
4. same, formed in 1966
5. area with numerous landslides, pre-1966
6. same, formed in 1966
7. major torrential inundation in valley plain
8. major inundations, without sedimentation
9. important deposition in valley plain
10. active torrents with dominant erosion
11. torrent with discontinuous erosion
12. lake
13. road

References

Benson, M.A., 1962. Factors influencing the occurrence of floods in a humid region of diverse terrain. USGS Water Supply Paper 1580 B.

Bird, J.F., 1980. Geomorphological implications of flood control measures. Austral. Geogr. Stud., 18(2): 169–182.

Blanck, J.P., 1960. Projet d'aménagement de la vallée du Niger entre Timbouctou et Labbezanga, Etude géomorphologique. Publ. C.G.A., Univ. de Strasbourg, France, 45 pp.

Burby, R.J., and French, S.P., 1981. Coping with floods, the land use management parameter. J. Amer. Planning Soc., 47(3): 289–300.

Burgess, C.N., 1967. Airphoto interpretation as an aid in flood susceptibility determination. Int. Conf. Water f. Peace. Washington D.C., 4: 867–881.

Burton, I., et al., 1969. The human ecology of coastal flood hazard in Megalopolis. Univ. Chicago, Dept. of Geogr. Res. Paper, 115, 196 pp.

Burton, I. et al., 1969. The shores of Megalopolis: coastal occupance and adjustment to flood hazard. Final report ONR Contract, 388–073, Elmer (New Jersey), 603 pp.

Cain, M., and Beatty, T., 1968. The use of soil maps in the delineation of flood plains. Water Res. Research, 4(1): 173–182.

Carter, V., et al., 1979. Wetland classification and mapping in western Tennessee. Ph. Eng. and R.S., 45(3): 273–284.

Castiglioni, G.B., et al., 1974. Note di commento alla Carta dell'alluvione del novembre 1966 nel Veneto e nel Trentino-Alto Adige, effetti morfologici e allagamenti 1:200,000. Publ. in: Atti del 21 Congr. Geogr. Ital. (Verbania, 1971) "Le calamità naturali nelle Alpi":269–291.

Cochrane, H.C., 1981. Flood loss simulation. Nat Res. Forum, 5(1): 31–67.

Deutsch, M. et al., 1974. Mapping of the 1973 Mississippi River floods from the Earth Resources Technology Satellite (ERTS)-1. Proc. Symp. Remote Sensing and Water Resources Management. Amer. Water Res. Ass., Urbanal (Ill.), USA: 39–55.

Environment Canada, 1977. Flood risk map Carman, Manitoba. Min. Supply and Services. Publ. 37–19, 21 pp.

Environment Canada, 1978. Cartes du risque d'inondations/ Flood risk maps. Rivière Chaudière, Quebec, du St. Denis à St. Joseph du Beaunes.

Ericksen, N.J., 1971. Human adjustment to flood in New Zealand. New Zealand Geogr., 27: 105–129.

Es, E. van, 1976. Levantamientos para fines hidrológicos de la zona inundabile de la Hoya del Río Magdalena. Revista CIAF, 3(1): 11–38.

Everitt, B.E., 1968. Use of the Cottonwood in an investigation of the recent history of a flood plain. Amer. J. Sci. 266: 417–439.

Eyre, L.A., 1980. The June 12, 1979 flood disaster in Jamaica: A satellite survey. Remote Sens. Quat., 2(2): 29–46.

Ghanem, H., 1973. A study of the morphological changes in part of the Zeroud Riverbasin, (Central Tunisia), induced by the flood of September/October 1969. M.Sc. thesis ITC, Enschede, 85 pp.

Gronet, R., 1980. Flood in the Narew and Bug valleys. Preliminary results of interpretation of Landsat-3 images of April 3, 1979. Proc. 3rd Symp. Remote Sensing, Warszawa, April 1979: Publ. P.A.N., Warszawa: 90–91.

Hanke, S.H., 1972. Flood losses–will they ever stop? J. Soil and Water Cons., 27: 242–243.

Hallberg, G.R., et al., 1973. Application of ERTS-1 imagery to flood inundation mapping. Proc. Symp. Significant results obtained from the ERTS-1 satellite. NASA SP-327, 1: 745–753.

Harker, G.R., 1974. The delineation of floodplains using automatically processed multispectral data. Techn. Rep. RSC-60, Remote Sensing Center, Texas A.& M. Univ., 227 pp.

Howells, D.H., 1977. Urban flood management: problems and research needs. J. Water Res. Planning and Management Div., ASCE, 103, Nov. 1977: 192–212.

Hughes, D.A., 1980. Floodplain inundation: processes and relationships with channel discharge. Earth Surface Processes, 5(3): 297–304.

Kates, R.W., 1962. Hazard and choice perception in flood plain management. Univ. Chicago, Dept. of Geogr., Res. Paper, 78, 157 pp.

Kates, R.W.,1964. Flood hazard evaluation. In: G.F. White (Editor), Choice of adjustment to floods. Univ. Chicago, Dept. of Geogr., Res. Paper, 93: 135–147.

Kates, R.W., 1965. Industrial flood losses, Univ. Chicago, Dept. of Geogr., Res. Paper, 98, 76 pp.

Kesik, A.B., 1981. Flood-affected areas in Canada: inventory, assessment and cartographic presentations: Synopsis. Paper IGU Working Group River and Coastal plains Seminar, Wageningen, 99 pp.

Lambert, R., and Vigneam, J.P., 1981. Les inondations catastropiques de juillet 1977 en Gascogne. Problèmes de prévision et de prévention. Ann. Géogr., 90 : 497–499.

Lee, G.B., et al., 1972. Development of new techniques for delineation of flood plain hazard zones, Part 2 by means of detailed soil surveys. Univ. Wisconsin Water Res. Center, Madison (Wis.), 77 pp.

Mensching, H., 1970. Die Hochwasserkatastrophe in Tunesien im Herbst 1969. UNESCO Publ., Paris: 1–44.

Monnet, G., 1959. La prévention de l'inondation, problème de géomorphologie. Rev. Géom. dyn., X: 150–155.

Nakano, T., et al., 1961. Report on the survey of the abnormal tidal waves, Tsunami, caused by the Chilian earth quakes on May 24, 1960. Publ. Geogr. Survey Institute, Min. Construction: 97–100

Nakano, T., et al., 1962. Landform classification for flood prevention using aerial photographs. Proc. First Symp. Comm. VII, I.S.P., Delft: 447–452.

Newson, M., 1980. The geomorphological effectiveness of floods–a contribution stimulated by two recent events in Wales. Earth Surface Processes, 6(1): 1–16.

Oya, M., 1955. A geomorphological survey map of the Kiso Riverbasin showing flood stricken areas

Oya, M., 1961. Die Ueberflutungen bei Nagoya (Japan) im Gefolge des Taifuns "Vera" (Isewan-Taifun). Geogr. Rundschau, 13(2): 376–379.

Oya, M., 1963. Topographical survey map of the Yoshino River basin, Shikoku, Japan, showing classification of flood areas. Publ. Res. Bur., Sci. and Techn. Agency, Tokyo: 38/12

Oya, M., 1964. Geomorphological map of Khulna (E. Pakistan) and its vicinity, showing classification of flood stricken areas

Oya, M., 1965. A comparative study on geomorphology and flooding among the plains of the Cho Shui Chi, Chao Phraya, Irrawaddy and Ganges. Geogr. Rev. Japan, 37, 1964: 20 pp. Also: Proc. UNESCO Symp. Sci. Problems on Humid Trop. Zone Deltas, Dacca 1964. UNESCO Publ., Paris: 23–28.

Oya, M., 1966. Topographical survey map of the Kano River basin, Japan, 1:50,000, showing classification of flood flow areas. Publ. Res. Bur., Sci. and Techn. Agency, Tokyo, 41/4

Oya, M., 1967. Geographical study of the flood immediately downstream from Pamong in the Mekong River. Report Comm. Coord. Investigations Lower Mekong Basin, ECAFE, 40 pp.

Oya, M., 1970. Land use control and settlement plans in the

flooded area of the city of Nagoya and its vicinity. Japan. Geoforum, 4: 27–35.
Oya, M., 1971. Geomorphological flood analysis on the Naktong River basin, southern Korea. Waseda Univ., Tokyo, 88 pp.
Oya, M., 1971. Geomorphological land classification map of the Neyagawa River basin (Osaka and the surrounding area), 1:25,000, indicating areas subject to will be flooding. Publ. Nat. Res. Centre f. Disaster Prevention, Sci. and Techn. Agency, Tokyo, 23 pp.
Oya, M., 1973. Relationship between geomorphology of alluvial plain and inundation. Asian Profile, 1(3): 479–538.
Oya, M., 1976. Map showing river geomorphology of the Nobi Plain, Japan, 1:50,000
Oya, M., 1977. Comparative study of the fluvial plain based on the geomorphological land classification. Geogr. Rev. Japan, 50(1): 1–31.
Oya, M., and Akagiri, T., 1968. Relationship between the geomorphology and flooding in the basin of the Kazuryu in the central part of Japan. Geogr. Series Hiroshima Geogr. Ass., 9: 1–13.
Oya, M., and Nakamura, S., 1979. Geographical study on the inundation caused by the local rainfall in the Neyagawa River near Osaka. Misc. Rep. Res. Inst. Nat. Res., 72: 13–32.
Parker, D., 1981. The value of hazard zone mapping. Water Authority Section Surveys in England and Wales. Disasters, 5(2): 120–124.
Pasotti, P., et al., 1976. Aerofoto interpretación de un Sector del Delta Entrerriano. Publ. 60 Inst. Fisiogr. y Geol. Dr. Alfredo Castellanos, Univ. Rosario, Argentina, 38 pp.
Penning Rowsell, E.C., and Chatterton, J.B., 1977. The benefits of flood alleviation. A manual of assessment techniques. Saxon House/Teakfield Press, Farnborough (U.K.), 291 pp.
Pirazzoli, P., 1973. Inondations et niveaux marins à Vénise. (Flooding and sea levels in Venice). Mém. Lab. Géom. E.P.H.E., Dinard, 22, 284 pp.
Platt, R.H., 1976. The national flood insurance program: some midstream perspectives. J. Amer. Inst. Planners, 42(7): 303–313.
Rango, A., and Anderson, A.T., 1974. Flood hazard studies in the Mississippi Riverbasin using remote sensing. Water Res. Bull, 10(5): 1060–1081.
Rango, A., and Salomonson, V.V., 1974. Regional flood mapping from space. Water Res. Research, 10(3): 473–484.
Roy. Soc., N.Z., 1976. Proc. Tsunami Research Symp., Wellington (N.Z.) 1974. Roy. Soc. N.Z./UNESCO Publ., 258 pp.
Reckendorf, F.F., 1973. Technique for identifying flood plains in Oregon. Ph.D. thesis Oregon State Univ., 344 pp.
Santema, P., 1966. The effect of tides,coastal currents, waves and storm surges on the natural conditions prevailing in deltas. Proc. UNESCO Symp. Sci. Problems on Humid Trop. Zone Deltas, Dacca, 1969. UNESCO, Paris: 109–113.
Santema, P., 1966. Influence of flood protection works on physical and biological environment. Proc. UNESCO Symp. Sci. Problems on Humid Trop. Zone Deltas, Dacca, 1964. UNESCO, Paris: 333–340.
Schick, A.P., 1971. A desert flood: physical characteristics, effects of Man, geomorphic significances, human adaptation–a case study of the southern Arava watershed, Jerusalem Stud. Geogr., 2: 91–155.
Schroeder Lanz, H., 1962. Zur Ermittlung von Sturmflutdeichschäden mit Hilfe von Luftbildern. Proc. First Symp. Comm. VII, I.S.P., Delft: 438–446.
Scheaffer, J.F., 1960. Flood Proofing: An element in a flood damage reduction program. Univ. Chicago, Dept. of Geogr. Res. Paper, 65, 190 pp.
Sigafoos, R.S., 1961. Vegetation relation to flood frequency near Washington D.C., USGS Prof. Paper, 424-C: 248–250.
Sigafoos, R.S., 1964. Botanical evidence of floods and floodplain deposits. USGS Prof. Paper, 485-A, 36 pp.
Smith, K., and Tobin, C.A., 1982. Human adjustment to flood hazard.(2nd Ed.) Longman Group Ltd., London, 130 pp.
Sollers, S.C., et al., 1978. Selecting reconnaissance strategies for flood plain surveys. Water resources Bull., Amer. Water Res. Ass., 14(2): 359–373.
Stuckmann, G., 1974. Geomorphologische Aspekte der Hochwasserkatastrophe im Juli 1972 auf den Philippinen. Erdkunde, 28(1): 48–54.
Tada, F., and Oya, M., 1968. Topographical survey map of the Kuzuryu Riverbasin, Japan, 1:50,000, showing classification of flood flow area. Publ. Res. Bur., Sci. and Techn. Agency: 43/3.
Tada, F., and Oya, M., 1968. Geomorphological survey map of the Yoshino Riverbasin, Shikoku, in the western part of Japan, showing classification of flood-stricken areas. Przeglad Geograficzny, 40: 289–292.
Tada, F., et al., 1977. A geomorphological survey map of the Tsugaru Plain, 1:50,000, indicating areas subject to flooding
Tag Eldeen, 1980. Predisaster physical planning: integration into physical planning–A case study in Tunisia. Disasters, 4(2): 211–222.
Tricart, J., 1960. Etude géomorphologique du projet d'aménagement du lac Faguibine (Rép. du Mali). African Soils, 5(3): 207–289.
Tricart, J., 1980. Mécanismes normaux et phénomènes catastrophiques dans l'évolution des versants du basin du Guil (Htes-Alpes, France) (I). Publ. C.G.A., Strasbourg.
Various, 1980. Various papers on flood problems. Disasters, 4(4)
Verstappen, H.Th., 1975. The coastal swamps. In: Development Plan for West Pasaman, Sumatra. Inst. Dev. Res./ A.D.P. German Techn. Ass.: 44–48.
Verstappen, H.Th., 1977. Remote sensing in Geomorphology. Elsevier Sci. Publ. Comp., Amsterdam, 214 pp.
Verstappen, H.Th., 1977. Remote sensing in coastal zone research. Proc. 13th session ESCAP/CCOP, Kuala Lumpur, part 2: 137–153.
Verstappen, H.Th., 1975. Landforms and inundations of the lowlands of South Central Java. ITC Journal 1975, 511–520. (Also in: Publ. Serayu Valley Project, Final Report, Vol 2, 1978: 65–73).
Viereck, L.A., 1973. Ecological effects of river flooding and forest firest on permafrost in the taiga of Alaska. In: Permafrost, the North American contribution to the 2nd int. conf., 13–28 July 1973. Irkutsk, U.S.S.R., U.S. Acad. of Sci.: 60–67.
White, G.F., 1945. Human adjustments to floods. Univ. Chicago, Dept. of Geogr. Res. Paper, 29, 236 pp.
White, G.F. (Editor), 1961. Papers on Flood Problems. Univ. Chicago, Dept. of Geogr. Res. Paper, 70, 234 pp.
White, G.F., (Editor), 1964. Choice of Adjustment to Floods. Univ. Chicago, Dept. of Geogr. Res. Paper, 93, 150 pp.
Williamson, A.N., 1974. Mississippi river flood maps from ERTS-1 digital data. Water Res. Bull., 10(5): 1050–1059.
Wisner, R.B., 1979. Flood prevention with mitigation in the P.R. Mozambique. Disasters, 3(3): 293–306.
Wolman, M.G., 1974. Evaluating alternative techniques of flood plain mapping. Water Res. Research, 7(6): 1383–1392.
Yanggen, D.A., et al., 1966. Use of detailed soil surveys for zoning. J. Soil and Water Conserv., 21: 123–126.

DROUGHT SUSCEPTIBILITY AND DESERTIFICATION SURVEYS

14.1 The Concepts of Drought and Desertification

14.1.1 Introduction

Although drought is invariably related to a deficiency in rainfall, it is by no means purely a climatological matter. The essence of the concept is not dryness in itself, but the effect of rainfall deficiency on nature and society. Sanford (1977) defines drought as 'the shortage of some commodity such as water, crops and grazing, brought about by low or maldistributed rainfall'. He argues, with justification, that there are as many forms of drought as there are types of desirable commodities whereby the abundancy is mainly determined by rainfall. Among the major types, one can list agricultural drought, cattle drought and wildlife drought. Agriculture is most vulnerable to drought: for the rains to be effective they have to be properly distributed both in space (the fields) and in time (the growing season). Cattle, being mobile, can go for water when the need arises; in this respect wild life is the most flexible of all but the animals may die off massively after the last pan dries out.

Within the field of agriculture one has to diversify since each crop has its own specific water needs. For each type of land utilization a critical drought threshold exists. Certain exceptional conditions may be disastrous for one type of land use or crop, whereas another is hardly affected because the threshold below which critical water stress for that plant occurs has not been reached (Sastri and Ramakrishna, 1980; Walker (Ed.), 1979; Shears, 1980).

In the context of drought it is important to underline when and where water is available. Small quantities may be of crucial importance if available at the proper place and time, whereas much larger quantities may be useless or even harmful if not provided in the right season and locality. With respect to distribution of rainfall, timewise, it should be stressed that optimal use and proper storage of the water available during a brief wet spell in a long-lasting drought may substantially decrease the adverse effects of the water deficit.

A consequence of the definition of drought given above is that no drought can occur in uninhabited areas and where few animals occur. These are areas which are so arid that they are unsuitable for human uses and unattractive to most animals. Since drought is not a problem in areas where adequate rainfall can always be relied upon, it is obvious that the basic problem is in the variability of the rainfall, which at times creates situations where water availability does not meet the demands which have gradually evolved in periods of normal or above-normal rainfall. The higher these demands become as a result of increasing human or animal population or economic development, the more drought-prone an area will become.

Desertification has been defined at the U.N. Conference on Desertification in Nairobi, 1978, as the deterioration of land, water and other natural resources under ecological stress. In most cases it is a slow and almost unperceivable process, resulting in the decline of vegetation, the reactivation of dunes and wind erosion in agricultural lands. Most characteristically it is developed in the semi-arid belt bordering deserts but it is also a widespread phenomenon in other parts of the world as a result of inappropriate land utilization. The desiccation of arid and semi-arid lands, which is especially felt in periods of drought, is in fact the cumulative effect of two entirely different factors or groups of factors which are respectively of anthroprogenous and climatic nature (Plan of operation, 1977; UN, 1977).

14.1.2. Anthropogenous and Climatic Factors

The anthropogenous factor boils down to inappropriate use of the land and becomes particularly devastating when the population rapidly increases and inappropriate technology is introduced. This may lead to intensified agricultural land use in marginal areas and to overgrazing of natural grass lands. Slashing of scrub and,in particular, burning practices are other important aspects which contribute to a gradual deterioration of the environment.

The climatic factor is basically associated with secular fluctuations in the average situation and seasonal displacements of the Inter Tropical Convergence Zone (ITCZ); in tropical areas most of the precipitation is concentrated here (Beard, 1969). During years of below-average rainfall in areas where man has caused environmental degradation, the effects of periods of drought become increasingly serious as a result of the cumulative effect of these two groups of factors. Fig. 14.1 is a synthesized graph illustrating this situation.

Anthropogenous and climatic factors are to a certain degree interrelated and it is thought by many investigators that they reinforce each other. The graph of fig. 14.1 is therefore a somewhat optimistic approximation. The mechanism of this interrelation is believed to operate along the following lines: A deterioration/decrease of the vegetative cover causes a raise in albedo of the land, resulting in subsiding air currents introduced or reinforced in the margins of the desert where the rainfall consequently decreases and the vegetation is further reduced. Investigations using NOAA satellite data indicate that a reversal of the process may also occur where the vegetation re-appears after a drought which has caused a massive death toll or

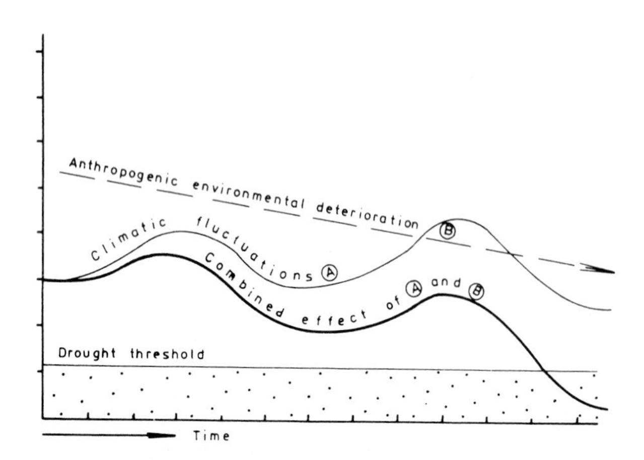

Fig. 14.1 Interaction of anthropogenic and climatological causative factors of drought.

a displacement of cattle. A natural increase in precipitation after the drought is then reinforced.

Leaving apart for the moment the extent to which the interrelation between anthropogenous and climatic factors increases the amplitude of the environmental changes brought about, it is evident that for all practical purposes of drought relief these two groups of factors are quite distinct and require different approaches.

The factor of climatic fluctuations is basically beyond our control; dry and wet periods alternate at irregular and usually short intervals. Dry years tend to occur in groups rather than at random. Human response to this situation is inevitably optimally bracing oneself against the adverse effects of a drought which is certain to come sooner or later. The measures to be considered include matters such as: an early-warning system, the localization of areas most likely to be seriously affected, the building-up of an infrastructure (roads, slaughter houses, food- and fodder stocks) and the preparation of a scenario for the emergency relief operations required.

In theory the anthropogenous factor can be influenced by concerted human effort and timely measures. However, a major operation covering many years is needed to stop environmental deterioration or to reverse the process. The problems involved are technologically complex and numerous. Decrease of humus content, loss of litter and the erosion by wind and water have affected the topsoil. Water losses result in increased surface run-off, infiltration and/or evaporation; storage capacity is reduced, river courses are filled with sand, etc., inundations during flash-floods and the covering of agricultural fields by wind-laid deposits are introduced.

More critical, however, are the socio-economic problems to be faced and the cultural attitudes which make it difficult to get new methods of tilling and water conservation accepted by the local population.

It is evident from the aforesaid that each of the following aspects merit full attention:

- immediate relief in case of drought and
- long-term programmes for combating desertification or improving the landscape ecological situation in already deteriorated areas.

They are, in fact hardly separable though requiring different approaches.

14.1.3 Use and Misuse of Groundwater Resources

It should be understood that the availability of large quantities of water is in itself not a panacea for problems of drought and desertification. Permanent availability by way of deep wells is not necessarily a proper solution to the problem. Herdsmen in the vicinity of abundant groundwater resources may be tempted to unduly increase the number of their cattle for which the grass resources may be insufficient. During wet years when the amount of rain water increases, the biomass will grow in accordance with the increased surface water and soil moisture situation. The subtle balance between water and grass resources, essential for well-planned cattle raising and herding is thus automatically maintained. However, this is not the case in areas where groundwater resources are used and unless the wells are properly managed and strict rules for the use of the water are enforced, overgrazing leading to rapid deterioration of the vegetal cover is almost inevitable. In areas of nomadic herdsmen the annual nomadic circuit may be replaced by permanent grazing around wells, leaving valuable grass resources unused while devastating the areas around the wells.

The dimensions of the devastated zones are such that they can be readily recorded from orbital altitudes. The Landsat image of fig. 14.2, picturing a part of northern Botswana, exemplifies this. A number of deep 'drought relief wells' have been constructed in Botswana following the disastrous drought of the nineteen-sixties which drastically reduced the number of cattle. It is understandable, though unfortunate, that these drought relief wells are not used solely in emergency situations, but throughout. As a result, a large number of cattle continuously graze around these wells and in a radius of about 10 km the grass resources are depleted to an alarming extent and only an occasional Acacia still survives.

Consequently, the albedo of the ground is substantially raised and the devastated areas appear as whitish discs, measuring approximately 1 cm in diameter on the 1:1,000,000 Landsat image.

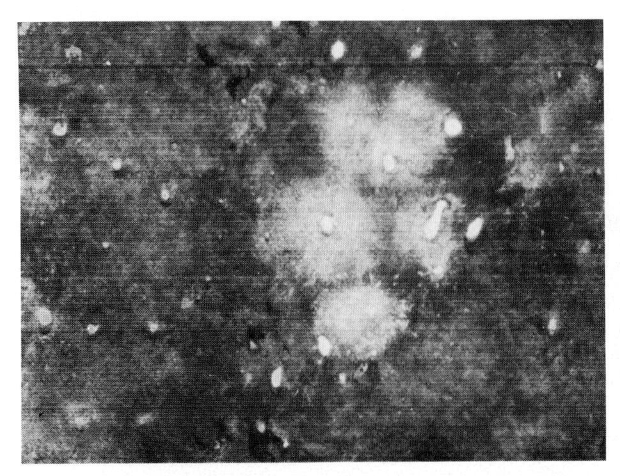

Fig. 14.2 Devastation of grass resources around drought relief wells in the Central Kalahari, Botswana. Landsat image, Band 5; 1:1,000,000. Bright white spots are pans. The whitish devastated areas measure 10-15 km across.

14.2 Human Adaptation to Drought Hazard

If a human group is to survive under such harsh, arid conditions, it should optimally adapt itself to the limited availability of water and select the sites of its settlements and its usually semi-nomadic way of life accordingly (Bharara, 1980). In this context of spatial distribution of water availability, environmental geomorphology is of great importance. Some tribes, such as the bushmen in the Kalahari Desert, have in the course of generations developed such a pragmatic knowledge about this subject that one may, with justification, incorporate it in their ethnoscience. They know where water can be found by shallow digging in an interdune depression. In many cases the quantities of water so obtained are so small that only a so-called 'zip well' is feasible.

The women then suck the water out of the sands using a straw and consequently store it in the shell of an ostrich egg. They also adapt themselves socially and divide themselves into smaller wandering groups when dryness comes to diminish their water requirements. In the last instance, it is not the wells that determine their itinerary and the location of their permanent or seasonal settlements, but areas where the green watermelon grows or can be cultivated. This variety of the watermelon accumulates considerable quantities of water and, contrary to the common, sweet watermelon, can be stored for several months (fig. 14.3).

The geomorphologist may follow-up this environmental approach to the search for water using sophisticated lines of research and discover that the depressions or pans utilized form a chain and mark an ancient drainage system that was in operation before the present aridity developed, which caused the invasion of the drainage system by wind-laid sands and dunes. The extended stereo pair (1:40,000)

Fig. 14.3 Green watermelon stored in village near Khutse, Botswana; a source of water in the dry season.

of fig. 14.4 gives an example of this from the Khutse Area, Central Kalahari, Botswana. A more efficient use of the shallow groundwater is thus possible.

Most oases in arid lands are situated at locations where groundwater is at shallow depth or where permanent water eyes exist. Such situations may occur where dry riverbeds or wadi break through a rock barrier, either a resistant geological formation or a dike. Although these riverbeds may occasionally be the scene of a flash flood, the flow is usually confined to the fill of sandy or gravelly sediments forming the valley bottom. Rock barriers buried by the valley fill, pond up the water in the coarse deposits, raising the groundwater at and immediately upstream of the barrier. Water eyes and shallow wells are suitable sites for growing date palms; some crops prevail and settlements are preferably located nearby.

An example is given in the photo mosaic of fig. 14.5 which pictures the Bardagué Valley, near the village of Bardai, situated at a height of about 1020 metres in the Tibesti Mountains, northern Chad, at a scale of 1:20,000. The Bardagué River is shown flowing from right to left in the lower portion of the photo. It is a dry wadi which has water only after an occasional rain shower. However, water percolates in the sands of the flood plain. It is at or near the surface at and immediately upstream of locations where the river traverses resistant rocks in a narrow gorge. The water in the floodplain is at shallow depth or even surfaces.

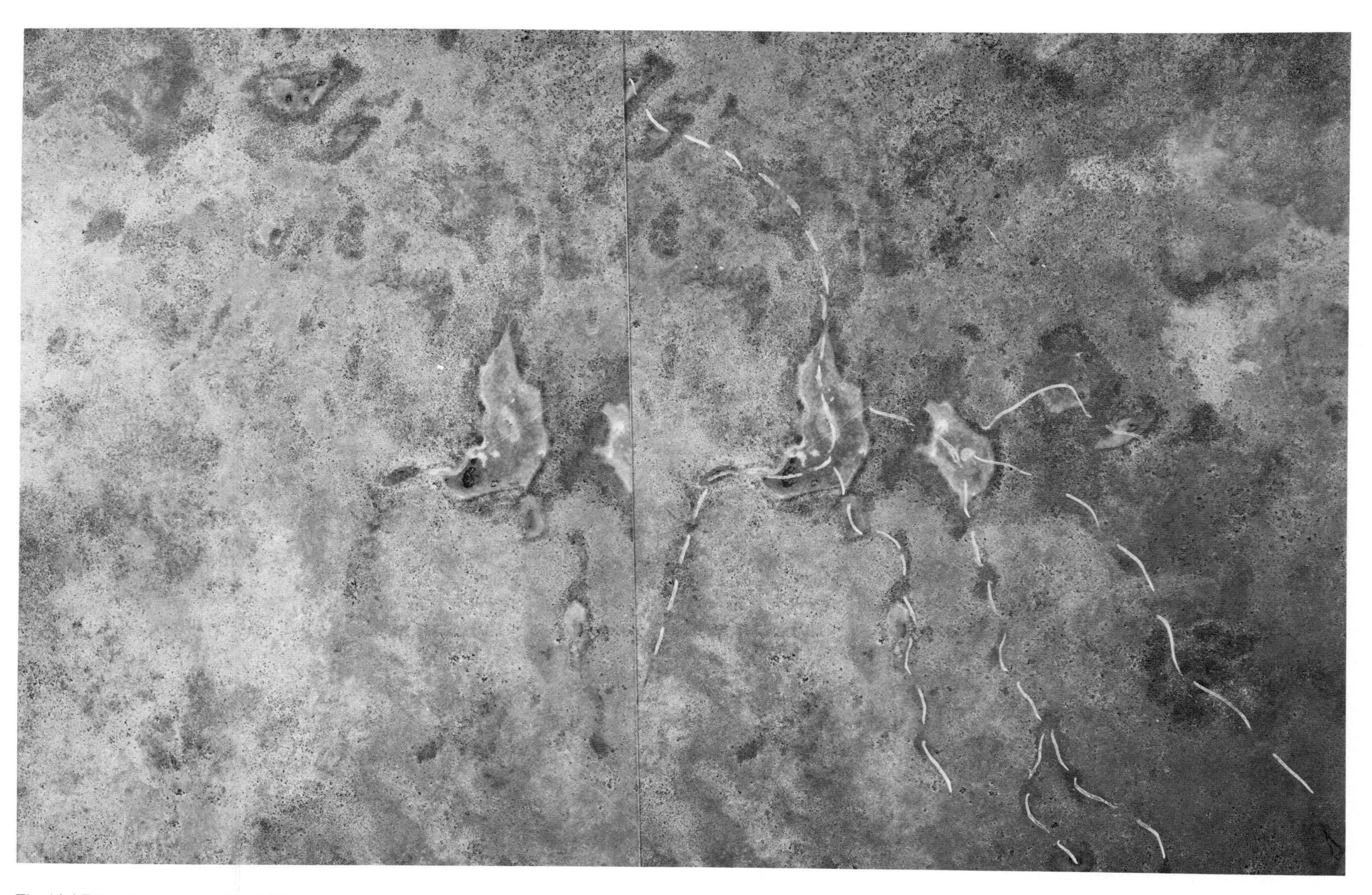

Fig. 14.4 Extended stereopair (1:40,000) of an ancient drainage system largely obliterated by invading wind-blown sands near Khutse, Central Kalahari, Botswana.

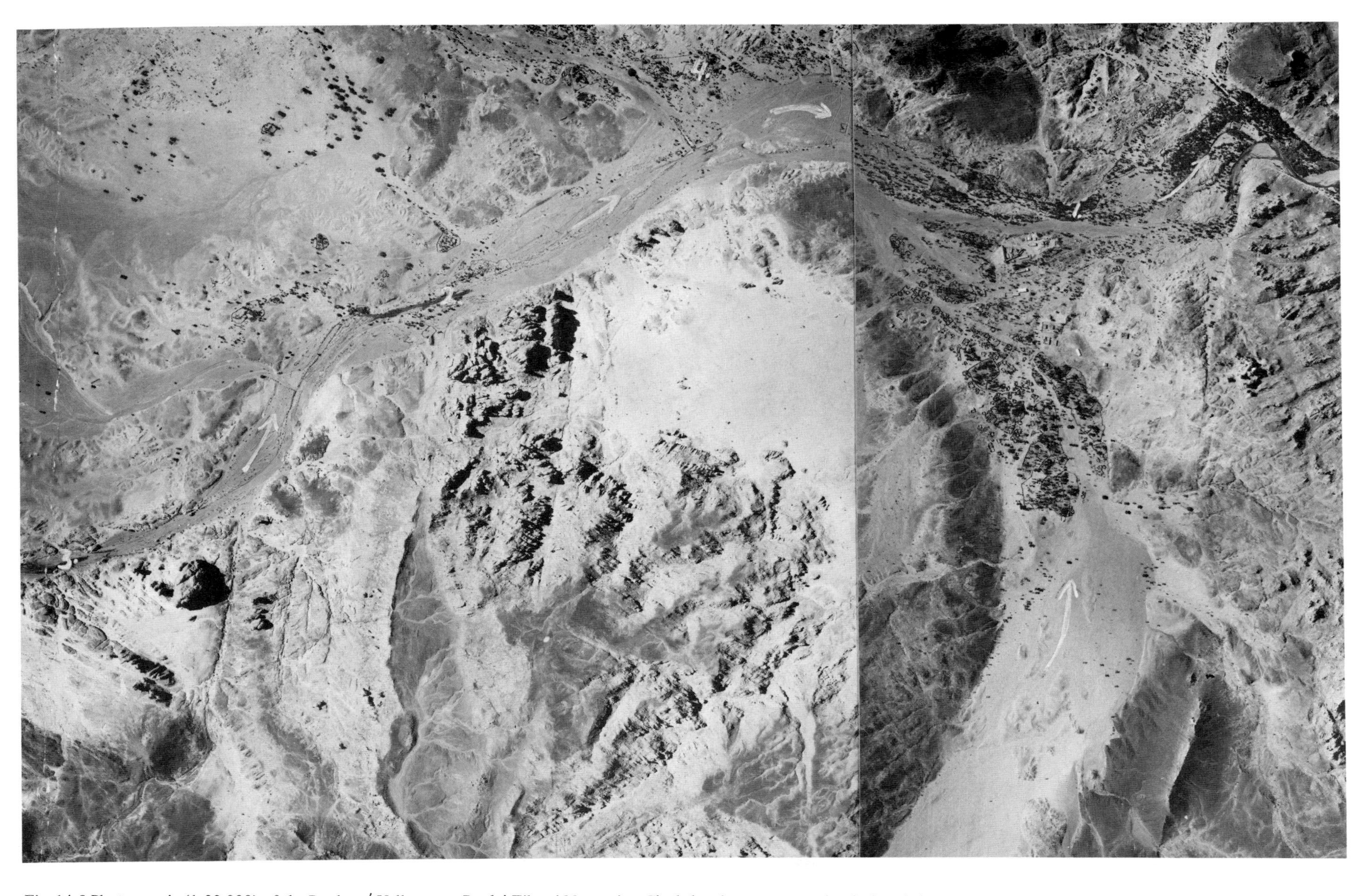

Fig. 14.5 Photomosaic (1:20,000) of the Bardagué Valley near Bardai, Tibesti Mountains, Chad, showing water eyes (guelpa) and short stretches of surface flow where rock barriers pond up the water percolating in the gravelly river bed (1, 2 and 3). The locations of Bardai, the houses nearby and the date plantations are dictated by the water availability.

Bardai, a district capital, is situated near the right-hand side of the mosaic at a river junction (1) upstream of a narrow gorge where the river breaks through a series of strongly jointed sandstones. The site of the settlements with the surrounding fields and the date palms in the gorge is explained by the effect of this barrier on the groundwater. Farther upstream (2) a similar situation exists but at a smaller scale. A permanent water eye, fields, and a few small villages depend on the water ponded up here. A third barrier (3) can be observed even farther upstream near the left edge of the mosaic. Some scattered houses (4) possibly depend on a barrier formed by the curved sandstone barrier occurring directly to the left (west) of Bardai although this needs further confirmation.

The area is extremely arid; the average annual precipitation amounts to 11,2 mm. Physical disintegration of rocks is a dominant process and the sandstones occupying the larger part of the image ultimately disintegrate into sand grains that build sandy wash plains at the foot of the near-vertical slopes maintained in the sandstones.

Disintegration is strongly governed by jointing and differential erosion between the various layers; the resulting landforms reflect this.

The curved, monoclinal sandstone ridge separating the village of Bardai from the sandy wash plain, and mentioned above as one of the barriers, is bound at the other side by a steep escarpment that marks the edge of the sandstone plateau and barrier from which the sands derive. Narrow ridges and fissures indicate the joint directions. Isolated sandstone stacks rising from the sandy wash plain are proof of the former larger extent of the sandstone plateau.

The darker areas in the northwestern parts of the wash plains are the slightly older, inactive parts; deflation has resulted here in a veneer of large rock fragments which serves as a protective cover against further wind work. These parts are now slightly higher than their surroundings and wash no longer affects them.

The sandstone plateau is rather dark in photographic tone as a result of the ferricrete covering it. No such coating occurs in the ravines, however, where occasional water work has destroyed it. The erosional glacis near the upper photo edge are also darker in tone as the rock fragments covering them have a marked desert varnish. The glacis are no longer active and are in the process of being dissected by ravines. The sandy wash plain is of a younger origin and is a 'live' form. A basalt dike can be seen in the lower left.

14.3 Surveying and Monitoring Techniques

14.3.1. General

Surveying and mapping of the present situation in (semi) arid lands and monitoring of the (temporal) dynamics of the landscape by recording the dry and wet season conditions and the secular changes occurring is evidently the first step towards the evaluation of the environment and an efficient planning of drought relief measures. International organisations such as FAO, UNESCO and WHO have prepared maps related to desertification on scales ranging from 1:5,000,000 to 1:25,000,000. Climatological data and vegetation zones are major items included in these maps, which give a concise picture of the global situation and are useful for delineating major problem areas. However, for operational purposes, more detailed maps are required. The type and amount of data needed should be carefully studied prior to the implementation of surveying and monitoring to avoid 'data cemeteries', the gathering of unnecessarily vast quantities of data which are not fully adequate and subsequently remain largely unused.

The factors to be included in these surveys are often diversified. It is important to avoid an unduly large number of sectorial maps and the data should, therefore, be skillfully combined to reflect the characteristics of the terrain configuration and land types.

It is clear that geomorphology plays an important role in this integrated evaluation of (semi) arid lands. Surveys will have to be carried out in close cooperation with governmental and other organisations engaged in combating the adverse effects of drought. It has been stated in the previous sub-section that the local population has learned the whims of the semi-arid environment by trial and error through the generations and has usually developed a very pragmatic mode of adaptation to the ecosystem. This precious experience should be used as a starting point for survey work and as a guideline for the evaluation of the land. Efficient use of the environmental data gathered can only be expected if optimum interaction exists between surveyors, government officials and local population and if the resulting data are coupled with information from socio-economic and cultural fields. The surveying and monitoring programmes should be geared to long-term environmental planning, on one hand, and at the same time to mitigation of the immediate effects of drought disaster. This dualism between short and long term benefits characteristic for surveying for drought and desertification has been discussed in the previous sub-section.

14.3.2. Multiphase Approach to Surveying using Satellite Imagery

It is superfluous to emphasize that for survey and monitoring operations to be successfully and speedily carried out, the use of remote sensing methods now available is indispensible (Fanale et al., 1978; Tricart, 1976).

The modes of remote acquisition of environmental data are diversified and a careful selection of the most appropriate technique for a specific purpose is essential for obtaining optimum results. When surveying and monitoring arid and semi-arid lands for the purpose of combating desertification, one has to consider the inherent characteristics of those areas and the related problems such as:

- the large extent of the areas to be mapped
- the spasmodic nature of short-lived phenomena such as floods and vegetation 'explosions' after rain
- the important seasonal changes in vegetation
- the limited scope for field observations due to the size of the area, transportation difficulties and the discontinuous aspect of natural processes
- the need for providing rapid and ready-for-use information, if possible by simple means
- the fact that the surveys should ultimately lead to directives about land use practices which are understandable and acceptable by the local population.

Since extensive and sparsely inhabited areas of low economic importance are involved, a multistage approach will be advisable which will allow for a gradual 'zooming in' from the global/continental context to the regional and,ultimately, the local situation. In this concept one can distinguish between:

1. explorative surveys at scales of approx. 1:1,000,000 to reveal major patterns of terrain configuration and the essentials of the landscape ecological dynamics. Landsat data, but also low-resolution satellite imagery such as that provided by NOAA and Meteosat are of use in this global/continental phase.
2. reconnaissance surveys at scales of approx. 1:250,000 to provide more detailed information on the same features of large critical areas. Blow-ups of Landsat images, aerial photographic mosaics, photo runs or photo blocks are useful tools in this regional phase.
3. (semi) detailed surveys at scales of approximately 1:25,000 - 1:50,000 of areas of particular interest by way of stereoscopic observation of pairs of vertical aerial photographs for obtaining appropriate information in the local phase.

The fieldwork associated with the three phases mentioned above is either by traverses, particularly in phase 1,or by site observation, particularly in phase 3. Observations and oblique 'pin-point' overboard photographs from small, low-flying aircraft also have proved to be valuable as a link between orbital and ground observations.

Data obtained from orbiting satellites have a vital place in surveying and monitoring for drought and desertification, particularly on the global/continental and regional levels. The four main reasons for this are the synoptic view they provide of large areas, the multispectral recording, the repeated coverage allowing for multi-temporal analysis and the digitized form in which the data are stored. Both the information on the 'static' configuration of the land and the information on the changing, dynamic features are adequately presented for mapping on exploratory and reconnaissance scales. Other tools, such as aerial photographs and ground truth gathering methods,should be efficiently incorporated in the work.

Low resolution, meteorological satellite data can be successfully used on the global/continental level. The thermal infrared images, for example produced by satellites such as ATS, Nimbus, NOAA and ESA satellites, give a clear picture of the cloud patterns, monitor the general meteorological situation and provide dynamic information on the position of the ITCZ and the related rains. A general picture of surface water resources including their seasonal variability can be obtained as demonstrated by fig. 14.6. Geomorphologically more rewarding are the computer print-outs of the onboard HRIR (infrared radiometers) which have a spatial resolution of a few kilometres. The data recorded are 'black body' temperatures in degrees Kelvin which are approximately comparable to actual temperatures when water bodies are concerned. The patterns produced over land areas are associated with the heat capacity of the surface features concerned, which is in most cases strongly affected by soil moisture conditions (Pouquet, 1969; Verstappen, 1977). Night recordings are preferably used; and areas recorded then as relatively warm may be so because of higher heat capacity due to soil moisture. Cool areas may be very dry or highly reflective (saline flats or dunes). Interesting indications for land evaluation and drought susceptibility classification may be obtained on the global/continental level.

High resolution multi-spectral Landsat in the 0.5-1.1μ range and (Landsat-3) in the thermal (11.5μ) infrared are of great importance at the regional level. Visual interpretation of these multi-spectral data can be done for each band separately (fig. 14.7), but preferably in combination using diazo transparancies, simulated false colour composites or a colour additive viewer such as the I^2S Model 6040. Digital interpretation has a much greater resolution and flexibility for proportionally combining the various bands, e.g., by way of principal component analysis (Verstappen, 1977).

14.3.3. Monitoring by Orbiting Satellites

Multi-temporal interpretation of sequential Landsat data,which is of the utmost importance for monitoring dynamic phenomena, can be done visually by super-

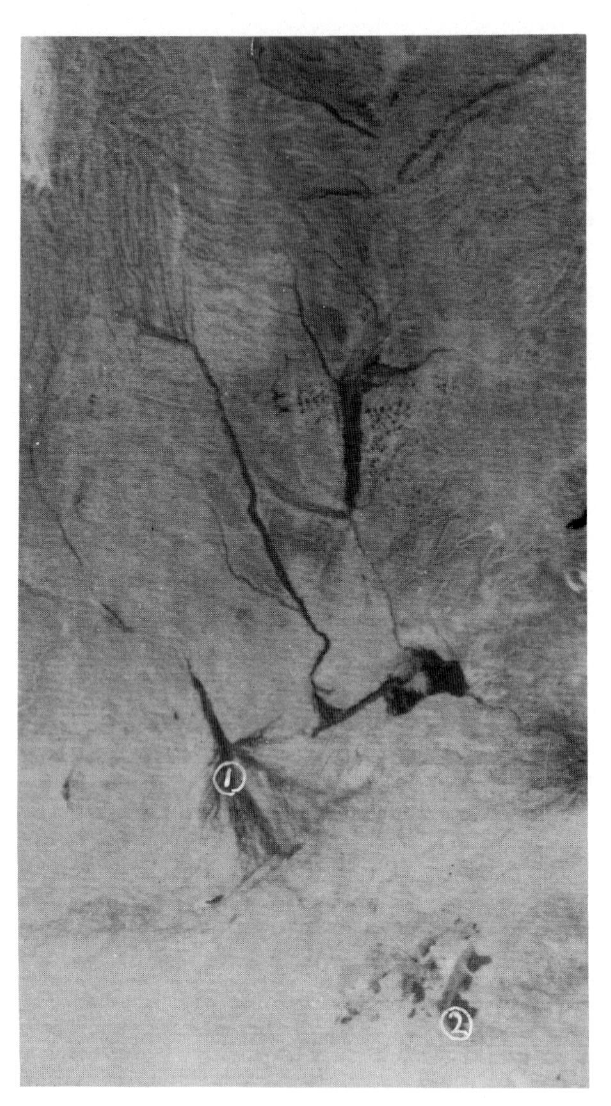

Fig. 14.6 Northern Botswana and adjacent areas as recorded by the NOAA satellite at a scale of 1:10,000,000. The Okavanga inland delta (1) and the Makgadikgadi pans (2) are clearly visible on this low-resolution, wet-season image and so is the SW-NE stretching fault structure, an incipient part of the east African rift zone, at the lower end of the delta. The Boteti River (fig. 14.7) can hardly be seen.

posing sequential images of the same band (e.g. 5 or 7) of a frame by diazo or I^2S techniques. If different coloured diazos are used or the positive of one pass is combined with the negative of another, changes in surface reflection can be recorded, which in band 5 relate to changes in vegetation (biomass) and in band 7 to changes in soil humidity, etc.

These studies can be carried out digitally provided that a geometric correction is applied to make sure that the pixels of identical terrain features are compared. Calibration for differences in atmospheric conditions and in sun angle is also required in order to make the radiance patterns of multi-temporal images fully reliable. This can be crudely done by assuming that,e.g., the radiance of reflection of rather stable features such as water surfaces and urban areas was identical during both passes and rotating the axes of the principal components to correct variable elements such as vegetation and land use. More reliable is the use of certain equations and computer programmes developed for the purpose (Otterman, 1976, 1977; Robinove and Chavez, 1978; Robinove et al., 1981). The estimation of biomass changes by way of visual interpretation of temporal Landsat Band 5 imagery can be substantially improved and even quantified by digital approaches. When the number of pixels of all radiance classes (Band 5) is monitored, the rise in radiance with the advent of the rains is a measure of chlorophyll, thus biomass (DHV, 1978; Pearson et al., 1976; Tucker et al. 1973; Tucker, 1977). The strongest correlation between reflectance,as measured with a spectro-radiometer, and green biomass has been found at 0,68 and 0,80 μ In fact, the measurement of albedo, either by band or in a broad spectral range using digital satellite data, and comparing it with photo meter measurements in the field and actual clip-plot figures is a promising field of study that provides an (indirect) measure of the process of desertification. If linked to vegetation/biomass, these data can be turned into a direct means of quantifying this process. At the present state of the art only relative values of albedo can be obtained,but with improvements to be realised particularly in the forthcoming Earth Radiation Budget Satellite System (ERBSS) absolute albedo values are within reach (Berry and Ford, 1977).

In summarizing, one may say that the monitoring of cloud cover, vegetation changes and albedo, but also matters such as dust storms (Rapp et al., 1976), is a capability of satellite observation unequalled by any other means of observation. For geomorphological purposes three aspects of surveying for drought and desertification are of prime importance, viz. drought susceptibility surveying, monitoring desertification in rural areas and morphodynamic mapping for fixation of dunes. In each of these fields aerospace technology plays an essential role as will be explained in the following three sub-sections.

14.4 Drought Susceptibility Surveys, the Concept

Drought susceptibility surveying is a new concept which is rooted in the consideration that geomorphological terrain factors and the related geological, pedological and vegetation characteristics have a much greater influence on the medium and large scale distribution of drought hazard than climatological factors alone. The importance of variability of rainfall has already been pointed out in sub-section 14.1 and,in connection herewith, one may define drought susceptibility as the spatial

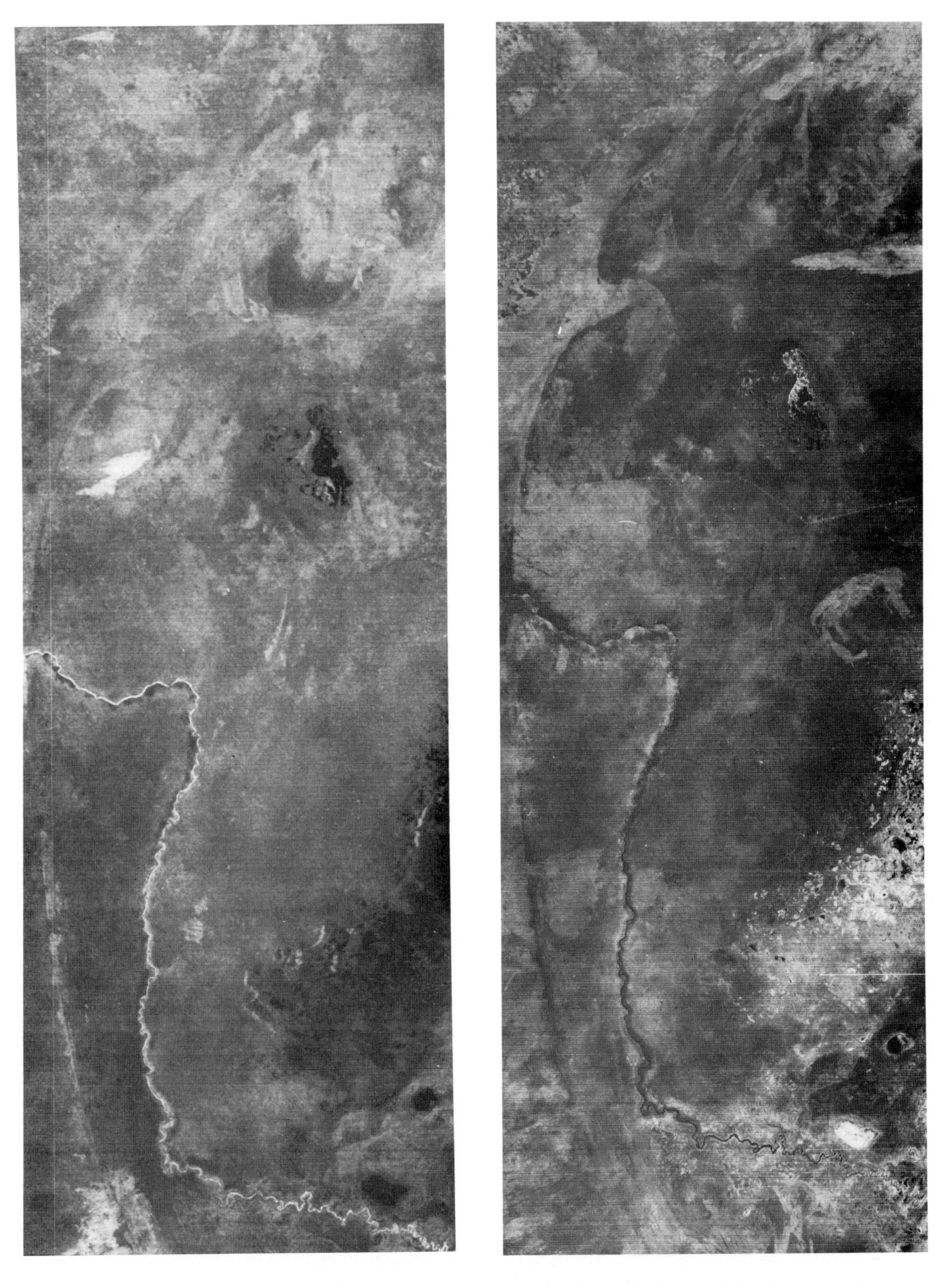

Fig. 14.7 High resolution Landsat images (Band 7; 1:1,000,000) of the Boteti River and part of the Makgadikgadi pans in northern Botswana from 14 September 1972 (left) and 12 March 1975 (right) respectively. On the left image the pans are wet (dark) but the riverbed is already dry, whereas on the right image this situation is reversed. Note the distinct barrier bar connected with the 945 metres phase of Lake Palaeomakgadikgadi and some, less distinct, lower and more recent shore features. Compare with figs. 14.6, 14.8 and 14.10.

and temporal variability of water resources. Thus drought susceptibility surveying amounts to the mapping of terrain units and their evaluation with respect to the quantity, quality and seasonal/secular variability of surface water, soil moisture and groundwater resources. The ultimate aim is a break-down of the land into drought susceptibility classes from which the areas of potential affliction can be defined and a plan of action can be formulated to be implemented whenever a drought occurs. An advantage of the approach is that the survey results can be used time and time again. The method is complementary to the monitoring aspects mentioned in sub-section 14.5.3.

Drought susceptibility may be regarded purely as a function of variations in a set of fundamental environmental factors - geomorphic site properties, climate, soil character and vegetation. A Drought Susceptibility Index based on such criteria could be regarded as a goal of research in this field. The human response could then be geared or planned in the light of the opportunity, or lack of it, so revealed. Such an index would be essentially predictive in nature, rather than purely descriptive.

The surface water situation and its variations can be directly mapped but soil moisture and groundwater conditions require indirect means of evaluation of a multidisciplinary nature. Geomorphological terrain classification has a key position in the surveys for reasons already mentioned. Among the factors to be considered, the following deserve special mention:

- General terrain configuration, because of its effect on the flow patterns of (ephemeral) surface water and the location of seasonal lakes and pans.
- Height above the (e.g., perched) watertable, due to its effect on water availability for domestic and other uses.
- Traces of old drainage lines dating from more humid periods of the past, because of their effect on the distributional patterns of shallow groundwater and soil moisture.
- Soil texture and related phenomena such as sealing, because of the effect on infiltration and groundwater recharge, soil moisture conditions and the evaporation, e.g., in pans.
- The distributional pattern of drought resistant vegetation, as it offers a concise way of evaluating the drought hazard in the various parts of the terrain.

In the meantime, the general concept of variability in water resources as a criterion in drought susceptibility may be further elaborated on. Under natural conditions where surface water and soil moisture are used, the grass resources (biomass) fluctuate in phase and in proportion with the water resources. Variability of the latter as a criterion is then adequate. If, however, sizeable quantities of water are made available throughout the year, e.g., by way of deep wells, and the water resource variability becomes theoretically zero, the grass resources continue to fluctuate with the season and become the only major limiting factor in the development of range lands. The discrepancy between water resources and grass resources so created often results in overgrazing near the wells where the cattle concentrate.

It is evident that solving the water problem only is not enough, as has been demonstrated by the Landsat image of fig. 14.2 showing devastated grazing land in Central Botswana, around the so-called drought relief wells. Under such conditions, both water and the resulting grass resource variabilities should be considered, using temporal interpretation of Landsat bands 5 and 7.

The data gathering required can be realised on the basis of satellite imagery and aerial photographs in combination with an efficiently devised limited field check. Landsat imagery is particularly important for the purpose, especially if images taken during the dry and wet seasons are compared. The distributional patterns of surface water and soil humidity and their seasonal fluctuations can be mapped on a reconnaissance level. Data on surface temperatures can be obtained from meteorological satellites such as NOAA-2 and the ESA meteo-satellite. Terrain reflectivity can also be measured using satellite data, aircraft and ground observations. It is important to realize that in this connection the albedo of the ground varies with humidity and vegetation cover.

When using Landsat band 7, changes in the extent of water bodies and soil humidity are optimally depicted, whereas the same procedure with band 5 gives a clear picture of the seasonal vegetation changes. Since the hazard of the drought rises with increasing variability of water (humidity) and vegetation patterns, useful information can thus be obtained.

Aerial photographs provide more detailed information on the spatial differentiation of the areas studied but lack the advantage of monitoring seasonal and secular variations. Sequential aerial photography taken at an interval of several decades is very useful in quantative studies on desertification (sub-section 14.8), though is of little use for drought susceptibility surveying. Ground truth gathering centering upon geomorphology, vegetation, sedimentology, geology and hydrology is an essentail part of the work. Since large areas are normally surveyed, a 'zooming in' from general to specific using a multi-phase approach, as outlined in sub-section 14.5.2, is essential.

The variability of water resources as a major factor in the concept of drought hazard implies that the areas with maximum variability fall into the high-risk category. Parts that are almost continuously dry and parts that are almost continuously moist both come into the low-risk category.

The important difference between the two is, however, that if - even a minor - fluctuation occurs in a usually moist zone, numerous people and large number of cattle will be affected, whereas if the same occurs in an almost consistently dry area this will be of concern only for the few living there.

The 'pressure on the water' by the human group is thus a second major criterion in drought susceptibility surveying together with the variability already mentioned. In rural (semi-) arid areas, data on the number of villages, inhabitants and heads of cattle provide reliable data of relatively easy access for the assessment of the magnitude of the water requirements and thus of the magnitude of the effect of a potential drought. One could, with justification, use the term drought susceptibility only for the system which uses purely natural environmental criteria, and the term drought hazard when the human factors are taken in consideration. Then the distinction already mentioned (agricultural drought, cattle drought, wildlife drought), and even further subdivision, e.g., by crop in the case of agricultural drought, may be required.

The effect of pressure on the water is clearly demonstrated by the following example. Shallow wells in semi-arid areas dry up in years of average rainfall during the wet season when the large number of cattle feeding on the extensive grasslands overdraw the wells; but when most of the animals are in cattle stations elsewhere during the dry season, the wells yield sufficient water for the few inhabitants of the villages and their cattle.

In fact, mobility of the user is considered a third major criterion in drought susceptibility, of equal importance as variability in water resources and the pressure on the water requirements. Basically, the drought susceptibility or hazard decreases with increasing mobility. This needs further elaboration. In case of agricultural land use, mobility is minimal: the useful rain should occur exactly in the right place (the field) at the right time (the growing season). Cattle can go to the water over a certain distance, provided that adequate grass resources are available. Still greater is the mobility of the nomadic herdsmen who usually have a routine annual itinerary. This, however, does not apply to other herding systems. Maximum mobility is found in wildlife that moves seasonally over many tens of kilometres (or more even). They cling to the shrinking surface water resources (pans and rivers) with the advance of the dry season and may die off massively in the event of drought. It is evident that for each of these three groups of water users, a separate drought susceptibility classification is required.

14.5 A Drought Susceptibility Survey in Northern Botswana

This survey was executed to serve as a pilot survey for drought susceptibility mapping by a multidisciplinary team of the Department of Geography, University College of Botswana, and the International Institute for Aerial Survey and Earth Sciences, Enschede, the Netherlands, in 1978 in the framework of a project sponsored by the Netherlands Universities Foundation for International Cooperation (NUFFIC). Northern Botswana was selected for the purpose because of the marked seasonal variations in surface water features, humidity and vegetation.

The area shows a variety of landforms and is of considerable geomorphological interest notwithstanding its minor relief: it is largely developed between the 910 and the 940 m contour lines. Young Quaternary deposits of aeolian, fluvial and lacustrine origin are dominant in most of the area. According to the 1:1,000,000 geological map, they form part of the (Tertiary to recent) formation of the Kalahari Sands. Older rocks, shales, mudstones, marls and basalts overlying sandstones of the Stormberg formation, and sandstones, shales and limestones of the Ghanzi formation also occur in the southeast although they are covered by Kalahari beds. Locally dolerite dikes form ridges and indicate major directions of faulting. Calcrete and silcrete formation is of common occurrence throughout the area. Climatic fluctuations of the and neotectonics have played an important role in the development of the landforms (fig. 14.8).

Forms of lacustrine origin comprise of ancient lake shores or barriers of the young Pleistocene Lake Palaeo-Makgadikgadi. Three phases of this ancient lake have been mapped, with lake levels of 945, 920 and 912 metres a.s.l., respectively. The highest lake level has left the most marked traces in the terrain: a sandy barrier ridge is accompanied to the western, landward side by a zone of wind blown sands in which a striation can be observed from the air, supposedly stretching in the direction of the local winds prevailing at the time of deposition. A comparable striated zone at places also accompanies the 920 m ridge although it is interrupted and much less distinct. It could not be traced along the shores of the 912 m lake level.

To the east the ridges are accompanied by a fairly monotonous and topographically somewhat lower zone that only emerged from the lake following a drop in its level. At several localities, fossil (at present dry) valleys can be observed which once originated where shallow groundwater emerged from the sands of the ridges. In the southeast, the former lake bottom shows distinct structural control and older geological formations here

Fig. 14.8 Geomorphological outline of the Gidikwe-Makgadikgadi area in northern Botswana based on Landsat images, aerial photographs and field observations.

Key: 1. barrier bar (Gidikwe ridge) of 945 m lake level; 2. same, 920 m lake level; 3. same, 912 m lake level; 4. ancient longitudinal dunes west of 1 and 2; 5. ancient lake bottom east of 1 and 2; 6. ancient lake bottom, structurally controlled; 7. ancient fluvial forms (D1, D2, D3: deltas); 8.recent/sub-recent fluvial forms (D4); 9. recent/sub-recent pans and dunes; 10. same, lowest parts.

occur at relatively shallow depth. Elsewhere the ancient lake bottoms are, at some distance from the ridges, often modified by more recent geomorphological processes and events. These parts are mainly areas where fluvial (Boteti) deposition has occurred after a drop of the lake level. Three distinct ancient deltas fall in this category. Recent and sub-recent fluvial deposits are mainly in a zone bordering the Boteti River.

The water/soil moisture and vegetation patterns are readily recorded by Landsat and the seasonal change detection has been optimised by superposing diazo-prints of dry and wet season images, using complementary colours. This procedure was applied to bands 5 and 7 to detect seasonal changes in vegetation and soil moisture/ water bodies respectively. The principle is explained by the graph of fig. 14.9. A no-change line runs diagonally across the graph. It is estimated that the results obtained at the exploratory and reconnaissance level in northern Botswana can be extrapolated to eastern Botswana at a later stage through the use of the land system maps of these areas prepared previously by D.O.S. Ideally, a drought susceptibility map of the whole of Botswana should be available as an operational tool in combating the adverse effects of drought (Breyer, 1979; Verstappen, 1979; Verstappen and Cooke, 1981).

Although the survey was centered on geomorphology, other environmental parameters related to soils, hydrology and vegetation were also taken into consideration. The ancient dunes, the old (Palaeo-Makgadikgadi) lacustrine features and other phenomena contributing to an understanding of humid and dry periods during the Quaternary were specifically studied.

All existing data were utilized and, excepting the aspects already stressed in the previous sub-section, special attention was paid to the climatic fluctuations that occurred in recent times using the meteorological data as a reference. As a matter of fact, drought is a recurrent phenomenon in Botswana and droughts have occurred in the mid-1870 s, the mid-1890 s, 1912-1916, the early 1920 s and 1930 s, 1947 and the early 1960 s and 1970 s.

The drought of the early 1960's was particularly severe: about 25% of the population received Government support against famine (Sanford, 1977) and 1.5 to 2 million head of cattle died off.

Following the exploratory mapping of the whole of northern Botswana, a more detailed, reconnaissance level survey was carried out using Landsat blow-ups at scale 1:250,000 as a base for plotting data obtained from the interpretation of aerial photographic mosaics and the scanning of stereopairs. Digital interpretation of Landsat data was also an important aspect in this phase. Fig. 14.10 gives an example of an area near Rakops.

Semi-detailed work based on thorough interpretation of indicative stereopairs was a third step in the multiphase approach to the survey. The fieldwork was carried out by traverses for the exploratory and reconnaissance phases and by site investigations in the semi-detailed phase of the work. Lastly, considerable attention was paid to the human adaptation to drought conditions by investigations in the villages during the fieldwork. An efficient combination of sophisticated aerospace technology and grassroot-level field research was thus achieved.

The data gathered on landform, lithology, soil types, vegetation, etc. were subsequently translated into hydrologically relevant terms, particularly of sub-surface water storage which gives a first assessment of the available water resources. This was substantiated by the local description of water resources. The water resources of the various geomorphological terrain units distinguished could therefore be evaluated on the basis of information gathered on the sites of more detailed investigation. The exploratory geomorphological terrain classification for water resource assessment (1:1,000,000) and the reconnaissance level survey (1:250,000) ultimately led to the maps of fig. 14.11 which give a general picture of the agricultural and the cattle drought hazards.

Three matters of interest are: first, some extensive, comparatively high, sandy areas, such as those near the left edge of the map to the west of Gidikwe Ridge, with only scanty sub-surface and no surface water resources have nevertheless been grouped in a rather low drought susceptibility class as they are almost continuously dry without substantial seasonal variability. The use of those areas is very extensive and the water requirements are thus negligible. Second, it can be observed that salination is an important factor particularly in the pan-areas. Digital Landsat data and particularly colour-coded principal component images have proved of particular value also for mapping water resources and salinization (Breyer, 1979). Density slicing has also been successfully applied for these purposes (fig. 14.12). Third, the valley bottom of the Boteti River has been assessed with a negative drought susceptibility value because the flood-retreat cultivation practised there is adversely affected by above-normal rainfall; the valley bottom runs dry too late or not at all under those conditions, whereas this type of cultivation is usually optimum in comparatively dry years. In fact, adjacent to the Boteti River an agricultural system has evolved which is remarkably well adapted to climatic fluctuations inherent to the area. In comparatively dry years a crop is obtained from the flood-retreat ('molapo') fields, whereas in years of abundant rains successful farming is feasible on the higher adjacent areas, mainly

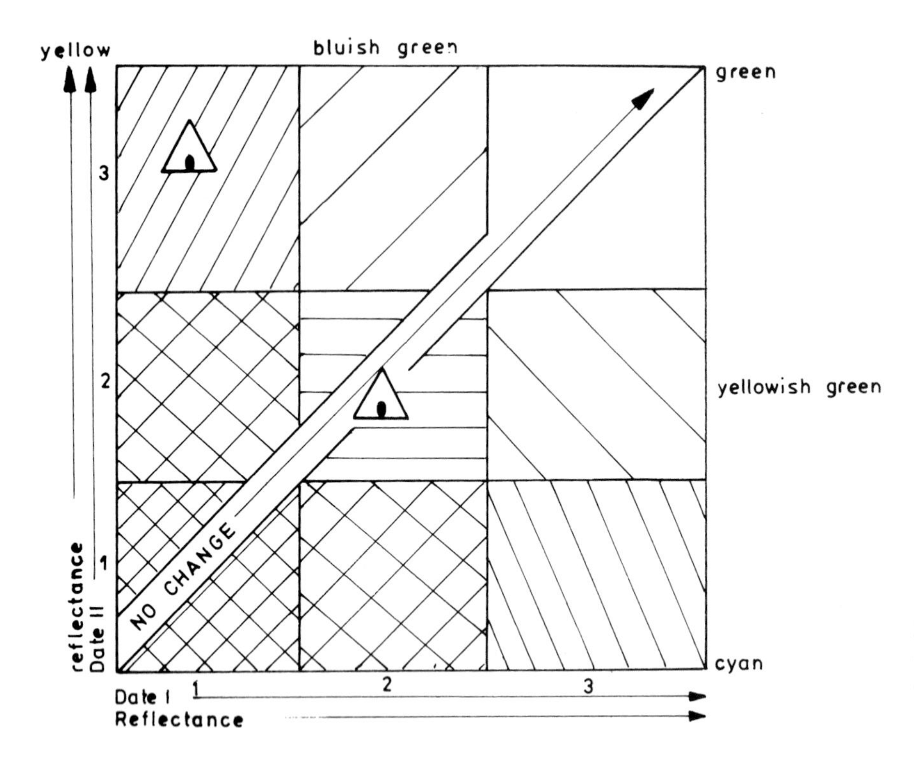

Fig. 14.9 Feature space, Landsat Band 7, of two different dates. The no-change line, green when diazos of complementary colours are superposed, runs diagonally across the picture.

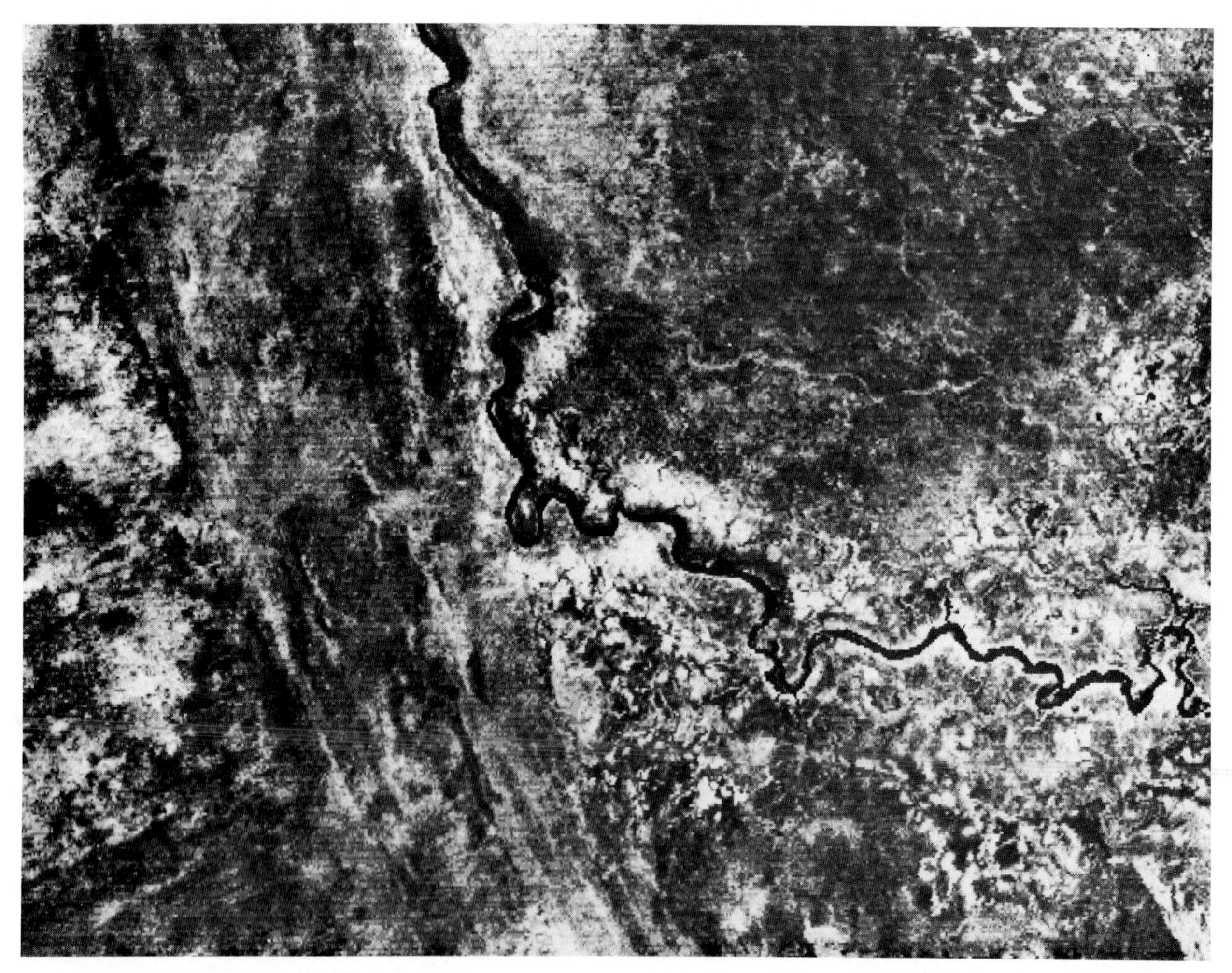

Fig. 14.10 Digital Landsat image (Band 7: 1:250,000) of the Boteti River and surroundings near Rakops, northern Botswana. Note the ancient lacustrine barrier bars stretching NNW-SSE to the west of the river and the longitudinal relief-less dune features in the SW corner of the picture.

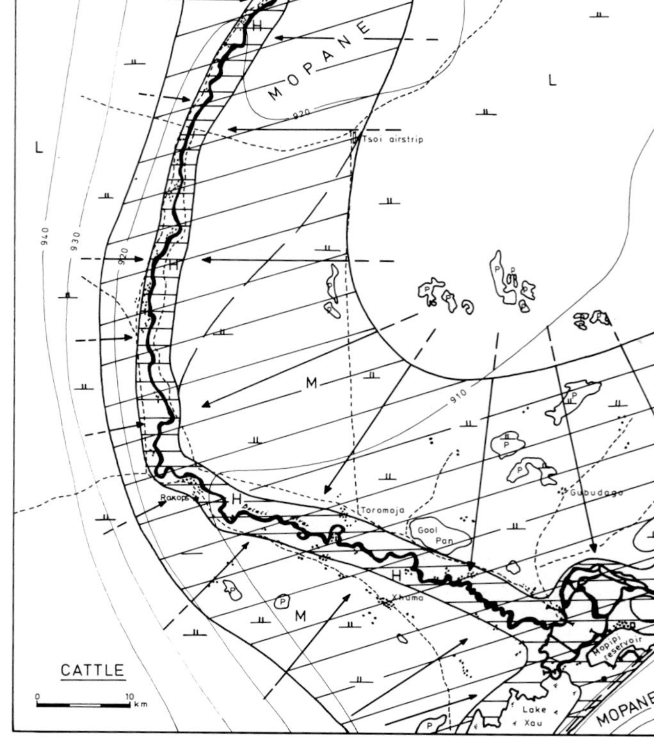

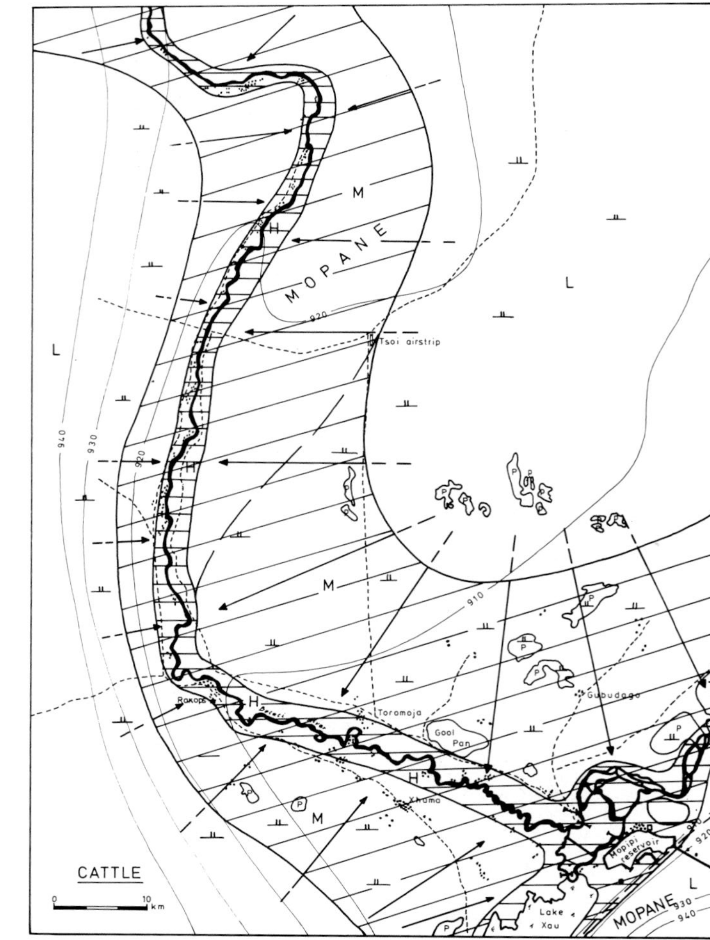

Fig. 14.11 Agricultural (left) and cattle (right) drought situation along part of the Boteti River in northern Botswana.

Fig. 14.12 Density slicing giving distribution of salt efflorence near Lake Xau, in northern Botswana (Breyer, 1977).

with silt loams or sandy loams, at least where no calcrete occurrences interfere with agricultural uses. The extended stereopair of fig. 14.13 shows the molapo fields alongside the Boteti River where planting is done immediately following the emergence of the field with the advent of the seasonal decrease in discharge. The oblique airphoto of fig. 14.14, taken from low flying aircraft, shows the fields on the adjacent higher terrain which were in use in the comparatively wet year of 1978. Flood retreat cultivation in all likelihood plays a similar stabilizing role also in other semi-arid parts of Africa affected by secular climatic fluctuations and drought.

14.6 Surveying Desertification in Rural Areas

14.6.1 Basic Survey Methods

Windwork adversely affects agricultural fields and grazing land in many parts of the world, particularly in arid and semi-arid lands (Ibrahim, 1978; Mensching and Ibrahim, 1976, 1978; Mabbut and Berkowicz, 1981; Meckelein, 1980; Orlova, 1979) and also in other climatic zones. It is often referred to as wind erosion although this is a rather abstract terminology, first because the phenomena described under this heading are especially of accelerated, anthropogenic nature. Second, not only the erosive effects of windwork, resulting in a removal of the topsoil and in the formation of blow holes and other depressions should be included but also the airborne transportation of fine soil particles,in the form of a dustbowl (Borcher, 1971) and sand, by saltation and creep, and the depositional features such as dunes and sand, dust or loess covers.

The gradual deterioration of the land by these aeolian processes as initiated or accentuated by its inappropriate use, is often referred to as desertification. The distributional patterns of the various phenomena involved are basically influenced by the following three groups of factors:

1. Climatic factors, particularly related to rainfall, air humidity and wind but also encompassing factors such as sunshine and temperature, mainly affect the broad distributional patterns and are important especially in small-scale mapping.
2. Terrain factors affecting the susceptibility to drought, elaborated upon already in sub-section 14.4, are a major causitive factor of the more specific distributional patterns and become increasingly important when surveying on semi-detailed and detailed scales is concerned. Apart from broad terrain classes, factors of apparently minor importance such as terrain roughness and soil moisture content may be of great effect (Biesel and Hsieb, 1966).
3. Intensity and type of land utilisation ultimately determine which parts of the terrain units and classes mapped will be most severely subjected to desertification.

The surveying of desertification should logically start with mapping the present state of deteoriation reached. However, for an understanding of the desertification patterns, the causative anthropogenic factors and the prevailing conditions of natural environment and their dynamics should also be incorporated. Aerial photographs are of great use in these studies as is evident from the example of a westward expanding sand sheet in an area of ancient dunes to the east of Lake Chad, shown in fig. 14.19. This example also illustrates that it is not always feasible to clearly separate between the mapping of the actual situation and the assessment of the future distributional patterns of desertification, in other words the desertification hazard. Sequential aerial photography and comparison of Landsat imagery with an earlier aerial photographic coverage of the area of study facilitates the assessment of this hazard by extrapolation in the future of the trends observed in the last few decades.

An example of desertification survey is given in fig. 14.15 from part of northern Darfur, Sudan (Ibrahim, 1980).

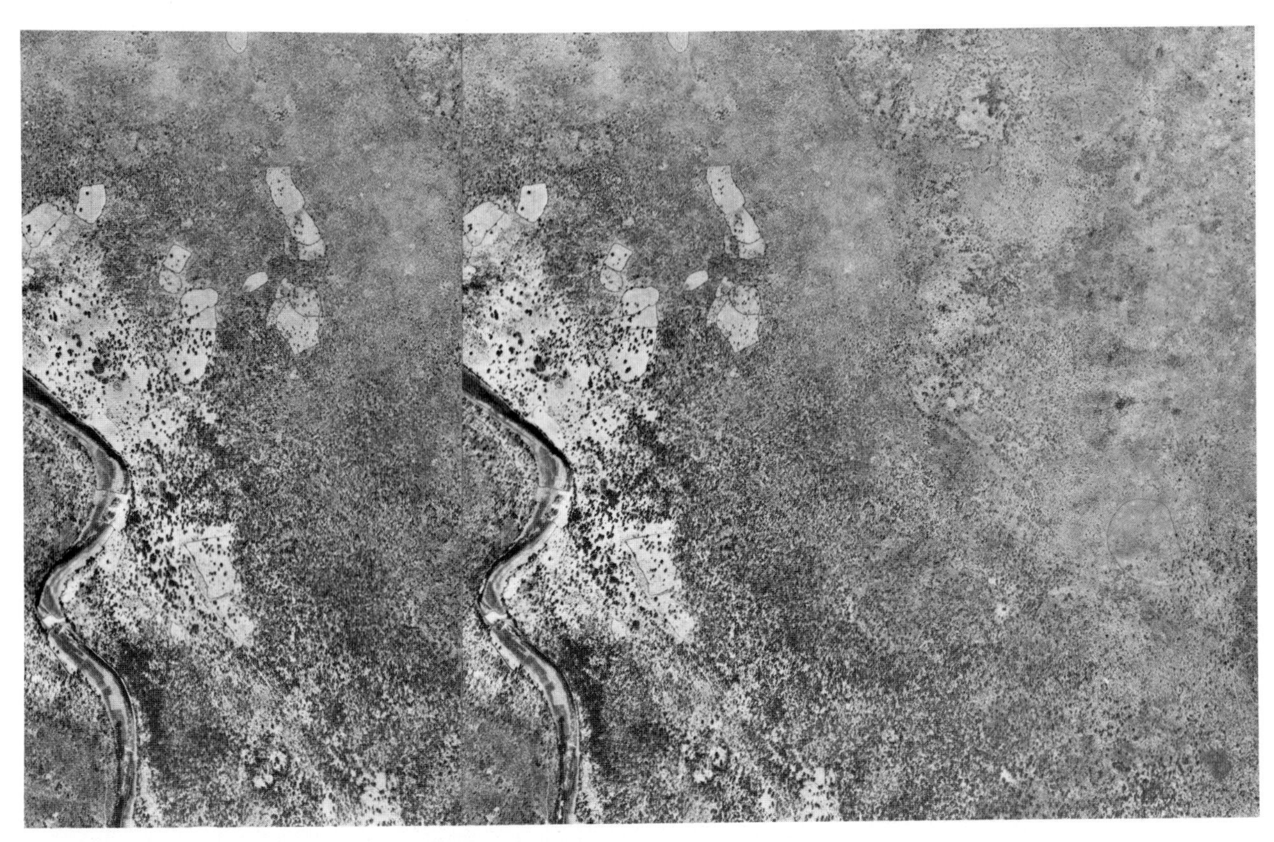

Fig. 14.13 Extended stereopair (1:50,000) showing flood retreat (molapo) cultivation in the valley bottom of the Boteti River in Botswana and larger, dry fields at some distance from the river.

Fig. 14.14 Low altitude oblique view of dry fields of the type pictured in fig. 14.13, near the Boteti River, Botswana.

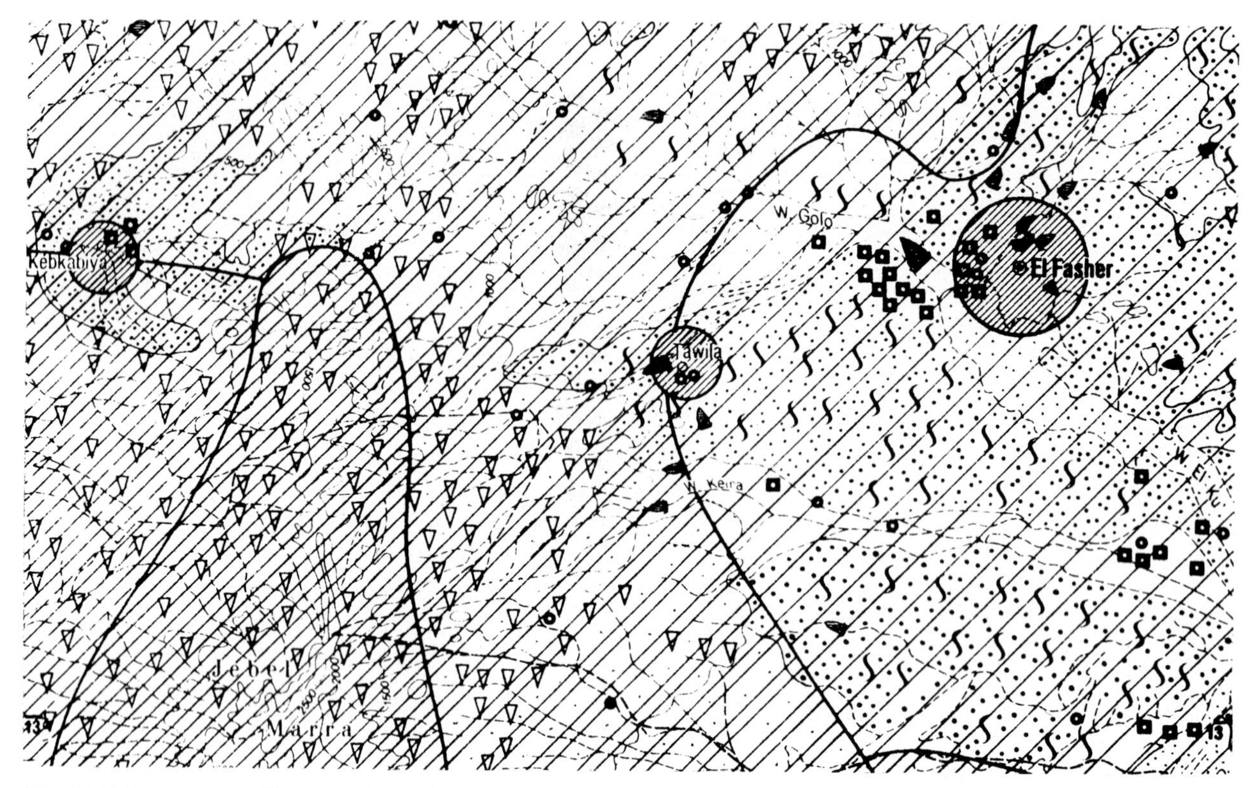

Fig. 14.15 Part of a desertification map of northern Darfur, Sudan (Ibrahim, 1980).

Key:

Degree of Desertification (Hazard)

- large settlement with severely affected surroundings
- smaller settlement with severely affected surroundings
- severely affected areas (rain-fed cultivation on old dunes; overgrazing and clearing near settlements and watering points
- moderately affected areas (rain-fed cultivation in south of area; overgrazing through all-year-round grazing)
- areas exposed to desertification hazards (seasonal grazing; clearing in wadis, depressions and highlands)
- areas less exposed to desertification hazards (seasonal grazing in more humid areas)

Vegetation-Climatic Zones

- semi-desert, thorn-scrub and grassland (11 arid months, 100-200 mm)
- *Acacia senegal* savanna (10 arid months, 200-400 mm)
- *Acacia mellifera* savanna (10 arid months, 200-400 mm)
- tree savanna of the pediplains (9 arid months, 400-700 mm)
- mountain woodland savanna (7-8 arid months, 600-1000 mm)

Morphodynamic Processes

- fluvial erosion on shallow soils of mountains and pediments reactivation of Goz dunes
- dune formation

Watering Points

- seasonal pond (rahad)
- dam (hafir)
- pumping station
- well

The concept, as evident from the legend, is based on mapping the present degree of desertification around settlements as well as in rural areas on one hand and the vegetative and climatological zones as an indication of desertification susceptibility of the environment on the other. Furthermore, some data on morphodynamics and on water resources are added.

Further elaboration of the geomorphological and pedological environmental factors may lead to more detailed information which is especially important for assessing future developments of the desertification problem in an area. For a complete insight into the hazard distribution, it is also necessary to indicate where the desertification processes have already caused such devastation making the land economically lost. The hazard for further desertification there is usually low: combating the desertification should be concentrated on those parts where the land is not yet too strongly eroded or covered by sediments and is therefore still a valid resource. Outlining the areal extent and degree of erosion, transportation and sedimentation of sand and/or finer textured materials should be included in the survey.

14.6.2 Desertification Processes and Soil Losses

The soil losses produced by windwork are, of course, maximum in dry years as has been clearly observed in the USA during the dry nineteen-thirties and seventies. Huge amounts of soil can become airborne even during one single storm (Goudie, 1978; Saarinen, 1969). It has been estimated that on one occasion in 1934 about 300 million tons of soil became airborne over the Great Plains and on this occasion some fields were at once stripped of a plow-deep soil cover. Depending on the annual rainfall, 400.000 to 15 million ha of land were damaged by wind erosion, i.e.,more than about 37 tons soil material lost per ha per year.

Adverse effects of windwork in rural areas have been studied in many countries. In the USA studies on the mechanism of aeolian sand transportation (Zingg, 1954; Task Committee, 1965), the physics of wind erosion in general (Chepil, 1945; Chepil and Woodruff, 1963), soil losses and soil productivity (Skidmore and Woodruff, 1968) are numerous. Fryberg and Lyles (1977) have recently outlined further required research. Borsy (1975) studied the problem in agricultural land of Hungary using both wind tunnel experiments and field observations on soil losses. See also: Akalan (1965); Anonymous (1978); Azizov (1977); Kuhlmann (1958).

The damage is mainly done by way of removal of fine fractions as a result of which the infertile coarser grains remain, which may move and form dunes. Pebbles and gravel may in time form a 'desert' pavement, checking the windwork, but the land, of course, is lost for agricultural uses.

The above mentioned annual soil loss of 37 ton/ha is used to estimate the acreage affected by desertification since it is considered to be visible by the eye. Soil removal/accumulation to a depth of 2.5 cm indicates an involved quantity of 125-400 ton/ha (Kimberlin et al., 1977). Evidentely,the limit of 37 ton/ha is a rather arbitrary one and considerably lower annual soil losses will, if continuous, also result in an appreciable degree of desertification in the course of the years.

Important damage to crops may also occur even if the soil loss is not considerable: the abrasive action of soil particles swept by the wind is sufficient to cause injuries to plants and to substantially reduce yields (Downes et al., 1977). The vulnerability to this phenomenon varies with the crops and considerable research has been done in this field (Jensen, 1954; Joshi, 1969; Lyles, 1975, 1977). If the topsoil is lost, decrease in plant nutrients and organic matter and also changes in the soil texture will be even stronger. Quantification of the effect of desertification on productivity of the land is not easy. Nevertheless, valuable attempts have been made in this direction by converting annual soil loss to average soil removal in cm depth and relating this to crop yield (Lyles, 1975). The (potential) soil loss rate can be estimated using a wind erosion equation developed by the Agricultural Research station of the US Dept. of Agriculture.

Much also depends on the agricultural practices used, such as stubble mulching, fallow years, etc. However, certain soils are much more susceptible to windwork than others (Layer, 1936; Bakhti and Ibrahim, 1982). In the above mentioned wind erosion equation, a wind erodibility factor is used and soils are classified in a number of wind erodibility groups for which a certain multiplification factor is used. Sandy soils and sandy loams are considered particularly vulnerable but one may wonder whether this vulnerability of relatively coarse textured soils has been overemphasized when compared to the finer textured soils (Chepil and Woodruff, 1963; Gillette, 1977).

In fact the emission of fine particles by windwork has been studied in much less detail. A major difference between the movement of finer and coarser materials is that the latter is limited by the maximum carrying capacity of the wind which depends on the vertical momentum flux resulting from wind speed and turbulence. Thus, after a certain distance,only as many grains can become airborne as are deposited, thereby limiting erosion. The fines, however, are carried off in suspension and are not confined to such volume control. The ratio of fine particle emission to total soil movement varies with wind speed, soil texture and mineralogical content.

The percentage of fine particles in dry parent soil was compared with fine particles moving at 0 - 1.3 cm above the ground and with wind-blown fines moving at 1 m above the surface; the grain size distribution of airborne dust collected by aircraft (fig. 14.16) was also given, from which data the strongly increased amount of fines in airborne transportation of soil particles is evident. Airborne fine particles are produced by the sand-blasing effect of saltation. Saltating grains release fines when bumping on the ground and tend to disaggregate fine particles. Salinity of the soil surface facilitates the production of fines and thus increases potential wind erosion.

One should be aware of the fact that almost any soil,

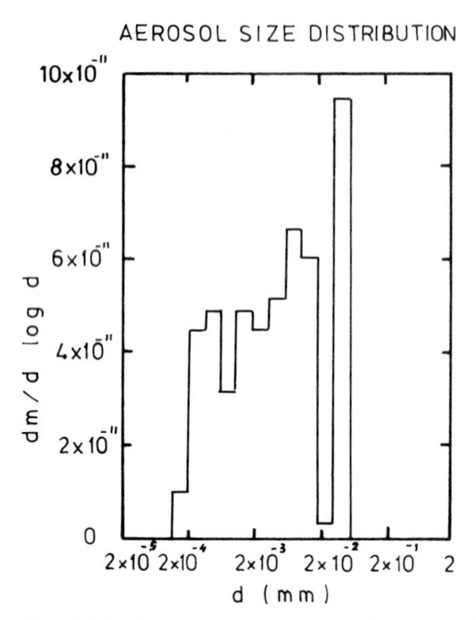

Fig. 14.16 Grain size distribution of airborne dust from dust bowl (Gillette and Walker, 1977).

if stripped of a protective cover and exposed to drought, is, to a certain extent, susceptible to desertification if inappropriate agricultural practices, overgrazing, etc. occur. Coarse, non-erodible surface elements play an important stabilizing role. They tend to reduce wind velocity and turbulence and thus reduce the vibration of soil particles when a threshold surface wind velocity is reached.

The potential soil losses of the land depend on a number of factors related to climate and terrain and vary considerably from one area to another. In the USA soil erodibility by windwork is expressed in tons/ha (or tons per acre). Soil is lost from flat, smooth large fields, not protected by vegetation, shelter, sealing, etc. and subject to the climate prevailing in the drier parts of the Great Plains (Chepil, 1956). Although the absolute values so obtained are not applicable to other (semi) arid areas where different climatic conditions occur, the picture so obtained about the relative erodibility of various soil types is certainly of general use.

It has been established that particles having a diameter of 0.84 mm and up can be considered non-erodible. It is not so much grain size distribution which matters, but rather the soil cloddiness (Chepil, 1955), resulting from natural, soil forming processes and agricultural practices. The larger, non-erodible particles or clods outcropping and slightly projecting above the surface tend to protect the fines to their leeward from the impact of saltating grains and thus reduce their rate of removal. Ultimately a (semi) stabilized surface is formed. (Schwab et al., 1971). The cohesion of soil particles (Smalley, 1970) is an important matter. The sealing of the soil surface caused by surface wash during the occasional rain showers may temporarily reduce wind erosion to less than 20% but the sealing cover is normally cracked soon after its formation and the fines forming it are then rapidly blown out leaving the soil again prone to wind erosion.

Terrain roughness is a second important factor in estimating potential soil losses in all soil erodibility classes. It merits full attention because of its effect on the aerodynamics of the wind flow. Substantial research has been carried out on the effect of knolly terrain (Chepil et al., 1964), low ridges (Armbrust at al., 1964) and other roughness factors (Lyles et al., 1974). The roughness may either increase or decrease the actual or potential wind erosion. Roughness of the soil surface generally decreases the soil losses.

Low ridges of about 5-15 cm height and perpendicular to the dominant wind increase the surface roughness and have a similar effect; they tend to decrease the wind velocity at the ground. Erodibility is distinctly reduced and furthermore airborne soil particles are trapped; this is a positive factor. Higher ridges tend to be eroded at their top and cause changes in the flow pattern of the wind which results in the same adverse affects as those inherent to knolly terain.

Larger terrain roughness features,such as knolls, increase the erodibility, probably because the rise of the ground to their windward side results in compression of the airflow and thus in an increase of the wind velocity. The effects of knolls increases with the angle of their windward slope and decreases with the length of this slope. A 2% slope of less than 150 metres length results in a 10% rise in erodibility and an 8% rise increases it to 350%. Windward slopes longer than 150 metres show no appreciable rise in erodibility, irrespective of their slope angle. Wind flow seems to be adjusted to relief and thus no longer results in increased erodibility. This also explains why the effect of major relief forms on wind erosion is less than one would assume.

Soil humidity (Chepil, 1956) is a third terrain factor to be considered in desertification surveys. Although increased soil humidity after rainfall may temporarily stop or reduce wind erosion on the wetted fields, this phenomenon is usually too short-lived in (semi)arid areas to be of relevance. However, certain parts of the terrain may be consistently somewhat more humid than others and these soil humidity patterns strongly affect the spatial distribution and intensity of wind erosion.

These three groups of terrain factors - soil conditions (particularly texture, structure, cloddiness, etc.), terrain configuration (including micro relief) and soil humidity - should receive full attention in desertification surveys and should be complemented with data on vegetative cover, sizes and pattern of fields, etc.

It should be clearly understood that the erodibility of the surface material is, in itself, not a reliable indicator for the assessment of actual soil losses by wind. The wind erosion hazard in potentially susceptible areas is, of course, low where the fertile soils have already been removed by windwork of the past. One should first define the zones of degradation and deflation and thereafter delineate those parts where substantial or potentially removable sands and/or fines are being or are threatened to be blown out.

Many features indicating accelerated windwork and terrain forms such as knolls, are clearly visible on aerial photographs of appropriate scales (Ahman 1974; Hurault, 1966) and thus have a key position in the systematic surveying of these phenomena (Bergsma, 1977). Geomorphological contributions to this field may relate either to the delineation of broad morpho-climatic zones (Meckelein, 1973) or to detailed mapping (Williams, 1968). One may also study matters such as the areas of provenance of the wind borne materials and the zones of sedimentation (Heine, 1970), the stabilizing role of vegetation on aeolian processes (Marshall, 1970) or the quantitative aspects of deflation (Babayev and Cheredinchenko, 1972).

14.6.3. Quantification of soil losses

Attempts at quantifying the soil losses due to desertification by the Soil Conservation Service (SCS) in the USA have resulted in the establishment of an empirical wind erosion equation (Niles, 1961; USDA, 1961; Woodruff and Siddoway, 1965) mentioned earlier.

The soils are classed in a number of soil erodibility classes and the potential erodibility in ton/ha estimated, further assumptions being made for the other factors listed (table 14.1). Subsequently, the annual soil loss is converted to cm soil depth/year and correlated to the percentage reduction in crop yield (USDA, 1961; Lyles, 1977; Simmons and Dotzenko, 1974). Gross soil losses over large areas can also be assessed. Corrections to the erodibility are applied for knoll erodibility and the climatic factor is composed of wind velocity and soil humidity components.

The equation relates the potential annual soil loss (E) to a number of factors, as follows:

$$E = f (I, C, L, L. V) \text{ in which}$$

E = potential annual soil loss in ton/ha
I = soil erodibility
C = climatic factor
K = soil surface-roughness factor
L = unsheltered field width in main wind direction
V = equivalent vegetative cover.

TABLE 14.1 Soil texture and erodibility

Texture class of surface soil	average erodibility ton/ha
very fine sand, fine sand and coarse sand	220
loamy sand (very fine-coarse) or sapric organic soil materials	134
sandy loam (very fine-coarse)	86
clay, silty clay, non-calc. clay loam or silty clay loam with 35% clay	86
calc. loam/silt loam or calc. clay loam with 35% clay	156
noncalc. loam/silt loam with 20% clay, or sandy clay loam, sandy clay or hemic organic materials	47
non-calc. loam/silt loam with 20% clay or non-calc. clay loam with 35% clay	38

(Soil Conservation Service, USDA, Report on wind erosion conditions in the Great Plains, 1976). Hayes (1972) gives somewhat different values.

An example of the application these principles to the assessment and classification of the wind erosion hazard as observed from erosional and depositional features in the field is given in an FAO land classification project in northern Sudan (v.d. Kevie, FAO/UNDP 3/76, Sudan).

Rating	Evidence of wind erosion/deposition	Texture class of surface soil
4 (low)	unstabilized dunes 100 cm high or continuous unstabilized sand sheet 20 cm thick	sand (very fine-medium)
3	many unstable hummocks 20-100 cm high and 20 m apart and/or A horizon partly/wholly eroded	loamy sand, loamy fine sand
2	unstabilized hummocks 20-100 cm high and 20 m apart or 20 cm high and 2m. apart; and/or A horizon partly eroded, tillage implements may reach underlying horizon	sandy loam; calc. loam/silt loam; calc. clay loam/ silty clay loam with 35% clay; clay/silty clay
1 (high)	no evidence of significant erosion and depositon; well developed A horizon	non-calc. loam/ silt loam; non-calc. clay loam/ silty clay loam with 35% clay; silt, sandy clay loam; sandy clay

14.7 Control Methods of Desertification on Agricultural Land

Soil losses by deflation can be checked or minimized by reducing the velocity of surface winds and by increasing the resistance of the surface material to the blasing effect of the wind. All wind erosion control methods are based on either of these two principles or on a combination of both. The control methods can be divided into three groups, namely those related to farming methods and tillage, those based on the application of chemical compounds and those based on the erection of wind barriers. The latter reduce the wind speed; chemical compounds, on the contrary increase the resistance of the surface to wind work, whereas farming methods and tillage affect both factors. Good reviews of the various methods in use are given by FAO (1960); Knottnerus and Peerlkamp (1972); Woodruff and Chepil (1956) and Hayes (1972).

Once the soil particles are air borne, only barriers or wind breaks can trap (at least part of) this material by reducing the transportation capacity of the wind. This trapping is most successful in the case of sandy grains moving mainly by saltation near the surface. It is much more difficult to efficiently trap aeolian dust as it also is transported at a greater height. This may be of considerable importance in combating dangerous visibility situations (Hagen and Skidmore, 1977). In areas of aeolian sedimentation other problems and control methods exist, as discussed in sub-section 14.9.

Wind breaks in the first place affect the air current. Part of it is diverted over their top and, particularly if of the open type (e.g., trees), another part is filtered through. They reduce the wind speed over a certain distance to the leeward side where the deflation is therefore reduced. The leeward effectiveness depends on type, density, height, etc., but it is usually limited to a maximum range of about 5-15 times their height. Care should be taken for gaps in the wind breaks, as wind will be funelled through and thus be increased instead of reduced. Dense wind breaks may be used, e.g., to protect farm houses, whereas more open types are sufficient for the protection of orchards and other fields. Completely closed mechanical wind breaks are uncommon for agricultural uses.

Immediately downwind of a wind break, crops may be adversely affected and some farmers may take an unfavourable view to them. The benefit of the wind break, however, is at some greater distance and largely outweighs the disadvantage immediately next to it. Wind breaks also have an effect on temperature, air humidity and surficial soil moisture content. A distinct decrease in desiccation may thus occur in many cases. Studies on the effect of optimum design and spacing of wind breaks and shelterbelts are made by Baltaxe (1967); Hagen (1972, 1976); Hagen and Skidmore (1971); Prakash (1959); Skidmore and Hagen (1977); Woodruff (1956), Woodruff et al., (1972).

Chemical sprays and/or windbreaks are, of course, only optimally effective if perpendicular to the abrasive winds. This poses a problem where winds from various directions may have harmful effects. The adhesives of rubber plast or petroleum derivates stabilize the soil surface and only exert a negligible influence on the airflow pattern by way of smoothing the surface and modifying the friction in the air-soil interface. A great variety of industrial products is on the market and Armbrust and Dickerson (1971) compared the cost and effectiveness of 34 different types. The sprays only give a temporarily protection and repeated application is thus required. The cost of these methods is generally prohibitively high for application over large areas. They may form part of soil stabilisation programmes, however (Gorke and Hulsman, 1971). See also ISRC (1971); Letey (1963); Lyles et al. (1969, 1974) and Weymouth (1967).

A wide range of possibilities for reducing desertification exists in the appropriate tillage methods and other agricultural practices. Plowing perpendicular to the abrasive winds leaving low (5-15 cm) ridges across the field results in a sizeable reduction of wind erosion. If tillage can be done when the soil is moist after the rains, many more non-erosive clods will be formed than when plowing in dry soil and also this way wind erosion is reduced.

Mulching also plays an essential role in combating desertification of agricultural lands. The plant residues left on the surface will partly cover the threatened soil surface, reduce the wind velocity over the field and trap airborne soil particles from upwind areas.

The application of cattle feed manure has similar effects. Standing stubble generally gives the best results, particularly when it is relatively long. If the mulch material is flattened on the surface considerably larger quantities are required to achieve adequate protection.

In any case, even small quantities of mulching material (e.g., 500 kg/ha) have distinct beneficial effects although to be fully effective several thousands of kg/ha may be required. The effectiveness generally depends on climatic, soil and terrain conditions prevailing in the area. The characteristics of the wind, such as turbulence, and its increased velocity with height near the ground are especially important. (Woodruff et al, 1974; Armbrust, 1977; Chepil, 1944; Chepil et al., 1960, 1963; Lyles and Alison, 1976; McCalla and Army, 1961; Siddoway et al., 1965). Among other agricultural practices, the use of grass barriers and other wind strips (Aase et al., 1976), the introduction of narrow field strips (Chepil, 1957; Hagen, 1972), the selection of appropriate crops and the use of tillage fallows (Fryrbear, 1963; Fenster and Wicks, 1977) should be mentioned.

It is evident that the geomorphologist engaged in the control of desertification of agricultural land should closely cooperate with agricultural engineers to ensure optimum results. His contributions lie in the field of investigating the dynamics of deflation and aeolian transportation of soil particles as affected by the various protective devices, in the breakdown of the land into classes of susceptibility to wind erosion and,whenever possible, to quantify his observations.

14.8 Surveying Active Dunes and Predicting their Future Development

It is evident from the previous sub-sections that quantification of the desertification process and measuring the rate of advance of the desert fringe are badly needed. This is not an easy matter because usually the situation is one of gradual deterioration of broad marginal zones of the desert rather than its advance with a knife-edge boundary. In the case of active dunes and related effects of wind work, however, the quantification aspect and the hazard evaluation of downwind areas is often more readily within reach. Aerial photography and other kinds of remote sensing imagery play an important role in surveys of this kind.

Even if only one aerial photographic coverage is available, this will in itself permit the precise mapping of the areal extent of the wind-blown areas and indicate the most endangered down-wind areas on the basis of the characteristics of the dune morphology in combination with existing wind data. One may also measure the dunes and thus arrive at a fairly precise quantitative assessment of the bulk of the dunes and the masses of sand involved in dune movement; however, the amount of through-flowing sand will be more difficult to estimate.

More important than mapping the momentary situation is the assessment of the aeolian dynamics which should form the basis for a prognosis of the situation to be expected at a given time in the future. Tendencies towards increasing (or occasionally decreasing) instability should receive particular attention in this connection (Verstappen, 1972). Several indicators for dune dynamics can be found in their geomorphological characteristics, such as:

1. Dominance of sand tails or shadows tapering out downwind points to greater instability than the prevailance of steep leeward slopes having the maximum angle of repose.
2. The velocity of the dunes downwind is related to their bulk.
3. Moist zones to the leeward of the wandering dunes may act as traps for blown-through sand grains and also tend to slow down the displacement of the dunes which then usually build up vertically.
4. Recently formed blow holes, the disruption of the curved parts of parabolic dunes and the formation of parallel longitudinal dune ridges from the then separated wings of the paraboles, point to increased instability.
5. The formation of low dunes forming a blanket over previously existing dunes of greater magnitude points to reactivation of aeolian activity, e.g., due to overgrazing or to agricultural practices which are inadequate from an environmental view point (Verstappen, 1957).

Although the information so obtained will be mainly of a qualitative nature, it is valuable for evaluating the rate of desertification and the magnitude of the hazard for the downwind areas.

Much more precise information can be given of those areas from which several aerial photographic coverages exist taken with an interval of,e.g., 5, 10, 20 or 30 years. Sequential interpretation of these images then leads to quantitative data about the sand encroachment during the period elapsed between the two aerial surveys. These temporal changes can also be extrapolated into the future and may then serve purposes of prognosis and planning of dune stabilization campaigns. The size or bulk of the dunes can then be correlated precisely to their rate of movement and a decreased size be immediately interpretated as increased movement and instability. Bulk transport rates and rates of divergence can also be calculated. An interesting study of this kind is by Lettau and Lettau (1969) concerning barchans of southern Peru. It was found that about 2 mm of sand is removed annually from the land surface and that the dunes have been formed in the last 100 years. It was also established how and to what extent sand from one barchan is trapped by another barchan situated farther downwind. It was estimated that each dune is completely renewed this way in about 14 years and has moved 1700 metres in that period.

An example of the utility of sequential aerial photographic interpretation for the purpose of monitoring dune development and planning stabilization programmes is given in the airphotos of figs. 14.17 and 14.18. The photographs date from 1954 (1:50,000) and 1958 (1:15,000) respectively.

The dune area depicted is located along the north coast of the Cap Vert peninsula to the east of the village of Camberene. Old, stabilized, longitudinal dunes dominate inland. They stretch in a NE-SW direction and were formed by the 'Harmattan' tradewind under dry continental conditions when the sea level was low. Because of their colour they are known as 'red dunes'. Their age is 21,000 - 15,000 B.C.

The sea invaded the more seaward parts of these dunes and formed embayments about 5500 B P during the peak of the Nouakchottian (Atlanticum) transgression. A beach barrier

Fig. 14.17 Blow-up (1:15,000) of a 1:50,000 airphoto, dating from 1954 of the coastal dunes near the village of Camberene, Cap Vert, Senegal. Compare with fig. 14.18.

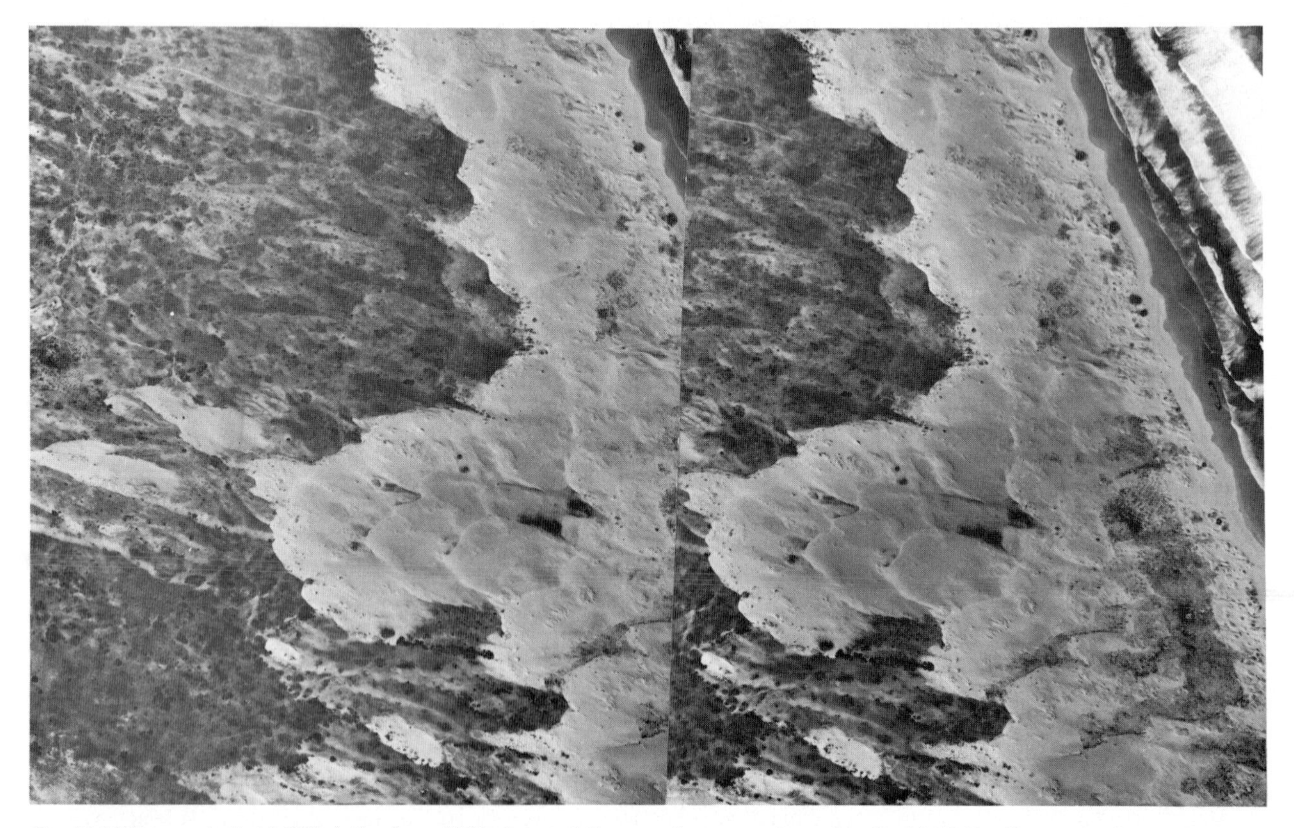

Fig. 14.18 Stereopair (1:15,000) dating from 1958 of part of the same dune area pictured in fig. 14.17. Reafforestation has gradually lead to a decrease in dune mobility.

was subsequently formed and closed off the embayments from the sea around 4000 B P (Michel, 1969).

A new series of dunes, the parabolic type, developed on the barrier between 3000 and 2000 B P under the influence of the marine trade winds when the sea level was high. The direction of these 'white dunes' is NNE-SSW and thus deviates somewhat from the earlier mentioned continental longitudinal 'red dunes'.

At least two periods of aeolian activity can be distinguished in the white dunes, viz. the sub-recent vegetated ones and the recent barren dunes.

Fixation of the dunes has been attempted by an ambitious reafforestation scheme of *Casuarina equisetifolia* on the beach barrier since the natural vegetation, mainly *Cenchrus biflorus* (cram-cram grass) is insufficient. *Euphorbia balsamifera* is locally used inland to create hedges for combating sand drift in the fields. The reafforestation blocks in the dunes are clearly visible on the aerial photograph of fig. 14.17. The photograph of fig. 14.18, taken four years later shows that reafforestation in this interval has been extended farther westward. The vegetation is, however, not yet completely effective in fixing the dunes which invade the plantations at various places at a rate of 5 m annually or even 30 m as a maximum. The inland creeping of the dunes can be accurately mapped by comparing the 1954 and the 1958 aerial photographs. It is evident from this example that sequential air photography is of great value in dynamic geomorphological studies.

Very often reactivation of dunes or accelerated encroachment of sand is caused by misuse of the land in one way or another. A striking example of this is given by Ibrahim (1978) from the Sudan using sequential aerial photography. The cause here was the cultivation of millet in zones of below-critical rainfall. The environmental deterioration so caused is,of course, particularly outspoken when a number of relatively dry years occur. The process of dune reactivation is often initiated under such conditions. The effect of man on dune reactivation and the role of sequential aerial photography to monitor it are also clearly demonstrated by the example of the coastal dunes near Parangtritis, Southern Java, Indonesia, given in section 6.

The formation of new dunes or a sand sheet superposed on older dunes that have been inactive or almost inactive for long periods of time, briefly referred to above, is illustrated by the single air photo of fig. 14.19 with the accompanying interpretation, showing a dune area near Kanem to the north of Lake Chad at a scale of 1:40,000

The area is characterized by stabilized transverse dunes, mostly about 10-15 metres high and formed by the NE 'Harmattan' winds prior to the existence of Lake Palaeo-Chad. The dunes are strongly subdued and the difference in steepness of windward and leeward slopes is only minor (6° and 12° respectively). They are covered by grasses and scrub and brown or reddish-brown sandy steppe soils have developed, except in the interdune depressions where hydromorphic and halomorphic soils occur (Pias, 1957). The original dune ridges are not easily seen in this picture and are only faintly indicated by the direction of the small lakes locally occurring in the interdune depressions. Some of these lakes are interconnected by a winding episodic channel.

Distinct, fresh looking and large, though low, parabolic dunes have been superposed on the old transverse dunes in more recent times and form a sand sheet which is partly responsible for their almost complete obliteration. Since the parabolic dunes occur below the 320 m shore line of Lake Palaeo Chad, they are in any case less than 5 000 years old, but probably much younger/recent.

Recent aeolian activity is also mentioned (Hurault, 1966) and results mainly from overgrazing. The rather great difference in age of the transverse and the parabolic dunes and the fact that only a thin veneer of actively moving sand is present, indicates that superposition and not transformation of these two dune types has occurred.

14.9 Controlling Drifting Sand and Dune Displacement

14.9.1 General Stabilisation of Areas of Wind-blown Sand

If an area is adversely affected by wind-blown sand or invaded by wandering dunes, it is obvious that,whenever possible, it should be attempted to locate the source area of the sand and to treat this subsequently by appropriate means of stabilisation. It may be, however, that the source area cannot be effectively treated because of low soil moisture content, depth of groundwater, remote situation, extensiveness, etc., in which case protective measures have to be taken in a downwind area which is affected by the blown-out sand. The sand may be derived from local sources, e.g., as the result of improper land utilization as discussed in sub-section 14.6. One should always bear in mind that the type and intensity of aeolian processes are strongly dependent on the often subtle equilibrium of the incoming and outgoing quantities of sand and that stabilisation in the source area resulting in smaller quantities of incoming sand in downwind areas may possibly increase the deflation there and thus result in a shifting of the source area.

The practical problems connected with processes of aeolian transportation and deposition of sand are diverse and any projection aiming at sand stabilisation will require a precise survey of the situation,taking sources area, trajectory and terrain configuration of the area to

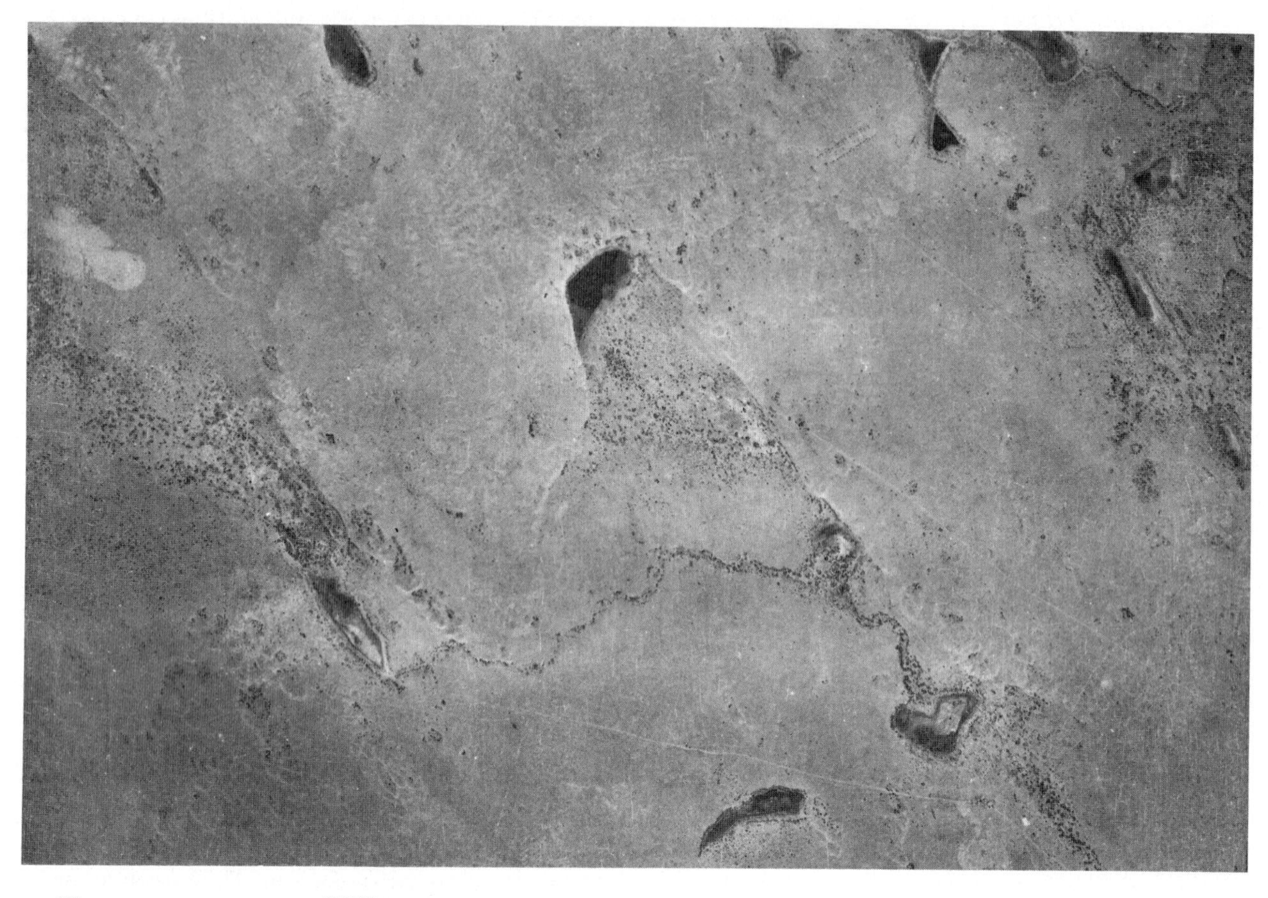

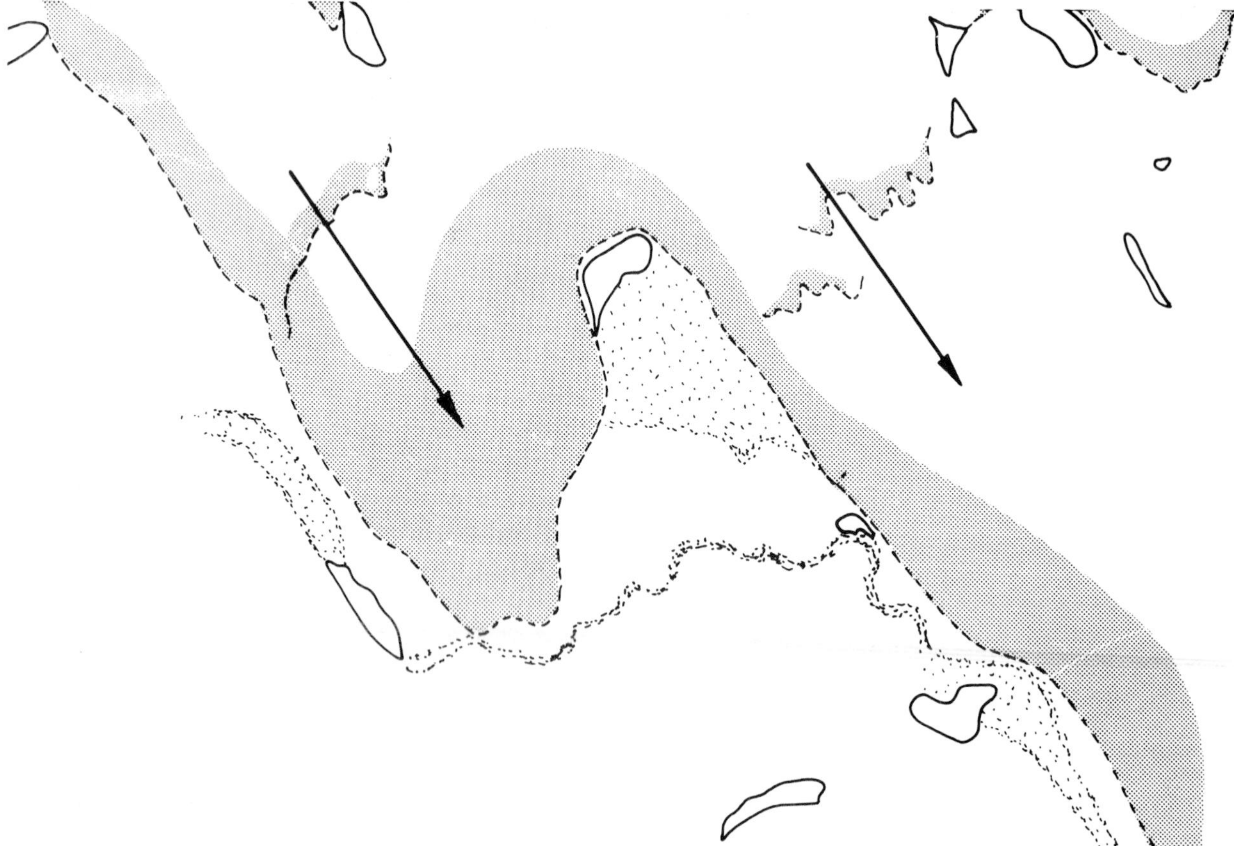

Fig. 14.19 Vertical airphoto (1:50,000) with interpretation (bottom) of an area near Bol to the east of Lake Chad, showing the gradual spread of a sand sheet over a field of ancient transverse dunes, demonstrating increased desertification due to overgrazing.

be protected into consideration,with special emphasis on the aeolian processes as generated by wind speed, direction and regularity and the friction and turbulence induced by the surface roughness of the terrain.

In some areas,the problem is mainly drifting sand, which may accumulate near obstructions or which may cause vertical accumulation in humid depressions,or where the wind velocity decreases for one reason or another and the amount of outgoing sand is less than amount of incoming sand. In other areas large quantities of sand, irregularly distributed over the area, may accumulate in the form of dunes. Objects may become deeply buried or overridden,to be uncovered years later when the dune has passed on to the leeward. In other areas the main problem is the windborne transportation of fine particles, causing dust bowl phenomena and deterioration of the land.

One has to distinguish between general stabilisation of larger or smaller tracts of land and localised stabilisation to protect certain objects of particular interest. The strategy of a sand or dune fixation project will be seriously influenced by this, because the cost of some means of stabilisation will be prohibitive if to be applied over large areas. Another matter that also affects the mode of approach is whether or not the areas downwind of those to be protected can absorb an increased sand drift without undue environmental or economic damage. In the affirmative case,one may, instead of attempting to stabilise the sand in and to the windward of the affected area, consider the possibility of promoting the through-flow of sand to the downwind zones. This particularly applies to localised protection of special objects. If through-flow of sand is not permissible, a decrease of incoming sand by accumulating and stabilising it, e.g., by fixation of dunes,should be effected before it reaches the area to be protected.

Sand stabilisation problems are a major issue in many arid and semi-arid areas. They may, however, also be of importance in other climatic zones and are often found in areas of coastal dunes at all latitudes (Chzao Sin-Chaun, 1953, 1959; Haas and Steer, 1969; Horilkawa and Shen, 1960; Ibrahim, 1969; Liao Yu Pin, 1964).

The means at our disposal to combat drifting sand and to control undesirable effects of wandering dunes can be divided into three categories, viz.: biological, chemical (including, e.g., petroleum derivates) and mechanical. It is generally agreed that the biological measures of reafforestation or regreening are the only ones that will be of long-term benefit and particularly so when extensive areas have to be treated. Chemical and/or mechanical devices may offer good temporary solutions until the vegetation has a chance to establish itself firmly and they are also frequently applied for localised problems such as the protection of buildings or other structures. In many cases a programme encompassing a combination of methods from these three categories has to be effectuated to reach optimal results. It is often only after several years that the effects of measures taken show effect (Lehotsky, 1972). In the next two sub-sections the relevance of methods of the three categories just mentioned will be discussed for purposes of general, over-all stabilisation and for the solution of localised problems, respectively. Considerable experience has been gained in the field of sand and dune stabilisation in various parts of the world, particularly in recent years. Bhimaya (1974); Bhimaya and Kaul (1960); Kaul (1970); Bhimaya et al. (1962) and Prakash and Chowdhary (1957) give examples from Rajasthan, India; Saccardy (1953), Lavouder (1928); Metro (1953); Messines (1952) from northern Africa; Loubser (1958) and Westhuyzen (1957) from southern Africa; Ali (1977) from Sudan; Dougrameji and Kaul (1972) from Iraq; Prego (1971) from Argentina; Reifenberg (1950) and Tsuriell (1974) from Israel; Woodhouse (1967) from the USA; Chzao Sin-Chann (1953) from China and Orlava (1940), and Petrov (1954, 1957, 1971) from the USSR.

It has been previously stated that the biological approach of regreening/reafforestation is by far the most realistic way of achieving long-lasting stabilisation of barren sandy areas. This does not imply, however, that it is exclusively the concern of foresters and botanists: the matter is far too complex to be remedied by simple monodisciplinary approaches. In the majority of cases, the (re)activation of sand drifts and dune movement has anthropogenic causes related to increasing pressure on the land which leads to overgrazing or to agricultural misuse of marginal lands,such as the millet cultivation in some climatologically critical zones in Sudan studied by Ibrahim (1978). The aeolian reactivation in such cases is usually initiated in exceptionally dry years or sequences of dry years.

Therefore, prior to launching a reafforestation programme, one should study the socio-economic and agricultural factors in relation to the climatological situation, particularly to the variability of annual rainfall, and also the geomorphological, hydrological and pedological characteristics of the land (Leontyev, 1962). The geomorphologist then has to provide information on the state of the wind work, quantifying the amount of sand involved and emphasizing the dynamics of the aeolian system prevailing in the area. On the basis of this study,the zones where reafforestation is most urgently required and where the best chances for success are present can be indicated. It is only after all aspects of the situation have been analysed and pieced together that the biologists can decide on either a programme of fencing to keep animals out of critical areas and thus to promote natural restoration of the vegetation, and/or for a programme of reafforestation, indicating what to plant and how to do it. The proper positioning of shelter belts can also be planned on this basis. Reafforestation should not be considered a panacea for all sand/dune fixation problems. Where excessive sand supply occurs, it may even be altogether impossible. In any case it is a laborious, expensive operation and requires watering of the plants during the initial stages.

Reafforestation in the early phases often has to be supported by applying chemical soil stabilizers or mechanical devices such as wind screens to prevent the young seedlings from being killed or unduly damaged by excessive sand drift and/or buried by sand deposition. The sands may also be poor in minerals and have a low water moisture holding capacity due to their texture. Planting in bricks is a practice successfully applied in several countries and mulching techniques have
useful in many cases. The selection of endemic drought resistant pioneer vegetation for purposes of regreening or reafforestation is a sound principle but the introduction of species from other and even slightly drier areas has also been successful.

Reafforestation may serve several purposes; the mode of planting and the species used vary accordingly. A main distinction is between protective reafforestation for purposes of sand stabilisation and reafforestation to increase the productivity of grazing land, agricultural fields, etc. Reafforestation may also be carried out to provide firewood and fodder, or shade along roads and around settlements. Overall cover of the land with tree or scrub vegetation or shelter belts, either in rows perpendicular to the dominant wind direction or in checkerboard patterns,may be aimed at. Shelter belts cause a decrease in wind speed to the leeward as they break the wind and divert it over them and as a result they protect grass cover strips 10-15 times wider than the windbreaks' height. (Raheja, 1963).

The geomorphologist surveying in the context of reafforestation projects should pay special attention to many different factors, the most important of which are:

. the delineation of the source areas of the drift sand
. the type and velocity of sand movement
. the average grain size of the sand and the related threshold wind velocity
. the quantity of incoming and outgoing drift sand
. the volume and velocity of individual dunes
. the dynamics of the wind work occurring
. the nature of the substratum (hard rock, clay, permeable material, etc.)
. the soil humidity conditions
. the occurrence of (ephemeral) surface and sub-surface water
. the degree of salination at the surface and in the root horizon
. the depth of the groundwater
. the occurrence of impervious layers (including pans, calcretes, etc.) possibly causing perched water tables
. the salinity of the groundwater
. the morphographic/genetic situation with special emphasis on sand relief

On the basis of all these factors,the geomorphologist should be able to arrive at a geomorphological breakdown of the land and a terrain classification geared to purposes of reafforestation. He should also endeavour to indicate zones most suitable for (re)afforestation, and anthropogenic modification of the dune relief such as flattening to facilitate planting and to bring the groundwater within reach of the roots. Monitoring the processes of dune migration and sand displacement (Lettau and Lettau, 1969; Simonett, 1975) is another important aspect of applied geomorphological studies in (semi) arid areas. The interpretation of sequential aerial photographs and Landsat imagery is most useful for these purposes as has been explained in sub-section 14.8.

Suitable places for (re)afforestation may be interdune depressions. Where these are too saline however, higher areas such as lunette dunes with a hardplan at a shallow depth causing a perched water table may be more suitable. Also where ephemeral surface water occurs the possibility may be more favourable than elsewhere. For the stabilisation of individual dunes,he may,e.g., suggest planting at the windward side of the dune to deprive the dune of incoming sand. In a second phase of the work the dune may be entirely vegetated. The mechanical and chemical means often used in combination with reafforestation will be discussed in the next sub-section on localised stabilisation problems.

The study may ultimately lead also to pure research, e.g., in the laboratory on the aerodynamic problems related to sand transportation and deposition and to close cooperation with civil and agricultural engineers when it comes to deciding on the details of the measures of mechanical and chemical stabilisation and,of course, with foresters and botanists for the further implementation of the reafforestation programme.

14.9.2 Localised Problems Caused by Sand Drifts and Wandering Dunes

Although biological approaches may also be of use when projects of limited areal extent geared to the protection of specific objects such as buildings, settlements, (rail)roads, etc.,are concerned, chemical and mechanical methods of control then gain relatively in importance. Chemical soil stabilisation products and also most mechanical sand stabilisation devices are too costly for general application over extensive areas and also require repeated application because of their limited period of effectiveness. Then are very useful, however, in an introductory phase of restoration of natural vegetation or reafforestation and also for the protection of small-sized objects of special importance. Their main disadvantage is that, contrary to reafforestation programmes, they do not improve the environmental conditions by way of accumulation of humus, soil formation, micro-climatological amelioration, etc.

Before a localised protection scheme is initiated, the geomorphologist should investigate whether there are realistic chances for success at justifiable cost. It may well be that the best or even the only solution is to live with the problem,which may even lead to the advice to abandon the site and to move to a more favourable one. It is very important in case of the erection of new structures to investigate the vulnerability of a proposed site to sand drift and dune invasion, prior to construction (Peters, 1964; Saini and Borrack, 1967). Where extensive areas of active wandering dunes occur upwind of the site, there is usually no hope to achieve stabilisation at acceptable cost. There may be situations, however, where the sand dunes pass just aside of the structure, oasis, etc., which then will in all likelyhood be safe for centuries (Stevens, 1974). Living with the problem may, when roads are concerned, also mean the construction of two, three or more alternate routes through the critical areas in such a way that at least one route will be free of wandering dunes and thus be motorable.

If the geomorphological investigations lead to the conclusion that the situation can be essentially improved or even (semi) permanent protection can be provided, the geomorphologist should, on the basis of his knowledge of the prevailing aeolian processes, be able to indicate what measures should be taken and at what locations.

The guiding principle is that the dune should either be stabilized or be destroyed ('killed') and that the drift sand be deposited or blown-through by way of skillfully regulating the amounts of incoming and outgoing sands by chemical, mechanical and/or biological measures.

The capacity of the downwind areas to absorb the sand blown out of the area to be protected should be carefully considered. The sand flow can be regulated by increasing/decreasing the velocity and/or turbulence of the near-surface wind or by diverting it sideways of the object to be protected. A few examples illustrate how this guiding principle can be put to practice.

A road traversing an area badly affected by drifting sand may require considerable maintenance to keep it motorable. If the sand-carrying winds blow from one main direction (e.g., trade winds or dry-monsoon winds) the cost of maintenance can be considerably reduced by giving a slight upwind inclination to its surface. The airstream will then be slightly compressed which causes a small increase in near-surface wind speed and thus in a slightly increased sand transporting capacity. Instead of sand being deposited on the road, the road will be blown clear of sand which will be carried farther downwind. Fig. 14.20 demonstrates the effect of the slope angle of the road surface on the sand accumulation at and near the road (Kerr and Nigra, 1952).

Another possibility for preventing undue sand accumulation on a road is by decreasing the surface roughness upwind which will result in a slight increase in wind speed and thus provoke a through flow of sand over the road instead of deposition. This decrease of surface roughness required can be realised by either mechanical means or, more likely, by applying chemicals (rubber or petroleum derivates) to the soil. If a (rail)road cannot be effectively protected by forcing the sand to blow over it, shelter belts should be constructed at its upwind side as to prevent the sand from reaching the (rail)road. Shelter belts may be laid out perpendicular to the wind, parallel to the (rail)road or in a checker board. Vegetation should be applied whenever possible, but frequently mechanical devices (reeds, branches, etc) are to be relied upon. Considerable experience has been gained in this field in the arid lands of Central Asia (Chzao Sin-Chaun, 1953; Ivanov, 1975; Khodzkhayev, 1947; Li Min-Ghan et al., 1960; Podryadov, 1958).

When wandering dunes are concerned comparable considerations apply. It is important, in any case, to effectuate remedial measures in an early stage because otherwise the structure to be protected will be buried by sand before the measures taken become effective. One may attempt to stop the advancement of a dune by constructing fences built from concrete, brick, metal or wood and measuring about 1-1½ metres in height perpendicular to the wind direction at some distance upwind of the dune. A set of three fences has been used in several cases. The first one of these collects the bulk of the sand and has to be repeatedly raised. The second one only collects the remaining overflowing sand and the third one, nearest to the dune, is to check the effectiveness of the two other fences and to provide a warning when the first, windward fence needs to be heightened (fig. 14.21).

The fences will deprive the dune from incoming sand and thus slow down the sand circulation of the sand within the dune and the movement of the dune involved. The

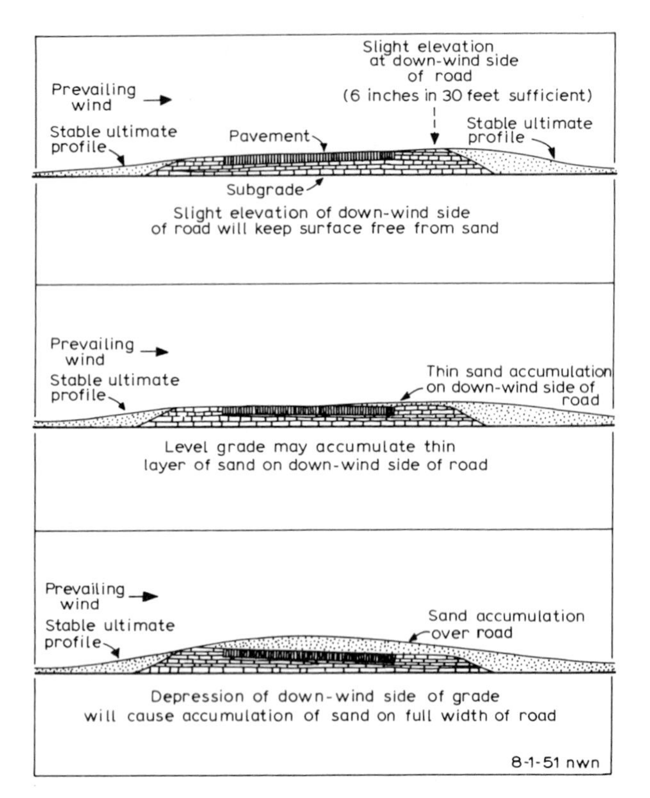

Fig. 14.20 The effect of the inclination of a road surface on the accumulation of sand (Kerr and Nigra, 1952).

dune may be subsequently stabilised by vegetation if circumstances permit. Instead of fences, vegetation upwind of the dune has a similar effect, the afforestation being extended to the dune itself after its stabilisation. Where strong winds occur the outgoing sand flow may continue but then at the expense of the dune itself which will rapidly decrease in size while moving on until it disappears. It may, however, also reach areas where the fence system is no longer effective and then re-establish itself.

Another possibility is to destroy the dune by causing turbulence and thus increasing the wind speed on the dune by placing patterns of wind screens in such a way that the wind is funneled through and erodes the dune. Partial fixation of the dune with chemicals or petroleum derivates may give the same results. One can also attempt to create turbulence by strewing boulders or pebbles on the dune. The size and spacing should, however, be precisely determined by experiments.

When the dune gradually decreases in volume, the pebbles (shells, etc.) tend to form a more or less coherent cover or 'desert pavement' and thus even partially protect the dune. In all these cases,the area downwind of the protected structure will have to support an extra amount of incoming sand which may either remain in transit or be deposited in the form of dunes or as a sand sheet.

A third interesting possibility is to stabilize the windward slope of the dune with petroleum derivates. This slope will then be stabilised instantly,but the incoming sand from upwind areas will blow over it and be deposited to the leeward. The shape of the,e.g., barchan,dune will then be modified and ultimately an oblong-shaped dune (fig. 14.22) will be formed, which is aerodynamically stable. From then onwards only through-flowing sand drift will occur. In the meantime, however, the leeward of the dune is greatly extended and may have buried the structure to be defended unless the stabilisation of the windward side is carried out when the distance from dune to structure is at least three times the width of the dune (Kerr and Nigra, 1952). If one wants to prevent this leeward growth, the treatment of the windward slope should be coupled with the construction of fences upwind of the dune and perpendicular to the wind direction to stop the flow of incoming sand as described above.

Small objects may be protected by fences making an acute angle with the dominant wind direction, either as a single line or V-shaped. Such fences will, at least partly, divert the wind around the structure to be defended. Turbulence and minor variations in wind direction, however, will cause accumulation of sand at their leeward side and ultimately the fences and the structure will be buried by sand unless the fences are repeatedly raised. The more acute the angle between the two sides of V-shaped diverting fences or between the one-line fence and the general wind direction, the longer they will be in operation but the smaller evidently the area protected will be.

The effect of chemical products and petroleum derivates (Compo, 1973; Bielfeldt and Taubner, 1966; Harpaz et al., 1965; Lyles et al., 1969, 1974; Letey et al., 1963; ISRC, 1971; Shell, 1969; Weymouth, 1967) is temporary only and they may also unfavourably affect the environment. They may, however, have an essential role in the early phases of reafforestation, as mentioned earlier. An extensive review of literature on dune stabilisation is given by Hagedorn et al., (1977). Paylore (1976) gives a more general bibliography on desertification. Hagedorn et al., also give a concise review of the constructional details of fences, the commercially available chemical soil stabilizers and petroleum derivates.

The importance of dune stabilisation for villages situated on the fringe of the desert and for oases surrounded by active, wind-blown sands is demonstrated by the two aerial photographs of fig. 14.23 which show encroaching sands and moving dunes in the oases of Aboud and El Barka in Algeria (scale 1:25,000). In the last mentioned oasis the villagers have repeatedly constructed fences at the windward side. Figs. 14.24 and 14.25 show ground views of comparable situations in Nefta, Tunesia and Faya Largeau, Chad, respectively.

Finally, it should also be mentioned here that in coastal areas one may want to construct a dune (ridge),

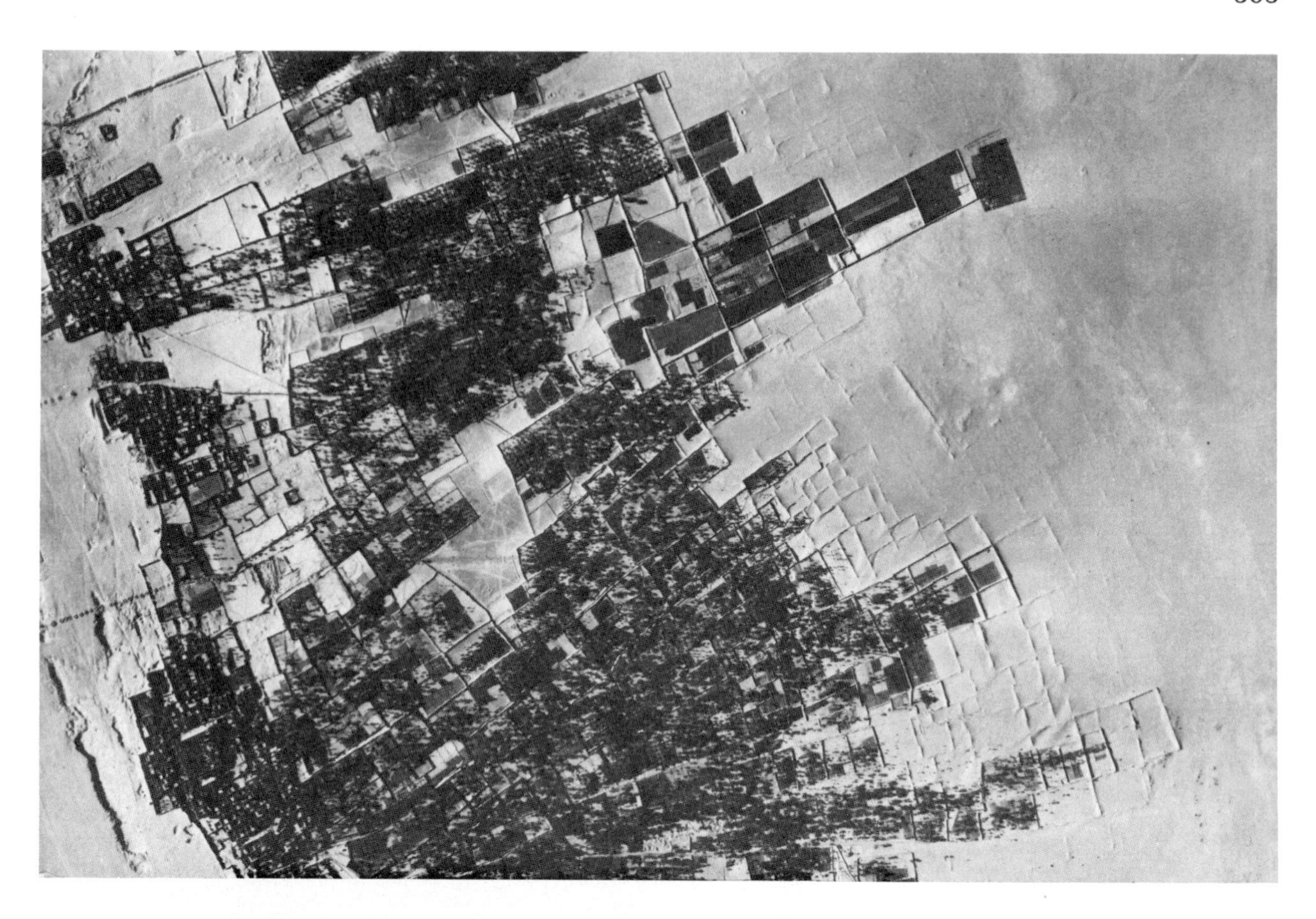

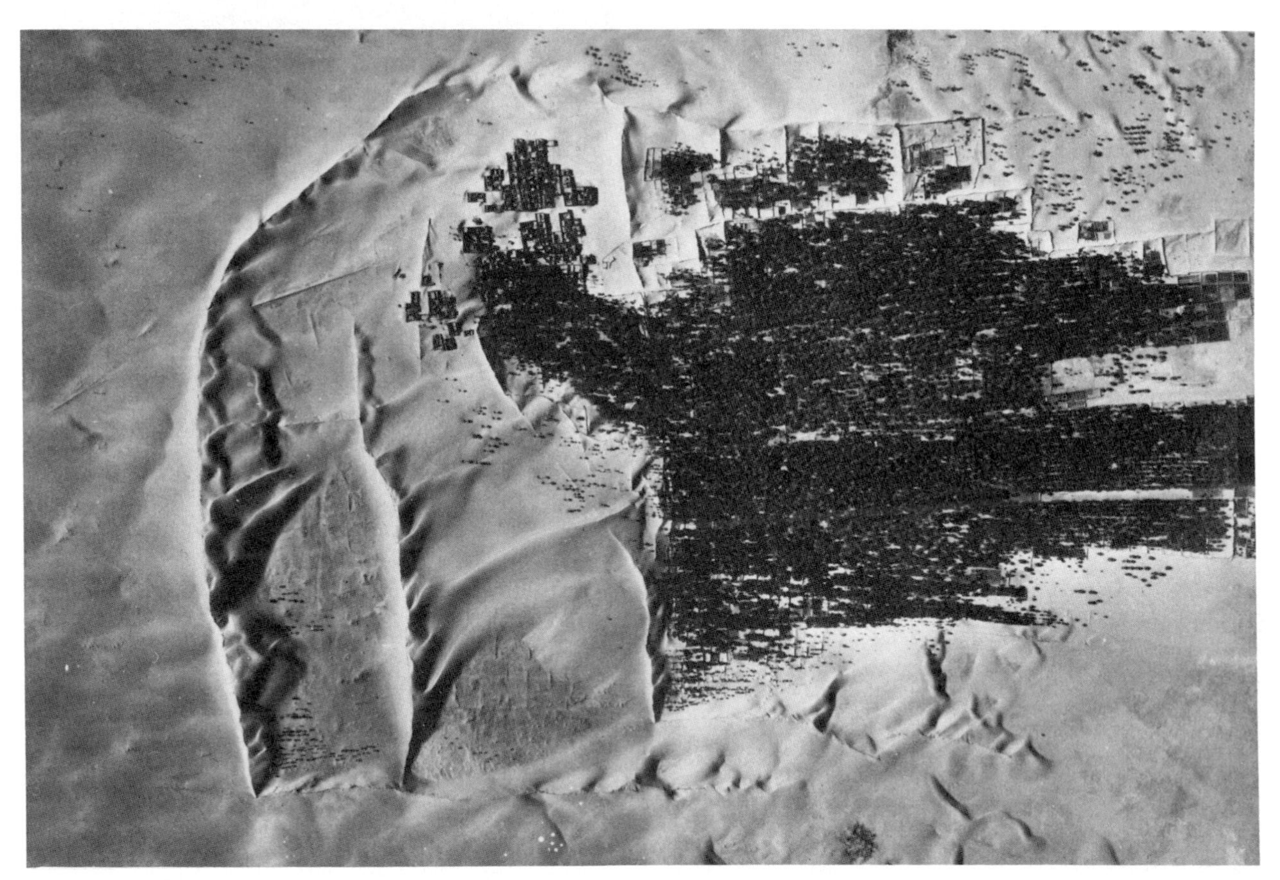

Fig. 14.23 Vertical air photos showing encroaching sand dunes and sand sheets in the oasis of Aboud (top) and El Barka (bottom) in Algeria. Scale 1:50,000.

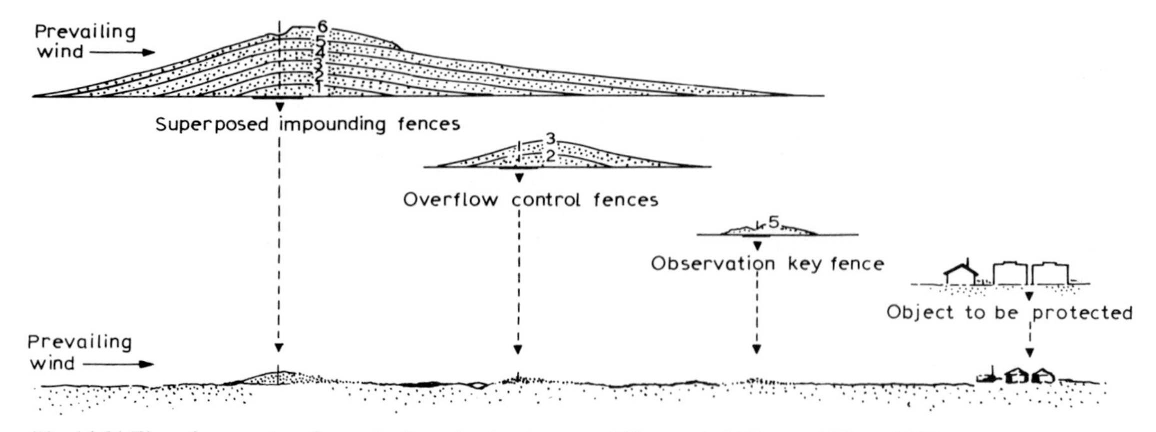

Fig. 14.21 Three-fence system for protection of extensive areas (villages, etc.) (Kerr and Nigra, 1952).

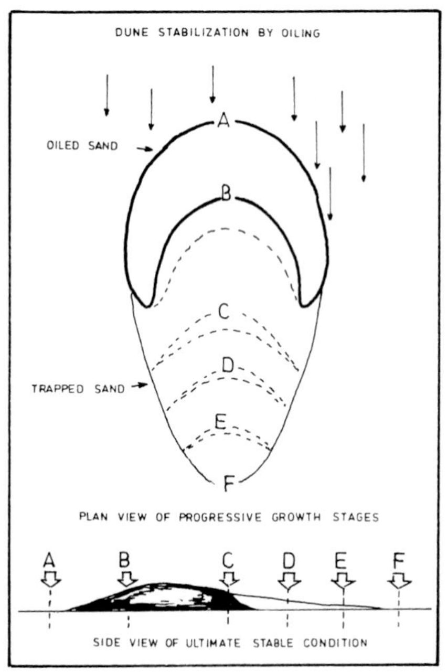

Fig. 14.22 Dune stabilisation by oiling (Kerr and Nigra, 1952).

Fig. 14.25 Dunes invading Faya Largean, Chad.

For fig. 14.23, see previous page.

Fig. 14.24 Dunes invading date palms in Nefta, Tunesia.

either a completely anthropogenic dune by means of bulldozing (which is, of cource, expensive if a large bulk of sand is to be worked) or a semi-natural one by reducing the wind speed, e.g., reed fences, and thus to create a dune to the leeward with the aim of achieving a more effective protection of the hinterland from the invasion by the sea. (Adriani and Terwindt, 1973; Mulder and Plutschouw, 1975).

References

Aase, J.K., et al., 1976. The use of perennial grass barriers for wind erosion control, snow management and crop production. Proc. Symp. Shelterbelts on the Great Plains. Denver: 69–78.

Adriani, M.J., and Terwindt, J.H.J., 1975. Sand stabilization and dune building. Rep. R.W.S., Deltadienst, Den Haag, 52 pp.

Ahman, R., 1975. Vinderosion i sydskane. Svensk Geografisk Arsbok 1974, 50: 232–240.

Akalan, I., 1965. Rüzgar erozyonu ve korunma çareleri (Wind erosion and its control). Ankara Univ. Ziraat Fakültesi Yilligi, 15(4).

Ali, A.D.,1977. The Sudan experience in the field of desert encroachment control and rehabilitation. Proc. Afr. Reg.

Conf. Desertification, Khartoum

Anonymus, 1978. Wind erosion and its control (1968–1978). Annotated bilbiography. Commonwealth Bur. Soils, Harpenden S.A, 72 pp.

Anonymus, 1978. Sand dune stabilization (1952/1976), Annotated bibliography, Commonwealth Bur. Soils, Harpenden SB, 12 pp.

Armbrust, D.V., et al., 1964. Effects of ridges on erosion of soil by wind. Soil Sci. Soc. Amer., 28(4)

Armbrust, D.V., and Dickerson J.D., 1971. Temporary wind erosion control: cost and effectiveness of 34 commercial materials. Soil and Water Conserv., 26(4): 154–157.

Armbrust, D.V., 1977. A review of mulches to control wind erosion. Trans. Amer. Soc. Agric. Eng., 20(5): 904–905/ 910.

Azizov, A., 1977. Influence of soil moisture on the resistance of soil to wind erosion. Soviet Soil Sci., 9(1): 102–105.

Babayev, A.G., and Cherednichenko, G., 1972. Method for determining the degree of deflation in shifting sands. Problemy Osvoenija Pustyn, 5: 41–45 (in russian).

Bakhit, A., and Ibrahim, F., 1982. Geomorphological aspects of the process of desertification in western Sudan. Geojournal 6(1): 19–24.

Baltaxe, R., 1967. Airflow patterns in the lee of model wind breaks. Archiv f. Meteor., Geophys. u. Bioklimatol, Allgem.u.biol.Klimatologie, B15(3), 26 pp.

Bates, C.G., 1920. Wind break as a farm asset. U.S. Dept. Agric., Farmers Bull., 1405.

Beard, J.S., 1969. Drought effects in the Gibson Desert. J. Roy. Soc. W. Austral., 51(2): 39–50.

Bergsma, E., 1977. Aerial photo-interpretation for accelerated wind erosion study and conservation planning. ITC lecture notes (unpubl.), 18 pp.

Berry, L., and Ford, R.B., 1977. Recommendations for a system to monitor critical indicators in areas prone to desertification. Publ. Clark Univ., Worcester (Mass.), 66 pp.

Bharara, L.P., 1980. Social aspects of drought perception in arid zone of Rajasthan. Annals of Arid Zone, 19(2): 154–167.

Bhimaya, C.P., 1974. Sand dune fixation. F.A.O. Rep. TA 3252, Rome.

Bhimaya, C.P., et al., 1962. Sand dune rehabilitation in western Rajasthan. Proc. 5th World Forestry Congress, 1960, Seattle: 358–363.

Bhimaya, C.P., and Kaul, R.N., 1960. Some afforestation problems and research needs in relation to erosion control in arid and semi-arid parts of Rajasthan. Indian For., 96: 453–468.

Bhimaya, C.P., and Chowdhary, M.D., 1961. Plantation of wind breaks in the Central Mechanized Farm, Suratgarh, an appraisal of techniques and results. Indian For., 87: 354–867.

Bielefeldt, H., and Taübner, K., 1966. Wirtschaftliche Verfahren zur Befestigung von Böschungsflächen. Bitumen, 4: 116–118.

Bisal, F., and Hsieh, J., 1966. Influence of moisture on erodibility of soil by wind. Soil Science, 102: 143–146.

Borchert, J.R., 1971. The Dust Bowl in the 1970s. Ann. Ass. Amer. Geogr., 61: 1–22.

Borsy, Z., 1975. Recent results ot wind erosion studies in hungarian blown sand areas. Földrajzi Ertesito, 23(2): 227–236.

Breyer, J., 1979. The application of remote sensing techniques for geomorphological terrain classification and terrain feature dynamics mapping in northern Botswana. M.Sc. thesis, ITC, Enschede, the Netherlands, 232 pp.

Chepil, W.S., 1944. Utilisation of crop residues for wind erosion control. Sci. Agric., 24(7)

Chepil, W.S., 1945. Dynamics of wind erosion: II. Initiation of soil movement, Soil Science, 60: 397–411.

Chepil, W.S., 1955. Factors that influence clod structure and erodibility of soil by wind: IV, Sand, Silt and Clay. Soil Science, 80: 155–162.

Chepil, W.S., 1956. Conversion of relative field erodibility to annual soil loss by wind. Proc. Soil Sci. Soc. Amer., 20: 143–145.

Chepil, W.S., 1956. Influence of moisture on erodibility of soil by wind. Proc. Soil Sci. Soc. Amer., 20: 288–292.

Chepil, W.S., 1957. Width of field strips to control wind erosion. Kansas Agric. Exper. Station, Techn. Bull., 92, 16 pp.

Chepil, W.S., et al., 1960. Anchoring vegetative mulches. Agric. Eng., 41(11): 754–755/759.

Chepil, W.S., et al., 1963. Mulches for wind and water erosion control. USDA/ARS Publ. 41–84, 23 pp.

Chepil, W.S., and Woodruff, N.P., 1963. The physics of wind erosion and its control. Advances in Agronomy, 15: 211–302.

Chepil, W.S., et al., Wind erodibility of knolly and level terrain. J. Soil and Water Conserv., 19(5): 179–181.

Chzao Sin-Chaun, 1953. Sand fixation and afforestation in Jullin. Scientia sinica, 9(2) (in chinese)

Chzao Sin-Chaun, 1959. Investigation and protection of railways in sand deserts. Peking, (in chinese)

Compo, 1973. Agrosil Anwendung. Compo Techn. Info.

D.H.V. Consulting Engineers, 1978. Botswana, country wide animal and range assessment project. Methodology Report B, 73 pp. Amersfoort, the Netherlands (in coop. with ITC, Enschede)

Dougrameji, J., and Kaul, R.N., 1972. Sand dune reclamation in Iraq–present status and future prospects. Annual of Arid Zone, 11(3–4): 133–144.

Downes, J.D., et al., 1977. Influence of wind erosion on growing plants. Trans. Amer. Soc. Agric. Eng., 20(5): 885–889.

Eimern, J. van, et al., 1964. Wind breaks and shelterbelts. WMO Techn. Notes, 59, 181 pp.

Fanale, R., et al.,1978. Remote sensing methods for the study and monitoring of desertification. Remote Sensing Div., SWCRC, Nat Park Service, Albuquerque (New Mexico) USA, 20 pp.

F.A.O., 1960. Soil erosion by wind and measures for its control on agricultural land. Agric. Dev. Paper, 71

Faure, H., and Gae, J.Y., 1981. Will the Sahelian drought end in 1985? Nature, 291, 5815

Fenster, C.R., and Wicks, G.A., 1977. Minimum tillage fallow systems for reducing wind erosion. Trans. Amer. Soc. Agric. Eng., 20(5): 906–910.

Ferber, A.E., 1969. Windbreaks for conservation. USDA Info. Bull., 339(334), USDA/SCS

Fryrear, D.W., 1963. Annual crops as wind barriers. Trans. Amer. Soc. Agric. Eng., 6: 340–343/352.

Fryrear, D.W., and Lyles, L., Wind erosion research accomplishments and needs. Trans. Amer. Soc. Agric. Eng., 20 (5): 916–918.

Gillette, D.A., 1977. Fine particulate emissions due to wind erosion. Trans. Amer. Soc. Agric. Eng., 20(5): 890–897.

Gorke, K., and Hulsman, J., 1971. Soil stabilization. Bull. Int. Soc. Soil. Sci., 38.

Goudie, A.S., 1978. Duststorms and their geomorphological implications. J. Arid. Env., 1(4), 291–311.

Haas, J.A., and Steers, J.A., 1964. An aid to stabilization of sand dunes: experiments at Scolt Head Isl., Geogr. J., 130 (2): 265–267.

Hagedorn, H., et al., 1977. Dune stabilization. A survey of literature on dune formation and dune stabilization. Deutsche Ges. Techn. Zusammenarbeit, Eschborn B.R.D.,

193 pp.
Hagen, L.J., and Skidmore, E.L., 1971. Turbulent velocity fluctuations and vertical flow as affected by windbreak porosity. Trans. Amer. Soc. Agric. Eng., 14(4): 634–637.
Hagen, L.J., et al., 1972. Designing narrow strip barrier systems to control wind erosion. J. Soil and Water Conserv. 27(6): 269–272.
Hagen, L.J., 1976. Wind break design for optimum wind erosion control. Proc. Symp. shelterbelts on the Great Plains, Denver: 31–36.
Hagen, L.J., and Skidmore, E.L., 1977. Wind erosion and visibility problems. Trans. Amer. Soc. Agric. Eng., 20(5): 898–903.
Harpaz, Y., et al., 1965. Effects of a synthetic rubber crust mulch on the emergence and early growth of three grasses on sand dunes. Israel J. Agric. Res., 15(3): 149–153.
Hayes, W.A., 1972. Designing wind erosion control systems in the Midwest Region. RTSC. Agron. Techn. Note LI-9, USDA/SCS
Heine, K., 1970. Einige Bemerkungen zu den Liefergebieten und Sedimentationsräumen der Lösse im Raum Marburg/Lahn auf Grund tonmineralogischer Untersuchungen. Erdkunde, 24(3): 180–194.
Hellden, U., and Stern, M., 1980. Monitoring land degradation in southern Tunisia. A test of Landsat data and digital data. Publ. Lund Univ. Fys. Geogr. Inst, 42 pp.
Horikawa, K., and Shen, H.W., 1960. Sand movement by wind action (on the characteristics of sand traps). Beach Erosion Board, Corps of Eng., Techn. Memo 119, 51 pp.
Hurault, J., 1966. Etude photo-aérienne de la tendance à la rémobilisation des sables éoliens au nord du lac Tchad (régions de Mao et Bol). Trans. 2nd Symp. Comm VII, I.S.P., Paris, IV, 7 pp.
Ibrahim, K.M., 1969. The control of drifting sands in the north coastal region of U.A.R. Pak. J. Forestry, 19(4): 456–471.
Ibrahim, F.N., 1978. Antrhopogenic causes of desertification in western Sudan. Geojournal 2(3): 243–254.
Ibrahim, F.N., 1980. Desertification in Nord Darfur. Dr. thesis Univ. Hamburg. Hamburger Geogr. Stud., 35, 175 pp.
ISRC, 1971. Soil stabilisation, the Unisol 91 stabilisation technique. Int. Synthetic Rubber Comp. Ltd., Southampton
Ivanov, A.P., 1975. Complex methods for the protection of highways from sand drifts in shifting sand conditions (in russian). Problemy Osvoejija Pustyn, 6: 87–89.
Jensen, M., 1954. Shelter effect: investigations into aerodynamics of shelter and its effects on climate and crops. Danish Techn. Press, Copenhagen, 264 pp.
Joshi, K.L., 1969. Problems of desert agriculture in Punjab, Haryana and West Rajasthan. Science and Culture, 35(10): 551–555.
Kaul, R.N. (Editor), 1970. Afforestation in arid zones. Junk N.V., Den Haag, 435 pp.
Kerr, R., and Nigra, J.O., 1952. Eolian sand control. Bull. Amer. Ass. Petrol. Geol., 36(8): 1541–1573.
Khodzkhayev, A.A., 1947. Borbas peschanymi zanosami na zheleznykh dorogakh (The protection of railways from sand drifts). M. Transzheldorizdat
Kimberlin, L.W., et al., 1977. The potential wind erosion problem in the United States. Trans. Amer. Soc. Agric. Eng., 20(5): 873–879.
Knottnerus, D.J.C., and Peerlkamp, P.K., 1972. Wind erosion. What can be done about it? (in dutch). Bedrijfsontwikkeling, 3(2): 175–179.
Kuhlmann, H., 1958. Quantitative measurements of aeolian sand transport. Geografisk Tidsskrift, 57: 51–72.
Lavauden, L., 1928. Reboisement et la fixation des dunes de Bizerte. Revue Eaux et Forêts, 12: 351–361.
Layer, C.G., 1936. Reclamation of drift sands. Farming in S. Africa, 11: 53–57.
Lehotsky, 1972. Sand dune fixation in Michigan thirty years later. J. Forestry, 70(3): 155–160.
Leontyev, A.A., 1962. Pesvcanye pustyni Svednej Azii i ih lesomeliorativnoe osvoenie (The sandy deserts of Soviet Central Asia and their reclamation by forests). Gosizdat U.S.S.R., Tashkent
Letey, J., et al., 1963. Wind erosion control with chemical sprays. Calif. Agr., 17(10): 4–5.
Lettau, L., and Lettau, H., 1969. Bulk transport of sand by the barchans of the Pampa de la Joya in southern Peru. Ztschr. f. Geom., 13(2): 182–195.
Liao Yu-Pin, 1964. Le contrôle des deserts en Chine. Acta Geographica, 51: 18–20.
Li Min-Ghan, et al., 1960. About the protection of the railway Baotou-Lanjour from sand drift. Lin-e-zsi-kan, 3, Peking.
Loubser, J.H., 1958. Fight against shifting dunes in the Kalahari sandveld. Farming in S. Afr., 34(2): 22–23.
Lyles, L., et al., 1969. Spray-on adhesives for temporary wind erosion control. J. Soil and Water Conserv., 24(5): 190–193.
Lyles, L., et al., 1974. How aerodynamic roughness elements control sand movement. Trans. Amer. Soc. Agric. Eng., 17(1): 134–139.
Lyles, L., et al., 1974. Commercial soil stabilizers for temporary wind erosion control. Trans. Amer. Soc. Agric. Eng., 17(6): 1015–1019.
Lyles, L., 1975. Possible effects of wind erosion on soil productivity. J. Soil and Water Conserv., 30(6): 279–283.
Lyles, L., and Allison, B.E., 1976. Wind erosion: the protective role of simulated standing stubble. Trans. Amer. Soc. Agric. Eng., 19(1): 61–64.
Lyles, L., 1977. Wind erosion: Process and effect on soil productivity. Trans. Amer. Soc. Agric. Eng., 20(5): 880–884.
Mabbutt, J.A., and Berkowicz, S.M., 1981. The threatened dry lands. Regional and systematic studies on desertification. Proc. Pre-Congress Symp. 24th IGU Congress, Japan, 1980. Publ. School Geogr., Univ. N.S. Wales, Kensington, Austral., 153 pp.
Marschall, J.K., 1970. Assessing the protective role of shrub-dominated rangeland vegetation against soil erosion by wind. Proc. IXth Int. Grassland Congr.: 19–23.
McCalla, T.M., and Army, T.J., 1961. Stubble mulch farming Advances in Agronomy, 13: 126–194.
Meckelein, W., 1973. Climatic-geomorphological zones and land utilisation in the coastal deserts of the north Sahara. In: D.H.V. Amiran and A.W. Wilson (Editors), Coastal deserts, their natural and human environments. Univ. Arizona Press. Tucson: 159–165.
Meckelein, W., 1980. Desertification in extremely arid environments. Stuttgarter Geogr. Stud., 95, 203 pp.
Mensching, H.G., and Ibrahim, F.N., 1976. Desertifikation im Zentral Tunesischen Steppengebiet. Nachr. Akad. Wiss. Göttingen, II, Math. Phys. Kl., 8, Göttingen
Mensching, H.G., and Ibrahim, F.N.,1981. Mapping desertification in the Sudan, a methodological approach. IGU Working Group Desertification, Fujinomija, Japan, August 1980. Univ. N.S. Wales Publ.: 19–28.
Messines, J., 1952. Sand dune fixation and afforestation in Lybia. Unasylva, 6(2): 50–58.
Metro, A.E., 1953. Techniques de fixation et de mise en valeur des dunes du littoral atlantique Marocain. Ann. Recherche forestrière au Maroc: 3–56.
Michel, P., 1969. Les bassins des fleuves Sénégal et Gambie, étude géomorphologique. Thèse Strasbourg, 1170 pp.

Mulder, W., and Plutschouw, W., 1975. Antropogene duinen in het deltagebied. Rep. R.W.S. Milieudienst, Middelburg
Niles, J.S., 1961. A universal equation for measuring wind erosion. ARS Spec. Report: 22-69
Nistri, P.F., 1972. L'erosione eolica nei paesi caldo aridi – Rivista Agric. Subtrop. e Tropicale, 66(4–6; 7–9): 111–121.
Orlova, M.A., 1940. Peski Astrakhanskoi polupustini, methody ikh ukrepleniya i khozyaistvennogo ispolzvaniya (Sands of the Astrakhan semi-desert, methods of their fixation and economic utilisation). Goslostekhizdat
Otterman, J., 1977. Monitoring surface albedo change with Landsat. Geoph. Res. Letters, 5(4), 441 pp.
Otterman, J., 1978. Effects on multispectral radiometry of reflection from adjacent terrain and subsequent atmospheric scattering over the object pixel and their elimination. Proc. 44th Annual Meeting Amer. Soc. Photogramm.: 363–372.
Orlova, M.A., 1979. Wind transport of salts. Soviet Soil Sci., 11(2): 196–206.
Orlova, M.A., 1979. Role of eolian factor in salt regime of desert solonchaks (in russian). Problemy osvoclya Pustyn. A.N.T. I.S.S.R. Akad. Nauk Turkmeniski
Paylore, P. (Editor), 1976. Desertification. A world bibliography. ISM Conf., Moscow
Pearson, R.L., et al., 1976. Spectral mapping of shortgrass prairie biomass. Ph. Eng. R.S. 42: 317–323.
Peters, J., 1964. Stabilization of sand dunes at Vandenberg Air Force Base. J. Soil. Mech. and Foundations Div., Amer. Soc. Civ. Eng., 90: 97–106.
Petrov, M.P., 1954. Podvizhnye peski pustin Soyuza SSR i borbas mimi. (Drifting sands of the U.S.S.R. deserts and their control). M. Geografgiz
Petrov, M.P., 1957. La phytoamélioration des déserts de sable en USSR. Annales de Géogr., 357(5)
Petrov, M.P., 1971. Sand stabilization methods in arid lands. Protection of agricultural and settlement areas. In: McGinnies, W.G., et al. (Editors), Food, Fiber and Arid Lands, Tucson: 355–368.
Pias, J., 1957. Etude pédologique des cuvettes lacustres de la bordure du Lac Chad. Publ. ORSTOM 57–13
Plan of Action, 1977. Plan of action to combat desertification. U.N. Conf. on Desertification, A/Conf. 74/L. 36, Nairobi.
Podryadov, N.A., 1958. Borba s peschanymi zanosami na zheleznykh dorogakh (The protection of railways from sand drifts). M. Transzheldorizdat,
Pouquet, J., 1969. Possibilities for remote detection of water in arid and sub arid lands derived from satellite measurements in the atmospheric window 3.5–4.2 μm. Proc. Conf. Arid Lands in a changing world, Tucson (Ariz.), 9 pp.
Prakash, M., and Chowdhary, M., 1957. Reclamation of sand dunes in Rajasthan. Indian Forester, 83(8): 492–496.
Prakash, M., 1959. Importance of shelterbelts to check wind erosion. J. Soil and Water Conserv.,7: 61–66.
Prego, A.J., et al., 1971. Stabilization of sand dunes in the semi-arid Argentine Pampas. In: McGinnies, W.G., et al. (Editors), Food, Fiber and Arid Lands, Tucson: 369–392.
Raheja, P.C., 1963. Shelterbelts in arid climates and special techniques for tree planting. Annals of Arid Zone, 2(1): 1–13.
Rapp, A., et al., 1976. Can desert encroachment be stopped. Ecol. Bull., Stockholm, 24, 21 pp.
Reifenberg, A., 1950. Man-made dune encroachment on Israel's coast. Proc. 4th Int. Soil. Sci. Congr., Amsterdam: 325–327.
Robinove, Ch. J., and Chavez, P.S.(Jr.),1978. Landsat albedo monitoring method for an arid region. Trans. A.A.A.S. Symp. Arid Region Plant Research, Texas. Nr. 8: 1–2.
Robinove, Ch.J., et al., 1981. Arid land monitoring using Landsat albedo difference images. Remote Sensing of Environment, 11(2): 133–156.
Saarinen, T.F., 1969. Perception of the drought hazard on the great plains. Univ. Chicago, Dept. of Geography, Res. Paper, 106, 183 pp.
Saccardy, L., 1953. Fixation des dunes sur le littoral méditerranéen de l'Afrique du Nord. Colloques Internat. CNRS, 35: 277–286.
Saini, B.S., and Borrack, G.C., 1968. Control of wind blown sand and dust by building design and town planning in the arid zone. Proc. 3rd Austral. Building Res. Conf., Melbourne: 221–224.
Sanford, S., 1977. Dealing with drought and lifestock in Botswana. Overseas Dev.Inst. London Publ., 114 pp.
Sastri, A.S., and Ramakrishna, Y.S., 1980. A modified scheme of drought classification applicable to the arid zone of western Rajasthan. Annals of Arid Zone, 19(2): 65–72.
Schab, G.O., et al., 1981. Soil and water engineering, 3rd Ed., Chapter VII, Wind erosion control, Wiley, New York, 525 pp.
Shell, 1969. Sandfix. Shell Bitumen Review, 3 pp.
Shears, P., 1980. Drought relief and agricultural rehabilitation. Disasters, 4(4): 469–474.
Siddoway, F.H., et al., 1965. Effect of kind, amount and placement of residue on wind erosion control. Trans. A.G. A.E.: 327–332.
Simmons, S.R., and Dotzenko, A.D., 1974. Proposed indices for estimating the inherent wind erodibility of soils. J. Soil and Water Conserv., 29(6): 275–276.
Simonett, D.S., 1975. Examples and potentials for monitoring dune sands and dune migration. F.A.O. Publ., Misc/75/11, Rome
Skidmore, E.L., and Woodruff, N.P., 1968. Wind erosion forces in the USA and their use in predicting soil loss. Agric. Handbook, 3463, 42 pp.
Skidmore, E.L., and Hagen, L.J., 1977. Reducing wind erosion with barriers. Trans. Amer. Soc. Agric. Eng., 20(5): 911–915.
Smalley, I.J., 1977. Cohesion of soil particles and the intrinsic resistance of simple soil systems to wind erosion. J. Soil Sci., 21: 154–161.
Stevens, J.H., 1974. Stabilization of aeolian sands in Saudi Arabia's Al Hasa oasis. J. Soil and Water Conserv., 29(3): 129–132.
Task Committee, 1965. Sediment transportation mechanics: wind erosion and transportation. Progress Report, Proc. Amer. Soc. Civ. Eng., J. Hydr. Div., 91, HY2: 267–287.
Tricart, J., et al., 1976. Apports de la télédétection à l'étude des régions arides et subarides. Journées d'étude C.N.L.A.-T., 67 pp.
Tsuriell, D.E., 1974. Sand dune stabilization in Israel. Report FAO, Danish Funds in Trust. FAO/DEN/TF 114, Rome.
Tucker, C.J., 1977. Resolution of grass canopy biomass classes. Ph. Eng. R.S., 43: 1059–1067.
Tucker, C.J., 1977. Spectral estimation of grass canopy variables. R.S. of Environm., 6: 11–26.
Tucker, C.J., et al., 1973. Measurement of the combined effect of biomass, chlorophyll and leaf water on canopy spectral reflectance of the short grass prairie. Proc. 2nd Ann. R.S. Earth Res. Conf., Univ. Tennessee: 1–34.
U.N., 1977. Desertification, its causes and consequences. Pergamom Press, Oxford, 448 pp.
U.S.D.A., 1961. A universal equation for measuring wind erosion. Agric. Res. Service Spec. Report: 22–69.
U.S.D.A., 1968. Wind erosion factors in the USA and their use in predicting soil loss. Agric. Handbook, 346 pp.

Verstappen, H.Th., 1957. Short note on the dunes near Parangtritis (Java). Tijdschr. Kon. Ned. Aardrijksk. Gen., 74: 1–6.
Verstappen, H.Th., 1972. On dune types, families and sequences in areas of unidirectional winds. Göttinger Geogr. Abh., 60, Hans Poser Festschrift: 341–354.
Verstappen, H.Th., 1977. Remote sensing in geomorphology. Elseviers Sci. Publ., Amsterdam, 214 pp.
Verstappen, H.Th., 1979. Drought susceptibility survey and the concept of monitoring landscape ecology. Proc. Symp. Drought in Botswana, Gaborone 1978. Botswana Soc., 10: 75–81.
Verstappen, H.Th., 1979. Les levés de sensibilité à la sècheresse et la notion de surveillance de l'écologie paysagique. Trav. Inst. Géogr. Reims, 39/40: 27–36.
Verstappen, H.Th., and Zuidam, R.A. van, 1970. Orbital photographs and the earth sciences: a geomorphological example from the Sahara. Geoforum, 2: 33–47.
Verstappen, H.Th., and Cooke, H.J. (Editors), 1981. A drought susceptibility pilot survey in northern Botswana. Final report project 527/6.3 NUFFIC/ITC, 237 pp.
Walker, R.H. (Editor), 1979. The management of semi-arid ecosystems. Elseviers Sci. Publ., Amsterdam, 398 pp.
Walls, J., 1980. Land, man and sand: desertification and its solution. MacMillan, New York, 335 pp.
Westhuyzen, J.J. van der, 1957. Combating sand dunes at Port Edward. Farming in S. Afr., 33(7): 37–39.
Weymouth, N., 1967. Soil stabilization. Rubber Plast. Age, 48(3): 253–255.
Williams, M.A.J., 1968. A dune catena on the clay pans of the west-central Gezira, Sudan. J. Soil Sci., 19: 367–370.
Woodhouse, W.W., 1967. Dune stabilization with vegetation on the Outer Banks of North Carolina, CERC Techn. Memo, 22, 45 pp.
Woodruff, N.P., 1956. The spacing interval for supplemental shelterbelts. J. Forestry, 54: 115–122.
Woodruff, N.P., and Chepil, W.S., 1956. Implements for wind erosion control Agric. Eng., 37: 751–754/758.
Woodruff, N.P., and Siddoway, F.H., 1965. A wind erosion equation. Proc. Soil Sci. Soc. Amer., 29: 602–608.
Woodruff, N.P., et al., 1972. How to control wind erosion. U.S.D.A. Info. Bull., 354: 1–22.
Woodruff, N.P., et al., 1974. Using cattle feedlot manure to control wind erosion. J. Soil and Water Conserv., 29(3): 127–128.
Zingg, A.W., 1954. The wind erosion problem in the Great Plains. Trans. Am. Geoph. Un., 35: 252–8.

SLOPE STABILITY AND EROSION SURVEYS

15.1 The Geomorphological Context

Slope processes of all kinds and intensities are an integral part of dynamic geomorphology and it is thus essential that all information related to them should be duly emphasized in analytical geomorphological surveying, as has been indicated in sub-section 15.3. Phenomena of accelerated erosion and/or imminent major mass movements are of particular importance in this respect. This means that not only the actual erosion processes and features are mapped, but also the data concerning the numerous factors which govern or influence them, such as slope gradient and form, lithology, unconsolidated materials and soils, etc. Analytical geomorphological surveying is thus a natural starting point for erosion surveys.

Usually, however, more detailed and diversified erosion-related information is required for erosion surveys than can be found on the average analytical geomorphological map of adequate scale. For example, exact data on the type, thickness and profile characteristics of unconsolidated materials and soils are necessary. Also information concerning type and percentage cover of vegetation, normally not included in analytical geomorphological maps,will often have to be indicated on erosion maps. Finally, there may be information included in the analytical geomorphological map but deserving more emphasis and thus a higher position in the hierarchical classification of the legend. Slope steepness, for example, may be visualized on the geomorphological map only by the spacing of contour lines whereas in erosion maps it may merit the use of the best cartographical means of expression: coloured area symbols. Therefore, for purposes of erosion survey,a specific, morpho-conservation map must be derived from the analytical geomorphological map with the additional data obtained from the erosion survey (see section 11).

No attempt will be made in this section to give a uniform legend for erosion surveys because several types of erosion surveys exist (as will be explained farther on). The map contents vary with the aim of the survey, the type of survey selected, the area/problem at hand and the mapping scale required (Monjuvent, 1973; Prochal, 1972; Rauschkolb, 1972; Rathjens, 1979; Stebelsky, 1974; Stehlik, 1975; St. Onge, 1974; Toy, 1977; Werner, 1966; Williams, 1978).

Evidently, linear erosion processes and their forms (ravines, gullies, rills) and also diffuse processes,such as surface wash and sheet erosion, should be emphasized above all. Special processes such as piping also merit full attention. Mass movements ranging from creep or solifluction to major slumping or catastrophic landsliding cannot be neglected, either. It is a fairly common situation that both groups of processes occur simultaneously in an area or are potentially present. One may then be in the delicate position in which combating erosion potentially increases the hazard of mass movements or vice versa. It is therefore essential to always evaluate all processes that occur or may occur on a slope, in other words, the dynamics and state of equilibrium on a slope must be properly appraised before deciding on matters of conservation and stabilization. Mass movements may be a problem especially where the geological strata dip towards the valley (fig. 15.1). Improved internal drainage may stop or reduce the slope instability (fig. 15.2).

Some authors prefer to speak of 'land erosion' rather than 'soil erosion' (Garland, 1979).

For many, the term slope (in)stability is strictly related to mass movements. From a geomorphological point of view it is quite defendable, however, to describe the general state of the slope regardless of whether its destruction is due to mass movements or processes of erosion by water (v. Asch, 1980). One may then consider erosion surveys to be a special type of slope stability surveys. However, since the matter of slope instability due to mass movements has already been elaborated upon in section 8 (on engineering) and also touched in section 13 (on river basin development in the context of flood susceptibility surveys) this section is mainly devoted to processes of erosion by water. Wind erosion is a distinct type of erosion survey which has been included in section 14 because it is considered more appropriate to discuss it in the context of drought susceptibility and desertification. Another reason for not incorporating wind erosion in this section is that it is not related to slopes and is particularly effective in plain lands.

Erosion surveys (in the sense of erosion by water) are of special importance in four major fields of application, namely: engineering, environmental planning, river basin development and agriculture. The development of deeper gullies and ravines by vertical cutting or headward erosion is of interest to the engineer because it may affect the siting, construction or maintenance of engineering works and poses problems which are distinct from those provoked by mass movements. The same applies to the environmental planner to whom it is significant whether erosion by water or mass movements has to be dealt with. Hydrologists,on the contrary, are above all interested in the quantity and type of material that reaches the river and is transported as bed load or in suspension. This ultimately affects the river in various ways or causes siltation of reservoir lakes. Except when slopes surrounding the latter are concerned,

Instable—main discontinuity dips out from slope, forming potential slip surface

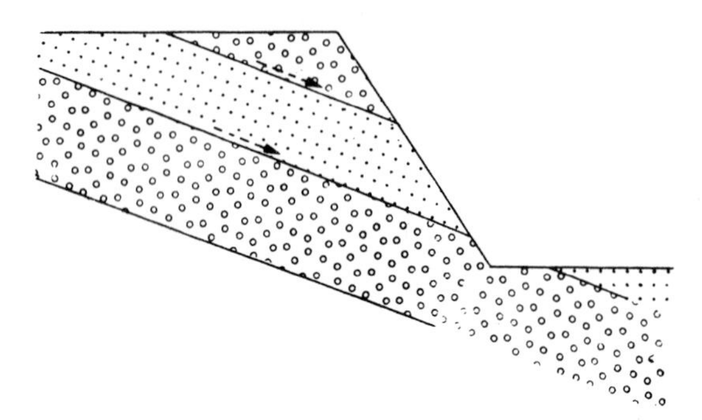

Stable—main discontinuity dips into slope

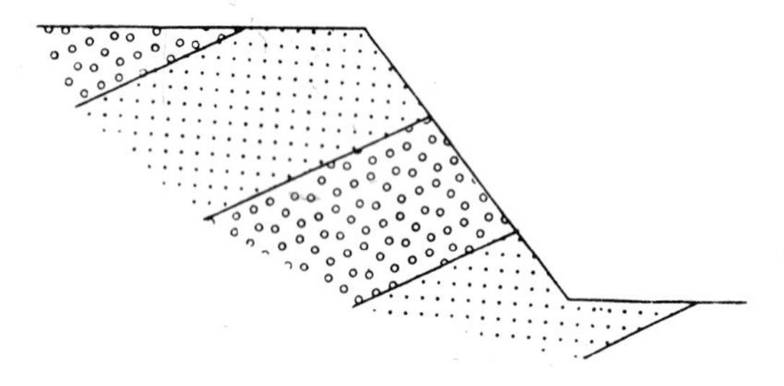

Fig. 15.1 Dip angle and slope instability.

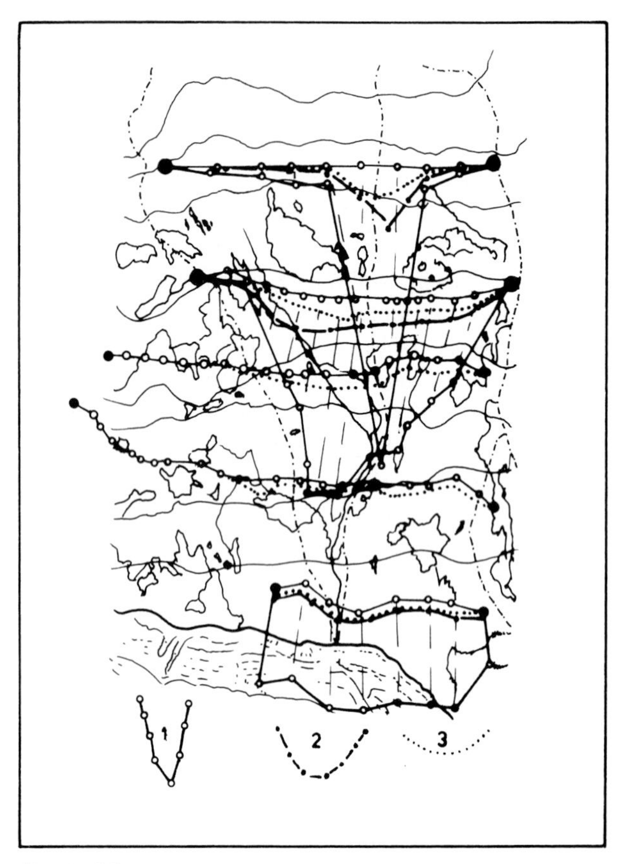

Fig. 15.2 Example of decreasing slope movements following stabilisation measures. 1: displacement along transects in 1974/1975; same 1975/1976; 1976/1977.

it is rather irrelevant to him whether mass movement or erosion by water is the producing agent of the material. The issue only becomes important if a reduction of the quantity of transported material is aimed at.

Since accelerated erosion by water is a serious and widespread threat in many agricultural fields, soil erosion studies are,with justification, a subject of study of long standing in soil science (FAO, 1977; v.d. Kevic, 1976; Kirkby and Morgan, 1980). The removal of the fertile top soil may drastically reduce productivity. Gullies and ravines, however, often reach considerably greater dimensions and cut not only through the entire soil profile but also remove the underlying unconsolidated materials and even soft or disintegrated bed rock. Likewise, mass movements are, in part, shallow and then mainly affecting the soil profile and in other cases deep-seated phenomena. Indications for creep, the formation of terracettes, colluvial processes leading, e.g., to agricultural or plow terraces,should receive due attention as well as landslide scars and deposits.

It is evident from the aforesaid that erosion surveys should not be confined to soil erosion by water only, but should take all processes into consideration including matters such as lateral sapping by rivers which may, e.g., initiate gullying in concave meander curves. They should be effectuated in a geomorphological context emphasizing also various terrain parameters (Carson and Kirkby, 1977; Chorley, 1978).

15.2 Terrain Parameters and Erosion Susceptibility

One cannot expect to attain a full understanding of the present situation of erosion (including its development in the past and in the future) if one restricts oneself to the study and mapping of the various kinds of geomorphological processes involved, although the outcome of such exercise may be useful. Fig. 15.3 gives an interesting example of a detailed map of slope processes in part of the Chilean Andes (Araya, 1964), in which processes such as diffuse run off, creep and terracette formation are indicated and a complete picture of the active dynamics of the area is given together with the deposits, breaks of slope, etc. Clearly these kinds of studies are a valid contribution to erosion studies, but even if one added further details about matters,such as the intensities of the various processes, including data about processes of the past, no complete picture would be obtained for lack of information on relief, lithology, soils, etc. Therefore, these data

Fig. 15.3 Slope processes in part of the Estanquez Valley, Chile, 1:25,000 (Araya, 1964). Processes mapped comprise of: diffuse runoff, creep, terracettes, badlands, sheet-solifluction, cup-solifluction, spoon-shaped slides, major slides, slumps. Deposits are classed as cone terraces, fans, torrent deposits, mud flow, etc. with indication of texture.

will have to be added in erosion surveys.

Slope Characteristics

Among the factors affecting processes of erosion by water, the morphological characteristics of the slopes concerned, such as gradient, length, profile, micro relief, configuration in plan and foreland,are of major importance (Stokking, 1972). On steep slopes the velocity of overland flow tends to be relatively high and the infiltration rate lower than on gentler slopes of the same material. This increases erodibility. Long slopes tend to build up large quantities of overland flow and consequently erosion, particularly near the foot of the slope, except when a high infiltration rate compensates for it. The profile characteristics of the slope also have to be considered: on convex slopes erosion generally increases downslope with increasing steepness, whereas on concave slopes the upper parts are usually the most susceptible to erosion. In the case of straight slopes the distribution of erosion is more complex: it may occur particularly in the - often slightly convex - top parts and/or in the lower parts where the build-up of overland flow is most outspoken. However, other factors, such as surface roughness, etc.,may also be dominant governing factors. Breaks of slope are of particular importance in evaluating the erosion susceptibility of a certain part of a slope. It is therefore often advisable to study the gradient of that part as well as the parts situated immediately upslope, because changes in gradient are often more indicative than the absolute general steepness of the slope as a whole.

When a slope has concave contours, a concentration of overland flow will occur which may easily lead to the formation of a gully or ravine in the depressed central portion. Where a convex configuration occurs, there will be a tendency to a diffusion of the water which makes the erosion susceptibility less. Slopes with straight contours take an intermediate position in this respect; the length of the slope in this case is often the leading factor in the distribution of gullies and ravines because of greater build-up of overland flow on long slopes.

If a flat or gently sloping foreland of adequate dimensions occurs in front of the eroding slope, the detached and subsequently downslope transported materials may be directly redeposited at its foot in the form of fans or gentle colluvial slopes.

New fields may be created or old ones fertilized, provided that the material deposited is not too coarse or deprived of nutritive minerals to reduce the productivity. If a river is situated in the immediate vicinity of the foot of the eroding slope, the situation is entirely different: the eroded material will then straight away contribute to the bedload and the foot of the slope may also be the scene of active gullying or ravination as a result of undercutting by lateral sapping or vertical cutting by the river.

Rock and Soil Type

The lithology of the eroding slopes is also a major factor which governs the distributional pattern of erosion by water. In the Chama Valley in the Venezuelan Andes, e.g., the Mesozoic,La Quinta, as well as the Cambrium, and Mucuchachi formations are by far the most vulnerable (see sub-section 15.3). It is therefore necessary in erosion studies to divide the terrain into litho-morphological units and to specify the susceptibility to the various processes of slope erosion of each of them.

Another important aspect in this connection is weathering. In deeply weathered parent materials, e.g., under humid tropical conditions or where mafic rocks occur, a deep layer of weathered material and soil is usually available which, if denuded or destabilized otherwise, may produce large quantities of debris. Soft clastic or weakly metamorphosed rocks may also produce considerable quantities of debris and are, generally speaking, more easily erodible than the mafic volcanic materials. Hard rocks, such as quartzites, usually have a thin cover of disintegrated materials and soils and, if eroded, will yield only small quantities of detached particles. Weathering depends, of course, on a number of factors, such as the mineral content of the materials concerned, the climatic conditions, the length of the weathering period, etc. In the Chama Valley, in the Venezuelan Andes, e.g., the intermediate river terraces are the ones least susceptible to erosion because they are already slightly compacted and have not yet been weathered to a considerable extent. The lower, younger terraces are composed of completely unconsolidated gravels and the older, higher ones are deeply weathered; both of them therefore are more vulnerable to erosion by water (sub-section 15.6).

Certain characteristics of the soil profile have a great effect on matters such as permeability, infiltration rate and storage capacity (Barnett et al., 1971; Dzhadan, et al., 1975; Fournier, 1972; Poelman, 1971). They are therefore important governing factors in the amount and type of overland flow and in the erosion so caused (fig. 15.4). Total depth, structural stability and porosity of the profile are of particular importance in this context. All related matters such as the texture, organic matter content, permeability of the subsoil, drainage conditions, salinity, etc., therefore, have to be given full attention. The susceptibility to surface sealing, resistance to splash erosion, depth of impervious layers are also vital in assessing the erosion hazard on slopes. Whereas the structural stability of the soil mainly depends on structural and textural characteristics of the soil profile, the infiltration rate is affected by soil-related factors such as porosity in combination with the meterological/climatological factors of the intensity and duration of rainfall. Together they determine the degree of saturation of the soil.

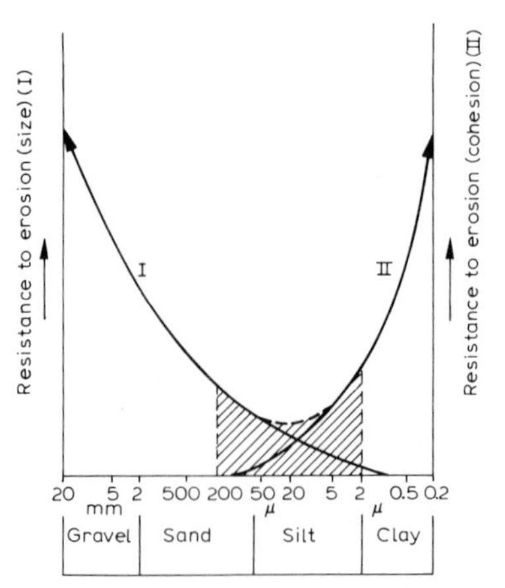

Fig. 15.4 The influence of particle size (1) and cohesion (11) on the erodability of soils. The zone of minimum resistance is indicated by shading.

Other Factors

The ultimately produced overland flow may be of various types and/or intensities and thus results in specific erosion processes. It may : 1) be entirely surficial and start directly at the top of the slope, 2) be delayed by absorption in swelling clays or infiltration in cracks, 3) operate at shallow depth, e.g., at plough depth, 4) start near the foot of the slope where a state of saturation has been reached, or 5) be absent altogether due to high infiltration rates (Kirkby and Chorley, 1967; Emmett, 1970). The vegetative cover of the slopes concerned may alter the situation of overland flow and erosion by water completely (O'Loughlin, 1974; Orlovskiy. 1975; Merseneau and Dyrness, 1972; Giardino, 1974). It, of course, intercepts the rain and tends to protect against the impact of the raindrops. Furthermore, it reduces the speed of the runoff and also contributes to the retention of the rain, thus delaying and decreasing peak runoff. It also extracts moisture from the soil by transpiration and thus facilitates infiltration and reduces overland flow in the early stages of the next rain. Agricultural practices likewise have an important effect on matters of overland flow and erosion by water (Black and Siddoway, 1971; Datiri, 1974; Garland, 1979; Quilty, 1973; Richardson, 1973; Rossi, 1979; Selby, 1972; Siemens and Oschwald, 1976; Sorby, 1974; Zapletin et al., 1969).

It is evident, therefore, that in surveys related to erosion by water all these factors should be taken into consideration and be mapped in one way or another as is elaborated upon in sub-sections 15.4 and 15.6. One should bear in mind, however, that the erosion of a certain part of the terrain is affected to a considerable degree by the situation upslope as well as downslope of it. It is therefore necessary to extend the survey over a larger zone than the area to be studied. One should also consider not only the aspect of detachment of soil material but also the transport and deposition aspects of accelerated erosion. Finally, distinction should be made between soil erosion affecting the producivity of the agricultural fields and deeper erosion which may be called form erosion. The latter affects matters such as communication lines and is thus of interest to planners and engineers. Hydrologists engaged in estimating the sediment delivery to the streams are mainly interested in form erosion as this produces the largest quantities of material. Although the distinction between soil and form erosion is not always significant, it is evident that where thin soils occur on hard rocks, soil erosion may have serious effects on agricultural productivity of the fields even though form erosion will be unimportant and the sediment delivery relatively low.

Examples

The aerial photographs of figs. 15.5 and 15.6 from Indonesia demonstrate the importance of lithomorphological units in the survey of erosion by water. The degree of erosion and the forms carved by it are quite distinct in the two units pictured, which are respectively volcanic tuffs of the Lake Toba Area, Sumatra, and silts and marls in the Palu Valley, Sulawesi. The extended stereopair of fig. 15.7 showing badlands in the Dakotas, USA, at scale 1:20,000 is even more elucidating. Three sub-horizontal beds can be distinguished, each having different photographic tones, dissection and slope form. Only minor patches of the thin dark-toned capping layer are left as the ravines have cut into the thicker underlying whitish layer and even into the lowest layer, which is again dark-toned. Advanced dissection and almost vertical slopes characterize the dark capping formation; deep dissection by a dense, intricate drainage pattern and fairly steep slopes occur in the whitish formation, whereas more subdued forms and less steep slopes have developed in the lower dark formation. The different forms, slope angles, etc., are clearly linked to lithology, the lower bed being more clayey than the silty loamy whitish formation. The rate of the erosion by water and the forms produced are thus associated with the litho-morphological situation.

15.3 The Slope Processes Involved

Slope processes form a major part of the morphodynamics to be studied in the context of any analytical geomorphological survey. They have therefore already been discussed in section 11. However, in an erosion survey there usually is a need for a considerably greater amount of detail as to the type, intensity and distributional patterns of these processes and they therefore merit some further consideration here (Arnoldus, 1974; Bell, 1968; Cambell, 1974; Ekern, 1953; Ellison, 1952; Imeson and Kwaad, 1980; Kinnell, 1974; Kronfelder-Krans, 1981; McIntyre, 1958; Mosley, 1974; Ologe, 1972; Yamamoto and Anderson, 1973; Zaborski, 1971).

The detection of accelerated erosion in many cases is an easy matter as abundant evidence is, unfortunately, available in many areas. The early phases of development, however, may escape attention and it is precisely in these phases that the situation can still be brought under control with comparatively simple means. Even when erosion processes are clearly visible on a slope, the diagnosis of the occurance of accelerated erosion may not be all that

Fig. 15.5 Aerial photograph, scale 1:10,000, showing severe gullying in volcanic tuffs near Lake Toba, Sumatra, Indonesia. Note steep valley sides and heads in poorly consolidated permeable material.

easy. In areas of certain poorly consolidated and non-resistant rocks, fast erosion may be due to natural causes and, in this case, the same kind of processes are acting and the same kind of erosion forms will be developed as those which result from accelerated erosion. The problem is aggrevated by the fact that in many parts of the world deforestation and expanding land use have quickened the processes of slope erosion for one or two centuries or even longer periods of time. As a result the natural rate of geological erosion cannot be established, thus depriving the erosion surveyor of a means of comparison.

Evidently in every area where man has diminished the vegetative cover of the slopes for agricultural or other purposes, the erosion has been accelerated to a certain degree. By practicing appropriate erosion control and careful farm management, one can only hope to keep the soil losses within acceptable limits. This is the aim of most conservation projects.

The intensity and role of the various processes involved in erosion by water vary with the three major phases of accelerated erosion that are distinguished by the author, namely the pre- or early erosion, soil erosion and form erosion phases. Splash caused by the impact of raindrops on the newly denuded slopes is a major element in the pre- or early phase, during which also an important change in type and increase in intensity of overland flow is produced. Although, as has been mentioned above, in this phase it is technically still rather easy to prevent lasting damage and

Fig. 15.6 Aerial photograph, scale 1:40,000 showing severely eroded area in SW Sulawesi (Celebes, Indonesia). Deforestation of the silty/marly deposits has led to major changes in landform and to economic and ecological deterioration. Note the abandoned lowland fields in the lower left corner.

to alleviate the situation, socio-economic conditions often prevent timely action.

In some cases ignorance, in others the pursuit of short-term profits may be the main reason for this, but more commonly perhaps the accelerated erosion is initiated in a period when, for one reason or another, the population has virtually no other option for survival than deliberately raping the land. A rare, extreme meteorological event may ultimately trigger the accelerated erosion once a critical situation has been created by man.

Fig. 15.7 Extended stereopair, scale 1:20,000 showing a badland area in the Dakotas, USA.

Sheet and rill erosion are dominant processes in the phase of soil erosion and lead to truncation of soil profiles and ultimately to the downslope transportation of all soil material, exposing the bare hard rock or unconsolidated overburden. Rills subsequently may become gullies. The phase of form erosion starts in fact when a number of sizeable gullies are formed and gradually develop into ravines and valleys. At the same time, of course, sheet and rill erosion are both still proceeding on the remaining but shrinking undissected interfluves.

Soil losses from agricultural lands are dominant in the phase of soil erosion and this is clearly the main field of interest of the soil scientist. In a survey on accelerated erosion it would be one-sided, however, to single out only this particular aspect. Considerably larger quantities of material may be removed during the phase of form erosion. One should thus study the whole process of erosion, transportation and deposition from a morphodynamic point of view. The stage of erosion reached by the gullies, ravines and valleys, and also their degree of activity, must then be carefully analysed. A certain degree of stabilization may be reached when, after a period of rapid expansion, the incision approaches its natural limitation, the divide line. Activity may also decrease as a result of conservation measures introduced by man.

The gross erosion of the watershed can be estimated by the survey and, provided that the deposition within the basin, e.g., at the foot of slopes, in the channel, etc., can be quantified, the sediment yield of the watershed can also be determined. These matters and also the sediment delivery ratio (the sediment yield expressed in % of gross erosion) are of particular interest for hydrologists and have therefore already been touched upon in sections 4 and 13 of this book, dealing with hydrological applications and flood susceptibility, respectively.

Splash or pluvial erosion is an effective process initiating the detachment of soil particles on bare slopes. Where the soil contains a considerable quantity of fines, such as in silty soils, these may fill the voids between the superficial particles. This sealing will decrease infiltration, increase overland flow and therefore favour erosion.

Overland flow is another effective process in the detachment of soil particles and it also governs the downslope transportation of the detached material towards the lines of concentrated run-off. Its velocity and thickness play a leading part in this connection. The type of overland flow is another important matter: it may be laminar, mixed laminar-turbulent, turbulent or subdivided (super turbulent) and may develop surge flow and rain wave trains. Its erosive force increases in this order. Where infiltration occurs, overland flow may be delayed, restricted to times and places where soil saturation occurs, or be absent, as has been mentioned in the previous sub-section.

Rain falling on dry soil, e.g., at the beginning of the rainy season, is of special importance. Loose, dry soil particles are readily washed away and the air between the grains may for a short time resist the infiltration of water. Sudden wetting of dry soils either by splach or by overland flow may cause the air inside the clods to be entrapped and subsequently forced out by the penetrating water. As a result the clod may explode and disintegrate. Infiltration of the run-off will decrease when long duration of the rain results in saturation of the soil. Therefore, climatological factors have to be taken into careful consideration.

It is difficult to separate sheet erosion from rill erosion and for practical purposes the two are often grouped under one heading. It operates by way of unconcentrated wash and braids and, although usually inconspicuous, is capable of carrying off large quantities of soil material because of the extensive surface areas over which it occurs. It is most effective where slopes have convex or straight contours and is a dominant process particularly under the savanna type of climate. As a result of micro relief (surface roughness) or patches of grass or other vegetation which cause concentration of flood, the depth and speed of the overland flow will increase and incision of rills will start. Rills may disappear seasonally due to plowing or a natural phenomenon such as frost work. Rills tend to develop in drainage systems by way of micro piracy and cross grading processes. The erosion may remain in this phase more or less continuously,without leading to the formation of gullies and ravines,where the combined effect of erosion susceptibility of the land, rainfall erositivity and farming practices is not too unfavourable. Slopes convex or straight in plan, of course, are most suitably shaped for this. An interesting nomogram has recently been developed by Boon and Savat (1981) to predict the chances for the development of rill erosion on the basis of unit-discharge, slope angle and median grain size. In the majority of river basins, sheet and rill erosion account for 70% or more of the gross erosion.

Quantification of the rate of sheet/rill erosion can be pursued in various ways:

1. estimation of the thickness of the truncated soil horizon (s). One may establish,for example, 4-7 erosion-depth classes
2. measuring the number and dimensions of rills formed since the obliteration of the rills of the previous season
3. measuring the turbidity of surface run-off and determining the quantity of material deposited in small reservoirs or upstream of gabions
4. applying the universal soil loss equation A = RKLSCP in which:

	values and variations	
A = average annual soil loss		
R = rainfall erosivity factor	20 - 600	30x
K = soil erodibility factor	0,05 - 0,7	35x
L = length of slope run-off factor		
S = steepness of slope run-off factor up to 22%	0,3 - 7,0	20x
C = cropping and management factor	0,4 - 8,0 (% or R-increments)	200x
P = supporting conservation practice factors	0,25 - 1,0	4x

(Wischmeier, 1958, 1976; Wischmeier et al., 1971; Wischmeier and Smith, 1962, 1978).

The vertical aerial photograph of fig. 15.8 shows accelerated erosion in some fields near Astoria, El Salvador, at a scale of 1:10,000. Sheet erosion is dominant and shows up as extensive, irregularly shaped, light-toned areas where crop growth is poor. Some rill and even gully erosion can also be seen and the deposition zones con-

Fig. 15.8 Vertical airphoto, scale 1:10,000 showing sheet erosion near Astoria in El Salvador. The process has effected particularly the irregularly shaped whitish areas.

nected with these are also clearly recognizeable. The erosion in these flat areas, which have a savanna climate, is concentrated in the period following the harvest when the land is barren and heavy showers occur. Geomorphologically the area is made up of two river terraces which are both under agricultural use and are separated by a scrub-covered terrace scarp which runs obliquely across the lower part of the stereopair. The slope of the terraces is faint and is approximately parallel to the irregular scarp. The latter is a form inherited from the past when the lower terrace still functioned as the flood plain and the river was sapping laterally. No signs of recent erosion can be seen at the scarp.

In most drainage basins gully erosion is quantitatively distinctly less important than sheet/rill erosion and generally produces less than 10% of the gross erosion. It is more easily recognized, however, and may at places have disastrous affects. In larger basins, of 250 sq. km or more, gully erosion is normally concentrated in a few relatively small parts of it, which together, however, may supply most of the sediment if they are in the phase of rapid development mentioned above. Gullies are usually characterised by an irregular long profile, collapsable sides and intermittent flow. In cross-section they may be V-shaped (particularly in sands and gravel), U-shaped (silts) or broader (clays). Their growth is often spasmodic and maximum growth rate is usually reached in the earlier stages of development. Even if stabilization has been reached, care has to be taken that no revival of the gullying activities is initiated. Major gullies occur mainly on steep slopes or scarps, where lateral sapping or vertical cutting results in oversteepening and on slopes which have concave contours. They may develop in many kinds of materials but favoured are among others: mixtures of sands, silts and clays, unconsolidated fine sands, fresh volcanic ashes and tuff deposits.

Gullies normally start as a deep 'master' rill where concentration of water occurs. They subsequently cut deeper into the soil profile and its substratum and extend upslope by way of headward erosion provoked by water entering from above. Large and small mass movements may contribute to the headward growth of gullies and ravines: oversteepening of the uppermost portion of the ravine causes instability which triggers mass movements. These in turn result in destruction of the vegetation, in the preparation of soil material for transportation and in an increase in size of catchment.

The photo of fig. 15.9 pictures terracettes on a slope in the Venezuelan Andes situated at the upslope end of a major ravine. Piping may also be an important mechanism, particularly where the soil characteristics lead to cracking in case of drying out. The photo of fig. 15.10 picturing piping phenomena on some steeply sloping potato fields in the Chama Valley, Venezuela, illustrates the process. Ultimately the ravines will reach dimensions and densities which render the land completely unsuitable for cultivation (Carrero, 1978). In situations such as that pictured in the photo of fig. 15.11, remedial works are extremely difficult to realize if not altogether impossible and,in any case, it is much too late to be beneficial to the local farmers. The photo of fig. 15.12 pictures a slope in the Venezuelan Andes where a variety of erosion processes occurs simultaneously,including sheet, rill and gully erosion aggravated by road construction and related roadside dumping of excavated materials. Terracetting as pictured in fig. 15.9 may contribute to the formation of the minor break of slope that marks the boundary of the gently sloping rounded top parts of the hill in the centre where erosion is considerably less severe. Field boundary gullies are a special variety common in agricultural lands. They have been studied by among others, Bergsma (1975).

Because of the variety of factors and processes involved in gullying, it is often difficult to assess the growth rate of individual gullies on a quantitative basis. It is, of course, possible, provided that aerial photographs of adequate scale are available, to measure the depth, width and length of the gullies and thus to calculate the volume of the eroded material,but if the year of origin of the gully is unknown it is difficult to determine the annual production rates. In the majority of cases the annual growth in length is limited to a few metres which makes it difficult to use sequential aerial photographs to measure this growth rate. The stereopairs of fig. 15.13 give an example in which this is marginally possible. They are at a scale of 1:20,000 and picture the configuration of some gullies near Roma, Lesotho, in 1951, 1961 and 1971,respectively. Although the area concerned is severely eroded, the growth rate of the gullies is rather unimportant. Most of the soil loss is due to sheet and rill erosion.

Mass movements have already been briefly mentioned above as a process involved in erosion by water (Cadiot, 1979; Imquist, 1980; Perrusset, 1981). This is, of course, largely limited to certain susceptible rock types and soils. Creep also comes under this heading. In fact the erosion by means of various mass movements is the most difficult to quantify with reasonable precision. Usually within a basin it is limited to a few well-defined areas. Meijerink (1975) mentions from the Serayu River Valley, Central Java, Indonesia that the critical Merawu formation outcrops in only about 10% of the surface area but yields about 15,000 tons/year per sq. km. For comparison it may be mentioned that the terraced volcanic slopes of other parts in that basin yield at the most 100 ton/year per sq. km. Mapping areas susceptible to landsliding in dif-

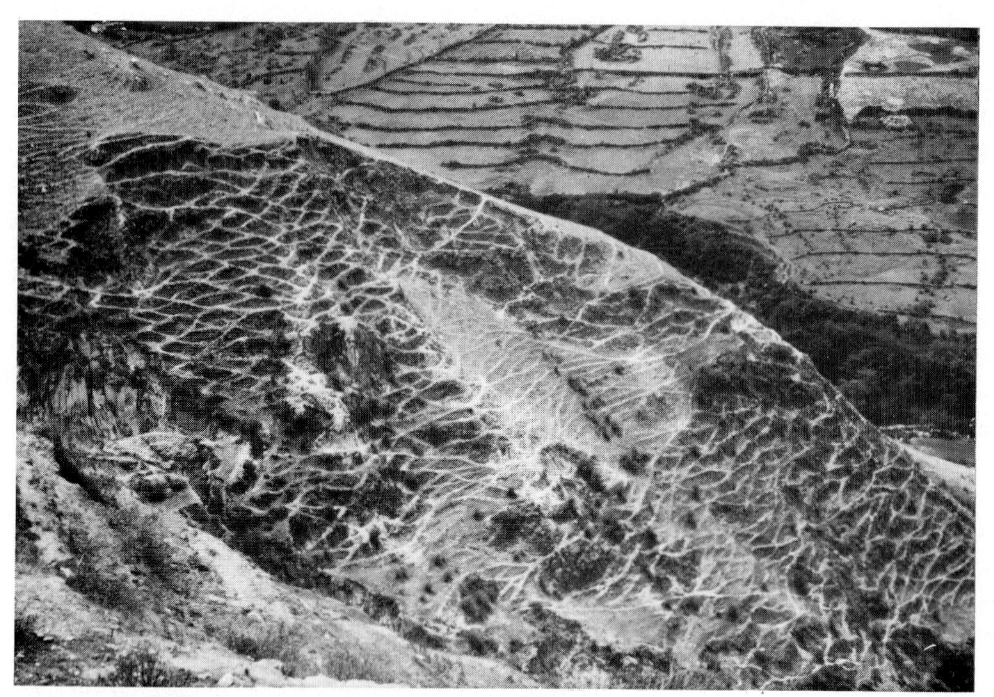

Fig. 15.9 Terracettes giving way to stronger forms of solifluction beneath a break of slope in the Chama Valley, Venezuela.
Note the agricultural terracing on the slopes in the distance.

Fig. 15.10 Piping phenomena, mostly along field boundaries, on a steep slope in the Venezuelan Andes.

ferent degrees is an aid in approximating the contribution to the gross erosion by mass movements. Creep, however, escapes such approaches and has to be determined in the first place in the field. One should realize, however that the observation of road cuts leads to an exaggerated estimate since the process there is accelerated by over-steepening. Careful measurement on experimental plots over adequate periods of time is essential. Colluviation terraces may assist in arriving at a qualitative assessment of colluviation processes along a slope, but for quantification experimental plots are the answer.

15.4 Types and Methods of Erosion Surveys

The types of erosion surveys are diversified as far as scales, aims or concepts are concerned. Therefore the survey methods and maps produced vary accordingly. With respect to scale, three main categories can be distinguished:

Detailed Surveys, which are based on extensive field observations and use of large scale aerial photographs or ortho-photographs (e.g. 1:1,000 - 1:5,000). These usually relate to specific areas of limited extent. Farm surveys as well as surveys for the protection of settlements, bridges or roads fall under this category.

Medium Scale Surveys. Because a catchment of a river is a natural hydrological unit, it is also a logical framework for the study of erosion and sedimentation processes. Erosion surveys on medium scales therefore preferably cover one or more complete drainage basins.

Fig. 15.11 Severe gullying disrupting the agricultural potential of a village in the Venezuelan Andes.

Fig. 15.12 Diversified erosion phenomena on largely terraced slopes in the Venezuelan Andes. Note the break of slope separating the gently sloping and comparatively less eroded top part from the remainder of the slope. Roadside erosion is noticeable in the left.

For practical reasons, however, they are sometimes limited to administrative units and part of the erosion system then remains unstudied. The scale of these surveys depends on the size of the catchments and on the aims of the project, however; 1:20,000 - 1:25,000 is rather common in use. As more intense erosion is often concentrated in certain parts of limited extent, it may be advisable to map the basin on a scale of 1:50,000 or 1:100,000 first, to decide on the areas for which a more detailed survey is required. Aerial photographs of medium scale are an essential prerequisite for an efficient implementation of these surveys.

Small Scale Surveys. These usually relate to an entire country or region, the aim being to provide a nation-wide inventory of the state of erosion for purposes of land use planning, conservation and hydrological development. A country-wide survey programme and a compilation of all existing data are essential in this case and should be directed by an efficient utilization of aerospace images. Small scale aerial photographs and photo-mosaics, e.g. in scales of 1:50,000 - 1:70,000, are useful tools in this context but their interpretation and the subsequent mapping may still amount to an almost prohibitive work load where very large areas are concerned. Therefore, the utilization of satellite imagery, either photographically enlarged to a scale of, e.g., 1:200,000 or 1:250,000,or digitally processed images of similar scales, is useful in the first phase of the work (Morrison and Cooley, 1973). Annotations made during a quick persual of aerial photographs in this same phase can be reduced to the same scale and superposed on the imagery. Once a general picture has been obtained, the problem can be approached step by step by way of small scale aerial photographs (e.g., 1:50,000), medium scale aerial photographs (e.g., 1:20,000), fieldwork, etc. In principle this 'zooming in' will expedite the survey. Attention can be concentrated on the most critical areas, while the field survey will increase in detail while zooming in.

Among the various aims or concepts of erosion surveys the following major types can be distinguished:

Surveys of Actual Erosion Processes

These surveys on the 'dynamics' of accelerated erosion by water should not be merely restricted to mapping the ongoing sheet/rill and gully erosion, but encompass all types of laminar and linear slope processes including matters such as diffuse wash of various intensities as well as shallow and deeper seated mass movements (creep, terracettes, slides/slumps/mudflows). Proper differentiation is essential. The effect on both denudation and deposition in the area should be carefully analyzed. It is usually not too difficult for an experienced geomorphologist to indicate the type(s) of processes occurring on a slope, but their intensity and rate of development is often more difficult to assess. The existing options for quantification range from detailed measurements on experimental plots to estimations and classifications into 3-5 classes (very slight, slight, moderate, severe, very severe). Sequential aerial photographs may be of help in this context. An additional problem exists where several processes occur simultaneously on a slope and the contribution of each of them to the soil loss and/or sediment delivery has to be specified. It would be short sighted to restrict this type of survey to mapping processes only without indicating the factors that have an effect on their efficiency. Therefore, normally, matters such as lithology, slope characteristics, hydrological features (springs, seepage, etc.) cover type and conservation practices should also be mapped. This is not always done, however, and in many cases only one of these factors, e.g., lithology, is given.

Surveys of the Present State of Erosion

These surveys are of a more static character and amount to an accurate mapping of the present extent of laminar (sheet and rill) and linear (gully, ravine) erosion features. Data for this survey can be collected in the field, e.g., by studying the decapitated and buried soil profiles and by measuring the length, width and depth of gullies and ravines. Aerial photographs facilitate these surveys to a considerable degree. The areas where decapitated soil profiles occur are often marked by a different (usually lighter-toned) grey tone or colour. Gullies and ravines are clearly pictured and can be counted and measured. The terrain can be rather easily subdivided into a number of erosion classes differentiated according to type and intensity of erosion.

Examples are from Rao (1975); Williams (1978); etc. The classification so obtained does not automatically imply that the active erosion is maximal in the most severely eroded parts of a basin or a slope. Where gravel or other coarse material has covered the surface during the phase of sheet/rill erosion, a protective cover has been formed which slows down,if not entirely precludes, the further progress of erosion. Also gullies do not necessarily grow most rapidly where they are most spectacularly developed. They may have reached the divide line or otherwise have achieved a certain degree of stabilization. Minor gullies on relatively gentle slopes and much less densely dissected terrain may at present cut and grow headward much more violently than areas that have already gone through this phase earlier. For a proper evaluation of the static erosion map so obtained, it is essential to include ample information on the geomorphological context of the mapped erosion patterns. Other relevant environmental data on lithology, land use, etc., must also be considered.

Surveys for Prediction of Erosion Hazard

Although the types of survey regarding erosion processes and the present state of erosion certainly yield data of considerable practical importance, the major interest in terms of surveys for development is in the forecasting of further developments of accelerated erosion that can be expected in the future, particularly in the near future. It is this information, combined with the assessment of the economic importance of the areas endangered,that governs the priority of various parts of a conservation project and the funds to be alotted. It ultimately determines the location and type of conservation measures to be taken.

Some confusion exists in this area where terminology is concerned. The term erosion hazard, often used in this context, is typically man-centered and relates to the dangers presented to man and to the works of man. It is affected on one hand by the erosion susceptibility - resulting from the erodability of the land and the erosivity of the precipitation - and on the other hand by the mode, intensity, duration and spatial distribution of land utilization. It was the latter that triggered the environmental degradation and then the accelerated erosion to which, subsequently, man himself fell victim. The term erosion susceptibility is thus distinct from erosion hazard in as much as it only relates to factors of the natural environment cf. flood hazard/susceptibility and drought hazard/susceptibility; sections 13 and 14). Some authors, however, use the term erosion susceptibility to indicate potential erosion and erosion hazard, thus the erosion that actually can be expected to occur in the near future regardless of whether or not it endangers man and his works.

Erosion hazard may take the form of initiating erosion in formerly unaffected areas by headward cutting of gullies, by branching or widening of ravines, etc. It may also amount to further degradation of areas already previously affected where, therefore, a further decline in productivity can be expected. Often both aspects occur simultaneously in an area. Quantification of the hazard in terms of soil losses to be expected or of future sediment delivery also can be successfully attempted and is of use in planning conservation works. Usually in deteriorating environments an increase in both erosion intensity and quantities of detached, transported and deposited material can be expected. However, one should bear in mind that the growth of gullies and ravines tends to diminish in the later stages of their development when a certain degree of stabilization has been reached and that the quantities of annually eroded material may then be gradually reduced notwithstanding the fact that the erosion is still progressing and the hazard still extending (Bergsma, 1974).

There are two main approaches to erosion hazard surveys which can be applied simultaneously so as to obtain optimum results. First, the hazard can be assessed on the basis of the survey of the present state of erosion and on that of the processes operating at present.

Sequential survey is important here in measuring the thickness of the soil horizon that has been removed within, say, a decade and in quantitatively establishing the development of the gullies and ravines occurring in the area of study. Sequential aerial photography is very useful in this connection and may, e.g., result in singling out particularly active gullies. Also the growth of areas affected by sheet/rill erosion may be studied this way, although differences in soil humidity, atmospheric conditions and illumination between the photographic flights used for the purpose may complicate the matter The sequential aerial photographs of figs. 15.13 and 15.14 give examples from Lesotho and Venezuela, respectively. The effect of conservation practices sometimes is noticeable. Nir and Klein (1974) describe the development of four gullies in the Hanal Shigma basin, Israel, using aerial photographs dating from 1955 and 1966 and using geomorphological criteria. Differences in photo scale, relief displacement, etc. also may cause problems. If the geomorphological situation of the most actively eroded areas is taken into consideration by assessing, e.g., the effect of nearby break of slopes or the amount of overland flow from upslope areas, and if geomorphological units are outlined, the assessment of the erosion can be considerably improved. A second approach to erosion hazard surveys is assessing the erosion susceptibility of the area concerned on the basis of the various environmental parameters and land utilization factors that affect the erosion processes on the slopes. Apart from geomorphological factors such as slope characteristics also lithology, soils and vegetative cover have to be considered.

It is sometimes useful to give a separate rating for the hazard of the main types of erosion (sheet/rill, gully, mass movements, etc.) and to sum these ratings at the end so as to arrive at a complete assessment of the erosion hazard. This parametric approach is facilitated if the various parameters are listed and rated in tabular form. The results of the exercise, of course, have to be considered with care as sheet/rill and gully erosion are quite distinct, and plain summation is thus in itself a rather crude approximation only. Nevertheless, useful results can be obtained, even if only a limited number of parameters are considered (Richter, 1965; Morgan, 1979). The variable factors related to land utilization and agricultural practices merit special attention (Obeng, 1972; Meade, 1969).

The Survey Methods Involved

In order to carry out an erosion survey efficiently a systematic approach is required in which the following steps and methods can be distinguished:

1. Mapping of all Erosion Features
   that can be traced on aerial photos and/or observed in the field. Line symbols will be used in this phase in order to map gullies and ravines of various dimensions; the aerial extent of sheet and/or rill erosion can also be indicated. Various degrees of erosion intensity of the features can also be distinguished in this phase, either for every type of erosion separately or for a combination of types. Shallow and also deeper seated mass movements should be plotted too, either each feature individually or the zones where they are generally occuring. Apart from the modes of degradation of the terrain the aggradational aspects, i.e., the zones of deposition of the eroded materials are to be mapped.
2. Mapping the Progress of Erosion
   in the recent past using sequential aerial photos and other successive data on accelerated erosion by water. This step will give further indications on the spatial distribution of the intensity of the various erosion processes present. Both lateral spread and changes in vertical dimensions of both linear and laminar processes

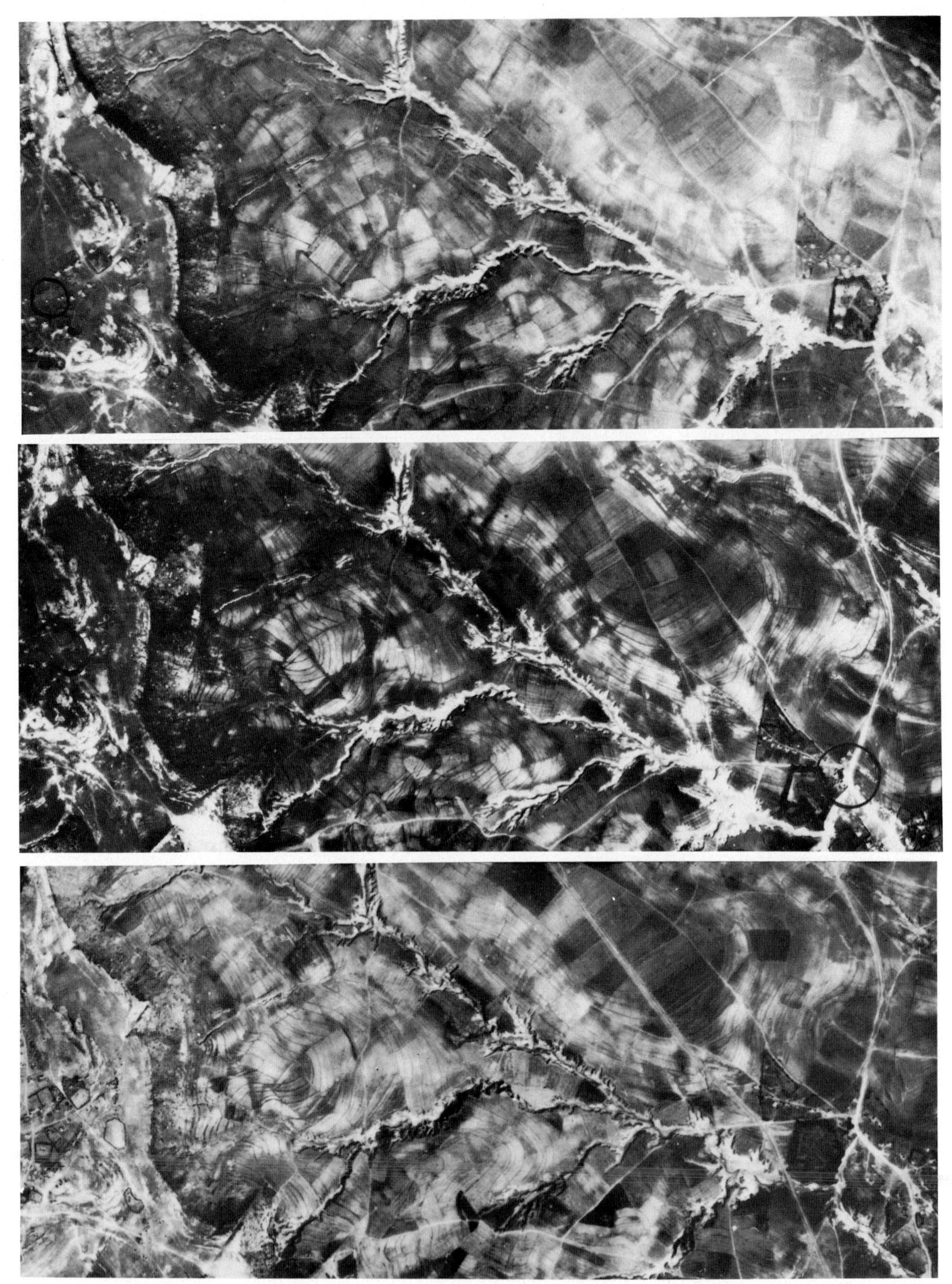

Fig. 15.13 Sequential aerial photographs showing gully development near Thaba Bosiu, Lesotho, in the years 1951 (top), 1961 (centre) and 1971 (bottom). Note that gullies already well-established in 1951 have not increased drastically but smaller ones (e.g., top part of photographs) went through the violent initial stages in this period. Note extensive sheet erosion and the conservation terraces that have been constructed as to stabilize the gullies and to stop further deterioration of the land. Scale 1:20,000.

Fig. 15.14 Sequential aerial photographs of an area near Lagunillas, Venezuelan Andes, showing various erosion and concervation features. Part of the area pictured in the left at a scale of 1:50,000 can also be seen in the central photograph which is at a scale of 1:20,000 and was taken 10 years later. The development of the ravines in the interval is minor; extensive conservation measures-terracing, reafforestation and the construction of gabions in major ravines have been implemented. The photo to the right shows the difference between protected and non-protected parts of the terrain.

should be given attention.

3. Subdivision of the Terrain into Erosion Units each of which is characterized by a well-defined degree of erosion of specified types. Since lithology is an important factor in accelerated erosion, it is usually possible to incorporate it in the hierarchic classification of units applied in the legend. Processes can be indicated by line symbols and a rather complete picture of the erosion can be given, e.g., within a drainage basin. It is even possible to utilize the data so obtained to arrive at an assessment of the contribution of the units distinguished in the gross erosion and sediment yield of the basin. If some field measurements are available on erosion of the land and/or on the load of the river, quantification of the problems is within reach.
4. Geomorphological Survey resulting in a breakdown of the land into geomorphological units. The erosion units just mentioned normally are clearly associated with the geomorphological situation even though this may not yet have been studied specifically for the purpose. Improved results can be expected once an analytical geomorphological survey is carried out. The state of erosion can be then integrated directly in the geomorphological situation which expedites the erosion survey. In fact one has two options at this stage: one may either attempt to arrive at a geomorphological breakdown of the land and establish the erosion in each of the geomorphological units distinguished, or alternatively, one may outline erosion units first which then have to be matched with the geomorphological units. The geomorphological units are considered as landscape ecological units as has been formulated in section 10. The diversified information so obtained gives a complete picture of the erosion susceptibility of the land and, taking the present distribution and rate of erosion and the land utilization into consideration, the erosion hazard can be adequately assessed. One may, therefore, if interested in matters of sediment yield, use the following procedures:
   a. assess the erosion processes and their activity in every geomorphological unit
   b. consult soils data to evaluate the erodability
   c. estimate sediment production, e.g., in 4 or 5 classes
   d. measure the percentage surface area that every class covers of the total catchment
   e. compare with known sediment yield from other, similar catchments and/or use available data, if any, from the basin studied
   f. give weighing factor to every class and calculate total sediment yield by multiplying surface area and weighing factor
5. Quantification by Parametric and Other Methods
   Since numerous environmental parameters act as a factor in matters of accelerated erosion, it is elucidating to assess the effect of each of them separately, e.g, in a tabular form. This also facilitates matters of quantification. Tables for rating each of these factors, such as

slope length, gradient and the various erosion processes, exist and contribute to a systematic, unbiased approach to the problem. Land use type, vegetative cover, deforestation, etc.,also should be included. A scoring system,for example, ranging from 1 (normal) to 5 (very severe) can be applied to the various parameters and also to the erosion hazard ultimately established and mapped. Tables 15.1 and 15.2 give examples (Morgan, 1980; Bergsma, 1975 and van Zuidam, 1979).

6. Production of a Map Showing Proposed Conservation Measures and Land Use Planning

This is the ultimate document to be used for combating accelerated erosion in the area of study.

15.5 Combating Erosion by Water: Conservation Practices

Erosion surveys of the kinds described in sub-section 15.4 will lead to the location of the areas which are in different degrees affected partially or fully by erosion processes of various types. They will also provide information about the intensity of the erosion and on the growth rate of the areas affected. Estimations about the soil losses from agricultural fields and the sediment yield of catchments can thus be made more precisely. Also the causes and the mechanism of the erosion processes can be better assessed. The ultimate aim, however, is in most cases not only to map and to analyse the state of erosion in the area of study, but to contribute to the stabilization or improvement of the situation through an adequate conservation scheme. It is mainly the scale of the survey which determines whether this contribution will be on the level of national planning, in intermediate phases or on the farm level.

The occurrence, development and distributional pattern of accelerated erosion is governed by the interplay of natural and cultural factors. Erosion control by conservation is thus a complex matter. The susceptibility to erosion of the natural environment is the result of the erodibility of the land (Bryan, 1969, 1974; Calinescu, 1967) on one hand and the erosivity of the rains on the other (Bergsma, 1981; Clement and Bonn, 1975; Joshua, 1977). There is little room for improvement in these spheres; one basically must be content with repairing or reducing the damage to the environment and to its vulnerability to erosion caused by inappropriate use of the land. Mapping and analysing the physical aspects of the matter in the context of an erosion survey and placing this survey in a valid geomorphological framework is important, but the efforts will only become really effective and yield practical results if the socio-economical, political, etc. factors which have given rise to the prevailing agricultural system, and thus to the accelerated erosion, form an integral part of the study. This applies to surveys at all scales; examples of striking differences in the degree of erosion between adjacent fields separated by a fence and subjected to different systems of farm management are numerous. The Landsat image of fig. 15.15 (1:1,000,000) picturing parts of Lesotho and of the Republic of South Africa, illustrates this but on a much smaller scale. The straight fence, erected between the two countries in the beginning of this century without relation to the natural environment, separates the severely eroded (light-toned) fields in Lesotho from the much better vegetated (dark-toned) lands in the RSA and indicates differences in population pressure, agricultural systems and political situation rather than anything else.

Three broad spheres can be outlined in the field of combating erosion by water:

1. the susceptibility of the natural environment
2. the socio-economic and political contexts
3. the technical solutions for the problems and the connected agricultural system.

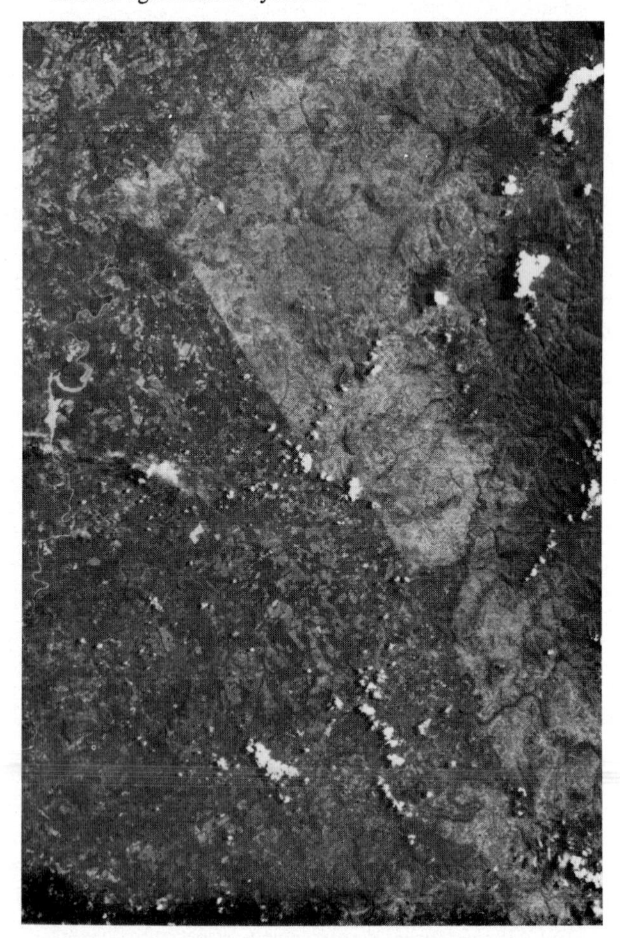

Fig. 15.15 Landsat Band 5 image showing differences in vegetal cover and in state of erosion at either side of the boundary between Lesotho and the Republic of South Africa: the considerably higher albedo of densely populated Lesotho as compared to the adjacent rural areas of the RSA illustrates the importance of socio-economic and political factors in matters of erosion.

The first sphere has been touched upon in the previous sub-sections and the second sphere is outside the scope of this book on applied geomorphology. This sub-section is devoted to the third sphere: conservation measures and their applicability.

Most conservation programmes relate to two distinct types of activities, namely stabilization of the gullies and ravines already formed and erosion control of the slopes in affected or threatened areas. Gully and ravine control concentrates on actively developing ones - by way of headward erosion, vertical cutting, gully side collapse, etc. - and that are thus centres from which the further degradation of the land spreads. Erosion control on slopes comprises a whole range of measures in the field of reafforestation and of agricultural practices such as tillage, mulching and crop rotation and also of the construction of conservation terraces of various types. A first and important application of erosion surveys is to locate the areas to be treated and the gullies and ravines to be controlled and to establish the priorities for each of them as to arrive at an optimally efficient conservation plan covering a number of years or decades. It should be stressed again, however, that conservation measures of the kinds just outlined will ultimately be successful only if they form part of an overall scheme of land use planning, including the creation of hydrological forest zones, etc.,preferably for the entire drainage basin (Bakanina, 1972; Barton, 1972; Beasley, 1972; Bennett, 1939; Blase and Timmons, 1965; Diamond and Kawamura, 1974; Dzyadevich, 1970; FAO, 1956; Gill, 1979; Harrison, 1972; Held et al., 1962; Held and Clawson, 1965; Hudson, 1971; Kawamura and Diamond, 1975; Meyer et al., 1971; Peasly, 1975; Posey, 1973; Richter, 1973; Shukhla, 1974; Thronson, 1973; Zaslavski, 1975).

The control of gullies and ravines is not an easy task, particularly when they are in the stage of rapid development. It involves mostly two or three types of action:

1. Stopping vertical cutting particularly in the lower reaches by the construction of a number of check dams or gabions of which the photo of fig. 15.16 gives an example. Also some detritus will be collected this way at their upstream side. A diversity of structures has been developed in the course of the years (wire bolsters, netting dams, drip structures, etc.) the efficiency of which varies with the conditions prevailing.
2. Stopping or slowing down the headward erosion in the upstream end and at incipient side gullies. This can be attempted by slowing down the velocity of the inflowing water (fig. 15.17), by stabilization using suitable types of scrub and grasses, by the construction of horse-shoe shaped diversion bunds to prevent the water from entering the gully, etc.
3. Stabilizing the entire course of the gully, by filling or shaping, by transforming it into a grassed water way or by paving its bed with blocks. It may also be required to pre-

Fig. 15.16 Checkdam or gabion in a ravine in southern Italy.

Fig. 15.17 Combating retrogressive erosion in the heads of some steep ravines in Burundi.

vent lateral sapping which might introduce or accelerate gullying at the affected localities. Groines then may have to be constructed (fig. 15.18). If at times, the discharge of the ravine is exceptionally large and damaging, the diversion of the water and its temporary storage or detention in small reservoirs at suitable sites may also be contemplated to spread the water, where the relief conditions and land utilization render this possible.

Morgan (1980), Flannery (1976) and Schwab et al. (1966) elaborate on the methodology and on the construction types which are in use for gully control.

The treatment of eroded or threatened slopes also is a complex matter and a range of measures exists in the field of reafforestation and conservation farming. The most appropriate selection of these has to be made for the problem at hand. Some of these conservation practices are simple and relatively cheap whereas others are technically more complex or elaborate and costly. They all come under two categories, namely agricultural practices geared to maintaining a stabilizing vegetative cover as well as possible and agricultural practices geared to reduction of soil losses by tillage practices, contouring, the construction of terraces, etc. Not all these practices are equally effective, however, and therefore sometimes expensive or elaborate measures are inevitable for adequate combating of erosion.

With regard to practices concerned with stabilizing vegetative cover, matters such as critical area planting, lifestock exclusion, the use of grasses and legumes in rotation, range seeding and pasture/hayland planting are highly effective, whereas controlled burning and brush and weed control have only a low efficiency in combating sheet, rill and gully erosion on slopes. Similarly, with respect to tilling practices and terracing, low efficiency is reached if only simple tillage practices or even contour farming methods are applied; contour strip cropping and terracing combined with cross-slope farming are of medium efficiency and only conservation bench terraces together with contour farming, field strip cropping or contour strip cropping (and also bench terraces) are really highly effective from a conservation point of view.

Conservation oriented tillage practices include, e.g., minimum tillage, mulch tillage and zone tillage. Green manuring and crop rotation are often used in connection. Contouring decreases the danger of plow lines to develop into rills and gullies but it is only effective on gentle to moderate slopes and provided that the soils are not too highly erodible. Strip cropping involves the planting (and rotation) of crops in strips, running either more or less precisely parallel to the contour lines (contour strip cropping) or approximately parallel to the contour lines but maintaining equal field width (field strip cropping).

Fig. 15.18 Groynes to prevent lateral sapping of a tributary of the Crati River, Calabria, Italy.

Sometimes strips are laid out especially on critical parts of the slope which are constantly kept under grasses and/or legumes (buffer strip cropping). The width of the strips depends on the slope gradient and ranges approximately from 20 to 50 metres.

Conservation terraces are basically of two types, namely bench terraces and broad base terraces. Bench terraces form level surfaces and result from both cutting and filling, thus involving considerable disturbance of the soil material. Nowadays they are constructed in some countries such as the USA only on gentle slopes (up to 3-4°) but in many other countries they have been constructed in the past and even are still constructed to date on considerably steeper slopes. The broad base terraces consist of a broad channel with an embankment or ridge at its downslope side. They leave the slope largely undisturbed and the farming machinery can drive across them. They may be given a slight gradient and then act as a diversion terrace that serves to drain surface run-off at low speed obliquely over the slope towards a protected drainage line such as a grassed waterway. The airphotos of fig. 15.13 show an example from Lesotho. They may also act as an interception terrace at locations where diversion is not urgently required. The most common, level, broad base terrace is meant to act as a water retention terrace. It is effective particularly in areas of low to moderate rainfall. The most critical are areas characterized by occasional heavy cloud bursts as is the case in many semi-arid countries. Overtopping of the ridges then becomes a serious hazard as can be seen e.g., in the aerial photographs of central Tunesia pictured in fig. 15.19. The so-called ‘tabia dams’ there, were disrupted at numerous places during the 1969 flood of the Zeroud and Merguelil rivers. Compartments are often established within these terraces to avoid the ac-

Fig. 15.19 Vertical aerial photograph, scale 1:50,000, showing carefully spaced so-called 'tabia dams' in Central Tunesia. Exceptional rains may cause overtopping and destruction of the earthen bunds or ridges.

cumulation of too much water at critical spots. Fig. 15.20 shows an example from Venezuela. Much longer compartments can be found in other areas with less relief. Where slopes are steep and particularly in semi-arid areas, the water retained may be insufficient and then a modified profile may be opted for which results in the so-called Zing bench conservation terrace. The upslope part of this is sloping and acts as a catchment for the horizontal downslope part of it. The vertical spacing of these conservation terraces of various types depends on the slope gradient upslope of them, on the permeability, erodibility and cover of the soils and on the rainfall characteristics of the area concerned.

Ven te Chow (1964) gives the following formula in this connection:

$VI = aS + b$ in which,

VI = vertical interval

a = constant depending on the region (0,2-0,6)

b = soil constant (2 for resistant, well covered soils)

S = slope gradient (in %) upslope of the terrace

It is evident that the geomorphologist in the context of erosion surveying and conservation schemes can make a contribution to the siting, shaping and spacing of terraces, grassed waterways and other structures required in relation to the terrain configuration, but he should do so in consultation with the agricultural engineer in charge of

the location of farm roads, the siting of buffer strips and hydrological forest zones. For drawing up and implementing a conservation plan, detailed information will be required specifically about:

1. The extent and activity of laminar (sheet/rill) erosion
2. The number of gullies and their location
3. A rating of gully activity specified for bed, walls and head of each gully and with indication of partial or full stabilization by vegetation and/or structures.
4. Dimensions of gullies (length, width, depth)
5. Land lost by gullying in square metres.
6. Sediment production in cubic metres.
7. Vegetation treatment (species to be planted, where, when and how; natural vegetation found in gullies).
8. Structural treatment (number, types and location).
9. Observations on behaviour of existing vegetation performance of existing structures and maintenance requirements.
10. Cost calculations on labour and material,cost of planting and of construction and of the maintenance required.

It is evident from the aforesaid that if timely action is taken, in many cases simple conservation measures may suffice and expensive measures such as terracing can be avoided. Even if the latter is not the case the efficiency of terraces and the like will be substantially increased if simple conservation practices are implemented in conjunction. This is examplified by investigations such as those by Aarstad and Miller (1973) who mention a drop in surface run-off from 40% to 1% following the creation of small basins between crop rows; and by Lal (1974) who elaborates on the effect of soil management practices on erosion and erosion risk in shifting cultivation areas. Anderson and Brooks (1975) studied the effect of seeding of grasses and legumes on the erosion of burned forest areas. Meyer (1971) emphasized the effect of mulches of crushed stone, gravel and woodchips for erosion control and Harrold and Edwards (1972) mention the effect of no-tillage practices in corn growing.

For proper implementation of these and other conservation measures, a thorough knowledge not only of the physical aspects of erodability of the land and erosivity of the rain should be available, but also the agricultural practices that give rise to, or favour,the erosion should be fully understood. The latter aspects have been emphasized by several authors such as Aghassy (1973) who mentions the effect of Bedouin ploughing techniques in past decades as a factor in parts of Israel; Servant (1974) who points to matters such as burning brush and market gardening as

Fig. 15.20 Narrow conservation terrace near Lagunillas, Venezuela, divided into compartments as to minimize the danger of concentration of rain water which may lead to overtopping and destruction. Compare with the aerial photographs of fig. 15.14.

factors in Tahiti; and Obeng (1972) who found clearing and burning practices, the introduction of mechanized farming techniques, inadequate tillage implements and injudicious lumbering to be major factors in Ghana.

15.6 Geomorphology and Integration of Erosion Survey Data

The information required in surveys on erosion by water and particularly for the assessment of the erosion hazard prevailing in the area of study and the conservation measures to be implemented is diversified, as will be evident from the previous sub-sections. This situation may first of all pose serious cartographic problems. To avoid overloading of the erosion map and in order to present the data in an organized manner, more than one map may be required. Furthermore, integration of the various groups of data, such as terrain form, process, lithology, cover type, etc., may be needed in tabular form to elucidate their interrelation and importance in the context of accelerated erosion (Boodt and Gabriels, 1980; Carrero, 1978; Chakela, 1981; Christianson, 1981; Ciccacci et al., 1980; Gregory and Walling, 1973; Holy, 1980; Mensching, 1979; Miholics, 1970; Rapp et al., 1972; Tufescu, 1967; van Zuidam-Cancelado, 1979).

Generally speaking, it can be said that these problems increase with decreasing scale. In large scale, e.g., farm surveys, the number of factors and the diversity of the land involved is comparatively limited in most cases and the surveyor can proceed almost directly to the study of specific gullies, badland areas or unstable zones. Quantification of the problem may, also in these large scale surveys, lead to the necessity of tabular presentation of data in order to facilitate their assessment. An erosion survey carried out by Bergsma (1975) in Central Java, Indonesia, may serve as an example. The survey was carried out using aerial photographs at a scale of 1:20,000 (in the absence of more detailed airphotos). The area could be divided into a number of simple geomorphological units in which slope characteristics (steepness) played an important part. Line symbols were used for mapping gullies and also the numerous scars of mass movements.

Fig. 15.21 illustrates the procedure and gives the legend used. After this phase, the attention could be concentrated directly on gully development and particularly to certain field-boundary gullies as given by fig. 15.21 (right) at a scale of 1:1,400. A quantitative analysis of the frequency of occurrence and surface area of gullies and landslide scars in the various geomorphological situations, as determined with the aid of the aerial photographs, was subsequently given in tabular form (see tables 15.1 and 15.2).

Where complete drainage basins have to be surveyed, the complexity of the situation usually is greater and so are the problems of cartographic and tabular presentation of data. Rao (1975) in an erosion survey of the comparatively small (60 sq. km.) Oliva River basin, Calabria, Italy, at a scale of 1:10,000, organized the data gathered according to the following schemes.

1. Lithology
2. Geomorphological units
3. Slope classes
4. Processes

(A) Erosion	(B) Slope instability
a) diffuse run-off without loss of soil	a) creep (laminar)
b) sheet erosion	b) solifluction in 'spoons'
c) concentrated run-off (no erosion)	c) terracettes
d) rill erosion	d) rock/debris slide (active/inactive)
e) gully erosion	e) slump/rotational slide (active/inactive)
f) linear erosion in valleys (none/light/severe)	f) rock/bedding slide (active/inactive)
g) headward erosion	g) landslide (active/inactive)
h) badlands	h) mud/earth flow (active/inactive)

i) lateral river erosion	i) incipient slide
j) downward river cutting	j) slide direction

% area affected

5. Soil

Three thickness classes: <25cm/25-75cm/>75cm
Three texture classes: fine/medium/coarse
Other aspects: (large/small gravel)
Moist horizon

% area affected

6. Cover type

a) non-vegetated	b) annual crops
c) grass	d) improved grasslands
e) scrub	f) forest: dispersed/dense

% area per sub-unit

7. Structural data	dip, strike, fault, fracture, schistosity, etc.
8. Topography	divide lines/ break of slope morphometry
9. Hydrographic data	springs, swamps, etc., alluvium (fixed/mobile)
10. Man-made features	afforestation, terraces, contouring, river training, excavation, fill and dump, etc.

All this information, except items 5, 9 and 10, subsequently was incorporated in two separate maps which gave the geomorphological units and features and the erosion classes (map 1) and slope classes and cover types (map 2) respectively. On map 1 the coloured area symbols were used for the geomorphological units and hachures for the erosion classes. Geomorphological features were pictured by line symbols. On map 2 the coloured area symbols were used for the slope classes and various screens were applied for mapping the cover types. Finally correlations between lithology, land use and various types of erosion were given in tabular form (table 15.3).

The importance of, particularly, lithology and cover type as factors in matters of run-off from slopes and sediment yield of catchments is evidenced in a very striking manner by the tables 15.4 and 15.5 based on the observations of various authors on the situation of the island of Java, Indonesia (Meijerink, 1975). The differences in run-off (expressed in % or rainfall) between well-covered slopes and areas where dry cultivation is practiced or which are barren are very outspoken and so are the differences in sediment yield between young volcanic and (mainly Tertiary) sedimentary rocks.

An interesting legend has been devised by the Instituto de Desarollo de los Recursos Naturales Renovables (INDERENA) in the framework of erosion surveys that cover large areas in the Colombian Andes. They give a clear picture of the multidisciplinary nature of the data required and of the integrated approaches to the problem which are therefore to be applied. The method is comprised of a map related to the present land use and the degree of protection against erosion and another, geomorphological map, which includes data on lithology, landforms and processes. The legends of these maps are given below.

Present Land Use and Degree of Protection against Erosion

Legend:

1. Forest

Primary forest
Secondary forest: regenerated
regenerating
Secondary forest with human influence:
without clearings
with clearings
Scrub: high without clearings - with clearings
low without clearings - with clearings

2. Grassland

Natural grassland (praderas) - without overgrazing
" with overgrazing
'Potreros' with natural grassland - without overgrazing
" with overgrazing
'Potreros' with improved grassland - without overgrazing
" with overgrazing

3. Agricultural Land/Crops

Permanent crops with 'sombrio'
Permanent crops without 'sombrio' - without grassland
" with grassland
Semi-permanent crops - without grassland
" with grassland
Annual crops 'limpios' - in flat areas
" in undulating areas
" on slopes

4. Various

Rocky outcrop
Strongly eroded area - slides - gullying - sheet erosion
Built-up areas

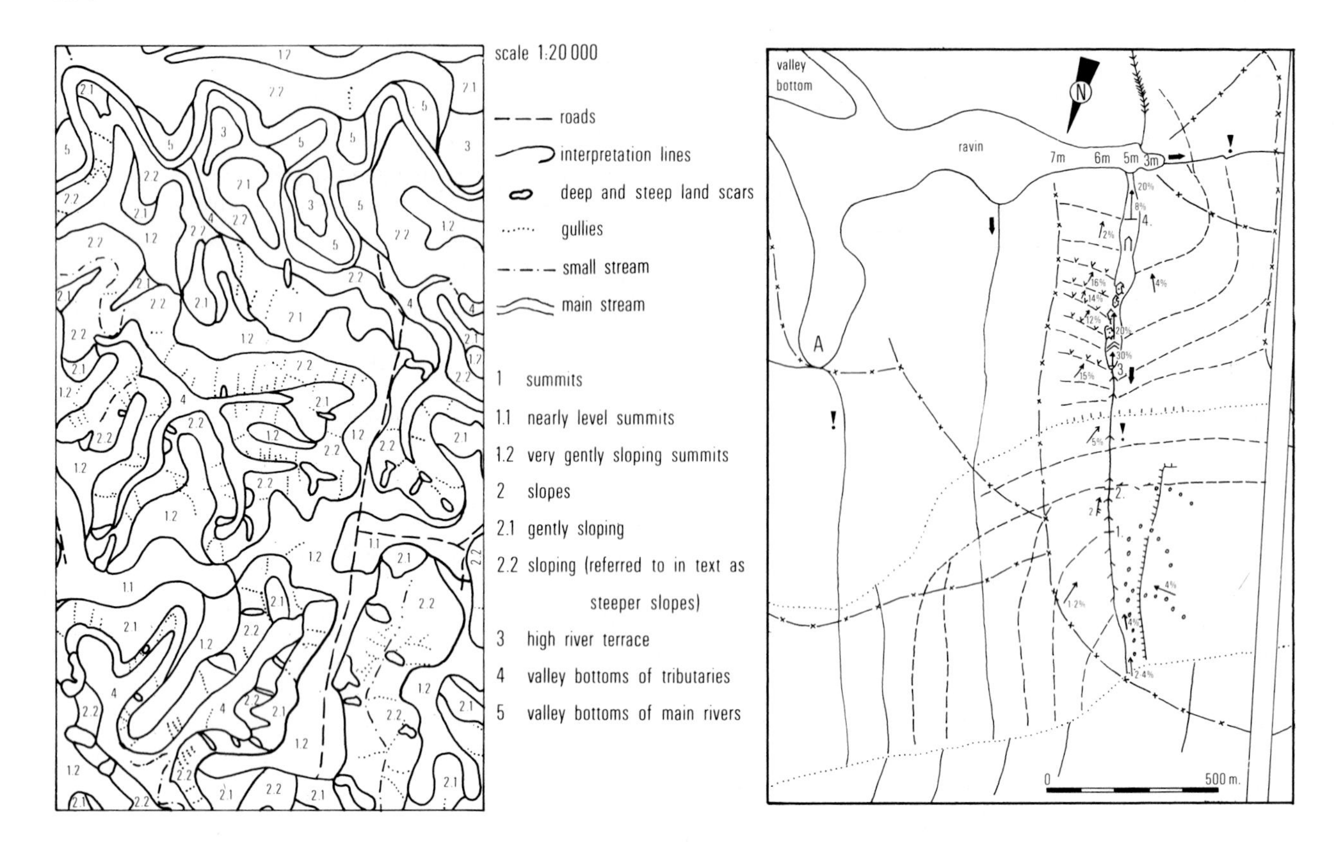

Fig. 15.21 Some details of erosion features in the Serayu Valley, Central Java, Indonesia. Left: break-down of the land into geomorphological units on the basis of aerial photograph interpretation with indication of (mostly field-boundary) gullies. Scale 1:20,000. Right: detail of a field boundary gully at a scale of 1:1,400. (Bergsma, 1975).

Table 15.1 Data on the occurrence of gullies (Bergsma, 1975).

Mapping unit	% area	length mm on AP scale	number	% length	% number
nearly level summits	8	0	0	–	–
very gently sloping sides of summits	24	31	19	6	5
gentle slopes	11	97	47	20	13
steep slopes	35	350	293	73	82
valley bottoms and terraces	22	0	0	–	–
total area of 910 ha	100	478	359	99	100

Table 15.2 Data on the occurrence of land scars (Bergsma, 1975).

Slope position of land scar	Surface area of land scars in mm^2 on airphotos of 1:20,000 — Number of land scars in each size class: 1	2	3	4	5	6	7	8	9	10	11	12	Number of scars: Gentle slope	Steep slope	Total	Area mm^2
Lower slope	19	17	6	2	1	–	–	1	–	–	1	–	3	44	47	103
Upper slope	9	10	5	1	1	3	–	–	–	2	1	–	5	27	32	102
Full slope	2	5	3	–	2	1	1	–	–	–	–	1	1	14	15	56
Total area 910 ha													9	85	94	10,4 ha

METAMORPHIC ROCKS	SEDIMENTARY ROCKS	IGNEOUS ROCKS
1 Phyllitic schist (soft metamorphics) 2 Schist & gneiss (hard metamorphics)	1 Sandstone 2 Silty Clay 3 Conglomerate 4 Limestone/dolomite	1 Granite

PRESENT LAND USE

	Sym-bol	Description	Area (hectares)	Percentage 10 20 30 40 50	Area (hectares)	Percentage 10 20 30 40 50	Area (hectares)	Percentage 10 20 30 40 50	Total Area	%
Cover Type		Olive	322.75		597.25		4.00		924.00	15.31
		Olive & cereals	60.75		94.25				155.00	2.57
		Crops productive	17.75		34.50				52.25	0.86
		Crops, erosion susceptible	187.75		58.00				245.75	4.07
		Scrub grass	142.50		581.25				723.75	11.99
		Broad leaf vegetation, dispersed	1004.00		720.00		69.50		1793.50	29.72
		Broad leaf vegetation, dense	1695.75		383.00		62.50		2141.25	35.48
									6035.50	100.00

EROSION

	Sym-bol	Description	% of occurrence in the whole area	Perc. of occurrence 10 20 30 40 50	% of occurrence in the whole area	Percentage of occurrence 10 20 30 40 50 60 70	% of occurrence in the whole area	Percentage of occurrence 10 20 30 40 50 60 70	% of total occurrence
Erosion		Diffuse run off without loss of soil	2.08		2.57				4.65
		Sheet erosion	19.4		25.95				45.35
		Gully erosion	1.6		1.2		0.37		3.17
		Headward erosion			0.23				0.23
		Bed erosion	1.53		1.8		0.14		3.47
Mass Movement		Slide	3.3		0.74		0.2		4.24
		Rock fall	0.04		0.97				1.01
		Creep	29.8		2.9		1.52		34.22
		Earth – mud flow	1.4		2.26				3.66
		Total	59.15	Total	38.62	Total	2.23	Total	100.00

Table 15.3 Relationships between lithology, land use and erosion in the Oliva Valley, Calabria, Italy (Rao, 1975).

Geomorphological Map

Legend:

I. Lithology

Volcanic rocks/deposits:

basic rocks - slightly or unweathered - strongly weathered

intermediate/acid rocks ” ”

tuffs/pyroclastics ” ”

coarse ashes and pumice ” ”

fine ashes ” ”

Igneous (plutonic) and metamorphic rocks

crystalline rocks ” ”

igneous-metamorphic complexes ”

metamorphic schists ” ”

Sedimentary rocks

clays resistant

silts ”

sands ”

conglomerates ”

lutitas intercalated resistant beds

lutitas with non-resistant

agglomerates

Quaternary alluvium

old and middle Quaternary

young Quaternary

Extra symbols/hachures

thin covers over other material: hachures over other symbol (color)

partly calcareous formations: open circles over other symbol (color)

II. Forms and Processes

A. Structural data

faults - certain/conjectural

fractures ” ”

dip ” ”

B. Hydrography and fluvial dynamics

streams - permanent, seasonal, ephemeral

gully

waterfall

rapid

meandering river

Catchment	Drainage area km2	Sediment yield ton/km^2/year	Source
Volcanic			
Ciliwung	130	250–375	Rutten
k Rambut	4.5	532	van Dijk
k Banyuputih	225	750–1000	Rutten
k Brantas	10000	875–1500	Rutten
Mainly volcanic			
Citarum	73000	800–1200	Modified after Soemarwoto
Cimanuk	3000	1000–2000	Rutten
Mixed volc – Sedimentary			
k Tandjum	210	750–1000	Rutten
Cilamaya	225	2500–3500	Rutten
k Lusi	860	2500–3500	Rutten
k Serayu	> 700	3500–4500	Rutten
k Cilitung		7500	Nedeco
k Solo			Soemarwoto
Sedimentary			
k Jragung	101	4000–6250	Rutten
k Cacaban	7.9	6600	Rutten
k Pengaron	41	9250–12500	van Dijk

Table 15.4 Influence of lithology on sediment yields, as observed for a one year period (Meyerink, 1975).

braided river
old river course
overflow channel
beach
'salida de madre'
sapping or undercutting
potential dam/reservoir site
site of potential sedimentation
swamp
lagoons - permanent - seasonal

C. Topographical forms
overhang
scarp
divide line
terrace
fan
glacis
scree - active - stabilized
torrent deposits - active - stabilized
texture of deposits (lettering system)
age of Quaternary formation

D. Water erosion
intense diffuse run-off - without loss of soil-sheet erosion
concentrated (linear) run-off - without erosion
" with considerable damage
" rills - gullies - riverbed erosion
headward erosion

E. Mass movements
terracettes - active - stabilized
solifluction - in sheets active - stabilized
" in 'lupas' (slices)
mudflow - active - stabilized
slide "
scree "
spoon-shaped erosion - active - stabilized
rock falls, etc. "

F. Critical areas
sub-watershed
other zone
endangered road section

If an erosion survey results in an overwhelming amount of data,even two maps may not suffice to present all information adequately. In this case one may, e.g., produce special maps for lithology and superficial formations, for the vegetative cover, and for the erosion processes. Ultimately a general erosion map (series) may then be produced which is the outcome of the assessment of the data collected first in the other maps produced. A suggested legend for each of these maps is given below.

Lithology and Surficial Formations

Lithology:
claystone
sandstone
silts/loams

Vegetation Land use	Station	elevation a s l	mean annual rain-fall	Runoff in % of rainfall during observation period min	max	average
Bare soil	4 stations, W Java					25–55%
	Janlappa	100	3000			32
	Monggot	150	2200			42
Dry cultivation	Janlappa	100	3000	13	19.5	16.2
	Ciwidej	1750	3200	4.7	17.3	10.6
	Klakah	200	3200	–	–	11.9
	Ngadisari	2000	1500	10.1	10.3	10.2
	Cobarrondo	1500	1800	–	–	2.0
Young forest plantation <2½ years old	Ciparaj	–	–	2.1	20.6	9.5
	Monggot	150	2200	9.0	10.7	9.5
	Monggot	150	2200	2.8	7.5	2.8
Grass, mainly alang-alang (Imperata)	Monggot	150	2200	2.5	8.9	5.4
	Klakah	200	3200	–	–	5.0
	Janlappa	100	3000	0.35	7.6	3.9
	Cobarrondo	1500	1800	0.2	3.0	1.6
	Ciwidej	1750	3200	0.1	1.7	0.5
	Arcamanik	1300	2500	0.28	0.28	0.28
Bamboo + grass	4 stations W Java	–	–	–	10–20	–
				–	5–10	–
Jungle	Janlappa	100	3000	1.0	4.3	2.6
	Klakah	200	3200	–	–	2.1
	Ciwidej	1750	3200	0.35	2.7	1.5
Forest						
Teak	Monggot	150	2200	–	8.0	–
Mahogany	Monggot	150	2200	2.4	4.6	3.6
Thinned Pinus	Arcamanik	1300	2500	3.2	3.2	3.2
Rain forest	Ciparaj	1000	4000	3.5	12.3	6.2
Rain forest	Bogor	–	–	–	–	2.4
Rain forest	Arcamanik	1300	2500	0.55	–	2.4
Rain forest	Ciwidej	1750	3200	0.4	2.4	1.5
					4.5	1.3

Table 15.5 Run-off from small plots under different vegetation on Java, data after Coster (1938)

clays with coherent sand/limestone layers (30% & 50%)
compact claystone
siltstone
sandstones
limestone
alluvial terraces
mudflow
proglacial plain (sandy)
fine-textured colluvium
fluvio lacustrine deposits
lake deposits

Surficial deposits:
weathered rock, soils, colluvial material, etc.

Some major soil characteristics:
thickness - <25cm - 25-75 cm >75 cm
texture - heavy - medium - light
percentage coverage surface deposits: circular graph

Complementary annotations:
presence in/near surface deposits of
large blocks <25% - 25-75% ->75%
small blocks/numerous boulders " "
concentration of gravel at surface " "
moist horizon " "

Vegetative cover:
Dominant vegetative cover:
non-vegetated
85% non-vegetated and 15% various
grassland & agricultural land
dispersed grass & barren
65% - 75% grass vegetation
scrub, etc.
trees

Percentage of various vegetation types:
circular graph, e.g., 50% cultivated; 10% grass & crops; 25% low grass; 5% high grass; 5% scrub; 5% trees

Erosional processes on analytical map :
Dominant process
normal diffuse run-off (no erosion; coherent soil; protection by vegetation)
pluvial (splash) erosion (with or without diffuse run-off on weak materials)
intense diffuse run-off - without rills - with rills (stabilized gullies)
numerous active gullies
badlands

Secondary processes
percentage area affected; circular graph

Additional annotations
depth of gullies - <1m; 1-3m; >3m
processes originating in other terrain unit

Major ravines and gullies
stabilized
active - in weak materials - with some resistant material - with cover of blocks
lateral erosion in main rivers
areas occupied by man with subsequent development of erosion

Factors slowing down erosion
cover with gravel
cover of boulders/blocks
outcrop of hard rocks
presence of hard rock in subsoil
resistant soil
gentle slope
vegetation

SUMMARY LEGEND OF GENERAL EROSION MAP

EROSION CLASS	STATIC ASPECT	DYNAMIC ASPECT
no erosion	everywhere topsoil preserved deeper than 25cm with humic horizon	diffuse run-off; protective vegetation
slight erosion	in 75% area topsoil preserved 25cm; remainder < 25cm	splash erosion diffuse run-off if soil or vegetation poor
moderate erosion	in 25%-75% area topsoil preserved >25cm deep	sheet erosion and some rills; important colluvium
strong erosion	less than 25% area topsoil preserved (thin)	at foot of slope intense sheet erosion & rills & gullies
very strong erosion	no topsoil left	large gullies & badlands

Althouth it is clear from the examples given that the information required in erosion surveys covers a field much wider than geomorphology only, it is also evident that a thorough geomorphological analysis of the terrain and the subsequent landscape ecological assessment of each of the geomorphological units distinguished is a valid approach to the study of accelerated erosion by water. A particularly elucidating case is presented by Schmitz (1980) from the Thaba Bosiu area in Lesotho where an erosion survey has been carried out by ITC, Enschede,and the National University of Lesotho (NUL) jointly (Schmitz and Verstappen, 1978).

The area of study is composed of subhorizontal sedimentaries of the Stormberg series, dating from the Triassic. The oldest beds belong to the Beaufort series which are covered by the Molteno beds, also the fairly soft Transition and Red beds and finally, by the scarp-forming Cave sandstones. The sequence is in the mountains and the foothills covered by a thick sequence of basaltic beds, the so-called Drakensberg beds. Near-vertical dolorite dykes are a common occurrence and are morphologically conspicuous. The profile of fig. 15.22 gives the geomorphological units distinguished within the Thaba Bosiu area, which have been mapped at a scale of 1:50,000 (fig. 15.23)*. These units appeared to be strongly correlated not only with the main soil series but particularly also with the erosion processes and the erosion features caused by them,as will be evident from the tables 15.6 and 15.7. The plateau is, in part, badly eroded and so is the accumulation glacis where rill and gully erosion spreads headward and sideways and deepens in mantles of colluvium. The aerial photographs selected for the study of gully growth using sequential aerial photographs (fig. 15.13) demonstrate this. Large gullies, locally known as 'dongas', can be traced in the depressions and valley fills. The widening of the dongas is effected more by headward retreat, branching and spur reduction than by massive wall retreat. Piping is important at donga heads. Sheet erosion is widespread locally, for example, on the higher erosion glacis. A particularly interesting phenomenon is the soil scarp collapse that occurs at places on sandstone plateaus and rocky shelves and to some extent also on basaltic slopes. Seepage water in-

*For fig. 15.23 see p. 398

Geomorphological units	Dominant type of process *erosive*	Resultant features
Steep basaltic slopes	soil slip	cattle steps, terracettes
	rilling	rills
Less steep basaltic slopes	soil slip	cattle steps, terracettes
Plateaux, rock shelves	soil scarp collapse	bare rock surfaces
Debris slopes	rilling	rills
	headward erosion	donga heads
Planation surfaces	sheet wash	truncated soils
		bare rock exposures on rims
	soil scarp collapse	bare rock exposures on rims
Footslope glacis	piping	tunnels, collapse holes
	gullying	discontinuous dongas
		badlands
	rilling	discontinuous rills
Valley glacis	piping	tunnels, collapse holes
	gullying	second generation dongas, strongly branched, badlands
Alluvial terrace edges	headward erosion	donga heads
	slumping	stepped, hummocky slopes
	accumulative	
Footslopes, depressions	colluviation	colluvial wedges
Floodplains	sedimentation	alluvial accumulations
Stream channels	sedimentation	sand banks, gravel banks
Dams	sedimentation	alluvial accumulations

Table 15.6 Processes and forms of accelerated erosion and sedimentation.

Geomorphological unit	Main associated soil series
Steep basaltic slopes	
upper slopes	Popa
middle slopes	Matsana
Footslopes glacis in basalt	Fusi
Planation surfaces in basalt	
undissected	Machache
very mildly dissected	Matsa'aba
mildly dissected	Ralebese
Valley glacis in basalt	Thabana
Sandstone plateaus	
bordering basaltic outcrops	Matela
undissected	Berea
upland areas	Ntsi
footslope glacis	Thoteng
valley glacis	Theko
Planation surfaces in sedimentary strata	
crests and upper slopes	Leribe
middle slopes	Rama
Accumulation glacis in sedimentary strata	
footslope glacis	Maliele
valley glacis	Sephula, Tsiki, Thotheng, Theko
Infilled valleys	Maseru, Phechela
Highes accumulation terrace (t_1)	Khabos
gravel mantled deposits	Kubu
Middle accumulation terrace (t_2)	Phechela, Khabos, Maseru

Table 15.7 Geomorphological units with main associated soil series (Schmitz, 1980).

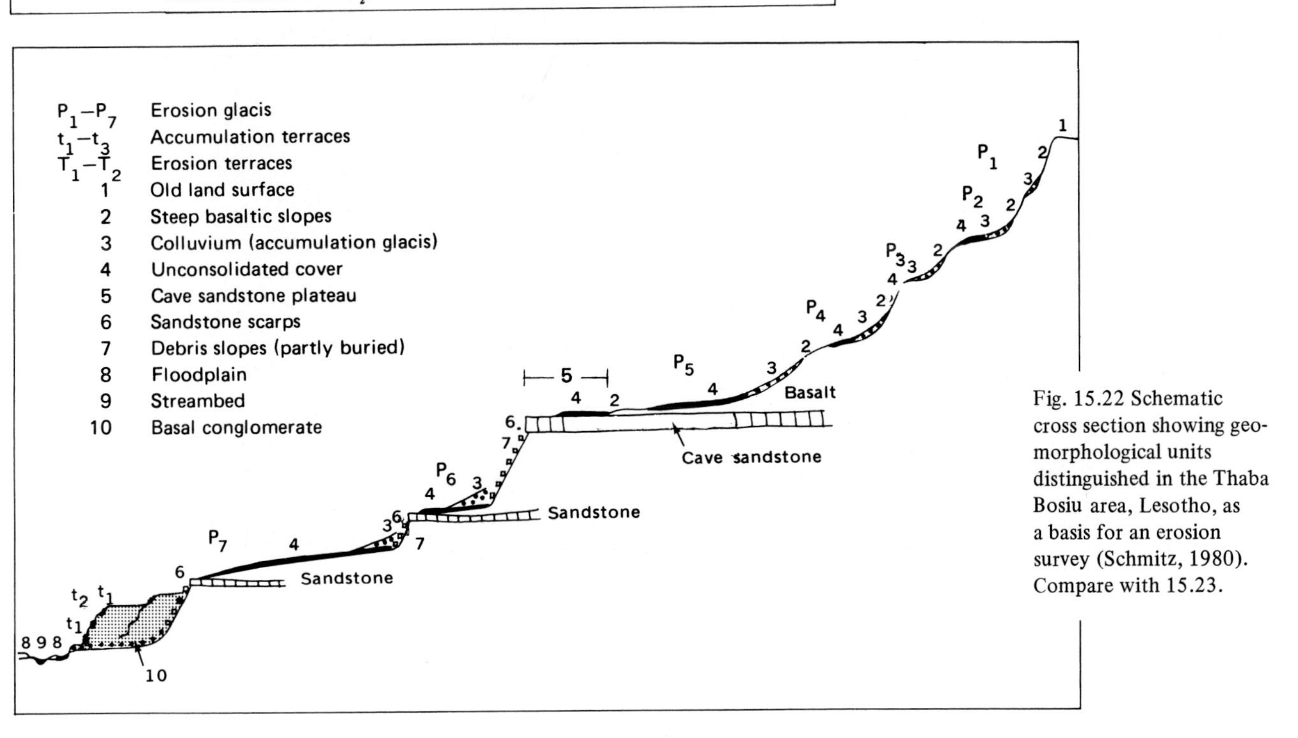

Fig. 15.22 Schematic cross section showing geomorphological units distinguished in the Thaba Bosiu area, Lesotho, as a basis for an erosion survey (Schmitz, 1980). Compare with 15.23.

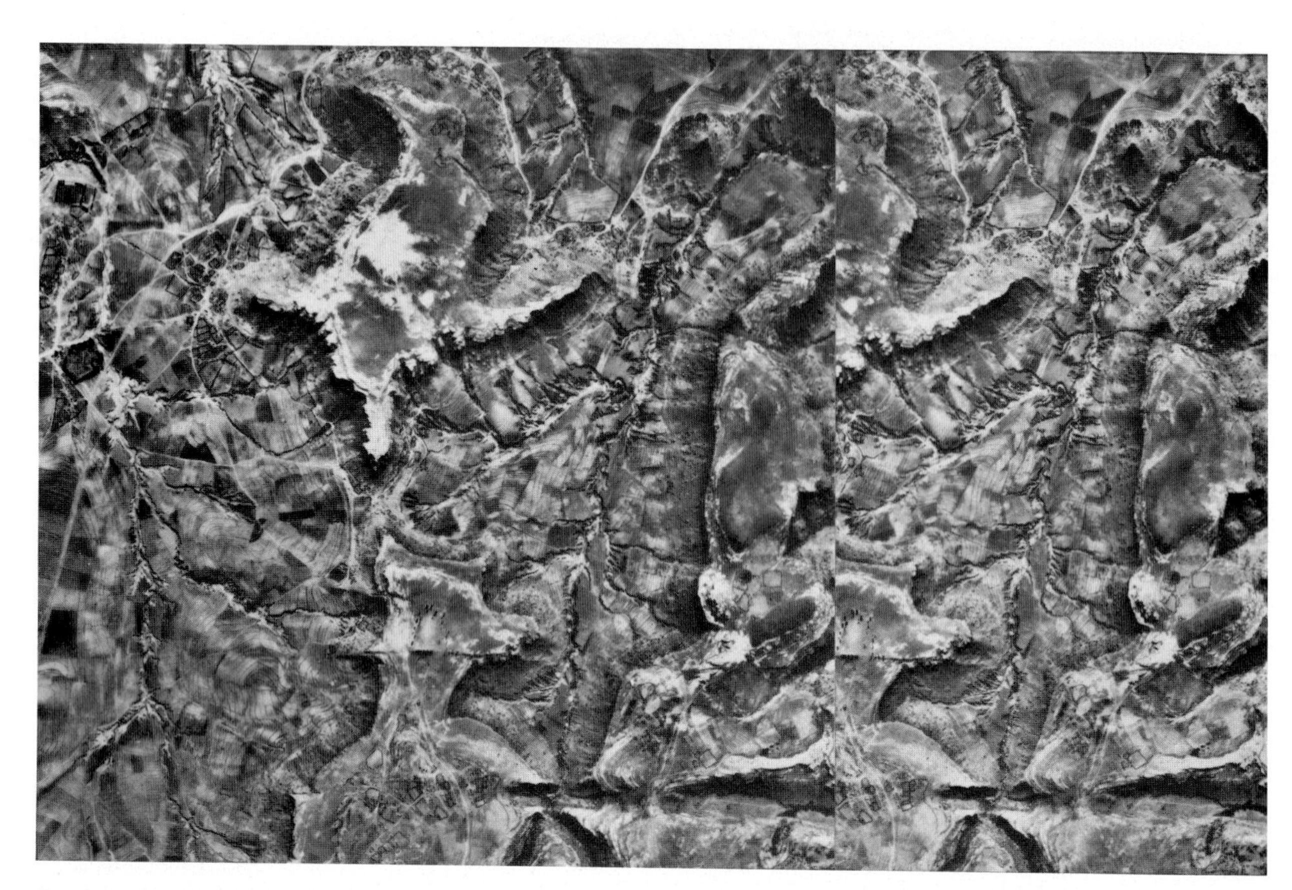

Fig. 15.24 Extended stereopair, scale 1:50,000 showing erosion features in the Thaba Bosiu area, Lesotho, and specifically the whitish areas affected by scarp-collapse erosion. Compare with fig. 15.25.

Fig. 15.25 Ground view of an area affected by the scarp-collapse erosion phenomena pictured in the aerial view of fig. 15.24. Only barren rock is left, wetted by emerging groundwater.

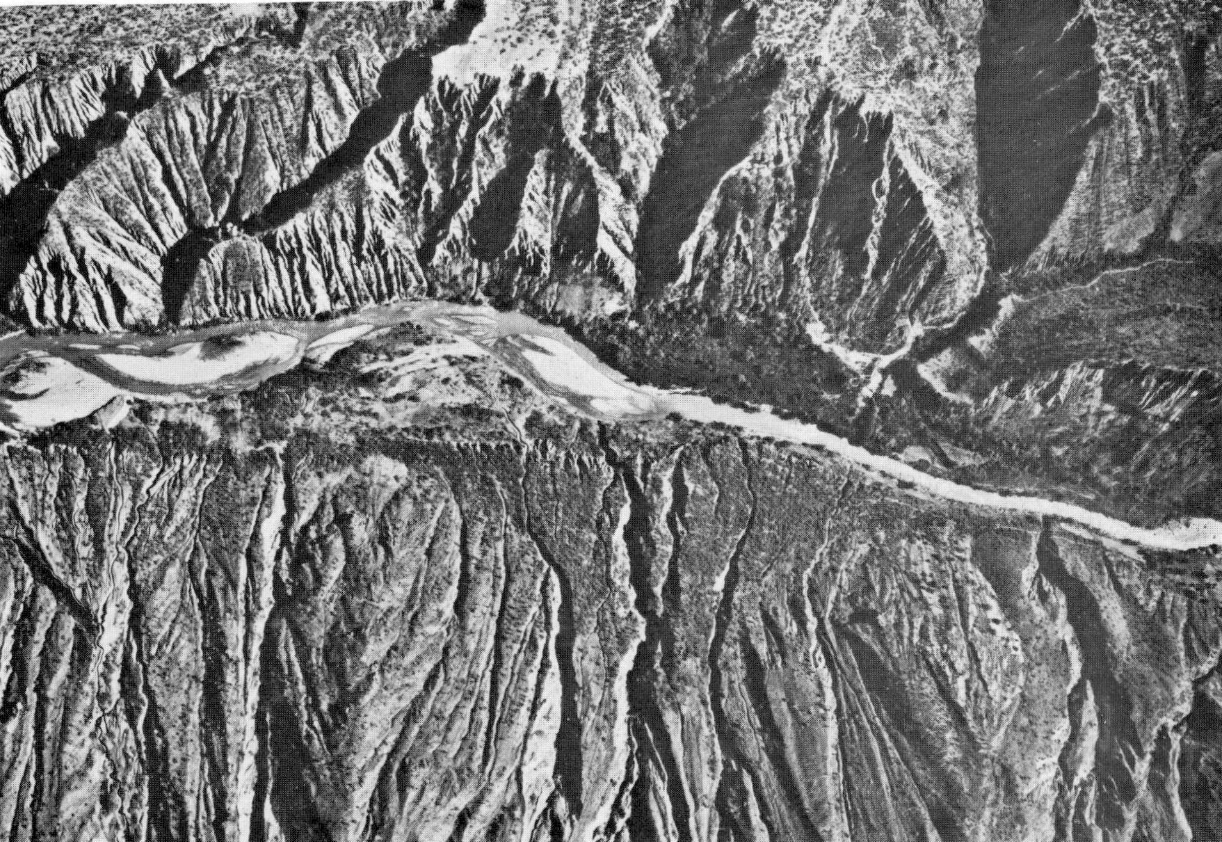

Fig. 15.26 Violent accelerated linear erosion in the Chama Valley, Venezuelan Andes. Top: active retrogressive erosion on steep slopes with slumping in the heads of ravines. Bottom: differences in gully erosion in gravelly terrace scarps (top) and the long slopes at the opposite side of the river. Scale 1:20,000.

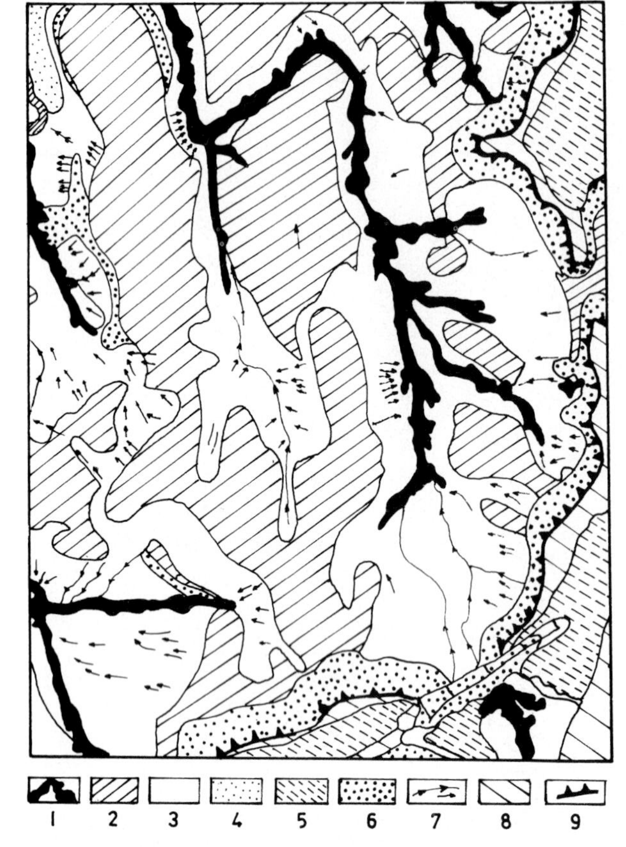

Fig. 15.23 Part of the geomorphological map of the Thaba Bosiu area. Compiled by Schmitz (1980) for purposes of erosion survey Compare with fig. 15.22
Key:
1. major ravines (donga)
2. planation surface in sedimentary strata
3. accumulation glacis
4. accumulation terrace, valley infill
5. higher sandstone plateau
6. debris slope
7. rills and gullies
8. lower sandstone plateau
9. escarpment

filtrates into the thin soil layer and a lateral subsurface flow is generated on the slopes when it reaches the underlying rocks. It emerges again at the lower end where a gradually upslope migrating scarp is formed. The extended stereopair of fig. 15.24 gives an aerial view and fig. 15.25 a ground view of this peculiar erosion process that has only been described by a few investigators.

The state of erosion may be judged from the combination of processes occurring within a unit and their relative importance. The activity of erosion may be judged from the percentage cover of vegetation in addition to other indications and e.g., in the case sequential aerial photography is not available.

Because of the variety of conditions under which accelerated erosion by water occurs, no standardized approach to the problem can be offered. The geomorphologist in charge will have to use his inventiveness to select the most appropriate survey methods for the area studied. This will be obvious from a comparison of the erosion features in the flat lands pictured in fig. 15.8 and the violent erosion processes on the steep slopes of the Chama Valley in the Venezuelan Andes pictured in fig. 15.26.

References

Aarstad, J.S., and Miller, D.E., 1973. Soil management to reduce runoff under center-pivot sprinkler system. J. Soil and Water Conserv., 28(4): 171–173.

Aghassy, J., 1973. Main-induced badlands topography. In: D. R. Coates (Editor), Environmental geomorphology and landscape conservation. Vol. 3, Non-urban regions. Benchmark Papers in Geology, Dowden, Hutchinson & Ross Inc., John Wiley & Sons Ltd.: 124–136.

Anderson, E.W., and Brooks, L.E., 1975. Reducing erosion hazard on a burned forest in Oregon by seeding. J. Range Management, 28(5): 394–397.

Araya, J.F., 1964. La cartografía de los procesos morfogenéticos actuales en Chile semi-árido-transicionale. Revista Geografica, 4/5, 11/13: 53–77.

Arnoldus, H.M.J., 1974. Soil erosion. A review of processes, assessment techniques. Proc. FAO/UNDP expert consultation on soil degradation. AGL: SD-74/7, Rome

Asch, Th.W.J. van, 1980. Water erosion and landsliding in a mediterranean landscape, Thesis, Utrecht State Univ., 236 pp.

Bakanina, F.M., 1972. Water-retaining dikes–an effective means of protecting soils from erosion (in russian), Gidrotekhnika i melioratsiya, 6: 44–46.

Barnett, A.P., et al., 1971. Erodibility of selected tropical soils. Trans. Amer. Sci. Agric. Eng., 14(3): 496–499.

Barton, K., 1972. The need for land stabilization in the Wabash Valley. Indiana State Univ., Dept. Geogr. Geol., Prof. Paper, 3: 3–23.

Beasley, R.P., 1972. Erosion and sediment pollution control. Iowa State Univ. Press, 320 pp.

Bell, G.L., 1968. Piping in the badlands of North Dakota. Proc. 6th Annual Eng. Geol. and Soils Eng. Symp., Boise Idaho (Boise, Idaho Dept. Highways): 242–257.

Bennett, H.H., 1939. Soil Conservation, McGraw-Hill, New York, 993 pp.

Bergsma, E., 1974. Soil erosion sequences on aerial photographs. ITC Journal, 3: 342–376.

Bergsma, E., 1975. Field boundary gullies in the Serayu River-basin, Central Java. ITC Journal, 3: 75–91.

Bergsma, E., 1981. Indices of rain erosivity. ITC Journal, 4: 460–484.

Black, A.L., and Siddoway, F.H., 1971. Tall wheatgrass barriers for soil erosion control and water conservation. J. Soil & Water Conserv., 26(3): 107–111.

Blase, M.G., and Timmons, J.F., 1965. Soil erosion control problems and progress. J. Soil and Water Conserv., 16: 1961: 157–162. Repr.. in: I. Burton and R.W. Kates (Editors), Readings in Resource Management and Conserv., Univ. Chicago Press, Chicago: 338–347.

Boodt, M. de, and Gabriels, D. (Editors), 1980. Assessment of erosion. John Wiley & Sons., Chichester, 536 pp.

Boon, W., and Savat, J., 1981. A nomogram for the prediction of rill erosion. Proc. Conf. Conserv., 80, Silsoe, U.K.: 303–319.

Bryan, R.B., 1968. The development, use and efficiency of indices of soil erodibility. Geoderma, 2: 5–26.

Bryan, R.B., 1974. A simulated rainfall for the prediction of

soil erodibility. Ztschr.f. Geom., Suppl., 21: 138–150.
Cadiot, B., 1979. Les mouvements de terrain en Ardèche: approche historique et essai de zonage régional du risque. Bull. B.R.G.M., 3(1): 49–58.
Călinescu, M., 1967. L'établissement d'un indice d'érosion en vue de l'amélioration des terrains dégradés de la Plaine de Transsylvanie. Symp. Int. Géom. Appl., Bucharest: 77–84.
Cambell, I.A., 1974. Measurements of erosion on badlands surfaces. Ztschr.f.Geom., Suppl., 21: 122–137.
Carrero, M.J.A., 1978. Procesos erosivos en la cuenca media del Rio Chama, Estado Merida, Venezuela. M.Sc.thesis, ITC Enschede, 180 pp.
Carson, M.A., and Kirkby, M.J., 1972. Hillslope form and process. Cambridge Univ. Press, 475 pp.
Chakela, Q.K., 1981. Soil erosion and reservoir sedimentation in Lesotho. Thesis Uppsala Univ. Naturgeogr. Inst. Report, 54, 150 pp.
Chorley, R.J., 1978. The hillsope hydrological cycle. In: Kirkby, M.J. (Editor), Hillslope Hydrology. Wiley Publ. Comp., Chichester, USA, 389 pp.
Christianson, C., 1981. Soil erosion and sedimentation in semi-arid Tanzania. Medd. Dept. Phys. Geogr. Univ. Stockholm, A 119, 208 pp.
Ciccacci, S., et al., 1980. Contributo dell analisi geomorfica quantitativa allo valutazione dell'entita dell'erosione nei bacini fluviali. Boll. Soc. Geol. Ital., 99: 455–516.
Clément, P., and Bonn, F., 1975. Réactions de surfaces terrestres aux facteurs climatiques. (Reactions of land surfaces to climatic factors). Geos: 5–7.
Datiri, B.T., 1974. Soil and water conservation practices particularly in areas of shifting cultivation. Soils Bull., F.A.O., 24: 237–241.
Diamond, S., and Kawamura, M., 1974. Soil stabilization for erosion control. Interim report November'71–May'74. Purdue Univ., Lafayette (Ind.). Joint Highway Research Project, 74–12
Dzhadan, G.I., et al., 1975. Effects of the degree of erodibility of soils on their agrochemical properties and grain crop yield. Soviet Soil Science, 7(5): 579–582.
Dzyadevich, I.A., 1970. Prevention and control of water erosion on irrigated land. Soviet Hydrol., Selected Papers, 5: 492–504.
Ekern, P.C., 1953. Problems of raindrop impact erosion. Agric. Eng., 34(1): 23–28.
Ellison, W.D., 1952. Raindrop erosion and soil erosion. Emp. J. Exp. Agric., 20(78): 81–97.
Ellwell, H.A., 1978. Soil loss estimation. Publ. Dept. Conserv. and Extension, Salisbury, Zimbabwe
Emmett, W.W., 1970. The hydraulics of overland flow on hill slopes. USGS Prof. Paper, 662–4
F.A.O., 1956. Soil erosion by water, some measures for its control on cultivated lands. Rome, 284 pp.
F.A.O., 1977. Assessing soil degradation. F.A.O. Soils Bull., 34, Rome, 146 pp.
Flannery, R.D., 1976. Handbook for gully control and reclamation. Lesotho Agric. College, Maseru, 152 pp.
Fournier, F., 1972. Aspects of soil conservation in the different climatic and pedologic regions of Europe. (Nature and Environment Series, Council of Europe and H.M.S.O., London), 194 pp.
Fournier, F., 1972. Rational use and conservation of soil. Geoforum, 10: 35–48.
Garland, G.G., 1979. Rural, non-agricultural land use and rates of erosion in the Natal Drakensberg. Envir. Conserv., 6(4): 273–276.
Garland, G.G., and Humphrey, B., 1980. Assessment and mapping of land erosion potential in mountainous recreational areas. Geoforum, 11(1): 63–70.
Giardino, J.R., 1974. When elephants destroy a valley. Geographical Magazine, 47(3): 174–181.
Gill, N., 1979. Watershed development with special reference to soil and water conservation. F.A.O. Soils Bull., 44, 257 pp.
Gregory, K.J., and Walling, D.E., 1973. Drainage basin form and process, a geomorphological approach. Arnold, London, 456 pp.
Harrison, E.A., 1973. Erosion control methodology. A bibliography with abstracts. U.S. Nat. Techn. Inf. Serv., Springfield (Va.), Report 1964, 73 pp.
Harrold, L.L., and Eswards, W.M., 1972. A severe rainstorm test of no-till'corn. J. Soil and Water Conserv., 27(1), 30 pp.
Held, R.B., et al., 1962. Soil erosion and some means for its control. Agric. and Home Ec. Expl. Station, Iowa State Univ. of Sci. and Techn. Spec. Report, 29, 32 pp.
Held, R.B., and Clawson, M., 1965. Soil conservation in perspective. John Hopkins Press. Baltimore, 344 pp.
Holy, M., 1980. Erosion and environment. Oxford Univ. Press, 225 pp.
Hudson, N., 1971. Soil Conservation. Batsford, London, 320 pp.
Imeson, A.C., and Kwaad, F.J., 1980. Gully types and gully prediction. Geogr. Tijdschr., 14(5): 430–441.
Inquist, R.C., 1980. Conceptual modelling of landslide distribution in space and time. Bull. Int. Ass. Eng. Geol., 21: 178–186.
Jones, A.C., 1976. Protected lands in New South Wales. J. Soil Conserv. Serv. N.S.W., 32(1): 30–33.
Joshua, W.D., 1977. Soil erosive effect of rainfall in the different climatic zones of Sri Lanka. Proc. Symp. Erosion and Soil Matter Transport in inland waters. Publ. IASM, 122
Kawamura, M., and Diamond, S.,1975. Stabilization of clay soils against erosion loss. Techn. paper, Purdue Univ., Lafayette (Ind.). Joint Highway Research Project, 75–4, 36 pp.
Kawamura, M., and Diamond, S., 1975. Stabilization of clay soils against erosion loss. Clay and Clay Minerals, 23(6): 444–451.
Kevie, W. van der, 1976. Manual for land suitability classification for agriculture. Part II, Guidelines for soil survey party chiefs. FAO/UNDP Wad Medani
Kirkby, M.J., and Chorley, R.J., 1967. Throughflow, overland flow and erosion. Bull. I.A.S.H., 12: 5–21.
Kirkby, M.J., and Morgan, R.P.C. (Editors), 1980. Soil erosion. John Wiley & Sons. Chichester, 326 pp.
Kinnell, P.I.A., 1974. Splash erosion: some observations on the splash-cup technique. Proc. Soil. Sci. Soc. Amer., 38 (4): 657–660.
Kronfelder-Kraus, G. (Editor), 1981. Beiträge zur Wildbacherosion und Lawinenforschung. Mitt. Forstl. Bundes-Versuchsanstalt, Wien, 138, 162 pp.
Lal, R., 1974. Soil erosion and shifting agriculture. F.A.O. Soils Bull., 24: 48–71.
McIntyre, D.S., 1958. Soil splash and the formation of surface crusts by raindrop impact. Soil Sci., 5(1): 57–74.
Meade, R.H., 1969. Errors in using modern stream-load data to estimate natural rates of denudation. Bull. Geol. Soc. Amer., 80(7): 1265–1274.
Meijerink, A.M.J., 1975. Hydrological reconnaissance survey of the Serayu Riverbasin. ITC Journal, 3: 25–53.
Mensching, H., 1974. Angewandte Geomorphologie,Beispiele aus den Subtropen und Tropen. Verh. 42, D. Geogr. Tag Göttingen: 25–34.
Mersereau, R.C., and Dyrness, C.T., 1972. Accelerated mass wasting after logging and slash burning in western Oregon.

J. Soil and Water Conserv., 27(3): 112–114.
Meyer, J.D., and Mannering, J.V., 1971. The influence of vegetation and vegetative mulches on soil erosion. In: Biological effects in the hydrological cycle, Proc. 3rd Int. Seminar Hydrol. Professors, 1971. Purdue Univ. (Ind.): 355–366.
Meyer, L.D., et al., 1972. Stone and woodchip mulches for erosion control on construction sites. J. Soil and Water Conserv., 27(6): 264–269.
Miholics, J., 1970. Some questions on the geomorphological investigation of soil erosion for practical purposes. Földrajzi Ertesito, 19(2): 135–144.
Monjuvet, G., 1973. L'érosion sur les Alpes françaises d'après l'exemple du Massif du Pelvoux. Rev. Géogr. Alpine, 61 (1): 107–120.
Morgan, R.P.C., 1979. Soil erosion. Topics in Applied Geography. Longman, London, 113 pp.
Morgan, R.P.C. (Editor), 1980. Soil conservation, problem and prospect. John Wiley & Sons, Chichester, New York, 576 pp.
Morrison, R.B., and Cooley, M.E., 1973. Application of ERTS-1 imagery to detecting and mapping modern erosion features and to monitoring erosional changes in S. Arizona. Progress Report 72–75. Goddard Space Flight Center, Greenbelt (Maryland)
Mosley, M.P., 1974. Experimental study of rill erosion. Trans. Amer. Soc. Agric. Eng., 17: 909–913.
Nir, D., and Klein, M., 1974. Gully erosion induced in land use in a semi-arid terrain (Hanal Shigma, Israel). Ztschr.f. Geom., Suppl., 21: 191–201.
Obeng, H.B., 1972/1973. Problems of soil erosion in cereal crop production in Ghana. In: O.A.U./S.T.R.C. Seminar on the environmental factors influencing the yield of tropical food crops, Dakar, July 1971. African Soils, 17, (1–3): 75–83.
Ologe, K.O., 1972. Gullies in the Zaria, 1(1): 55–66.
O'Loughlin, C.O., 1974. The effect of timber removal on the stability of forest soils. J. Hydrol., New Zealand, 13(2): 121–134.
Orlovskiy, V.B., 1975. Increasing the effectiveness of erosion control by forest plantations (in russian). Pochvovedeniye, 8: 119–122.
Perrusset, A.Chr., 1981. Glissements de terrain et aménagement du territoire: principes de réalisation d'une carte prévisionnelle de risques quantifiés. Bull. Int. Ass. Eng. Geol., 23: 7–11.
Peasly, B.A., 1975. Soil conservation in the Narrabri area. J. Soil Cons. Service N.S.W., 31(1): 9–18.
Poelman, J.N.B., 1971. Erosie van lössgronden (Water erosion of dutch loess soils). Boor en Spade, 17: 177–187.
Posey, J., 1973. Erosion-proofing drainage channels. J. Soil and Water Conserv., 28(2): 93–95.
Prochal, P., 1972. Soil erosion in the hill and mountain regions of Poland (in polish). Zeszyty Problemowe Postepów Nauk Rolniczych, 138: 73–84.
Quilty, J.A., 1973. Soil conservation structures for marginal arable areas. A field study. J. Soil Cons. Service N.S.W., 29(3): 119–129.
Rao, D.P., 1975. Applied geomorphological mapping for erosion surveys: the example of the Oliva Basin, Calabria. ITC Journal, 3: 341–350.
Rapp, A., et al., 1972. Studies on soil erosion and sedimentation in Tanzania. Geografiska Annaler, 59A(3/4), 379 pp.
Rathjens, C., 1979. Die Formung der Erdoberfläche unter den Einfluss des Menschen. Teubner Studienbücher d. Geographie. Stuttgart, 160 pp.
Rauschkolb, R.S., 1971. Land degradation. F.A.O. Soil Bull., 13
Richardson, C.W., 1973. Runoff, erosion, and tillage efficiency on graded-furrow and terraced watersheds. J. Soil and Water Cons., 28(4): 162–164.
Richter, G., 1965. Bodenerosion. Schäden und gefährdete Gebiete in der Bundesrepublik Deutschland. Bundesanstalt f. Landesk. u. Raumforschung, Bad Godesberg. Publ. 152, 592 pp.
Richter, G., 1973. Schutz vor Bodenerosion–ein wichtiger Bestandteil des Umweltschutzes. Geogr. Rundschau, 10: 377–386.
Rossi, G., 1979. L'érosion à Madagascar: l'importance des facteurs humains. Cahiers d'Outre Mer, 32(128): 355–370.
Schwab, G.O., et al., 1981. Soil and Water Conservation Engineering. Wiley, New York, 525 pp.
Selby, M.J., 1972. The relationships between land use and erosion in the central North Island, New Zealand. J. Hydrol (New Zealand), 11(2): 73–87.
Servant, J., 1974. Un problème de géographie physique en Polynésie Française: l'érosion, exemple de Tahiti. Cahiers ORSTOM Sér. Sci. Humaines, 11(3–4): 203–210.
Shukla, P.N., 1974. Soil conservation and its beneficiary effects in district Etawah (U.P.). Geogr. Observer, 10: 59–65.
Siemens, J.C., and Oschwald, W.R., 1976. Erosion for corn tillage systems. Trans. Amer. Soc. Agric. Eng., 19(1): 69–72.
Sorby, E.D., and Young, N.S.W., 1974. Erosion and its control on orchards. J. Soil Cons. Services N.S.W., 30(4): 174–184.
Schmitz, G., 1980. A rural development project for environmental management in Lesotho. ITC Journal, 2: 349–363.
Schmitz, G., and Verstappen, H.Th. (Editors), 1978. Some methods in resources surveys, with examples, from Lesotho. Report, NUFFIC/ITC–6. N.U.L. Project, 178 pp.
Stebelsky, I., 1974. Environmental deterioration in the central Russian black earth region: the case of soil erosion. Canadian Geographer, 18(3): 232–249.
Stehlik, O., 1975. The problem of soil erosion in the formation of environment on the territory of the Czech Socialist Republic. Stud. Geom. Carpatho-Balcanica, 9: 159–167.
Stocking, M.A., 1972. Relief analysis and soil erosion in Rhodesia using multivariate techniques. Ztschr.f.Geom., 16(4): 432–433.
St. Onge, D.A., and Lengelle, J., 1974. Erosion susceptibility maps, Swan Hills, Alberta, Canada. Geol. Surv., Open File Report, 169/240
Thronson, R.E., 1973. Comparative costs of erosion and sediment control construction activities. Eng. Sci. Inc. Berkely (Calif.), 211 pp.
Toy, T.J. (Editor), 1977. Erosion: research techniques, erodibility and sediment delivery. Geo-abstracts, Norwich, 86 pp.
Tufescu, V., La géomorphologie au service de l'amélioration des terrains dégradés en Roumanie. Symp. Int. Géom. Appl., Bucharest, 3 pp.
Ven Te Chow (Editor), 1964. Handbook of applied hydrology. McGraw-Hill, New York, 1500 pp.
Werner, D., 1966. Entwicklung, Leistungspotential und Erosions disposition abtragungsgeschädigter Böden in Thüringen sowie Möglichkeiten ihrer Verbesserung. Forschungsabschlussbericht 3504 60/4 – 05/1. Inst.f. Miliorationswesen Univ. Jena
Williams, D.F., 1978. The identification and location and intensity of gully erosion, Basilicata Province, southern Italy. Unpubl. Ph.D. Thesis, Reading Univ.
Wischmeier, W.H., 1958. Evaluation of factors in soil loss equation. Agric. Eng., 39: 435–462.

Wischmeier, W.H., 1976. Use and misuse of the universal soil loss equation. J. Soil and Water Conservation, 31(2)
Wischmeier, et al., 1971. A soil erodibility nomograph for farmland and construction sites. J. Soil and Water Conserv., 26: 189–198.
Wischmeier, W.H., and Smith, D.D., 1962. Soil loss estimation as a tool in soil and management planning. Int. Ass. Sci. Hydrol., Publ. 59: 138–159.
Wischmeier, W.H., and Smith, D.D., 1978. Predicting rainfall erosion losses: a guide to conservation planning. U.S. Dept. Agric. Handbook 537
Yamamoto, T., and Anderson, H.W., 1973. Splash erosion related to soil erodibility indexes and other forest soil properties in Hawaii. Water Res. Research, 9(2): 336–345.
Zaborski, B., 1971(1972). On the origin of gullies in loess. Acta Geogr. Debrecina, 10: 109–112.
Zapletin, V.Y., et al., 1969. Anti-erosion organization of territory and use of land under the conditions of the Central Chernozem Zone and problems of its anti-erosion organization (in russian). Zapiski Voronezhskogo Sel' skokhozyoystvennogo Inst., 39, 184 pp.
Zaslavski, M.N., 1975. Methodological problems in constructing a general system of erosion control practices for the U.S.S.R. (in russian). Pochvovedeniye, 9: 115–122.
Zuidam, R.A. van, and Zuidam-Cancelado, F.I., van, 1979. Terrain analysis and classification using aerial photographs, a geomorphological approach. ITC Textbook VIII.6, 310 pp.

Chapter 16

AVALANCHE MAPPING

16.1 The Avalanche Hazard, General

Avalanches are a major natural hazard in many populated mountainous areas and particularly in alpine mountains at middle and high latitudes. In some more densely populated mountain ranges such as the Alps in Europe, the avalanche danger has increased considerably since extensive deforestation during the eighteenth and nineteenth centuries has in many places substantially lowered the timber line. The control of and the protection against avalanches has gained enormously in importance as a result of the growing popularity of skiing which brings thousands of people on snow-covered and avalanche-prone slopes during the winter season and in early spring time. The construction of roads and railways in high mountains has also added to the increased awareness of the avalanche hazard. With respect to this situation, avalanche studies have become a focus of attention in alpine areas and several countries have been prompted to organize a monitoring and warning system with the aim of reducing the number of accidents and loss of human life and to substantially mitigate material damage.

Characteristically, the factors to be considered in evaluating the avalanche hazard are of two types, namely variable and non-variable factors. The variable factors relate to the meteorological conditions during and after the period(s) of deposition of the snow cover and to the hydrological conditions of the snow metamorphosis which affects the density, plasticity and firmness of the snow cover (Shen and Roper, 1970; Sommerfeld et al. 1976; Sulakvelidze and Dolor, 1973). The non-variable factors relate to the terrain configuration and are thus of geomorphological significance. The only (non-geomorphological) terrain factor which is liable to change with time is the vegetative cover of the slopes concerned. It is obvious that the continuous observation of the variable snow-related, climatological and hydrological factors is the hard core of avalanche monitoring and early warning (Bader et al., 1954; Bradley, 1973; Calembert, 1968; Curtis and Smith, 1973; Langmuir, 1970; Miller and Miller, 1974; Roch, 1954, 1955). The geomorphologist may, for reasons of pure science, be interested in avalanches as a geomorphological agent causing the down-slope transportation of debris and contributing to slope development in alpine mountains (Gray, 1974; Pippan, 1974). His contribution to the practical problems of minimizing the avalanche hazard, however, is mainly in the study, classification and evaluation of terrain forms with respect to the downslope movement of snow. It is this aspect that will be highlighted in this section. As an introduction, it is necessary to briefly describe the main types of avalanches and their characteristics; only thereafter can their relationship to the terrain configuration be explained (Salome, 1979).

16.2 Main Types of Avalanches

The characteristics of avalanches are diverse and depend on the condition of the snow involved, the weather prevailing before and at the moment of their detachment, downslope movement and deposition. Continuous observation of the type of snow and the gradual metamorphosis of the snow crystals, e.g., using a snow penetrometer, is important to estimate imminent avalanche danger. The formation of cup crystals, by constructive metamorphism and the occurrence in the snow profile of surface heat, which develop when moist air above the snow is warmer than the snow surface should be particularly observed as these layers are potential sliding planes for overlying snow. Several avalanche types can be distinguished on the basis of certain criteria and each of these types has its own specific hazard aspects. Some are of importance mainly for skiers and others in the higher parts of alpine mountains during the snow season, whereas other types are of particular importance for villagers living below the forest line, for traffic on mountain highways, etc.,as will be explained below.

The main criteria are the following (cf. Fig. 16.1):

1. The Compaction of the Snow Involved

Loose snow avalanches usually detach from a single point, whereas slab avalanches of compacted snow normally start in a broader zone and leave behind a distinct and steep snow wall at their upper limit.

2. The Type of the Sliding Surface

In case of full-depth or 'ground' avalanche, the ground acts as a sliding plane and as a result boulders and other debris may become embodied in the avalanche. A surface avalanche moves over underlying snow layers and the total volume of snow tends to be less than that of a ground moraine of comparable areal extent.

3. The Shape or Cross Section of the Avalanche Track

Depending on the topographic situation, the snow of an avalanche will either be channeled in ravines or move unconfined over portions of a slope. Most large avalanches gain sufficient momentum to reach the lower portions of the slopes and below the forest line along well-established tracks. The most damaging avalanches are among these channeled avalanches because the large masses of snow involved reach inhabited areas where serious damage may occur. The unconfined avalanches are more common on

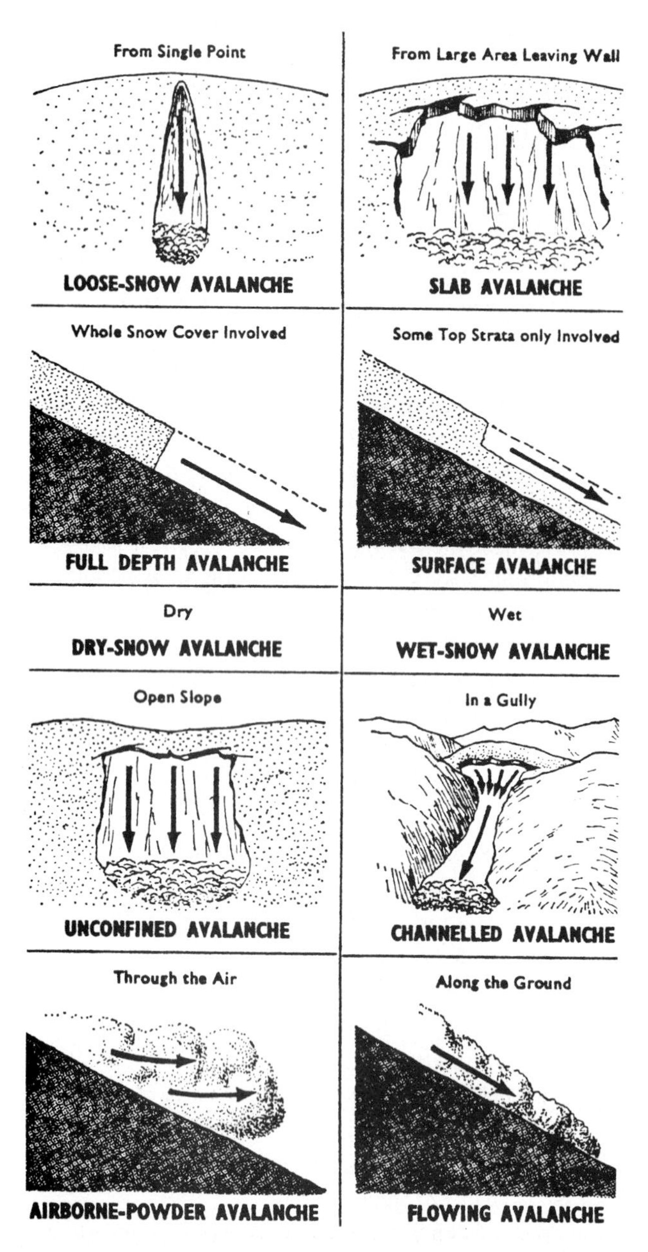

Fig. 16.1 Avalanche classification system by Haefeli and de Quervain (1955) according to type of breakaway (top), position of sliding surface, humidity of snow, the form of the track in cross section and the form of movement (bottom).

the higher slopes, although channeled avalanches are also a common occurrence there.

4. The Humidity of the Snow

In dry-snow avalanches,no free water is present, whereas in wet-snow avalanches this is distincly the case. An intermediate form is the damp avalanche, which is sometimes mentioned. The water content of avalanches has a profound effect on the weight and mode of movement (see ad. 5), causing snow flows or slides.

5. The Type of Movement

Most avalanches are ground-borne (flowing avalanches), particularly when the snow is wet or damp. In case of dry snow, however, a portion of the snow may become airborne and then form 'powder avalanches', characterised by a fast-moving cloud of snow over the avalanche track. Severe damage to the slope opposite the one in question may occur, due to the blast of the pressure wave caused by the fast moving, snow-laden air.

Dry avalanches occur predominantly in early and midwinter, whereas wet snow avalanches are common when the snow melts with the rising temperatures of late winter and early spring. It is evident that the traditional classification of avalanches into powder avalanches (Staublawinen), fig. 16.2 and ground avalanches (Grundlawinen) fig. 16.3, used for centuries in the Alps, has been considerably refined, although its essentials still persist in the modern snow avalanche classification system by Haefeli and de Quervain (1955), referred to in the above lines (see also de Quervain, 1973). It should be understood, however, that most avalanches are of a mixed nature and classification is not always an easy task. This applies particularly to certain larger or even catastrophic avalanches. Ice avalanches are a closely related phenomenon; they are almost exclusively due to glacier movement and often labelled glacier avalanches. Large masses of ice may be involved when parts of a steeply sloping glacier suddenly break away along pre-established crevasses and drop or slide downslope (fig. 16.4). These events are usually confined to the higher parts of alpine mountains where population is absent or scarce. If this is not the case, rare but catastrophic events may occur, particularly as substantial masses of earth and rock material may also be involved.

This was the case with the ill-famed disasters of the Santa Valley, Peru, where two avalanches descending from the slopes of Mt. Huascaran (6786 m) in 1962 and 1970, are still fresh in mind. In 1962 a mass of ice of approximately 2.5 - 3 million cubic metres detached itself at an altitude of about 6300 m. It fell 1000 m down a steep cliff and subsequently hurtled downslope, releasing snow avalanches and incorporating rock and soil masses until its volume was augmented to an estimated 13 million cubic metres and reached a maximum velocity of 65-90 km/hr. It finally came to rest 15 km downslope killing 4000 people and destroying six villages. The town of Hungay was left unharmed on that occasion, but this was not the case when a similar event of even considerably greater magnitude occurred in the valley in 1970. About 7 million cubic metres of ice fell on that occasion, triggering snow avalanches of about 4-5 million cubic metres and involving 30 million cubic metres of rock and earth. This ice/snow mixture reached a velocity of 400 km/hr and killed 20,000 people in Hungay, almost completely destroying this town (Lliboutry, 1971).

Fig. 16.2 Dry snow (airborne powder) avalanche (Flaig, 1955).

Fig. 16.3 Wet snow avalanche tunneled through to end road blockage, Aare Valley, Switzerland.

Comparable catastrophic glacial mud flows, though of a somewhat different nature, may result when an ice-marginal or a supra-glacial lake suddenly empties in a glacier crevasse, when falling seracs or parts of hanging glaciers drop into morainic lakes or in cases of surging glaciers. Major avalanches are sometimes induced by earthquakes in much the same way as landslides in other environments. The huge avalanche that occurred in the Puget Bay area, Alaska, caused by the earthquake of 27 March, 1964, is an example of this (Hoyer, 1971). About 1.8 million cubic metres of snow, rock and soil were involved. Basically, the destructive force of avalanches results from the bulk of material involved, its high velocity and, in cases of powder avalanches, of the air pressure produced. The velocity of powder avalanches is particularly high and may easily reach values in the order of 200-300 km/hr. The wind velocities inside the snow emulsion of the cloud may be even twice as much. The devastation to houses, forests, etc., caused by these avalanches is thus understandable.

Fig. 16.4 Ice (slab) avalanche beneath hanging glacier.

Fig. 16.5 shows a distinct avalanche track at the Polish side of the Tatra Mountains. Wet snow and particularly full-depth avalanches carrying rock and soil material have formed debris ridges, the so-called avalanche moraines, at either side of the ravine, thus facilitating the identification of the avalanche track. Fig. 16.6 shows the effect of a powder snow avalanche in the northern Caucasus, where repeated avalanching contributed substantially to the formation of the gravel fans in the far distance and a recent one (1975) destroyed a telephone line, bridge railing and young forest at the opposite side of the valley. Figs. 16.7 and 16.8 show an avalanche track and damage to forest in the Alps near Grenoble. For general information on avalanches, their dynamics, classification and effects, reference is made to Flaig (1955); Fraser (1966); Krasser (1964) and Paulcke (1938); new research results and protective devices can be found in Anon (1972); Int. Glacialological Society (1980) and Proceedings (1965); a good bibliography is by Knapp (1972).

16.3 Protection from Avalanche Hazards

In order to minimize damage to villages, houses and other structures in avalanche prone areas, it is essential to carefully select building sites. They should be outside the main danger zones such as the common avalanche tracks

Fig. 16.5 Typical avalanche track with avalanche 'moraines' at either side in the Tatra Mountains, Poland.

Fig. 16.6 Destruction of telephone pole and bridge railing by airblast of avalanches that came down from the distant slopes and talus cones. Caucasus, USSR.

Fig. 16.7 Avalanche track near Grenoble in the French Alps showing destruction of forest, coarse deposits and hazard to secondary road.

Fig. 16.8 Forest damage due to avalanche activity in the French Alps near Grenoble.

and preferably under the natural protection of rocks, bulges or other favourable terrain features (Voellmy, 1955). In fact, the site of many settlements in the Alps, for example, were carefully selected in the distant past and the present problems often arise from the gradual growth of these settlements,resulting in construction outside the safer places,or from extensive deforestation decreasing the safety of a site. Isolated houses and sheds in the higher parts of the Alps are usually skillfully situated, e.g., propped against or immediately below steep rocks rising from the slope from which avalanches may come down.

When settlements or isolated structures have to be protected from avalanches, measures are often taken immediately upslope of the endangered house(s). This is the oldest type of avalanche defense and protection,and several modern devices and principles of this kind go back to practices established by mountain people centuries ago and by trial and error. It aims at the protection of areas of limited extent by comparatively simple means (Coaz, 1910). Its main zone of implementation is the lower inhabited parts of the slopes and the sides of the valley bottoms, thus the downslope sections of the avalanche tracks including the parts where the avalanche snow originating from the higher slopes comes to rest. Newer methods of a different and larger scope than the localized avalanche defense measures downslope are developed for combating the avalanche hazard where it originates, namely the higher slopes,especially those above the forest line. This approach requires protective measures of a more generalized nature, covering sloping surfaces of considerable areal extent (Anon., 1972; Bucher, 1956; Fraser, 1974; Frutiger, 1966; Haefeli, 1960; Haefeli and Bucher, 1939; Oechslin, 1955; Schaerer, 1969; USDA, 1968).

In the case of localized avalanche protection of settlements and isolated structures, one has three basic options: letting the snow pass over the structure, diverting the avalanche away from the object to be protected or breaking the force of the avalanche and thus diminishing the velocity, impact and length of track, as a result of which it will, one hopes, not reach the object to be protected. Transitions between these three types may also occur. The first principle is applied to houses built into the slope having the roof at the same height or slightly lower than the terrain upslope. Sometimes an earthen fill is applied to the backwall of the house in order to match the height of the roof with the slope. Fig. 16.9 illustrates the various possibilities. The principle also finds application in the tunnels, galleries and simple snowsheds which presently protect many roads and railways in alpine mountains (fig. 16.10).

The traditional way of diverting an avalanche away

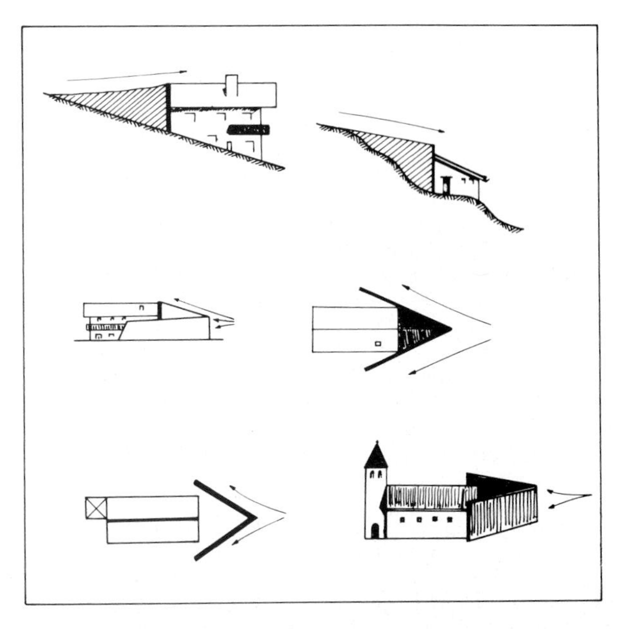

Fig. 16.9 Various ways of protection of structures on avalanche-exposed sites based on overpassing or deflection of snow.

Fig. 16.10 Road protection from avalanches and drifting snow.

from a house or church is the construction of a splitting wedge, V-shaped in plan and pointing upslope. Combinations and/or transitions between an earth fill matching the roof of a building with the slope and a splitting wedge are also found. In old churches one may come across a V-shaped backwall serving the purpose of avalanche protection. Evidently, much stronger structures of heavily anchored, reinforced concrete are now in use; a diversion dam, placed at a slight angle with the avalanche course may also be effective (fig. 16.11) and small splitting wedges may be successfully used to protect transmission

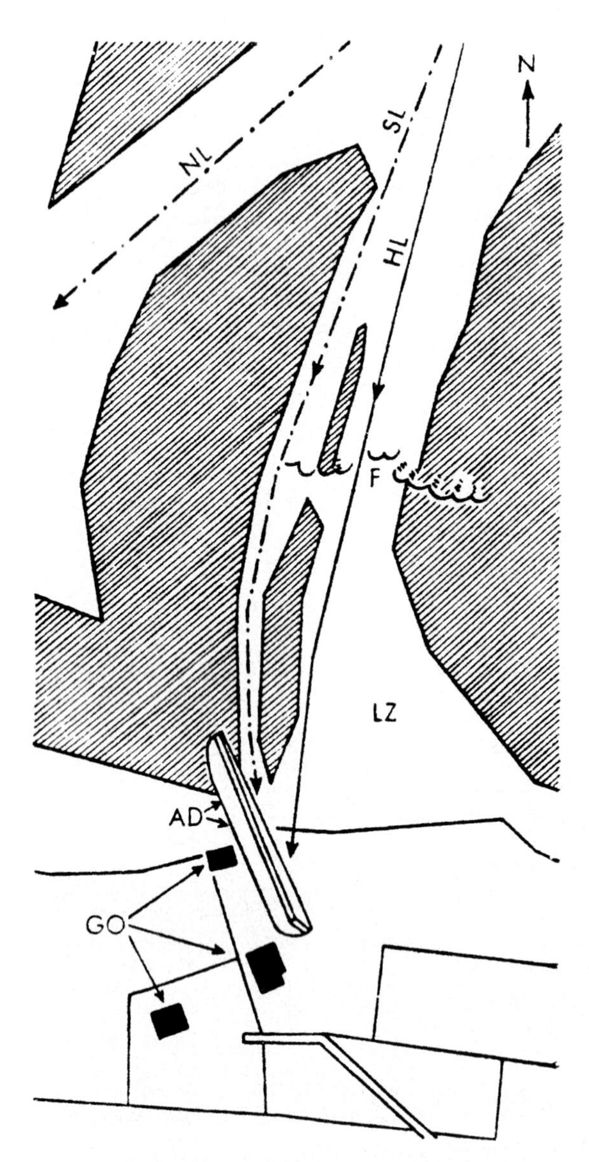

Fig. 16.11 Diversion dam for the protection of structures on the fringe of an avalanche track.

Fig. 16.12 Modern avalanche splitting wedge of reinforced concrete.

line pilings (fig. 16.12). The third principle, that of breaking the force of the avalanche, may be effectuated by a system of splitting wedges causing lateral movements in the snow masses,though it is often more practical to create a network of stone-covered earthen mounds, e.g., by bulldozer, in the lower portion of the avalanche track where the natural tendency towards deposition of the avalanche snow already exists.

Reafforestation plays an important role in generalized avalanche protection, since dense, mature and healthy forest can support and absorb large masses of snow. In order to get the forest installed (and not immediately devastated), particularly in the most critical zones (the avalanche tracks) it is necessary to erect snow-supporting structures in the break away zone of the avalanches. These structures are also known as stabilization barriers, being the counterpart of the diversion barriers mentioned earlier. Initially, vertical walls, both free and with an up-slope fill, were used (fig. 16.13). Later it proved to be more useful to give these walls a slight mountainward inclination to improve their strength. Nowadays, in most areas, the walls are succeeded by fence-type snow support structures, either snow bridges with horizontal bars or snow rakes with vertical shafts (fig. 16.14). They are given a downslope inclination of about 105° to achieve maximum strength and to make optimum use of the height of the structures. Avalanche nets, developed in Austria and France, also have come into use.

In addition to snow-support structures in or below the break-away zone of a full-fledged avalanche protection scheme, snow or drift fences are sometimes used upslope of this zone to promote snow accumulation, thereby decreasing the build-up of snow cover in and below the break-away zone and thus the chances of an avalanche occurring. If it does occur, its volume is decreased. Snow baffles, placed at certain selected locations to prevent deposition of snow, e.g., where dangerous cornices might be otherwise formed, are another type of structure used.

A different approach to the mitigation of avalanche hazards often practiced among other places, in ski resorts is the controlled release of avalanches using explosives (Mellor, 1968, 1973). Snow covered slopes can be made safe at will and avalanches can be released before the snow volume becomes too large to cause avalanches of devastating dimensions (Gardner and Judson, 1970).

16.4 Terrain Factors and Avalanches

In avalanche studies, the geomorphologist will have to cooperate with meteorologists monitoring the weather conditions and hydrologists monitoring the metamorphosis of the snow and thus its stability. When avalanche defense matters are concerned, important input can be expected from specialists such as foresters and civil engineers. The real geomorphological contribution to avalanche studies is focused on the study of the mechanism of the avalanche process, including its areal extent, frequency and particularly the terrain configuration as a major factor in the degree and distributional pattern of the avalanche hazard (Bjornsson, 1980; LaChapelle and Leonard, 1971). These terrain factors are elaborated upon in this section.

The slope angle is a major terrain factor to be considered in this context. It is unusual for an avalanche to start on gentle slopes or on very steep slopes. The former are comparatively avalanche free and stable because the pressure on the ground exerted by the weight of the snow has only a small down-slope component at these gradients. The latter are comparatively safe because the snow slips before large masses of it have the chance to accumulate. In fact it is rare for a major avalanche to be initiated at slopes of less than 22° and of more than 50°. Smaller masses of snow may avalanche on considerably steeper slopes, and in fact do so, but the smaller masses involved are seldom devastating. The most dangerous slopes normally have gradients of 30-40°. One should be aware, however, that these values are indicative rather than rigid limits, much depends on the weather, snow conditions and terrain configuration factors other than straightforward slope gradient. It is also clear that relatively small masses of snow (or icicles, etc.) falling from steep slopes may trigger larger avalanches on less steep (e.g., 30-40°) slopes farther down, and that large avalanches, due to the momentum gained, may invade slopes having a gradient of less than 22° (in fact, even flow upslope over short distances) and that the air blast build-up by a fastly moving powder snow avalanche may be extremely devastating also on the opposite side of a valley.

Other slope characteristics to be considered are diverse and can be classified according to the influence they exert on the accumulation, metamorphosis and movement of the snow. Irregularities of a slope are generally a favourable factor since they tend to increase the friction at the base of the snow. Ridges and rock terraces caused by outcropping resistant beds, but also large boulders and even minor matters such as terracettes, are effective in this respect, especially when the thickness of the snow cover increases in mid- and late winter. However, only ground avalanches and not the more common surface avalanches can be efficiently impeded. The vegetative cover may also affect the friction: slopes covered with long grass are particularly hazardous, whereas dense scrub and forest are stabilizing factors. Under some conditions, however, slope irregularities may introduce increased vulnerability for avalanches. For example, when a steep rock, cliff or a cornice rises above a snow covered slope, snow or ice breaking away from the cornice and icicles, rocks or large blocks of ice detached from a slope glacier may trigger avalanches. Isolated trees or a forest may have a similar effect when swaying from the wind; the impact of their stems against the snow cover and of the snow falling down from their branches are triggering factors on avalanche-prone slopes.

The general form of the slope profile and especially the occurrence of major convexities and concavities, merits attention. This applies specifically when, some time after the snow fall, the gradual metamorphosis of this snow has resulted in some degree of compaction, some increase of density and particularly in greater cohesion. Under these conditions the process of snow creep, brought about by both settling of the snow and gravity, will cause tension, as a result of which slabs of snow may

Fig. 16.13 Various types of stabilization barriers for avalanche control in the break away zone: with upslope fill (left); free (centre) and inclined (right).

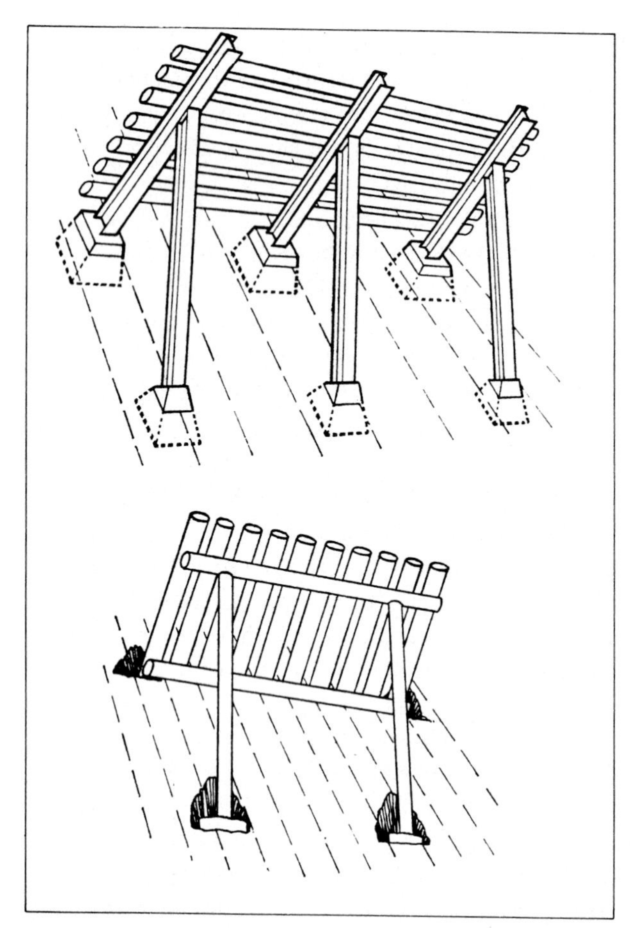

Fig. 16.14 Common structures for avalanche protection in the break away zone: the snow bridge (top) and the snow rake (bottom).

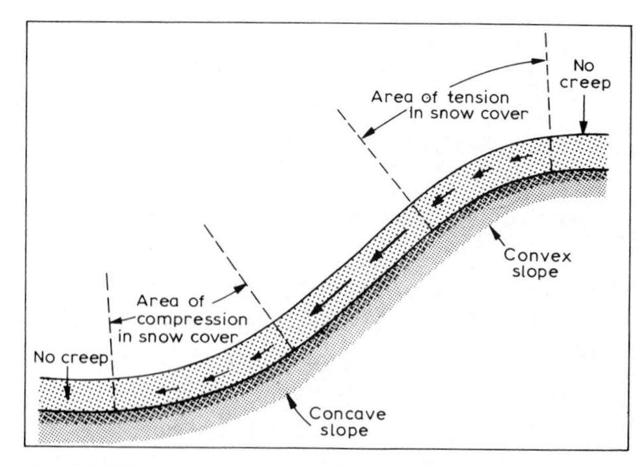

Fig. 16.15 Tension and compression of snow cover: the combined effect of snow creep and slope profile.

slide down over the ground, giving rise not only to an interesting type of soil erosion but ultimately also leading to cracks in the snow which are the potential break aways of slab avalanches. This tension is strongest and the occurrence of cracks and break aways is therefore most likely on convex parts of the slope where the intensity of creep increases downslope with increasing slope gradient. On concave parts of a slope, on the contrary, the downslope decrease in gradient slows down the creep movement and thus compaction will occur. Normally, the snow in these parts does not develop tension cracks and its greater strength even gives some support to the snow formed on the steeper gradients upslope (fig. 16.15).

The aspect or orientation of the slope with respect to the sun influences the mode and degree of snow metamorphosis and therefore the stability of the snow cover as a whole. Generally speaking, metamorphosis is less on the cooler, north-exposed slopes where the sun rays extend only during short periods of time and under a slanted angle. As a result of this and also because these sites are favourable for deep hoar development, dry avalanches are frequent on north-facing slopes, particularly in early and mid-winter. South facing slopes, however, become increasingly dangerous in late winter and with the advent of spring, when snow melt, accelerated by rising temperatures and more intense insolation, renders them more liable to wet snow avalanches.

It will be clear that the distinction between north and south facing slopes is a rough indication only; much depends on altitude, general temperature level, general exposure to nocturnal cooling, etc.,of the slopes concerned. Exposure with respect to the prevailing wind direction also has to be considered, although this is in many cases a variable factor. Windward slopes do not normally favor thick snow accumulation, particularly when they are steep. Furthermore, the snow deposited there tends to be fairly compact and covered by a hard crust which adds to the stability of such snow slopes. Ayzenberg et al. (1970) state that in many parts of the Carpathians, wet-snow avalanches in spring time are a major form of avalanching resulting in a volume of 10.10^3 - 300.10^3 cbm of snow cones annually. They are found mainly on the northern slopes due to the prevailing direction of snow transport. Leeward slopes are generally more favourable for the accumulation of thick layers of only slightly compacted snow slabs or loose snow. They are thus particularly avalanche prone. The difference in snow accumulation on windward and leeward slopes is clearly demonstrated by fig. 16.16.

The cross profile of a slope or its configuration affects avalanches in various ways. Where concavities in plan occur, such as at the confluence of several ravines in the higher parts of a slope, a concentration of avalanching snow occurs and a considerable mass of snow may be channeled into major ravines farther downslope. Distinct avalanche tracks may thus be pointed out where avalanches come down from time to time. Although it should be stressed that not all avalanches are confined to such tracks, the majority certainly is in most mountainous areas. One should also be aware of the fact that even where small or

Fig. 16.16 The effect of aspect or orientation of slope on snow accumulation: one-sided snow cover of a steep ridge on the Aiguille du Chardonnet, Switzerland.

medium-sized avalanches are confined to well-established avalanche tracks, it may be that large masses of snow of catastrophic avalanches escape from the track where sudden curves in its course occur and thereafter take an unusual path causing damage at unexpected locations. Also in the lower part of the slopes where the avalanches terminate and the snow is accumulated, the terrain configuration in plan is important as it affects the direction of flow and thus the distributional pattern of the avalanche risk in the inhabited parts of the valleys.

Terrain factors are, of course, also of major concern when decisions are to be made regarding the siting of houses, transmission line pilings and other structures and about the most efficient siting of avalanche protection structures such as deflecting walls and splitting wedges. Snow or drift fences are successfully applied where plateaus of moderate relief occur above steep valley slopes, e.g., in Scandinavia (Ramsli, 1974). They may also efficiently modify the depositional pattern of snow near mountain crests and thus reduce the development of dangerous cornices. The spacing of avalanche support fences (snow rakes and snow bridges) is calculated using formulae in which the slope gradient is a main parameter. The steeper the slope, the closer the spacing between the structures should be in order to provide adequate support to the snow cover.

16.5 Surveying and Mapping for Avalanche Control

It will be clear from the previous sections that avalanche risk results from fairly complex relationships between various terrain factors,on one hand,and matters of weather conditions and snow stratigraphy,on the other (Armstrong, 1974; Jail and Vivian, 1970; Korolev and Koroleva, 1970; Kotliakov and Touchinski, 1974; Mears, 1975). The detailed study of the relief is thus a logical starting point for geomorphological surveys related to avalanches. Aerial photographs and other remote sensing imagery taken during the snow-free summer season may be used for the purpose (Otake, 1980). All matters of terrain configuration specified in the preceding section may be effectively studied and an ultimate classification of the terrain geared to the avalanche risk can be established. If data about the location and frequency of avalanches are also available, calculations on the statistical probability of their occurrence in each of the terrain classes distinguished is feasible (Takahashi et al., 1968).

The location of common avalanche tracks can be mapped with reasonable accuracy from aerial photographs taken in the snow-free season, provided the ground truth gathering, essential in any type of photo interpretation, is adequately carried out. The terrain configuration, the scars in the alpine forest and the so-called avalanche moraines formed at either side of the avalanche tracks by the earth and boulders transported by the avalanches (figs. 16.5 and 16.7) are reliable indicators. Estimations about velocity, kinetic energy per unit volume of snow and the impact on the ground may also be ventured, using the topography, induced starting point of the avalanche and the downslope gradient (and its variations) as major parameters (Bojo, 1968). In many cases avalanche mapping is more or less limited to the plotting of avalanche tracks with the amount of detail given depending on the scales used. In case of generalized mapping at scales in the order of 1:50,000, the main avalanche tracks can only be indicated by line symbols. More important for practical purposes are usually the detailed maps in scales of 1:5,000 on which the areal extent of the avalanches of the past can be mapped and their situation with respect to houses, villages, etc.,indicated (Ramsli, 1974). Main hazard areas can then be pointed out, defense structures planned and cost-benefit matters evaluated. In case of avalanche studies for road protection, special attention should be paid to the zones where avalanche tracks cross the road to decide on adequate types of defense. The more money spent on the latter, the shorter the periods of closure and the lower the cost of snow clearance tend to be (Schaerer, 1969, La Chapelle and Brown, 1971, 1972). Examples of avalanche track maps are given in figs 16.17 and 16.18. Complete hazard zoning maps should be the ultimate aim (Cazabat, 1970; Fruticer, 1980; Hestnes, 1980; Kienholz, 1977; Tesche, 1977).

Further data about avalanche tracks and the frequency of their use can be obtained in various ways. Mapping of the three main parts of an avalanche, starting zone, track and deposition or run out area (Perla, 1971) is important. Tree-ring dating in avalanche tracks may contribute to the estimation of avalanche frequency and thus to the re-occurrence rate of damage in inhabited parts downslope. The age of the trees in the avalanche zone is only a rather inaccurate criterion. More reliable are changes in the growth-ring pattern from concentric to eccentric caused by tilting,as well as the search for datable scars on these trees (Potter, 1969). The geomorphologist looks for further details on avalanche moraines which are particularly reliable avalanche indicators particularly above the forest line, but in many cases also in other parts of the track. The linear features or tongues of earth and fine debris stretching in the slope direction,which are often found on top of these boulder ridges downslope of major blocks and resulting from the drag caused by recent avalanches, are particularly important indicators for frequently used avalanche tracks.

A full-fledged avalanche survey should not only indicate the position of avalanche tracks but should also incorporate matters:

1. Slope Classification. Several gradient classes should be distinguished, the limits of which are significant in the avalanche context (sub-section 15.4). One might, e.g., distinguish slopes of $<22^o$, $22-30^o$, $30-40^o$ (highest risk), $40-50^o$, and $>50^o$. Apart from the slope gradient,also the regularity of the slope which affects the friction and thus the susceptibility to sliding, the general slope profile, the configuration in plan, the vegetative cover, etc. should be considered when arriving at an overall risk evaluation.
2. Avalanche Tracks. These should be mapped in detail, indicating the limits between the three major parts of the avalanche: catchment-track-area of deposition. Data on frequency and age of avalanches, type and velocity, volume, etc.,may be added either to the map or in an accompanying report. Details on the processes and their violence may also be elucidating.
3. Special Terrain Features. Certain terrain types or sites which are of particular interest in relation to avalanches should be accurately mapped and emphasized by adequate symbols. Such features are to be found in both the upslope areas where the avalanches have their origin, as well as along their tracks and in the low, usually more gently sloping areas where they come to rest. They include:
   a. high-risk zones for avalanche initiation, usually in the $30-40^o$ gradient zone.
   b. major convexities in the downslope profile where tension may produce cracks resulting in break aways of slab avalanches
   c. steep rocks and cliffs rising from steep snow fields from where snow and ice falling down from cornices or falling rocks may trigger avalanches
   d. sudden curves in avalanche tracks where major avalanches may suddenly divert from the track
   e. natural avalanche protection offered by the terrain configuration such as rocks rising from avalanche-prone slopes or ridges which may divert avalanches, etc.
4. Vegetation, in particular the distribution of dense high forest which may prevent the initiation of avalanches, although possibly not offering full protection downslope from avalanches falling from above. Zones where reafforestation is needed also may be added.
5. Snow Protection Structures already present and the erection of those required. Endangered houses and other structures in dangerous positions are to be specifically indicated although one should never forget that the unpredictability of avalanches is such that a completely successful result cannot be guaranteed.
6. Areas Affected by Previous Avalanches. This information should be based on documents, interviews and field indications. Frequency of occurrence is also a matter to be assessed.

16.6 The Study of a Damaging Avalanche, Using Aerial Photographs as an Example

Three major mixed avalanches came down from the ENE and S slopes of Piz Mondin (3120 m) and Piz Alpetta (2975 m) and descended the steep slopes of the upper Inn valley near Finstermünz, in eastern Switzerland on 18 February, 1962. They were about 3,6 km long and since they were only partly confined to tracks and rushed down along a rather broad track of slope (1,5 km), an unusually large surface area was affected by them. Approximately 90-100 ha mature forest, including numerous 120-150 year old trees,were destroyed, totaling about 20,000 cbm. The hamlet of Vinadi miraculously escaped disaster, the damage to three houses was only minor and its 17 inhabitants were rescued. Serious damage also occurred on the opposite side of the valley in Austria, mainly

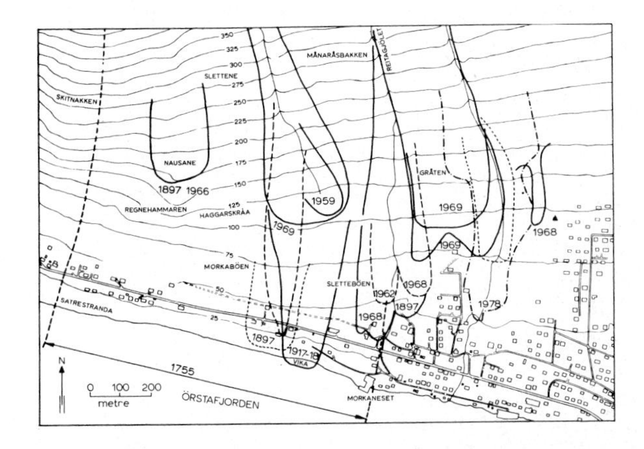

Fig. 16.17 Avalanche track registration map of an area in Norway, simplified from a 1:5,000 map with contour interval of 5 m.

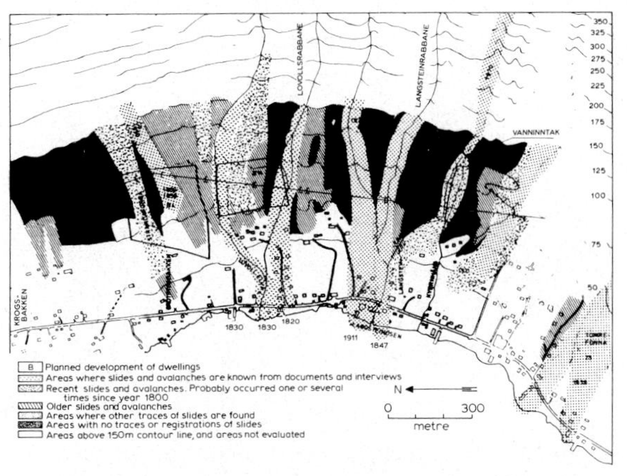

Fig. 16.18 Example of a geomorphological hazard map of avalanches and slides, Norway, simplified from a 1:5,000 map with contour interval of 5 m.

by the air blast caused by the avalanche cloud. The stereopair of fig. 16.19 at an approximate scale of 1:27,000 indicates the situation. The hamlet of Vinadi can be seen at the road junction near the right edge and in the lower part of the picture.

Ruedi and Bischoff (1962) studied avalanches and the damage caused by them using vertical aerial photographs taken before catastrophic avalanches and comparing them with those taken immediately afterwards. Single obliques, terrestrial photographs and 1 :10,000 cadastral and forest inventory maps were also used. The aerial photographic approach proved to be very efficient for arriving at a rapid determination of the affected surface area for the planning of debris and snow clearing operations and for estimating the volume of timber involved. The areal extent of the avalanches, including their break away zones, main avalanche tracks and deposition cones could easily be mapped. Also the different modes of movement of the avalanche snow could be recognized (channeled, diffuse over slopes, air borne powder, etc.) and the areas affected by each of them mapped,whereafter a comparison with the type and intensity of forest damage associated with them could be made.

Most trees snapped off near the ground or were uprooted by any type of avalanche motion. In air blast zones some trees were also broken several metres above their bases. On the opposite slope, a spruce and pine forest was stripped of branches without the trunks being felled. These matters can be seen in the stereopair of fig. 16.19 in the lower left where bridges of avalanche snow occur over the River Inn. The cause of the major damage affected by these avalanches was not an excessive thickness of the snow cover but rather the broad catchment area from where the snow - at least partly - forced its way through narrow and deep ravines such as those of the Val Mondin, Val Zipla, Val Fontana and the narrow furrows near Finadi. The main road (lower one in the photograph) was blocked by avalanche snow over a considerable distance but it had been cleared by the time the photographs were taken.

The terrain configuration had a profound effect on the behaviour of the avalanches and on the damage caused, especially in certain well-defined localities. Where ravines changed direction or widened, the avalanche snow left the tracks normally in use by smaller avalanches and moved on,both ground- and airborne,affecting areas normally not visited by avalanches. Where more gentle slopes or short steep slope reversals gave way to steeper gradients farther downslope,parts of the avalanches were shot into the air and the resulting airblast of the snow-air mixture caused considerable extra damage. Proper, channeled ground moraines only occurred in the ravines already mentioned and in the end phase of the avalanche process. It was such a comparatively minor channeled ground avalanche, laden with shattered timber, that caused some slight damage in the otherwise unaffected hamlet of Vinadi.

The stereopair also shows some avalanche protection structures built in connection with the road system. Some avalanche sheds and galleries by which the snow passes over the road can be easily recognized.

It is evident that after the occurrence of such a rare, catastrophic avalanche causing substantial damage to the forest cover, a new evaluation of the avalanche risk in this zone has to be made because the violence and frequency of the average size avalanche may have been substantially increased by the deforestation. Planned reafforestation and avalanche protection programmes could also make good use of aerial photographic approaches, combined with other research methods in the field.

Fig. 16.19 Stereo pair of an avalanche that has occurred in the Upper Inn Valley near Finstermünz, Switzerland, on 18 February 1962. For explanation see text.

References

Armstrong, R.L., 1974. Avalanche hazard evaluation and prediction in the San Juan mountains of southwestern Colorado. Proc. Symp. Snow and ice resources. Monterey (Cal.), Dec. 2–6, 1973. U.S. Nat. Acad. Sci.: 346–355.

Armstrong, R.L., et al., 1974. Development of methodology for evaluation and prediction of avalanche hazard in the San Juan mountain area of southwestern Colorado. Inst. Arctic a. Alpine Res., Univ. Colorado, Occas. Paper, 13, 141 pp.

Anon., 1972. Lawinenschutz in der Schweiz. Beiheft 9 zum Bündnerwald. Ztschr.d.Bündner Forstwesen, 225 pp.

Ayzenberg, M.M., et al., 1970. Snow avalanches in the Chernogora Massif (Ukrainian Carpathians) and urgent tasks in their study (in russian). Trudy Sredneaziatskogo Nauchnoissledovatel' skogo Gidrometeorologicheskogo Instituta, 51(66): 145–150.

Bader, H., et al., 1954. Snow and its metamorphism. U.S. Army Corps of Eng. Snow, Ice and Permafrost Research Establ. Transl. 15

Bjornsson, H., 1980. Avalanche activity in Iceland, climatic conditions and terrain features. J. Glaciol., 26(94): 13–23.

Bojo, C., 1968. Snow avalanche analysis on aerial photographs. Japan Soc. Photogramm., 7(3): 125–129.

Bradley, J.C., et al., 1973. Snow pack stability indices relative to the climax avalanche, Final report 1, Jan 1971–1 July 1973. Montana State Univ., 23 pp.

Bucher, E., 1956. Beitrag zu den theoretischen Grundlagen des Lawinenverbaues. Geologie d. Schweiz, Geotechn. Ser. Hydrologie, 6, 1948, Bern. Engl. transl.: Contributions to the theoretical foundations of Avalanche Defense Construction. U.S. Army Corps of Eng. Snow, Ice and Permafrost Research Establ. Transl. 18

Calembert, L., 1968. Glissements et avalanches catastrophiques. Bull. Séances Acad. Roy. Outre-Mer Belge: 692–703.

Cazabat, C., 1970. Les cartes de localization probable des avalanches. Bull. I.G.N., Paris, 11: 20–27.

Chapelle, E.R. La, and Brown, C.B., 1971. Avalanches on the North Cascades Highway SR-20. Summary report. Washington State Dept. Highways, Interim Rep. April '70. -Sept. '71, 74 pp.

Chapelle, E.R. La, and Brown, C.R., 1972. Avalanche studies, 1971–1972. Report on methods of avalanche control on Washington mountain highways. Washington State Dept. Highways. Second Ann. Rep., 55 pp.

Chapelle, E.R. La, and Leonard, R., 1971. North Cascades Highway SR-20 Avalanche Atlas. Washington Univ., Seattle, 177 pp.

Coaz, J., 1910. Statistik und Verbau der Lawinen in den Schweizeralpen. Bern

Curtis, J.O., and Smith, F.W., 1973. Material property and boundary condition effects on stresses in avalanche snowpacks. Proc. Symp. Adv. North Amer. Avalanche Technol. 1972. USDA Forest Service, Gen. Techn. Rep. RM-3: 14–23.

Flaig, W., 1955. Lawinen. 2nd Ed. Brockhaus, Wiesbaden, 251 pp.

Fraser, C., 1966. The Avalanche Enigma. John Murray, London, 301 pp.

Fraser, C., 1974. Avalanches–lessons in defence and control. Unasylva, 26(105): 23–29.

Frutiger, H., 1966. A Manual for Planning Structural Control of Avalanches. USDA, Forest Service Res. Paper RN 19, 68 pp.

Frutiger, H., 1980. Swiss avalanche hazard maps. J.of Glaciol., 26, 94: 518–519.

Gardner, N.C., and Judson, A., 1970. Artillery control of avalanches along mountain highways. U.S. Dept. Agric. Service. Res. Pap. RM 61, 26 pp.

Gray, J.T., 1974. Geomorphic effects of avalanches and rockfalls on steep mountain slopes in the central Yukon Territory. Proc. 3rd Guelph Symp. Geomorph. (1973): 107–117.

Haefeli, R., and Bucher, E., 1939. Recherches récentes en matière de lutte contre les avalanches. L'Annuaire de l'Ass. Suisse de Clubs de Ski, 35

Haefeli, R., 1960. Zur Geschichte der Bremsverbauung von Lawinen. Schweiz. Ztschr.f.Forstwesen, 111(8): 420–427.

Haefeli, R., and Quervain, M. de, 1955. Gedanken und Anregungen zur Benennung und Einteilung von Lawinen. Die Alpen, 31(2): 72–77.

Hestnes, E., 1980. Skredfarevurdering (with english summary). Norges Geotekn. Inst., Oslo, Publ., 132: 61–81.

Hoyer, C., 1971. Puget Peak avalanche, Alaska. Geol. Soc. Amer. Bull., 82(5): 1267–1284.

Int. Glaciological Society, 1980. Proc. Symposium on snow in motion, Colorado, 1979. J. of Glaciol., 26, 94, 527 pp.

Jail, M., and Vivian, R., 1971. Les glissements de terrain et les éboulements dans les Alpes françaises du Nord en 1970. Revue Géogr. Alpine, 59(4): 473–502.

Kienholz, H., 1977. Kombinierte geomorphologische Gefahrenkarte 1:10,000 von Grindelwald. Catena, 3(3/4): 265–294.

Knapp, L. (Editor), 1972. Avalanches, including debris avalanches: a bibliography. Office of Water Res. Research, Water Res. Sci. Inf., 72–216, 91 pp.

Korolev, A.I., and Koroleva, L.M., 1970. Concerning avalanche processes (in russian). Trudy Sredneaziatskogo Nauchno-issledovatel'skogo Gidrometerologicheskogo Instituta, 51(66): 135–139.

Kotlianov, V.M., and Touchinski, G.K., 1974. Le régime actuel des glaciers et des avalanches dans le Caucase. Rev. Géogr. Phys. Géol. dyn., 16: 299–312.

Krasser, L., 1964. Grundzüge der Schnee und Lawinenkunde. Bregenz.

Langmuir, E., 1970. Snow profiles in Scotland. Weather, 25 (5): 205–209.

Lliboutry, L., 1971. Les catastrophes glaciaires. La Recherche, 2(12): 417–425.

Mears, A.I., 1975. Dynamics of dense-snow avalanches interpreted from broken trees. Geology, 3(9): 521–523.

Mellor, M., 1968. Avalanches,Cold Regions Science and Engineering. U.S. Army Material Command. Cold Regions Res. and Eng. Lab. Hannover (N.H.), 215 pp.

Mellor, M., 1973. Controlled release of avalanches by explosives. Symp. Advances in North American Avalanche Technology, 1972. USDA Forest Service, Gen. Techn. Rep. RM-3: 37–49.

Miller, L., and Miller, D., 1974. The computer as an aid in avalanche hazard forecasting. Proc. Symp. Snow and Ice resources, Montery (Cal.), Dec. 2–6, 1973. U.S. Nat. Acad. Sci.: 356–362.

Oechslin, M., 1955. Der Kampf gegen die Lawinen. Die Alpen, 37(2): 88–93.

Otake, K., 1980. Snow survey by aerial photographs. Geojournal, 4(4): 367–369.

Paulcke, W., 1938. Praktische Schnee und Lawinenkunde. Springer Verlag, Berlin, 218 pp.

Perla, R.I., 1972. Snow avalanches of the Wasatch Front. In: Environmental Geology of the Wasatch Front, 1971, Utah Geol. Ass. Publ. 1: 01–025.

Pippan, T., 1974. Bedeutung der Lawinentätigkeit für gegenwärtige geomorphologische Prozesse im Hochgebirge vom Salzburg. Abh. Akad. Wiss. Göttingen, 29: 301–312.

Potter, N. Jr., 1969. Tree-ring dating of snow avalanche

tracks and the geomorphic activity of avalanches, northern Absaroka Mountains. Geol. Soc. Amer. Spec. Paper, 123 : 141-165.

Proceedings, 1965. International Symposium on scientific aspects of snow and avalanches research. Schweiz. Inst. f. Schnee u. Lawinenforschung, Davos

Quervain, M. de, 1973. Eine internationale Lawinenklassifikation. Ztschr.f.Gletscherk.u.Glazialgeol., 9(1/2)1: 189-206.

Ramsli, F., 1974. Avalanche problems in Norway. In: Natural Hazards (G.F. White, Editor). Oxford Univ. Press, London, Toronto, Chapter 22: 175-180.

Roch, A., 1951. On the study of avalanches. Sierra Club. Bull., 36(5): 88-93.

Roch, A., 1955. Le méchanisme du déclenchement des avalanches. Die Alpen, 31(2): 94-105.

Ruedi, H., and Bisschoff, N., 1962. Determination of the extent of forest damage caused by an avalanche. Mitt. Schw. Anst. Forstwesen, 38(1): 213-224.

Salome, A.I., 1979. Sneeuwlawinen in de Alpen. Geogr. Tijdschr., 13(4): 286-298.

Schaerer, P.A., 1969. Planning avalanche defence works for the Trans-Canada Highway at Rogers Pass. (B.C.). In: J.G. Nelson & M.J. Chambers, (Editors), Geomorphology, Process and method in Canadian geography. Methuen Ltd.: 219-234.

Shen, H.W., and Roper, A.T., 1970. Dynamics of snow avalanches. Bull. Int. Ass. Sci. Hydrol., 15(1): 7-26.

Sommerfeld, R.A., et al., 1976. A correction factor of Roch's stability index of slab avalanche release. J. of Glaciol., 17(75): 145-147.

Sulakvelidze, G.K., and Dolov, M.A., 1967. Physics of snow, avalanches and glaciers. USDA Forest Service, Washington D.C., 585 (translated from russian).

Takahashi, S., et al., 1968. The probability of occurrence of the avalanche. Japan Soc. Photogramm. 7(3): 117-124.

Tesche, T.W., 1977. Avalanche zoning, current status, obstacles and future needs. Proc. Western Snow Conf., Albuquerque (N.M.): 41-44.

U.S.D.A., 1968. Snow avalanches, a handbook of forecasting and control measures. U.S.D.A. Forest Service, Agric. Handb. 194, 84 pp.

Voellmy, A., 1955. Ueber die Zerstörungskraft von Lawinen. Schweizer Bauzeitung, 73: 159-165; 212-217; 246-249; 280-285.

NATURAL HAZARDS OF ENDOGENOUS ORIGIN

17.1 Volcanic Hazard Survey and Risk Zoning

17.1.1 Introduction

Volcanic disasters rank high among natural hazards in countries where active volcanoes occur in densely populated areas. The assessment of these hazards then becomes a major issue in environmental management with the aim of preventing or at least mitigating the disastrous effects of volcanic eruptions (Howard and Dickinson, 1978; Ney, 1975; Tazieff, 1967; Sparks, 1981).

Continuous surveillance of potentially dangerous volcanoes and monitoring of solfatara and fumarole temperatures, (micro) seismicity, general thermal state, topographic changes, crustal deformation and of any other signs of increased or forthcoming volcanic activity has become general practice. Thermal infrared surveys,either ground-based or,preferably, airborne,have become of increasing importance in recent years (Friedman and Williams, 1968; Friedman et al., 1969; Moxham, 1970) and also the methods of seismic observation (Unger and Decker, 1970; Unger and Mills, 1973; Tilling et al., 1975; Koyanagi et al., 1975) are now being generally applied. A complete review of geophysical, geochemical and other types of monitoring methods for purposes of prediction is given by Gorshkov et al. (1971).

Since volcanic activity is to a considerable degree unpredictable in most cases as far as time and magnitude of eruption is concerned and because the type and centre of eruption are subject to variations, it is evident that monitoring, to be effective, has to be supported by other means of appraising the potential volcanic hazards existing in an area. Even then one has to face the fact that satisfactory protection can only be achieved by an early warning system based on customary patterns of eruptions, realising that the effects of extraordinary volcanic events are more difficult to assess and may at any time cause unforseen disaster. These types of events are infrequent, however, and when all possible measures have been taken, this small remaining risk factor simply has to be accepted. More critical is the proper assessment of time and magnitude of the eruption, particularly when evacuation of the population from endangered areas is concerned. Evacuation should be effectuated timely,but on the other hand,not too early, unnecessarily or in the wrong areas.

The types of volcanic hazards are diverse and vary from one volcano to another (Duncan et al., 1981; Healy, et al., 1978; Hoblitt, 1980; Mullineaux and Peterson, 1979; Murton and Shimabukuro, 1974; Preusser, 1973). Airborne tephra products are a dominant hazard when explosive eruptions are concerned. The wind direction at the time of eruption is an important element in their distribution. Hot, pyroclasic flows and cold volcanic mudflows (lahar) are mostly concentrated in valleys and ravines. An example has been given in section 6 (fig. 6.15) of the ill-famed mudflows endangering the west and southwest slopes of the Merapi volcano, central Java, Indonesia. The water of the mudflows in this case is derived from the wet monsoon rains. A second example of volcanic hazards affecting rural development, given in chapter 6, relates to the Kelut volcano, east Jave, Indonesia. The source is in this case the water of the crater lake being blown out as a result of the eruptions (fig. 6.43). The lake was drained after the disastrous eruption of 1919, thereby greatly reducing the hazard (Van IJzendoorn, 1953). In areas such as Iceland and the Antarctic,melting snow and ice on the volcanic slope may provide the water generating mudflows (Baker, 1969). Lava flows (Walker, 1973) associated with strato volcanoes often emerge in the lower parts of the volcanic cones. The vertical aerial photograph of fig. 17.1, showing the NE slope of the Ciremai volcano, western Java, Indonesia,at the scale of 1:40,000, gives an example. Rock avalanches are usually associated with active volcanic plugs or other steep top parts of volcanic cones. Accompanying glowing clouds may be particularly dangerous (Davies et al., 1978; Fisher, 1980).

Additional hazards include factors such as eruption-induced tsunami (Suwa, 1980) that have caused a high death toll, e.g., during the 1883 eruption of the Krakatau volcano in the Sunda Straits, Indonesia. Starvation due to prolonged or lasting nonproductivity of the land was the cause of 82,000 out of the 92,000 deaths from the disastrous eruption of the Tambora volcano, Indonesia, in 1815 (Suwa, 1980). Nowadays this is not a major concern because of the better means of communication and thus the timely availability of emergency aid. Also the techniques of conditioning volcanic materials for agricultural uses have been improved (Muljadi et al., 1972); they include the application of moist butimenous micelles to improve the water retention capacity.

17.1.2. Methods in Volcanic Hazard Assessment

It is evident from the previous sub-section that many factors are involved in volcanic hazard assessment,making this a rather complex matter. The studies should cover the following grounds:

1. Monitoring the volcanic activity and related phenomena as previously outlined.
2. Analysis of data on type, frequency, magnitude, etc. of historical eruptions and particularly on the distributional patterns of the areas affected by them. The aim of this approach is to extrapolate these data to future eruptions and thus to arrive at a better appraisal of

Fig. 17.1 Vertical aerial photograph, scale 1:40,000 showing numerous lava flows that have emerged from the lower north-easterly slopes of Mt. Ciremai, W. Java, Indonesia,during the early part of the growth of this strato volcano.

the inherent hazards.

3. Mapping the sequence of volcanic deposition. The stratigraphic succession of different volcanic layers reflects changes with time in the type of volcanic eruption and the areas affected over a considerably longer period of time than that covered by historical records, mentioned under heading 2. Absolute dating of the chronology so obtained, e.g., by 14c dating methods is possible in a number of cases. It is of particular importance in the context of these studies to determine the genesis of the deposits concerned in terms of the causative geomorphological processes: airborne tephra, volcanic mudflow, avalanche, etc. Grandell and Waldron (1969), Anon (1974) and Grandell and Mullineaux (1975) elaborate on the matter.
4. Studying the denudational history of the volcanic body with alternating periods of construction. These factors are responsible for the present configuration of the volcano which governs, to an important extent, the spreading of the volcanic products resulting from present and future eruptions. Relics of old calderas, scarps formed by extinct craters, old lava flows and toloids, rising above the slope of the volcanic cone are often an effective protection for the areas farther downslope. Deep ravines radiating from the top area of a volcano, on the contrary, may act as major channels for the transportation of pyroclastics. The aerial photograph (1:35,000) of fig. 17.2, showing the top area of the slender but extinct cone of Mt. Kiematabu (1,730 m), the strato volcano dominating the island of Tidore, Indonesia, gives an example and pictures a breach of the crater rim with a major barranco which would primarily direct the volcanic products to the north if ever a central eruption of minor/medium magnitude would occur. Fisher (1980) states that the pyroclastic flows and associated ash clouds of the ill-famed 1902 eruption of Mt. Pelee, Martinique, that resulted in the total destruction of the town of St. Pierre (28,000 casualities) have been topographically controlled by the Rivière Blanche Valley and the subsequent glowing clouds have been controlled by the presence of the plug.
5. Volcanic hazard appraisal proper and risk zoning. This is a matter of considerable importance to which affected countries should direct their full attention (Escher, 1920; Neumann v. Padang, 1958; Furuya, 1978; Kusumadinata, 1979; Suwa, 1980, Westercamp, 1981). It is obvious from the surveying aspects mentioned above that geomorphological considerations rank high in the assessment of volcanic hazard and in the ultimate breakdown of the land in risk zones for specific types of volcanic disaster. Nevertheless,some of the modes of risk zoning applied at present tend to take the geomorphological situation of the volcanic environment insufficiently into consideration and also do not or only partly distinguish properly between different types of hazard provoked by the various processes involved.
6. Establishing an efficient early-warning system on the basis of the hazard appraisal and taking the location of possibly endangered settlements, the distribution of rural population, the road system available for evacuation and emergency rescue, etc.,into account (Sörensen and Gersmehl, 1980).

Fig. 17.2 Aerial photograph showing the top part of the Mt. Kiematabu strato volcano that dominates the island of Tidore, Indonesia. Note the breach in the crater rim.

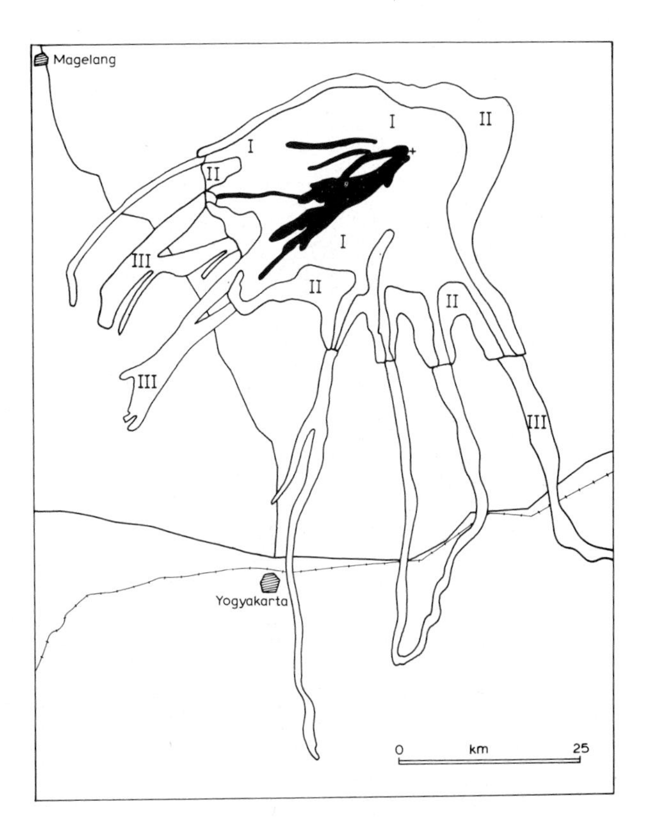

Fig. 17.3 Volcanic hazard map of the Merapi volcano, Java, Indonesia, showing a closed zone (I), where glowing cloud risk is always present, and two further hazard zones (II and III). In zone III rain-fed volcanic mud flows (cold lahars) are the danger type. Reduced and simplified from a hazard map made by the Volcanological Survey, Bandung, Indonesia. Black: lava and glowing cloud deposits.

7. Implementing protective measures such as reafforestation of devastated slopes, construction of check dams in ravines affected by mudflows, etc. (Kadomura and Yamamoto, 1980; Kadomuro et al., 1980). Artificial drains for emptying dangerous crater lakes, such as mentioned for the Kelut volcano, east Java, Indonesia (Van IJzendoorn, 1953; Smart, 1981),is also among the measures to be contemplated. The planning of optimal use of the land, the proper (re)location of settlements, major structures, bridges, etc., are other matters that should be guided by the results of the volcanic hazard appraisal and the resulting maps, as a part of the management of volcanic regions. Engineering works for combating the adverse effects with respect to irrigation, navigability, etc.,of excessive sedimentation in the rivers near active volcanic cones comes under the same heading (Thal Larsen, 1926; Verstappen, 1963).

The map of fig. 17.3 gives a broad volcanic hazard zoning of the Merapi volcano, central Java, Indonesia. Distinction has been made between the closed zone - in which various pyroclastic flows are indicated - and two further hazard classes. The second one stretches along the major ravines radiating from the top area of the volcano and relates to volcanic mud flows (lahar). Since they traverse the densely populated lower slopes of the volcano, the danger is concentrated there. The most active flows in the last decades, occur in the SSW where settlements and bridges,in particular along the road Magelang-Yogyakarta,are affected. These flows are not separated from those in the south which have been inactive since the beginning of this century. The old crater rim protecting the areas to the north and east of the top area is not indicated. Nevertheless, the map gives a good general picture of the hazard problems (Cool, 1931; Escher, 1920).

17.1.3 Volcanic Hazard Zoning of Ternate: an Example

The volcanic hazard map and risk zoning of the island of Ternate, northern Moluccas, Indonesia, given in fig. 17.4,serves to exemplify the role of geomorphology in these studies. It is based on field observations by the author (Verstappen, 1964) that were preceded by the interpretation of aerial photographs (fig. 17.5).

The island consists entirely of the Gamalama volcano, also known as the 'Peak of Ternate',which rises to 1,715 m

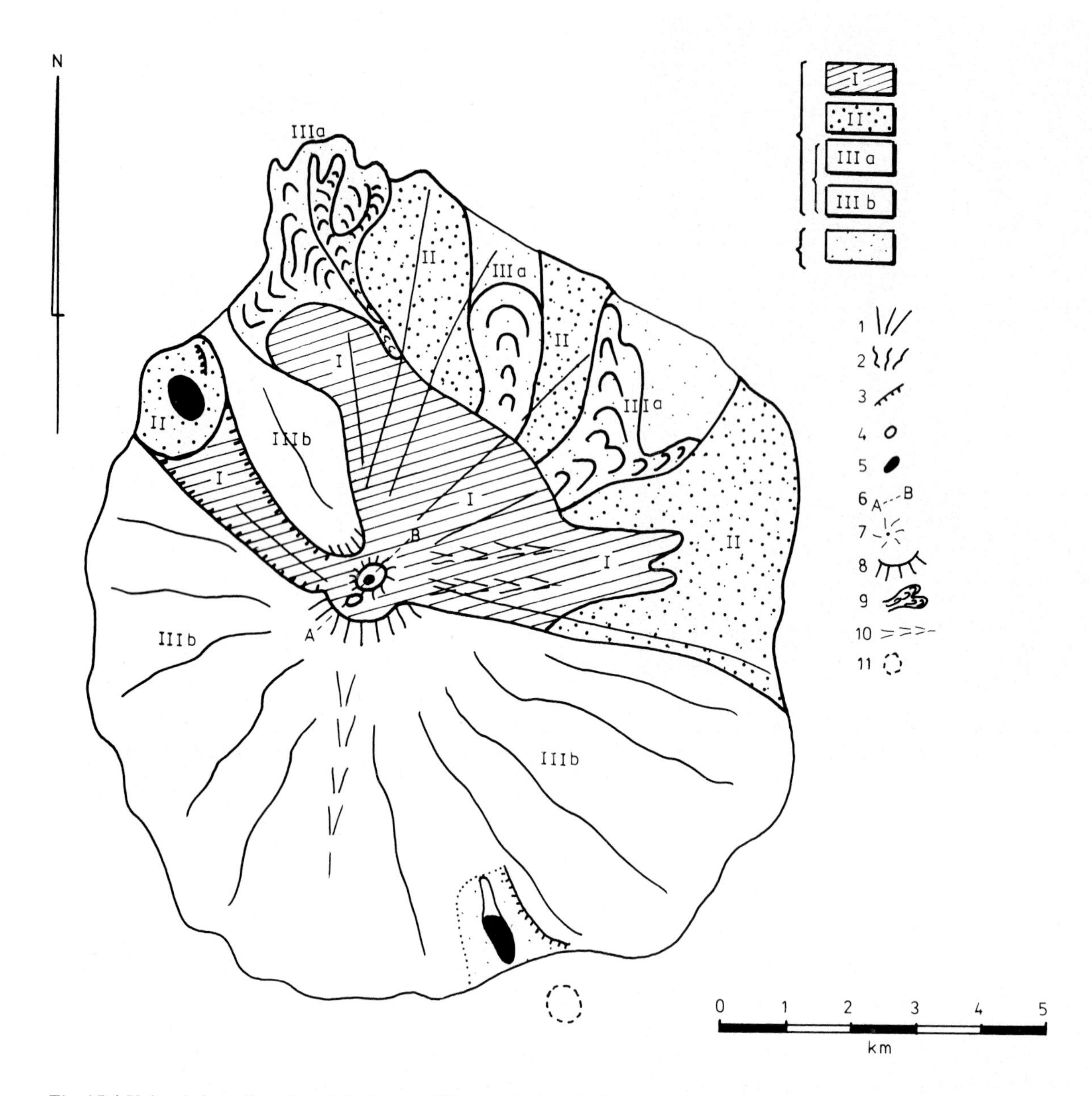

Fig. 17.4 Volcanic hazard zoning of the island of Ternate, Indonesia, based on geomorphological survey and airphoto interpretation.
Key: Hazard classification for central eruption:
Hazard zone I : high slopes of young Arfat cone
Hazard zone II : lower slopes of same without old lava flows
Hazard zone IIIa: lower slopes of same protected by old lava flows
Hazard zone IIIb: old Somma slopes
Fine dots: Areas potentially subjected to rare parasitic eruptions
Geomorphological phenomena:
1. young Arfat cone
2. old Somma slopes (Gamalama)
3. major radial rift
4. aligned crater pits
5. main crater pit in Mt. Arfat crater
6. alignment of crater pits
7. Mt. Arfat cone
8. relics of Somma crater
9. lava flows
10. major barrancos
11. submarine parasitic eruption point

a.s.l. It is a comparatively active strato-volcano, the most recent eruption of which dates from the end of 1962. A number of parasitic eruption points, both of explosive and effusive nature, adorn its lower slopes. The eruptions are usually at the main crater, located in the top area. A cinder cone, Mt. Arfat,occurs here containing four crater pits, three of which are situated in an older, larger eruption point,whereas the fourth lies immediately to the southwest of this. These crater pits are aligned along a SW-NE stretching fissure.

To the south of the young Mt. Arfat cone, a part of an old 'Somma' crater rim is preserved and a smaller relic also occurs in the northwest. The dissection of the slopes of the volcano reflects this situation: those of the north-

Fig. 17.5 The northern slopes of Mt. Arfat, mostly composed of volcanic tuffs and the lava flows occurring at the northern tip of the island of Ternate. Aerial photograph 1:40,000, compare with the map of fig. 17.4.

eastern sector are the least dissected and result from Mt. Arfat volcanic activity. Those of the other parts of the island are distinctly more intensely dissected, with the exception of the rift valley separating the northwestern relic from the remainder of the Somma slopes.

Two explosion craters, with lakes, are situated on the lower slopes in the NW rift zone. An old report makes mention of an eruption here in 1775,but this has not been confirmed. Another, oblong-shaped, explosion lake occurs along a supposed radial fault in the lower SSE slopes. The dip of the volcanic ashes at its seaward side suggests that another, submarine, explosion crater is present immediately off-shore. Lava flows are confined to the lower northern and northeastern Mt. Arfat slopes. Outflows in historical times (18th century) have been reported.

Summarizing, the following observations are important for protection against volcanic disaster. The western, southwestern, southern and southeastern inactive slopes of the island and a narrow strip in the northwest are relatively well-protected by the Somma crater. The young ashes of Mt. Arfat are transported downward along the northern, northeastern and eastern slopes and to some extent also along the northwestern rift. On several occasions mud flows and stone avalanches have been reported from these slopes. Two barrancos on the eastern slopes are particularly important in this respect because they reach the sea not far from the town of Ternate. The northern-most one is the site of the mudflows of 1897. At present, however, this side of the mountain is rather well vegetated and bare ashes are found on the northern slopes. The aerial photograph of fig. 17.5 indicates the situation there and shows the ashes, the northern lava flows and the NW rift with the explosion lakes.

The hazards of the rare parasitic eruptions are difficult to assess. Only surveillance can lead to timely measures. Explosive eruptions will, in all likelihood be situated in the NW rift and along the supposed radial fault SSE. Effusive eruptions are most likely on the lower northern and northeastern slopes. In these parts,there is a risk of rare effusive events, but much more important is the protective effect of the existing lava flows as a result of their higher topographic situation. The areas downslope and covered by them are safe in case volcanic mud flows from Mt. Arfat may come down in that direction.

In case a central eruption might necessitate rapid evacuation of the population within the island, it is recommendable for the inhabitants of the town of Ternate to seek refuge at the old NE lava flow (where also the villagers from the northeastern part of the island could flee). Evacuation of the NW rift zone would also be advisable. The map of fig. 17.4 gives the volcanic zoning of the island, together with a few relevant geomorphological phenomena. Three hazard classes can be distinguished, based on the assumption of a central eruption. The highest risk exists on the higher and middle slopes in the NE sector, in the NW rift and particularly near the barrancos in the east. The lowest risk exists on the old Somma slopes and also on and downslope of the lava flows in the north and northeast. The areas most likely affected by rare and rather unpredictable parasitic eruptions are indicated separately. The map illustrates the role of geomorphology in volcanic hazard appraisal.

17.2 Earthquake Hazard Survey and Risk Zoning

17.2.1 Various Approaches

Earthquakes are a permanent natural hazard in many tectonically unstable areas and threaten society there with sudden disasters of unknown dimensions (Eiby, 1980; Gates, 1970; Steinbrugge et al., 1973; Tricart, 1973). The long and irregular recurrence interval of major quakes combined with the - until recently completely - unpredictability of the place, the time and magnitude of their occurrence are characteristic for this type of hazard, resulting, on one hand, in fatalistic attitudes of the affected population; on the other hand, they are a challenge for the responsible authorities, scientists and engineers (Nazarian and Hadjian, 1980).

There are several ways and means to reduce the loss of life and the material damage caused by earthquakes and thus to mitigate the disasters. They can be briefly summarized as follows:

Predicting and even preventing earthquakes

These attempts are still in their infancy. Since, however, an earthquake was successfully predicted in the Haicheng area, China, in 1975, this matter has received increased attention. The traditional approaches of studying spatial regularities, time distribution and return period of major earthquakes in themselves are, of course, inadequate for a reliable prediction. They are, however, at present supplemented by geodetic, geophysical and other approaches which may in the future lead to methods of predicting the location, time and magnitude of major earthquakes with the high reliability required for the authorities to issue an early warning and for the population to trust the warning and to follow instructions given to them (Anon, 1969; Rikatake, 1975). One should not forget that such a warning will inevitably have serious social, economic, psychological and other consequences and result in hardships for those concerned. It should therefore, be based on certain evidence (Roberts, 1979).

The geodetic approaches include distance measurement of fault movements, e.g., by geodimeter, measurement of elevation changes, tilt and other anomalous crustal activities in potential epicentral areas (Plafker and Rubin, 1967; Okusa and Anma, 1980). The geophysical approaches relate to precursors such as (micro) seisms, dilatancy-related electrical resistivity changes of rocks, changes in seismic velocities (variations in travelling time of transverse and longitudinal waves), acoustic emissions, long term changes in strain as measured by deep electromagnetic sounding and/or gravity measurements, etc. (Pakiser, 1970; Scholz et al., 1973). In addition to these approaches, there are diverse environmental indications such as changes in groundwater level, geochemical changes in the water of hot springs (Rinehart and Murphy, 1969), fluctuations in oil flow and even peculiar animal behaviour. The ultimate aim of all these approaches combined should be to in the (distant) future either give an early warning or to release the strain artificially before it has built up to a dangerous level that might cause disaster. Although such possibilities are as yet remote, the efforts should be concentrated on the prediction and mitigation of the quakes themselves. Obviously, the prediction of earthquakes is largely outside the field of specialisation of the geomorphologist, although he may be involved in neotectonic studies related to it (Inouchi and Sato, 1975). He can render good services, however, for the mitigation of earthquake disasters.

Study of the damage (and its distribution) caused by previous earthquakes

Thorough stock-taking of the damage caused by an earthquake in a certain area and assessment of the data obtained may in several respects be a lesson for the future. Such inductive studies of past, historical earthquakes may thus assist in mitigating future earthquake disasters assuming that the same patterns will then also apply, notwithstanding possible differences in epicenter, magnitude, etc., of the quake. A considerable similarity in damage patterns caused by subsequent earthquakes has indeed been proved on several occasions. The percentage of damage to houses of various construction types (wooden, brick, concrete, single or multi-storey, etc) is a much used parameter and the percent-damage maps produced, giving spatial distribution, act as a useful tool in assessing the effects of future earthquakes in the area. The matter has been dealt with by various authors (Tada et al., 1951; Kanakubo and Tanioka, 1980; Murayama, 1980). These maps indicate the vulnerability of the various construction types and may contribute to better designs by engineers and building codes imposed by the authorities. Insurance companies will also be interested in these data on the distributional patterns of damage for a proper assessment of the risk and the corresponding premium (Northey, 1973). Apart from damage to houses, the damage caused to other structures, such as roads, railways and dikes, may also be considered. It should be borne in mind, however, that in order to assess the loss of life and property of a future earthquake of a certain magnitude, information is required about the density of housing, the number of inhabitants, etc., in addition to the percentage-damage data.

Assessing the land in terms of earthquake risk

The geomorphological lay of the land, including matters such as soil and sub-soil conditions and slope stability, has an important effect on the distributional patterns of earthquake damage. This matter has received attention in many countries, particularly after severe earthquakes (USGS, 1964; Gupta and Virdi, 1975; Mareus, 1964; Post, 1967; Nat. Res. Council, 1968, 1969). The map of fig. 7.9, showing the situation of the town of Quetta, Pakistan, may serve as an example. The severe earthquake that struck the town in 1935 caused a very high death toll in the old, densely populated part of the town located on the alluvial plain which, being saturated with water, transmitted the shocks in full vigor. Much less damage was done in the higher parts of the town situated on gravel fans made up of coarse, angular and dry material. On other occasions well-defined geomorphological situations, such as narrow zones near terrace edges and slopes formed in water-saturated glacial till, have proved to be particularly susceptible. Earthquake hazard or risk zoning is therefore a realistic means of mitigating earthquake disaster by applying deductive methods. The geomorphologist can play an active part in it.

The information obtained from the distributional pattern of earthquake risk is of considerable importance for land use planning,particularly in densely populated urban and suburban areas. Urban extensions should preferably be directed to areas that will supposedly be least effected by earthquakes. Building codes may vary with the hazard zone. The percentage-damage or vulnerability maps of earlier earth quakes often give a clear indication about the parts of the terrain that are particularly susceptible to earthquake damage. These inductive studies are a useful aid in the deductive studies of environmental zoning of the affected areas and, in fact, are often executed in this context; they may also contribute to the quantitative assessment of potential disaster if data on the concentration of population and structures are added.

Understandably, earthquake hazard zoning has received considerable attention in the countries and areas affected. Grant-Taylor et al. (1974) have studied the matter in Wellington, New Zealand; Iliev (1974) reports on the situation in Bulgaria; Matsuda (1980) did work in the densely populated Kanto Plain, Japan,where Nakano (1980) specifically investigated the hazards faced by the Tokyo metropolitan area. In the USA the Office of Emergency Preparedness deals with the subject (Gates, 1970). Hart (1974) studied matters of zoning surface fault hazards in California where Borcherdt (1975) and Steinbrugge (1968) have investigated the hazards occurring in the vulnerable San Francisco Area (see also Proceedings, 1978). The matter will be elaborated upon in the next sub-section.

Mitigation by planning and preparedness

The data gathered on earthquake risk and particularly the hazard zoning maps are essential information for the authorities in earthquake zones. If maps on the density of population and houses, larger structures, etc.,are added, it can be attempted to define and even quantify the disaster in terms of casualities and material losses for earthquakes of specific magnitude. Computer simulation of earthquake hazard is a useful tool in this connection. Friedman (1970) gives an example of the San Francisco area using variables such as ground conditions (Kanai, et al., 1963), earthquake intensity and epicentral locations of previous quakes. Apart from matters such as land use planning (Schmoll et al., 1975; Steinbrugge, 1968, 1970) and, particularly in densely populated urban areas, building codes, the planning of emergency operations and the problems related to secondary dangers such as fire outbreaks, ruptures in water supply (Paxton, 1972), epidemics, etc. have to be considered in the context of earthquake emergency preparedness.

17.2.2 Geomorphology in Earthquake Hazard Zoning

The subject of terrain as an important factor in the distributional patterns of earthquake damage, already touched upon in the previous sub-section, merits further elaboration in this context as it is of considerable interest to the geomorphologist (Karmick and Algermissen, 1978). To this aim, the various causes of earthquake damage have to be analysed. The major causes can be listed as follows:

1. vibration
2. deformation
3. liquefaction or settlement in (almost) flat land
4. slope failure of various kinds in areas of relief
5. flooding
6. fire.

Of these, only the first two causes are directly related to crustal movements. The remaining four major causes are indirect results of the earthquake, but are often responsible for a major part of the damage. Geomorphological aspects play a part in all six major causes listed, though in varying degrees.

The effects of vibration depend, to a considerable degree, on the conditions of the land. In soft ground and particularly where fine-textured and water-saturated material occurs,such as back-swamp clays, peat formations, moist lacustrine or fine-sandy fluvial deposits, the vibrations and the consequential acceleration of the shock waves is much stronger,and it has been found on many occasions that the damage in such soft ground terrain is 5-10 times greater than in adjacent hard rock areas. The inertia of structures of yielding material may cause them to settle into the foundation medium. Rigid structures often fare

Fig. 17.6 Pile foundation of a 3-storey house in Leoni, south Italy, that sank in the water-saturated lacustrine deposits of the site on the occasion of the earthquake of November, 1980. The first floor almost reached the ground (foreground).

Fig. 17.7 A 4-storey house in Leoni, south Italy, built on lacustrine deposits with high groundwater level, destroyed during the earthquake of November, 1980, notwithstanding the adapted 'flat-foot' foundation.

better than non-rigid ones. Figs. 17.6 and 17.7 give examples from buildings in Leoni, south Italy, constructed on lacustrine sediments and severely damaged by the earthquake of November, 1980. Boore et al. (1972); Grand-Taylor et al. (1974) and Kachodoorian (1968) give examples. Within the soft ground areas the damage varies with the ground conditions and borings up to 30 metres may be required for proper assessment. It is the slowest, surficial (L) waves of the earthquake that are responsible for the rocking of the ground and thus for the damage caused (Yeats, 1981).

The effects of deformation include horizontal and/or vertical movements along fault lines, tilt, etc. (Nakano, 1966; Research Group, 1980). Precise mapping of fault lines in tectonically unstable area is important for mitigating earthquake damage because buildings situated across fault lines are often much more severely affected than similar buildings just slightly to either side of the fault. Indications of neotectonics thus need careful study. Fig. 17.8 shows recent faultscarps in volcanic tuffs near Lake Toba, Sumatra, Indonesia,and fig. 17.9 shows active faults in an area of urbanized fluvial terraces in Japan (Research Group, 1980). Radbruch (1969) gives an interesting example of active faults in the Hayward fault zone, California, USA, where slow neotectonic movements cause damage to structures straddling zones of differential movements. The damage may be either sudden, during earthquakes, or result from continuous strain and slow movements. Similar examples are known from numerous parts of the world (Garcia, 1966). Cracking and deformation of buildings straddling faultlines in areas of soft ground are often marked in the terrain by a minor faultscarp in the alluvial deposits and aggravated also by the effect of the fault on the groundwater situation. Regional tectonic subsidence accompanying earthquakes may also result in major destruction (McCulloch and Bonilla, 1970).

Liquefaction and settlement are among the indirect causes of severe damage when earthquakes occur. Earthquakes affecting areas of soft ground may result in sudden changes in the state of saturated sands. Liquefaction may occur when loose sands are in a critical state of equilibrium.

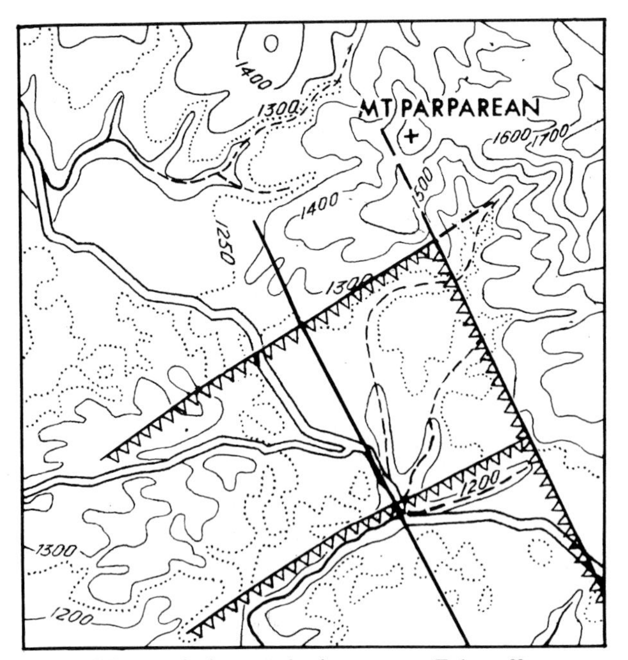

Fig. 17.8 Recent fault scarps in the youngest Toba tuffs, Sumatra, Indonesia: distinct indications of neotectonic movements. Scale 1:100,000.

Cyclic mobility is another process that may be triggered by earthquakes in sands of any kind. The susceptibility to this phenomenon of saturated materials is often difficult to assess by testing because of the inevitable disturbance during sampling and testing (Castro, 1975). This phenomenon may cause ground failure even in flat terrain but may also trigger dangerous landslides in hilly land (Bolton Seed, 1969). The ground failures may range from flow-landslides to lateral-spread and quick condition failures (Youd, 1973). Sand jets may also be ejected on such occasions (Nishimura et al., 1965; Waller, 1968).

Slope failure triggered by the vibration or liquefaction, etc.,related to earthquakes ranges from landslides of various types and rock falls to avalanches of snow and glacier ice. These phenomena often account for much of the damage when densely populated mountainous areas are struck by a major earthquake. It is thus of considerable

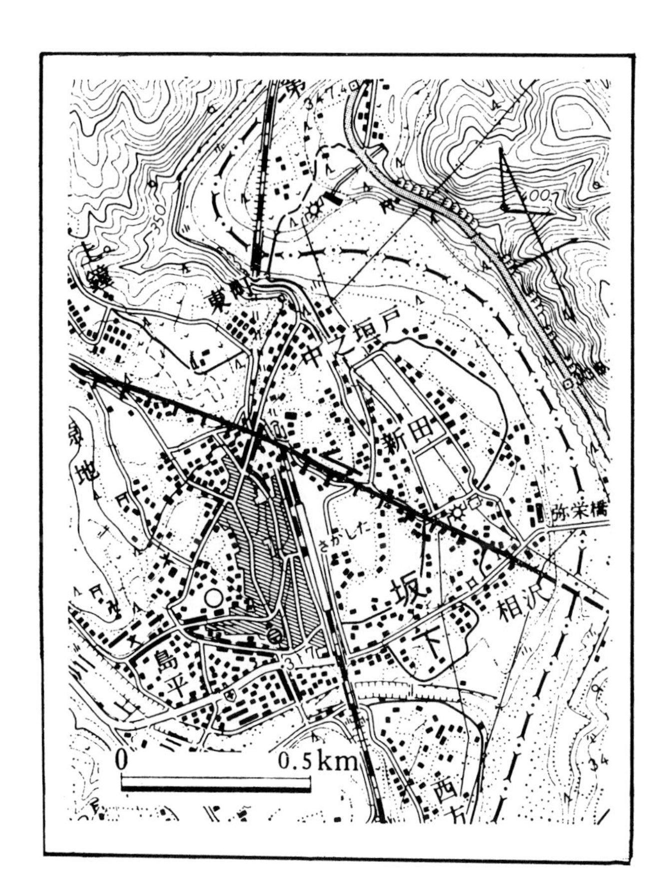

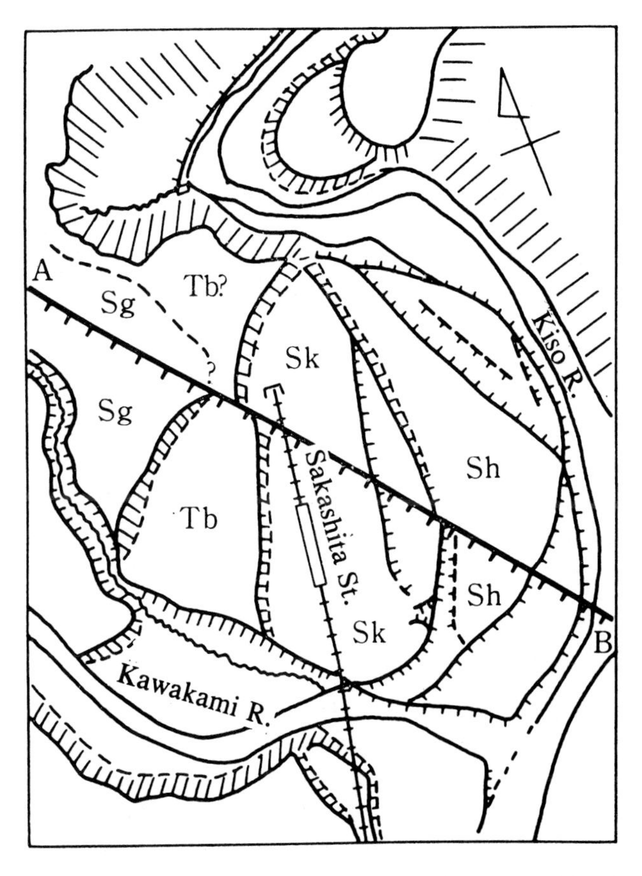

Fig. 17.9 Active fault in Atera area, Honshu, Japan. Left: the urbanisation pattern, right: the geomorphological situation (Sg, Tb, Sk and Sh are terraces of the Kiso River). After: Research Group, 1980.

importance to properly assess the susceptibility to slope failure in such areas, taking into consideration matters such as material, slope angle and water content.

Hogg (1980) describes and maps the effects of a rockfall-type earthquake disaster of 1976 in the region of Friuli, situated north of Udine, in the surroundings of Venzone, north Italy. The main shock occurred on 6 May 1976 (Richter 6.4) but the seismic event was characterized by a series of after-shocks, the epicenters of which are indicated in the map, together with the landslides and rockfalls triggered by the seism. These indirect effects of the earthquake were responsible for most of the damage. They were all associated with potentially unstable slopes. The lithology of the area - breccias and limestones - is also indicated. The transmissivity of the unconsolidated, alluvial deposits of the valley bottom in this case did not cause major damage: only the sites directly below instable slopes were affected.

Other cases of earthquake-related rockfall disaster are mentioned by Morton (1971), Clapperton and Hamilton (1971) and Cooke and Townsend (1970). The two last-mentioned publications relate to the ill-famed huge avalanche, rockfall and fluid debris avalanche of 1970 in Huascaran, Peru. The role of earthquakes in triggering avalanches (La Chapelle, 1968) and their effects on glaciers (Field, 1968; Tuthill, 1968) were studied especially after the Great Alaskan Earthquake of 1964. Eckel (1970) describes a variety of landslides that occurred on that occasion.

Earthquake-triggered landslides are one of the most studied phenomena in the field of earthquake disasters (Gams, 1979). Both reactivation of existing slides and the the formation of new, major slides may occur. The stability of slopes under earthquake conditions depends, apart from the strength of the quake, also on matters such as number of effective pulses, steepness of slope, type of rock, degree of fracturing and, in case of sedimentary beds, on their inclination with respect to the slope direction (Kobayashi, 1968; Morton, 1971). Furthermore, changes in soil properties, e.g., liquefaction, may be produced and intercalations of weak rocks also have to be considered.

Iliev (1974) divides slopes into three classes of seismic resistance: slopes in resistant rocks, slopes formed in material of medium resistance and slopes in clayey and sandy sediments. These matters, of course, affect planning in earthquake zones and post-earthquake reconstruction. The measures involved may aim either at protection by stabilization of slopes and scarps or at removing structures or even entire settlements from endangered localities (Farhoudi, 1975; Garcia, 1966 . See also: Bolton Seed, 1967; 1970; Parry, 1974; Wahler, 1973). Artificial fills are often particularly susceptible to instability and settlement. The bay fills of San Francisco are notorious in this respect (Bolton Seed, 1969).

During the earthquake that struck Sendai, Japan, on 12 June 1978, the new residential areas situated on artificially levelled, rolling terrain developed in soft upper-Tertiary beds, proved to be particularly vulnerable when situated on fills. Most damage, in fact, occurred at the transition between cut and fill zones where differential settling of the ground was a major cause. Fig. 17.9 shows the general situation and fig. 17.10 demonstrates the relationship between damage and filled-up terrain in the new residential areas to the north of the centre (after: Tamura, 1980; see also Murayama, 1980). Substantial damage also occurred in the lowlands, but the old parts of the town situated on the gravels of Pleistocene terraces were only slightly affected.

Flooding is another potential cause of earthquake-induced damage. It may result from failure of river banks or from slides ponding up rivers and thus forming temporary lakes which may be suddenly emptied when the swelling waters break through the landslide material. Flooding may also be related to tsunami provoked in the sea and/or major lakes as a result of sub-aquous volume changes. The waves formed can be divided in back-fill waves rushing toward the land to fill the void produced by the sinking ground and far-shore waves that hit the opposite shore (McCulloch, 1966). The tsunami may travel thousands of kilometres across the oceans and cause great damage in distant places (see section 13). Emptying of existing natural, e.g., glacier-dammed lakes (Marcus, 1968),or of storage lakes due to dam failure (Bolton Seed, 1975; Nishimura et al., 1969; Okusa and Anma, 1980) may also give rise to additional earthquake disaster.

Fire caused, e.g., by the rupture of gas pipes is another serious indirect earthquake hazard, particularly in urban areas. Responsible authorities should be aware of the areas liable to fire outbreak and those of extreme fire hazard, beforehand. On the basis of likely ignition points and duration of the fires, their spread can be estimated. The only contribution a geomorphologist can possibly make in this context is assessing the accessibility of the areas concerned under post-earthquake emergency conditions (flooding, etc.). Serious damage was caused in Tokyo after the Kanto earthquake because certain burning parts of the town could not be reached by the fire brigade.

The diversity of factors to be taken into consideration in the geomorphological assessment and zoning of the land for earthquake hazard will be clear from the preceding pages. The problem will vary from one region to another depending on the environmental situation. Aerial photographs are an almost indispensable tool for such surveys (Oswal, 1968; Farhoudi, 1975; Kanakubo and Tanioka, 1980). The maps produced are, in the first place, those correlating the potential damage to land conditions. Furthermore, maps correlating the damage to ground surface changes can be made and specific maps for counter measures (% houses destroyed, etc.) may also be prepared. The survey methods have been studied specifically by Kadomura (1967, 1968) for the Shizuoka-Shimuzu Area in central Honshu, Japan. Aerial photograph interpretation, geomorphological field studies, systematic borings and terrain classification ultimately lead to adequate earthquake hazard zoning maps. The deductive approach of assessing the susceptibility of the land is supported by the inductive method of studying the spatial distribution of the damage caused by previous earthquakes. A flow diagram of the research work is given in fig. 17.12 and the various maps of fig. 17.13 further elucidate the approach selected for these specific geomorphological surveys which ultimately lead to the hazard zoning. Fig. 17.14 shows the damage percentage of houses caused by the earthquake that hit the area in 1935: the correlation with the hazard zoning map of fig. 17.11 is evident.

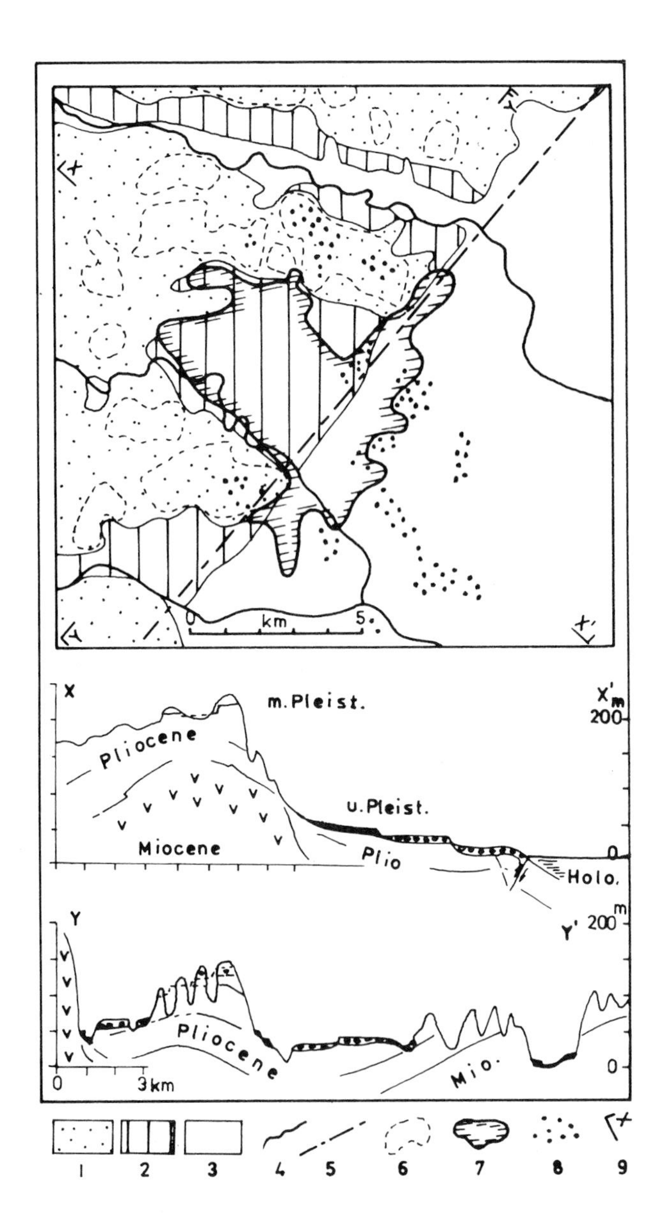

Fig. 17.10 Geological - geomorphological situation of the town of Sendai, Honshu, Japan, with two cross-sections (X-X and Y-Y). Note the artificial fills in the zone of urban extension and the distributional pattern of the damaged houses (Tamura, 1980).
Key:
1. hilly area
2. flat uplands
3. lowlands
4. river
5. active fault line (not in 1978)
6. artificial fill
7. built-up area around 1960
8. severely damaged area (1978 earthquake)
9. location of profiles

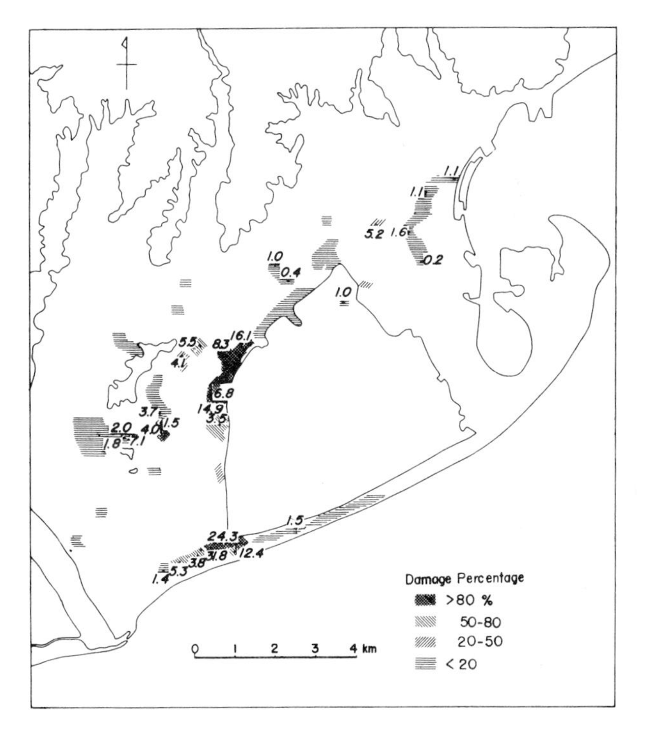

Fig. 17.14 Distributional pattern of damage percentage of houses in the Shizuoka-Shimizu Area, Japan, due to the earthquake of July, 1935. The figures indicate the percentage of totally collapsed houses in each village. Compare with fig. 17.13, lower right.

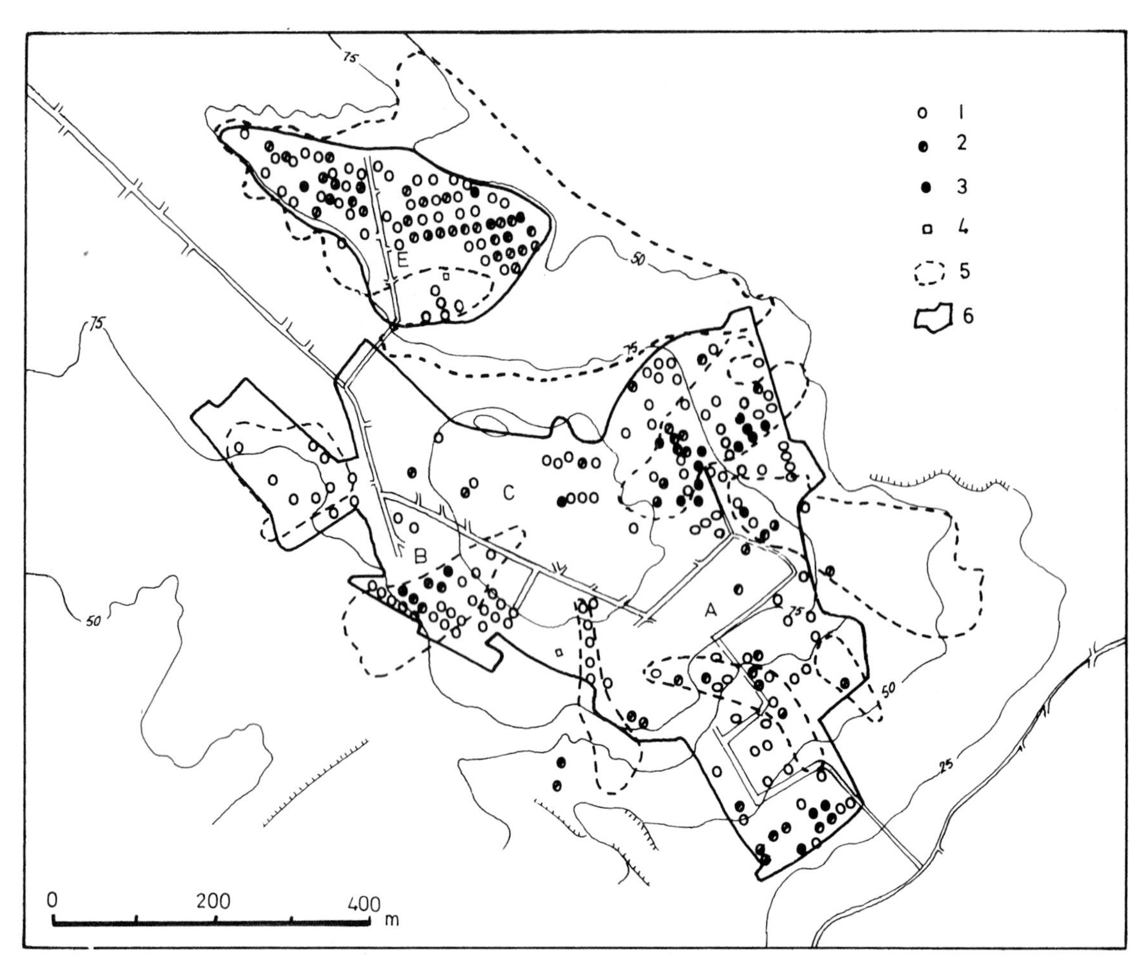

Fig. 17.11 Detailed damage survey in the Midorigaoka area (Sendai) of the 1978 earthquake (Tamura, 1980).
Key:

1. slightly damaged house
2. moderately damaged house
3. severely damaged house
4. rupture of water supply pipe
5. artificial fill
6. survey area

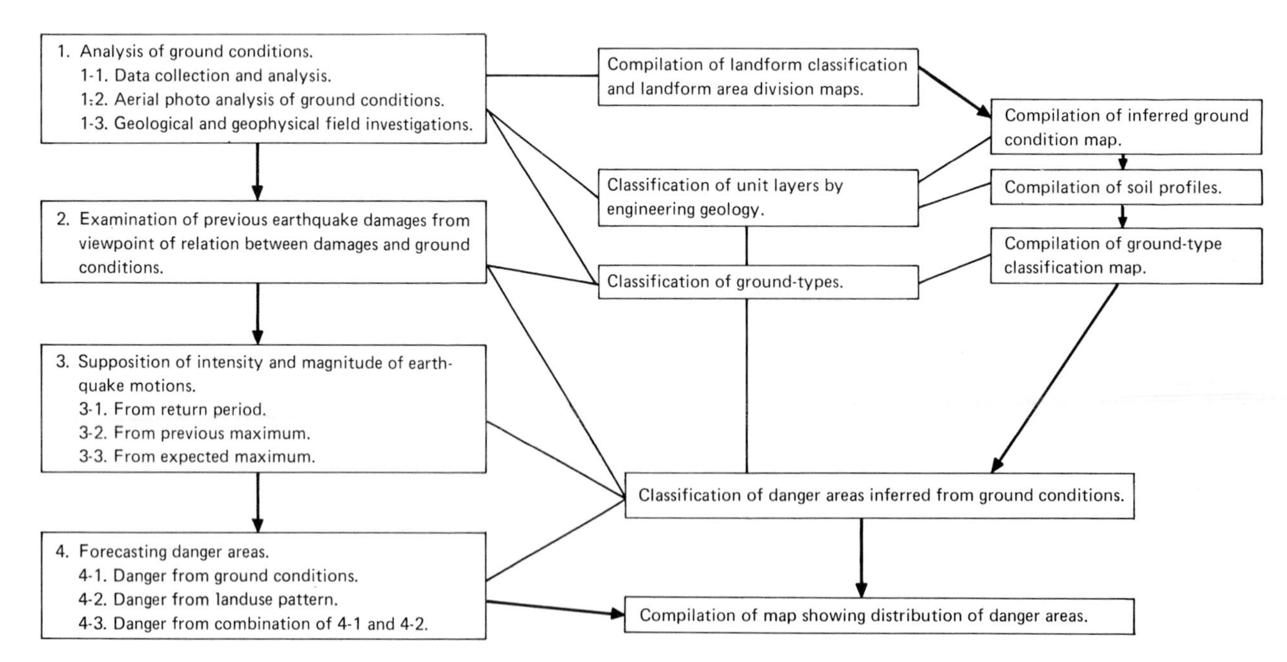

Fig. 17.12 Flow diagram of earthquake hazard zoning research according to Kadomura (1968). Aerial photograph interpretation and the assessment of soft ground conditions play a major role.

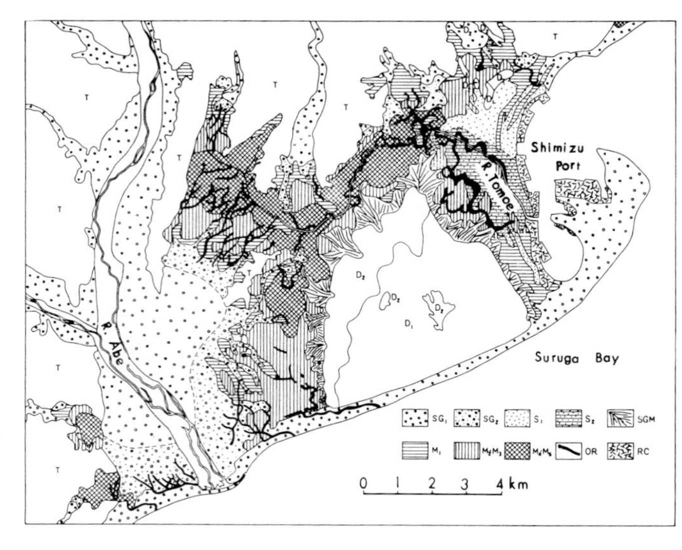

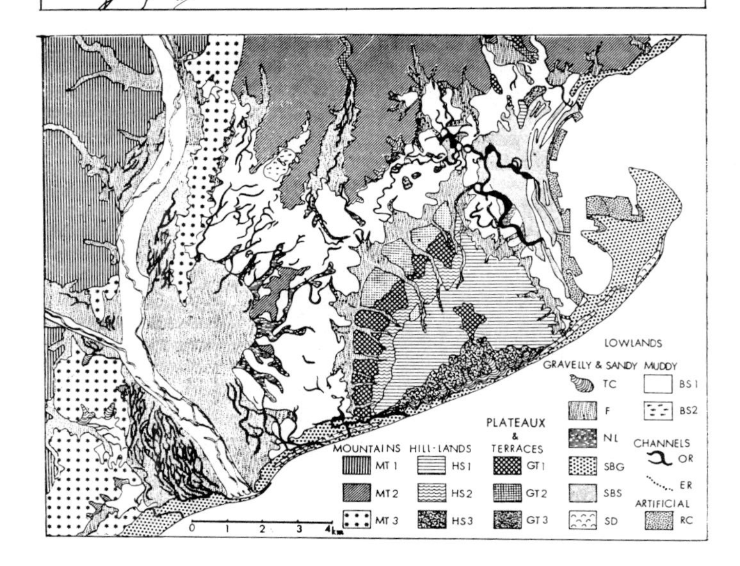

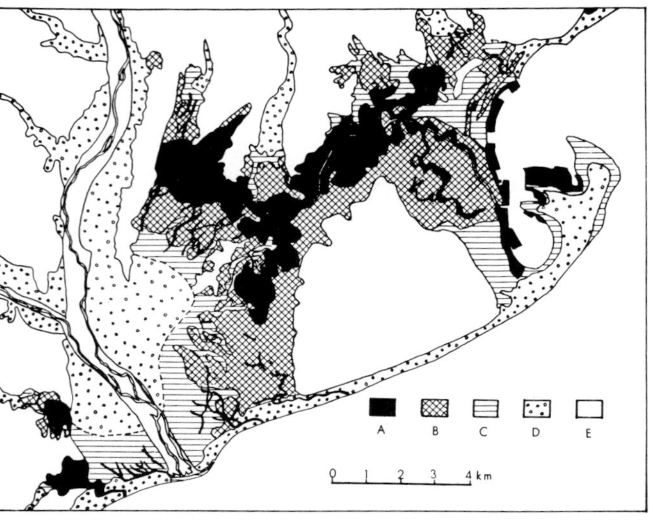

Fig. 17.13 Four types of maps made of the Shizuoka, Shimizu Area, Japan to assess the earthquake hazard (Kadomura, 1968). Upper left: main landform units. Lower left: landforms and surface geology, compiled from aerial photographs. Upper right: Distribution of ground types (texture/thickness) G: gravel, S: sand, M: mud. Lower right: Earthquake hazard zoning A: high.

Key: MT-mountains (1. Palaeogene; 2. Neogene, 3. basaltic). HS-hills (1. Pleistocene, gravelly, 2. Pleistocene, muddy, 3. severely eroded). GT-gravelly plateau terrace (1. upper, 2. middle, 3. lower). Lowlands: TC-talus cone, F-fan, NL-natural levee, SBG-shingle bar, SBS-sand-bank, SD-sand dune, BS1-backswamp, etc., BS2-Asahata-numa marsh, OR-old channel, ER-elevated river, RC-reclaimed land.

References

Volcanic hazard

Anon., 1974. Appraising volcanic hazards of the Cascade Range of the NW United States. Earthquake Inform. Bull. USGS, 6(5): 3–10.

Baker, P.E., 1969. A volcano erupts beneath the antarctic ice. Geogr. Mag., 42(2): 115–126.

Cool., W., 1931. Het Merapi-gebeuren in Midden-Java, bij de jaarwisseling 1930/31. De Ingenieur, 46(37): A 341–357.

Crandell, D., and Waldron, H.H., 1970. Volcanic hazards in the Cascade Range. In: Olson, R.A., and Wallace, M.M. (Editors), Geologic hazards and public problems. Proc. Conf. San Francisco., 1969. U.S. Govt. Printing Office, Washington, D.C.: 5–18.

Crandell, D.R., and Mullineaux, D.R., 1975. Technique and rational of volcanic hazard appraissals in the Cascade Range, N.W. United States. Environm. Geol., 1(1): 23–32.

Davies, D.K., et al., 1978. Glowing avalanches from the 1974 eruption of the volcano Fuego, Guatemala. Bull. Geol. Soc. Amer., 89(3): 369–384.

Duncan, A.M., et al., 1981. Mount Etna Volcano: environmental impact and problems of volcanic prediction. Geogr. J., 147(2): 164–179.

Escher, B.G., 1920. Gloedwolken en lahars, vulkanische catastrophes in Nederlandsch Indië. Tropisch Nederland, 3: 291–320.

Escher, B.G., 1920. Over het vulkanisme van Java in verband met de uitbarsting van den Merapi. De Ingenieur, 46(37): 357–369.

Fischer, R.V., 1980. Destruction of St. Pierre, Martinique, by ash-cloud surges, May 8 and 20, 1902. Geology, 8(10): 472–482.

Friedman, J.D., and Williams, R.S. Jr., 1968. Infrared sensing of active geological processes. Proc. 5th Symp. Remote Sensing of Environment, Ann Arbor: 787–815.

Friedman, J.D., et al., 1969. Infrared surveys in Iceland. Preliminary report. USGS Prof. Paper 650–C: 89–105.

Furuya, T., 1978. Preliminary report on some volcanic disasters in Indonesia. SE Asain studies (Kyoto), 15(4): 591–597.

Gorshkov, G.S., et al., 1971. The surveillance and prediction of volcanic activity. A review of methods and techniques. UNESCO Earth Science Series, 8, 166 pp.

Healy, J., et al., 1978. The eruption of Ruapehu, New Zealand on 22 June 1969. DSIR Bull., 244, 80 pp.

Hoblitt, R.P., et al., 1980. Mount Helens eruptive behaviour during the past 1,500 yrs. Geology, 8(11): 555–560.

Howard, A.D., and Dickinson, W.R., 1978. Volcanic environments. In:Howard, A.D., and Remson, I. (Editors), Geology in environmental planning. McGraw-Hill Book Comp., New York: 246–273.

Kadomura, H., and Yamamoto, H., 1980. Usu volcanic disasters and human adjustment. Guidebook Meeting IGU Working Gr. Perception of Environment, Hokkaido: 109–127.

Kadomura, H., et al., 1980. Erosion and mudflows at Usu Volcano after the 1977–1978 eruption. Envir. Sci., Hokkaido, 3(2): 155–184.

Koyanagi, E., et al., 1975. Reawakening of Mauna Loa Volcano, Hawaii: a preliminary evaluation of seismic evidence. Geoph. Res. Letters, 2(9): 405–408.

Kusumadinata, K., et al., 1979. Data dasar Gunung api Indonesia. Publ. Dir. Vulkanologi, Bandung, Indonesia, 820 pp.

Moxham, R.M., 1970. Thermal features at volcanoes in the Cascade Range as observed by aerial infrared surveys. Bull. Volcanol., 34: 77–106.

Mullineaux, D.R., and Peterson, D.W., 1974. Volcanic hazards on the island of Hawaii. USGS Reston (Va.) Publ., U.S. Govt. Repr. Ann., 25, 61 pp.

Muljadi, D., et al., 1972. Soil conditioning of volcanic debris. Results obtained on Java and Bali. Pédologie, 22(1): 100–124.

Murton, B., and Shimabukuro, S., 1974. Human adjustment to volcanic hazard in Puna District, Hawaii. In: G.F. White (Editor), Natural Hazards, Oxford, New York: 151–159.

Neumann von Padang, M., 1958. Enkele maatregelen om de bevolking tegen de gevolgen van vulkanische uitbarstingen in Indonesië te beschermen. Erts, 10: 74–79, 111–117.

Ney, C.S., 1975. Volcanoes in our lives. Geoscience Canada, 2(4): 188–192.

Preusser, H., 1973. Vulkanausbruch auf Heimaey, Island. Erdkunde, 104(3/4): 193–199.

Smart, G.M., 1981. Volcanic debris control, Gunung Kelud, E.Java. Int. Ass. Sci. Hydrol., Publ., 132: 604–623.

Sorensen, J.H., and Gersmehl, P.J., 1980. Volcanic hazard warning system: persistence and transferability. Envir. Management, 4(2): 125–136.

Sparks, R.S.J., 1981. Triggering of volcanic eruptions. Nature, 290(5806), 448 pp.

Suwa, A., 1980. The surveillance and prediction of volcanic activities in Japan. Geojournal, 4(2): 153–160.

Tazieff, H., 1967. The menace of extinct volcanoes. UNESCO Courier: 4–13.

Thal Larsen, J.H., 1926. Bestrijding van de zandbezwaren aan den voet van werkzame vulkanen op Java. De Ingenieur, 41(51): 1009–1023.

Tilling, R.I., et al., 1975. Rockfall seismicity-correlation with field observations; Makopuhi Crater, Kilauea Volcano. Hawaii. J. of Research, USGS, 3(3): 345–361.

Unger, J.D., and Decker, R.W., 1970. The micro earthquake activities of Mt. Rainier, Washington. Bull. Seismol. Soc. Amer., 60: 2023–2035.

Unger, J.D., and Mills, K.F., 1973. Earthquakes near Mount Helens, Washington. Bull. Geol. Soc. Amer., 84(3): 1065–1067.

Verstappen, H.Th., 1963. Some observations on Indonesian volcanoes. Tijdschr. Kon. Ned. Aardrijksk. Gen., 81(3): 237–251.

Verstappen, H.Th., 1964. Some volcanoes of Halmahera (Moluccas) and their geomorphological setting. Tijdschr. Kon. Ned. Aardrijksk. Gen., 81(3): 297–316.

Walker, G., 1973. Diminishing prospects for Heimaey. New Scientist, 57(835): 477–478.

Westercamp, D., 1981. Cartographie du risque volcanique à la Soufrière de Guadeloupe: rétrospective et technique actuelle. Bull. Int. Ass. Eng. Geol., 23: 25–32.

IJzendoorn, M.J. van, 1953. The eruption of Gunung Kelud on August 31, 1951, proved the utility of the Kelud tunnelworks. Berita Gunung Berapi, 3/4, 15 pp.

Earthquake hazard

Anon., 1969. Können Erdbeben vorausgesagt werden? Ztschr. f.Wirtschaftgeogr., 13(8): 271–272.
Bolton Seed, H., 1967. Soil stability problems caused by earthquakes. Cal. Univ. Dept. Civil Eng., Berkeley, 20 pp.
Bolton Seed, H., 1969. Landslides during earthquakes due to soil liquefaction. J. Soil Mech. and Found. Div. A.S.C.E., SM4, 95, 11–23.
Bolton Seed, H., 1969. Seismic problems in the use of fills in San Francisco Bay. In: Geol. and Eng. aspects of San Francisco Bay fill, Cal. Div. Mines and Geol. Spec. Rep. 97: 87–99.
Bolton Seed, H., 1970. Earth Slope stability during earthquakes. In: R.L. Wiegel (Editor), Earthquake engineering. Englewood Cliffs. N.J., Prentice-Hall: 383–401.
Bolton Seed, H., et al., 1975. Dynamic analysis of the slide in the lower San Fernando Dam during the earthquake of February 9th, 1971. J. Geotechn. Eng. Div., ASCE, GT9, 101: 889–911.
Boore, D.M., et al., Ground motion values for use in the seismic design of the trans-Alaska pipeline system. USGS, Circular 672, 23 pp.
Borcherdt, R.D., 1975. Studies for seismic zonation of the San Francisco Bay region. USGS Prof. Paper, 941–A, 102 pp.
Castro, G., 1975. Liquefaction and cyclic mobility of saturated sands. J. Geotechn. Eng. Div., ASCE, 101 (GT6): 551–569.
Chapelle, E.R. la, 1968. The character of snow avalanching induced by the Alaska earthquake. In: The Great Alaska Earthquake of 1964. Hydrology. Nat. Acad. Sci., Washington D.C.: 355–361.
Clapperton, C.M., and Hamilton, P., 1971. Peru beneath its eternal threat. The Geogr. Mag., 43(9): 632–639.
Cooke, R.U., and Townshend, J.R.G., 1970. Pattern of Peru's great earthquake. The Geogr. Mag., 42: 765–766.
Eckel, E.B., 1970. The Alaska earthquake, March 27th, 1964, lessons and conclusions. USGS Prof. Paper, 546, 57 pp.
Eiby, G.A., 1980. Earthquakes. Heinemann, Auckland, London, 209 pp.
Farhoudi, G., 1975. Luftbild Schiras, Iran – Gefährdung moderner Stadtplanung durch tektonische Aktivitäten. Erdkunde, 106(1/2): 1–9.
Field, W.O., 1968. The effect of previous earthquakes on glaciers. In: The Great Alaska Earthquake of 1964. Hydrology. Nat. Acad. Sci., Washington D.C.: 252–265.
Friedman, D.G., 1970. Computer simulation of the earthquake hazard. In: R.A. Olson and M.M. Wallace (Editors), Geologic hazards and public problems. U.S. Office Emergency Preparedness, Region 7, Proc. Conf. San Francisco, (Calif.,)1969, Washington C.D., U.S. Govt. Pr. Off.: 153–181.
Gams, I., 1979. On the mass movements triggered by the earthquake in 1979 and neotectonics in Montenegro (S. Yugoslavia). Proc. 15th meeting IGU Comm. Geom. Survey and Mapping, Modena, Italy: 117–127.
Garcia, W., 1966. Observaciones sobre los daños producidos por el sismo del 24 de setiembre de 1963, en los pueblos de Mañas, Gorgor y Rajanya, prov. de Cajatambo, dep. de Lima. Bol. Com. Carta Geol. Nac., Peru, 13: 111–117.
Garcia, W., 1966. Movimientos sismicos ocurridos en Abancay en diciembre 1963 y inero, febrero 1964, prov. de Abancay, dep. de Apurimac . Bol. Com. Carta Geol. Nac., Peru, 13: 143–151.
Gates, G.O., 1970. Earthquake hazards. In: R.A. Olson and M.M. Wallace (Editors), Geologic hazards and public problems. U.S. Office Emergency Preparedness, Region 7, Proc. Conf., San Francisco, Calif., 1969, Washington D. C., U.S. Govt. Printing Office: 19–52.
Grant-Taylor, T.L., et al., 1974. Microzoning for earthquake effects in Wellington, New Zealand. New Zealand Dept. Sci. and Ind. Res., Bull. 212, 62 pp.
Gupta, V.J., and Virdi, N.S., 1975. Geological aspects of the Kinnaur earthquake, Himachal Pradesh. J. Geol. Soc. India, 16(4): 512–514.
Hart, E.W., 1974. Zoning for surface fault hazards in California: The new special studies zones maps. Cal. Geol., 27(10): 227–230.
Hiramatsu, T., 1980. Urban earthquake hazards and risk assessment of earthquake prediction. J. Phys. Earth., 28: 59–101.
Hogg, S.H., 1980. Reconstruction following seismic disaster in Venzone, Friuli. Disasters, 4(2): 173–186.
Iliev, I., 1974. Influence of the geological structure of slopes on their seismic resistance (in bulgarian). Izv. Geol. Inst. BAN. Ser. Inzen. Geol. i Hidrogeol., 23: 149–159.
Iliev, I., 1974. Pecularities of landslides caused by earthquakes (in bulgarian). Izv. Geol. Inst., Ser. Inzen. Geol. i Hidrogeol., 23: 161–166.
Inouchi, N., and Sato, H., 1975. Vertical crustal deformation accompanied with the Tonankai earthquake of 1944. Geogr. Surv.Inst. Bull. Tokyo, 21(1): 10–18.
Kachadoorian, R., 1968. Effects of the earthquake of March 27, 1964, on the Alaska highway system. USGS Prof. Paper, 545–C, 66 pp.
Kadomura, H., 1967. Basic concepts of photogeomorphological analysis of soft ground conditions. Geogr. Rep. Tokyo Metrop. Univ., 2: 237–254.
Kadomura, H., 1968. Predicting the areas of danger from natural disaster due to soft ground conditions. An application of aerial photo analysis. Geogr. Rep. Tokyo Metrop. Univ., 3: 11–29.
Kanai, K., et al., 1963. Relation between earthquake damage and nature of the ground II, Case of Nagaoka earthquake. Bull. Earthquake Res. Inst., 41(1): 271–277.
Kanakubo, T., and Tanioka, S., 1980. Natural hazard mapping. Geojournal, 4(4): 333–340.
Karmick, V., and Algermissen, S.T., 1978. Seismic zoning. In: The assessment and mitigation of earthquake risk. UNESCO, Paris
Kobayashi, H., 1968. Mechanism of earthquake damage to embankments and slopes (in japanese). Zisin, Series 2, 21(3): 178–189.
Marcus, M.G., 1968. Effects on glacier-dammed lakes in the Chugach and Kenai mountains. In: The Great Alaska Earthquake of 1964, Hydrology. Nat. Acad. Sci., Washington D.C.: 329–347.
Matsuda, I., and Mocmizuki, I., 1980. A method of predicting distribution of intensity for seismic zonation of the Kanto Plain. Abstracts 24th IGU Congres, Tokyo, 4: 86–87.
McCulloch, D.S., 1968. Slide-induced waves, seiching and ground fracturing at Kenai Lake. In: The Great Alaska Earthquake of 1964, Hydrology. Nat. Acad. Sci., Washington D.C.: 47–81. Also: USGS Prof. Paper, 543–A.
McCulloch, D.S., and Bonilla, M.G., 1970. Effects of the earthquake of March 27, 1964, on the Alaska Railroad. USGS Prof. Paper, 545–D, 161 pp.
Morton, D.M., 1971. Seismically triggered landslides above San Fernando Valley. Cal. Geol., 24(4/5): 81–82.
Murayama, Y., 1980. Damages of residential area in Sendai and its vicinity caused by the Miyagi-ken-oki earthquake (in japanese, with english summary). Ann. Tohoku Geogr. Ass., 32(1), 10 pp.

Nakano, T., 1966. Systematic photo-geomorphological analysis of land deformation due to earthquake. Geogr. Rep. Tokyo Metrop. Univ., 1: 221–228.
Nakano, T., 1980. Earthquake damage, damage prediction and counter measures in Tokyo. Abstracts 24th IGU Congress Tokyo, Vol. 4: 74–75.
Nat. Res. Council, 1968. The Great Alaska Earthquake of 1964, Hydrology. Committee on the Alaska Earthquake. Div. Earth Sciences, Nat. Res. Council. Publ. 1603, Nat. Acad. Sci., Washington D.C., 441 pp.
Nat. Res. Council, 1969. Towards reduction of losses from earthquakes. Conclusions from the Great Alaska Earthquake of 1964. Committee on the Alaska Earthquake, Div. Earth Sciences, Nat. Acad. Sci., Washington D.C., 34 pp.
Nazarian, H.N., and Hadjian, A.N., 1980. Earthquake-induced lateral soil pressures on structures. J. Geotechn. Eng. Div., 106(11)
Nishimura, K., et al., 1965. Geomorphological accidents caused by the Niigata Earthquake. Sci. Report Univ., Series 7 (Geography), 14(3): 147–165.
Nishimura, K., et al., 1969. Geomorphological accidents caused by the Tokachi-oki Earthquake. Sci. Reports Tohoku Univ., Series 7 (Geography), 18(1): 41–58.
Northey, R.D., 1973. Insurance claims from earthquake damage in relation to soil pattern. Geoderma, 10(1/2): 151–159.
Okusa, S., and Anma, S., 1980. Slope failures and tailings: dam damage in the 1978 Izu-Okshima-Kinkai Earthquake. Eng. Geol., 16(3/4): 195–224.
Oswal, H.L., 1968. The airphoto as a tool in some aspects of earthquake research. Third Symp. on earthquake engineering, Roorkee, India, 6 pp.
Pakiser, L.C. Jr., 1970. Earthquake prediction and modification research in progress. In: R.A. Olson and M.M. Wallace (Editors), Geologic hazards and public problems. U.S. Office Emergency Preparedness, Region 7, Proc. Conf. San Francisco, Cal., 1969. U.S. Govt. Printing Off. Washington D.C.: 297–303.
Parry, W.T., 1974. Earthquake hazards in sensitive clays among the central Wasatch Front, Utah. Geology, 2(11): 559–560.
Paxton, W.H., 1972. Salt Lake City water supply system and earthquake damage. In: Environmental geology of the Wasatch Front, Utah Geol. Ass. Publ., 1: P1–P8.
Plafker, G., and Rubin, M., 1967. Vertical tectonic displacements in south central Alaska during and prior to the great 1964 earthquake. Proc. 11th Pacific Sci. Congr. Tokyo, 1966, Symp. 19, J. Geosciences, 10: 53–66.
Post, A.S., 1967. Effects on glaciers. In: The Great Alaska Earthquake of 1964, Hydrology. Nat. Acad. Sci., Washington D.C., 1968: 266–308. Also: USGS Prof. Paper, 544–D, 1967.
Proceedings, third congres IAEG Madrid, 1978. Vol. 1–2 Regional Planning/Terrain evaluation of regional/urban development. (16 papers on natural hazards).
Radbruch, D.M., 1969. Aerial and engineering geology of the Oakland East quadrangle, California. USGS Quad. Map, GQ/769
Research Group, 1980. Active faults in and around Japan: the distribution and the degree of activity. J. Nat. Disaster Sci., 2/2: 61–99.
Rikitake, T., 1975. Earthquake precursors. Bull. Seismol. Soc. Amer., 65.
Rikitake, T., 1975. Earthquake prediction. In: Dev. in Solid Earth Geophysics. Elsevier Publ. Comp., Amsterdam, 357 pp.
Rinehart, J.S., and Murphy, A., 1969. Observations on pre- and post-earthquake performance of Old Faithful geyser. J. Geoph. Res., 74(2): 574–575.
Roberts, J.L., 1979. Reflections on environmental hazard. Nature and Resources, 15(4): 24–30.
Schmoll, H.R., et al., 1975. Geologic consideration for redevelopment planning of Managua, Nicaragua, following the 1972 earthquake. USGS Prof. Paper 914, 23 pp.
Scholz, C.H., et al., 1973. Earthquake prediction: a physical basis. Science, 181 (4102): 803–810.
Steinbrugge, K.V., 1968. Earthquake hazards in the San Francisco Bay area a continuing problem in public policy. Univ. Cal. Inst. Govt. Stud., 80 pp.
Steinbrugge, K.V., 1970. Earthquake hazard abatement and land use planning directions towards solutions. In: R.A. Olson and M.M. Wallace (Editors), Geologic hazards and public problems. U.S. Office Emergency Preparedness, Region 7, Proc. Conf. San Francisco, Cal., U.S. Govt. Printing Off.: 143–152.
Steinbrugge, K.V., et al., 1973. Earthquake hazard and public policy in California, Engineering Issues. J. Prof. Activities, ASCE, 99(PP4), Proc. Paper 10105: 513–519.
Tada, F., et al., 1951. The relation between the percentage of houses destroyed by the Tonankai Earthquake and the configuration of the ground. Misc. Dep. Inst. Nat. Res., 19/21: 93–103 (in japanese with english summary).
Takasaki, M., et al., 1966. The relation of damage caused by the Niigata Earthquake to the topographical conditions in Niigata plain. Rep. Co-op. Res. Disaster Prev., 11: 13–18 (in japanese with english summary).
Tamura, T., 1980. The earthquake of 1978 in the Sendai area. Proc. IGU Conference Tokyo.
Tricart, J., 1973. Un problème de géomorphologie appliquée: le choiz des sites d'habitat dans une région sismique (Andes centrales, Pérou). Ann. de Géogr., 82(449): 8–27.
Tuthill, S.J., et al., 1968. Post-earthquake studies at Sherman and Sheridan glaciers. In: The Great Alaska Earthquake of 1964, Hydrology. Nat. Acad. Sci., Washington D.C.: 318–328.
U.S.G.S., 1964. The Alaska Earthquake, March 27, 1964. USGS Prof. Papers: 541–546.
Vogt, J., 1981. Présentation de la carte sismotectonique de la France, 1:1,000 000. Bull. Int. Ass. Eng. Geol., 23: 11–12.
Wahler, W.A., 1973. Earthquake research applied to slope stability. J. Construction Div., ASCE, 99 (C01). Proc. Paper 9837, 4 pp.
Waller, R.M., 1968. Water sediment ejections. In: The Great Alaska Earthquake of 1964. Hydrology. Nat. Acad. Sci., Washington D.C.: 97–116.
Yeats, R.S., et al., 1981. Active fault hazard in southern California: ground rupture versus seismic shaking. Bull. Geol. Soc., Amer., 92(4): 189–196.
Youd, T.L., 1973. Liquefaction. Earthquake Information Bull., 5(6): 11–17.

Subject Index